Advanced Structured Materials

Volume 191

Common engineering materials are reaching their limits in many applications, and new developments are required to meet the increasing demands on engineering materials. The performance of materials can be improved by combining different materials to achieve better properties than with a single constituent, or by shaping the material or constituents into a specific structure. The interaction between material and structure can occur at different length scales, such as the micro, meso, or macro scale, and offers potential applications in very different fields.

This book series addresses the fundamental relationships between materials and their structure on overall properties (e.g., mechanical, thermal, chemical, electrical, or magnetic properties, etc.). Experimental data and procedures are presented, as well as methods for modeling structures and materials using numerical and analytical approaches. In addition, the series shows how these materials engineering and design processes are implemented and how new technologies can be used to optimize materials and processes.

Advanced Structured Materials is indexed in Google Scholar and Scopus.

Aleksander N. Guz · Holm Altenbach · Viacheslav Bogdanov · Vladimir M. Nazarenko
Editors

Advances in Mechanics

Current Research Results of the NAS of Ukraine

Springer

Editors
Aleksander N. Guz
S. P. Timoshenko Institute of Mechanics
National Academy of Sciences of Ukraine
Kyiv, Ukraine

Viacheslav Bogdanov
S. P. Timoshenko Institute of Mechanics
National Academy of Sciences of Ukraine
Kyiv, Ukraine

Holm Altenbach
Otto-von-Guericke-Universität Magdeburg
Magdeburg, Germany

Vladimir M. Nazarenko
S. P. Timoshenko Institute of Mechanics
National Academy of Sciences of Ukraine
Kyiv, Ukraine

ISSN 1869-8433 ISSN 1869-8441 (electronic)
Advanced Structured Materials
ISBN 978-3-031-37312-1 ISBN 978-3-031-37313-8 (eBook)
https://doi.org/10.1007/978-3-031-37313-8

This Springer imprint is published by the registered company Springer Nature Switzerland AG
The registered company address is: Gewerbestrasse 11, 6330 Cham, Switzerland

Paper in this product is recyclable.

Preface

The publication of this book is timed to coincide with the 145th anniversary of Stephen P. Timoshenko (1878–1972)—a prominent, world-renowned scientist in mechanics, reputable science administrator, and university teacher. He entered the history of Ukrainian science as a co-founder of the Ukrainian Academy of Sciences (now the National Academy of Sciences of Ukraine), the founder and the first director of the Institute of Mechanics of the National Academy of Sciences of Ukraine, now named after him.

Stephen Timoshenko (1878–1972)

Stepan Prokofievich Timoshenko (later known as Stephen Timoshenko) was born on 23 December 1878 in the village of Shpotovka, Uyezd of Konotop in the Chernigov Governorate, which at that time was a territory of the Russian Empire (today in Konotop Raion, Sumy Oblast of Ukraine) into the family of land surveyor Prokofi Timoshenko (1847–1932) and his wife Jozefina Sarnavska (1854–1922), a daughter of a military retiree. In 1896, he finished a vocational school and entered the Institute of Railway Engineers in St. Petersburg, where teaching was provided by mathematician D. A. Grave, civil engineer D. I. Zhuravsky and railway and civil engineer N. A. Belelubsky, who specialized in mechanics, O. A. Brandt, an expert in thermodynamics, and other distinguished professors. After graduation, S. P. Timoshenko worked as an assistant in this institute's mechanical laboratories. In 1901, pursuing his scientific interests, he visited France. Later, in 1903, he started working at the strength of materials laboratory of the St. Petersburg Polytechnic Institute, which sent him to Germany (1904–1906) to study the advances of leading scientific schools in the field of mechanics. There he attended lectures and trained at the Munich and Göttingen universities under the guidance of outstanding professionals in mechanics A. Föppl (1854–1924) and L. Prandtl (1875–1953).

During his work at the St. Petersburg Polytechnic Institute, Timoshenko met Victor Kirpichov (1845–1913)—a well-known expert in strength of materials, the founder and first rector of the Kharkov Polytechnic Institute of Emperor Alexander III (now the National Technical University "Kharkiv Polytechnic Institute") and the Kiev Polytechnic Institute of Emperor Alexander II (now the National Technical University of Ukraine "Igor Sikorsky Kyiv Polytechnic Institute")—who, according to the reminiscences of Timoshenko himself (Timoshenko 1968), made the decisive influence on his career, research, and teaching. V. L. Kirpichov advised Timoshenko to participate in the competition for a vacant professor position at the Kiev Polytechnic Institute. Elected to the position in 1906, Timoshenko moved to Kyiv, where he soon headed the Chair of Strength of Materials, and in 1909 became the dean of the Mechanics and Structural Engineering departments (1909–1911). In 1907 he defended his thesis and got an associate degree in applied mechanics.

The years at the Kiev Polytechnic Institute were very fruitful for the scientist. He radically changed the teaching courses in the theory of elasticity and strength of materials, wrote and published fundamental textbooks and tutorials in oscillation theory, theory of deformable system stability, mechanical engineering, applied dynamics, theory of structures, and theory of plates and shells. The result of his long and indefatigable work was the monograph *Stability of Elastic Systems* (1910), which was awarded D. I. Zhuravsky Prize "For outstanding works in structural mechanics."

After Professor S. P. Timoshenko, in January 1911, signed the protest condemning police violence against students and teachers, he was dismissed, together with several other professors. He returned to St. Petersburg, where he worked for a year and a half as a teacher in the Electrotechnical and Polygraphy Institutes. In 1912, Timoshenko went on a scientific trip to Great Britain. In 1912–1917 he again worked in St. Petersburg, held professorships at the Polytechnic, Electrotechnical Institutes, and the Institute of Railway Engineers, and was involved as a scientific consultant in the construction of warships. In those years, Timoshenko wrote textbooks *Course in Strength of Materials* (1911) and two-volume *Theory of Elasticity* (1914–1916), which became engineering classics.

Dismissed deans of the Kiev Polytechnic Institute (from left to right): S. P. Timoshenko, O. V. Nechayev, and K. G. Schindler (1911)

After the Bolshevik coup late in 1917, Timoshenko returned to Kyiv and, at the invitation of the Polytechnic Institute's Academic Council, resumed his professor ship at this institute. In July–October 1918, he actively participated in the work of the Commission for elaborating a draft law on establishing the Ukrainian Academy of Sciences (below the Commission), which was led by Professor Vladimir Vernadsky, a mineralogist and geochemist. Its objective was to elaborate a concept of the Academy's research activities, as well as its organizational and personnel structure (Paton 2018). In the Commission, S. P. Timoshenko was among those few representatives of natural and engineering branches who laid the foundation of the future Academy, proposed conceptual principles of natural science development, and headed the first Academy institutes and departments.

In July 1918, at a meeting of the Commission, Timoshenko made a policy-setting presentation regarding the organization of research in applied natural science at the Ukrainian Academy of Sciences (UAS). In the speech, he noted, in particular, that the time when sciences and engineering were going different ways was over and that the powerful tools provided by mathematics and mechanics were often used for dealing with purely technical tasks. He also referred to developed countries, where there were numerous research institutions in which representatives of "pure" science and applied knowledge worked hand in hand at the solution of scientific problems that were of practical importance (The History of Academy of Sciences of Ukraine 1918–1923). According to Timoshenko, Ukraine began to be independent in challenging economic conditions, so, first and foremost, it should develop research to raise labor productivity and explore natural resources, along with the methods to use them. He proposed that in the future Academy, a class of applied natural science be organized, joining that way science and technology to pursue major tasks: to advance science in Ukraine, explore natural resources, develop and improve methods of using them, participate in the training of professional personnel and professors for the institutions of higher education. He also expressed the idea about the advisability of granting the Academy the right to train scientific personnel of the highest qualification and awarding academic degrees, the right to award doctoral degrees for outstanding research activities, which earlier had been the prerogative of universities.

The work of the Commission was finalized by adopting on 14 November 1918 the "Law of the Ukrainian State on the Foundation of the Ukrainian Academy of Sciences in Kyiv," which was signed by Hetman of Ukraine Pavlo Skoropadsky. On the same day, by his order, the first 12 full members (academicians) of the UAS were appointed, Timoshenko being among them. On 27 November 1918, the Academy began its activities with the first Constituent Assembly. This meeting, under the chairmanship of the oldest academician O. I. Levytsky, with S. P. Timoshenko as the youngest academician doing secretarial work, unanimously elected Academician V. I. Vernadsky the President (Head) of the Academy. Academician M. T. Kashchenko was elected the Head of the Department of Physics and Mathematics, and Academician S. P. Timoshenko became Department's Secretary, whose functions he faithfully performed till his departure for emigration early in 1920.

Due to his inexhaustible organizational energy and personal qualities, the Institute of Technical Mechanics was founded among the first institutions of the Academy. On 30 November 1918, the session of its General Assembly approved Timoshenko as the director of this institute. The assembly also elected him the head of the Chair of Applied Mechanics and the leader of the Construction Materials Section.

The Institute of Technical Mechanics started its work with studies of the strength of reinforced concrete beams under changing loads and the strength of a spherical film under uniform pressure, which S. P. Timoshenko and E. K. Garf carried out. Timoshenko wrote his works *On the Strength of Airplanes* and *Calculations of Arches* at that time. He also started publishing *Proceedings of the Institute of Technical Mechanics*. Furthermore, developing the ideas of the integration of science and technology, scientific research and the implementation of joint applied studies, Timoshenko initiated and headed a specialized council at the laboratory of construction materials, which was responsible for approving sample norms for various construction materials, the programs of respective works and reports of the sampling station.

A special page in active science-administration activities of Timoshenko at the Academy was his work as the deputy of Vernadsky in the Standing Commission for Exploring Natural Resources of Ukraine. It united academicians in natural sciences and leading UAS professionals in nature exploration and started publishing collected works *Natural Resources of Ukraine*. Several sections were set up under this Commission, particularly the construction materials section headed by Timoshenko.

So, summing up the dedicated work of Academician Timoshenko in the Ukrainian Academy of Sciences, it can be stated that he was one of the like-minded partners of Vernadsky and, together with him, laid the foundation for future achievements of the National Academy of Sciences of Ukraine. Due to his active participation, the Department of Physics and Mathematics acquired top priority. Furthermore, he took the helm of the first research institution in the Academy—the Institute of Technical Mechanics (Bogdanov 2018; Guz 2008).

Because of the political and economic situation that developed after Bolsheviks seized Kyiv in 1919, the activities of the UAS were suspended, forcing S. P. Timoshenko to leave Ukraine. In March 1920, he and his family arrived in Serbia. As he could not find a job in Belgrade, from April 1920, he headed the Department of Strength of Materials at the Zagreb Polytechnic Institute (at the time, Croatia was a part of the Kingdom of Serbs, Croats, and Slovenes) (Soderberg, 1878–1972).

In 1922, the scientist accepted the invitation to move to the USA, where from 1923 to 1927, he worked in Philadelphia—at first as an engineer and then as a scientific consultant for the research department of the *Westinghouse* company. From 1927, Timoshenko held the Chair of Applied Mechanics at the University of Michigan (in Ann Arbor), within which, in 1929, he opened a summer school of mechanics for certified specialists, which was of great importance for the advancement of these science and training professionals in the USA. There he provided specialized lecture courses where hundreds of engineers enriched their knowledge. In addition, quite a few doctoral theses were defended under his guidance. Concurrently, Timoshenko organized the Section of Mechanics at the American Society of Mechanical Engineers.

S. P. Timoshenko in the laboratory of engineering mechanics at Stanford University (1939)

From 1936, for a quarter of a century, the scientist lived in the town of Palo Alto in California and taught at Stanford University, heading the Chair of Mechanics till 1943 and holding a professorship till 1960. In addition, S. P. Timoshenko devoted more than 60 years to the development of engineering science in various countries.

After retiring in 1960 from 1964, he lived with his daughter in Wuppertal (Germany). He died on 29 May 1972 and was buried beside his wife in Palo Alto (California, USA).

S. P. Timoshenko (in the center) with his brothers Serhiy (left) and Volodymyr (right) in Palo Alto (1947)

S. P. Timoshenko revisited Ukraine twice (in 1958 and 1967), where he was received with honors at leading scientific institutions and universities of Kyiv and Kharkiv. He recollected his 1958 visit to the Institute of Technical Mechanics, which had been founded by him (at that time, it was the Institute of Structural Mechanics of the Academy of Sciences of the Ukrainian SSR): “My tour of the laboratory gave me much satisfaction. The basic idea of marrying science to engineering, which had so inspired me when I organized the mechanics department at the Kiev Academy of Sciences, was a living reality here. This laboratory that I had planned was now actively participating in solving the country’s important engineering problems” (Timoshenko 1968).

S. P. Timoshenko in the Strength of Materials Department of the Kiev Polytechnic Institute (1958)

Due to his bright scientific and engineering talent and phenomenal single-minded dedication to his work, Timoshenko made a tremendous contribution to world science. His scientific works became the foundation for the development of many areas of mechanics. For many years the Ukrainian scientist took a leading place among U.S. experts in the field of mechanics and was the founder of mechanical engineering in that country. It was not by chance that when opening the Twelfth International Congress of Applied Mechanics, which took place in August 1968 at Stanford University, the president of the Congress, Prof. N. J. Hoff noted that S. P. Timoshenko "did so much for applied mechanics, particularly in the United States."

The main areas of the scientist's research were fundamental works and further development of highly relevant problems in the mechanics of solids: strength, stability and oscillations of mechanical systems, structural mechanics, and the theory of structures. He made an especially significant contribution to advancing the applied theory of elasticity, the theory of stability of shell and plate elastic systems, including those reinforced with stiffeners. He solved a number of problems on stress concentration near holes and the strength of railway rails. In the theory of thin-walled systems, the so-called Timoshenko model is widely used, which allows considering the effect of shear strains. S. P. Timoshenko's studies dealing with bending, torsion, oscillation, and impact in structural members are of great importance. Making novel achievements in science and engineering, S. P. Timoshenko carried out fundamental works in the strength of materials, applied elasticity theory, and oscillation theory, which were ahead of their time, yet found broad practical application in developing modern aerospace equipment, engineering structures, and in shipbuilding. Professor Timoshenko intensely studied the history of the science of strength of materials and significantly contributed to the formation of engineering education. He created a series of unique monographs on the main areas of engineering mechanics (Timoshenko 1925; Timoshenko 1930; Timoshenko 1933; Timoshenko 1935; Timoshenko 1937; Timoshenko 1940; Timoshenko 1945; Timoshenko 1948; Timoshenko 1951;

Timoshenko 1953; Timoshenko 1954; Timoshenko 1959), on which many generations of specialists in the field of mechanics around the world were educated and continue to be educated.

The scientific achievements of S. P. Timoshenko were acknowledged all over the world. He was elected a member of the Ukrainian (1918), Polish (1935), French (1939), Italian (1948) Academies of Sciences, the USSR Academy of Sciences (1958), and London Royal Society (1944). The title of doctor *honoris causa* was conferred on him by Lehigh University (USA, 1936), University of Michigan (USA, 1938), Zurich Institute of Technology (Switzerland, 1947), University of Bologna (Italy, 1954), Zagreb Polytechnic (Yugoslavia, 1956), and Turin Polytechnic (Italy, 1960). In 1957, he became the first recipient of the S. P. Timoshenko Medal, instituted by the American Society of Mechanical Engineers "for the invaluable contribution and personal example as the leader of new era in applied mechanics."

In 1993, the outstanding contribution of academician Stephen P. Timoshenko to Ukrainian and world science was honored by the National Academy of Sciences of Ukraine (NAS) by bestowing his name to the NAS Institute of Mechanics, whose founder and first director he was, and by establishing S. P. Timoshenko Prize for major achievements in theoretical and applied mechanics in 1999.

The S. P. Timoshenko Institute of Mechanics of the NAS of Ukraine was founded on 30 November 1918 by a resolution of the General Meeting of the Ukrainian Academy of Sciences (UAS), and it was the first institution conducting research in engineering at the UAS (now it is the National Academy of Sciences of Ukraine).

Monument to S. P. Timoshenko on the territory of the National Technical University "Igor Sikorsky Kyiv Polytechnic Institute"

Changes in the Institute's structure and research areas over the years of its existence are evidenced by the changes in its name (in 1918–1929, it was referred to as the Institute of Technical Mechanics, in 1929–1959—as the Institute of Structural Mechanics, and since 1959 it has been the Institute of Mechanics).

Besides well-known works by Timoshenko, the major achievements of the first years of its activities were the new scientific area created by N. M. Krylov and N. N. Bogolyubov—the asymptotic theory of non-linear oscillations and establishing the world-recognized Kyiv school of non-linear mechanics. In structural mechanics, problems of the strength of engineering structures and structure materials, particularly the strength of aircraft engines, were studied (Paton 2018; Guz 2008; Guz 2019).

Important accomplishments of the Institute in the following years were the development of methods for computing rocketry structural elements, including those that considered the concentration of stresses, numerical methods for investigating complex rotational shells, and the theory of materials thermoplasticity under non-isothermal deformation. Among top-priority research results, one should mention the development of the three-dimensional theory of deformable body stability; investigations of the stability of structural elements made of composite materials and the stability of mine workings on its basis; advancing the theory of elastic wave propagation in bodies with initial (residual) stresses and using it to create a method for determining biaxial and triaxial stresses in non-destructive ultrasonic testing; the development of analytical and numerical methods of non-stationary hydroelasticity of shells; the theory of elastic wave diffraction for the case of multiply connected bodies; methods to predict physicomechanical properties of composites of various structures, as well as the theory of multiphase media; the method of matrix-valued Lyapunov functions in the theory of stability of mechanical system motion (Paton 2018; Guz 2008; Guz 2019).

S. P. Timoshenko with scientists of the Institute of Mechanics

In recent years, the recognition of the world scientific community has been won by the research results obtained by associates of the S. P. Timoshenko Institute of Mechanics of the NAS of Ukraine in the fields of composite materials mechanics, including nanocomposite mechanics; non-classical problems of fracture mechanics (in particular, fracture in composites under compression along reinforcing element); brittle fracture of materials containing cracks, taking into account the initial (residual) stresses acting along cracks; fracture under compression along parallel cracks; fracture of bodies containing cracks under the action of dynamic loads (with taking into account interactions of crack faces; fracture of thin-walled bodies containing cracks under tension in the case of previous loss of stability; long-term fracture of viscoelastic materials); mechanics of coupled fields in materials and structure elements, methods for analyzing the dynamics and functional stability of complex non-linear systems (Paton 2018; Guz 2019). Institute's scientists have developed further the theory of anisotropic layered rotational arbitrarily-shaped shells with layers of variable stiffness and elaborated methods for numerical solutions of a wide class of problems on their stress state under non-axisymmetric loading, the theory of thermo-viscoplasticity of materials in the processes of complex loading at elevated temperatures, the theory of impact interaction of solid and deformable bodies with liquid and elastic medium, the theory of deformation and damage of homogeneous and composite materials of various structures. They have proposed new analytical and numerical methods of studying stationary and non-stationary thermoplasticity problems, solving contact problems for elastic bodies with initial (residual) stresses, and calculating shells with holes under physically and geometrically non-linear deformations. The dynamics of solid bodies, liquid drops, and elastic bodies in compressible viscous fluid, the theory of waves in mixtures, and the model of finite fibers in the three-dimensional theory of composite materials stability have also been developed.

Over the years of existence of the S. P. Timoshenko Institute of Mechanics of NAS, its scientists have published nearly 500 monographs, including 8 summarizing multi-volume editions covering various areas of mechanics (Guz 2018). Since 1955 the Institute has been publishing the international scientific journal *Applied Mechanics*, which is translated and published as *International Applied Mechanics* by Springer Publishers. It publishes research and reviews articles presenting new scientific results in mechanics of deformable solids, structure elements, and mechanical engineering. Just in 2000–2009, the journal published about 170 generalizing review articles authored by renowned scientists from 26 countries.

The S. P. Timoshenko Institute of Mechanics promotes scientific cooperation with foreign science centers and universities. In recent years, it has conducted joint research and published co-authored monographs and papers with scientists of the University of Bologna (Italy), University of Aberdeen (Scotland), Heriot-Watt University (Scotland), Otto-von-Guericke Universität Magdeburg (Germany), Sofia University' St. Kliment Ohridski' (Bulgaria), Technische Universität Berlin (Germany), Harbin Institute of Technology (China), Institute of Fundamental Technological Research of Polish Academy of Sciences (Poland).

Today's structure of the National Academy of Sciences of Ukraine has the Department of Mechanics, which, besides the S. P. Timoshenko Institute of Mechanics, includes 5 more institutions: the G. S. Pisarenko Institute for Problems of Strength, the M. S. Polyakov Institute of Geotechnical Mechanics, the Institute of Technical Mechanics, the Institute of Hydromechanics, and the Institute of Transport Systems and Technologies (Paton 2018). The principal areas of their research are mechanics of deformable solid bodies, mechanics of fluid, gas, and plasma, general mechanics, mechanics of solids and rocks, mechanics of spacecraft and aircraft, and mechanics of ground transport systems.

Some research in the field of mathematical problems of mechanics, mechanics of materials, dynamics, and structural strength is also carried out at the institutes of other departments of the Academy, such as the Institute of Mathematics (Kyiv), the Ya. S. Pidstrigach Institute of Applied Problems of Mechanics and Mathematics (Lviv), the Institute of Applied Mathematics and Mechanics (Cherkasy), the V. M. Glushkov Institute of Cybernetics (Kyiv), the E. O. Paton Institute of Electric Welding (Kyiv), the G. V. Karpenko Physico-Mechanical Institute (Lviv), and the A.M. Pidgorny Institute of Mechanical Engineering Problems (Kharkiv). Investigations in various fields of mechanics are also carried out at leading universities of Ukraine, in particular, the T. Shevchenko National University of Kyiv, the National Technical University of Ukraine "Igor Sikorsky Kyiv Polytechnic Institute," the O. Honchar National University of Dnipro, and the I. Franko National University of Lviv.

This book presents articles by scientists of S. P. Timoshenko Institute of Mechanics and other institutes of the NAS of Ukraine, which address numerous sections of modern mechanics, in particular, mechanics of composite materials, fracture mechanics, the strength of materials and structures, thermoviscoelasticity and plasticity, mechanics of shell structures, contact mechanics, theory of wave propagation, dynamics of mechanical and hydromechanical systems.

Kyiv, Ukraine — Aleksander N. Guz
Magdeburg, Germany — Holm Altenbach
Kyiv, Ukraine — Viacheslav Bogdanov
Kyiv, Ukraine — Vladimir M. Nazarenko

References

Bogdanov VL, Dubrovina LA (2018) The role of academician S.P. Timoshenko in founding the Ukrainian Academy of Sciences and the Institute of Technical Mechanics of the UAS. Her NAS Ukraine 12:67–78

Guz AN (ed) (2008) S.P. Timoshenko Institute of Mechanics of the NAS of Ukraine (1918–2008). History. Structure. Informational Aspects. Litera, Kyiv

Guz AN, Rushchitsky JJ (2018) For the 100th anniversary of the S. P. Timoshenko Institute of Mechanics of the NASU: Books (Monographs and Textbooks) Published by the Institute. Int Appl Mech 54(2):121–142

Guz AN (2019) For the 100th anniversary of the S. P. Timoshenko Institute of Mechanics of the National Academy of Sciences of Ukraine (NASU). Int Appl Mech 54(1):3–33

Paton BE (ed) (2018) The national academy of sciences of Ukraine (1918–2018). The 100th anniversary of its foundation. Akademperiodyka, Kyiv
Soderberg, C.R.: Stephen P. Timoshenko (1878–1972). Biographical Memoir. National Academy of Sciences, Washington D.C. (1982)
The History of Academy of Sciences of Ukraine. 1918–1923. Documents and materials. Kyiv (1993) [in Ukrainian]
Timoshenko SP (1925) Applied elasticity. Westinghouse Mechanical Night School Press, East Pittsburg
Timoshenko SP (1930) Strength of materials. D. van Nostrand Company, Princeton
Timoshenko SP (1933) Theory of elasticity. McGraw-Hill Book Company Inc., New York
Timoshenko SP (1935) Elements of strength of materials, Part I & II. D. van Nostrand Company, Princeton
Timoshenko SP (1937) Theory of elastic stability. McGraw-Hill Book Company Inc., New York
Timoshenko SP (1940) Theory of plates and shells. McGraw-Hill Book Company Inc., New York
Timoshenko SP (1945) Theory of structures. McGraw-Hill Book Company Inc., New York
Timoshenko SP (1948) Advanced dynamics. McGraw-Hill Book Company Inc., New York
Timoshenko SP (1951) Theory of elasticity. McGraw-Hill Book Company Inc., New York
Timoshenko SP (1953) History of strength of materials. McGraw-Hill Book Company Inc., New York
Timoshenko SP (1954) Collected papers. McGraw-Hill Book Company Inc., New York
Timoshenko SP, Woinowsky-Krieger S (1959) Theory of plates and shells. McGraw-Hill Book Company Inc., New York et al.
Timoshenko SP (1968) As I remember. D. van Nostrand Company, Princeton

Contents

Chapter 1
A Brief Review of the Development of Mechanics in the National Academy of Sciences of Ukraine

Holm Altenbach, Viacheslav Bogdanov, Anatolii Bulat, Aleksander Guz, and Vladimir Nazarenko

1.1 Introduction

The origin of scientific research in the field of mechanics at the National Academy of Sciences of Ukraine (NAS of Ukraine, Academy) is rightfully associated with the name of an outstanding scientist, organizer of science, and teacher, one of the founders of the Academy Stepan Prokofievich Timoshenko (later known as Stephen Timoshenko, 1878–1972). He was the first academician in mechanics in the National Academy of Sciences of Ukraine (at the time of its creation in 1918, it was called the Ukrainian Academy of Sciences) and the founder of the Academy's first institute of a technical profile—the Institute of Technical Mechanics (later this institute was renamed into the Institute of Structural Mechanics, and then the Institute of Mechanics). Although Timoshenko worked at the Academy for less than two years (from

H. Altenbach
Lehrstuhl Technische Mechanik, Institut für Mechanik, Fakultät für Maschinenbau, Otto-von-Guericke-Universität Magdeburg, Magdeburg, Germany
e-mail: holm.altenbach@ovgu.de

V. Bogdanov (✉) · A. Guz · V. Nazarenko
S.P. Timoshenko Institute of Mechanics, National Academy of Science of Ukraine, Kyiv, Ukraine
e-mail: bogdanov@nas.gov.ua

A. Guz
e-mail: guz@nas.gov.ua

V. Nazarenko
e-mail: vmnazarenko@nas.gov.ua

A. Bulat
M.S.Polyakov Institute of Geotechnical Mechanics, National Academy of Sciences of Ukraine, Kyiv, Ukraine
e-mail: bulat@nas.gov.ua

A. N. Guz et al. (eds.), *Advances in Mechanics*, Advanced Structured Materials 191,
https://doi.org/10.1007/978-3-031-37313-8_1

the time of its founding in November 1918 until March 1920, when he went abroad before the capture of Kyiv by the Bolsheviks), his contribution to the formation of the Academy and its institutions was very significant. This is discussed in more detail in the preface to this book, as well as in Bogdanov and Dubrovina (2018), Guz (2008), Paton (2018). Timoshenko himself briefly but very vividly describes this period of his creative life in his well-known book of memoirs Timoshenko (1968).

The results of research conducted at the Academy in the 1930s by N.M. Krylov and N.N. Bogolyubov on asymptotic theory of nonlinear oscillations received worldwide recognition (Krylov and Bogolyubov 1947). These studies laid the foundations for a new scientific direction—nonlinear mechanics, which is still actively developing in many scientific centers of mechanics and mathematics (Martynyuk et al. 2009). In the same years, at the Academy, A.N. Dynnik, N.V. Kornoukhov, M.A. Lavrent'yev, G.N. Savin, S.V. Serensen and their students carried out important research in the field of strength of engineering structures and structural materials, dynamics and strength of machine parts and mechanisms, primarily aircraft engines and mine mechanisms, mining mechanical engineering.

In the first years after World War II, the Academy researchers developed structural analysis methods in rocket engineering, analytical methods for studying stress concentration near holes and cavities, and methods for structural analysis of plates and shells of variable thickness under thermomechanical loading. They also studied vibrations of mechanical systems taking into account energy dissipation, created and applied fundamentals of the theory of cumulative fatigue damage, and built the general theory of inertial navigation systems.

In the 60s and 70s of the last century, research in the field of mechanics expanded significantly at the Academy, which was due both to the formation and development of new directions in mechanics (computational mechanics, mechanics of polymers and composite materials, fracture mechanics, mechanics of coupled fields, etc.), and the country's needs in the scientific and technical support of important industries (aerospace engineering, naval architecture, heavy engineering, mining). In addition to the Institute of Structural Mechanics (since 1959—the Institute of Mechanics) and the Institute of Hydrology (since 1964—the Institute of Hydromechanics), which have been previously formed in Kyiv, the Institute for Problems of Strength (Kyiv, 1966), the Institute of Geotechnical Mechanics (Dnipropetrovsk, 1967) and the Institute of Technical Mechanics (Dnipropetrovsk, 1980) were established. On the basis of these institutes, the Department of Mechanics of the National Academy of Sciences of Ukraine (Department of Mechanics) was formed in 1983. More details about these institutes' research directions will be discussed below.

The chapter presents information about the scientific institutes of the National Academy of Sciences of Ukraine doing research in the field of mechanics, their main areas of research, and scientific and technical cooperation with industrial enterprises of Ukraine.

1.2 Department of Mechanics of the NAS of Ukraine

In February 1983, within the structure of the Academy, the Department of Mechanics was established, uniting five research institutes, five special design and technology bureaus, and pilot manufacturing. The Department of Mechanics organized and coordinated the research in the field of mechanical system analysis (statics, dynamics, and stability of mechanical systems with lumped and distributed parameters), material and structural strength, hydrodynamics, hydromechanics, hydraulic engineering, mining engineering (mechanical design engineering for mining in complex geological conditions). Academicians A.N. Guz (1983–1988), V.T. Troshchenko (1988–1993), V.V. Pilipenko (1993–2004), and A.F. Bulat (from 2004 to the present) worked as Academicians-secretaries of the Department of Mechanics. Some research in the field of mathematical problems of mechanics, mechanics of materials, dynamics, and structural strength is also carried out at the institutes of other departments of the Academy, such as the Institute of Mathematics (Kyiv), the Ya.S. Pidstrigach Institute of Applied Problems of Mechanics and Mathematics (Lviv), the Institute of Applied Mathematics and Mechanics (Cherkasy), the V.M. Glushkov Institute of Cybernetics (Kyiv), the E.O. Paton Electric Welding Institute (Kyiv), the G.V. Karpenko Physico-Mechanical Institute (Lviv), and the A.M. Pidgorny Institute of Mechanical Engineering Problems (Kharkiv).

The institutes of the Department of Mechanics of the National Academy of Sciences of Ukraine carry out research in the following scientific areas (Bulat 2013):

- Mechanics of composite and inhomogeneous media;
- Mechanics of fatigue and fracture mechanics;
- Dynamics and stability of mechanical systems;
- Dynamics and aerothermodynamics of mechanical and hydromechanical systems, power plants, aircraft, space vehicles, and their subsystems;
- Assessment of the limit state and formulation of criteria for the strength of materials and structures;
- Reliability and optimization of mechanical systems, survivability of structures;
- Development of the scientific foundations of mining and technical processes, equipment, and technologies for the extraction and processing of minerals.

Institutes of the Department of Mechanics of the National Academy of Sciences of Ukraine develop scientific cooperation with the universities of Ukraine, which conduct research and train personnel in mechanics. These are the Taras Shevchenko National University of Kyiv, the Ivan Franko National University of Lviv, the Oles' Honchar Dnipro National University, the National Technical University of Ukraine "Igor Sikorsky Kyiv Polytechnic Institute," the National Technical University "Kharkiv Polytechnic Institute," the National Technical University "Dnipro Polytechnic," the Lviv Polytechnic National University and many others. The graduates of these universities usually join the scientific staff of the institutes of the Academy's Department of Mechanics.

The results of basic and applied scientific research in the institutes of the Department of Mechanics have found application in many sectors of the national economy, such as mechanical engineering (energy, chemical, transport, agricultural), metallurgical and mining industries, nuclear power, oil refining, production of space hardware, manufacturing of marine and aviation equipment. Institutes belonging to the Department of Mechanics have constant scientific and technical contacts with large industrial and design enterprises of Ukraine, such as the *Antonov* State Enterprise (aircraft industry), the O.G. Ivchenko Zaporizhzhia Machine-Building Design Bureau *Progress* (manufacturing of gas turbine engines for aviation and industrial applications), the M.K. Yangel Design Bureau *Pivdenne* (rocket engineering), the Ukrainian Energy Machines Joint-Stock Company *Turboatom* (creation of steam and hydraulic turbines), the National Nuclear Energy Generating Company *Energoatom* (nuclear power reactors). The institutes also sustained contact with enterprises in the mining and processing industry.

Prominent mechanical scientists worked at the Academy. Among them are O.K. Antonov, N.G. Bondar, V.S. Budnik, Ya.M. Grigorenko, A.N. Dynnik, O.G. Ivchenko, A.Yu. Ishlinsky, N.A. Kilchevsky, A.D. Kovalenko, V.O. Kononenko, S.M. Konyukhov, N.V. Kornoukhov, A.S. Kosmodamiansky, M.A. Lavrent'yev, V.A. Lazaryan, G.V. Logvinovich, V.O. Lotarev, V.I. Mossakovsky, V.V. Pilipenko, G.S. Pisarenko, M.S. Polyakov, V.M. Poturaev, G.N. Savin, S.V. Serensen, V.T. Troshchenko, V.F. Utkin, Yu.N. Shevchenko, M.K. Yangel. Currently, 12 academicians work in the Department: V.L. Bogdanov, A.F. Bulat, V.T. Grinchenko, A.N. Guz, D.S. Kiva, V.D. Kubenko, V.V. Matveev, A.A. Martinyuk, V.M. Nazarenko, O.V. Pilipenko, V.V. Kharchenko, and V.P. Shevchenko. The Department is represented by 18 corresponding members of the NAS of Ukraine: A.P. Alpatov, B.O. Bluss, M.I. Bobyr, A.O. Borisyuk, M.D. Borisyuk, G.O. Voropaev, O.Ya. Grigorenko, Ya.O. Zhuk, O.P. Korostilyov, I.F. Kravchenko, O.P. Krukovskyi, E.I. Nikiforovich, O.Ya. Oliynik, J.J. Rushchitsky, Yu.M. Savchenko, V.I. Timoshenko, I.S. Chernyshenko, and O.V. Shymanovs'kyj. The Academy's Department of Mechanics also has 5 foreign members: Holm Altenbach (Germany), Satya N. Atluri (USA), Herbert Mang (Austria), Stefan Markus (Slovakia), and Michael Yarymovych (USA). In total, 614 researchers work at the institutes of the Department of Mechanics of the National Academy of Sciences of Ukraine, including 121 Doctors of Science (DSc) and 269 Candidates of Science (PhD). Below we present more detailed information on the scientific institutions belonging to the Department of Mechanics.

1.3 Institutes of the Department of Mechanics of NAS of Ukraine

1.3.1 S.P. Timoshenko Institute of Mechanics of the NAS of Ukraine

The S.P. Timoshenko Institute of Mechanics is the oldest institute of the Department of Mechanics of the National Academy of Sciences of Ukraine and has more than 100 years of history. Its goal was to solve fundamental problems of mechanics, train highly qualified scientists and researchers, and generalize scientific research results with the following implementation in engineering practice. Detailed information about the history and modern research of the Institute is presented in the works Guz et al. (1989), Guz (1998, 2008, 2018a, b).

The Institute was established in 1918 as the Institute of Technical Mechanics, being part of the Ukrainian Academy of Sciences and the country's first technical institute. In 1929 it was transformed into the Institute of Structural Mechanics. In 1959, the Institute of Structural Mechanics was renamed to the Institute of Mechanics of the Academy of Sciences of the Ukrainian SSR, and in 1993 it was named after the founder and the first director of the Institute, Stepan Prokopovich Timoshenko.

The Institute was headed by well-known scientists in mechanics academicians S.P. Timoshenko (1918–1920), D.A. Grave (1921), K.K. Siminsky (1921–1932), S.V. Serensen (1932–1940), N.V. Kornoukhov (1940–1944), F.P. Beliankin (1944–1958), G.N. Savin (1958–1959), A.D. Kovalenko (1959–1965), V.O. Kononenko (1965-1975), and A.N. Guz (1976–2022). Since 2022, the institute has been headed by Academician V.M. Nazarenko.

Well-known Institute's scientists, who were members of the Academy in various periods, are N.N. Bogolyubov, N.M. Krylov, A.N. Dynnik, E.O. Paton, Yu.A. Mitropolsky, G.S. Pisarenko, G.V. Karpenko, N.N. Davidenkov, K.K. Khrenov, G.Yo. Sukhomel, R.F. Ganiev, B.N. Gorbunov, B.D. Grozin, A.M. Pen'kov, O.G. Ivakhnenko, S.N. Kozhevnikov, V.M. Maizel', A.F. Ulitko, I.Ya. Shtajerman, Ya.M. Grigorenko, L.P. Khoroshun, Yu.N. Shevchenko, Yu.N. Nemish, and N.A. Shul'ga (Guz et al. 1989; Guz 1998, 2018a, b).

In the first decades after the establishment of the Institute, scientific research was carried out in the following areas: development of standards for structural materials, determination of the physical and mechanical properties of rocks, elaboration of granite strength models, development of non-destructive methods for assessing the mechanical properties of wood, assessment the strength of bridge elements.

In the 1930s, the study of the strength of structural steels and machine parts, depending on mechanical factors and processing conditions, progressed. The laws of similarity under fatigue were established, and the role of the three-dimensional stress state was estimated; a statistical theory of fatigue failure was proposed, and a three-axis testing machine was created. Based on the study of the ultimate bearing capacity of steel and wooden structures, a new limiting state method for structural analysis was developed.

During the Second World War, the Institute was evacuated to Ufa (Bashkiria, Russian Federation), where its research was concentrated on the dynamic strength of military structure elements, aviation, and transport engineering.

After returning to Kyiv in 1944–1958, a study was made of the stability of engineering structures within and beyond the limits of elasticity, and methods for determining the strength, stability, and deformability were developed. In the field of strength of engineering structures, the theory and methods for structural analysis of torsion plates and rotational shells were developed, the energy dissipation during vibrations in the material of turbine blades was studied, a new phenomenon of adsorption fatigue of metals was revealed, a method of mechanical testing of hardened and brittle estimates of the wear resistance of steels was proposed according to a complex of physical and mechanical characteristics of the surface layer.

Since the late 1950s, the Institute, along with scientific developments in traditional areas in the theory of elasticity and thermoelasticity, shell theory, stress concentration, and investigation of fatigue, strength, and plasticity of materials and structural elements, has received significant development in new scientific areas: deterministic and stochastic mechanics of composite materials structures, three-dimensional theory of stability and wave dynamics of deformed bodies, theory of coupled mechanical and physical fields in structural elements, thermoplasticity, numerical methods of shell theory, analytical mechanics of polyaggregate systems, nonlinear theory of spatial vibrations of solids.

Among the major results of scientific research of the Institute during this period are the establishment of a three-dimensional theory of the stability of deformed bodies and the study on its basis of the stability of structural members made of composite materials and mine workings, the development of the theory of propagation of elastic waves in bodies with initial (residual) stresses and the elaborating using this method the methodology of determining biaxial and triaxial stresses in non-destructive ultrasonic studies, the development of analytical and numerical methods for non-stationary hydroelasticity of shells, the theory of diffraction of elastic waves for the case of multiply connected bodies, methods for predicting the physical and mechanical properties of composites of different structures and the theory of multiply connected media, the method of matrix-valued Lyapunov functions in the theory of stability of motion of mechanical systems (Guz et al. 1989; Guz 1998, 2008).

The Institute of Mechanics was the creator of new scientific institutions of the Academy. So, in 1945, based on the Laboratory of Agricultural Mechanics of the Institute of Structural Mechanics, the Laboratory of Mechanical Engineering and Problems of Agricultural Mechanics of the Academy of Sciences of the Ukrainian SSR was founded as an independent research institution, transformed in 1950 into the Institute of Mechanical Engineering and Problems of Agricultural Mechanics of the Academy of Sciences of the Ukrainian SSR (since 1996—Physical-technological Institute of Metals and Alloys of the National Academy of Sciences of Ukraine). In 1964, based on the Laboratory of Hydraulic Machines, the Kharkiv Branch of the Institute of Mechanics of the Academy of Sciences of the Ukrainian SSR was established, the main scientific directions of which were a comprehensive study of processes in thermal and hydraulic machines. In 1970, the Kharkiv branch of the Institute of Mechanics of the Academy of Sciences of the Ukrainian SSR was transformed into the Kharkiv branch of the Institute of Technical Thermophysics of the Academy of Sciences of the Ukrainian SSR, and then into the A.M. Pidgorny Institute of Mechanical Engineering Problems NAS of Ukraine. Based on the Dnipropetrovsk branch of the Institute established in 1968, the Institute of Technical Mechanics was established in 1980.

During the years of Ukraine's independence (since 1991), scientists of the S.P. Timoshenko Institute of Mechanics of the NAS of Ukraine received recognized results in the field of mechanics of composite materials, including the mechanics of nanocomposites, non-classical problems of fracture mechanics (in particular, fracture in composites under compression along reinforcing elements, brittle fracture of materials with cracks, taking into account the action of initial (residual) stresses along cracks, fracture under compression along parallel cracks, destruction of bodies with cracks under dynamic loading, taking into account the interaction of crack edges, destruction of thin-walled bodies with cracks in tension in the case of preliminary buckling, long-term destruction of viscoelastic materials), mechanics of coupled fields in materials and structural members, methods for analyzing the dynamics and stability of the functioning of complex nonlinear systems (Guz et al. 1989; Guz 1998, 2008, 2018a, b; Paton 2018; Zagorondy 2021). Institute scientists have developed a theory of anisotropic layered shells of revolution of arbitrary shape with layers of variable stiffness and developed methods for numerically solving a wide class of problems about their stress state under nonaxisymmetric loading, a theory of thermo-viscoplasticity of materials in complex loading processes at elevated temperatures, a theory of shock interaction of solid and deformable bodies with liquid and elastic medium, the theory of deformation and damage of homogeneous and composite materials of different structure. They developed new analytical and numerical methods for studying stationary and non-stationary problems of thermoelasticity, solving contact problems for elastic bodies with initial stresses, and calculating shells with holes under physically and geometrically nonlinear deformations. Also developed are the dynamics of solids, liquid drops, and elastic bodies in a compressible viscous liquid, the theory of waves in mixtures, and the model of short fibers in the three-dimensional theory of the stability of composite materials.

To date, the main scientific areas of research of the S.P. Timoshenko Institute of Mechanics of the NAS of Ukraine are mechanics of composite and inhomogeneous media; mechanics of shell systems; mechanics of coupled fields in materials and structural members; fracture mechanics and fatigue; dynamics and stability of mechanical systems. The structure of the Institute includes 11 scientific departments, in which 91 researchers are working, including 35 Doctors of Science (DSc) and 41 Candidates of Science (PhD).

The *Department of Dynamics and Stability of Continua* studies the problems of three-dimensional stability of deformable bodies, the theory of wave propagation and diffraction in deformable bodies, mechanics of composite materials and structural members, aerohydroelasticity, mechanics of a compressible viscous fluid, rock mechanics; investigates non-classical failure and fracture problems, contact problems for elastic bodies with initial stresses; analyzes the stress concentration around holes in shells; develops analytical and numerical methods for solving nonlinear problems in the theory of composite shells, and for calculating non-circular cylindrical shells.

The *Department of Rheology* performs theoretical and computer analysis of nonlinear waves with different initial profiles that propagate in materials; develops fundamentals of nanomechanics of materials and mechanical models of materials of different structural levels (macro-, micro-, nano-levels); elaborates rheological models of soil layers concerning the analysis of seismic waves; develops nonlinear approaches to study the deformation of new auxetic materials.

The *Department of Computational Mechanics and Techniques* is engaged in the development of methods for studying the stress state and non-classical fracture processes of structural elements made of highly elastic and composite materials, edge effects in composites; conducts research on stochastic composites of various structures with physically nonlinear components; studies the problems of stability, nonlinear deformation and supercritical behavior of anisotropic shells, taking into account the initial geometric and structural imperfections.

The main research areas of the *Department of Computational Methods* are the development of the theory of thin and non-thin layered shells assembled from anisotropic layers of variable thickness; the development of discrete-continuum numerical methods for solving problems of the stress state of systems consisting of shells of rotation of an arbitrary shape, under the action of distributed and local loads, the dynamics of layered non-circular cylindrical shells; the study of stationary and non-stationary dynamic processes in piezoceramic inhomogeneous bodies of spherical and cylindrical shape; the development of application packages for the study of shells using the developed methods; the study of biomechanical processes in the field of surgical, orthopedic and orthodontic dentistry.

The *Department of Structural Mechanics of Thin-Walled Structures* investigates natural and forced vibrations of layered shells of revolution, taking into account design features (the presence of added masses, discreteness and inhomogeneity of the filler, asymmetry of layering); studies transient processes in trash-layered shells of revolution under combined non-stationary loading; develops problems of the dynamics of trash-layered asymmetric shells of revolution with a discrete-symmetric filler under explosive loading.

The *Department of Thermoplasticity* investigates the problems of thermoviscoplasticity of isotropic and anisotropic materials under deformation along arbitrary trajectories and taking into account the type of stress state under variable loading processes; carries out modeling of deformation processes in structural members in the form of shells and spatial bodies under thermal force loading; develops methods for calculating the stress-strain state of shell structural members under variable thermoradiation loading and methods for solving problems of the theory of thermoplasticity for three-dimensional bodies, plates, and shells made of materials with shape memory; formulates multi-parameter equations of state that describe the processes of elastic-plastic deformation of materials, taking into account the type of stressed state of the material.

The *Department of Thermoelasticity* conducts research on the problems of thermomechanics of inelastic materials and structural members, taking into account the interaction of mechanical, thermal, and electromagnetic fields; develops numerical-analytical methods for studying the durability and electrothermomechanical behavior of inelastic spatial and thin-walled structural members with piezoelectric inclusions, taking into account dissipative heating; elaborates models and methods for studying active damping of forced harmonic oscillations of current-wall structural members using piezoelectric inclusions; creates thermomechanical models of modern additive technologies.

The *Department of Creep Mechanics* develops a nonlinear three-dimensional theory of hereditary viscoelasticity, including the construction of a system of constitutive equations, the substantiation of the structure and the identification of heredity nuclei, the solution of problems of creep and stress relaxation in structural members made of polymer and composite materials; carries out construction of models and criteria for the long-term destruction of metal, polymer, and composite materials due to creep and high-cycle fatigue under conditions of a complex stress state; formulates the main hypotheses of the nonlinear mechanics of continuum damage and models of damage accumulation due to creep and high-cycle fatigue.

The *Department of Fracture Mechanics of Materials* deals with the modeling of linear viscoelastic composite materials; develops models of the cohesive zone; explores the problems of long-term fracture of anisotropic viscoelastic solids and contact fracture mechanics; studies the processes of destruction of nonlinear elastic materials.

The *Processes Stability Department* develops methods for analyzing the technical (practical) stability of movement; integral methods for analyzing the stability of the motion of systems with a small parameter; generalization of the direct method for studying the stability of motion based on a new class of auxiliary functions, matrix-valued Lyapunov functions; construction of the theory of stability of dynamic equations on time scales; developing of new methods for analyzing the stability and boundedness of the motion of polynomial systems and systems with a fractional-like derivative of the state vector.

The *Department of the Theory of Vibrations* deals with the dynamics of the interaction of structural members with the environment; optimal motion control of spacecraft with electric and combined propulsion systems; contact interaction of piezoelectric bodies under force and temperature loading.

Theoretical and experimental results obtained at the S.P. Timoshenko Institute of Mechanics of the NAS Ukraine for many years have found application in rocket and space, aviation, shipbuilding, and other industries. The Institute's developments are used in the engineering practice of leading research and design organizations and enterprises in Ukraine and other countries to assess the strength, reliability, and durability of materials and standard structures.

The staff of the Institute published about 500 scientific monographs (see more information in Guz and Rushchitsky (2018), Guz and Rushchitsky (1998), Guz (2008)), including 8 generalizing multi-volume monographs (Guz 2011, 1983, 2003, 1989, 1982, 2018, 1993, 1986). The Institute publishes the international scientific journal *Applied Mechanics* (founded in 1955). Since 1965, this journal has been translated into English in the USA (New York) as *Soviet Applied Mechanics*, first by Faraday Press, and since 1968 by Plenum Publishing Corporation. Since 1992 *Applied Mechanics* has been translated into English and published as *International Applied Mechanics* (first by Kluwer Academic Publishers and since 2004 by Springer).

S.P. Timoshenko Institute of Mechanics of the NAS of Ukraine develops scientific cooperation with foreign scientific centers and universities. Thus, in recent years, joint research has been carried out, and monographs and articles have been published with scientists from the University of Bologna (Italy), the University of Aberdeen (Scotland), Heriot-Watt University (Scotland), Otto-von-Guericke-Universitat Magdeburg (Germany), Sofia University 'St. Kliment Ohridski' (Bulgaria), Technische Universitat Berlin (Germany), Harbin Institute of Technology (China), Institute of Fundamental Technological Research of Polish Academy of Sciences (Poland).

1.3.2 Institute of Hydromechanics of the NAS of Ukraine

The history of the Institute of Hydromechanics of the National Academy of Sciences of Ukraine began in 1926 (Bulat 2013; Grinchenko 2002; Paton 2018) when the Research Institute of Water Management of Ukraine was organized based on the Department of Hydrogeology of the Kyiv Polytechnic Institute. In 1936, this institute was transferred to the Academy of Sciences of the Ukrainian SSR, and in 1938, it was reorganized into the Institute of Hydrology of the Academy of Sciences of the Ukrainian SSR.

During World War II, the Institute was evacuated to Ufa (Bashkiria, Russian Federation) as a department of hydraulic structures of the Institute of Structural Mechanics of the Academy of Sciences of the Ukrainian SSR. In 1944, the Institute resumed its activities in Kyiv under the name of the Institute of Hydrology and Hydraulic Engineering of the Academy of Sciences of the Ukrainian SSR, which in 1964 was reorganized into the Institute of Hydromechanics of the Academy of Sciences of the Ukrainian SSR.

At the time of its creation, the activities of the Institute were aimed at solving important applied problems of hydraulic engineering construction in Ukraine. Employees of the Institute took part in substantiating water management issues of such important projects as the construction of the Dnipro Hydroelectric Station and the creation of a complex of industrial facilities and transport systems in the Donbas and the Dnipro region. The hydrological communication service established at the Institute was the first scientific center in the Soviet Union, where methods of long- and short-term hydrological forecasts were developed for the conditions of the Dnipro River.

During the years of its existence, the Institute was headed by Academician E.V. Oppokov (1926–1928), Professor A.V. Ogievsky (1928–1940), Academician

G.I. Sukhomel (1940–1958), PhD M.M. Didkovsky (1958–1966), Academician G.V. Logvinovich (1966–1972), Corresponding Member O.Ya. Oliynik (1972–1981), Corresponding Member O.D. Fedorovsky (1981–1987), Academician V.T. Grinchenko (1987–2021). Since 2021, the Institute has been headed by the Corresponding Member of the NAS of Ukraine G.O. Voropaev.

At present, the research activity of the Institute of Hydromechanics of the NAS of Ukraine is aimed at developing a wide range of scientific areas of modern hydraulic engineering and hydromechanics. The Institute is a leading center in the field of fluid mechanics in Ukraine. His research is focused on the development of new methods of analysis and control of hydrodynamic processes to create new technologies in ecology, medicine, and mining; improvement and creation of new theories and methods for calculating hydromechanical processes and developing recommendations based on them to improve the efficiency of using new models of water transport; development of scientific foundations for the prevention of environmental disasters associated with the displacement of slopes and erosion of the shores of water basins and the effective use of the shelf zone.

The Institute of Hydromechanics of the NAS of Ukraine employs 103 researchers, including 22 Doctors of Science (DSc) and 51 Candidates of Science (PhD). Scientific schools have been created here in the following areas: wave and vortex hydrodynamics, covering research in the field of acoustics of liquid and gas flows, medical acoustics, hydrodynamics of moving objects, the interaction of waves and flow with engineering structures; high-speed hydromechanics, covering research in the field of high-speed hydromechanics, supercavitation, reduction of resistance to movement of bodies in the water, hydrodynamic technologies; hydrothermodynamics and heat and mass transfer in gas-liquid and cellular media and systems "liquid-solid particles", covering research in the field of physicochemical hydrodynamics, energy efficiency, hydro- and pneumatic transport, hydraulic engineering, and land reclamation.

A characteristic feature of the Institute's activities is the combination of theoretical and experimental research using the existing experimental base, which allows for solving fundamental problems of hydromechanics and hydraulic engineering. The experimental complex for hydrodynamic studies includes a multi-purpose high-speed hydrodynamic tube, an experimental pool, and a high-speed hydrodynamic basin. The high-speed multi-purpose hydrodynamic tube is designed to study the hydrodynamic characteristics of bodies by the reverse flow method. The experimental capabilities of the hydrodynamic tube cover the range of velocity values from close to zero (creeping modes of motion) to the speed of sound propagation in water—1450 m/s. The experimental pool was created as a specialized laboratory for the experimental study of the hydrodynamic characteristics of ships and their propulsion and stern systems. The high-speed hydrodynamic basin was designed as a specialized laboratory for experimental research in the field of aero- and hydrodynamics of high-speed vessels with dynamic support principles (gliders, hydrofoils, ground-effect vehicles, etc.).

The traditional topics of the Institute of Hydromechanics research include the study of a wide sill spillway, unidentified currents and channel processes in rivers and behind hydraulic structures, turbulent flow characteristics in the near-bottom area and near the free boundary, water resistance in earthen channels, as well as flow dynamics at the turn of an open channel.

In a new scientific area—the hydraulics of stratified flows, a new method has been developed for calculating heat transfer to the atmosphere from the surface of cooling reservoirs, which makes it possible to assess both the cooling properties of the water area and the pollution of the atmosphere by thermal emissions.

The study of the patterns of flow around moving objects is one of the main scientific areas of research of the Institute. The results of studying the static stability of ships, the hydrodynamic theory of ship motion in canals and rivers under supercritical conditions, and the hydrofoil theory are obtained. Under the leadership of Academician G.V. Logvinovich, a new scientific direction was formed—hydro-jet propulsion, boundary layer control, and hydrobionics, the theory of supercavitation, high-speed movement of objects near the boundaries. The systematic study of cavitation flows laid the foundation for the theory of the spatial movement of free boundaries under the influence of perturbation factors, the development of new methods for modeling and optimizing cavitation flows, methods for organizing cavitation movements of objects in water at high speed and achieving the speed of sound by objects in motion.

One of the topical scientific areas of modern mechanics is the study of the laws governing the interaction of a fluid with elastic structures. The processes of energy exchange between flows and waves in liquids and deforming bodies are very difficult for mathematical modeling, and assessments of quantitative characteristics and understanding of interaction processes are determining factors in the course of creating efficient structures. Seismic waves are one of such characteristic types of impact on structures. The methods developed at the Institute for assessing the consequences of seismic disturbances on hydraulic structures were widely used in the construction of hydroelectric power stations.

The Institute of Hydromechanics of the NAS of Ukraine researches the patterns of blood flow in the vessels and researches breathing noises. Physical and mathematical models of processes, methods, and means of experimental measurement of breathing noise characteristics have been developed. Based on research on the generation and propagation of sound in complex inhomogeneous media, an environmentally friendly medical complex has been developed, created, and certified for the diagnosis and monitoring of human broncho-pulmonary diseases, which is used in medical institutions.

1.3.3 Institute of Technical Mechanics of the NAS of Ukraine and State Space Agency of Ukraine

The Institute of Technical Mechanics of the National Academy of Sciences of Ukraine studies scientific and technical problems associated with the creation of launch vehicles and spacecraft (Bulat 2013; Paton 2018). Its history began with a small scientific subdivision—the sector of problems of technical mechanics, organized in 1966 on the initiative of Academician M.K. Yangel, the Chief Designer of the Separate Design Office #586 (now the M.K. Yangel State Enterprise Design Bureau *Pivdenne*), one of the founders of rocket technology. In 1968, this sector was transformed into the Dnipropetrovsk branch of the Institute of Mechanics of the Academy of Sciences of the Ukrainian SSR, headed by Academician V.A. Lazaryan. In 1980, based on the Department, the Institute of Technical Mechanics of the Academy of Sciences of the Ukrainian SSR was established, headed by Academician V.V. Pilipenko. In 1995, the functions of the main institute of the rocket and space industry of Ukraine were entrusted to the Institute. Since 2003, the Institute of Technical Mechanics of the NAS of Ukraine has been headed by Academician O.V. Pilipenko.

The main scientific directions of the Institute are the dynamics of mechanical and hydromechanical systems, systems of launch vehicles, railway, and road transport; aerothermodynamics of power plants, aircraft, and spacecraft and their subsystems; strength, reliability, and optimization of mechanical systems, launch vehicles, and spacecraft; mechanics of the interaction of a solid body with an ionized medium and electromagnetic radiation; systematic analysis of trends and prospects for the development of rocket and space technology.

During its activity, the Institute has become a leading research institution in the field of scientific and methodological support for the creation and operation of promising objects of rocket and space technology in Ukraine and has achieved certain success in the field of fundamental and applied research in the development of efficient technologies in energy, industry, railway, and automobile transport.

The scientists of the Institute carried out large-scale research in the field of cavitation self-oscillations in hydraulic systems, developed the scientific and technical foundations of a new original scientific direction in the field of dynamics of liquid rocket propulsion systems and liquid launch vehicles and the theory of low-frequency cavitation self-oscillations in pump power systems of liquid rocket propulsion systems; a linear theory of longitudinal stability of liquid-propellant launch vehicles has been developed, taking into account the phenomena of cavitation in pumps of liquid-propellant rocket engines; a nonlinear theory of longitudinal oscillations of liquid-propellant launch vehicles was created. These results were used to calculate the longitudinal stability of the launch rockets *Zenith*, *Dnepr*, *Cyclone-4*, and *Cyclone-4M*, and the development and experimental testing of a thermohydrodynamic damper on a cryogenic liquid to ensure the longitudinal stability of the *Energia* rocket.

A significant place in the subject of the Institute belongs to the work on the aerogasdynamics of launch vehicles, engines, and spacecraft. Methods for numerical simulation of problems of aerogasdynamics of aircraft at the stage of their entry into the atmosphere and during flight at high supersonic speed were developed.

A complex of scientific equipment for the space experiment *Potential* for the diagnostics and monitoring of the parameters of the polar ionospheric plasma has been developed and manufactured. The effectiveness and information content of the devices created at the Institute was confirmed during operation on board the Ukrainian spacecraft *Sich-2*.

The employees of the Institute have developed mathematical models and effective methods for calculating the main parameters of promising space systems.

New spacecraft control tools, models, and algorithms for controlling space-based solar power plants are proposed; models and methods for choosing the design parameters of space manipulators and hexopods, models of spacecraft that are designed in Ukraine; a theory of mobile control of space vehicles has been created, which makes it possible to increase their reliability and use vehicles in some emergencies.

Currently, 101 researchers work at the Institute of Technical Mechanics of the NAS of Ukraine, including 12 Doctors of Science (DSc) and 40 Candidates of Science (PhD).

The Institute carries out basic and applied research in the field of mechanical engineering, thermal power engineering, railway transport, rolling production, and mining. Thus, the Institute created a fundamentally new air suspension of a car with quasi-zero stiffness and separate damping of low-frequency vibrations of the sprung and high-frequency vibrations of the unsprung mass of the car; a series of works on the use of the cavitation phenomenon to obtain high-frequency high-amplitude oscillations of the liquid, caused by periodic separation, has been completed, a new cavitation-pulse technology for obtaining finely dispersed water-coal has been proposed; scientific and methodological support has been created for performing all the necessary stages of aerodynamic design of compressor lattices and calculating spatial turbulent gas flows in multistage compressors of aircraft gas turbine engines, which is used in the design of gas turbine engines for aircraft and power plants; a comprehensive modernization of freight car bogies is proposed, which allows several times to increase the resource of the elements of the friction damping system of bogies, significantly reduce the wear of the contact surfaces of wheels and rails, reduce energy consumption for traction due to a decrease in the forces of interaction of wheels with rails; designs have been developed and a number of new devices for reducing the sound level of a shot for small arms manufactured by NATO member countries have been developed.

The Institute of Technical Mechanics of the NAS of Ukraine publishes the scientific journal *Technical Mechanics*. The journal publishes scientific articles highlighting new results of theoretical and experimental research in the field of kinematics, dynamics, motion control, motion stability of elastic-viscous systems and systems with liquid masses, hydro- and gas-dynamics, heat and mass destruction, strength and reliability of structures, spacecraft mechanics, mechanical aspects of space research.

1.3.4 G.S. Pisarenko Institute for Problems of Strength of the NAS of Ukraine

The G.S. Pisarenko Institute for Problems of Strength of the National Academy of Sciences of Ukraine was established in 1966 based on the Sector of Strength Studies at the Institute for Problems in Materials Science of the Academy of Sciences of the Ukrainian SSR (Bulat 2013; Paton 2018; Troshchenko et al. 2016). Since 2002, the Institute has been named after its founder and first director, Academician G.S. Pisarenko.

Academician G.S. Pisarenko (1966–1988), Academician V.T. Troshchenko (1988–2011), Academician V.V. Kharchenko (2011–2021), and Corresponding Member A.P. Zinkovsky (2021–2022) were the directors of the Institute. Since 2022, the Institute has been headed by Doctor of Technical Sciences O.Yu. Chirkov.

The name of the founder of the Institute G.S. Pisarenko is associated with outstanding results in the development of the theory of oscillations of dissipative systems in a nonlinear formulation, with the study of the damping properties of mechanical systems and the development of methods for improving the vibration reliability of highly stressed structural members; with the study of various aspects of the strength of materials and structural members, in particular, at high and low temperatures under a complex stress state of impulse loads, under the influence of an aggressive gas environment and neutron irradiation; on the development of criteria for the limiting state and on the justification of the possibility of using new classes of structural materials in technology.

At the time of establishment, the Institute for Problems of Strength of the Academy of Sciences of the Ukrainian SSR was tasked with theoretical and experimental studies aimed at formulating criteria for the strength and bearing capacity of materials and structural members, as well as increasing their strength following the requirements of the latest branches of technology, taking into account structural and technological factors, the type of stress state and real modes of power and thermal overload in a wide range of temperatures.

Today, the activities of the Institute for Problems of Strength of the NAS of Ukraine are aimed at the development of basic and applied research in the field of experimental mechanics of deformable solids and the strength of materials and structural members. The main scientific research directions are the limit state and criteria of strength of materials and structures, calculation and experimental methods for studying the stress-strain state, fracture mechanics and survivability of structures, and oscillations of non-conservative mechanical systems.

Since 1969, the Institute has been publishing the international scientific journal *Problems of Strength*, which, starting from the first issue, is translated into English and published by the Springer publishing house as journal *Strength of Materials*. The journal publishes the results of experimental and theoretical research in the field of strength of materials and structural members and mechanics of deformable solids.

The Institute has designed and produced more than 150 original test facilities and benches to study the basic mechanical characteristics of a wide range of metallic structural materials and alloys, non-metallic composite materials in a wide temperature range, and types of mechanical stress, taking into account the impact of aggressive media, and radiation.

The Institute traditionally performs research aimed at solving the problems of ensuring the strength and durability of structural members for various purposes: nuclear and thermal power plants, main oil and gas pipelines, oil refining and chemical industries, gas turbines, aircraft structures, and railway transport. Thus, fundamental and applied results were obtained at the Institute, these which were introduced into the practice of calculations to justify the over-design life of power units of nuclear power plants in Ukraine.

A methodology and a computer expert system have been created for the current monitoring of the technical condition and ensuring the integrity of the main pipelines to increase their service life and optimize the frequency and volume of diagnostic and repair work. The system was implemented at the Ukrainian State Enterprise *Ukrhimtransammiak* and the State Enterprise *Ukrtransnafta*. A method for rapid assessment of the dynamic stability of rotor blades of axial compressors for aircraft gas turbine engines is proposed, the effectiveness of which is confirmed by the results of predicting the dynamic stability of compressor blades of impellers for engines produced by the O.G. Ivchenko Zaporizhzhia Machine-Building Design Bureau *Progress*.

For now, the staff of the G.S. Pisarenko Institute for Problems of Strength of the NAS of Ukraine is 97 researchers, including 19 Doctors of Science (DSc) and 54 Candidates of Science (PhD).

1.3.5 M.S. Polyakov Institute of Geotechnical Mechanics of the NAS of Ukraine

The M.S. Polyakov Institute of Geotechnical Mechanics of the National Academy of Sciences of Ukraine was established in 1967 based on the Dnipropetrovsk branch of the Institute of Mechanics of the Academy of Sciences of the Ukrainian SSR (this branch, in turn, was formed in 1964 by transforming the Department of Mining Problems of the Institute of Electrical Engineering of the Academy of Sciences of the Ukrainian SSR, which dealt with the extraction of mineral raw materials, organization of underground transport and lifting raw materials from a considerable depth) (Bulat 2013, 2017; Paton 2018).

When establishing the Institute, the following main goals of its activity were defined: study the problems of physics and mechanics of rocks; development of new effective methods of destruction of rocks; elaboration of control methods for aerogasdynamics processes; investigation of the physical foundations and scientific substantiation of progressive methods for the mineral resources mining.

The founder and first director of the Institute in 1967–1975 was Academician M.S. Polyakov. During this period, the scientists of the Institute carried out fundamental research in the field of petrophysics and mechanics of rocks and massifs, new methods of their destruction, the theory of mining machines and working processes, mine aerogasdynamics, dynamic manifestations of rock and gas pressure.

From 1975 to 1992, the Institute worked under the guidance of Academician V.M. Poturaev. During this period, the studies started earlier were continued, and new scientific directions were formed to study the destruction of rocks, the normalization of thermal conditions during underground coal mining, and the development and implementation of vibrotransport machines and units for the mining industry. Research on the mechanics of deformation and destruction of rubber elements of machines under cyclic loads in the atmosphere at various temperatures, as well as in aggressive environments, has been widely developed. Scientific foundations and methods for controlling the geomechanical and gas-dynamic processes of a rock mass, which is shifting as a result of underground mining of coal seams, have been developed. The geomechanical foundations of rock pressure dynamics and non-traditional hydrodynamic methods for preventing coal and gas outbursts have been developed.

For the mining industry, the scientific foundations of open pit mining methods and technologies have been developed using controlled caving, displacement of soft rocks in technological processes, as well as the mechanics of explosive and gravitational destruction and movement of hard rocks. The fundamentals of the mechanics of vibration transportation and pipeline vibration-pneumatic transport of bulk materials have been developed, and several vibration-pneumatic machines have been designed on their basis. Research has been carried out on the creation of technical means and technologies for the extraction of solid minerals from the bottom of the World Ocean based on the use of vibration effects.

Since 1992, the Institute has been headed by Academician A.F. Bulat. During this period, the Institute developed the scientific foundations for the technology of active control of the state of the rock mass, in which an analytical description of the stress-strain state of the rock mass near the stope was given, taking into account free gas in the reservoir, geological disturbances, and boundary deformation of rocks. With the use of these results, technologies have been developed for clearing operations and for carrying out horizontal mine workings in mines with steep and sloping coal seams. The mechanisms of unloading and degassing of the reservoir were studied, the parameters of the effective impact of downhole vibration sources on the reservoir were determined, and downhole pneumatic vibrators were created.

The Institute has developed the concept of complex degassing and utilization of mine gas methane, which is based on the organizational division in time and underground space of the processes of extraction of two energy carriers—coal and methane. The program of work, compiled based on this concept, for the first time in Ukraine, was fully implemented at the O.F. Zasyad'ko coal mine.

Today, the Institute's scientists study the properties of rocks and massifs, the processes of their destruction and the control of the stress-strain state; investigate scientific foundations of mining processes, equipment, and technology for the extraction and processing of minerals; analyze physical, technical and geological foundations of coal mine methane production technologies; examine energy saving and reliability of mining operations.

The M.S. Polyakov Institute of Geotechnical Mechanics of the NAS of Ukraine employs 169 researchers, including 29 Doctors of Science (DSc) and 66 Candidates of Science (PhD).

1.3.6 Institute of Transport Systems and Technologies of the NAS of Ukraine

The Institute of Transport Systems and Technologies of the National Academy of Sciences of Ukraine "Transmag" was established in 1995 (Bulat 2013; Paton 2018). Its director is a Doctor of Technical Sciences, Professor V.O. Dzenzersky.

To date, research at the Institute is developing under the main scientific areas of work: physical and technical problems of creating magnetic levitation transport systems and devices, their controls and power supply; problems of mechanics and aerodynamics of vehicles, including those that levitate over profiled supporting surfaces; problems of creation and operation of high-energy on-board power sources for vehicles.

The Institute of Transport Systems and Technologies of the NAS of Ukraine employs 53 researchers, including 4 Doctors of Science (DSc) and 17 Candidates of Science (PhD).

References

Bogdanov VL, Dubrovina LA (2018) The role of academician S.P. Timoshenko in founding the Ukrainian Academy of Sciences and the Institute of Technical Mechanics of the UAS. Herald NAS Ukraine 12:67–78

Bulat AF (2013) Department of Mechanics of the NAS Ukraine. Herald NAS Ukraine 11:21–29

Bulat AF (2017) Several pages of the history and present activity of the M.S. Polyakov Institute of Geotechnical Mechanics of the National Academy of Sciences of Ukraine. Herald NAS Ukraine 6:77–89

Grinchenko VT (ed) (2002) Institute of Hydromechanics of the National Academy of Sciences of Ukraine. Intergraphik, Kyiv

Guz AN, Nemysh YuN, Gumenyuk BP (1989) Institute of Mechanics (1919–1989). Naukova Dumka, Kyiv

Guz AN, Rushchitsky JJ (2018) For the 100-th anniversary of the S.P. Timoshenko Institute of Mechanics of the NASU: books (Monographs and textbooks) published by the Institute. Int Appl Mech 54(2):121–142

Guz AN, Rushchitsky JJ (1998) Main Ukrainian historical and modern sources on mechanics: focus on the S.P. Timoshenko Institute of Mechanics. Appl Mech Rev 40(10):199–208

Guz AN (ed) (2005–2011) Advances in mechanics, in 6 vols, 7 books. A.S.K. (vols 1–3), Litera Ltd, (vols 4–6), Kyiv

Guz AN (ed.) (1982–1983) Mechanics of composite materials and structural members, in 3 vols. Naukova Dumka, Kyiv

Guz AN (ed-in-chief) (1993–2003) Mechanics of composites, in 12 vols Naukova Dumka (vols 1–4), A.S.K. (vols 5–12), Kyiv

Guz AN (ed) (1987–1989) Mechanics of coupled fields in structural members, in 5 vols. Naukova Dumka, Kyiv

Guz AN (ed-in-chief) (1980–1982) Methods for shells analysis, in 5 vols. Naukova Dumka, Kyiv

Guz AN (ed) (2016–2018) Modern problems of mechanics, in 3 vols. Litera Ltd, Kyiv

Guz AN (ed) (1990–1993) Non-classical problems of fracture mechanics, in 4 vols, 5 books. Naukova Dumka, Kyiv

Guz AN (ed) (1998) S.P. Timoshenko Institute of Mechanics (dedicated to the 80-th anniversary). ASK, Kyiv

Guz AN (ed) (2008) S.P. Timoshenko Institute of Mechanics of the NAS of Ukraine (1918–2008). History. Structure. Informational Aspects. Litera, Kyiv

Guz AN (ed) (1984–1986) Spatial problems of the theory of elasticity and plasticity, in 6 vols. Naukova Dumka, Kyiv

Guz AN (2018a) For the 100-th anniversary of the S.P. Timoshenko Institute of Mechanics of the National Academy of Sciences of Ukraine (NASU). Int Appl Mech 54(1):3–33

Guz AN (2018b) To 100-th anniversary of the S.P. Timoshenko Institute of Mechanics of the NASU (National Academy of Sciences of Ukraine). Litera Ltd, Kyiv

Krylov NM, Bogolyubov NN (1947) Introduction to nonlinear mechanics. Princeton University Press, Princeton

Martynyuk AA, Mishchenko EF, Samoilenko AM, Sukhanov AD (2009) Academician Nikolai Nikolaevich Bogolyubov (for the 100-th anniversary of his birth). Int Appl Mech 45(7):693–698

Paton BE (ed) (2018) The National Academy of Sciences of Ukraine (1918–2018). The 100-th anniversary of its foundation. Akademperiodyka, Kyiv

Timoshenko SP (1968) As I Remember. D. van Nostrand Company, Princeton

Troshchenko VT, Matveev VV, Kharchenko VV, Zinkovsky AP (2016) Ensuring the strength of materials and structures is the key to creating highly reliable equipment. To the 50th anniversary of the founding of the G.S. Pysarenko Institute of Strength Problems of the National Academy of Sciences of Ukraine. Herald NAS Ukraine 10:68–73

Zagorondy AG (ed) (2021) National Academy of Sciences of Ukraine in 1991–2021. To the 30-th anniversary of the independence of Ukraine. Akademperiodyka, Kyiv

Chapter 2
Thermoviscoplasticity Equations of Isotropic Material with Stress Mode Dependence

Maya Babeshko, Alexander Galishin, Vitalii Savchenko, and Mykola Tormakhov

2.1 Introduction

In modern technology and industry, many elements of structures work under the action of fields of various nature (mechanical, thermal, radiation) and materials are deformed both within and beyond elasticity limit.

To establish the optimal conditions for the operation and manufacturing of structural members, it is necessary to determine the elastoplastic stress–strain state of the body taking into account the loading history of some of its elements and the factors accompanying the nonisothermal loading. To this aim is intensively developed fracture mechanics investigation, which deals with the final stage of fracture, i.e., crack development in brittle materials. The initial stage of fracture, i.e., damage of materials during deformation, is much less investigated. However, a crack nucleates and grows in a material prepared for fracture during a period when a variety of irreversible physical and mechanical processes occur in it at micro- or macroscales. In this process, the initial structure of the material changes, pores nucleate, grow, and coalesce, and microdefects are formed. Defects weaken the cross section, reduce the effective area over which stresses are distributed, thus contribute to further fracture. Therefore, fracture under combined loading can be considered to occur in several stages. At the first stage, damages that are much smaller than some typical size of the structure are accumulated. This process dominates until microcracks form, which lead to the formation of a main crack, followed by the fracture of the material. Therefore, the development of adequate theory and methods for determining the elastoplastic stress–strain state and damage of solids under nonisothermal loading is an important task of solid mechanics.

M. Babeshko (✉) · A. Galishin · V. Savchenko · M. Tormakhov
S.P. Timoshenko Institute of Mechanics, National Academy of Science of Ukraine, Kyiv, Ukraine
e-mail: bab41@ukr.net

V. Savchenko
e-mail: savchenko_vitalij@ukr.net

A. N. Guz et al. (eds.), *Advances in Mechanics*, Advanced Structured Materials 191,
https://doi.org/10.1007/978-3-031-37313-8_2

Various isolated studies on elastoplastic stress–strain state of solids under non-isothermal loading began to be published in the late 1950–60s. These studies used the model of perfectly plastic body, the theory of small elastoplastic deformations regardless of the loading history, the theory of flow with isotropic hardening, and, later, the theory of flow with anisotropic hardening. However, the validity of the constitutive equations used in these studies was not analyzed.

The thermoviscoplastic state of solids began to be regularly studied in the 1970–80s using experimentally validated constitutive equations, taking into account the loading history, and analyzing the applicability of these equations to specific processes of nonisothermal loading (Grigorenko et al. 2002; Shevchenko 1970; Savchenko 1986a; Shevchenko et al. 1980, 1992; Shevchenko and Savchenko 1987). In fact, these monographs provided the basis for the new thermoviscoplasticity as division of solid mechanics, which studies the occurrence and development of displacements, strains, and stresses nonisothermally loaded beyond on the elastic limit in some regions of the material solids.

The main task of thermoviscoplasticity is the construction of constitutive equations that describe the laws of material behavior beyond the elastic limit, and the development of methods for determining displacements, strains, and stresses that occur in solids subjected to the combined action of external nonuniform heating, surface and volume forces that causes irreversible deformation of some elements of the body.

Generally, the solutions of boundary-value problems of thermoviscoplasticity are strongly dependent on the validity of constitutive equations based on one thermoviscoplasticity theory or another. It is well to bear in mind that modern solid mechanics uses many models of the behavior of solids at various stages of deformation. Hence, there are many constitutive equations describing various deformation processes in solids.

Despite the great variety of constitutive equations, the deformation processes they describe are, in fact, classed as simple or nearly simple deformation or combined deformation along paths of small curvature. The applicability of constitutive equations to the description of the deformation processes in a structural member is not even questioned. The S.P. Timoshenko Institute of Mechanics of the National Academy of Sciences of Ukraine was the first to formulate and resolve this issue using the shape of deformation paths.

Note that all the above thermoviscoplastic equations are based on the assumption that the relationship between the stress and strain intensities is independent of the stress mode, i.e., these equations are the same for tension, compression, and torsion. Actually, the majority of structural steels are moderately sensitive to the stress mode (Ohashi et al. 1975; Shevchenko et al. 2007b). The tensile (compressive) or torsional stress–strain curves at small strains of these materials differ by less than 10%. However, the difference of the tensile (compressive) and torsional stress–strain curves for D16T aluminum alloy reaches 40% (Bondar' 2004), i.e., its mechanical characteristics depend on the third deviatoric stress invariant (stress mode). The materials sensitive to the stress mode include various grades of cast iron whose tensile, torsional, and compressive stress–strain curves differ substantially.

Stress mode is characterized by angle that indicates orientation of shear stress in the octahedral plane related to the principal stresses projection onto this plane (Kachanov 2004; Shevchenko and Tormakhov 2010). This angle has simple relationship with the Lode parameter (Lode 1926) and expressed in terms of the second and third deviatoric stress invariants. Rather Lode parameter is expressed in terms of the principal stresses. Commonly the stress state of structural elements is analyzed using constitutive equations that disregard the stress mode. For the first time, the stress mode was considered for the plasticity problems in Shevchenko et al. (2006, 2007a, b), Shevchenko and Tormakhov (2012a) where constitutive equations for simple deformation (Shevchenko and Tormakhov 2012a) and deformation along paths of small curvature (Shevchenko et al. 2006) were derived. These equations relate the components of the engineering stress and strain tensors, assuming that the strains have elastic and inelastic components. The constitutive equations describing deformation along paths of small curvature are based on the assumption that the directional tensors of stresses and inelastic-strain increments coincide, while the constitutive equations describing deformation along straight paths are based on the assumption that the directional stress and strain tensors coincide. The equations include two nonlinear experimentally found functions. One of these functions relates the first invariants of the stress and strain tensors, while the other function relates the second invariants of the respective deviators. These functions are individualized in two series of reference tests on tubular specimens under proportional loading at several constant values of the stress mode angle and several temperatures.

The constitutive equations from Shevchenko et al. (2006), Shevchenko and Tormakhov (2012a), which allow for the stress mode, were experimentally validated in Shevchenko et al. (2006, 2007a, b), Shevchenko and Tormakhov (2010, 2012b) and are widely used to solve boundary-value problems.

With a linear connection between the first invariants of the stress and strain tensors and independence from the stress mode of the connection between the second invariants of the corresponding deviators, the constitutive equations in Shevchenko et al. (2006) and Shevchenko and Tormakhov (2012a) are transformed into traditional relations (Shevchenko et al. 1992; Shevchenko and Savchenko 1987; Shevchenko and Terekhov 1982) of the theory of deformation processes along the trajectories of small curvatures and the theory of simple processes, which are based on the relationship between the Prandtl–Reuss theory (Prandtl 1923; Reuss 1930) and the Hencky theory (Hencky 1924), respectively.

Let us formulate general constitutive equations that can be used to study various processes of nonisothermal loading of various isotropic structural elements with and without regard to the damage to material structure, stress mode, and loading history.

2.2 Thermoviscoplastic Problem Formulation

We will present a general statement of problems of thermoviscoplasticity, which in each individual case, when solving specific problems, is adapted to the geometry of the object under investigation, its material and condition of thermo-force loading.

Consider a compound body made of isotropic inelastic materials. To describe it, we will use an orthogonal coordinate system q_i ($i = 1, 2, 3$). The body is subject to volume and surface forces and nonuniform heating that cause small strains in its elements. Assume that the material characteristics depend on temperature and stress mode. The body is loaded and heated so that its elements undergo simple (or nearly simple) deformation or deformation along paths of small curvature, accompanied by inelastic deformation and unloading. The components of the body are made of dissimilar materials and were joined without tension at initial temperature so that they are in perfect mechanical and thermal contact. The stress–strain state of such solids is analyzed by determining the temperature (non-stationary heat conduction problem) and the displacement, strains, and stresses at fixed time points (thermoviscoplastic problems).

Methods for solving the problem of non-stationary thermal conductivity have been developed both for spatially axisymmetric bodies and for shells of revolution. They are described in detail in the works (Galishyn et al. 2012; Grigorenko et al. 2002; Shevchenko and Savchenko 1987).

In order to solve the problem of thermoviscoplasticity, static equations, the relationship between deformation components and displacements, and constitutive equations are used in the previous case.

2.3 Constitutive Equations

The strain tensor is represented as the sums of elastic, plastic and creep strains tensors. Inelastic changes of first invariant deformation tensor are also assumed. The variation in the elastic strains with stresses follows the generalized Hooke's law.

The impact of damage as a result of material creep on body deformation processes will be modeled using the damage parameter, which characterizes the reduction of the effective volume where the corresponding stress components act.

To account for the deformation history, the whole process of loading and heating is divided into rather small time intervals (steps) so that their endpoints are as close as possible to the onsets of unloading of elements of the body. The history is traced by solving the problem at each step.

The relations between the components of the stress tensor σ_{ij} and the components of the strain tensor ε_{ij} (constitutive equations) have the same form, irrespective of the stage of deformation. Let us represent these relations in the form of the generalized Hooke's law with additional terms:

$$\sigma_{ij} = 2G'\varepsilon_{ij} + \left(K' - 2G'\right)\varepsilon_0\delta_{ij} - \sigma_{ij}^{*}. \tag{2.1}$$

The coefficients of these equations G', K' and additional terms σ_{ij}^{*} depend on the theories of plasticity and the method of linearization.

If we use the equations of thermoviscoplastic deformation theory along paths of small curvature, linearized by the method of additional stresses, $G' = G$, $K' = K = 2G\,(1+\nu)\,/(1-2\nu)$, where G is the shear modulus and ν is Poisson's ratio.

The additional stresses are

$$\sigma_{ij}^{*} = 2G\left[e_{ij}^{(n)} + \frac{1+\nu}{1-2\nu}\left(\varepsilon_T + \varepsilon_0^{(p)} + \varepsilon_0^{(c)}\right)\delta_{ij}\right], \tag{2.2}$$

where $\varepsilon_T = \alpha_T\,(T - T_0)$, α_T is the coefficient of linear thermal expansion; G, ν and α_T depend on temperature T; T_0 is the initial temperature; ${e_{ij}}^{(n)}$ are the components of the unelastic strain deviator;

$$\varepsilon_0^{(p)} = \frac{1}{3}\varepsilon_{ii}^{(p)}, \quad \varepsilon_0^{(c)} = \frac{1}{3}\varepsilon_{ii}^{(c)}$$

are the first invariants of the plasticity strain tensor and creep strain tensor, respectively.

As it was noted above, when performing calculations using (2.1), the loading process must be divided into stages. At the end of the N th stage, the components of the deviator of inelastic deformation are determined as the sum of their increments

$$e_{ij}^{(n)} = \sum_{k=1}^{N} \Delta_k e_{ij}^{(n)}. \tag{2.3}$$

We will use the dependence between the first invariants of the stress $\sigma_0 = \sigma_{ii}/3$ and strain $\varepsilon_0 = \varepsilon_{ii}/3$ tensors to determine $\varepsilon_0^{(p)}$ and $\varepsilon_0^{(c)}$:

$$\sigma_0 = F_1\left(\varepsilon_0^{*}, T, \omega_\sigma\right), \tag{2.4}$$

$$\varepsilon_0^{*} = \varepsilon_0 - \varepsilon_T - \varepsilon_0^{(c)}, \tag{2.5}$$

$$\varepsilon_0^{(c)} = \sum_{k=1}^{N} \Delta_k \varepsilon_0^{(c)}, \tag{2.6}$$

$$\omega_\sigma = \frac{1}{3}\arccos\left[-\frac{3\sqrt{3}}{2}\frac{I_3\,(D_\sigma)}{S^3}\right] \quad (0 \leqslant \omega_\sigma \leqslant \pi/3)\,; \tag{2.7}$$

ω_σ is the angle of the stress mode (Babeshko and Shevchenko 2010; Kachanov 2004), $I_3\,(D_\sigma) = |s_{ij}|$ is third invariant of the stress deviator D_σ, $s_{ij} = \sigma_{ij} - \sigma_0\delta_{ij}$ are the stress deviator components, and S is the intensity of the tangential stresses

$$S = \left(s_{ij}s_{ij}/2\right)^{1/2}. \tag{2.8}$$

In some works (Bondar' 2004; Savchenko 2008), instead of ω_σ, the angle

$$\vartheta = \frac{1}{3} \arccos \left(\frac{3\sqrt{3} I_3 (D_\sigma)}{2S^3} \right)$$

is used.

The increment $\Delta_k e_{ij}^{(n)}$, an arbitrary stage of loading, is determined by the expression

$$\Delta_k e_{ij}^{(n)} = \left\langle \frac{S_{ij}}{S} \right\rangle_k \Delta_k \Gamma^{(n)}, \tag{2.9}$$

where $\Delta_k \Gamma^{(n)}$ is the increment of the intensity of inelastic shear deformations,

$$\Delta_k \Gamma^{(n)} = \Delta_k \Gamma^{(p)} + \Delta_k \Gamma^{(c)}, \tag{2.10}$$

$\Delta_k \Gamma^{(p)}$ and $\Delta_k \Gamma^{(c)}$ are the increments of intensities of instantaneous accumulated shear deformations and creep deformations, respectively. The angle brackets in (2.9) mean the average value for the stage. To determine $\Delta_k \Gamma^{(p)}$, we will assume existence of dependence

$$S = F_2 \left(\Gamma^*, T, \omega_\sigma \right), \tag{2.11}$$

where Γ^* is the intensity of instantaneous shear deformations,

$$\Gamma^* = \frac{S}{2G} + \Gamma^{(p)}, \quad \Gamma^{(p)} = \sum_{k=1}^{N} \Delta_k \Gamma^{(p)}. \tag{2.12}$$

The functions F_1 in (2.4) and F_2 in (2.11) are calculated on the results of the first series of above-mentioned basic experiments on the proportional load of tubular samples, as described in Shevchenko et al. (2006, 2007a, b), Shevchenko and Tormakhov (2010, 2012b). To determine $\Delta_k \varepsilon_0{}^{(c)}$ and $\Delta_k \Gamma^{(c)}$, we use the results of above-mentioned second series of basic creep experiments. The curves obtained in these experiments can be used directly.

Authors of Shevchenko and Tormakhov (2012b) use the following approximating expressions:

$$\dot{\Gamma}^{(c)} (S, T, \omega_\sigma) = \exp \left(c_2 \ln (c_1 S) + c_3 + c_4 T + c_5 \omega_\sigma + c_6 \omega_\sigma^2 \right), \tag{2.13}$$

$$\dot{\varepsilon}_0^{(c)} (\sigma_0, T, \omega_\sigma) = \exp \left(d_2 \ln (d_1 \sigma_0) + d_3 + d_4 T + d_5 \omega_\sigma + d_6 \omega_\sigma^2 \right), \tag{2.14}$$

where c_i and d_i $(i = 1, \ldots, 6)$ are coefficients calculated from condition of best approximation by expressions (2.13) and (2.14) of experimental data. Then, we have

$$\Delta_k \varepsilon_0^{(c)} = \dot{\varepsilon}_0^{(c)} \Delta_k t, \quad \Delta_k \Gamma^{(c)} = \dot{\Gamma}^{(c)} \Delta_k t \tag{2.15}$$

$(\Delta_k t = t_k - t_{k-1}$ is the stage duration).

Note that the given constitutive equations can be used in a simplified version (Babeshko et al. 2010), when instead of (2.4), a linear dependence between the first invariants of the stress and strain tensors is assumed.

If we use the equations of the theory of small elastoplastic deformations allowing for the loading history and linearized by the method of elastic solutions, the coefficients in the constitutive equations (2.1), $G' = G_0$ and $K' = K_0$, the additional terms are as follows:

$$\sigma_{ij}^{*} = 2G_0\omega e_{ij} + 2G^{*}e_{ij}^{1n} + \left[K_0\omega_1\varepsilon_0 + K^{*}\left(\varepsilon_0^{1n} + \varepsilon_T\right)\right]\delta_{ij}, \tag{2.16}$$

where G_0 and K_0 are the shear and bulk modules of the material at temperature T_0,

$$\omega = 1 - \frac{G^{*}}{G_0}, \quad \omega_1 = 1 - \frac{K^{*}}{K_0}, \quad 2G^{*} = \frac{S}{\Gamma}, \quad K^{*} = \frac{\sigma_0}{\varepsilon_0 - \varepsilon_T}, \tag{2.17}$$

e_{ij}^{1n} and ε_0^{1n} are the irreversible deviatoric strains and the mean strain at the time of unloading of a body element, respectively. If the process is loading, they are equal to zero. If the process is unloading, then

$$e_{ij}^{1n} = e_{ij}^{*} - \frac{s_{ij}}{2G^{1}}, \quad \varepsilon_0^{1n} = \varepsilon_0^{1} - \varepsilon_T^{1} - \frac{\sigma_0^{1}}{K^{*}}, \tag{2.18}$$

where the superscript "1" refers to the moment of unloading.

When the equations of the theory of small elastoplastic deformations are linearized by the method of variable parameters, coefficients in the constitutive equations (2.1), $G' = G^{*}$, $K' = K^{*}$ and σ_{ij}^{*} are expressed as

$$\sigma_{ij}^{*} = 2G^{*}e_{ij}^{1n} + K^{*}\left(\varepsilon_0^{1n} + \varepsilon_T\right)\delta_{ij}. \tag{2.19}$$

The nonlinear constitutive equations are linearized by the method of successive approximations using the instantaneous stress–strain surfaces (2.4) and (2.11) and the creep curves (or expresses (2.13) and (2.14)) corresponding to the temperature at the current step and the stress mode angle calculated in the previous approximation:

$$\sigma_0 = F_1\left(\varepsilon_0^{*}, T_m, \omega_{\sigma k-1}\right), \quad S = F_2\left(\Gamma^{*}, T_m, \omega_{\sigma k-1}\right), \tag{2.20}$$

where m is the step number, k is the approximation number. We determine σ_{ij}^{*} in (2.1) using (2.20).

With the use of the constitutive equations (2.1), a decoupling system is formulated for determining the thermoviscoplastic state of spatial bodies and shells. For spatial bodies of revolution, the variational Lagrange equation and the finite element method are used. The problem is reduced to solving in each approximation at each load stage the system algebraic equations, the coefficients, and right-hand parts of which are calculated based on the results of the previous approximation. For shells of revolution when using the hypotheses of Kirchhoff–Love or rectilinear element, the problem is reduced to solving in each approximation at each load stage the systems of ordinary differential equations, the coefficients and free terms of which are calculated based on the results of the previous approximation. With non-axisymmetric loading, the given loads and decoupling functions are presented in the form of trigonometric series along the circular coordinate. Methods of constructing the process of successive approximations are described in Babeshko and Savchenko (1998, 2018); Babeshko and Shevchenko (2010, 2014), Savchenko (2004a, 2006, 2008), Savchenko and Shevchenko (2004b); Shevchenko and Savchenko (2016), Savchenko and Babeshko (2018).

Note that the writing of the constitutive equations in the general form (2.1) facilitates the development of algorithms and appropriate software tools for solving boundary-value problems of thermoviscoplasticity.

The methods developed using (2.1) have found their application in the study of the thermally stressed and deformed state of structural elements in the form of 3-D bodies (Savchenko 2008, 2012) and shells of revolution (Babeshko et al. 2015; Babeshko and Shevchenko 2010, 2014; Babeshko and Savchenko 2018; Galishin and Shevchenko 2010) made of materials whose properties depend on the stress mode.

Elaborating upon the studies (Altenbach 2003; Il'yushin 1967; Kachanov 1974; Khazhinskii 1971; Khoroshun and Shikula 2002a, b; Namestnikov and Shesterikov 1985), a method for analyzing the stress–strain state of compound solids of revolution subject to damage was developed.

At each step of loading, the effect of creep damage of a material on the deformation of the body is modeled using a damage parameter ω^p to characterize the decrease in the effective volume in which the corresponding stresses act, and true stresses rather than engineering stresses are used (Rabotnov and Mileiko 1970):

$$\tilde{\sigma}_{ij} = \frac{\sigma_{ij}}{1-\omega^p}, \tag{2.21}$$

where σ_{ij} are engineering stresses, i.e., loads per undamaged area elements of the body. This damage parameter characterizes the change in the initial structure of the material, the nucleation, growth, and coalesce of microdefects during deformation, which decrease the effective areas over which the stresses are distributed. If we use the equations of theory of small curvature linearized by the method of additional stresses, the relationship between stresses σ_{ij} and strains ε_{ij} (2.1) remains in the form

$$\sigma_{ij} = 2G\varepsilon_{ij} + (K-2G)\,\varepsilon_0\delta_{ij} - \sigma_{ij}^*,$$

where

$$\sigma_{ij}^* = \omega^p\left[2G\varepsilon_{ij} + (K-2G)\,\varepsilon_0\delta_{ij}\right] \\ +2G\left(1-\omega^p\right)\left[e_{ij}^{(n)} + \frac{1+\nu}{1-2\nu}\left(\varepsilon_T + \varepsilon_0^{(p)} + \varepsilon_0^{(c)}\right)\delta_{ij}\right].$$

At the end of the m th step of loading, the nonlinear strains ε_{ij}^n (sum of plastic $\varepsilon_{ij}{}^p$ and creep $\varepsilon_{ij}{}^p$ strains) are defined by (2.9) and (2.10) for $\varepsilon_0^n = 0$ and are calculated from the instantaneous thermomechanical surface (2.11), and creep curves $\varepsilon^c = \varepsilon^c(\sigma, T, t)$ until the fracture of the specimen. The scalar damage parameter ω^p determined from a kinematic equation where the rate of variation in damage is related to some equivalent stress, using stress rupture curves, also obtained in uniaxial-tension tests. The equivalent stress is determined from a failure criterion. It is assumed that the damage parameter is a functional of the loading process. A method for determining it is detailed in Savchenko (2004a), Shevchenko and Mazur (1986). Another, simpler approach is to use the ratio of current creep strain to maximum creep strain as a damage parameter. The onset of fracture of a body at a given point of its element can be estimated by either using a damage parameter or comparing the equivalent stresses to the ultimate strength.

There are other forms of representation of the damage parameter. For example, in Galishin (2008), Galishin and Sklepus (2019), Shevchenko and Galishin (2003); Savchenko (2008), the parameter of damage was introduced to describe the third unstable sections of the creep curves.

The process of successive approximations in solving boundary-value problems of thermoviscoplasticity is detailed in Grigorenko et al. (2002), Savchenko and Shevchenko (2004b), Savchenko et al. (2006), Shevchenko and Savchenko (1987).

The study of the thermally stressed state of the bodies of revolution, that takes into account damage to the material during loading, is presented in Galishin and Sklepus (2019), Savchenko (2003, 2008), Savchenko and Babeshko (2018), Shevchenko and Mazur (1986), and the shells of revolution are studied in Galishin (2008), Galishin and Sklepus (2019), Shevchenko and Galishin (2003).

2.4 Numerical Results

Using methods based on Eq. (2.1), a number of boundary-value problems were solved and the impact of taking into account the dependence of material properties on the stress mode and damage was analyzed. We will present some results of solving specific problems.

Using the method of taking into account material damage, the stress–strain state and strength of the turbine disk of a fighter's jet engine were analyzed considering the system of forces (Savchenko 2014) that act on the disk during the period from takeoff to landing (see Fig. 2.1 for a design model). The disk, which is initially at a temperature of 20 °C, is subjected to the body forces caused by the rotation of the disk and to surface forces uniformly distributed over the disk rim and generated by the repelled blades. The surface forces depend on the speed as follows: $t_{nr} = 174(n/n_{\max})^2$ MPa, where n and $n_{\max}$ are, respectively, the current speed and the maximum speed (13 and 300 rpm). How the speed varies during one cycle from takeoff to landing is shown in Fig. 2.2. The temperature field for the same time interval and for several values of the radius was measured with thermocouples. The

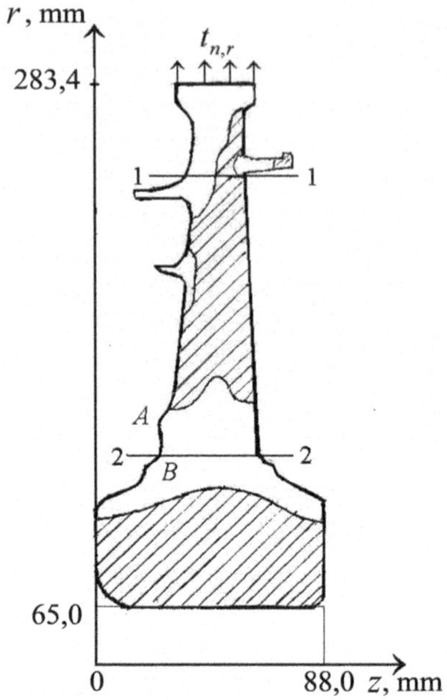

Fig. 2.1 Calculation scheme of the disk

temperature for the other values of the radius was found by linear interpolation. It was assumed that the temperature is constant throughout the thickness of the disk.

The boundary conditions specified on the surface AB (Fig. 2.1) prohibited displacements of the disk along the axis of rotation (the axial displacements and the shear stresses were set at zero). The period from takeoff to landing is divided into 48 stages, according to the disk speed change program. At the last stage, the speed dropped to zero, the temperature of the disk being quite high. The residual stresses and displacements were determined after cooling of the engine at $n = 0$ and $T = 20\,°\text{C}$. The stages are numbered in Fig. 2.2.

The calculations show that plastic strains in the disk occur at the instant denoted by "*a*" in Fig. 2.2 in Section *1–1* (Fig. 2.1), near the right lateral surface of the disk. The plastic zone grows into the disk with time. In Fig. 2.1, the portion of the disk's cross section that undergoes plastic deformation throughout the entire cycle, i.e., to the 48th stage, is hatched. The shape of the plastic area suggests that the rotation and surface load cause bending of the disk. The temperature of the disk and the time of operation of the engine during one cycle are such that creep strains hardly develop. As indicated in Rabotnov and Mileiko (1970), the creep strains should be taken into account when they exceed the instantaneous strains by more than 10%.

Some of the calculated results are presented in Fig. 2.3. Figure 2.3 shows the radial and hoop stresses in Sections *1–1* (*a*) and *2–2* (*b*) at the instants denoted by *b*,

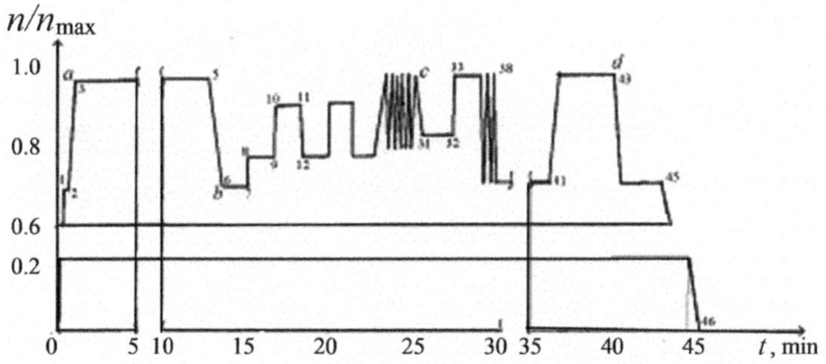

Fig. 2.2 Program of changes in time of the number of revolutions of the disk during the cycle

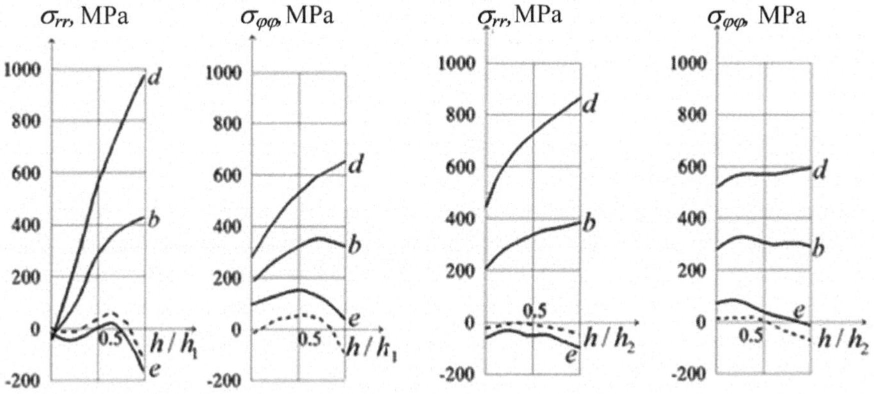

Fig. 2.3 Radial and circular stresses in Sections 1-1 (**a**) and 2-2 (**b**)

d, and *e* in Fig. 2.2. The dashed lines represent the residual stresses after full cooling of the disk in the first cycle.

The calculated results have shown that the cyclic nature of loading is manifested only in the second and third cycles. In the fourth cycle, the accumulated plastic strains have no effect.

The calculated results suggest that the stress–strain state near the rim and hub of the disk displays strongly pronounced spatial behavior. Therefore, using plate theory or shell theory to analyze the state of the disk would lead to significant errors. Failure to account for the changes in the disk speed during flight would result in an even larger error. However, if the disk is subject to repeated loading, i.e., if the engine operates following the same program, the stress–strain state will stabilize after the third cycle. The strength of the disk subject to such a loading program was assessed as the ratio of the maximum to allowed value of the shear-stress intensity that does not exceed 0.6–0.7. This is indicative of an adequate safety factor for the disk. All calculations are described in detail in Savchenko (2014).

Using the given constitutive equations, a mathematical model was proposed (Shevchenko et al. 2015) for studying the thermoplastic stress–strain state and

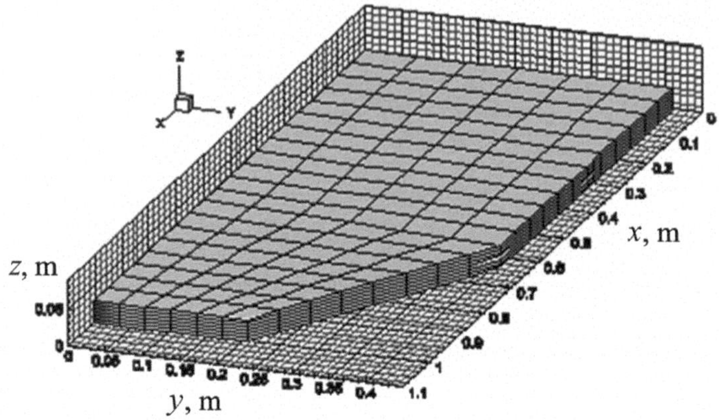

Fig. 2.4 Finite element mesh of the plate

strength of elements of the launch table of the rocket, which allows estimating the resource of the structure.

A general statement of the problem was formulated and a methodology was developed for assessing the strength and determining the resource of the structure, which includes many elements of different geometries, in the process of repeated force and thermal loading. The assessment of the strength of the structure and its service life is based on the results of a numerical study of the temperature and stress–strain state of the most stressed structural elements.

As an example of the application of the developed methodology for assessing the strength and resource of the researched structure, the results of the investigation of the plate element of the starting table are given.

Let us determine stress–strain state and estimate the residual service life of plate element having a constant thickness used for the protection of electronic equipment during the rocket launch. During the rocket launch, the plate undergoes an intense thermal mechanical load. The analysis of input data has showed that the mechanical load on the plate can be neglected as compared with the thermal one. Therefore, hereafter, the research deals with thermoplastic deformation of the plate under cyclical thermal load. The geometry of a plate quarter with a grid consisting of finite number of elements is given in Fig. 2.4. At the crossing $x = 0$ and $y = 0$ the symmetry conditions are set. Beyond the contour, the plate is deemed thermally insulated. On the inner surface $z = 0$, the conditions of convective heat exchange with environment are established. Initial temperature of the plate is $T_0 = 308$ K. The load cycle consists of heating and cooling processes. At the first stage, the outer surface $z = 0.03$ m is heating during 13.2 s. Temperature of external environment, heat exchange coefficient, and specific radiant heat flux vary with time reaching their maximums at $2.8s \leqslant t \leqslant 3.2s$. Having been heated the plate is cooling down to the initial temperature T_0 during 6000 s. The plate is made of 10XCHD steel. This material is deemed to harden linearly with ideal Bauschinger effect.

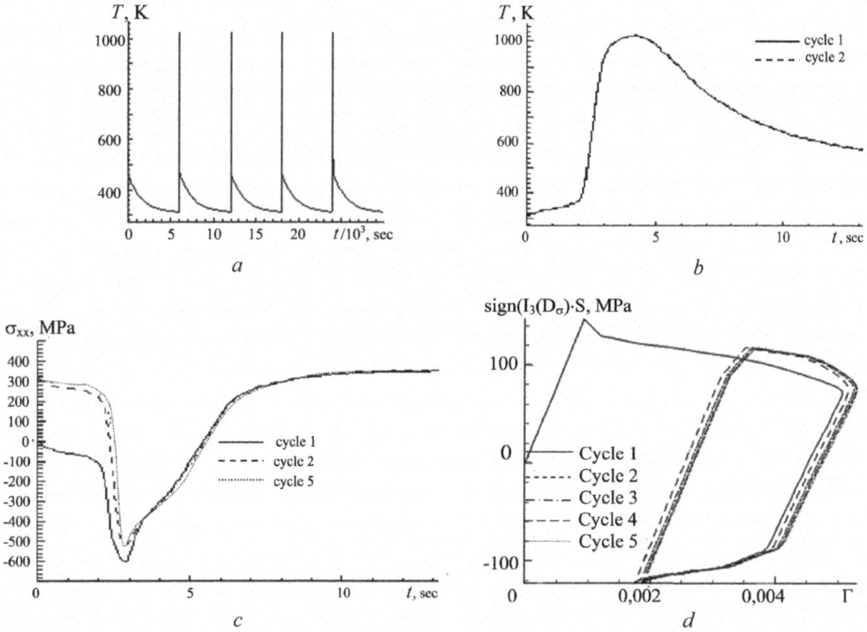

Fig. 2.5 Temperature distribution over time T for a point P_0

The calculations have showed that while heating and cooling the plate, the temperature T and stress σ_{ij} distributions vary insignificantly at the crossings parallel to Oxy plane. The shear stresses can be neglected as compared with the normal ones, while stresses σ_{xx} and σ_{yy} differ slightly from each other and exceed stress σ_{zz} by an order of magnitude. It has been established that at $t = 2.8$ s significant temperature gradients appear and cause compressive stresses near the plate surface. As the heating rate decreases, temperature is distributed uniformly across the plate thickness, with the stresses reversing their sign due to a plastic deformation in the area of maximum temperature gradients, near the plate surface.

In Fig. 2.5a, b, one can see temperature T dependence with time at the point P_0 with coordinates $x = y = 0, z = 0.03$ m. Figure 2.5a shows temperature dependence for 5 cycles of heating and cooling; Fig. 2.5b illustrates change in temperature during heating for the first and second cycles. One can see that after the cooling, temperature reaches its initial value T_0, with the heating for different cycles varying insignificantly. In Fig. 2.5c, σ_{xx} stress distribution is given at the point P_0 during the heating for the first, second, and fifth cycles. Figures 2.5b, c show that stresses σ_{xx} reach their maximums at the moment of maximum warm-up of the plate. Having compared stresses at the end of heating of the first cycle and at the beginning of the second and fifth cycles, one can see that during the cooling the stresses do not vary significantly. Figure 2.5d features dependence sign $(I_3(D_\sigma))S \sim \Gamma$ at the point P_0 for 5 cycles of load (intensity of shear stresses S multiplied by sign of the third stress deviator

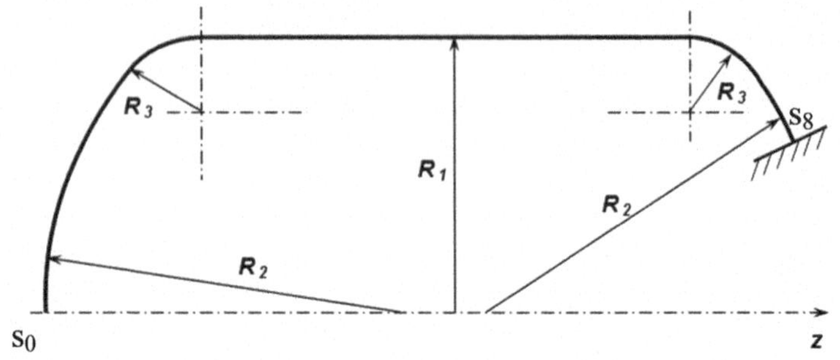

Fig. 2.6 Meridian of the shell simulating the rocket engine housing

$I_3(D_\sigma)$ for determining the sign of load). As one can see from Fig. 2.5d, the hysteresis loop is stabilized at the 5th load cycle, with the amplitude of total strain being $\Delta\varepsilon = 2/\sqrt{3}\Delta\Gamma \sim 0.34\%$. The residual life of this structure under repeated thermal load is estimated using the Coffin–Manson-type fatigue criterion (Birger et al. 1975).

The number of cycles before failure of the structure under consideration has been established to be $N \geqslant 3000$. To estimate the residual service life of this structure more accurately is impossible unless reliable data on cyclical plastic deformation and thermal fatigue strength of the material used are available. The detailed calculations are given in Shevchenko et al. (2015).

Let us determine the critical load for a shell of revolution modeling a motor housing subject to increasing internal pressure. The inside surface of the shell is chosen to be the coordinate one. The meridian (Fig. 2.6) consists of the following concatenated segments: one spherical (1), one toroidal (2), four cylindrical (3–6), one toroidal (7), and one spherical (8). The following dimensions are indicated in Fig. 2.6: $R_1 = 6.63$ c, $R_2 = 8.5$ c, $R_3 = 2.5$ c. The values of the meridional coordinate $s_0 = 0$, $s_8 = 57.16$ c. The thickness of the shell is constant and equal to 0.3 c within segments 1 and 2, varies linearly from 0.3 c to 0.15 c along segment 3, and is constant and equal to 0.15 c within segments 4–8.

The boundary conditions for solving the boundary-value problem were given in the following form: under conditions for $s_0 = 0$ a spherical pole (Grigorenko and Vasilenko 1981); with $s = s_8$—hard pinching.

The shell is made of the Kh18N10T alloy, for which the dependences of the material properties at different values of the angle ω_σ (2.7) are given in Babeshko and Shevchenko (2014), Babeshko et al. (2010). The limiting values of the intensity of tangential stresses for values $\omega_\sigma = 0, \pi/6, \pi/3$ equal to 324 MPa, 272 MPa, and 326 MPa, respectively.

The shell is subject to uniform internal pressure q_ς increasing from $q_\varsigma = 4$ MPa. The loading period is divided into steps so that the step size decreases with increase in the load. The computations demonstrate that plastic deformation occurs first within segment 7 of the shell under $q_\varsigma = 7$ MPa and then, as the load is increased, within all the other segments, except for the vicinity of the pole of segment 1. The loading

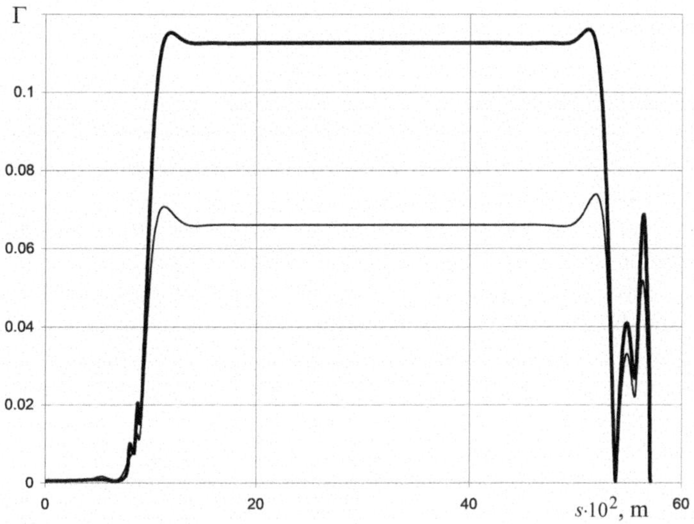

Fig. 2.7 Changes in the intensity of shear deformations along the shell meridian

process is active, i.e., no unloading occurs. Initially, while the deformation is elastic, the meridional stresses that arise in the toroidal segment near the end $s = s_8$ are maximum, yet not critical. With further increase in the load and development of plastic deformation, the circumferential stresses in the cylindrical segment of the shell become maximum. This means that the shell will fail longitudinally. The chosen failure criterion indicates that failure will occur in the cylindrical segment of the housing on the outside surface of the shell in the neighborhood of the point $s = 51.7$ cm, when $\omega_\sigma = 0.42$. The corresponding breaking stress $\sigma_n = S_n\sqrt{3} = 485$ MPa is reached at $q_\varsigma = 11.5$ MPa.

When the stress mode is disregarded, failure occurs in the same place, but the breaking stress $\sigma_n = 565$ MPa when the critical pressure $q_\varsigma = 13.5$ MPa, which exceeds the value obtained with allowance for the stress mode by 17%.

The breaking stresses calculated with and without regard to the stress mode differ a little, whereas the strains differ substantially. The maximum values of the shear-strain intensity within the cylindrical segment of the shell calculated with and without regard to the stress mode differ by more than 60%. Figure 2.7 demonstrates how the shear-strain intensity calculated with (heavy lines) and without (thin lines) regard to the stress mode vary along the meridian s (in centimeters) of the outside surface of the shell. The detailed calculations are given in Babeshko et al. (2015).

2.5 Conclusions

The problem of thermoviscoplasticity is stated and the general form of the constitutive equations describing the processes of inelastic deformation of an isotropic material is given, taking into account the dependence of its properties on temperature, stress mode, load history, and damage to the structure during deformation. The general form of the equations is preserved when using different theories of thermoplasticity and linearization methods. Using the given form of the constitutive equations, the authors have developed methods for solving the corresponding boundary-value problems for spatial bodies and shells of revolution. The stress–strain state of bodies and shells is determined step by step in the process of successive approximations. With non-axisymmetric loading, the specified loads and solving functions are represented by trigonometric series in the circular coordinate. The finite element method was used to solve the spatial problem, and the Runge–Kutta method with discrete orthogonalization of partial solutions was used to solve the problem of the theory of shells. It is worth noting that writing the constitutive equations in a general form facilitates the development of algorithms and corresponding software tools. A numerical study of the stress–strain state was performed with an assessment of the strength of a number of bodies and shells simulating the responsible elements of structures of various purposes. Some results of calculations obtained using methods and software tools developed on the basis of the given equations are given in graphic form.

References

Altenbach H (2003) Creep theory present-day problems and applications. Int Appl Mech 39(6):631–655

Babeshko ME, Savchenko VG (1998) Method of successive approximation in boundary-value problems of thermoplasticity. Int Appl Mech 34(3):232–238

Babeshko ME, Savchenko VG (2018) Allowing for the third deviatoric stress invariant in analyzing the deformation of thin shells. Int Appl Mech 54(2):163–171

Babeshko ME, Shevchenko YuN (2010) Method of successive approximations for solving boundary-value problems of plasticity with allowance for the stress mode. Int Appl Mech 46(7):744–752

Babeshko ME, Shevchenko YuN (2014) Studying the axisymmetric thermoviscoelastoplastic deformation of layered shells taking into account the third deviatoric stress invariant. Int Appl Mech 50(6):615–626

Babeshko ME, Shevchenko YuN, Tormakhov NN (2010) Approximate description of the inelastic deformation of an isotropic material with allowance for the stress mode. Int Appl Mech 46(2):139–148

Babeshko ME, Galishin AZ, Semenets AI, Shevchenko YuN (2015) On effect of a kind of stress state on strength of high pressure vessels. Int Appl Mech 51(3):319–325

Birger IA, Shorr BF, Demianushko IV et al (1975) Thermal strength of machine parts. Mashinostroenie, Moscow [in Russian]

Bondar' VS (2004) Inelasticity: versions of the theory. Fizmatlit, Moscow [in Russian]

Galishin AZ (2008) Axisymmetric thermoviscoelastoplastic state of thin flexible shells with damages. Int Appl Mech 44(2):158–166

Galishin AZ, Shevchenko YuN (2010) Determining the axisymmetric elastoplastic state of thin shells with allowance for the third invariant of the stress deviator. Int Appl Mech 46(8):869–876
Galishin AZ, Sklepus SN (2019) Prediction of the time of failure of axisymmetrically loaded hollow cylinders under conditions of creep. J Math Sci 240(2):194–207
Galishyn AZ, Steblianko PA, Shevchenko YuN (2012) determining the non-stationary temperature fields in thin layered shells of revolution under axisymmetric heating. In: Trans. Dniprodzerzhinsk State Technical University, Issue 2(19) (Mathematical Problems of Engineering Mechanics), DDTU, Dniprodzerzhinsk, pp 3–12 [in Russian]
Grigorenko YaM, Shevchenko YuN, Vasilenko AT et al (2002) Numerical methods, Vol. 11 of the 12-volume series “Mechanics of composite materials”. A.S.K., Kyiv [in Russian]
Grigorenko YaM, Vasilenko AT (1981) The theory of shells of variable stiffness, Vol. 4 of the five-volume series “Methods of calculation of shells”. Nauk. Dumka, Kyiv [in Russian]
Hencky H (1924) zur theorie plastischer deformationen und der hierdurch im material hervorgerufenen nachspannungen. ZAMM 4(4):323–334
Il’yushin AA (1967) A theory of long-term strength. Izv. AN SSSR, MTT 3:21–35
Kachanov LM (1974) Fundamentals of fracture mechanics. Nauka, Moscow [in Russian]
Kachanov LM (2004) Fundamentals of the theory of plasticity. Dover, New York
Khazhinskii GM (1971) Theory of creep and stress rupture of metals. MTT 6:29–36
Khoroshun LP, Shikula EN (2002a) To the theory of short-time microdamage of composite materials of the granular structure under a temperature action. Int Appl Mech 38(1):60–67
Khoroshun LP, Shikula EN (2002b) Short-time microdamage of layered materials under the temperature action. Int Appl Mech 38(4):432–439
Lode W (1926) Versuche uber den Versuche über den Einfluß der mittleren Hauptspannung auf das Fließen der Metalle Eisen. Kupfer und Nickel Z Phys 36:913–939
Namestnikov IV, Shesterikov SA (1985) Vector representation of the damage parameter. In: Deformation and fracture of solids. MGU, Moscow, pp 43–52 [in Russian]
Ohashi Y, Tokuda M, Yamashita H (1975) Effect of third invariant of stress deviator on plastic deformation of mild steel. J Mech Phys Solids 23(4):295–323
Prandtl L (1923) Anwendungsbeispile zu einem Henckyschen satz uber das plastische gleichgewicht. ZAMM 3(6):401–406
Rabotnov YN, Mileiko ST (1970) Transient creep. Nauka, Moscow [in Russian]
Reuss A (1930) Berucksichtigung der elastischen formanderung in der plastizitatstheorie. ZAMM 10(3):266–274
Savchenko VG (2003) A method of describing damage in analyzing the nonaxisymmetric thermostressed state of orthotropic solids of revolution. In: Systems technologies, Issue 2(25), Dnepropetrovsk, pp 134–138 [in Ukrainian]
Savchenko VG (2004) Numerical nonaxisymmetric thermostress analysis of compound solids of revolution with damage. Int Appl Mech 40(3):275–282
Savchenko VG (2006) Stress state of compound solids of revolution made of damaged orthotropic materials with different tensile and compressive moduli. Int Appl Mech 42(11):1246–1255
Savchenko VG (2008) A method to study the nonaxisymmetric plastic deformation of solids of revolution with allowance for the stress mode. Int Appl Mech 44(9):975–981
Savchenko VG (2012) Analysis of the axisymmetric inelastic deformation of solids of revolution taking into account the stress mode. In: Trans. Dniprodzerzhinsk State Technical University, Issue 2(19) (Mathematical Problems of Engineering Mechanics), DDTU, Dniprodzerzhinsk, pp 13–17 [in Russian]
Savchenko VG (2014) Thermoviscoelastic state of bodies of revolution subject to creep damage under nonaxisymmetric thermomechanical loading. In: Trans. Dniprodzerzhinsk State Technical University, Issue 1(24), DDTU, Dniprodzerzhinsk, pp 170–174 [in Russian]
Savchenko VG, Shevchenko YuN (2004) Nonaxisymmetrical thermal stressed state of laminated bodies of revolution of orthotropic materials under nonisothermic loading. Mech Compos Mater 40(6):731–751

Savchenko VG, Shevchenko YuN (2006) Three-dimensional problems of thermoviscoplasticity, Vol. 1 of the six-volume series "Advances in mechanics". ASK, Kyiv, pp 625–660 [in Russian]
Savchenko VG, Babeshko ME (2018) Thermostressed state of layered bodies of revolution damaging under deformation. Int Appl Mech 54(3):287–305
Shevchenko YuN (1970) Thermoplasticity under variable loading. Nauk, Dumka, Kyiv [in Russian]
Shevchenko YuN (1986) Numerical methods for solving applied problems, Vol. 2 of the six-volume series "Three-dimensional problems in elasticity and plasticity". Nauk,Dumka, Kyiv [in Russian]
Shevchenko YuN, Terekhov RG (1982) Constitutive equations of thermoviscoplasticity. Nauk, Dumka, Kyiv [in Russian]
Shevchenko YuN, Mazur VN (1986) Solution of plane and axisymmetric boundary-value problems of thermoviscoplasticity with allowance for creep damage to the material. Sov Appl Mech 22(8):695–704
Shevchenko YuN, Savchenko VG (1987) Thermoviscoplasticity, Vol. 2 of the five-volume series "Mechanics of coupled fields in structural members". Nauk, Dumka, Kyiv [in Russian]
Shevchenko YuN, Galishin AZ (2003) Determination of thermoviscoelastoplastic state of laminates shells of revolution, taking into account the damage under creep. In: Thermal stresses '03. The fifth international congress on thermal stress and related topics. Virginia Polytechnic Institute and State University. Vol.: - P. TA -8-3-1–TA-8-3-4
Shevchenko YuN, Tormakhov NN (2010) Constitutive equations of thermoplasticy including the third invariant. Int Appl Mech 46(6):613–624
Shevchenko YuN, Tormakhov NN (2012a) Constitutive equations of thermoplasticity describing processes of proportional loading and incorporating the stress mode. In: Mathematical problems of engineering mechanics, Issue 2 (19), Dniprodzerzhinsk, DDTU, pp 69–73 [in Russian]
Shevchenko YuN, Tormakhov NN (2012b) Thermoviscoplastic deformation along paths of small curvature: constitutive equations including the third deviatoric stress invariant. Int Appl Mech 48(6):688–699
Shevchenko YuN, Savchenko VG (2016) Three-dimensional problems of thermoviscoplasticity: focus on Ukrainian research (review). Int Appl Mech 52(3):217–271
Shevchenko YuN, Babeshko ME, Piskun VV, Savchenko VG (1980) Three-dimensional problems of thermoplasticity. Nauk, Dumka, Kyiv [in Russian]
Shevchenko YuN, Babeshko ME, Terekhov RG (1992) Thermoviscoelastoplastic processes of combined deformation of structural members. Nauk, Dumka, Kyiv [in Russian]
Shevchenko YuN, Terekhov RG, Tormakhov NN (2006) Constitutive equations for describing the elastoplastic deformation of elements of a body along small-curvature paths in view of the stress mode. Int Appl Mech 42(4):421–430
Shevchenko YuN, Terekhov RG, Tormakhov NN (2007a) Elastoplastic deformation of elements of an isotropic solid along paths of small curvature: constitutive equations incorporating the stress mode. Int Appl Mech 43(6):621–630
Shevchenko YuN, Terekhov RG, Tormakhov NN (2007b) Linear relationship between the first invariants of the stress and strain tensors theories of plasticity with strain hardening. Int Appl Mech 43(3):291–302
Shevchenko YuN, Andrushko NF, Banyas MV, Galishin AZ et al (2015) Procedure of forecasting operation and extremal state of critical systems of the rocket technique under repeated thermoforce loading. Sci Innov 11(5):25–36

Chapter 3
Influence of Finite Initial Deformations on Velocities of Generalized Lamb Waves in an Incompressible Elastic Layer Interacting with a Layer of an Ideal Fluid

Olexandr Bagno

3.1 Introduction

At present, Lamb waves are increasingly widely used in fundamental theoretical research and practical applications. Papers (Akbarov et al. 2005; Akbarov and Ozaydin 2001; Akbarov and Ozisik 2003; Ottenio et al. 2007; Guz 2004, 2016a, b; Guz et al. 1976; Guz and Bagno 2018a; Guz 2002; Guz and Bagno 2018b; Guz 2014; Guz et al. 2016; Zhuk 1980a; Guz 1980b) are focused on studying the propagation of Lamb waves in prestressed bodies. The mentioned publications obtained qualitative and quantitative results mainly for rigid (metals, organic glass) compressible bodies that allow small initial deformations. However, unlike compressible, incompressible bodies (rubber-like, caoutchouc-like materials) are highly elastic—their initial deformation can be significant. In this regard, it is of particular theoretical and practical interest to study the influence of large (finite) initial deformations of an incompressible layer on Lamb waves' speed when interacting with a layer of an ideal compressible fluid.

The propagation of acoustic waves along the interface between a prestrained incompressible elastic half-space and a half-space of viscous fluid was analyzed in Ottenio et al. (2007). At the same time, it is of particular theoretical and practical interest to study the influence of large (finite) initial strains of an incompressible layer on the velocity of generalized Lamb waves during its interaction with a layer of an ideal compressible fluid.

3.2 Problem Statement. Basic Equations

In this chapter, to study the propagation of waves in a hydroelastic waveguide consisting of an incompressible elastic layer and a liquid layer, we use a prestressed body model, a compressible ideal fluid model, and the three-dimensional linearized

O. Bagno (✉)
S.P. Timoshenko Institute of Mechanics, National Academy of Science of Ukraine, Kyiv, Ukraine
e-mail: alexbag2016@gmail.com

A. N. Guz et al. (eds.), *Advances in Mechanics*, Advanced Structured Materials 191,
https://doi.org/10.1007/978-3-031-37313-8_3

Euler equations for the fluid and the three-dimensional linearized equations of finite deformations for the elastic body. The fluid is assumed to be at rest. We will use a problem formulation and a method based on the general solutions of the equations of motion of a compressible ideal fluid and a prestressed incompressible elastic body proposed in Guz (1980b, 2002, 2004, 2009, 2016a, b).

We will consider such dynamic processes in the hydroelastic system that the additional strains induced, i.e., strain perturbations, are much lower than the prestrains. The wave processes are assumed to be harmonic and low-amplitude. The elastic body is assumed to be in the initial stress-strain state. Note that unlike elastic bodies described in Lagrangian coordinates, fluids are described in Eulerian coordinates in their natural state. In the hydroelastic problem, the initial stress-strain state of the elastic body is a natural state relative to the fluid and the system as a whole. Since we will consider the propagation of small perturbations, the Eulerian and Lagrangian approaches to the description of the behavior of the media coincide. Therefore, the differences between the Lagrangian and Eulerian coordinates are ignored, eliminating the difficulties associated with formulating boundary conditions typical for nonlinear problems. Within the framework of the accepted models, the main relations for the system prestressed incompressible elastic body–ideal compressible fluid take the following form Guz (1980b, 2002, 2004, 2009, 2016a, b):

1. incompressible elastic body

$$\left(\tilde{\kappa}_{ij\alpha\beta}\frac{\partial^2}{\partial z_i \partial z_\beta} - \delta_{j\alpha}\rho\frac{\partial^2}{\partial t^2}\right)u_\alpha + \tilde{q}_{ij}\frac{\partial p}{\partial z_i} = 0, \quad z_k \in V_1, \tag{3.1}$$

$$\tilde{\kappa}_{ij\alpha\beta} = \lambda_i\lambda_\beta\kappa_{ij\alpha\beta}, \quad \tilde{q}_{ij} = \lambda_i q_{ij}, \quad \lambda_1\lambda_2\lambda_3 = 1, \qquad \tilde{q}_{ij}\frac{\partial u_j}{\partial z_i} = 0, \quad z_k \in V_1, \tag{3.2}$$

$$\tilde{Q}_j \equiv \left(\tilde{\kappa}_{1j\alpha\beta}\frac{\partial u_\alpha}{\partial z_\beta} + \tilde{q}_{1j}f\right)\cos\chi + \left(\tilde{\kappa}_{2j\alpha\beta}\frac{\partial u_\alpha}{\partial z_\beta} + \tilde{q}_{2j}f\right)\cos\eta + \left(\tilde{\kappa}_{3j\alpha\beta}\frac{\partial u_\alpha}{\partial z_\beta} + \tilde{q}_{3j}f\right)\cos\gamma, \quad z_k \in S; \tag{3.3}$$

2. ideal compressible fluid

$$\frac{\partial v_1}{\partial t} + \frac{1}{\rho_0}\frac{\partial p}{\partial z_1} = 0, \quad \frac{\partial v_2}{\partial t} + \frac{1}{\rho_0}\frac{\partial p}{\partial z_2} = 0, \quad \frac{\partial v_3}{\partial t} + \frac{1}{\rho_0}\frac{\partial p}{\partial z_3} = 0,$$
$$\frac{1}{\rho_0}\frac{\partial \rho^*}{\partial t} + \frac{\partial v_1}{\partial z_1} + \frac{\partial v_2}{\partial z_2} + \frac{\partial v_3}{\partial z_3} = 0, \quad \frac{\partial p}{\partial \rho^*} = a_0^2, \quad a_0 = \text{const}, \quad z_k = V_2, \tag{3.4}$$

$$p_{ij} = -\delta_{ij}p, \quad \tilde{P}_1 = -\,p\cos\chi, \quad \tilde{P}_2 = -p\cos\eta, \quad \tilde{P}_3 = -p\cos\gamma, \quad z_k \in V_2, \tag{3.5}$$

where δ_{ij} is the Kronecker symbol.

At the same time, the specifics of the interaction of elastic and liquid media are reflected by dynamic $\tilde{Q}_j = \tilde{P}_j$, $z_k \in S$ and kinematic $\partial \mathbf{u}/\partial t = \mathbf{v}$, $z_k \in S$ boundary conditions specified on the contact surface of solids and liquids S. Introduced tensors $\tilde{\kappa}_{ij\alpha\beta}$ and $\tilde{q}_{ij}$ depend on the type of the initial state and the type of elastic potential of the solid material. Expressions for calculating the components of these tensors are given in Guz (2016b). Simplifications for various variants of the theory of small initial strains are also proposed.

Here we have introduced the following notation: u_i are the components of the displacement vector $\mathbf{u}$ of the elastic layer; ρ is the density of the elastic body; λ_i are the elongations of the elastic body along the coordinate axes; v_i are the components of the velocity perturbation vector $\mathbf{v}$ of the fluid; ρ^* and p are the density and pressure perturbations in a fluid; ρ_0 and a_0 are the density of and the speed of sound in the fluid at rest; $\tilde{P}_j$ and $\tilde{Q}_j$ are the components of stress vector in a fluid and an elastic body, respectively; $\mathbf{N}^0 = (\cos\chi; \cos\eta; \cos\gamma)$ is the normal vector to the outer surface of the body; V_1 and V_2 are the volumes occupied by an elastic body and a fluid; S are the contact surface of elastic and liquid media.

Equations (3.1)–(3.3) describe the behavior of an elastic body. Small fluctuations of an ideal compressible fluid at rest describe relations (3.4) and (3.5).

Let us now consider an isotropic nonlinear elastic body whose elastic potential is an arbitrary twice continuously differentiable function of the components of the Green strain tensor. The body occupies a volume ($-\infty < z_1 < \infty$, $-h_2 \leqslant z_2 \leqslant 0$, $-\infty < z_3 < \infty$) and contacts with a layer of compressible ideal fluid occupying a volume ($-\infty < z_1 < \infty$, $0 \leqslant z_2 \leqslant h_1$, $-\infty < z_3 < \infty$). The external forces acting on these media are uniformly distributed along the Oz_3-axis. Therefore, the problem is plane, and we may restrict the study of wave propagation to the plane Oz_1z_2. Consequently, this problem is reduced to solving the system of equations of motion of an elastic body and a fluid under the following dynamic and kinematic boundary conditions:

$$\tilde{Q}_1\Big|_{z_2=0} = 0, \quad \tilde{Q}_2\Big|_{z_2=0} = \tilde{P}_2\Big|_{z_2=0}, \quad \tilde{Q}_1\Big|_{z_2=-h_2} = 0, \quad \tilde{Q}_2\Big|_{z_2=-h_2} = 0, \tag{3.6}$$

$$\tilde{P}_2\Big|_{z_2=h_1} = 0; \quad v_2\Big|_{z_2=0} = \frac{\partial u_2}{\partial t}\Big|_{z_2=0}. \tag{3.7}$$

The following notations are introduced here: h_1 is the thickness of the liquid layer, h_2 is the thickness of the elastic layer.

3.3 Problem-Solving Method

We will use the statements of hydroelasticity problems for prestressed bodies and ideal fluid and the general solutions proposed in Guz (1980b, 2002, 2004, 2009, 2016a, b). Wave processes in incompressible prestrained elastic bodies interacting with a compressible ideal fluid whose initial state is homogeneous are further studied.

The general solutions for the homogeneous stress-strain state in the plane case are as follows:

1. for an incompressible elastic layer

$$u_1 = -\frac{\partial^2 \chi_1}{\partial z_1 \partial z_2}, \quad u_2 = \lambda_1 q_1 \lambda_2^{-1} q_2^{-1} \frac{\partial^2}{\partial z_1^2} \chi_1, \quad q_i = \lambda_i^{-1}, \quad \lambda_1 \lambda_2 = 1,$$

$$p = \lambda_1^{-1} q_1^{-1} \Big\{ \lambda_1^2 \left[\lambda_1^2 a_{11} + s_{11}^0 - \lambda_1 \lambda_2 q_1 q_2^{-1} (a_{12} + \mu_{12}) \right] \frac{\partial^2}{\partial z_1^2}$$

$$+ \lambda_2^2 \left(\lambda_1^2 \mu_{12} + s_{22}^0 \right) \frac{\partial^2}{\partial z_2^2} - \rho \frac{\partial^2}{\partial t^2} \Big\} \frac{\partial}{\partial z_2} \chi_1;$$

2. for a compressible ideal liquid layer

$$v_1 = \frac{\partial^2 \chi_2}{\partial z_1 \partial t}, \quad v_2 = \frac{\partial^2 \chi_2}{\partial z_2 \partial t},$$

where the functions χ_i are solutions of the equations:

1. for an elastic layer of incompressible material

$$\Bigg[\frac{\partial^4}{\partial z_1^4} + \frac{\lambda_2^4 q_2^2 \left(\lambda_1^2 \mu_{12} + s_{22}^0 \right)}{\lambda_1^4 q_1^2 \left(\lambda_2^2 \mu_{12} + s_{11}^0 \right)} \frac{\partial^4}{\partial z_2^4} - \frac{\rho}{\lambda_1^2 \left(\lambda_2^2 \mu_{12} + s_{11}^0 \right)} \frac{\partial^4}{\partial z_1^2 \partial t^2}$$

$$+ \frac{q_1 q_2^{-1} \left(\lambda_2^2 a_{22} + s_{22}^0 \right) + q_1^{-1} q_2 \left(\lambda_1^2 a_{11} + s_{11}^0 \right) - 2 \lambda_1 \lambda_2 (a_{12} + \mu_{12})}{\lambda_1^2 \lambda_2^{-2} \left(\lambda_2^2 \mu_{12} + s_{11}^0 \right) q_1 q_2^{-1}} \frac{\partial^4}{\partial z_1^2 \partial z_2^2}$$

$$- \frac{\lambda_2^2 q_2^2 \rho}{\lambda_1^4 q_1^2 \left(\lambda_2^2 \mu_{12} + s_{11}^0 \right)} \frac{\partial^4}{\partial z_2^2 \partial t^2} \Bigg] \chi_1 = 0, \quad q_i = \lambda_i^{-1}, \quad \lambda_1 \lambda_2 = 1;$$

2. for a layer of an ideal compressible fluid

$$\left[\left(\frac{\partial^2}{\partial z_1^2} + \frac{\partial^2}{\partial z_2^2} \right) - \frac{1}{a_0^2} \frac{\partial^2}{\partial t^2} \right] \chi_2 = 0.$$

The following notations are used above: a_{ij} and μ_{ij} are quantities determined from the equations of state and depending on the type of elastic potential (Guz 2004, 2016a), s_{ii}^0 are the initial stresses.

Note that the equations for perturbations in elastic bodies are linear, but the initial stress-strain state is determined from the general nonlinear equations. In this connection, even though the general statement of hydroelastic problems for prestressed bodies in coordinates z_1 is similar to the statement of linear problems of classical hydroelasticity, there are substantial differences in the equations and the boundary conditions in these problem statements.

To analyze the propagation of harmonic perturbations, candidate solutions of the system of equations are in the class of traveling waves

$$\chi_j = X_j(z_2)\exp[i(kz_1 - \omega t)] \quad (j = 1, 2),$$

where k is the wave number, ω is the circular frequency, i is imaginary unit ($i = \sqrt{-1}$).

Note that the class of harmonic waves chosen in this paper, being the most straightforward and most convenient in theoretical studies, does not limit the generality of the results obtained since, as is known, a linear wave of an arbitrary shape can be represented by a set of harmonic components.

Next, applying the Fourier method, we arrive at two eigenvalue problems for the equations of motion of the elastic body and the fluid. Solving them, we find the eigenfunctions. Substituting the general solutions into the boundary conditions (3.6) and (3.7), we obtain a homogeneous system of linear algebraic equations for the arbitrary constants. From the condition for the existence of a nontrivial solution, we derive the dispersion equation. For a hydroelastic system consisting of an elastic layer and a liquid layer, the dispersion equation has the form

$$\det\left\| e_{lm}\left(c, \rho, \mu, \lambda_i, a_{ij}, \mu_{ij}, s_{ii}^0, \rho_0, a_0, \frac{\omega h_1}{c_s}, \frac{\omega h_2}{c_s}\right)\right\| = 0$$

$$(l, m = 1, 2, \ldots, 6), \tag{3.8}$$

where A is the phase velocity of normal waves in a prestressed layer, c_s is the velocity of a shear wave in the nonstressed elastic body ($c_s^2 = \mu/\rho$), μ is the shear modulus of the elastic body.

Note that the dispersion equations (3.8) do not depend on the type of elastic potential and represent arbitrary incompressible elastic bodies with high (finite) initial strains. It has the most general form from which it is possible to derive equations for many special cases examined in Guz et al. (1976, 2016), Zhuk (1980a). If we put $\lambda_i = 1$ ($s_{ii}^0 = 0$), then we obtain equations for Raleigh and Scholte–Stoneley and Lamb waves studied in detail using the classical theory of elasticity Guz et al. (2016).

3.4 Numerical Results and Their Analysis

The dispersion equation (3.8) is solved numerically in what follows. In this case, the calculations are carried out for a hydroelastic system consisting of a real elastic body and a fluid. It consists of an incompressible elastic layer and a layer of water. As a material for the elastic layer, we choose highly elastic rubber, which elastic properties of which are described by the Treloar elastic potential. In this case, we choose the following mechanical parameters of the hydroelastic system (Koshkin 1988): elastic layer—$\rho = 1200\,\mathrm{kg/m^3}$, $\mu = 1.2 \cdot 10^6$ Pa, liquid layer—$\rho_0 = 1000\,\mathrm{kg/m^3}$,

$a_0 = 1459.5\,\mathrm{m/s}$, $\bar{a}_0 = a_0/c_s = 46.153442$. In this waveguide, the material of the elastic body (rubber) is pliable and soft. In addition, when solving, the initial stress state is assumed to satisfy the relations $s_{11}^0 \neq 0$ and $s_{22}^0 = 0$. As shown in Guz (2004), with such a load, there is no analogy between problems in the linearized and linear formulations. Therefore, we cannot obtain the results for bodies with initial stresses from the solutions of the corresponding linear problem.

Note that we derived (3.8) without any additional requirements for the form of elastic potential—it takes place for elastic potential of arbitrary shape.

Figures 3.1, 3.2, 3.3, 3.4, 3.5 and 3.6 present the obtained results. Graphs in Fig. 3.1 illustrate the effect of fluid on the wave characteristics of a hydroelastic

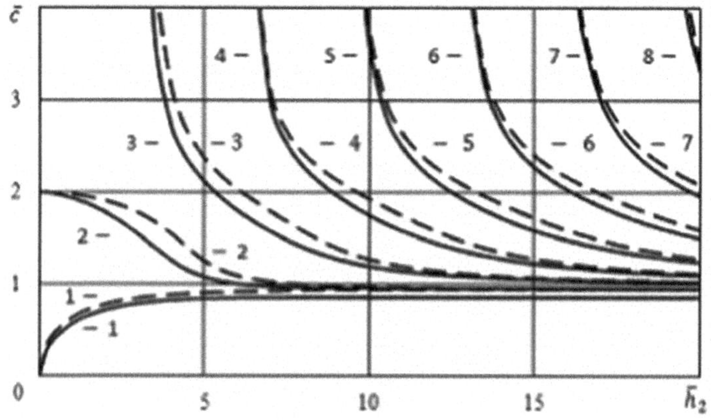

Fig. 3.1 Dependences of the dimensionless values of the phase velocities of the Lamb modes $\bar{c}$ on the dimensionless value of the thickness of the elastic layer $\bar{h}_2$ (solid lines correspond to $\rho_0 \neq 0$ and for the dashed lines $\rho_0 = 0$)

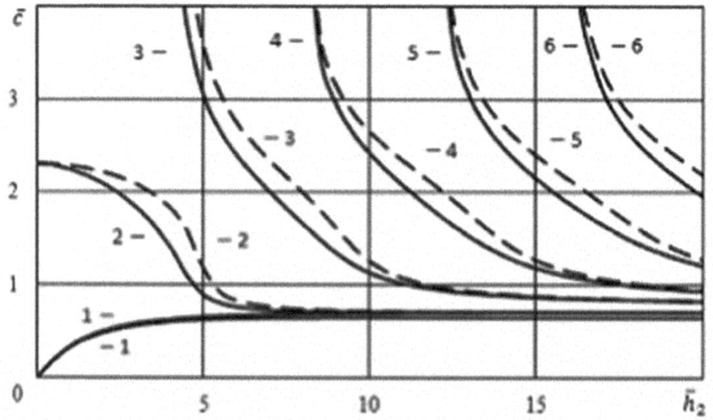

Fig. 3.2 Dependences of the dimensionless values of the phase velocities of the Lamb modes $\bar{c}$ on the dimensionless value of the thickness of the elastic layer $\bar{h}_2$ when $\rho_0 = 0$ (solid lines correspond to $\lambda_1 = 0.8$ and dashed lines correspond to $\lambda_1 = 1$)

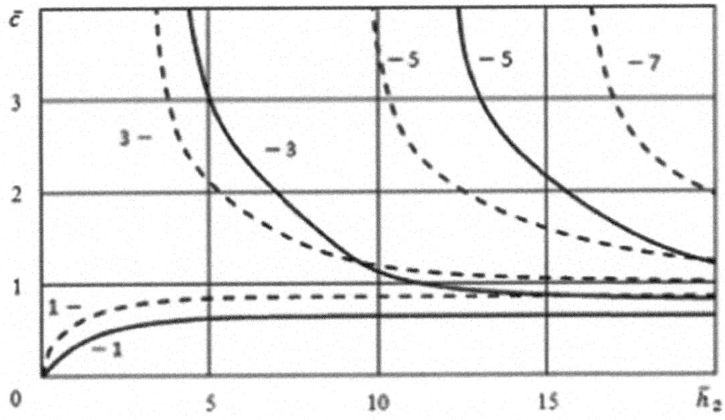

Fig. 3.3 Dependences of the dimensionless values of the phase velocities odd of the generalized Lamb modes $\bar{c}$ on the dimensionless value of the thickness of the elastic layer $\bar{h}_2$ when $\rho_0 \neq 0$ (solid lines correspond to $\lambda_1 = 0.8$, and dashed lines are obtained for $\lambda_1 = 1$)

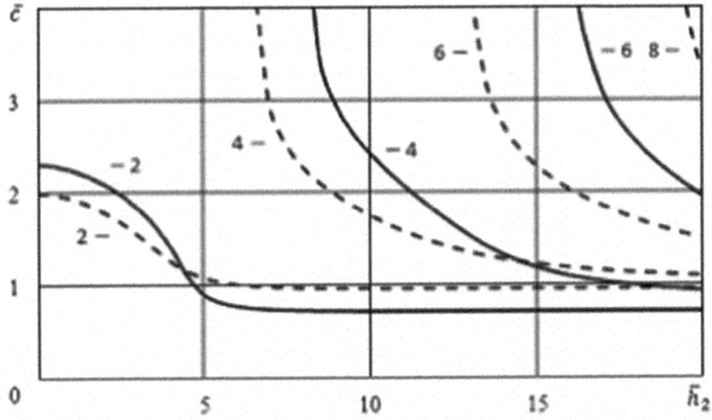

Fig. 3.4 Dependences of the dimensionless values of the phase velocities even of the generalized Lamb modes $\bar{c}$ on the dimensionless value of the thickness of the elastic layer $\bar{h}_2$ when $\rho_0 \neq 0$ (solid lines correspond to $\lambda_1 = 0.8$ and dashed lines correspond to $\lambda_1 = 1$)

system. At the same time, Fig. 3.1 shows the dispersion curves reflecting the dependences of the dimensionless values of the phase velocities of the Lamb modes $\bar{c}$ ($\bar{c} = c/c_s$) on the dimensionless value of the thickness of the elastic layer (frequency) $\bar{h}_2$ ($\bar{h}_2 = \omega h_2/c_s$). Solid curves are obtained for an elastic layer interacting with a thick liquid layer, the layer thickness of which is $\bar{h}_1$ ($\bar{h}_1 = \omega h_1/c_s$) is equal to 20. The dashed lines correspond to the case of an elastic layer that does not interact with the liquid. The graphs in this figure are obtained for elastic layers without initial deformations ($\lambda_1 = 1$).

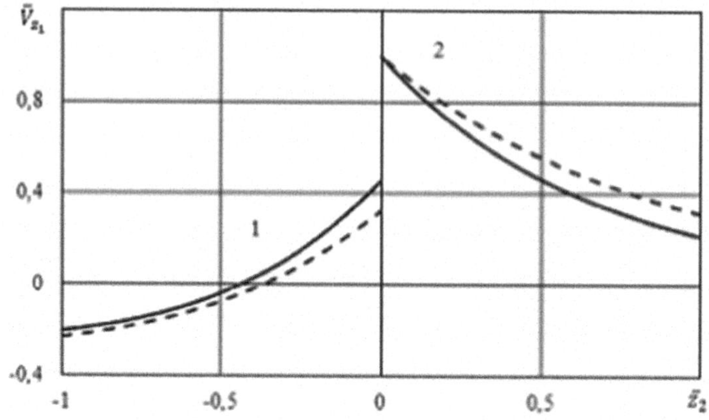

Fig. 3.5 Dependences of the dimensionless values of the longitudinal displacement amplitudes $\bar{V}_{z1}$ on the dimensionless transverse coordinate $\bar{z}_2$

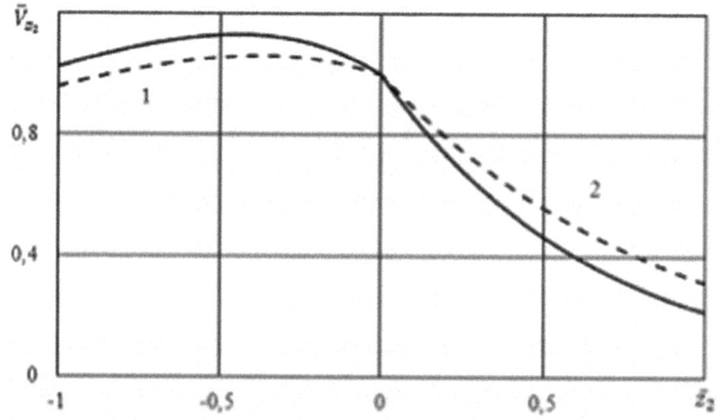

Fig. 3.6 Dependences of the dimensionless values of the transverse displacement amplitudes $\bar{V}_{z2}$ on the dimensionless transverse coordinate $\bar{z}_2$

Presented in Fig. 3.1 graphs illustrate the effect of liquid on the dispersion properties of the wave process. It follows from them that in a hydroelastic waveguide (solid lines) with an increase in the thickness of the elastic layer (frequency) $\bar{h}_2$ the velocities of the first modes tend to the speed of the Stoneley wave $\bar{c}_{st}$ ($\bar{c}_{st} = c_{st}/c_s = 0.859257$ at $\lambda_1 = 1$ and $\bar{c}_{st} = 0.650184$ at $\lambda_1 = 0.8$) from below, and the velocities of the second modes—to the speed of the Rayleigh-wave $\bar{c}_R$ ($\bar{c}_R = c_R/c_s = 0.955318$ at $\lambda_1 = 1$ and $\bar{c}_R = 0.709558$ at $\lambda_1 = 0.8$) above. For a purely elastic waveguide (dashed lines), the velocities of these Lamb modes tend to the speed of the Rayleigh-wave $\bar{c}_R$. In this case, the first modes tend to the velocity of the surface wave $\bar{c}_R$ ($\bar{c}_R = 0.9553303$ at $\lambda_1 = 1$ and $\bar{c}_R = 0.709558$ at $\lambda_1 = 0.8$) from below, and the velocities of the second modes, respectively, to $\bar{c}_R$ ($\bar{c}_R = 0.9553303$ at $\lambda_1 = 1$ and

$\bar{c}_R = 0.709558$ at $\lambda_1 = 0.8$) above. The speeds of all high-order modes tend to the speed of the shear wave in the material of the elastic body $\bar{c}_s$. At the same time, as the frequency (thickness) increases, they are dominated by transverse displacements, the amplitude of which on the layer surfaces tends to zero compared to their amplitudes in the thickness of the layer, that is, motions in high-order modes are displaced from the surface into the layer and are localized in its thickness.

Figure 3.2 shows the dispersion curves $\bar{c} = f\left(\bar{h}_2\right)$, reflecting the effect of initial strains on the phase velocities of the Lamb modes. The solid lines correspond to the precompressed elastic layer ($\lambda_1 = 0.8$). The dashed lines indicate the dispersion curves obtained in the absence of initial deformations ($\lambda_1 = 1$).

From the graphs shown in Fig. 3.2, preliminary deformations cause a change in the frequencies of the generation of Lamb modes and a shift in their dispersion curves. It is easy to see that the initial compression ($\lambda_1 = 0.8$) (solid lines) leads to a shift of critical frequencies and dispersion curves to the long-wavelength part of the spectrum.

The nature of the influence of initial deformations on the velocities of normal waves in an elastic-fluid system is illustrated by the graphs in Figs. 3.3 and 3.4. Here, the dependences of the dimensionless values of the phase velocities of the generalized Lamb modes are presented $\bar{c}$ on the dimensionless value of the thickness of the elastic layer $\bar{h}_2$. Solid curves are obtained for a system whose elastic layer is subjected to initial compression ($\lambda_1 = 0.8$). The dashed lines indicate the dispersion curves corresponding to the hydroelastic waveguide without initial deformations ($\lambda_1 = 1$). At the same time, Fig. 3.3 shows graphs for odd and Fig. 3.4 for even generalized Lamb modes. These figures show curves for hydroelastic waveguides whose fluid layer thickness $\bar{h}_1 = 20$.

From the graphs shown in Figs. 3.3 and 3.4, preliminary deformations in an elastic-liquid waveguide also cause a change in the frequencies of generation of generalized Lamb modes and a shift in their dispersion curves. It is easy to see that the initial compression ($\lambda_1 = 0.8$, solid lines), in contrast to a purely elastic layer, leads to a shift of the critical frequencies and dispersion curves to the short-wavelength part of the spectrum.

The graphs presented in Figs. 3.5 and 3.6 reflect the distribution of displacement amplitudes (velocities $\partial u_i/\partial t$ and v_i) in the near-contact region of the hydroelastic waveguide in the interval of coordinate change $\bar{z}_2$ equal to $-1 \leqslant \bar{z}_2 \leqslant 0$ in the elastic layer and $0 \leqslant \bar{z}_2 \leqslant 1$ in a layer of an ideal fluid. They show the dimensionless longitudinal, $\bar{V}_{z_1}$ (Fig. 3.5) and transverse $\bar{V}_{z_2}$ (Fig. 3.6), respectively, versus the dimensionless transverse coordinate $\bar{z}_2$ for lowest mode 1 (Figs. 3.1 and 3.3). The graphs shown in these figures are obtained for a hydroelastic waveguide with an elastic layer thickness of 20 and a liquid layer thickness of 20. In these figures, the dashed lines show the dependencies $\bar{V}_{z_1} = f(z_2)$ and $\bar{V}_{z_2} = f(z_2)$ for a hydroelastic waveguide in the absence of initial stresses ($\lambda_1 = 1$). Similar dependencies for an elastic-liquid waveguide, in which the elastic layer is under initial compression ($\lambda_1 = 0.8$), are represented by solid lines in the figures.

The phase velocity and profile of a Stoneley wave in the interface between solid and liquid half-spaces are known to depend on the mechanical parameters of the hydroelastic system and are determined by the ratio between the acoustic-wave velocity in the fluid and the Rayleigh-wave velocity in the solid half-space. The mechanical parameters of the "rubber-water" hydroelastic system are such that the speed of sound $\bar{a}_0$ ($\bar{a}_0 = 46.153442$) in the fluid is higher than the Rayleigh-wave velocity $\bar{c}_R$ ($\bar{c}_R = 0.955318$ for $\lambda_1 = 1$ and $\bar{c}_R = 0.709558$ for $\lambda_1 = 0.8$). According to Figs. 3.5 and 3.6, with such a ratio of the mechanical parameters of the components of the system, mode 1 of propagating along the interface of the media is localized near the surface of both the liquid and the elastic layer. In the short-wave portion of the spectrum, this quasi-surface (Stoneley-type) wave penetrates deeper into the elastic body than into the liquid. As follows from the graphs shown in Figs. 3.5 and 3.6, initial compression reduces the depth of penetration of the lowest mode into the fluid. As follows from the graphs shown in Figs. 3.5 and 3.6, initial compression ($\lambda_1 = 0.8$) reduces the depth of penetration of the lowest mode into the fluid.

3.5 Influence of the Liquid Layer on the Wave Process

The graphic material presented in Figs. 3.1 and 3.2 makes it possible to obtain information about the influence of the liquid layer on such wave characteristics of hydroelastic waveguides as phase velocities and dispersion of generalized Lamb modes. It follows that the liquid layer, interacting with the elastic layer, causes a change in the mode nucleation frequencies and their dispersion curves. In this case, the critical frequencies decrease, and the dispersion curves (solid lines) shift to the long-wavelength part of the spectrum. Thus, we can conclude that in a hydroelastic waveguide, the generalized Lamb modes' phase velocities become smaller than the corresponding Lamb modes' phase velocities in a purely elastic waveguide.

3.6 Influence of Finite Initial Strains on the Wave Characteristics of Generalized Lamb Modes

In Guz et al. (1976), for an elastic layer of a compressible material, it was shown for the first time that, in an elastic waveguide, the initial stresses cause a change in the frequencies of the generation of Lamb modes and a shift in their dispersion curves. Thus, we can conclude that, in the vicinity of critical frequencies, the phase velocities of the modes in the preliminarily deformed layer can be either less or greater than the phase velocities of the corresponding modes in the body without initial stresses. It causes the appearance in the spectrum of an elastic waveguide of frequencies (thicknesses) at which the initial stresses do not affect the phase velocities of a series of normal Lamb waves. From the graphs presented in Figs. 3.3 and 3.4, it follows that this regularity also takes place for an elastic layer of an incompressible material, both for non-interacting and interacting with the liquid layer. It is easy to see that for all generalized Lamb modes, except for the first one, there are elastic layers of specific thicknesses (certain frequencies) $\bar{h}_2$, at which the phase velocities do not depend on the initial compression λ_1. This pattern first revealed for compressible bodies and described in Guz et al. (1976) is general and is inherent in the frequency spectra of elastic waveguides not only from different materials (compressible, incompressible) but also elastic-liquid waveguides.

In addition, as follows from the graphs shown in Figs. 3.5 and 3.6, initial compression ($\lambda_1 = 0.8$) reduces the depth of penetration of the lowest mode into the liquid.

3.7 Conclusions

Thus, the analysis shows that in the elastic-liquid system "rubber (pliable material)–water" at $\bar{a}_0 > \bar{c}_R$, in the high-frequency part of the spectrum, the lowest mode penetrates the solid body and, like high-order modes, propagates in the elastic layer. In this case, the elastic layer is decisive in forming the wave field. It is the main waveguide through which wave disturbances propagate and most of the wave energy is transferred.

The proposed approach and the results obtained make it possible to establish the limits of the application of models of wave processes based on various versions of the theory of small initial deformations and the classical theory of elasticity.

References

Akbarov SD, Ozaydin O (2001) The effect of initial stresses on har-monic stress fields within the stratified half-plane. Eur J Mech A Solids 20(3):385–396

Akbarov SD, Ozisik M (2003) The influence of the third order elastic constants to the generalized Rayleigh wave dispersion in a pre-stressed stratified half-plane. Int J Eng Sci. 41(17):2047–2061

Akbarov SD, Emiroglu I, Tasci F (2005) The Lamb's problem for a half-space covered with the pre-stretched layer. Int J Mech Sci 47(9):1326–1349

Guz AN (1980) Aerohydroelasticity problems for bodies with initial stresses. Int Appl Mech 16(3):75–190

Guz AN (2002) Elastic waves in bodies with initial (residual) stresses. Int Appl Mech 38(1):23–59

Guz AN (2004) Elastic waves in bodies with initial (residual) stresses. A.S.K, Kyiv [in Russian]

Guz AN (2009) Dynamics of compressible viscous fluid. Cambridge Scientific Publishers, Cambridge

Guz AN (2014) Ultrasonic nondestructive method for stress analysis of structural members and near-surface layers of materials: focus on Ukrainian research (review). Int Appl Mech 50(3):231–252

Guz AN (2016a) Elastic waves in bodies with initial (residual) stresses. Part 1. General problems. Waves in infinite bodies and surface waves. LAP LAMBERT Academic Publishing, Saarbrucken [in Russian]

Guz AN (2016b) Elastic waves in bodies with initial (residual) stresses. Part 2. Waves in partially bounded bodies. LAP LAMBERT Academic Publishing, Saarbrucken [in Russian]

Guz AN, Bagno AM (2018a) Effect of prestresses on quasi-lamb waves in a system consisting of a compressible viscous liquid half-space and an elastic layer. Int Appl Mech 54(6):617–627

Guz AN, Bagno AM (2018b) Effect of prestresses on Lamb waves in a system consisting of an ideal liquid half-space and an elastic layer. Int Appl Mech 54(5):495–505
Guz AN, Zhuk AP, Makhort FG (1976) Waves in layer with initial stresses. Nauk, Dumka, Kyiv [in Russian]
Guz AN, Zhuk AP, Bagno AM (2016) Dynamics of elastic bodies, solid particles, and fluid parcels in a compressible viscous fluid (review). Int Appl Mech 52(5):449–507
Koshkin NI, Shirkevich MG (1988) Handbook of elementary physics. Nauka, Moscow [in Russian]
Ottenio M, Destrade M, Ogden RW (2007) Acoustic waves at the interface of a pre-stressed incompressible elastic solid and a viscous fluid. Int J Non-Linear Mech 42(2):310–320
Zhuk AP (1980) Stonley waves in a medium with initial stresses. Appl Mech 16(1):113–116 [in Russian]

Chapter 4
Fracture of Materials Loaded Along Cracks: Approach and Results

Viacheslav Bogdanov, Aleksander Guz, and Vladimir Nazarenko

4.1 Introduction

Technological processes of manufacturing structural materials and assembling structure elements made of them often generate fields of initial (residual) stresses and strains in such materials. Those initial stresses must be taken into account in the calculations of product strength and durability, especially if crack-like defects emerge in such products in the processes of their manufacturing and operation. In the situation when initial stresses act along crack surfaces (and this situation is typical, e.g., in laminar or unidirectional fibrous composites (Dvorak 2000; Malmeister et al. 1980; Shul'ga and Tomashevskii 1997), materials with thermal insulation or anticorrosion coatings) (Ainsworth et al. 2000), the approaches of classical fracture mechanics (Cherepanov 1979; Kassir and Sih 1975) prove to be inapplicable. This results from the fact that such initial stresses are not involved in the expressions for stress intensity factors, J-integral and the values of crack opening, hence, they do not influence material's fracture parameters in the framework of Griffiths–Irwin, Cherepanov–Reiss fracture criteria, critical crack openings or their generalizations (Guz 1991, 2021; Guz et al. 2020).

In the situations when initial stresses are significantly larger as compared to additional (operational) stresses, for investigating problems of such kind, in Guz (1980, 1991) the applicability of the approach within the linearized mechanics of deformable solid bodies (Guz 1999) was justified, while in Guz (1982, 1991) energy- and force-based criteria of brittle fracture of materials with initial (residual) stresses

V. Bogdanov (✉) · A. Guz · V. Nazarenko
S.P. Timoshenko Institute of Mechanics, National Academy of Science of Ukraine, Kyiv, Ukraine
e-mail: bogdanov@nas.gov.ua

A. Guz
e-mail: guz@nas.gov.ua

V. Nazarenko
e-mail: vmnazarenko@nas.gov.ua

A. N. Guz et al. (eds.), *Advances in Mechanics*, Advanced Structured Materials 191,
https://doi.org/10.1007/978-3-031-37313-8_4

were formulated. The results of studying some problems of the fracture mechanics of materials with initial (residual) stresses, which were obtained with reliance on this approach, were presented in Bogdanov (2007, 2010, 2012), Bogdanov et al. (2015), Guz (1991, 2021).

Another group of non-classical problems of fracture mechanics is the fracture of bodies compressed along the parallel cracks they contain, when fracturing process is initiated by the local loss of stability in the part of material adjacent to the crack (Bolotin 1994, 2001; Guz et al. 1992, 2020; Kachanov 1988; Kienzler and Herrmann 2000; Wu 1979). Under such loading mode, singular parts in corresponding exact solutions of the problems of linear theory of elasticity are absent and, hence, all stress intensity factors are equal to zero, due to which classical fracture criteria are not applicable (Guz 2021; Guz et al. 2020). Many engineering problems related to the calculations of products with predetermined defects are reduced to the force-based scheme of compression along crack-like defects. Problems of this kind are rather typical in modeling the action of tectonic forces in mountainous terrain (model of fissured-layered massif), in calculating various supports, and in evaluating the strength and durability of concrete structure members. Various approaches to determining critical compression parameters, which correspond to the abovementioned local loss of stability, were analyzed in detail in Guz (2021), Guz et al. (2020). The results of investigating some problems on body compression along both isolated and interacting cracks with the use of the approach in the framework of the three-dimensional linearized theory of stability of deformable bodies are presented in Bogdanov and Nazarenko (1994), Guz (2014, 2021), and Guz et al. (1992, 2020).

It should be noted that although in terms of research subject the problems on the fracture of pre-stressed bodies under the action of initial stresses along cracks and the problem on the compression of materials along cracks are different, in the formulation of those problems there is an essential common point, viz., the presence of load components directed in parallel to cracks, whose influence, in fact, cannot be taken into account with the methods of classical fracture mechanics. This permits the two abovementioned groups of non-classical problems of fracture mechanics to be united and considered as problems of materials fracture under the action of forces directed along cracks. As it will be shown below, they can be investigated jointly, using the methodology based on relations of linearized mechanics of deformable solid bodies.

4.2 Approach to Studying the Problems

Below, brief information about the procedure used to investigate problems on the fracture of cracked bodies under the action of loads directed along cracks, and about the general formulation of corresponding boundary value problems is given.

As noted above, starting with the works (Guz 1980, 1991), to investigate problems of fracture mechanics for pre-stressed materials, when initial stresses act along the cracks the material contains and these initial stresses are significantly larger than operational stresses, an approach within the linearized mechanics of deformable solid bodies started to be applied consistently. The key factor in substantiating this method is that the application of linearized relations for investigating the abovementioned class of fracture mechanics problems, on the one hand, permits broad use of the advantages of the linear model of the deformable body and, on the other, qualitative and quantitative description (as opposed to the classical procedures) of the main phenomenon related to the influence of the stress components acting along crack surfaces on fracture parameters.

In Guz (1981), Guz et al. (1992), Wu (1979), it was shown that under compression of bodies along parallel cracks they contain, the beginning (start) of the fracture process is caused by the loss of material's stability in local areas near the cracks, when compressive forces achieve the values critical for the given material and the geometry of cracks location. Here, to determine the critical values of the compressive forces mentioned, the relations of the three-dimensional linearized theory of deformable bodies stability (Guz 1999) can be used, since with the involvement of the abovementioned criterion of the fracture process beginning (start), the possibility of the transition of a part of the material in the vicinity of cracks into adjacent forms of equilibrium under small (as compared to the main values of the initial states) perturbations of stresses and displacements is analyzed.

It should be noted that until recently the abovementioned two classes of fracture mechanics problems, viz., the problems on the fracture of materials with initial stresses acting along cracks and the problems on the fracture of bodies under compression along cracks were considered separately, but at the same time, taking into account that there are common features in the formulations of and approaches to these two classes of problems, namely, the presence of the load component acting along the cracks and the use of linearized relations for problems solving, in Bogdanov et al. (2017), Guz et al. (2013), the applicability of the unified approach within the linearized mechanics of deformable solid bodies was substantiated for investigating fracture mechanics problems on pre-stressed cracked materials and the problems on the fracture of bodies compressed along cracks (the information about this method can also be found in Guz et al. (2020)).

This approach is simpler and more effective for determining critical (limit) loading parameters in the problems on bodies compression along the cracks they contain, since there is no need of individual investigations of eigenvalue problems within the 3D linearized theory of stability. The parameters mentioned are calculated in solving corresponding boundary value problems of the mechanics of fracture of pre-stressed materials, when under the continuous change of loading parameters, we determine the initial compressive stresses which, when achieved, lead to a resonance change in the amplitude values (of stresses and displacements) near crack tips. The initial loading parameters determined in this way will correspond to the eigenvalues of corresponding eigenvalue problems on bodies compression along cracks.

Besides, an important positive feature of this approach is the possibility to conduct investigations in a single general form for compressible and incompressible isotropic or transverse isotropic elastic bodies with arbitrary structures of elastic potential as applied to the theory of finite (large) initial strains, as well as various variants of the theory of small initial strains. There, the specification of material's model (e.g., the use of elastic potential of one type or another) is only carried out at the final stage of the investigation—in the numerical analysis of the characteristic equations, the resolving integral equations, etc., obtained in the general form. It should also be noted that when considering composite materials in this work, it is assumed that crack sizes are significantly larger than the sizes of composite's structural elements, while the cases of cracks location in the interfaces of composite's components are not considered. With such assumptions, following, e.g., Broutman and Krock (1974), Dvorak (2000), Malmeister et al. (1980), we will use the continuum model of the composite with the reduced (averaged) characteristics of the transverse isotropic body.

Now, we present the principal relations of the linearized mechanics of deformable solid bodies, which will be used for solving particular problems. Figure 4.1 shows schematically an unbounded body with tensile or compressive initial (residual) stresses S_{11}^0 acting along cracks located in parallel planes $y_3 = \text{const}$. It should be noted that to carry out the investigation of the stress-strain state of pre-stressed bodies is more convenient in the Lagrangian coordinates y_j $(j = 1, 2, 3)$, which are related to the initial state caused by initial stresses S_{11}^0. These coordinates can be presented via the Cartesian coordinates of the non-deformed (natural) state of the body x_j $(j = 1, 2, 3)$ by the following relations

$$y_j = \lambda_j x_j, \quad j = 1, 2, 3, \tag{4.1}$$

where $\lambda_j = \text{const}$ are coefficients of elongation (contraction) along coordinate axes Oy_j, which are caused by initial stresses S_{11}^0. Besides, operational stresses (additional to the initial ones) also act on the body (normal stresses Q'_{33} are shown as an example).

Under the action of initial (residual) stresses S_{11}^0 only, a homogeneous stress-strain state emerges in the material (both isotropic and transverse isotropic). (It is assumed for transverse isotropic material that cracks are located in the planes of material's properties symmetry and, thus, initial stresses are directed along the sym-

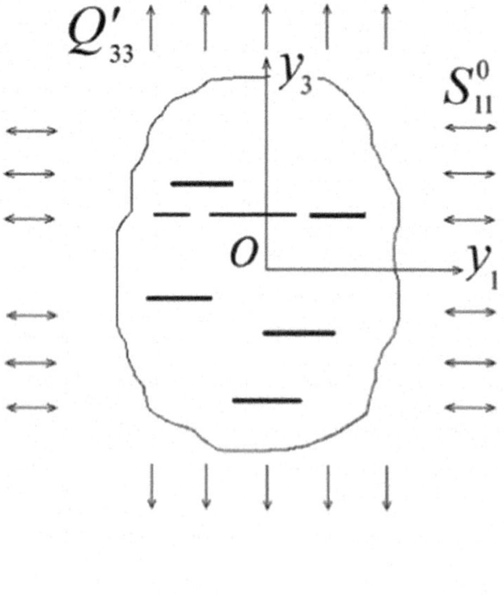

Fig. 4.1 Cracked body with initial stresses

metry axes of material's properties.) This stress-strain state is characterized by such expressions for components of the tensor of initial stresses $\tilde{S}^0$ and the vector of initial displacements $\mathbf{u}^0$:

$$S_{11}^0 = \text{const}, \; S_{22}^0 = \text{const}, \; S_{33}^0 = 0, \; S_{ij}^0 = 0, \; i \neq j; \; u_j^0 = \lambda_j^{-1} \left(\lambda_j - 1 \right) y_j. \tag{4.2}$$

For compressible bodies, linearized equilibrium equations in displacements are of the form Bogdanov et al. (2017), Guz (1999), Guz et al. (2020)

$$\omega'_{ij\alpha\beta} \frac{\partial^2 u_\alpha}{\partial y_i \partial y_\beta} = 0, \tag{4.3}$$

where u_α are displacements caused by the action of initial and operational stresses. Boundary conditions in stresses on a part of S_1 surface are presented as

$$N_i^0 Q'_{ij} = P'_j, \tag{4.4}$$

where N_i^0 are components of the ort of the normal to the surface of the body in the initial state (the body state caused by initial stresses S_{11}^0), while boundary conditions in displacements on a part of S_2 surface are of the form

$$u_j = f'_j. \tag{4.5}$$

The components of the fourth rank elasticity tensor $\tilde{\omega}'$, which are involved in (4.3) and in the linearized elasticity relations

$$Q'_{ij} = \omega'_{ij\alpha\beta} \frac{\partial u_\alpha}{\partial y_\beta}, \tag{4.6}$$

are given by expressions

$$\omega'_{ij\alpha\beta} = \frac{\lambda_i \lambda_j \lambda_\alpha \lambda_\beta}{\lambda_1 \lambda_2 \lambda_3} \left[\delta_{ij} \delta_{\alpha\beta} A_{i\beta} + \left(1 - \delta_{ij}\right) \left(\delta_{i\alpha} \delta_{j\beta} + \delta_{i\beta} \delta_{j\alpha} \right) G_{ij} \right] + \frac{\lambda_i \lambda_\beta}{\lambda_1 \lambda_2 \lambda_3} \delta_{i\beta} \delta_{j\alpha} S_{\beta\beta}^0, \tag{4.7}$$

where δ_{ij} is Kronecker symbol, A_{ij} are elasticity constants, G_{ij} are shear moduli, $S_{\beta\beta}^0$ are initial stresses, and λ_m are coefficients of elongation (contraction) along coordinate axes Oy_m, that is caused by these initial stresses.

The dependence between components of Piola–Kirchhoff non-symmetric stress tensor $\tilde{Q}'$ and Lagrange symmetric stress tensor $\tilde{S}$ is given by relations (Bogdanov et al. 2017; Guz 1999; Guz et al. 2020)

$$Q'_{ij} = \frac{\lambda_i \lambda_j}{\lambda_1 \lambda_2 \lambda_3} S_{ij} + \frac{\lambda_i}{\lambda_1 \lambda_2 \lambda_3} S_{in}^0 \frac{\partial u_j}{\partial y_n}.$$

Representations of the general solutions of linearized equilibrium equations (4.3) via harmonic potentials were constructed in Bogdanov et al. (2017), Guz (1999), Guz et al. (2020). By assuming that the axis of material's isotropy coincides with axis Oy_3 of the coordinate system, and conditions $\lambda_1 = \lambda_2 \neq \lambda_3$, $S_{11}^0 = S_{22}^0$, $S_{33}^0 = 0$ are satisfied, we have such representations of the general solutions for the circular cylindrical coordinate system (r, θ, y_3) obtained from the Cartesian one (Bogdanov et al. 2017; Guz 1999; Guz et al. 2020):

in the case of non-equal roots of the characteristic equation $(n_1 \neq n_2)$

$$\begin{aligned}
u_r &= \frac{\partial(\varphi_1+\varphi_2)}{\partial r} - \frac{1}{r}\frac{\partial\varphi_3}{\partial\theta},\\
u_\theta &= \frac{1}{r}\frac{\partial(\varphi_1+\varphi_2)}{\partial\theta} + \frac{\partial\varphi_3}{\partial r},\\
u_3 &= m_1 n_1^{-1/2}\frac{\partial\varphi_1}{\partial z_1} + m_2 n_2^{-1/2}\frac{\partial\varphi_2}{\partial z_2},\\
Q'_{33} &= C_{44}\left(d_1 l_1 \frac{\partial^2\varphi_1}{\partial z_1^2} + d_2 l_2 \frac{\partial^2\varphi_2}{\partial z_2^2}\right),\\
Q'_{3r} &= C_{44}\left(d_1 n_1^{-1/2}\frac{\partial^2\varphi_1}{\partial r\partial z_1} + d_2 n_2^{-1/2}\frac{\partial^2\varphi_2}{\partial r\partial z_2} - n_3^{-1/2}\frac{1}{r}\frac{\partial^2\varphi_3}{\partial\theta\partial z_3}\right),\\
Q'_{3\theta} &= C_{44}\left(d_1 n_1^{-1/2}\frac{1}{r}\frac{\partial^2\varphi_1}{\partial\theta\partial z_1} + d_2 n_2^{-1/2}\frac{1}{r}\frac{\partial^2\varphi_2}{\partial\theta\partial z_2} + n_3^{-1/2}\frac{\partial^2\varphi_3}{\partial r\partial z_3}\right),\\
&\qquad z_j = n_j^{-1/2} y_3, \quad j = 1, 2, 3;
\end{aligned} \tag{4.8}$$

in the case of equal roots of the characteristic equation $(n_1 = n_2)$

$$\begin{aligned}
u_r &= -\frac{\partial\varphi}{\partial r} - z_1\frac{\partial F}{\partial r} - \frac{1}{r}\frac{\partial\varphi_3}{\partial\theta},\\
u_\theta &= -\frac{1}{r}\frac{\partial\varphi}{\partial\theta} - z_1\frac{1}{r}\frac{\partial F}{\partial\theta} + \frac{\partial\varphi_3}{\partial r},\\
u_3 &= (m_1 - m_2 + 1)\, n_1^{-1/2} F - m_1 n_1^{-1/2}\Phi - m_1 n_1^{-1/2} z_1 \frac{\partial F}{\partial z_1},\\
Q'_{33} &= C_{44}\left[(d_1 l_1 - d_2 l_2)\frac{\partial F}{\partial z_1} - d_1 l_1 \frac{\partial\Phi}{\partial z_1} - d_1 l_1 z_1 \frac{\partial^2 F}{\partial z_1^2}\right], \quad \Phi \equiv \frac{\partial\varphi}{\partial z_1},\\
Q'_{3r} &= C_{44}\left\{n_1^{-1/2}\frac{\partial}{\partial r}\left[(d_1 - d_2)F - d_1\Phi\right] - n_1^{-1/2} d_1 z_1 \frac{\partial^2 F}{\partial r\partial z_1} - n_3^{-1/2}\frac{1}{r}\frac{\partial^2\varphi_3}{\partial\theta\partial z_3}\right\},\\
Q'_{3\theta} &= C_{44}\left\{n_1^{-1/2}\frac{1}{r}\frac{\partial}{\partial\theta}\left[(d_1 - d_2)F - d_1\Phi\right] - n_1^{-1/2} d_1 z_1 \frac{1}{r}\frac{\partial^2 F}{\partial\theta\partial z_1} + n_3^{-1/2}\frac{\partial^2\varphi_3}{\partial r\partial z_3}\right\},
\end{aligned} \tag{4.9}$$

where the roots of the characteristic equations take the form

$$\begin{aligned}
n_{1,2} &= c' \pm \sqrt{c'^2 - \frac{\omega'_{3333}\omega'_{3113}}{\omega'_{1111}\omega'_{1331}}}, \quad n_3 = \frac{\omega'_{3113}}{\omega'_{1221}},\\
c' &= \frac{\omega'_{1111}\omega'_{3333} + \omega'_{3113}\omega'_{1331} - \left(\omega'_{1133} + \omega'_{1313}\right)^2}{2\omega'_{1111}\omega'_{1331}}.
\end{aligned} \tag{4.10}$$

In (4.8) and (4.9), the potentials $\varphi_j\left(r,\theta,z_j\right)$, $\varphi\left(r,\theta,z_j\right)$, and $F\left(r,\theta,z_j\right)$ $(j = 1,2,3)$ satisfy Laplace's equations; the values C_{44}, m_i, d_i, and l_i $(i = 1,2)$ are determined by the choice of material's model and are linked with components of elasticity tensor $\tilde{\omega}'$ (4.7) (Bogdanov et al. 2017; Guz 1999; Guz et al. 2020).
in the case of non-equal roots of the characteristic equation $(n_1 \neq n_2)$

$$C_{44} = \omega'_{1313}, \quad m_i = \frac{\omega'_{1111}n_i - \omega'_{3113}}{\omega'_{1133} + \omega'_{1313}}, \quad d_i = 1 + m_i,$$
$$l_i = \frac{\omega'_{3333}m_i - \omega'_{1133}n_i}{n_i d_i \omega'_{1313}}, \quad i = 1,2; \tag{4.11}$$

in the case of equal roots of the characteristic equation $(n_1 = n_2)$ parameters C_{44}, m_1, d_1, d_2, l_1 are determined from (4.11), and parameters m_2, l_2 take the from

$$m_2 = \frac{\omega'_{1133} - \omega'_{1313}}{\omega'_{1133} + \omega'_{1313}}, \quad l_2 = \frac{\omega'_{3333}(m_1 + m_2 - 1) - \omega'_{1133}n_1}{n_1 d_2 \omega'_{1313}}. \tag{4.12}$$

In the case of axisymmetric linearized problems, the potential function f_3 in (4.8) and (4.9) should be set equal to zero, while the potential functions φ_1, φ_2 , φ, F are to be considered independent of coordinate θ.

Taking into account representations (4.8) and (4.9), the general statement of linearized problems (4.3)–(4.5) can be re-formulated in terms of harmonic potential functions $\varphi_j\left(r,\theta,z_j\right)$, $j = 1,2,3$ (in the case of non-equal roots) and $\varphi\left(r,\theta,z_1\right)$, $F\left(r,\theta,z_1\right)$, $\varphi_3\left(r,\theta,z_3\right)$ (in the case of equal roots). For the spatial boundary value problems on pre-stressed bodies containing circular cracks (which are also referred to as penny-shaped cracks), considered in this work, we will present the potential functions mentioned as Henkel integral transform in radial coordinate, reduce the problems to paired (dual) integral equations and then to Fredholm integral equations of the second kind, which will be investigated numerically.

4.3 Formulation of the Problems

Consider spatial problems on pre-stressed half-bounded body with a near-surface circular crack and those on an unbounded body with initial (residual) stresses, containing two parallel coaxial circular cracks. It should be noted that the former geometric scheme permits the analysis of the influence of initial stresses as well as the effect of the interaction of cracks and the free surface of the body on stress intensity factors in the vicinity of crack contours and on the critical compression parameters, which, when achieved, lead to the local loss of material's stability near cracks. The latter geometric scheme permits the evaluation of the influence of parallel cracks interaction on those parameters.

4.3.1 *Initially Stressed Half-Space with a Near-Surface Circular Crack*

Consider an elastic body occupying half-space $y_3 \geqslant -h$. There are initial stresses $S_{11}^0 = S_{22}^0$ acting along a near-surface crack of radius $r = a$, located in $y_3 = 0$ plane centered on axis Oy_3: $\{0 \leqslant r \leqslant a, 0 \leqslant \theta < 2\pi, y_3 = 0\}$ (Fig. 4.2). We assume that additional (with respect to initial stresses) fields of normal and shear forces Q'_{33} and Q'_{3r} act on crack faces, while the half-space boundary is free of loads. Boundary conditions of such non-axisymmetric problem are of the form

$$\begin{gathered} Q'_{33} = -\sigma(r,\theta),\ Q'_{3r} = -\tau_r(r,\theta),\ Q'_{3\theta} = 0\ (y_3 = (0)_\pm,\ 0 \leqslant r \leqslant a), \\ Q'_{33} = 0,\quad Q'_{3r} = 0,\quad Q'_{3\theta} = 0\quad (y_3 = -h,\ 0 \leqslant r < \infty). \end{gathered} \tag{4.13}$$

(Here and further $0 \leqslant \theta < 2\pi$, and subscripts "+" and "–" denote the upper and lower crack faces respectively).

Besides, the conditions of the attenuation of displacement vector and stress tensor components at infinity must be satisfied as

$$u_j \to 0,\quad Q'_{ij} \to 0\quad (r \to +\infty,\ y_3 \to +\infty). \tag{4.14}$$

Further, in constructing the solution of the problem examined, it is convenient to divide the half-space $y_3 \geqslant -h$ into two domains: domain "1" is the half-space $y_3 \geqslant 0$ and domain "2" is the layer $-h \leqslant y_3 \leqslant 0$. All the values relating to each of these domains will be marked with superscripts "(1)" and "(2)". For such subdivision into two domains, on the domain boundary (when $y_3 = 0$) outside the crack, it is necessary that the conditions of continuity for displacement and stress vectors be satisfied. Then the boundary conditions (4.13) can be written as

$$\begin{gathered} Q_{33}^{'(2)} = -\sigma(r,\theta),\ Q_{3r}^{'(2)} = -\tau_r(r,\theta),\ Q_{3\theta}^{'(2)} = 0\ (y_3 = 0,\ 0 \leqslant r \leqslant a), \\ Q_{33}^{'(2)} = 0,\ Q_{3r}^{'(2)} = 0,\ Q_{3\theta}^{'(2)} = 0\ (y_3 = -h,\ 0 \leqslant r < \infty), \\ u_3^{(1)} = u_3^{(2)},\ u_r^{(1)} = u_r^{(2)},\ u_\theta^{(1)} = u_\theta^{(2)}\ (y_3 = 0,\ a < r < \infty), \\ Q_{33}^{'(1)} = Q_{33}^{'(2)},\ Q_{3r}^{'(1)} = Q_{3r}^{'(2)},\ Q_{3\theta}^{'(1)} = Q_{3\theta}^{'(2)}\ (y_3 = 0,\ 0 \leqslant r < \infty). \end{gathered} \tag{4.15}$$

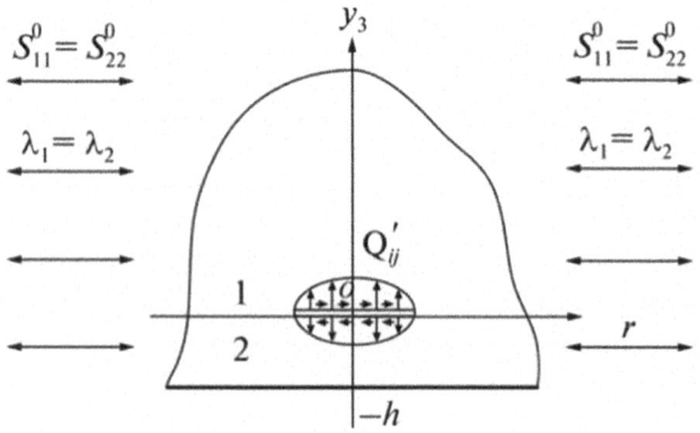

Fig. 4.2 Pre-stressed semi-infinite body with a circular near-surface crack

By using the representations of general solutions in terms of potential harmonic functions of form (4.8) for non-equal roots and form (4.9) for equal roots, from (4.15) we obtain the problem formulation in terms of harmonic potential functions $\varphi_i{}^{(k)}(r, \theta, z_j)$, $k = 1, 2$, $i, j = 1, 2, 3$ (in the case of non-equal roots) and $\varphi^{(k)}(r, \theta, z_1)$, $F^{(k)}(r, \theta, z_1)$ and $\varphi_3{}^{(k)}(r, \theta, z_3)$, $k = 1, 2$ (in the case of equal roots).

The formulations of axisymmetric problems for the pre-stressed half-space with mode I, mode II, or mode III cracks are carried out in the similar way, if in the corresponding boundary conditions, it is set that $u_\theta{}^{(k)} = 0$, $Q'_{3\theta}{}^{(k)} = 0$, $k = 1, 2$, while other components of displacement vector and stress tensor are considered to be independent of angular component θ.

4.3.2 Pre-stressed Body with Two Parallel Circular Cracks

Consider an unbounded elastic body with initial stresses $S^0_{11} = S^0_{22}$ that contains two circular cracks of the same radius $r = a$, which are located in parallel planes $y_3 = 0$ and $y_3 = -2h$ with centers on the Oy_3 axis (Fig. 4.3). Additional stresses Q'_{33} and Q'_{3r} (with respect to the initial ones $S^0_{11} = S^0_{22}$) acting on crack faces and the boundary condition are

$$
\begin{aligned}
&Q'_{33} = -\sigma(r, \theta),\ Q'_{3r} = -\tau_r(r, \theta),\ Q'_{3\theta} = 0\ (y_3 = (0)_\pm,\ 0 \leqslant r \leqslant a),\\
&Q'_{33} = -\sigma(r, \theta),\ Q'_{3r} = -\tau_r(r, \theta),\ Q'_{3\theta} = 0\ (y_3 = (-2h)_\pm,\ 0 \leqslant r \leqslant a),
\end{aligned}
\tag{4.16}
$$

where $0 \leqslant \theta < 2\pi$, and subscripts "+" and "–" denote the upper and lower cracks faces respectively.

For the case considered, the symmetry of the geometric and force-based schemes of the problem in respect of plane $y_3 = -h$ exists. Due to that, given (4.16), it can be re-formulated as a mathematically equivalent problem on the half-space $y_3 \geqslant -h$

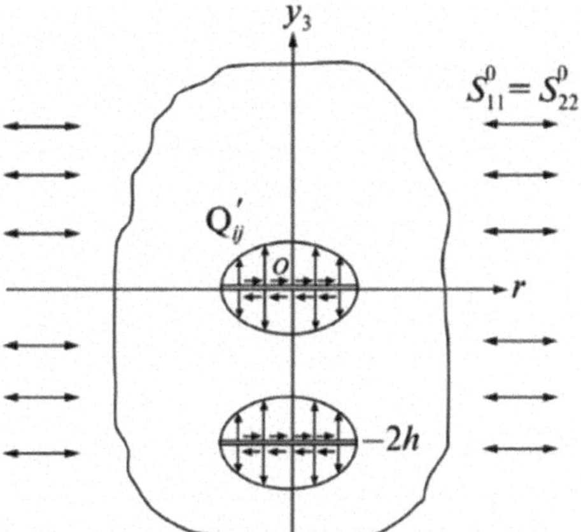

Fig. 4.3 Pre-stressed body with two parallel coaxial circular cracks

with a single mode I or mode II crack located in the plane $y_3 = 0$, with the following boundary conditions on its faces and on the half-space boundary:

for *mode I crack*

$$\begin{aligned} Q'_{33} = -\sigma(r,\theta),\quad Q'_{3r} = 0,\quad Q'_{3\theta} = 0 \quad (y_3 = (0)_\pm,\ 0 \leqslant r \leqslant a), \\ u_3 = 0,\quad Q'_{3r} = 0,\quad Q'_{3\theta} = 0 \quad (y_3 = -h,\ 0 \leqslant r \leqslant a); \end{aligned} \tag{4.17}$$

for *mode II crack*

$$\begin{aligned} Q'_{33} = 0,\quad Q'_{3r} = -\tau_r(r,\theta),\quad Q'_{3\theta} = 0 \quad (y_3 = (0)_\pm,\ 0 \leqslant r \leqslant a), \\ u_r = 0,\quad u_\theta = 0,\quad Q'_{33} = 0 \quad (y_3 = -h,\ 0 \leqslant r \leqslant a). \end{aligned} \tag{4.18}$$

Besides, the conditions of the attenuation of the displacement vector and stress tensor components at infinity (4.14) must be satisfied.

As can be seen from (4.17) and (4.18), the formulation of the problem on a space with two parallel coaxial cracks (4.16) is mathematically equivalent to the problems on the half-space $y_3 \geqslant -h$ with a mode I crack or a mode II crack that is parallel to its surface when the half-space boundary rests on a smooth rigid foundation.

We will confine ourselves to the axisymmetric problem and will consider individually the cases when normal stresses Q'_{33} (mode I cracks), radial shear stresses Q'_{3r} (mode II cracks) or tangential torsional stresses $Q'_{3\theta}$ (mode III cracks) act on crack faces.

Mode I cracks. Taking into account the symmetry of the geometric and force-based schemes of the problem in respect of plane $y_3 = -h$, we have the following boundary conditions specified on the cracks faces and on the half-space boundary and crack faces

$$\begin{aligned} Q'_{33} = -\sigma\,(r),\quad Q'_{3r} = 0 \quad (y_3 = (0)_\pm,\ 0 \leqslant r \leqslant a), \\ u_3 = 0,\quad Q'_{3r} = 0 \quad (y_3 = -h,\ 0 \leqslant r \leqslant \infty). \end{aligned} \tag{4.19}$$

The half-space $y_3 \geqslant -h$ is conditionally divided into two domains: “1”—the half-space $y_3 \geqslant 0$ and “2”—the layer $-h \leqslant y_3 \leqslant 0$. All the values relating to each of the domains mentioned are marked by superscripts “(1)” and “(2)”. Taking into account the conditions of the continuity of stresses and displacements on the boundaries of the domains, new boundary conditions of the problem are obtained from (4.17)

$$\begin{aligned} Q'^{(2)}_{33} = -\sigma(r),\ Q'^{(2)}_{3r} = 0 \quad (y_3 = 0,\ 0 \leqslant r \leqslant a), \\ u^{(2)}_3 = 0,\ Q'^{(2)}_{3r} = 0 \quad (y_3 = -h,\ 0 \leqslant r \leqslant \infty), \\ u^{(1)}_3 = u^{(2)}_3,\ u^{(1)}_r = u^{(2)}_r \quad (y_3 = 0,\ a < r < \infty), \\ Q'^{(1)}_{33} = Q'^{(2)}_{33},\ Q'^{(1)}_{3r} = Q'^{(2)}_{3r} \quad (y_3 = 0,\ 0 \leqslant r < \infty). \end{aligned} \tag{4.20}$$

Mode II cracks. The tangential radial stresses of the intensity $t_r(r)$ are applied to crack faces antisymmetrically in respect of the planes of cracks location. Taking into account the symmetry of the geometric and force-based schemes of the problem with respect to the plane $y_3 = -h$, equidistant from the cracks and dividing the half-space

$y_3 \geqslant -h$ into two parts, namely, the half-space $y_3 \geqslant 0$ (domain "1") and the layer $-h \leqslant y_3 \leqslant 0$ (domain "2"), we have such boundary conditions of the problem on mode II cracks:

$$\begin{gathered}
Q_{33}^{\prime(2)}=0,\ Q_{3r}^{\prime(2)}=-\tau_r(r) \quad (y_3=0,\ 0\leqslant r\leqslant a),\\
u_r^{(2)}=0,\ Q_{33}^{\prime(2)}=0 \quad (y_3=-h,\ 0\leqslant r\leqslant\infty),\\
u_3^{(1)}=u_3^{(2)},\ u_r^{(1)}=u_r^{(2)} \quad (y_3=0,\ a<r<\infty),\\
Q_{33}^{\prime(1)}=Q_{33}^{\prime(2)},\ Q_{3r}^{\prime(1)}=Q_{3r}^{\prime(2)} \quad (y_3=0,\ 0\leqslant r<\infty).
\end{gathered} \tag{4.21}$$

By using the representations of general solutions for displacements and stresses via potential functions (4.8) (for non-equal real roots) and (4.9) (for equal roots), from boundary conditions (4.20) and (4.21) the formulations of problems for mode I crack and mode II crack are obtained in terms of potential functions $\varphi_i{}^{(k)}\left(r, z_j\right)$, $i, j, k=1,2$ (in the case of non-equal roots) and $\varphi^{(k)}\left(r, z_1\right)$, $F^{(k)}\left(r, z_1\right)$, $k=1,2$ (in the case of equal roots).

Mode III cracks. The tangential torsional stresses of the intensity $\tau_\theta(r)$ are applied to crack faces antisymmetrically in respect of cracks location planes. With this loading scheme, only the components of displacement vector u_θ and stress tensor $Q'_{3\theta}$ will be non-zero, and owing to the axisymmetric nature of the problem they do not depend on the angular coordinate θ. The boundary conditions of the problem on mode III cracks can be written as follows

$$\begin{gathered}
Q_{3\theta}^{\prime(2)}=-\tau_\theta(r) \quad (y_3=0,\ 0\leqslant r\leqslant a),\\
u_\theta^{(2)}=0 \quad (y_3=-h,\ 0\leqslant r\leqslant\infty),\\
u_\theta^{(1)}=u_\theta^{(2)} \quad (y_3=0,\ a<r<\infty),\\
Q_{3\theta}^{\prime(1)}=Q_{3\theta}^{\prime(2)} \quad (y_3=0,\ 0\leqslant r<\infty).
\end{gathered} \tag{4.22}$$

The representations of general solutions in this case, given (4.8) and (4.9), are of the same form for the cases of both equal and non-equal roots, specifically

$$u_\theta=\frac{\partial}{\partial r}\varphi_3\left(r, z_3\right), \quad Q'_{3\theta}=C_{44}n_3^{-1/2}\frac{\partial^2}{\partial r\,\partial z_3}\varphi_3\left(r, z_3\right). \tag{4.23}$$

By using representations (4.23) we obtain the formulation of the problem for elastic body with mode III cracks in terms of the potential harmonic function $\varphi_3\left(r, z_3\right)$

$$\begin{gathered}
C_{44}n_3^{-1/2}\frac{\partial^2\varphi_3^{(2)}}{\partial r\,\partial z_3}=-\tau_\theta\left(r\right) \quad \left(y_3=0,\ 0\leqslant r\leqslant a\right),\\
\frac{\partial\varphi_3^{(2)}}{\partial r}=0 \quad \left(y_3=-h,\ 0\leqslant r<\infty\right),\\
\frac{\partial\varphi_3^{(1)}}{\partial r}=\frac{\partial\varphi_3^{(2)}}{\partial r} \quad \left(y_3=0,\ a<r<\infty\right),\\
\frac{\partial^2\varphi_3^{(1)}}{\partial r\,\partial z_3}=\frac{\partial^2\varphi_3^{(2)}}{\partial r\,\partial z_3} \quad \left(y_3=0,\ 0\leqslant r<\infty\right).
\end{gathered} \tag{4.24}$$

4.4 Fredholm Integral Equations

The mixed boundary value problems on harmonic potential functions, which were formulated in the previous section will be solved by first reducing them to paired (dual) integral equations and then—to systems of Fredholm integral equations of the second kind. For that, harmonic potentials will be represented via Fourier series (for non-axisymmetric problems) and Hankel integral expansions in radial coordinates. The process of solving non-axisymmetric problems will be exemplified by the problem on a half-space with a near-surface crack, and the solution of axisymmetric problems—by that on a body containing two parallel mode I cracks.

4.4.1 Half-Space with a Near-Surface Circular Crack

We will show calculation procedures for the case of equal roots of the characteristic equation ($n_1 = n_2$); the procedures for non-equal roots ($n_1 \neq n_2$) are carried out in the similar way.

The external loads on the crack faces (the right-hand parts of the first two expressions in (4.15)) are presented as Fourier series in coordinate θ, assuming that they are even functions in this coordinate

$$\sigma\,(r,\theta) = \sum_{n=0}^{\infty} \sigma^{(n)}(r)\cos n\theta\ , \quad \tau_r\,(r,\theta) = \sum_{n=0}^{\infty} \tau_r{}^{(n)}(r)\cos n\theta\ , \tag{4.25}$$

where coefficients in the expansions are of the form:

$$\begin{aligned}
\sigma^{(0)}\,(r) &= \frac{1}{\pi}\int_0^{\pi} \sigma\,(r,\theta)\,d\theta,\\
\tau_r^{(0)}\,(r) &= \frac{1}{\pi}\int_0^{\pi} \tau_r\,(r,\theta)\,d\theta,\\
\sigma^{(n)}\,(r) &= \frac{2}{\pi}\int_0^{\pi} \sigma\,(r,\theta)\cos n\theta\,d\theta,\\
\tau_r^{(n)}\,(r) &= \frac{2}{\pi}\int_0^{\pi} \tau_r\,(r,\theta)\cos n\theta\,d\theta, \quad n = 1, 2, \ldots.
\end{aligned} \tag{4.26}$$

In the case when $\sigma\,(r,\theta)$ and $\tau_r\,(r,\theta)$ are odd functions in θ, their transforms into Fourier series will be similar if cosines are changed to sines in (4.25); in the general case, when loads are arbitrary functions, the superposition of solutions should be used.

We present potential functions $\varphi^{(k)}\,(r,\theta,z_1)$, $F^{(k)}\,(r,\theta,z_1)$ and $\varphi_3^{(k)}\,(r,\theta,z_3)$ $(k = 1, 2)$ as Fourier series in coordinate θ with coefficients in the form of Hankel integral equations in radial coordinate r of the order corresponding to the harmonic in θ

$$\varphi^{(1)}(r,\theta,z_1) = -\sum_{n=0}^{\infty}\cos n\theta \int_0^{\infty} B_n(\lambda)e^{-\lambda z_1} J_n(\lambda r)\frac{d\lambda}{\lambda},$$
$$F^{(1)}(r,\theta,z_1) = \sum_{n=0}^{\infty}\cos n\theta \int_0^{\infty} A_n(\lambda)e^{-\lambda z_1} J_n(\lambda r)d\lambda,$$
$$\varphi_3^{(1)}(r,\theta,z_3) = \sum_{n=1}^{\infty}\sin n\theta \int_0^{\infty} C_n(\lambda)e^{-\lambda z_3} J_n(\lambda r)\frac{d\lambda}{\lambda},$$
$$\varphi^{(2)}(r,\theta,z_1) = \sum_{n=0}^{\infty}\cos n\theta \int_0^{\infty} \left[B_n^{(1)}(\lambda)\sinh\lambda(h_1+z_1) + B_n^{(2)}(\lambda)\cosh\lambda(h_1+z_1)\right] J_n(\lambda r)\frac{d\lambda}{\lambda\sinh\lambda h_1}, \tag{4.27}$$
$$F^{(2)}(r,\theta,z_1) = \sum_{n=0}^{\infty}\cos n\theta \int_0^{\infty} \left[A_n^{(1)}(\lambda)\cosh\lambda(h_1+z_1) + B_n^{(2)}(\lambda)\sinh\lambda(h_1+z_1)\right] J_n(\lambda r)\frac{d\lambda}{\sinh\lambda h_1},$$
$$\varphi_3^{(2)}(r,\theta,z_3) = \sum_{n=1}^{\infty}\sin n\theta \int_0^{\infty} \left[C_n^{(1)}(\lambda)\cosh\lambda(h_3+z_3) + C_n^{(2)}(\lambda)\sinh\lambda(h_3+z_3)\right] J_n(\lambda r)\frac{d\lambda}{\lambda\sinh\lambda h_3},$$
$$h_j = n_j^{-1/2}h, \quad j = 1, 3.$$

In expressions (4.27), A_n, B_n, C_n, $A_n^{(k)}$, $B_n^{(k)}$, and $C_n^{(k)}$ $(k = 1, 2)$ are unknown functions that are to be determined. It should be noted that the presentation of potential functions as (4.27) ensures that conditions (4.14) are satisfied.

Substitute expressions (4.26) and (4.27) into boundary conditions (4.15). Then, from the conditions presented in the second and fourth lines of (4.15), which are set on all planes $y_3 = -h$, $y_3 = 0$, we obtain six relations linking nine functions A_n, B_n, C_n, $A_n{}^{(i)}$, $B_n^{(i)}$, and $C_n^{(i)}$ $(i = 1, 2)$

$$
\begin{aligned}
B_n^{(1)}(\lambda) &= \mu_1 A_n^{(2)}(\lambda) + \left(1 - \frac{d_2}{d_1}\right) A_n^{(1)}(\lambda), \\
B_n^{(2)}(\lambda) &= \mu_1 A_n^{(1)}(\lambda) + \left(1 - \frac{d_2 l_2}{d_1 l_1}\right) A_n^{(2)}(\lambda), \quad C_n^{(2)}(\lambda) = 0, \\
A_n(\lambda) &= \left[\frac{\mu_1}{k}(1 + \coth\mu_1) - 1\right] A_n^{(1)}(\lambda) + \left[\frac{\mu_1}{k}(1 + \coth\mu_1) + 1\right] A_n^{(2)}(\lambda), \\
B_n(\lambda) &= \left[\left(1 - \frac{d_2 l_2}{d_1 l_1}\right)\frac{\mu_1}{k}(1 + \coth\mu_1) - \left(1 - \frac{d_2}{d_1}\right) - \mu_1 \coth\mu_1\right] A_n^{(1)}(\lambda) \\
&\quad + \left[\left(1 - \frac{d_2 l_2}{d_1 l_1}\right)\frac{\mu_1}{k}(1 + \coth\mu_1) + \left(1 - \frac{d_2 l_2}{d_1 l_1}\right) - \mu_1\right] A_n^{(2)}(\lambda), \\
C_n(\lambda) &= -C_n^{(1)}(\lambda), \quad \mu_1 = \lambda h_1, \quad k = \frac{d_2(l_1 - l_2)}{d_1 l_1}.
\end{aligned}
\tag{4.28}
$$

From the remaining boundary conditions (the first and the third lines in (4.15)), taking into account the following relations (Watson 1995)

$$
\begin{aligned}
\frac{2n}{\lambda r} J_n(\lambda r) &= J_{n-1}(\lambda r) + J_{n+1}(\lambda r), \\
2\frac{\partial J_n(\lambda r)}{\partial(\lambda r)} &= J_{n-1}(\lambda r) - J_{n+1}(\lambda r),
\end{aligned}
$$

and equating to zero the relations at $\cos n\theta$, $\sin n\theta$, we obtain (individually for each nth harmonic in coordinate θ) the system of paired integral equations

$$
\begin{aligned}
&\int_0^\infty \left\{n_1^{-1/2} d_1 \left[\mu_1 A_n^{(1)} + (k + \mu_1 \coth\mu_1) A_n^{(2)}\right] - n_3^{-1/2} C_n^{(1)}\right\} \\
&\qquad \times J_{n+1}(\lambda r)\lambda d\lambda = -\frac{1}{C_{44}}\left[\tau_r^{(n)}(r) + \tau_\theta^{(n)}(r)\right], \quad r \leqslant a, \\
&\int_0^\infty \left\{n_1^{-1/2} d_1 \left[\mu_1 A_n^{(1)} + (k + \mu_1 \coth\mu_1) A_n^{(2)}\right] + n_3^{-1/2} C_n^{(1)}\right\} \\
&\qquad \times J_{n-1}(\lambda r)\lambda d\lambda = \frac{1}{C_{44}}\left[\tau_r^{(n)}(r) - \tau_\theta^{(n)}(r)\right], \quad r \leqslant a, \\
&\int_0^\infty \left[(k - \mu_1 \coth_1) A_n^{(1)} - \mu_1 A_n^{(2)}\right] J_n(\lambda r)\lambda d\lambda \\
&\qquad = -\frac{\sigma^{(n)}(r)}{C_{44} d_1 l_1}, \quad r \leqslant a,
\end{aligned}
\tag{4.29}
$$

$$
\begin{aligned}
&\int_0^\infty X_1 J_{n+1}(\lambda r)\, d\lambda = 0, \quad r > a, \\
&\int_0^\infty X_2 J_{n-1}(\lambda r)\, d\lambda = 0, \quad r > a, \\
&\int_0^\infty X_3 J_n(\lambda r)\, d\lambda = 0, \quad r > a,
\end{aligned}
$$

where the following notations are used

$$
\begin{aligned}
X_1 &= \left(1-\frac{d_2 l_2}{d_1 l_1}\right)(1+\coth\mu_1)\left[\frac{\mu_1}{k}A_n^{(1)}+\left(1+\frac{\mu_1}{k}\right)A_n^{(2)}\right] \\
&\qquad - C_n^{(1)}(1+\coth\mu_3), \\
X_2 &= \left(1-\frac{d_2 l_2}{d_1 l_1}\right)(1+\coth\mu_1)\left[\frac{\mu_1}{k}A_n^{(1)}+\left(1+\frac{\mu_1}{k}\right)A_n^{(2)}\right] \\
&\qquad + C_n^{(1)}(1+\coth\mu_3), \\
X_3 &= 2\left(1-\frac{d_2 l_2}{d_1 l_1}\right)\left[\left(1-\frac{\mu_1}{k}\right)A_n^{(1)}-\frac{\mu_1}{k}A_n^{(2)}\right](1+\coth\mu_1).
\end{aligned}
\tag{4.30}
$$

To solve the system of paired equations (4.29), in accordance with the substitution method (Uflyand 1977), we present X_1, X_2, and X_3, given by (4.30) in the form that provides the identical satisfaction of those parts of the system of paired equations which are specified in the range $r > a$. Now, we introduce new unknown functions $\varphi(t)$, $\psi(t)$, and $\omega(t)$, which are continuous along with their first derivatives in the segment $[0, a]$, and represent via these functions the expressions X_j $(j = 1, 2, 3)$ as

$$
\begin{aligned}
X_1 &= \sqrt{\frac{\pi}{2}}\lambda^{3/2}\int_0^a \sqrt{t}\varphi(t)\, J_{n+1/2}(\lambda t)\, dt \\
&= \sqrt{\frac{\pi\lambda}{2}}\int_0^a \tilde{\varphi}(t)\left[a^{-n+1/2}J_{n-1/2}(\lambda a)-t^{-n+1/2}J_{n-1/2}(\lambda t)\right]dt, \\
X_2 &= \sqrt{\frac{\pi\lambda}{2}}\int_0^a \sqrt{t}\psi(t)J_{n-1/2}(\lambda t)dt, \\
X_3 &= \sqrt{\frac{\pi\lambda}{2}}\int_0^a \sqrt{t}\omega(t)J_{n+1/2}(\lambda t)dt \\
&= \sqrt{\frac{\pi}{2\lambda}}\int_0^a \tilde{\omega}(t)\left[a^{-n+1/2}J_{n-1/2}(\lambda a)-t^{-n+1/2}J_{n-1/2}(\lambda t)\right]dt, \\
\tilde{\varphi}(t) &\equiv \frac{d}{dt}\left[t^n\varphi(t)\right], \quad \tilde{\omega}(t) \equiv \frac{d}{dt}\left[t^n\omega(t)\right].
\end{aligned}
\tag{4.31}
$$

By using Weber–Schafheitling discontinuous integral (Bateman and Erdelyi 1953)

$$
\int_0^\infty \sqrt{\lambda}J_{n+1/2}(\lambda a)\, J_n(\lambda t) = \begin{cases} 0, & 0 \leqslant a < t \\ \sqrt{\dfrac{2}{\pi}}\dfrac{t}{a^{n+1/2}\sqrt{a^2-t^2}}, & 0 < t < a \end{cases}
\tag{4.32}
$$

and differentiation formulas for Bessel functions (Watson 1995)

$$
t^{-n}\frac{d}{dt}\left[t^n J_n(\lambda t)\right] = \lambda J_{n-1}(\lambda t), \quad t^n\frac{d}{dt}\left[t^{-n}J_n(\lambda t)\right] = -\lambda J_{n+1}(\lambda t),
\tag{4.33}
$$

it can be shown that the three last equations in system (4.29) (for the range $r > a$) are satisfied identically. Then, from the remaining three equations in (4.29) (for the range

$r \leqslant a$) we obtain Fredholm integral equations of the second kind (the procedure is shown in more detail in Bogdanov et al. (2017)):

$$\begin{aligned}
&(sk+q)\, f_1(\xi) + (sk-q)\, f_2(\xi) + \frac{4}{\pi}\int_0^1 f_1(\eta)\, K_{11}(\xi,\eta)\, d\eta \\
&+\frac{4}{\pi}\int_0^1 f_2(\eta)\, K_{12}(\xi,\eta)\, d\eta + \frac{4}{\pi}\int_0^1 f_3(\eta)\, K_{13}(\xi,\eta)\, d\eta \\
&= \frac{8\xi}{\pi}\int_0^{\pi/2} \nu_1'(\xi\sin\theta) d\theta, \quad 0 \leqslant \xi,\eta \leqslant 1, \\
&(sk-q)\, f_1(\xi) + (sk+q)\, f_2(\xi) + \frac{4}{\pi}\int_0^1 f_1(\eta)\, K_{21}(\xi,\eta)\, d\eta \\
&+\frac{4}{\pi}\int_0^1 f_2(\eta)\, K_{12}(\xi,\eta)\, d\eta + \frac{4}{\pi}\int_0^1 f_3(\eta)\, K_{23}(\xi,\eta)\, d\eta \\
&= \frac{8\xi}{\pi}\int_0^{\pi/2} \nu_2'(\xi\sin\theta)\, d\theta, \quad 0 \leqslant \xi,\eta \leqslant 1, \\
&skf_3(\xi) + \frac{4}{\pi}\int_0^1 f_1(\eta)\, K_{31}(\xi,\eta)\, d\eta + \frac{4}{\pi}\int_0^1 f_2(\eta)\, K_{32}(\xi,\eta)\, d\eta \\
&\frac{4}{\pi}\int_0^1 f_3(\eta)\, K_{33}(\xi,\eta)\, d\eta = -\frac{8\xi}{\pi}\int_0^{\pi/2} u'(\xi\sin\theta) d\theta, \\
&0 \leqslant \xi,\eta \leqslant 1, \quad s = \frac{n_1^{-1/2} d_1^2 l_1}{d_1 l_1 - d_2 l_2}, \quad q = n_3^{-1/2}.
\end{aligned} \tag{4.34}$$

The following dimensionless variables and functions are introduced in (4.34):

$$\begin{aligned}
\xi &= \frac{x}{a}, \quad \eta = \frac{t}{a}, \quad \beta = \frac{h}{a}, \\
f_1(\xi) &= a^{-n-1}\tilde{\varphi}(x) \\
&= a^{-n-1}\frac{d}{dx}\left[x^n \varphi(x)\right], \\
f_2(\xi) &= a^{-n-1} x^n \psi(x), \\
f_3(\xi) &= a^{-n}\tilde{\omega}(x) \\
&= a^{-n}\frac{d}{dx}\left[x^n \omega(x)\right], \\
u(\xi) &= \frac{\xi^n}{C_{44} l_1 \sqrt{n_1}}\sigma^{(n)}(a\xi), \\
\nu_1(\xi) &= \frac{\xi^{2n}}{C_{44}}\int_0^{\xi} \rho^{-n}\tau_r^{(n)}(a\rho)\, d\rho, \\
\nu_2(\xi) &= \frac{1}{C_{44}}\int_0^{\xi} \rho^{n}\tau_r^{(n)}(a\rho)\, d\rho.
\end{aligned} \tag{4.35}$$

The kernels in (4.34) are of the form Bogdanov et al. (2017):

$$K_{22}(\xi,\eta) = 2skn\beta_1\xi^{n-1}\eta^{-n-1}S_n(z_{11}) + \frac{4s}{k}n\beta_1^3\xi^{n-2}\frac{\eta^{-n-2}}{z_{11}^2-1}$$
$$\times\left\{\left[\left(\frac{8}{z_{11}^2-1}+n(n-1)+6\right)\frac{4\beta_1^2}{\xi\eta}-6z_{11}\right]S_n(z_{11})\right.$$
$$\left.+(n-1)\left[3\left(z_{11}^2-1\right)+\frac{16\beta_1^2 z_{11}}{\xi\eta}\right]P_n(z_{11})\right\}$$
$$-2sn\beta_1\xi^{n-2}\eta^{-n-2}\left[\left(\xi\eta-\frac{8\beta_1^2 z_{11}}{z_{11}^2-1}\right)S_n(z_{11})+4\beta_1^2(n-1)P_n(z_{11})\right]$$
$$+2qn\beta_3\xi^{n-1}\eta^{-n-1}S_n(z_{13}), \tag{4.36}$$

etc., where

$$\beta_j = \beta n_j^{-1/2} = \frac{h}{a}n_j^{-1/2} = \frac{h_j}{a},\quad z_{1j} = \frac{4\beta_j^2+\xi^2+\eta^2}{2\xi\eta},\quad j=1,3,$$
$$S_n(z) = \frac{Q_n(z)-zQ_{n-1}(z)}{4\left(z^2-1\right)},\quad P_n(z) = \frac{Q_{n-1}(z)}{4\left(z^2-1\right)},$$

$Q_n(z)$ is Lagrange function of the second kind. The geometric parameter $\beta = ha^{-1}$ is the dimensionless distance from the crack to the boundary surface of the body.

In the similar way, axisymmetric problems on a semi-infinite body containing near-surface mode I, mode II, and mode III cracks can be reduced to Fredholm integral equations of the second kind (Bogdanov et al. 2017; Nazarenko et al. 2000).

So, for *the axisymmetric problem on mode I crack* in a semi-infinite pre-stressed body, when normal stresses of $\sigma(r)$ intensity act on crack faces, in the case of equal roots we obtain such system of Fredholm integral equations of the second kind

$$\begin{aligned} f(\xi)+\frac{4}{\pi k}\int_0^1 f(\eta)K_{11}(\xi,\eta)\,d\eta-\frac{4}{\pi k}\int_0^1 g(\eta)K_{12}(\xi,\eta)\,d\eta &\\ = -\frac{4}{\pi k}\int_0^{\pi/2} s(\xi\sin\theta)\,d\theta,&\\ g(\xi)+\frac{4}{\pi k}\int_0^1 f(\eta)K_{21}(\xi,\eta)\,d\eta-\frac{4}{\pi k}\int_0^1 g(\eta)K_{22}(\xi,\eta)\,d\eta = 0,&\\ s(\xi) = \frac{\xi}{C_{44}d_1l_1}\sigma(a\xi),& \end{aligned} \tag{4.37}$$

with the kernels

$$K_{11}(\xi,\eta) = -\left[\frac{k}{2} I_1(2\beta_1,\eta) + \beta_1 I_2(2\beta_1,\eta) + \frac{\beta_1^2}{k} I_3(2\beta_1,\eta)\right],$$

$$K_{12}(\xi,\eta) = \frac{\beta_1^2}{k}\left[\eta^{-1} I_2(2\beta_1,\eta) - I_2(2\beta_1,1)\right],$$

$$K_{21}(\xi,\eta) = -\frac{\beta_1^2}{k}\xi I_4(2\beta_1,\eta), \tag{4.38}$$

$$K_{22}(\xi,\eta) = \begin{aligned}&\xi\left\{\frac{k}{2}\left[\eta^{-1} I_1(2\beta_1,\eta) - I_1(2\beta_1,1)\right]\right.\\ &\quad - \beta_1\left[\eta^{-1} I_2(2\beta_1,\eta) I_2(2\beta_1,1)\right]\\ &\quad \left. + \frac{\beta_1^2}{k}\left[\eta^{-1} I_3(2\beta_1,\eta) - I_3(2\beta_1,1)\right]\right\}.\end{aligned}$$

The following notations are introduced in the expressions for the kernels (4.38):

$$I_1(\beta,\eta) = \frac{\beta}{2\xi\eta\left[\zeta^2(\eta) - 1\right]},$$

$$I_2(\beta,\eta) = I_1(\beta,\eta)\left[4\zeta(\eta) I_1(\beta,\eta) - \frac{1}{\beta}\right],$$

$$I_3(\beta,\eta) = 4I_1^2(\beta,\eta)\left\{2\left[3\zeta^2(\eta) + 1\right] I_1(\beta,\eta) - \frac{3\zeta(\eta)}{\beta}\right\}, \tag{4.39}$$

$$I_4(\beta,\eta) = \begin{aligned}&12 I_1^2(\beta,\eta)\left\{16\zeta(\eta)\left[\zeta^2(\eta)+1\right] I_1^2(\beta,\eta)\right.\\ &\quad \left. - \frac{4}{\beta}\left[3\zeta^2(\eta)+1\right] I_1(\beta,\eta) + \frac{\zeta(\eta)}{\beta^2}\right\},\end{aligned}$$

$$\zeta(\eta) = \frac{\beta^2 + \xi^2 + \eta^2}{2\xi\eta}.$$

For the *axisymmetric problem on mode II crack* in a semi-infinite pre-stressed body, when radial shear stresses of $\tau_r(r)$ intensity act on crack faces, in the case of equal roots we obtain such system of Fredholm integral equations of the second kind:

$$f(\xi) + \frac{4}{\pi k}\int_0^1 f(\eta) K_{11}(\xi,\eta)\, d\eta - \frac{4}{\pi k}\int_0^1 g(\eta) K_{12}(\xi,\eta)\, d\eta = 0,$$

$$g(\xi) + \frac{4}{\pi k}\int_0^1 f(\eta) K_{21}(\xi,\eta)\, d\eta - \frac{4}{\pi k}\int_0^1 g(\eta) K_{22}(\xi,\eta)\, d\eta = -\frac{4\xi}{\pi k}\int_0^{\pi/2} q'(\xi\sin\theta)\, d\theta, \tag{4.40}$$

$$q(\xi) = \frac{\sqrt{n_1}\xi}{C_{44} d_1}\tau_r(a\xi),$$

where the kernels are of the form (4.38).

For the *axisymmetric problem on mode III crack* in a semi-infinite pre-stressed body, when tangent torsional loads of $\tau_\theta(r)$ intensity are applied to the crack faces antisymmetrically in respect of the plane of crack location, we obtain Fredholm integral equation of the second kind:

$$f(\xi) - \frac{1}{\pi}\int_0^1 f(\eta) K(\xi, \eta) d\eta = \frac{4\xi}{\pi}\int_0^{\pi/2} t'(\xi \sin\theta) d\theta,$$
$$t(\xi) = \frac{\sqrt{n_3}\xi}{C_{44}}\tau_\theta(a\xi), \tag{4.41}$$

where

$$K(\xi, \eta) = 8\beta_3\xi^2\left[\frac{1}{(4\beta_3^2 + \xi^2 + \eta^2)^2 - 4\xi^2\eta^2} - \frac{1}{(4\beta_3^2 + \xi^2 + 1)^2 - 4\xi^2}\right]. \tag{4.42}$$

4.4.2 Body with Two Parallel Circular Cracks

Now, we will show the results for the case of non-equal roots of the characteristic equation ($n_1 \neq n_2$); the procedures for non-equal roots ($n_1 = n_2$) are carried out similarly.

By performing procedures similar to those presented in the previous subsection, for the non-axisymmetric problem on a pre-stressed body containing two parallel coaxial circular mode I cracks, we obtain such system of Fredholm integral equations of the second kind

$$\left(s\frac{k}{k_1} + q\right) f_1(\xi) + \left(s\frac{k}{k_1} - q\right) f_2(\xi) + \frac{4}{\pi}\int_0^1 f_1(\eta) K_{11}(\xi, \eta) d\eta$$
$$+ \frac{4}{\pi}\int_0^1 f_2(\eta) K_{12}(\xi, \eta) d\eta + \frac{4}{\pi}\int_0^1 f_3(\eta) K_{13}(\xi, \eta) d\eta = 0,$$
$$\left(s\frac{k}{k_1} - q\right) f_1(\xi) + \left(s\frac{k}{k_1} + q\right) f_2(\xi) + \frac{4}{\pi}\int_0^1 f_1(\eta) K_{21}(\xi, \eta) d\eta$$
$$+ \frac{4}{\pi}\int_0^1 f_2(\eta) K_{22}(\xi, \eta) d\eta + \frac{4}{\pi}\int_0^1 f_3(\eta) K_{23}(\xi, \eta) d\eta = 0, \tag{4.43}$$
$$s\frac{k}{k_2} f_3(\xi) + \frac{4}{\pi}\int_0^1 f_1(\eta) K_{31}(\xi, \eta) d\eta + \frac{4}{\pi}\int_0^1 f_2(\eta) K_{32}(\xi, \eta) d\eta$$
$$+ \frac{4}{\pi}\int_0^1 f_3(\eta) K_{33}(\xi, \eta) d\eta = \frac{8\xi}{\pi}\int_0^{\pi/2} u'(\xi \sin\theta) d\theta,$$

where

$$u(\xi) = \frac{k_1 \xi^n}{C_{44} k_2} \sigma^{(n)}(a\xi), \quad s = \frac{n_2^{-1/2} d_1 d_2 l_1}{d_1 l_1 - d_2 l_2}, \tag{4.44}$$
$$q = n_3^{-1/2}, k_1 = l_1 \sqrt{n_2}, \quad k_2 = l_2 \sqrt{n_1}, \quad k = k_1 - k_2$$

The kernels in (4.43) are of the form Bogdanov et al. (2017)

$$K_{12}(\xi, \eta) = 2n\xi^{n-1}\eta^{-n-1}\left[-\frac{sk_2}{k_1}\beta_1 S_n(z_{11}) + s\beta_2 S_n(z_{12}) - q\beta_3 S_n(z_{13})\right]$$
$$+ \sqrt{\pi}\frac{\Gamma(n+1)}{\Gamma(n+\frac{1}{2})}\xi^{2n}\left[-\frac{sk_2}{k_1}R_n(2\beta_1, \eta) + sR_n(2\beta_2, \eta) - qR_n(2\beta_3, \eta)\right], \tag{4.45}$$

etc., where

$$\beta_j = \beta n_j^{-1/2} = \frac{h}{a} n_j^{-1/2} = \frac{h_j}{a}, \quad z_{1j} = \frac{4\beta_j^2 + \xi^2 + \eta^2}{2\xi\eta}, \quad j = 1, 2, 3,$$
$$S_n(z) = \frac{Q_n(z) - zQ_{n-1}(z)}{4(z^2 - 1)}, \quad R_n(b, \eta) = \frac{b}{4(b^2 + \eta^2)^{n+1}},$$

$Q_n(z)$ is Legendre function of the second kind, and $\Gamma(n)$ is gamma function. Here the geometric parameter $\beta = ha^{-1}$ is the dimensionless half-distance between the cracks.

The procedure of solving axisymmetric problems will be exemplified by the problem on a body containing two mode I cracks, for which boundary conditions are of the form (4.20). The harmonic potential functions involved in (4.8) will be presented as Hankel integral expansions

$$\varphi_1^{(1)}(r, z_1) = \int_0^\infty A(\lambda) e^{-\lambda z_1} J_0(\lambda r) \frac{d\lambda}{\lambda},$$
$$\varphi_2^{(1)}(r, z_2) = \int_0^\infty B(\lambda) e^{-\lambda z_2} J_0(\lambda r) \frac{d\lambda}{\lambda},$$
$$\varphi_1^{(2)}(r, z_1) = \int_0^\infty [C_1(\lambda)\cosh\lambda\,(z_1 + h_1) + C_2(\lambda)\sinh\lambda\,(z_1 + h_1)]\, J_0(\lambda r) \frac{\partial\lambda}{\lambda \sinh\mu_1}, \tag{4.46}$$
$$\varphi_2^{(2)}(r, z_2) = \int_0^\infty [D_1(\lambda)\cosh\lambda\,(z_2 + h_2) + D_2(\lambda)\sinh\lambda\,(z_2 + h_2)] J_0(\lambda r) \frac{\partial\lambda}{\lambda \sinh\mu_2},$$

where A, B, C_k, and D_k $(k = 1, 2)$ are unknown functions that are to be determined; $\mu_k = \lambda h_k = \lambda h n_k^{-1/2}$.

Substitute expressions (4.46) into boundary conditions (4.20). Then, from the conditions presented in the second and fourth lines of (4.20), which are set on all planes $y_3 = -h$, $y_3 = 0$, we obtain four relations linking six functions A, B, C_k, and D_k $(k = 1, 2)$

$$
\begin{aligned}
A(\lambda) &= \frac{1}{k}\left[(k_2 + k_1 \coth\mu_1)\, C_1(\lambda) + \frac{d_2 l_2}{d_1 l_1} k_1 (1 + \coth\mu_2)\, D_1(\lambda)\right], \\
B(\lambda) &= -\frac{1}{k}\left[\frac{d_1 l_1}{d_2 l_2} k_2 (1 + \coth\mu_1)\, C_1(\lambda) + (k_1 + k_2 \coth\mu_2)\, D_1(\lambda)\right], \\
C_2(\lambda) &= 0, \quad D_2(\lambda) = 0.
\end{aligned} \tag{4.47}
$$

From the remaining boundary conditions (4.20), the following system of paired (dual) integral equations is obtained

$$
\begin{aligned}
&\int_0^\infty [d_1 l_1 \coth\mu_1\, C_1(\lambda) + d_2 l_2 \coth\mu_2\, D_1(\lambda)]\, J_0(\lambda r)\, \lambda d\lambda = -\frac{\sigma(r)}{C_{44}}, \\
&\qquad\qquad r \leqslant a, \\
&\int_0^\infty \left[n_1^{-1/2} d_1 C_1(\lambda) + n_2^{-1/2} d_2 D_1(\lambda)\right] J_1(\lambda r)\, \lambda d\lambda = 0, \quad r \leqslant a, \\
&\int_0^\infty X_1 J_0(\lambda r)\, d\lambda = 0, \quad r > a, \\
&\int_0^\infty X_2 J_1(\lambda r)\, d\lambda = 0, \quad r > a,
\end{aligned} \tag{4.48}
$$

where

$$
\begin{aligned}
X_1 &= \frac{d_1 l_1}{d_2 l_2} (1 + \coth\mu_1)\, C_1(\lambda) + (1 + \coth\mu_2)\, D_1(\lambda), \\
X_2 &= \frac{d_1}{d_2}\sqrt{\frac{n_2}{n_1}} (1 + \coth\mu_1)\, C_1(\lambda) + (1 + \coth\mu_2)\, D_1(\lambda).
\end{aligned}
$$

Functions X_1 and X_2 are presented in the form permitting two last equations in (4.48) (for the range $r > a$) to be satisfied identically, viz.,

$$
\begin{aligned}
X_1 &= \sqrt{\frac{\pi\lambda}{2}} \int_0^a \sqrt{t}\varphi(t)\, J_{1/2}(\lambda t)\, dt \\
&= \int_0^a \varphi(t) \sin\lambda t\, dt, \\
X_2 &= \sqrt{\frac{\pi\lambda}{2}} \int_0^a \sqrt{t}\psi(t)\, J_{3/2}(\lambda t)\, dt,
\end{aligned} \tag{4.49}
$$

where $\varphi(t)$ and $\psi(t)$ are unknown functions continuous along with their first derivatives in the segment $[0, a]$. In that case, from the first two equations in (4.48) we obtain the system of Fredholm integral equations of the second kind

$$
\begin{aligned}
f(\xi) + \frac{2}{\pi k}\int_0^1 f(\eta)\, K_{11}(\xi,\eta)\, d\eta + \frac{2}{\pi k}\int_0^1 g(\eta)\, K_{12}(\xi,\eta)\, d\eta \\
= -\frac{4}{\pi}\int_0^{\pi/2} s(\xi\sin\theta)\, d\theta, \\
g(\xi) + \frac{2}{\pi k}\int_0^1 f(\eta)\, K_{21}(\xi,\eta)\, d\eta + \frac{2}{pk}\int_0^1 g(\eta)\, K_{22}(\xi,\eta)\, d\eta = 0,
\end{aligned} \tag{4.50}
$$

where

$$f(\xi) \equiv a^{-1}\varphi(a\xi), \quad g(\xi) \equiv a^{-1}\frac{d}{d\xi}[\xi\psi(a\xi)],$$
$$s(\xi) = \frac{\xi}{C_{44}d_2l_2}\sigma(a\xi). \tag{4.51}$$

The kernels are of the form:

$$\begin{aligned} K_{11}(\xi,\eta) &= k_1 I_1(2\beta_1,\eta) - k_2 I_1(2\beta_2,\eta), \\ K_{12}(\xi,\eta) &= k_1\{[I_0(2\beta_1,1) - I_0(2\beta_2,1)] \\ &\quad - \eta^{-1}[I_0(2\beta_1,\eta) - I_0(2\beta_2,\eta)]\}, \\ K_{21}(\xi,\eta) &= -k_2\xi[I_2(2\beta_1,\eta) - I_2(2\beta_2,\eta)], \\ K_{22}(\xi,\eta) &= -\xi\{[k_2 I_1(2\beta_1,1) - k_1 I_1(2\beta_2,1)] \\ &\quad - \eta^{-1}[k_2 I_1(2\beta_1,\eta) - k_1 I_1(2\beta_2,\eta)]\}, \end{aligned} \tag{4.52}$$

where

$$I_0(\beta,\eta) = \frac{1}{4}\ln\frac{\zeta(\eta)+1}{\zeta(\eta)-1},$$

$I_1(\beta,\eta)$, $I_2(\beta,\eta)$, and $\zeta(\eta)$ are determined from (4.39), and k_1, k_2, and k—from (4.44).

By performing similar procedures for the axisymmetric problem on mode II cracks in an unbounded body the following system of Fredholm integral equations of the second kind is obtained

$$\begin{aligned} f(\xi) - \frac{2}{\pi k}\int_0^1 f(\eta)K_{11}(\xi,\eta)\,d\eta - \frac{2}{\pi k}\int_0^1 g(\eta)K_{12}(\xi,\eta)\,d\eta &= 0, \\ g(\xi) - \frac{2}{\pi k}\int_0^1 f(\eta)K_{21}(\xi,\eta)\,d\eta - \frac{2}{\pi k}\int_0^1 g(\eta)K_{22}(\xi,\eta)\,d\eta & \\ = \frac{4}{\pi}\xi\int_0^{\pi/2} q'(\xi\sin\theta)\,d\theta, & \end{aligned} \tag{4.53}$$

where

$$q(\xi) = \frac{\sqrt{n_2}\xi}{C_{44}d_2}\tau_r(a\xi), \tag{4.54}$$

and the kernels are determined from (4.52).

For the axisymmetric problem on mode III cracks in an unbounded body, such Fredholm integral equation of the second kind is obtained

$$f(\xi) + \frac{1}{\pi}\int_0^1 f(\eta)K(\xi,\eta)\,d\eta = \frac{4\xi}{\pi}\int_0^{\pi/2} t'(\xi\sin\theta)\,d\theta,$$
$$t(\xi) = \frac{\sqrt{n_3}\xi}{C_{44}}\tau_\theta(a\xi), \tag{4.55}$$

where the kernel is of the form (4.42).

4.5 Stress Intensity Factors

Now we analyze the asymptotic distribution of stresses in the vicinities of crack edges in the investigated pre-stressed bodies containing cracks and determine stress intensity factors (*SIFs*), which like those in classical fracture mechanics (Cherepanov 1979; Kassir and Sih 1975) are coefficients at singularities in the stress distributions mentioned when approaching crack edges.

4.5.1 Half-Space with a Near-Surface Circular Crack

The procedure of determining stress intensity factors for the non-axisymmetric problem on a body with a near-surface circular crack will be considered in more detail.

From the representations of stresses via harmonic potential functions (4.9), given (4.27), (4.28), and (4.30), we obtain expressions for stress tensor components Q'_{33}, Q'_{3r}, and $Q'_{3\theta}$ in the domain $y_3 = 0, r > a$ (i.e., in the plane of crack location, outside its contour, in subdomain "2"). For Q'_{33} we have

$$Q_{33}^{\prime(2)}(r,\theta,0) = \tfrac{1}{4}C_{44}skl_1\sqrt{n_1}\sum_{n=0}^{\infty}\cos n\theta\left\{\int_0^{\infty}X_3J_n(\lambda r)\lambda d\lambda\right.$$
$$\left.-\frac{2}{k^2}\int_0^{\infty}\left[\mu_1^2(X_1+X_2)+\left(\frac{k^2}{2}+\mu_1^2+\mu_1 k\right)X_3\right]e^{-2\mu_1}J_n(\lambda r)\lambda d\lambda\right\}. \tag{4.56}$$

The analysis of expression (4.56) implies that the singularity when $r \to a$ only involves the first integral in the braces since in the second integral in the braces, as follows from the corresponding formulas of Bessel function integrals (Prudnikov et al. 1986b), this singularity is absent. In this connection, the first integral in braces in (4.56) will be analyzed in more detail. Given expressions (4.31) and formulas (4.33), performing integration by parts and taking into account the value of discontinuous integral (4.32), we obtain

$$\int_0^{\infty}X_3J_n(\lambda r)\lambda d\lambda$$
$$=\sqrt{\frac{\pi}{2}}\int_0^{\infty}\left[\int_0^a\sqrt{t}\omega(t)\lambda J_{n+1/2}(\lambda t)dt\right]J_n(\lambda r)\sqrt{\lambda}d\lambda$$
$$=-\frac{a^n\omega(a)}{r^n\sqrt{r^2-a^2}}+\int_0^a\frac{\tilde{\omega}(t)dt}{r^n\sqrt{r^2-t^2}}. \tag{4.57}$$

The integral

$$\int_0^a \frac{\tilde{\omega}(t)\,dt}{r^n\sqrt{r^2-t^2}}$$

does not have singularities when $r \to a$ (Prudnikov et al. 1986a). Then from (4.56) and (4.57), taking into account the expression

$$\omega(t) = t^{-n}\int_0^t \tilde{\omega}(t)\,dt,$$

we obtain

$$Q_{33}^{\prime(2)}(r,\theta,0) = -\tfrac{1}{4}C_{44}skl_1\sqrt{n_1}\sum_{n=0}^{\infty}\cos n\theta \times \left[\int_0^a \tilde{\omega}(t)\,dt\right]\frac{r^{-n}}{\sqrt{(r-a)(r+a)}} + O(1), \quad (4.58)$$

where symbol $O(1)$ denotes regular components that do not have singularities when $r \to a$.

Performing a similar analysis for other stress tensor components in the plane of crack location, we obtain

$$Q_{3r}^{\prime(2)}(r,\theta,0) = \tfrac{1}{4}C_{44}sk\sum_{n=0}^{\infty}\cos n\theta\,\frac{r^{-n+1}}{\sqrt{(r-a)(r+a)}} \times \left[\frac{\tilde{\varphi}(a)}{a} + a^{n-1}\psi(a)\right] + O(1), \quad (4.59)$$

$$Q_{3\theta}^{\prime(2)}(r,\theta,0) = \tfrac{1}{4}C_{44}q\sum_{n=1}^{\infty}\sin n\theta\,\frac{r^{-n+1}}{\sqrt{(r-a)(r+a)}} \times \left[\frac{\tilde{\varphi}(a)}{a} - a^{n-1}\psi(a)\right] + O(1). \quad (4.60)$$

Expressions (4.58)–(4.60) can be written as

$$\begin{aligned} Q_{33}^{\prime(2)}(r,\theta,0) &= \frac{K_I}{\sqrt{2\pi(r-a)}} + O(1),\\ Q_{3r}^{\prime(2)}(r,\theta,0) &= \frac{K_{II}}{\sqrt{2\pi(r-a)}} + O(1),\\ Q_{3\theta}^{\prime(2)}(r,\theta,0) &= \frac{K_{III}}{\sqrt{2\pi(r-a)}} + O(1). \end{aligned} \quad (4.61)$$

In (4.61), stress intensity factors (*SIFs*) are expressed by the following relations

$$
\begin{aligned}
K_I &= -\tfrac{1}{4}C_{44}skl_1\sqrt{n_1}\sqrt{\frac{\pi}{a}}\sum_{n=0}^{\infty}\cos n\theta\left[a^{-n}\int_0^a \tilde{\omega}(t)\,dt\right] \\
&= -\tfrac{1}{4}C_{44}skl_1\sqrt{n_1}\sqrt{\pi a}\sum_{n=0}^{\infty}\cos n\theta\int_0^1 f_3(\eta)\,d\eta, \\
K_{II} &= \tfrac{1}{4}C_{44}sk\sqrt{\frac{\pi}{a}}\sum_{n=0}^{\infty}\cos n\theta\left\{a^{-n+1}\left[\frac{\tilde{\omega}(a)}{a}+a^{n-1}\psi(a)\right]\right\} \\
&= \tfrac{1}{4}C_{44}sk\sqrt{\pi a}\sum_{n=0}^{\infty}\cos n\theta\,\left[f_1(1)+f_2(1)\right], \\
K_{III} &= \tfrac{1}{4}C_{44}q\sqrt{\pi a}\sum_{n=1}^{\infty}\sin n\theta\left\{a^{-n+1}\left[\frac{\tilde{\omega}(a)}{a}-a^{n-1}\psi(a)\right]\right\} \\
&= \tfrac{1}{4}C_{44}q\sqrt{\pi a}\sum_{n=1}^{\infty}\sin n\theta\,\left[f_1(1)-f_2(1)\right],
\end{aligned}
\tag{4.62}
$$

where functions $f_1(\xi)$, $f_2(\xi)$, and $f_3(\xi)$ are determined by solving the system of Fredholm integral equations (4.34).

Expressions (4.61) and (4.62) imply that the order of singularity in stresses distribution in the vicinity of a near-surface crack edge in a semi-infinite pre-stressed body is $-1/2$, i.e., it coincides with the order of singularity in stresses distribution near the crack edge in a body free of initial stresses (Kassir and Sih 1975). In addition, it follows from (4.62) that the mutual influence of a near-surface crack and material's free surface causes qualitative changes in the asymptotic distribution of stresses near the crack edge, viz., non-zero values of K_{II} and K_{III} in the case of loading crack faces only by normal forces (when $\sigma(r,\theta)\neq 0$, $\tau_r(r,\theta)=\tau_\theta(r,\theta)=0$) (in the problem on a body containing an isolated mode I crack $K_I\neq 0$, $K_{II}=0$, and $K_{III}=0$ (Bogdanov et al. 2017)) and non-zero values of K_I in the case when only tangent shear forces $\tau_r(r,\theta)$ act on crack faces (for such scheme of loading the faces of an isolated crack in an unbounded body, it was $K_I=0$, $K_{II}\neq 0$, and $K_{III}=0$ (Bogdanov et al. 2017)). Besides, it can be seen from expressions (4.62) that all three *SIFs* depend on initial stresses, since parameters C_{44}, s, k, q, l_1, and n_1 depend on the elongation (contraction) coefficient λ_1, which, in turn, is determined by the action of initial stresses $S_{11}^0 = S_{22}^0$.

Analyze the limit case of mode I crack location, when the distance between the crack and half-space boundary tends to infinity. It follows from the analysis of expressions (4.36) for the kernels of integral equations (4.34) that when $\beta\to\infty$, all the kernels in the limit become zero. Then (4.34) implies

$$
\begin{aligned}
f_1^{\infty} &= f_2^{\infty} = 0, \\
f_3^{\infty} &= -\frac{8}{\pi sk}\xi\int_0^{\pi/2} u'(\xi\sin\theta)\,d\theta, \\
f_j^{\infty} &\equiv \lim_{\beta\to\infty} f_j\,,\quad j=1,2,3.
\end{aligned}
\tag{4.63}
$$

From (4.63), taking into account (4.35) and performing the change of variables $\eta = \xi \sin\theta$, we obtain

$$f_3^\infty = -\frac{8}{\pi C_{44} s k l_1 \sqrt{n_1}} \frac{d}{d\xi} \int_0^\xi \frac{\eta^{n+1} \sigma^{(n)}(a\eta)}{\sqrt{\xi^2 - \eta^2}} d\eta.$$

Then, we have from (4.62)

$$\begin{aligned} K_I^\infty &\equiv \lim_{\beta\to\infty} K_I \\ &= 2\sqrt{\frac{a}{\pi}} \sum_{n=0}^{\infty} \cos n\theta \int_0^1 \frac{\eta^{n+1}\sigma^{(n)}(a\eta)}{\sqrt{1-\eta^2}} d\eta \\ &= \frac{2}{\sqrt{\pi a}} \sum_{n=0}^{\infty} \frac{\cos n\theta}{a^n} \int_0^a \frac{t^{n+1}\sigma^{(n)}(t)}{\sqrt{a^2-t^2}} dt, \\ K_{II}^\infty &= 0, \\ K_{III}^\infty &= 0, \end{aligned} \tag{4.64}$$

where Fourier coefficients $\sigma^{(n)}(x)$ $(n = 0, 1, 2, \dots)$ are determined from relations (4.26) via the normal load acting on the crack faces.

As can be seen, in this case *SIFs* do not depend on initial stresses and entirely coincide with the values obtained in the non-axisymmetric problem on a mode I crack in an infinite pre-stressed body (Bogdanov et al. 2017). In particular, when normal loads of the form

$$\sigma\,(r, \theta) = \sigma_1 \cos\theta \tag{4.65}$$

are applied to crack faces, we obtain

$$K_I^\infty = \tfrac{1}{2}\sqrt{\pi a}\sigma_1 \cos\theta\,, \quad K_{II}^\infty = 0, \quad K_{III}^\infty = 0. \tag{4.66}$$

It should be noted that the values of *SIFs* obtained by solving the problem on a pre-stressed body containing a near-surface mode I crack in the limit case of crack location, when the distance between it and the half-space boundary tends to infinity (those *SIF* values are given by expressions (4.64) and (4.66)) also entirely coincide with the values of *SIFs* which were obtained in the non-axisymmetric problem on an infinite body with a penny-shaped mode I crack within fracture mechanics of materials free of initial stresses (Kassir and Sih 1975).

By performing similar procedures in the case of axisymmetric problem on a half-space containing a near-surface mode I crack (Bogdanov et al. 2017), we obtain such expressions for stress tensor components near the crack edge:

$$\begin{aligned} Q_{33}^{\prime(2)}(r,0) &= \frac{K_I}{\sqrt{2\pi(r-a)}} + O(1), \\ Q_{3r}^{\prime(2)}(r,0) &= \frac{K_{II}}{\sqrt{2\pi(r-a)}} + O(1), \\ Q_{3\theta}^{\prime(2)}(r,0) &= 0. \end{aligned} \tag{4.67}$$

In (4.67), *SIFs* are determined from the expressions

$$\begin{aligned} K_I &= -\tfrac{1}{2}C_{44}kd_1l_1\sqrt{\pi a}\, f(1), \\ K_{II} &= -\tfrac{1}{2}C_{44}kd_1n_1^{-1/2}\sqrt{\pi a}\int_0^1 g(\eta)\,d\eta, \\ K_{III} &= 0, \end{aligned} \tag{4.68}$$

where functions f and g are determined by solving the system of Fredholm integral equations (4.37). It can also be shown that in the limit case of crack location, when the distance between the crack and the half-space boundary tends to infinity, we have

$$\begin{aligned} K_I^{\infty} &= 2\sqrt{\frac{a}{\pi}}\int_0^1 \frac{\eta s(\eta)}{\sqrt{1-\eta^2}}d\eta = \frac{2}{\sqrt{\pi a}}\int_0^a \frac{t\sigma(t)}{\sqrt{a^2-t^2}}dt, \\ K_{II}^{\infty} &= 0, \\ K_{III}^{\infty} &= 0. \end{aligned} \tag{4.69}$$

In particular, when uniform normal pressure of the form

$$\sigma(r) = \sigma = \text{const}, \tag{4.70}$$

acts on crack faces, we have from (4.69)

$$K_I^{\infty} = 2\sigma\sqrt{\frac{a}{\pi}}. \tag{4.71}$$

In the case of the axisymmetric problem on a half-space containing a near-surface mode II crack, *SIFs* are of the form (4.68), where functions f and g are determined by solving the system of Fredholm integral equations (4.40). In the limit case of crack location, when the distance between the crack and the half-space boundary tends to infinity, we have

$$\begin{aligned} K_I^{\infty} &= 0, \\ K_{II}^{\infty} &= 2\sqrt{\frac{a}{\pi}}\int_0^1 \frac{\eta^2 q(\eta)}{\sqrt{1-\eta^2}}d\eta = \frac{2}{a\sqrt{\pi a}}\int_0^a \frac{t^2\tau_r(t)}{\sqrt{a^2-t^2}}dt, \\ K_{III}^{\infty} &= 0. \end{aligned} \tag{4.72}$$

In particular, when a uniform shear load of the form

$$\tau_r(r) = \tau = \text{const}, \tag{4.73}$$

acts on the crack faces, we have from (4.72)

$$K_{II}^{\infty} = \frac{\tau}{2}\sqrt{\pi a}. \tag{4.74}$$

In the case of the axisymmetric problem on a half-space containing a near-surface mode III crack, *SIFs* are of the form Bogdanov et al. (2017):

$$K_I = 0, \quad K_{II} = 0, \quad K_{III} = \tfrac{1}{2}C_{44}n_3^{-1/2}\sqrt{\pi a}\int_0^1 f(\eta)\,d\eta, \tag{4.75}$$

where function f is determined by solving Fredholm integral equation (4.41). In the limit case of crack location, when the distance between the crack and the half-space boundary tends to infinity, we have

$$\begin{aligned} K_I^{\infty} &= 0, \\ K_{II}^{\infty} &= 0, \\ K_{III}^{\infty} &= 2\sqrt{\frac{a}{\pi}}\int_0^1 \frac{\eta^2 t(\eta)}{\sqrt{1-\eta^2}}d\eta = \frac{2}{a\sqrt{\pi a}}\int_0^a \frac{t^2\tau_\theta(t)}{\sqrt{a^2-t^2}}dt. \end{aligned} \tag{4.76}$$

In particular, when a uniform torsional load of the form

$$\tau_\theta(r) = \tau = \text{const} \tag{4.77}$$

acts on the crack faces, we have from (4.76)

$$K_{III}^{\infty} = \frac{\tau}{2}\sqrt{\pi a}. \tag{4.78}$$

It should be noted that from the analysis of the asymptotic distribution of stresses in the vicinity of the near-surface crack edge we can make conclusions concerning the order of singularities near the crack edges, the influence of initial stresses on stress intensity factors, as well as the effect of crack interaction with the body boundary, which are similar to those made in the consideration of the non-axisymmetric problem.

4.5.2 Body with Two Parallel Circular Cracks

For the non-axisymmetric problem on a pre-stressed body containing two parallel coaxial mode I cracks, in the case of non-equal roots ($n_1 \neq n_2$) we obtain the asymptotic distribution of stress tensor components as (4.61) with *SIFs* of the forms

$$
\begin{aligned}
K_I &= C_{44}\frac{sk}{4k_1}\sqrt{\pi a}\sum_{n=0}^{\infty}\cos n\theta \int_0^1 f_3(\eta)\,d\eta,\\
K_{II} &= C_{44}\frac{sk}{4k_1}\sqrt{\pi a}\sum_{n=0}^{\infty}\cos n\theta\ [f_1(1)+f_2(1)],\\
K_{III} &= \tfrac{1}{4}C_{44}q\sqrt{\pi a}\sum_{n=1}^{\infty}\sin n\theta\ [f_1(1)-f_2(1)],
\end{aligned}
\tag{4.79}
$$

where functions $f_1(\xi)$, $f_2(\xi)$, and $f_3(\xi)$ are determined by solving the system of Fredholm integral equations (4.43). It is seen from (4.79) that the effect of mutual influence of two parallel coaxial cracks in a pre-stressed body is evident in the appearance of non-zero values of $K_{II}{}^{\infty}$ and $K_{III}{}^{\infty}$ only under the action of a normal load on crack faces. It can also be shown that when the distance between cracks tends to infinity, in the limit we obtain the values of *SIFs* $K_I{}^{\infty}$, $K_{II}{}^{\infty}$, and $K_{III}{}^{\infty}$ as (4.64) (while for the special case of a non-axisymmetric normal load acting on crack faces (4.65) the values of *SIFs* are of the form (4.66)), which corresponds to physical considerations.

For the axisymmetric problem on two parallel coaxial mode I cracks located in a pre-stressed body, we obtain expressions for stress tensor components in the vicinities of cracks as (4.67), where *SIFs* are presented by the expressions

$$
\begin{aligned}
K_I &= -\tfrac{1}{2}C_{44}d_2l_2 f(1),\\
K_{II} &= \tfrac{1}{2}C_{44}d_2n_2^{-1/2}\int_0^1 g(\eta)\,d\eta,\\
K_{III} &= 0,
\end{aligned}
\tag{4.80}
$$

while functions f and g are determined by solving the system of Fredholm integral equations (4.50). In the limit case of cracks location, when the distance between them tends to infinity, we arrive at the values of *SIFs* of the form (4.69) (and in the special case of the load acting on cracks faces as (4.70), $K_I{}^{\infty}$ is of the form (4.71)).

By solving the axisymmetric problem on two parallel coaxial mode II cracks located in a pre-stressed body we obtain expressions for *SIFs* in the form of (4.79), where functions f and g are determined from the solution of the system of Fredholm integral equations (4.53). In the limit case of cracks location, when the distance between them tends to zero, we arrive at the values of *SIFs* as (4.72) (while in the special case of the load on cracks faces (4.73), $K_{II}{}^{\infty}$ is of the form of (4.74)).

Finally, when considering the axisymmetric problem on a pre-stressed body containing two parallel mode III cracks, we arrive at the values of *SIFs* as (4.75), where functions f and g are determined from the solution of Fredholm integral equation (4.55). In the limit case of cracks location, when the distance between them tends to infinity, we obtain the values of *SIFs* as (4.76) (while in the special case of the torsional load of the form (4.77) acting on cracks faces, the value of $K_{III}{}^{\infty}$ is of the form (4.78)).

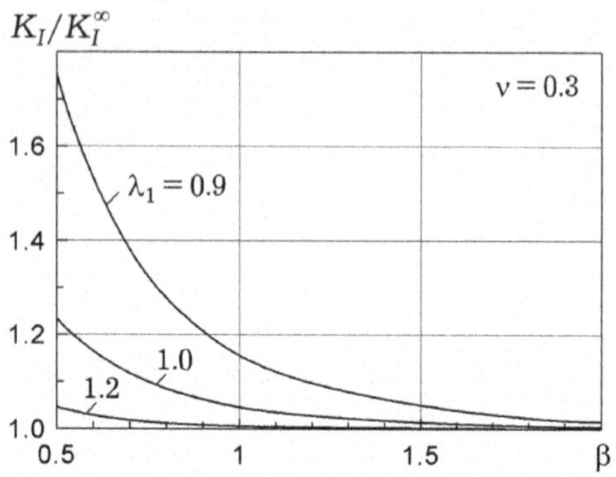

Fig. 4.4 Dependence of *SIFs* ratios K_I/K_I^∞ on the dimensionless distance between the crack and the boundary surface of the body $\beta = ha^{-1}$ for the harmonic-type potential

4.6 Numerical Results

Below we present the results of numerical investigation for some highly elastic materials and composites. The parameters of those materials, which are involved in the resolving Fredholm integral equations of the second kind, and expressions for stress intensity factors are given, e.g., in Bogdanov et al. (2017), Guz et al. (2020).

Highly elastic material with the elastic potential of harmonic type (a compressible body, equal roots) (John 1960). Consider the results of numerical calculation for the non-axisymmetric problem on a body containing a near-surface mode I crack, when crack faces are under a normal tensile load of (4.65) form.

Figure 4.4 shows the dependence of the stress intensity factors (*SIFs*) ratios K_I/K_I^∞ on the dimensionless distance between the crack and the half-space boundary $\beta = ha^{-1}$ for the value of Poisson coefficient $\nu = 0.3$. Here, K_I^∞ is determined from (4.66) and corresponds to the *SIF* value in the problem on an isolated mode I crack in an infinite pre-stressed body (this value, as shown in Sect. 4.5.1, coincides with the *SIF* value in the problem on a mode I crack in a body free of initial stresses). The dependences are given for the values of $\lambda_1 = 0.9$ (initial compression), $\lambda_1 = 1.2$ (initial tension) and $\lambda_1 = 1.0$ (no initial stresses). It can be seen that the interaction of the crack and the free body boundary increases substantially when the distance between them decreases. E.g., for $\lambda_1 = 0.9$ the value of $K_I/K_I{}^\infty$ when $\beta = 0.5$ is higher than the corresponding value of $K_I/K_I{}^\infty$ when $\beta = 2.0$ by a factor of 1.7. On the other hand, with the increase in the distance between the crack and the half-space boundary this mutual influence weakens rapidly, and the respective values of *SIFs* tend to the values obtained for an isolated crack in an infinite body. The precision acceptable for practical calculations, the mutual influence between the crack and the free surface can be neglected when the distance between them exceeds 2 crack radii.

Figure 4.5 illustrates the dependence of $K_I/K_I{}^\infty$ on the parameter of initial stresses λ_1 for different values of Poisson coefficient ν when $\beta = 0.5$. As can be seen from the figure, the compressibility of the material with harmonic-type potential, which is characterized by Poisson coefficient, noticeably influences the values of *SIFs*. E.g., when $\lambda_1 = 0.95$, $\beta = 0.5$, the value of $K_I/K_I{}^\infty$ for $\nu = 0.5$ exceeds

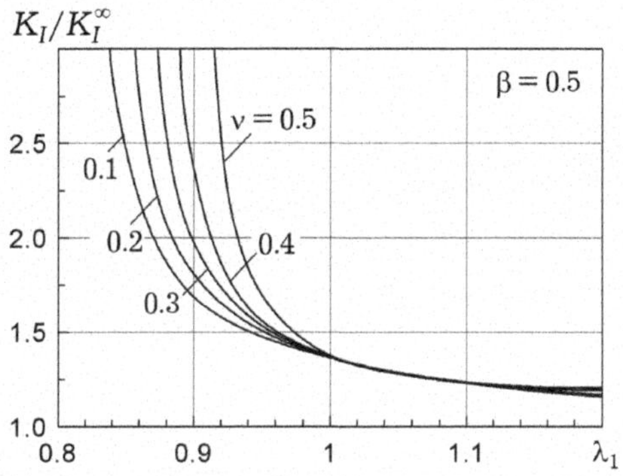

Fig. 4.5 Dependence of *SIFs* ratios K_I/K_I^∞ on elongation (or contraction) ratio λ_1 for different values of Poisson's ratio (the harmonic-type elastic potential)

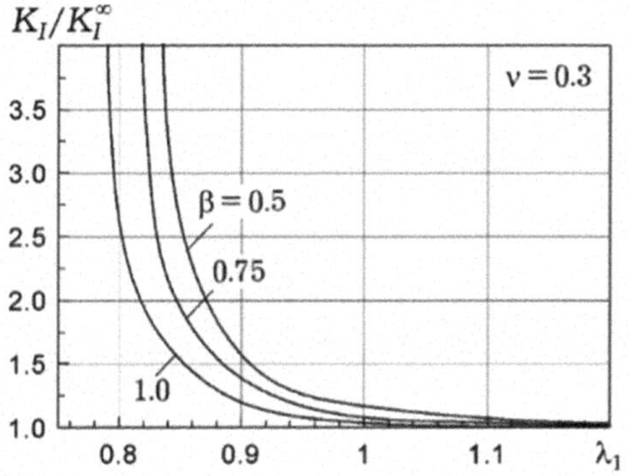

Fig. 4.6 Dependence of *SIFs* ratios K_I/K_I^∞ on elongation (or contraction) ratio λ_1 for the harmonic-type potential

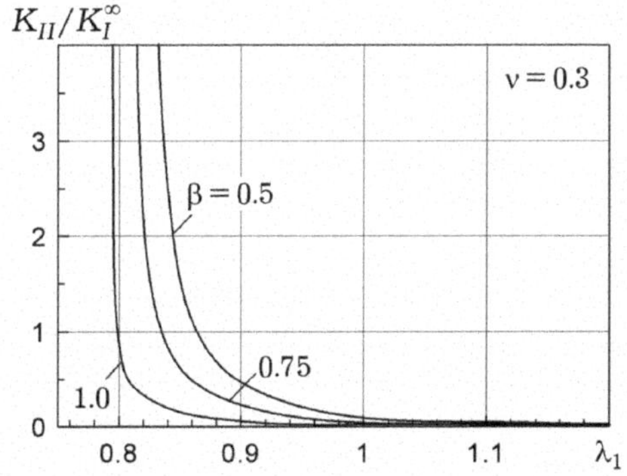

Fig. 4.7 Dependence of *SIFs* ratios K_{II}/K_I^∞ on elongation (or contraction) ratio λ_1 for the harmonic-type potential

the value of $K_I/K_I{}^\infty$ for $\nu = 0.1$ by 12%, while for $\lambda_1 = 0.9$, $\beta = 0.5$ these values differ by a factor of 2.2.

Figures 4.6, 4.7, and 4.8 show, respectively, the dependences of $K_I/K_I{}^\infty$, $K_{II}/K_I{}^\infty$, and $K_{III}/K_I{}^\infty$ on the parameter of initial elongation (contraction) λ_1 at different values of geometric parameter $\beta = ha^{-1}$ for the value of Poisson coefficient $\nu = 0.3$. As the figures imply, *SIFs* considerably depend on initial stresses, with the influence of contractive initial stresses being higher than that of tensile stresses.

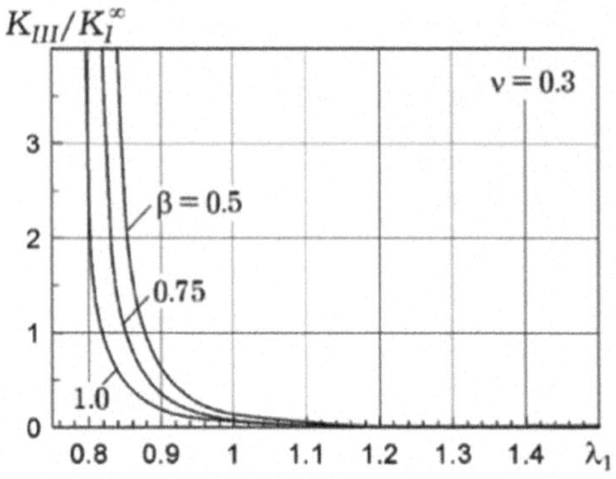

Fig. 4.8 Dependence of *SIFs* ratios K_{III}/K_I^∞ on elongation (or contraction) ratio λ_1 for the harmonic-type potential

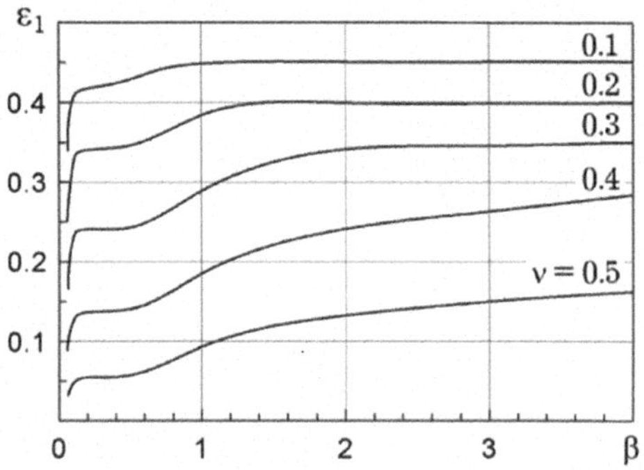

Fig. 4.9 Dependence of the critical values of relative contraction ε_1 on the geometric parameter β for the harmonic-type potential (non-axisymmetric form of stability loss)

The curves in Figs. 4.6, 4.7, and 4.8 have vertical asymptotes corresponding to a sharp (resonance) increase of the stress intensity factors at certain values of the initial contraction parameter $\lambda_1 < 1$. According to the unified approach within the linearized mechanics of deformable solid bodies, described in Sect. 4.1, this effect permits determining the critical (limit) values of contraction parameters, which, when achieved, cause local loss of material's stability in the vicinity of the crack.

Figure 4.9 shows for different values of Poisson coefficient the dependences of the values of relative critical contraction $\varepsilon_1 = 1 - \lambda_1$ corresponding to the local loss of material's stability in the vicinity of near-surface crack in the non-axisymmetric form (the first harmonic in coordinate θ) of the geometric parameter $\beta = ha^{-1}$. The figure implies that the mutual influence of the crack and the half-space boundary leads to a substantial decrease in the values of ε_1 and, respectively, in the critical contraction stresses as compared to the case of a single isolated crack in an unbounded body (in this case for the harmonic-type potential, critical contractions corresponding to the non-axisymmetric form of stability loss are calculated by the formula $\varepsilon_1 = (1 - \nu)/2$ (Guz et al. 1992, 2020)). At the same time, with an increase in the distance between the crack and the half-space boundary this influence becomes weaker, and corresponding critical parameters tend to the values obtained for the case of a single crack in a body.

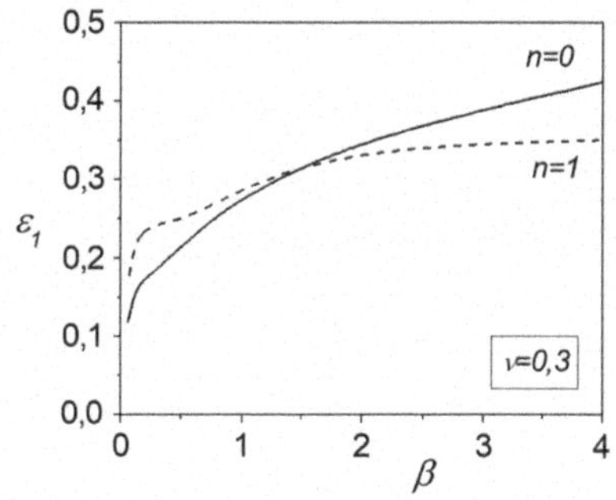

Fig. 4.10 Comparing the critical values ε_1 in the cases of the axisymmetric and non-axisymmetric forms of stability loss for the harmonic-type potential

Figure 4.10 compares for the same material the dependences of ε_1 on β that were obtained from the solution of the axisymmetric problem (the axisymmetric form of stability loss, solid line) and from the solution of the non-axisymmetric problem (the non-axisymmetric form of stability loss, dashed line). It should be noted here that the critical contractions corresponding to the axisymmetric form of stability loss are calculated for the harmonic-type potential with the formula $\varepsilon_1 = 1/(2+\nu)$ (Guz et al. 1992, 2020).

Highly elastic material with Bartenev–Khazanovich elastic potential (an incompressible body, equal roots) (Bartenev and Khazanovich 1960). The results of investigating the axisymmetric problems on a pre-stressed body containing two parallel coaxial cracks for this material are given here.

Figures 4.11 and 4.12 illustrate for mode I cracks, when forces of the form (4.70) act on crack faces, the dependences of the ratios of stress intensity factors $K_I/K_I{}^{\infty}$ and $K_{II}/K_I{}^{\infty}$, respectively, (here $K_I{}^{\infty}$ is determined from (4.71)) on the parameter of initial stresses λ_1 for different values of the dimensionless half-distance between the cracks $\beta = ha^{-1}$. It can be seen from the figures that *SIFs* K_I, K_{II} significantly depend on initial stresses. The curves shown in Figs. 4.11 and 4.12 have vertical asymptotes that correspond to the effect of resonance nature, when the initial contraction stresses (and, correspondingly, the parameter of initial contraction $\lambda_1 < 1$) achieve the values at which the local loss of material's stability occurs (in the form symmetric with respect to the plane $y_3 = -h$) in the vicinity of cracks under contraction along the cracks.

Figures 4.13 and 4.14 for mode II cracks, when forces of (4.73) form act on crack faces, show, respectively, the dependences of the ratios of the *SIFs* $K_{II}/K_{II}{}^{\infty}$ and $K_I/K_{II}{}^{\infty}$ (where $K_{II}{}^{\infty}$ is determined from (4.74)) on the parameter of initial stresses λ_1 for different values of the dimensionless half-distance between the cracks) $\beta = ha^{-1}$. The figures demonstrate the significant influence of initial stresses on *SIFs* K_I and K_{II}.

In the domain of compressive initial stresses ($\lambda_1 < 1$), the curves have vertical asymptotes corresponding to the resonance *SIF* change occurring when the values of initial compressive stresses tend to the values at which the local loss of material's stability in the vicinity of cracks occurs (in the antisymmetric in respect of plane $y_3 = -h$, or bending, form) under compression by forces directed along the cracks.

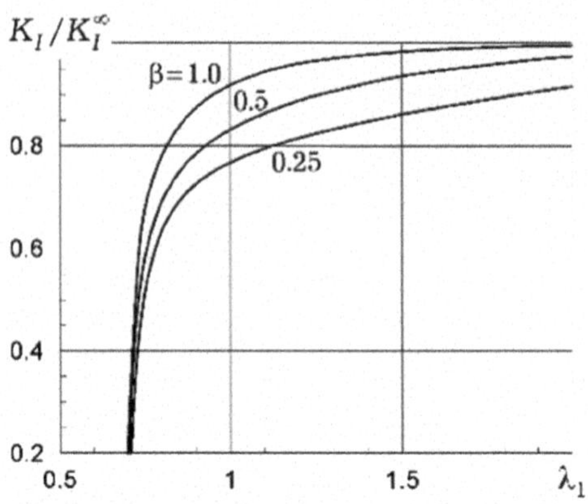

Fig. 4.11 Dependence of *SIFs* ratios K_I/K_I^∞ on elongation (or contraction) ratio λ_1 for Bartenev–Khazanovich potential

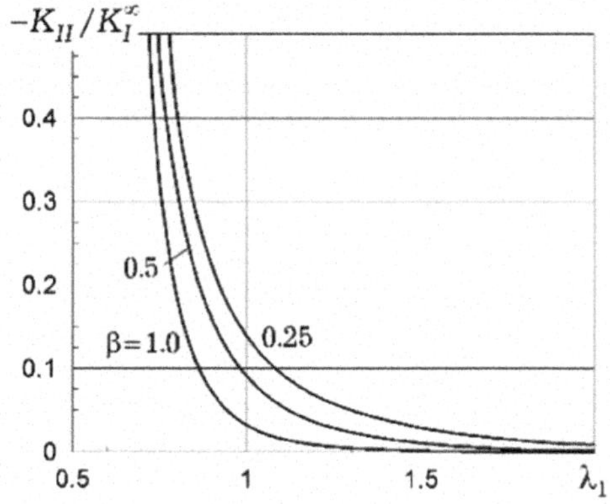

Fig. 4.12 Dependence of *SIFs* ratios K_{II}/K_I^∞ on elongation (or contraction) ratio λ_1 for Bartenev–Khazanovich potential

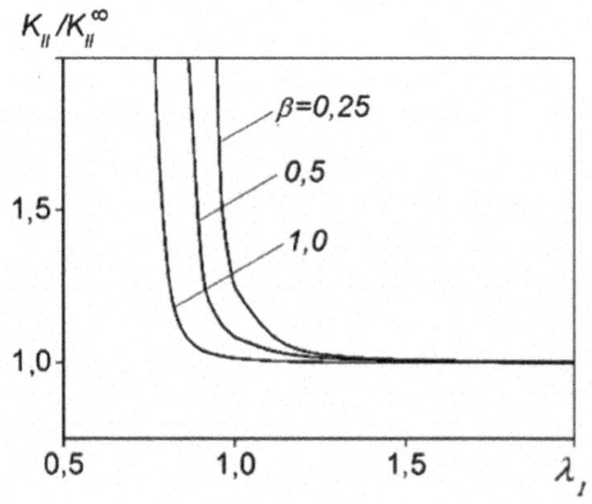

Fig. 4.13 Dependence of *SIFs* ratios K_{II}/K_{II}^∞ on elongation (or contraction) ratio λ_1 for Bartenev–Khazanovich potential

Here, it should be noted that the critical (limit) values of contraction parameters $\lambda_1 < 1$ for the antisymmetric (bending) form of stability loss are larger (and the critical (limit) compressive stresses, correspondingly, smaller) than the critical values for the symmetric form of stability loss which were obtained above (see Figs. 4.11

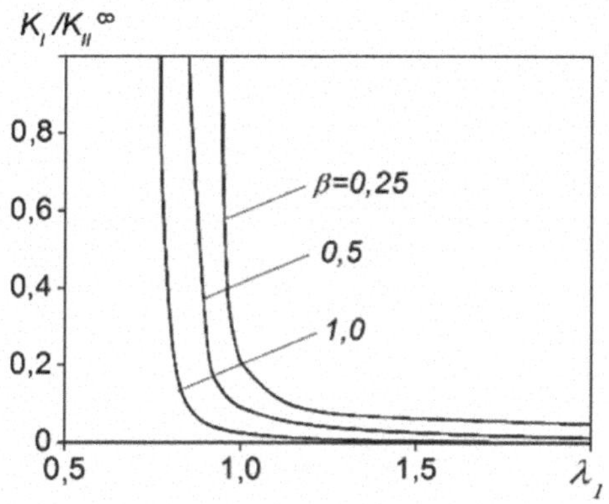

Fig. 4.14 Dependence of *SIFs* ratios K_I/K_{II}^∞ on elongation (or contraction) ratio λ_1 for Bartenev–Khazanovich potential

Table 4.1 Critical values of relative contraction $\varepsilon_1 = 1 - \lambda_1$ for Bartenev–Khazanovich potential

β	1/8	1/4	1/2	1	10
$\varepsilon_1^{(1)}$	0.304	0.307	0.307	0.307	0.307
$\varepsilon_1^{(2)}$	0.035	0.089	0.168	0.242	0.306
$\varepsilon_1^{(3)}$	0.010	0.039	0.117	0.224	0.306

and 4.12). This is clearly demonstrated in Table 4.1, which gives the values of relative critical (limit) contraction parameters $\varepsilon_1{}^{(1)} = 1 - \lambda_1{}^{(1)}$ and $\varepsilon_1{}^{(2)} = 1 - \lambda_1{}^{(2)}$ at which the local loss of material's stability occurs under compression along two parallel coaxial cracks (values $\varepsilon_1{}^{(1)}$ correspond to the symmetric form of stability loss, and $\varepsilon_1{}^{(2)}$—to the antisymmetric (bending) form of stability loss). As can be seen, in the entire range of β change, the values $\varepsilon_1{}^{(2)} < \varepsilon_1{}^{(1)}$, i.e., stability loss for this material takes place according to the bending form. It is also seen that at small distances between cracks their mutual influence results in a significant decrease of critical compression parameters. Yet, with increasing distance between cracks, the relative critical contraction parameters tend to the value of $\varepsilon_1 = 0.307$, which for Bartenev–Khazanovich potential corresponds to the critical (limit) contraction parameter in the case of a single isolated crack in an infinite body (Guz et al. 1992, 2020). Besides, this table shows the values of $\varepsilon_1{}^{(3)}$, which are relative critical contraction parameters obtained from the solution of the axisymmetric problem on compression of a semi-bounded body containing a near-surface crack.

For mode III cracks, when the crack faces are under load (4.77), Fig. 4.15 shows the dependences of the ratios of stress intensity factors $K_{III}/K_{III}{}^\infty$ (where $K_{III}{}^\infty$ is determined from (4.78)) on initial stress parameters λ_1 for different values of geometric parameter β, which proves a significant influence of initial stresses on the *SIF* K_{III}. In this case, however, there are no effects of the resonance change of the stress intensity factor, as opposed to the problems on mode I and mode II cracks, since, evidently, under compression of the material containing two parallel cracks there is no stability loss corresponding to the torsion problem.

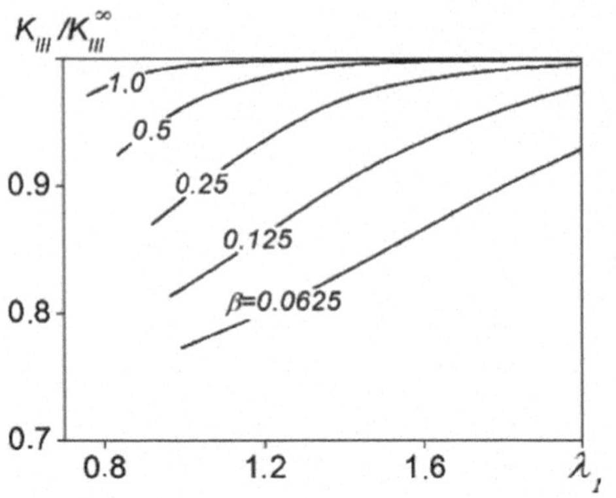

Fig. 4.15 Dependence of *SIFs* ratios $K_{III}/K_{III}{}^{\infty}$ on elongation (or contraction) ratio λ_1 for Bartenev–Khazanovich potential

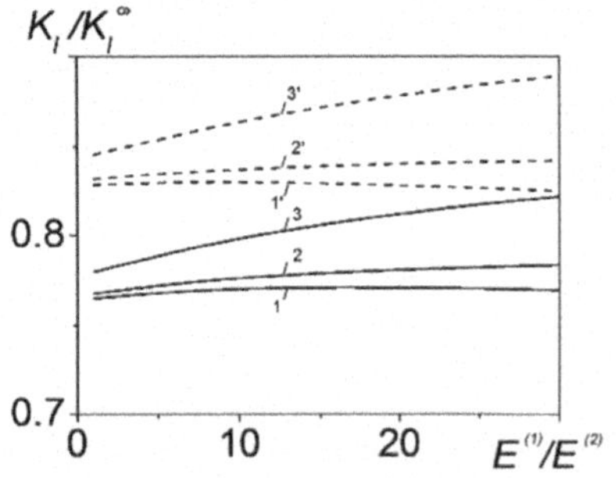

Fig. 4.16 Dependence of *SIFs* ratios $K_I/K_I{}^{\infty}$ on the ratio of elastic moduli $E^{(1)}/E^{(2)}$ for a laminated composite

Laminated two-component composite with isotropic layers (in macrovolumes, that is a transversely isotropic medium (Khoroshun et al. 1993), *a compressible body, non-equal roots).* For this material, the results of studying the axisymmetric problem of a pre-stressed body containing two parallel coaxial mode I cracks are presented.

Figure 4.16 shows that the ratios of stress intensity factors $K_I/K_I{}^{\infty}$ increase monotonously with the increase in the ratios of elastic moduli of the materials containing composite layers $E^{(1)}/E^{(2)}$ (the materials of the layers have identical Poisson's ratios $= 0.3$). Besides, it can be seen that for the values of dimensionless half-distance between the cracks $\beta = 0.25$ the corresponding values of $K_I/K_I{}^{\infty}$ (solid lines) are smaller than for $\beta = 0.5$ (dashed lines).

Figure 4.17 illustrates the dependence of the ratio $K_{II}/K_I{}^{\infty}$ on $E^{(1)}/E^{(2)}$. In Figs. 4.16 and 4.17, lines *1* and *1'* correspond to $\lambda_1 = 0.99$ (compressive initial stresses), lines *2* and *2'*—to $\lambda_1 = 1.0$ (no initial stresses), lines *3* and *3'*—to $\lambda_1 = 1.05$ (tensile initial stresses).

Figure 4.18 shows the dependence of the $K_I/K_I{}^{\infty}$ ratios on the glass concentration factor c_1 for different values of initial stress parameters λ_1, demonstrating the influence of initial stresses and mechanical characteristics of the composite on the values of stress intensity factors in the composition of aluminum/boron/silicate glass layers with those of epoxy/maleic resin.

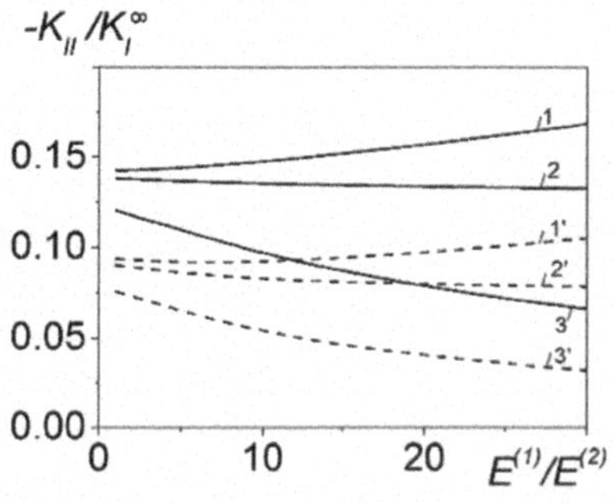

Fig. 4.17 Dependence of *SIFs* ratios $K_{II}/K_I{}^\infty$ on the ratio of elastic moduli $E^{(1)}/E^{(2)}$ for a laminated composite

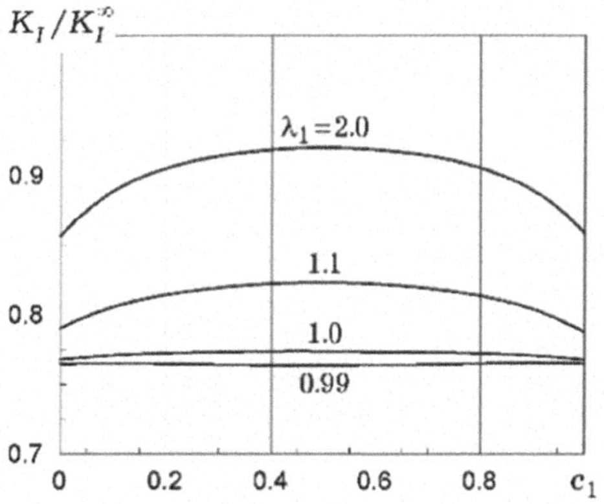

Fig. 4.18 Dependence of *SIFs* ratios $K_I/K_I{}^\infty$ on the glass concentration factor c_1 for aluminum/boron/silicate glass in epoxy/maleic resin

4.7 Conclusions

The results obtained in the research of the stress-strain state of pre-stressed materials containing near-surface cracks and two parallel coaxial cracks suggest the following conclusions:

- the order of singularity in the distribution of stresses in the vicinity of near-surface crack edge in a pre-stressed semi-bounded body and near the edges of parallel coaxial cracks in a pre-stressed unbounded body is equal to $-1/2$, i.e., it coincides with the order of singularity in the distribution of stresses near crack edges in the bodies free of initial stresses (Kassir and Sih 1975);
- in all the problems considered (with the exception of problems on torsion) a dramatic resonance change of stress intensity factors occurs when initial compressive forces approach the values corresponding to the local loss of material's stability in the vicinities of cracks, which permits the critical (limit) compression parameters to be determined directly from the solutions of corresponding non-homogeneous problems of the fracture mechanics of pre-stressed materials;
- the mutual influence between the crack and the half-space boundary (a near-surface crack) or between the cracks (two parallel cracks) causes a quantitative change (especially significant for small distances between cracks or between the crack

and the half-space boundary) in the values of stress intensity factors as compared to those obtained for an isolated crack in an infinite body. On the other hand, with an increase of the distance between the cracks (or the crack and the half-space boundary) the abovementioned influence gradually becomes weaker, and the values of stress intensity factors tend to the corresponding values obtained in the case of an isolated crack in an infinite material;

- the mechanical characteristics of materials produce a significant influence on the values of stress intensity factors;
- the critical (limit) compression parameters corresponding to the local loss of material's stability in the vicinities of cracks significantly depend on the geometric parameters of the problems (crack radii, distances between cracks, or those between the crack and material's boundary) and on the mechanical characteristics of materials.

References

Ainsworth RA, Sharples JK, Smit SD (2000) Effects of residual stresses on fracture behaviour - experimental results and assessment methods. J Strain Anal Eng Des 35(4):307–316

Bartenev GM, Khazanovich TN (1960) The law of high elastic deformation of network polymers. Vysok Soedinyeniya 2(1):21–28

Bateman H, Erdelyi A (1953) Higher transcendental functions. McCrew-Hill, New York

Bogdanov VL (2007) On a circular shear crack in a semiinfinite composite with initial stresses. Mater Sci 43(3):321–330

Bogdanov VL (2010) Influence of initial stresses on fracture of composite materials containing interacting cracks. J Math Sci 165(3):371–384

Bogdanov VL (2012) Influence of initial stresses on the stressed state of a composite with a periodic system of parallel coaxial normal tensile cracks. J Math Sci 186(1):1–13

Bogdanov VL, Guz AN, Nazarenko VM (2015) Nonclassical problems in the fracture mechanics of composites with interacting cracks. Int Appl Mech 51(1):64–84

Bogdanov VL, Guz AN, Nazarenko VM (2017) Unified approach in non-classical problems of fracture mechanics. LAP LAMBERT Academic Publishing, Saarbrücken [in Russian]

Bogdanov VL, Nazarenko VM (1994) Study of the compressive failure of a semi-infinite elastic material with a harmonic potential. Int Appl Mech 30(10):760–765

Bolotin VV (1994) Stability problems in fracture mechanics. Wiley, New York

Bolotin VV (2001) Mechanics of delaminations in laminate composite structures. Mech Compos Mater 37(5):367–380

Broutman LJ, Krock RH (1974) Mechanics of composite materials. In: Sendeckyj GP (ed) Composite materials, vol 2. Academic Press, New York

Cherepanov GP (1979) Mechanics of brittle fracture. McGraw-Hill, New York

Dvorak GJ (2000) Composite materials: Inelastic behaviour, damage, fatigue and fracture. Int J Solids Struct 37(1–2):155–170

Guz AN (1980) Theory of cracks in elastic bodies with initial stress - formulation of problems, tear cracks. Sov Appl Mech 16(12):1015–1024

Guz AN (1981) A criterion of solid body destruction during compression along cracks (two-dimensional problem). Dokl AN SSSR 259(6):1315–1318

Guz AN (1982) On the criterion of brittle fracture of materials with initial stresses. Dokl AN SSSR 262(2):285–288

Guz AN (1991) Brittle fracture of materials with initial stresses. In: Guz AN (ed) Nonclassical problems of fracture mechanics, vol 2. Nauk. Dumka, Kyiv [in Russian]
Guz AN (1999) Fundamentals of the three-dimensional theory of stability of deformable bodies. Springer, Berlin-Heidelberg-New York
Guz AN (2014) Establishing the foundations of the mechanics of fracture of materials compressed along cracks (review). Int Appl Mech 50(1):1–57
Guz AN (2021) Eight non-classical problems of fracture mechanics. In: Advanced structure materials, vol. 138. Springer, Cham
Guz AN, Bogdanov VL, Nazarenko VM (2020) Fracture of materials under compression along cracks. In: Advanced structure materials, vol. 138. Springer, Cham
Guz AN, Dyshel MS, Nazarenko VM (1992) Fracture and stability of materials with cracks. In: Guz AN (ed) Nonclassical problems of fracture mechanics, vol. 4, book 1. Nauk. Dumka, Kyiv [in Russian]
Guz AN, Nazarenko VM, Bogdanov VL (2013) Combined analysis of fracture under stresses acting along cracks. Arch Appl Mech 83(9):1273–1293
John F (1960) Plane strain problems for a perfectly elastic material of harmonic type. Commun Pure Appl Math 13(2):239–296
Kachanov LM (1988) Delamination buckling of composite materials. Kluwer Academic Publisher, Boston
Kassir MK, Sih GC (1975) Mechanics of fracture, vol 2. Three-dimensional crack problems. Noordhoff International Publishing, Leyden
Khoroshun LP, Maslov BP, Shikula EN, Nazarenko LV (1993) Statistical mechanics and the effective properties of materials. In: Guz AN (ed) Mechanics of composite materials, vol 3. Nauk. Dumka, Kyiv [in Russian]
Kienzler R, Herrmann G (2000) Mechanics in material space with applications to defect and fracture mechanics. Springer, Berlin
Malmeister AK, Tamuzh WP, Teters GA (1980) Strength of polymer and composite materials. Zinatne, Riga [in Russian]
Nazarenko VM, Bogdanov VL, Altenbach H (2000) Influence of initial stress on fracture of a half-space containing a penny-shaped crack under radial shear. Int J Fract 104:275–289
Prudnikov AP, Brychkov YA, Marichev OI (1986) Elementary functions. In: Integrals and series, vol. 1. Gordon and Breach Science Publisher, New York
Prudnikov AP, Brychkov YA, Marichev OI (1986) Special functions. In: Integrals and series, vol. 2. Gordon and Breach Science Publisher, New York
Shul'ga NA, Tomashevskii VT (1997) Process-induced stresses and strains in materials. In: Guz AN (ed) Mechanics of composite materials, vol 6. ASK, Kyiv [in Russian]
Uflyand YS (1977) Integral transformations in problem of the theory of elasticity. Nauka, Leningrad [in Russian]
Watson GN (1995) A treatise on the theory of Bessel functions. Cambridge University Press, Cambridge
Wu CH (1979) Plane-strain buckling of a crack in harmonic solid subjected to crack-parallel compression. J Appl Mech 46:597–604

Chapter 5
Evolution of Electron Structure of the Methane-Coal Sorption System Components and Properties

Anatolii Bulat, Oleksandr Burchak, Volodymyr Trachevskyi, and Andrey Tokar

5.1 Introduction

Carbonized fossil organic matter and, in particular, hard coal, is a complex multicomponent, heterogeneous, metastable composition of substances, whose mechanisms of sorption capacity formation are the subject of numerous studies. According to modern concepts, sorption interaction is caused by dispersion forces or by occurrence of instantaneous non-zero moments (electrical potentials) in molecules and atoms. The effectiveness of such interaction depends on the correlation of the frequencies of fluctuations of the electron density in molecules, i.e., in the first approximation, on the resonance of vibration frequencies of the sorbent molecules and individual fragments of the sorbate matrix.

Certain difficulties in studying sorption mechanisms are caused by the absence of a matrix architecture hierarchy (Alexeev 2012; Bulat et al. 2010) due to the metastability (Bulat et al. 2010) of coal and the variability of structures. Therefore, it is

A. Bulat · O. Burchak (✉)
N.S. Polyakov Institute of Geotechnical Mechanics of the National Academy of Sciences of Ukraine, Dnipro, Ukraine
e-mail: gvrvg@meta.ua

A. Bulat
e-mail: igtmnanu@ukr.net

V. Trachevskyi
Laboratory for Synthesis and Diagnostics of Nanosystems and Nanomaterials of the Analytical Research Department of the Technical Center of the National Academy of Sciences of Ukraine, Kyiv, Ukraine
e-mail: trachev@imp.kiev.ua

A. Tokar
Dnipro State Agrarian and Economic University of the Ministry of Education and Science of Ukraine, Dnipro, Ukraine
e-mail: atokar_2004@ukr.net

A. N. Guz et al. (eds.), *Advances in Mechanics*, Advanced Structured Materials 191,
https://doi.org/10.1007/978-3-031-37313-8_5

difficult to specify the very concept of the sorption center. Under these conditions, a statistically averaged behavior of the macrosystem is determined by the manifestation of the properties of collectivized functional groups in the composition.

In works (Blumenfeld et al. 1962; Burchak et al. 2016), by using the EPR spectroscopy method in combination with chemical studies, a physical nature of the adsorption interaction was proven and the relationship between the sorption capacity of the carbonized organic matter relative to oxygen molecules and the content of fragments with conjugated bonds in the structure of carbon matrices (aromatic and aliphatic components) was established. Further studies (Alexeev 2012; Bulat et al. 2010) showed that concepts about the course of physical sorption interaction of coal extend to the systems with other gases, with methane in particular.

The main problem in describing the specifics of the process of interaction between the aromatic component of carbonized organic matter and methane molecules under the influence of dispersion forces is high symmetry of the CH_4 molecule (centered tetrahedron), whose vector sum of dipole moments of the bonds (m = 0.4D) is zero ($m_{zv} = 0$). However, according to the classical theory this means that such a molecule should not interact with the sorption center of coal. But practical experience obtained from experiments with a change in pressure unquestionably proves the reality of sorption of methane molecules by coal (Alexeev 2012; Bulat et al. 2010, 2021). Under these circumstances, there is a need to go beyond the generally accepted ideas about configuration of methane molecule and about evolution of electron state in the carbon atoms under the conditions of sorption interaction. At the same time, the assumption about the resonance of the vibration frequencies of bonds with different degrees of hybridization requires additional research.

The relevance of studying the evolution of electron state in the carbon atoms under the conditions of sorption interaction is determined by the high importance of obtaining reliable information about the mechanisms of interphase interaction for creating adequate models of the processes in the organic matter-gas system and developing an algorithm for sophisticated control of the transformations.

The idea of the work was to study the relationship between external conditions and the electron state on the carbon atoms in the composition of methane molecules and in fragments of solid hydrocarbon matrices under conditions of sorption interaction.

The purpose of the work was to determine the range of changes in the operating conditions of interphase interaction mechanisms, to visualize the evolution of electron state in the methane carbon atoms during sorption, and to determine the driving forces of structural transformations in hydrocarbons.

5.2 Research Methods and Objects

The NMR ^{13}C method (NMR spectrometer AVANCE 400, Bruker), which allows studying the evolution of architecture of macromolecular compositions with the identification of interacting objects. The identification of interacting objects made it possible to study structural and functional transformations of carbonized organic matter and methane molecules by changing pressure in the coal matrix-gas system.

Based on the data of the IRS method (IR-Fourier spectrometer Nicolet iS10, Thermo Scientific), which allowed to identify functional groups and fragments with double and triple carbon-carbon bonds, changes in the intensity of vibrations in the corresponding absorption bands of the solid carbonized organics were assessed before and after mechanical compression.

Models of complexes describing the most likely contributions to interactions between non-polar molecules were built by the quantum-chemical methods. Calculations at the *ab initio* level of theory were performed by using the Gaussian 03, Revision E.01 software package with further visualization of the results in GaussView 3.0.

5.3 Experimental Studies

The occurrence of dipole moment in the methane molecules is possible due to changes in the geometry of the compound under the influence of external forces, for example, at the transition from tetrahedral structure to pyramidal (disphenoidal) structure. As regard to the coal-gas system, it can be assumed that on the trajectory to the combined system of the aromatic cluster, carbon atom of the methane molecule changes its electron structure (from the sp^3 state to the sp^2 state), according to L. Pauling's assumption for the entropic nature of hybridization. That is, hybridization of electron orbitals of carbon atoms in the methane molecule is determined by the conditions characterizing the state of the heterogeneous system and requires objective experimental confirmation.

One can be convinced of the realism of this hypothesis by recording the NMR ^{13}C spectrum of a methane-coal mixture with constant stoichiometry at different pressures (Bulat et al. 2021). During the research, the authors studied energy absorption regions characteristic for the methane molecules ($\delta \approx 0.0$ ppm) and aromatic ($\delta \approx 188$ ppm) fragments of the structure of solid hydrocarbons (Fig. 5.1).

The presence of changes in the spectral characteristics of methane and aromatic fragments also indicates a change in the hybridization of carbon atoms during the interaction of CH_4 with the solid coal matrix. In particular, increasing the pressure in the measured NMR-cell by ≈ 0.8 bar changed the ratio of the integral intensities of lines with the chemical shift of $\delta \approx 27.1$ ppm and $\delta \approx 188$ ppm. The presence of this difference proves the participation of the growing number of methane molecules under pressure in the interaction with the structural fragments of the solid matrix. The observed change in the shape of the line and the increase in absorption intensity within the region characteristic for sp2-hybridized carbon atoms in the solid matrix can be related to the transformation of methane molecules with a decrease of symmetry under the conditions of condensation interaction with the coal aromatic structures (Bulat et al. 2021).

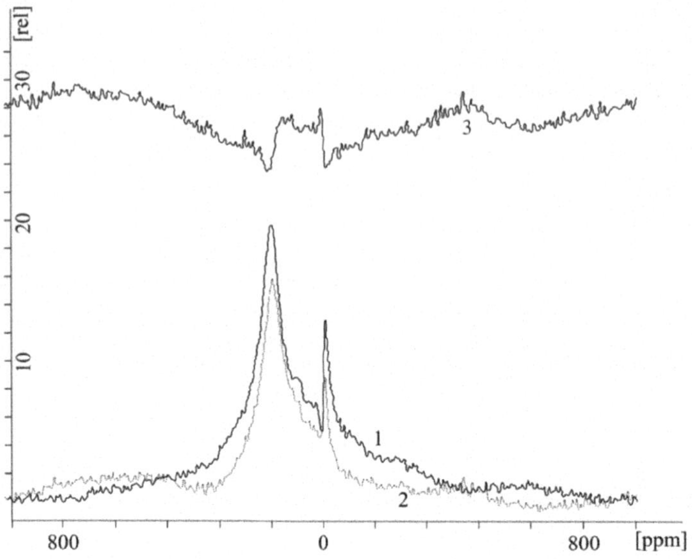

Fig. 5.1 Influence of pressure on the NMR spectrum of the coal-$^{13}CH_4$ system (1—initial state of the system, 2—increased pressure, 3—the difference between the spectra)

The information obtained by the NMR ^{13}C method allows to visualize the evolution of electron state in methane carbon atoms and to propose a possible mechanism of sorption interaction in the solid hydrocarbon matrix-methane system. In addition, with taking into account the obtained experimental results based on a qualitative assessment of the redistribution pattern of carbon atoms between the sp^3 and sp^2 states in specific thermodynamic conditions, it is possible to specify a change in the internal energy and reactivity of adsorbed methane molecules. Given that the overlapping of orbitals, which initiates spin exchange, occurs when the distance optimal for the organization of physical contact of partners is reached, the redistribution of integral intensities of NMR signals should be considered as a source of information and a way to understand the overall mechanism of the sorption (Bulat et al. 2021).

In order to verify the thesis regarding the possibility of the linear change of the bond vibration frequency (length) due to the change of thermodynamic conditions in the system, the influence of mechanical pressure on the state of the molecular structure of the coal matter was simulated by the IR-Fourier spectrometry method (Fig. 5.2).

In the course of the experiment, the IR spectrum of the coal was recorded in its three states: natural, compressed, and broken (crushed) after compressing; the ratio of the intensities of the vibration bands of methyl and methylene functional groups was evaluated.

The IR spectra (Fig. 5.2) show a change in the ratio of intensities of the bands of the characteristic valence vibrations of the CH_3/CH_2 groups, which is an indicator of the intramolecular redistribution of hydrogen atoms between the types of bonds in

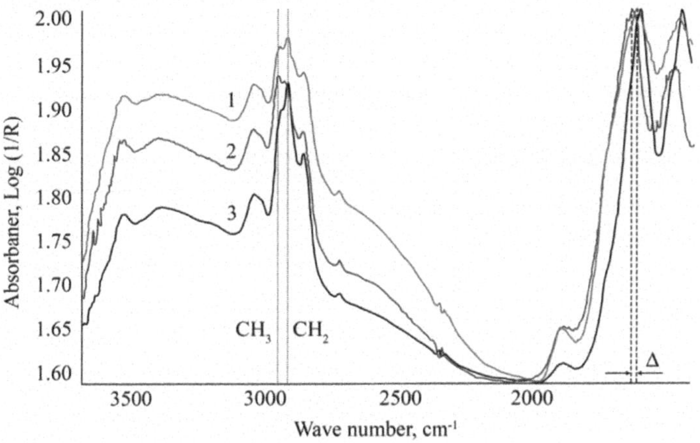

Fig. 5.2 Influence of mechanical factor on the IR spectra of coal (1—carbon matter in natural state ($CH_3/CH_2 = 0.697$), 2—carbon matter in compressed state ($CH_3/CH_2 = 1.129$), 3—carbon matter in broken state after compressing ($CH_3/CH_2 = 0.768$))

the coal structure. The effect of disproportionation ($\text{-}CH_2\text{-} \rightarrow \text{-}CH_3\text{+-}CH\text{=}$) of fragments of the aliphatic component was also recorded—it was initiated by mechanical pressure, reorganization of distribution according to the energy state and reactivity of conformers of hydrocarbon chains (Burchak and Balalaev 2010).

An increase of the energy potential of the matter was also detected—shifts of the absorption bands in the direction of an increase in the frequency of vibrations of interatomic bonds were recorded; so the shift of the D_{1600} absorption band by $\Delta \approx 14$ cm^{-1} corresponds to the accumulation of energy in the atomic-molecular architecture of the carbon matter of ≈ 1.68 J/mol. Besides, mechanical activation of the matter takes place under the influence of loads, that is, transformation of mechanical energy into chemical energy. As a result of reorientation, some fragments of the structure of the organic coal matter transfer into energetically more favorable conformational states, which are fixed by the character of the intra- and intermolecular interaction, and the energy barriers during the relaxation of the ascending situation can be so great that the caused changes become irreversible.

The obtained information on the directed evolution of the solid hydrocarbon matrix with increased mechanical load makes it possible to assume that the internal energy accumulated during the structural and functional reorganization caused by an external factor (mechanochemical activation) largely determines physical and chemical properties of the coal matter, in particular, sorption capacity of coal.

The picture compiled as a result of the analysis of the data set can serve as an argumentative proof of the reverse participation of alkanes in "aromatization" under the mechanical load. Accordingly, it can be assumed that during the interaction of alkyl groups (radicals) with aromatic structures in extreme conditions, the frequency of their characteristic vibrations converges, as the course of coarctate reactions predicts. That is, after the initial physical interaction (sorption) with the electron-deficient fragments of the coal matrix, methane molecules are transformed and participate in the formation of aromatic structures (Magdesieva 2013).

Thus, the cumulative effect of the mechanical load on the coal matrix consists in the expenditure of energy to disrupt the hierarchical heterogeneous architecture, structural and functional structure due to the macro- and micro-scale changes of:

- nature of interparticle and interphase interaction;
- interfragment contacts and conformational rearrangements;
- generation of highly reactive particles, the totality of which determines the evolution of the space-time organization of the matter under the action of external force loads.

Accordingly, the conducted research can give, in future, the algorithms for understanding the ways of forming the properties of carbon matrices and effective mastering of mechanochemical transformations caused by plastic deformation and developed against the background of absorption and transformation of elastic energy by a solid matter.

5.4 Theoretical Background of the Concept

One of the tasks of the research was to form a theoretical basis for the multi-stage mechanism of interaction of methane molecules and aromatic structures in the coal matter. To solve it, the nature and features of intermolecular interaction were investigated at the level of models with functional groups and individual molecules and with involvement of modern computational methods of quantum chemistry and vibrational spectroscopy.

At the initial stage of the research, of fundamental importance becomes the problem of choosing the model compounds, which would provide the possibility of adequate reproducing the features of the structure of the aromatic fragments of carbonized organic matter and the interphase interaction in the coal-methane system, and, at the same time, would be sufficiently simple from the point of view of conducting quantum-chemical calculations. The results of molecular modeling of the interaction between the benzene ring and methane molecule under standard conditions (without increasing the pressure) are shown in Fig. 5.3. Here, one of the possible vibratory states of such a system with the corresponding displacement vector is shown.

It is clearly seen in Fig. 5.3 that the elongated C-H bond is directed to the double bond of the ring, and not to the center of the benzene molecule. Accordingly, it should be assumed that the sorption interaction is evolutionarily accompanied by a change in the vibration frequencies (length) of the C-H bond in the methane molecule (dispersion forces), and is not related to the action of the magnetic field induced in the middle of the aromatic ring. Therefore, the development of conjugation systems is a structural characteristic of coal associated with interphase interaction.

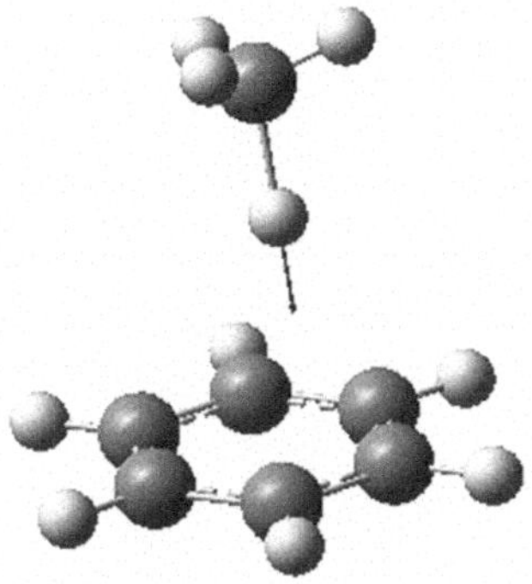

Fig. 5.3 Visualization of the results of molecular modeling of the interaction of the benzene ring with the methane molecule

Further, the quantum-chemical calculations were used for studying a sorption capacity of the cyclic structures with a system of conjugated bonds. By using the non-empirical generalized gradient approximation **PBE1PBE/6-311++G(d,p)**, the features of the intermolecular interactions that appeared in the model system composed of two completely non-polar molecules (benzene and methane) were investigated. It should be noted that interactions of this type are of great interest, first of all, from the point of view of theoretical explanation of their nature, and they can serve as a starting point for the preliminary assessment of some energetic and spectral characteristics of these systems. As for the related interactions between the purely aromatic fragments of molecules and manifestations of the so-called stacking effects, the situation is sufficiently good understood (Hill et al. 2006; Munshi and Guru 2005; Zhikol et al. 2005). It is well known that such interactions play a decisive role in ensuring the stability of many macromolecules, including biopolymers containing regions with heterocyclic structure, in particular proteins and nucleic acids.

In Fig. 5.4, critical points (3, −1) correspond to the areas with the highest electron density between the C_6H_6 and CH_4 molecules, which are converging, while the point in the center of the benzene ring has the lowest density. At the same time, the molecular graph indicates an interaction not with the bond of the benzene ring, but specifically with one of the equivalent carbon atoms contained in it.

When determining the energy parameters of intermolecular interactions of the $\mathbf{C_6H_6 \cdot CH_4}$ type, in addition to electron energies, special corrections for zero-point vibrations and basis set superposition errors of were also taken into account according to the Boys-Bernardi calculation procedure (Sordo 2001). The obtained results indicate the energetic effects of interactions, which do not exceed 1.9 kJ/mol and occur at distances of 3.159–3.174 Å and can be identified as weak dispersion forces. Calculation within the Bader's AIM-theory (Kolandaivel and Nirmala 2004) of values of the electron density $\rho(r) = 0.0035$ e/Å^3 and the Laplacian of the electron density $\nabla^2\rho(r) = 0.0024$ e/Å^5 at the single (3, −1) critical point between molecules (Fig. 5.4) also confirms this assumption, since such interactions are at least one order of magnitude lower than the energy of typical intramolecular bonds (Tokar and Chigvintseva 2021; Tokar et al. 2021; Tsirelson 2017).

Fig. 5.4 Molecular graph of the $C_6H_6 \cdot CH_4$ system

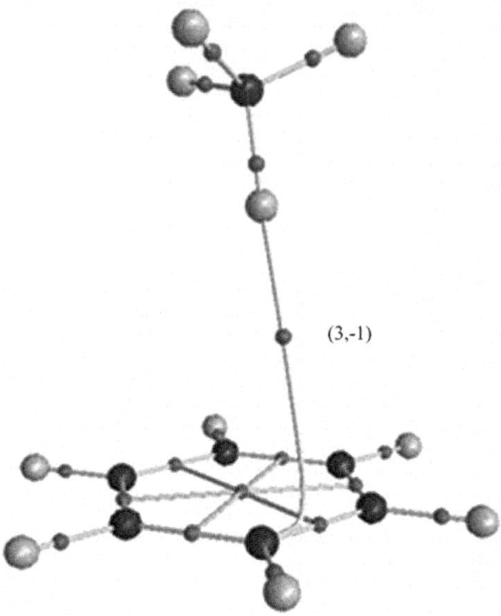

The determination of some thermodynamic parameters related to the processes of formation and destruction of the studied complex of molecules deserves special attention. For interactions with stoichiometry of the type $A + B = C$, special corrections to the quantities ΔH^0, ΔS^0 and ΔG^0, as it is known, are equal to the products $R \cdot T$, $R \cdot T \times (1 + \ln R' \times T)$ and $-R \cdot T \times \ln R' \cdot T$ (where R' corresponds to the value of the universal gas constant expressed in l·atm/mol·K), which under standard conditions are +2.5 kJ/mol, +34.9 J/mol·K and –7.9 kJ/mol, respectively (Benson 1976). The calculated values of the enthalpy, entropy, and Gibbs free energy of the formation of the $\mathbf{C_6H_6 \cdot CH_4}$ complex are +2.7 kJ/mol,—43.3 J/mol·K and +15.6 kJ/mol. At the same time, an insignificant contribution of the enthalpy factor is well consistent with the low energy intensity of the studied interactions, while the more significant contribution of the entropy component is easily explained by significant limitations in translational, rotational, and vibrational degrees of freedom of individual molecules under binding conditions. The latter becomes especially noticeable when analyzing the basic spectral characteristics of the studied system.

As it is known, harmonic vibrational frequencies calculated by the methods of quantum chemistry are usually higher than the similar parameters obtained experimentally. At the same time, the main source of errors, according to the authors (Merrick et al. 2007), is neglecting of effects of anharmonicity, as well as not complete description of electron correlations due to the use of limited sets of basic functions. To improve the accuracy of the calculated data in reproduction of the main spectral characteristics of the studied systems, one should use the auxiliary values of the scaling factors, which, in the case of the hybrid PBE functional, were 0.9944 and 0.9948 for the basis sets 6-311+G(d,p) and 6-311 ++G(d,p) (Butyirskaya 2011).

The results obtained in parallel by the IRS method indicate that the most intense vibration bands are located in the 3207 cm^{-1} and 689 cm^{-1} spectrum area and correspond to valence and deformation vibrations of the benzene ring. Of particular interest are the absorption bands in the area 3163 cm^{-1} and 1324 cm^{-1}, which characterize similar fluctuations in the "bound" methane molecule. At the same time, the most likely resonance frequency of vibrations, which corresponds to the direct interaction of molecules in the $\mathbf{C_6H_6 \cdot CH_4}$ complex under standard conditions, is ~3170 cm^{-1}.

The partial displacement of one of the hydrogen atoms of the methane molecule toward the aromatic system leads to barely noticeable changes in the symmetry of the absolute tetrahedron, which, in turn, is accompanied by insignificant changes in the initial hybridization states of some carbon atoms both in the aryl fragment and in the CH_4 molecule itself. In particular, the calculated data obtained within the framework of the Weinhold's NBO theory (Weinhold 2012; Weinhold and Landis 2012) convincingly testify in favor of the $s^{0,95}p^{1,79}$/ $s^{1,10}p^{2,52}$ hybrid state of the carbon atoms of the benzene ring against the $s^{1,16}p^{3,24}$-hybridization of the carbon atom of the methyl group, which is fully consistent with generally accepted concepts about the peculiarities of the structure of these systems in the basic energy state. But under the conditions of the transition to an excited or activated state (increased pressure in the system), it is probably worth expecting more significant changes in geometric parameters and energy characteristics of the studied molecules, which will certainly require further theoretical substantiation of the mechanism of such transformations.

Thus, the given results of quantum-chemical calculation studies convincingly testify to the substantial possibility of applying the methods and techniques of quantum chemistry to study the conditions and mechanisms of interphase interaction at the level of the aromatic component of a solid body—methane molecules. The results of the calculations are in good agreement with the data of spectroscopy (NMR, IRS) and can be of fundamental importance for determining the mechanisms of sorption interaction in the coal-methane system.

5.5 Conclusions

It is shown that the spread of values of the parameters of the NMR ^{13}C spectra of the carbon atoms in the sorption system and the shift of the absorption band of the IR spectrum under mechanical loading allow us to state that the frequency of electron density fluctuations due to dispersion interaction in the system of bonds of electron-deficient fragments of the composition does not have a discrete value, the metastable state of the carbon matter is a consequence of the adaptive ability of carbon atoms to change the degree of hybridization depending on external thermodynamic conditions.

Based on the data of spectroscopic measurements and quantum-chemical modeling of the evolution of the distribution of carbon atoms of the reacting components between the states with different degrees of hybridization, the mechanisms and driving forces of self-organization of the heterophase coal-gas system were determined.

The visualized evolution of electron state in the carbon atoms of the sorption system under mechanical loading makes it possible to assess the contribution of the obtained results to the laws and mechanisms of transformations of organic compounds as fundamental, which will serve as a basis for the development of new-generation technologies of nanosynthesis processes.

On the basis of the revealed regularities a possibility is considered for formulating in the future the relationship between the forces of interatomic interaction with such mechanical and rheological properties of a solid matter as strength, plasticity, and elasticity, which describe the reverse reaction of the solid body to mechanical loads.

Directional control of the coal matter state will allow to initiate the processes of intensive artificial generation of methane in coal seams and organize effective, safe extraction of gaseous hydrocarbon energy carriers directly from the rock stratum. The structural and functional destructive transformation of natural solid hydrocarbons is exhausted down to the molecular level and is associated with significant (up to industrial indicators) beneficiation of the remains with nanoparticles of compounds of iron, sulfur, phosphorus, and acutely deficient non-ferrous metals and rare earth elements.

References

Alexeev AD (2012) Physics of coal and mining processes. CRC Press, Boca Raton

Benson SW (1976) Thermochemical kinetics. John Wiley & Sons, New York-London-Sydney-Toronto

Blumenfeld LA, Voevodsky VV, Semenov AG (1962) The use of epr in chemistry. Publishing House of SO AN USSR, Novosibirsk [in Russian]

Bulat AF, Skipochka SI, Palamarchuk TA, Antsiferov VA (2010) Methangeneration in coal layers. Lyra LTD, Dnepropetrovsk (in Russian)

Bulat AF, Bogdanov VL, Trachevsky VV, Burchak OV, Serikov YuA (2021) Research of mechanisms and driving forces of the self-organization of the matrices of natural solid hidrocarbons. Dopov Nac Akad Nauk Ukr 3:26–32 [in Ukrainian]

Burchak AV, Balalaev AK (2010) The effect of changing the parameters of the IR spectra of coals in the series of metamorphism under mechanical pressure. Geotechnical mechanics: Interdepartmental collection of scientific works 87:190–198 [in Russian]

Burchak OV, Trachevsky VV, Balalaev OK (2016) The coal aromaticity coefficient comparison during carbonization. Geotechnical mechanics: Interdepartmental collection of scientific works 124:208–215 [in Ukrainian]

Butyirskaya EV (2011) Computational chemistry: bases of theory and work with the programs of Gaussian and GaussView. Solon-Press, Moscow (in Russian)

Hill J, Platts JA, Werner H-J (2006) Calculation of intermolecular interactions in the benzene dimer using coupled-cluster and local electron correlation methods. Chem Phys Phys Chem

Kolandaivel P, Nirmala V (2004) Study of proper and improper hydrogen bonding using Bader's atoms in molecules (AIM) theory and NBO analysis. J Mol Struct

Magdesieva TV (2013) New types of combined pericyclic reactions. Russ Chem Rev 82:228–247 (in Russian)

Merrick JP, Moran D, Radom L (2007) An evaluation of harmonic vibrational frequency scale factors. J Phys Chem A
Munshi P, Guru Row TN (2005) Charge density based classification of intermolecular interactions in molecular crystals. CrystEngComm
Sordo JA (2001) On the use of the Boys-Bernardi function counterpoise procedure to correct barrier heights for basis set superposition error. J Mol Struct
Tokar A, Chigvintseva O (2021) The quantum-chemical and spectral criteria for hydrogen bonding efficiency in structural analysis of aramides. Chem Chem Technol
Tokar A, Kabat O, Chigvintseva O, Belošević S (2021) Intermolecular interactions in complex systems "Polyamide-Silica Gel": The quantum-chemical interpretation. In: Karabegović I (eds) New technologies, development and application IV. Lecture Notes in Networks and Systems, vol. 233. Springer, Cham, pp. 875–882
Tsirelson VG (2017) The quantum chemistry. Molecules, molecular systems and solids. Laboratory of Knowledge, Moscow [in Russian]
Weinhold F (2012) Natural bond orbital analysis: A critical overview of relationships to alternative bonding perspectives. J Comput Chem
Weinhold F, Landis C (2012) Discovering chemistry with natural bond orbitals. Wiley, New Jersey
Zhikol O, Shishkin OV, Lyssenko KA, Leszczynski J (2005) Electron density distribution in stacked benzene dimers: A new approach towards the estimation of stacking interaction energies. J Chem Phys

Chapter 6
Stress-Strain State of a Two-Layered Half-Space With Initial Stresses Under the Influence of a Moving Load

Yuriy Glukhov, Stepan Babich, and Victoria Kornienko

6.1 Introduction

One of the problems of considerable scientific and practical interest is the problem of the propagation of waves in bodies under the action of moving loads applied to their boundaries. Problems of this kind arise in the calculation of airfield and road pavements, in the calculation of plates and shells in a moving medium or under the action of moving stamps, and in many other cases.

It should be noted that from both a mathematical and a physical point of view, a moving load is a very general concept. Therefore, the main regularities of the action of a moving load can be studied regardless of the physical process resulting in these problems.

At present, a number of scientific directions are being developed in the dynamics of elastic bodies with initial (residual) stresses, one of which is the study of the dynamics of materials under moving loads (the review of publications on this topic is partially presented in Guz et al. (2015)).

In this chapter, within the framework of the formulations of Guz et al. (2015), using the integral Fourier transform, the solution to the problem of the dynamic response of an elastic two-layer half-space with initial stresses under the action of a moving surface load is obtained in a general form for compressible and incompressible materials and various conjugation conditions for the elements of a layered medium.

Y. Glukhov (✉) · S. Babich · V. Kornienko
S.P. Timoshenko Institute of Mechanics, National Academy of Science of Ukraine, Kyiv, Ukraine
e-mail: gluchov.uriy@gmail.com

S. Babich
e-mail: babich_sy@ukr.net

V. Kornienko
e-mail: vf_kornienko@ukr.net

A. N. Guz et al. (eds.), *Advances in Mechanics*, Advanced Structured Materials 191,
https://doi.org/10.1007/978-3-031-37313-8_6

6.2 Problem Statement

Consider a layer of thickness of $2h$, lying on a half-space, the initial stress-strain state of which is determined by the following components of the displacement vector and the generalized stress tensor:

$$u_j^0 = \delta_{ij} (\lambda_i + 1) x_i, \quad \sigma_{ii}^{*0} \neq 0 \quad (i, j = 1, 2, 3), \tag{6.1}$$

where λ_i are extensions ($\lambda_i =$ const) along the axes of the Lagrangian coordinate system x_i, which coincides in the natural state with the Cartesian coordinate system. Along with the Lagrangian coordinates, we introduce the Cartesian coordinates ξ_i of the initial deformed state, related to the coordinates x_i by relation $\xi_i = \lambda_i x_i$.

To a free boundary layer moving with a constant speed $\mathbf{v}$ load, independent of the coordinates ξ_3, is attached. Such a load causes a plain deformed state in this layered medium.

To solve the problem, we use the relations of the linearized elasticity theory for bodies with initial stresses (Guz 2004).

Assuming that the pattern of deformations is invariant with respect to time in the coordinate system (y_1, y_2) moving together with the load, where $y_1 = \xi_1 - vt$ and $y_2 = \xi_2$, the equation of the steady motion of the half-space in terms of the function $\chi(y_1, y_2)$ can be written in the form

$$\left(\eta_1^2 \frac{\partial^2}{\partial y_1^2} + \frac{\partial^2}{\partial y_2^2}\right)\left(\eta_2^2 \frac{\partial^2}{\partial y_1^2} + \frac{\partial^2}{\partial y_2^2}\right) \chi^{(j)} = 0, \quad j = 1, 2. \tag{6.2}$$

Roots η_1 and η_2 of the equation are determined by the following equation

$$\eta^4 + 2A\eta^2 + A_1 = 0, \tag{6.3}$$

where coefficients A and A_1 in case of compressible material are formed from the correlations

$$\begin{aligned} 2A\tilde{\omega}_{2222}\tilde{\omega}_{2112} &= \tilde{\omega}_{2222}\left(\tilde{\omega}_{1111} - \tilde{\rho}v^2\right) \\ &\quad + \tilde{\omega}_{2112}\left(\tilde{\omega}_{1221} - \tilde{\rho}v^2\right) - \left(\tilde{\omega}_{1122} + \tilde{\omega}_{1212}\right)^2, \\ 2A_1\tilde{\omega}_{2222}\tilde{\omega}_{2112} &= \left(\tilde{\omega}_{1111} - \tilde{\rho}v^2\right)\left(\tilde{\omega}_{1221} - \tilde{\rho}v^2\right), \quad \tilde{\rho}\lambda_1\lambda_2\lambda_3 = \rho, \end{aligned} \tag{6.4}$$

and in case of incompressible material are formed from the correlations

$$\begin{aligned} 2A\tilde{q}_{22}^2\tilde{\varkappa}_{2112} &= \tilde{q}_{11}^2\tilde{\varkappa}_{2222} + \tilde{q}_{22}^2\left(\tilde{\varkappa}_{1111} - \tilde{\rho}v^2\right) - 2\tilde{q}_{11}\tilde{q}_{22}\left(\tilde{\varkappa}_{1122} + \tilde{\varkappa}_{1212}\right), \\ 2A_1\tilde{q}_{22}^2\tilde{\varkappa}_{2112} &= \tilde{q}_{11}^2\left(\tilde{\varkappa}_{1221} - \tilde{\rho}v^2\right), \quad \tilde{q}_{ij} = \delta_{ij}\lambda_i q_i, \quad \tilde{\rho} = \rho. \end{aligned} \tag{6.5}$$

In (6.4) and (6.5), ρ is the density of the material of the half-space in the natural state, $q_t{}^{\{s\}}$, $\tilde{\varkappa}^{\{s\}}$ and $\tilde{\omega}^{\{s\}}$ are the parameters which characterize the material of the elements of the layered medium (Guz 2004).

Let us assume that the motion of the layer can be described by a system of equations from the theory of plates, taking into account the influence of rotational inertia and transverse shear. For a plate under the influence of transverse and tangential surface forces, the corresponding equations are given in Achenbach et al. (1967). In the coordinate system (y_1, y_2), the equations of plate theory can be written as

$$
\begin{aligned}
2h\left(\frac{2G_1}{1-\nu_1}-\rho_1 \mathrm{v}^2\right)\frac{\partial^2 u}{\partial y_1^2}-\tau &= P_1,\\
2h\left(\varkappa G_1-\rho_1 \mathrm{v}^2\right)\frac{\partial^2 w}{\partial y_1^2}-2\varkappa G_1 h\frac{\partial \phi}{\partial y_1}-q &= P_2,\\
\frac{2h^2}{3}\left(\frac{2G_1}{1-\nu_1}-\rho_1 \mathrm{v}^2\delta_0\right)\frac{\partial^2 \phi}{\partial y_1^2}+2\varkappa G_1\left(\frac{\partial w}{\partial y_1}-\phi\right)-\tau &= 0.
\end{aligned}
\tag{6.6}
$$

In (6.6), G_1, ν_1, and ρ_1 are shear modulus, Poisson's ratio and the density of the plate material respectively; u and w are the displacements of the middle surface of plate ($y_2 = 0$), δ_0 is a constant, which takes a value of 1 or 0, depending on whether the plate rotation inertia is taken into account or neglected when deriving (6.6); ϕ is the angle of rotation of the plate cross-section; $\varkappa$ is the Timoshenko shift coefficient; q and τ are normal and tangential stresses forcing to the surface of the partition plate and the half-space; P_1 and P_2 are tangential and normal components of the load on the free surface of the plate. The bending moment in the plate is determined by the formula

$$
M=\frac{4}{3}\frac{G_1 h^3}{1-\nu_1}\frac{d\phi}{dy_1}. \tag{6.7}
$$

Let us consider two cases of contact between the plate and a half-space at $y_2 = -h$:

- rigid contact

$$
\tilde{Q}_{21}=\tau,\quad \tilde{Q}_{22}=q,\quad u_2=w,\quad u_1=u+h\phi; \tag{6.8}
$$

- non-rigid contact

$$
\tilde{Q}_{21}=0,\quad \tau=0,\quad \tilde{Q}_{22}=q,\quad u_2=w. \tag{6.9}
$$

Thus, the problem is reduced to solving the equations of motion (6.2) and (6.6) under the boundary conditions (6.8) or (6.9).

Using the equations of motion of the plate (6.6) and conditions (6.8) and (6.9), the boundary conditions can be written in the general form

$$\delta_1\theta_1\left(\frac{d^2u_1}{dy_1^2}-h\frac{d^2\varphi}{dy_1^2}\right)-\tilde{Q}_{21}=\delta_1P_1,$$
$$\theta_3\frac{d^2u_2}{dy_1^2}-2\varkappa hG_1\frac{d\phi}{dy_1}-\tilde{Q}_{22}=P_2, \tag{6.10}$$
$$\theta_2\frac{d^2\phi}{dy_1^2}+2\varkappa G_1\left(\frac{du_2}{dy_1}-\phi\right)-\delta_1\tilde{Q}_{21}=0.$$

Here the following notations are introduced

$$\theta_1=2h\left(\frac{2G_1}{1-\nu_1}-\rho_1\mathrm{v}^2\right),\quad \theta_2=\frac{2h^2}{3}\left(\frac{2G_1}{1-\nu_1}-\delta_0\rho_1\mathrm{v}^2\right),$$
$$\theta_3=2h\left(\varkappa G_1-\rho_1\mathrm{v}^2\right).$$

Parameter δ_1 equals 1 for rigid contact and 0 – for non-rigid contact.

The values of the functions $\eta_1^2\,(\mathrm{v})$ and $\eta_2^2\,(\mathrm{v})$ determine the form of equations of motion (6.2) and, accordingly, the choice of the form for solving the considered equations. The influence of the load movement speed on the value of the roots of Eq. (6.3) for a compressible and incompressible half-space is studied in detail in Guz et al. (2015). Let us write the solution of the problem in general form for equal and unequal roots of Eq. (6.3).

The stresses, displacements, and velocities of displacements in a half-space are determined by formulas presented in Guz (2004) using the functions $\chi^{(j)}$:

$$\tilde{Q}_{ij}=\left(\alpha_{ij}^{(12)}\frac{\partial^2}{\partial y_1^2}+\alpha_{ij}^{(22)}\frac{\partial^2}{\partial y_2^2}\right)\frac{\partial\chi^{(2)}}{\partial y_{2-\delta_{ij}}}$$
$$+\left(\alpha_{ij}^{(11)}\frac{\partial^2}{\partial y_1^2}+\alpha_{ij}^{(21)}\frac{\partial^2}{\partial y_2^2}\right)\frac{\partial\chi^{(1)}}{\partial y_{1+\delta_{ij}}},\quad i,j=1,2, \tag{6.11}$$

$$u_i=-\beta_{i1}^{(i)}\frac{\partial^2\chi^{(i)}}{\partial y_1\partial y_2}+\left(\beta_{i1}^{(j)}\frac{\partial^2}{\partial y_1^2}+\beta_{i2}^{(j)}\frac{\partial^2}{\partial y_2^2}\right)\chi^{(j)};\ i,j=1,2;\ i\neq j; \tag{6.12}$$
$$\dot{u}_i=\mathrm{v}\left[\beta_{i1}^{(i)}\frac{\partial^3\chi^{(i)}}{\partial y_1^2\partial y_2}-\left(\beta_{i1}^{(j)}\frac{\partial^2}{\partial y_1^2}+\beta_{i2}^{(j)}\frac{\partial^2}{\partial y_2^2}\right)\frac{\partial\chi^{(j)}}{\partial y_1}\right];$$
$$i,j=1,2;\ i\neq j; \tag{6.13}$$

where, in the case of compressible material,

$$
\begin{aligned}
&\alpha_{ii}^{(11)} = \tilde{\omega}_{ii22}\left(\tilde{\omega}_{1111} - \tilde{\rho}\mathrm{v}^2\right) - \tilde{\omega}_{ii11}\left(\tilde{\omega}_{1212} + \tilde{\omega}_{2211}\right), \quad \alpha_{ii}^{(12)} = \tilde{\omega}_{ii11}\left(\tilde{\omega}_{1221} - \tilde{\rho}\mathrm{v}^2\right),\\
&\alpha_{ii}^{(21)} = \tilde{\omega}_{ii22}\tilde{\omega}_{2112}, \quad \alpha_{ii}^{(22)} = \tilde{\omega}_{ii11}\tilde{\omega}_{2222} - \tilde{\omega}_{ii22}\left(\tilde{\omega}_{1122} + \tilde{\omega}_{2121}\right),\\
&\alpha_{ij}^{(11)} = \tilde{\omega}_{ij21}\left(\tilde{\omega}_{1111} - \tilde{\rho}\mathrm{v}^2\right), \quad \alpha_{ij}^{(22)} = \tilde{\omega}_{ij12}\tilde{\omega}_{2222},\\
&\alpha_{ij}^{(12)} = \tilde{\omega}_{ij12}\left(\tilde{\omega}_{1221} - \tilde{\rho}\mathrm{v}^2\right) - \tilde{\omega}_{ij21}\left(\tilde{\omega}_{1122} + \tilde{\omega}_{2121}\right),\\
&\alpha_{12}^{(21)} = \tilde{\omega}_{ij21}\tilde{\omega}_{2112} - \tilde{\omega}_{ij12}\left(\tilde{\omega}_{1212} + \tilde{\omega}_{2211}\right),\\
&\beta_{11}^{(1)} = \beta_{21}^{(2)} = \beta = \tilde{\omega}_{1212} + \tilde{\omega}_{2211},\\
&\beta_{i2}^{(j)} = \tilde{\omega}_{2jj2}, \quad \beta_{i1}^{(j)} = \tilde{\omega}_{1jj1} - \tilde{\rho}\mathrm{v}^2, \quad i, j = 1, 2, \quad i \neq j,
\end{aligned}
$$

and in the case of incompressible material,

$$
\begin{aligned}
&\alpha_{ii}^{(ii)} = (-1)^i \tilde{q}_{jj}^{-1}\tilde{\varkappa}_{1212} - \delta_{j2}\tilde{\rho}\mathrm{v}^2\tilde{q}_{11}^{-1}, \quad \alpha_{ij}^{(12)} = -\tilde{q}_{22}^{-1}\tilde{\varkappa}_{ij21}, \quad \alpha_{ij}^{(22)} = \tilde{q}_{11}^{-1}\tilde{\varkappa}_{ij12},\\
&\alpha_{jj}^{(ii)} = \tilde{q}_{jj}\tilde{q}_{ii}^{-2}\left(\tilde{\varkappa}_{iiii} - \delta_{j2}\tilde{\rho}\mathrm{v}^2\right) + \tilde{\varkappa}_{jjjj}\tilde{q}_{jj}^{-1} - \tilde{q}_{ii}^{-1}\left(2\tilde{\varkappa}_{1122} + \tilde{\varkappa}_{1212}\right),\\
&\alpha_{ij}^{(11)} = \tilde{q}_{22}^{-1}\tilde{\varkappa}_{ij21}, \quad \alpha_{ij}^{(21)} = -\tilde{q}_{11}^{-1}\tilde{\varkappa}_{ij12}, \quad i, j = 1, 2, \quad i \neq j,\\
&\alpha_{22}^{(12)} = \tilde{q}_{22}^{-1}\left(\tilde{\varkappa}_{1221} - \tilde{\rho}\mathrm{v}^2\right), \quad \alpha_{11}^{(12)} = \tilde{q}_{11}\tilde{q}_{22}^{-1}\alpha_{22}^{(12)}, \quad \alpha_{11}^{(21)} = \tilde{q}_{11}^{-1}\tilde{\varkappa}_{2112},\\
&\alpha_{22}^{(21)} = \tilde{q}_{22}\tilde{q}_{11}^{-1}\alpha_{11}^{(21)},\\
&\beta_{11}^{(1)} = \beta_{12}^{(2)} = \beta = \tilde{q}_{11}^{-1}, \quad \beta_{21}^{(2)} = \beta_{21}^{(1)} = \tilde{q}_{22}^{-1}, \quad \beta_{11}^{(2)} = \beta_{22}^{(1)} = 0.
\end{aligned}
$$

Taking into account (6.11) and (6.12), the boundary conditions (6.10) can be written as

$$
\begin{aligned}
&\left[\delta_1\theta_1\frac{\partial^2}{\partial y_1^2}\left(\beta_{11}^{(2)}\frac{\partial^2}{\partial y_1^2} + \beta_{12}^{(2)}\frac{\partial^2}{\partial y_2^2}\right) - \frac{\partial}{\partial y_2}\left(\alpha_{21}^{(12)}\frac{\partial^2}{\partial y_1^2} + \alpha_{21}^{(22)}\frac{\partial^2}{\partial y_2^2}\right)\right]\chi^{(2)}\\
&\quad - \left[\delta_1\theta_1\beta_{11}^{(1)}\frac{\partial^3}{\partial y_1^2\partial y_2} + \left(\alpha_{21}^{(11)}\frac{\partial^2}{\partial y_1^2} + \alpha_{21}^{(21)}\frac{\partial^2}{\partial y_2^2}\right)\right]\frac{\partial\chi^{(1)}}{\partial y_1} - \delta_1\theta_1 h\frac{\partial^2\varphi}{\partial y_1^2} = \delta_1 P_1,
\end{aligned}
$$

$$
\begin{aligned}
&- 2\varkappa hG_1\frac{\partial\phi}{\partial y_1} - \left[\theta_3\beta_{21}^{(2)}\frac{\partial^3}{\partial y_1^2\partial y_2} + \left(\alpha_{22}^{(12)}\frac{\partial^2}{\partial y_1^2} + \alpha_{22}^{(22)}\frac{\partial^2}{\partial y_2^2}\right)\right]\frac{\partial\chi^{(2)}}{\partial y_1}\\
&+ \left[\theta_3\frac{\partial^2}{\partial y_1^2}\left(\beta_{21}^{(1)}\frac{\partial^2}{\partial y_1^2} + \beta_{22}^{(1)}\frac{\partial^2}{\partial y_2^2}\right) - \frac{\partial}{\partial y_2}\left(\alpha_{22}^{(11)}\frac{\partial^2}{\partial y_1^2} + \alpha_{22}^{(21)}\frac{\partial^2}{\partial y_2^2}\right)\right]\chi^{(1)} = P_2,
\end{aligned}
$$

$$
\begin{aligned}
&\theta_2\frac{\partial^2\phi}{\partial y_1^2} - 2\varkappa G_1\phi - \left[\left(2\varkappa G_1\beta_{21}^{(2)} + \delta_1\alpha_{21}^{(12)}\right)\frac{\partial^2}{\partial y_1^2} + \delta_1\alpha_{21}^{(22)}\frac{\partial^2}{\partial y_2^2}\right]\frac{\partial\chi^{(2)}}{\partial y_2}\\
&+ \left[\left(2\varkappa G_1\beta_{21}^{(1)} - \delta_1\alpha_{21}^{(11)}\right)\frac{\partial^2}{\partial y_1^2} + \left(2\varkappa G_1\beta_{22}^{(1)} - \delta_1\alpha_{21}^{(21)}\right)\frac{\partial^2}{\partial y_2^2}\right]\frac{\partial\chi^{(1)}}{\partial y_1} = 0.
\end{aligned}
\tag{6.14}
$$

Thus, the problem of the steady motion of a two-layer half-space under the influence of a moving load is reduced to finding the functions $\chi^{(j)}$ and ϕ from the boundary conditions (6.14).

6.3 Solution of the Problem in the Image Area

The problem's solution will be found using the integral Fourier transform with respect to y_1 and the corresponding inversion formula. Applying the Fourier transform to Eq. (6.2), we obtain

$$\left(\frac{d^2}{dy_2^2}-k^2\eta_1^2\right)\left(\frac{d^2}{dy_2^2}-k^2\eta_2^2\right)\chi^{(j)F}=0,\quad j=1,2. \tag{6.15}$$

Let us define the problem's solution in a general form for the cases of unequal and equal roots, for various conditions of conjugation of the layer and half-space, and for any load speed (subsonic, transonic, and supersonic).

Boundary conditions (6.14) in the space of Fourier images have the form

$$\begin{aligned}
&\left(-\alpha_{21}^{(22)}\frac{d^3}{dy_2^3}-k^2\delta_1\theta_1\beta_{12}^{(2)}\frac{d^2}{dy_2^2}+k^2\alpha_{21}^{(12)}\frac{d}{dy_2}+k^4\delta_1\theta_1\beta_{11}^{(2)}\right)\chi^{(2)F}\\
&\quad-ik\left(\alpha_{21}^{(21)}\frac{d^2}{dy_2^2}-k^2\delta_1\theta_1\beta_{11}^{(1)}\frac{d}{dy_2}-k^2\alpha_{21}^{(11)}\right)\chi^{(1)F}+k^2\delta_1\theta_1 h\varphi^F=\delta_1 P_1^F,\\[2ex]
&-2ik\varkappa hG_1\phi^F+ik\left(-\alpha_{22}^{(22)}\frac{d^2}{dy_2^2}+k^2\theta_3\beta_{21}^{(2)}\frac{d}{dy_2}+k^2\alpha_{22}^{(12)}\right)\chi^{(2)F}\\
&\quad-\left(\alpha_{22}^{(21)}\frac{d^3}{dy_2^3}+k^2\theta_3\beta_{22}^{(1)}\frac{d^2}{dy_2^2}-k^2\alpha_{22}^{(11)}\frac{d}{dy_2}-k^4\theta_3\beta_{21}^{(1)}\right)\chi^{(1)F}=P_2^F, \qquad (6.16)\\[2ex]
&\left(k^2\theta_2+2\varkappa G_1\right)\phi^F-\left[k^2\left(2\varkappa G_1\beta_{21}^{(2)}+\delta_1\alpha_{21}^{(12)}\right)-\delta_1\alpha_{21}^{(22)}\frac{d^2}{dy_2^2}\right]\frac{d\chi^{(2)F}}{dy_2}\\
&\quad+ik\left[k^2\left(2\varkappa G_1\beta_{21}^{(1)}-\delta_1\alpha_{21}^{(11)}\right)-\left(2\varkappa G_1\beta_{22}^{(1)}-\delta_1\alpha_{21}^{(21)}\right)\frac{d^2}{dy_2^2}\right]\chi^{(1)F}=0.
\end{aligned}$$

The solution of the transformed Eq. (6.15), taking into account damping at infinity, is sought in the form

$$\begin{aligned}
\chi^{F(j)}&=\left[1-\delta_{j2}(1-\delta_{\eta_1\eta_2})\right]\\
&\quad\times\left\{C_1^{(j)}e^{k_1k\eta_1(y_2+h)}+\left[\delta_{\eta_1\eta_2}(y_2+h)+1-\delta_{\eta_1\eta_2}\right]C_2^{(j)}e^{k_2k\eta_2(y_2+h)}\right\},
\end{aligned} \tag{6.17}$$

where $C_m^{(j)}$ $(j,m=1,2)$ are integration constants,

$$\gamma_j=k_j\eta_j,\quad j=1,2,\quad \delta_{\eta_1\eta_2}=\begin{cases}0, & \eta_1\neq\eta_2\\ 1, & \eta_1=\eta_2\end{cases},\quad \delta_{j2}=\begin{cases}0, & j=1\\ 1, & j=2\end{cases}.$$

Let us introduce the integration constants:

$$C_1^{(1)} = iC_1, \quad C_2^{(1)} = iC_2, \quad C_1^{(2)} = C_1, \quad C_2^{(2)} = C_2. \tag{6.18}$$

Substituting (6.17) and (6.18) into (6.16), we obtain the following system of algebraic equations for the unknowns C_1, C_2 and ϕ^F:

$$\begin{aligned} k\left(a_{11}^{(1)} + ka_{11}^{(2)}\right)C_1 + \left(a_{12}^{(1)} + ka_{12}^{(2)} + k^2 a_{12}^{(3)}\right)C_2 + a_{13}\varphi^F &= k^{-2}\delta_1 P_1^F \\ k^2\left(a_{21}^{(1)} + ka_{21}^{(2)}\right)C_1 + k\left(a_{22}^{(1)} + ka_{22}^{(2)} + k^2 a_{22}^{(3)}\right)C_2 + a_{23}\phi^F &= -ik^{-1}P_2^F \\ k^3 a_{31}C_1 + k^2\left(a_{32}^{(1)} + ka_{32}^{(2)}\right)C_2 + \left(a_{33}^{(1)} + k^2 a_{33}^{(2)}\right)\phi^F &= 0, \end{aligned} \tag{6.19}$$

where

$$\begin{aligned} &a_{11}^{(1)} = -\gamma_{21}^{(11)} + \delta_{\eta_1\eta_2}\gamma_1\gamma_{21}^{(21)}, \quad a_{11}^{(2)} = \delta_1\theta_1\left(\delta_{\eta_1\eta_2}\theta_1^{(21)} - \beta_{11}^{(1)}\gamma_1\right), \quad a_{13} = \delta_1\theta_1 h, \\ &a_{12}^{(1)} = \delta_{\eta_1\eta_2}\left[\gamma_{21}^{(22)} + 2\gamma_2\left(\alpha_{21}^{(21)} - \alpha_{21}^{(22)}\gamma_2\right)\right], \quad a_{12}^{(3)} = -\delta_1\theta_1\beta_{11}^{(1)}\gamma_2\left(1 - \delta_{\eta_1\eta_2}\right), \\ &a_{12}^{(2)} = -\left[\delta_1\delta_{\eta_1\eta_2}\theta_1\left(\beta_{11}^{(1)} + 2\beta_{12}^{(2)}\gamma_2\right) + \left(1 - \delta_{\eta_1\eta_2}\right)\gamma_{21}^{(12)}\right], \quad a_{33}^{(1)} = -2\varkappa G_1, \\ &a_{21}^{(1)} = \gamma_1\gamma_{22}^{(11)} + \delta_{\eta_1\eta_2}\gamma_{22}^{(21)}, \quad a_{21}^{(2)} = \theta_3\left(\theta_2^{(11)} + \delta_{\eta_1\eta_2}\beta_{21}^{(2)}\gamma_1\right), \quad a_{23} = -2\varkappa h G_1, \\ &a_{22}^{(1)} = \delta_{\eta_1\eta_2}\left[\gamma_{22}^{(12)} - 2\gamma_2\left(\alpha_{22}^{(22)} + \alpha_{22}^{(21)}\gamma_2\right)\right], \quad a_{22}^{(3)} = \theta_3\theta_2^{(12)}\left(1 - \delta_{\eta_1\eta_2}\right), \\ &a_{22}^{(2)} = \delta_{\eta_1\eta_2}\theta_3\left(\beta_{21}^{(2)} - 2\beta_{22}^{(1)}\gamma_2\right) + \left(1 - \delta_{\eta_1\eta_2}\right)\gamma_2\gamma_{22}^{(12)}, \quad a_{33}^{(2)} = -\theta_2, \\ &a_{31} = 2\varkappa G_1\left(\theta_2^{(11)} + \delta_{\eta_1\eta_2}\gamma_1\beta_{21}^{(2)}\right) + \delta_1\left(\delta_{\eta_1\eta_2}\gamma_1\gamma_{21}^{(21)} - \gamma_{21}^{(11)}\right), \end{aligned}$$

$$\begin{aligned} &a_{32}^{(1)} = \delta_{\eta_1\eta_2}\left\{2\varkappa G_1\left(\beta_{21}^{(2)} - 2\gamma_2\beta_{22}^{(1)}\right) + \delta_1\left[\gamma_{21}^{(22)} + 2\gamma_2\left(\alpha_{21}^{(21)} - \alpha_{21}^{(22)}\gamma_2\right)\right]\right\}, \\ &a_{32}^{(2)} = \left(1 - \delta_{\eta_1\eta_2}\right)\left(2\varkappa G_1\theta_2^{(12)} - \delta_1\gamma_{21}^{(12)}\right), \\ &\theta_m^{(kj)} = \beta_{m1}^{(k)} - \beta_{m2}^{(k)}\gamma_j^2, \quad \gamma_{mk}^{(nj)} = \alpha_{mk}^{(1n)} - \alpha_{mk}^{(2n)}\gamma_j^2, \quad j, k, m = 1, 2. \end{aligned}$$

The solution of the system (6.19) can be written in the form

$$C_j = \frac{\delta_1 P_1^F U_1^{(j)} + iP_2^F U_2^{(j)}}{\Delta(k)}, \quad j = 1, 2, \quad \phi^F = \frac{\delta_1 P_1^F U_1 + iP_2^F U_2}{\Delta(k)}, \tag{6.20}$$

where

$$\begin{aligned} \Delta(k) &= k^2\left(b_0 + kb_1 + k^2 b_2 + k^3 b_3 + k^4 b_4 + k^5 b_5\right), \\ U_j^{(1)} &= k^{-1}\left(b_{10}^{(j)} + kb_{11}^{(j)} + k^2 b_{12}^{(j)} + k^3 b_{13}^{(j)} + k^4 b_{14}^{(j)}\right), \\ U_j^{(2)} &= -\left(b_{20}^{(j)} + kb_{21}^{(j)} + k^2 b_{22}^{(j)} + k^3 b_{23}^{(j)}\right), \\ U_j &= k^2\left(b_{30}^{(j)} + kb_{31}^{(j)} + k^2 b_{32}^{(j)}\right), \quad j = 1, 2, \end{aligned}$$

$$
\begin{aligned}
b_0 &= a_{33}^{(1)}\left(a_{11}^{(1)}a_{22}^{(1)} - a_{12}^{(1)}a_{21}^{(1)}\right), \quad b_5 = a_{33}^{(2)}\left(a_{22}^{(3)}a_{11}^{(2)} - a_{12}^{(3)}a_{21}^{(2)}\right),\\
b_1 &= a_{33}^{(1)}\left(a_{11}^{(2)}a_{22}^{(1)} + a_{22}^{(2)}a_{11}^{(1)} - a_{12}^{(1)}a_{21}^{(2)} - a_{12}^{(2)}a_{21}^{(1)}\right) + a_{23}\left(a_{31}a_{12}^{(1)} - a_{11}^{(1)}a_{32}^{(1)}\right),\\
b_2 &= a_{33}^{(1)}\left(a_{22}^{(2)}a_{11}^{(2)} + a_{22}^{(3)}a_{11}^{(1)} - a_{12}^{(2)}a_{21}^{(2)} - a_{12}^{(3)}a_{21}^{(1)}\right) + a_{33}^{(2)}\left(a_{11}^{(1)}a_{22}^{(1)} - a_{12}^{(1)}a_{21}^{(1)}\right)\\
&\quad + a_{23}\left(a_{31}a_{12}^{(2)} - a_{11}^{(1)}a_{32}^{(2)} - a_{11}^{(2)}a_{32}^{(1)}\right) + a_{13}\left(a_{21}^{(1)}a_{32}^{(1)} - a_{31}a_{22}^{(1)}\right),\\
b_3 &= a_{33}^{(2)}\left(a_{11}^{(2)}a_{22}^{(1)} + a_{22}^{(2)}a_{11}^{(1)} - a_{12}^{(1)}a_{21}^{(2)} - a_{12}^{(2)}a_{21}^{(1)}\right) + a_{33}^{(1)}\left(a_{22}^{(3)}a_{11}^{(2)} - a_{12}^{(3)}a_{21}^{(2)}\right)\\
&\quad + a_{13}\left(a_{21}^{(1)}a_{32}^{(2)} + a_{21}^{(2)}a_{32}^{(1)} - a_{31}a_{22}^{(2)}\right) + a_{23}\left(a_{31}a_{12}^{(3)} - a_{11}^{(2)}a_{32}^{(2)}\right),\\
b_4 &= a_{33}^{(2)}\left(a_{11}^{(2)}a_{22}^{(2)} + a_{11}^{(1)}a_{22}^{(3)} - a_{12}^{(2)}a_{21}^{(2)} - a_{12}^{(3)}a_{21}^{(1)}\right) + a_{13}\left(a_{21}^{(2)}a_{32}^{(2)} - a_{22}^{(3)}a_{31}\right),\\
b_{10}^{(1)} &= a_{22}^{(1)}a_{33}^{(1)}, \quad b_{11}^{(1)} = a_{22}^{(2)}a_{33}^{(1)} - a_{23}a_{32}^{(1)}, \quad b_{12}^{(1)} = a_{22}^{(3)}a_{33}^{(1)} + a_{33}^{(2)}a_{22}^{(1)} - a_{23}a_{32}^{(2)},\\
b_{13}^{(1)} &= a_{33}^{(2)}a_{22}^{(2)}, \quad b_{14}^{(1)} = a_{33}^{(2)}a_{22}^{(3)}, \quad b_{10}^{(2)} = a_{12}^{(1)}a_{33}^{(1)}, \quad b_{11}^{(2)} = a_{12}^{(2)}a_{33}^{(1)},\\
b_{12}^{(2)} &= a_{12}^{(3)}a_{33}^{(1)} + a_{33}^{(2)}a_{12}^{(1)} - a_{13}a_{32}^{(1)}, \quad b_{13}^{(2)} = a_{33}^{(2)}a_{12}^{(2)} - a_{13}a_{32}^{(2)}, \quad b_{14}^{(2)} = a_{33}^{(2)}a_{12}^{(3)},\\
b_{20}^{(1)} &= a_{21}^{(1)}a_{33}^{(1)}, \quad b_{21}^{(1)} = a_{21}^{(2)}a_{33}^{(1)} - a_{23}a_{31}, \quad b_{22}^{(1)} = a_{21}^{(1)}a_{33}^{(2)}, \quad b_{23}^{(1)} = a_{21}^{(2)}a_{33}^{(2)},\\
b_{20}^{(2)} &= a_{11}^{(1)}a_{33}^{(1)}, \quad b_{21}^{(2)} = a_{11}^{(2)}a_{33}^{(1)}, \quad b_{22}^{(2)} = a_{11}^{(1)}a_{33}^{(2)} - a_{13}a_{31}, \quad b_{23}^{(2)} = a_{11}^{(2)}a_{33}^{(2)},\\
b_{30}^{(1)} &= a_{21}^{(1)}a_{32}^{(1)} - a_{22}^{(1)}a_{31}, \quad b_{31}^{(1)} = a_{21}^{(1)}a_{32}^{(2)} + a_{21}^{(2)}a_{32}^{(1)} - a_{22}^{(2)}a_{31},\\
b_{30}^{(2)} &= a_{11}^{(1)}a_{32}^{(1)} - a_{12}^{(1)}a_{31}, \quad b_{31}^{(2)} = a_{11}^{(1)}a_{32}^{(2)} + a_{11}^{(2)}a_{32}^{(1)} - a_{12}^{(2)}a_{31},\\
b_{32}^{(1)} &= a_{21}^{(2)}a_{32}^{(2)} - a_{22}^{(3)}a_{31}, \quad b_{32}^{(2)} = a_{11}^{(2)}a_{32}^{(2)} - a_{12}^{(3)}a_{31}.
\end{aligned}
$$

Let us apply the Fourier conversion to (6.7), (6.11) and (6.13):

$$
\begin{aligned}
\tilde{Q}_{jm}^{F} &= \left(-k^2\alpha_{jm}^{(12-\delta_{jm})} + \alpha_{jm}^{(22-\delta_{jm})}\frac{d^2}{dy_2^2}\right)\frac{d\chi^{(2-\delta_{jm})F}}{dy_2}\\
&\quad + ik\left(-k^2\alpha_{jm}^{(11+\delta_{jm})} + \alpha_{jm}^{(21+\delta_{jm})}\frac{d^2}{dy_2^2}\right)\chi^{(1+\delta_{jm})F}, \quad j,m = 1,2,\\
\dot{u}_j^F &= -ik\mathrm{v}\left(-k^2\beta_{j1}^{(m)} + \beta_{j2}^{(m)}\frac{d^2}{dy_2^2}\right)\chi^{(m)F} - k^2\mathrm{v}\beta_{j1}^{(j)}\frac{d\chi^{(j)F}}{dy_2},\\
&\qquad j,m = 1,2, \quad i \neq m,\\
M^F &= \frac{4}{3}\frac{ikG_1h^3}{1-\nu_1}\phi^F.
\end{aligned}
\tag{6.21}
$$

Taking into account (6.17), (6.18) and (6.20), Eq. (6.21) can be represented in the form

$$
\begin{aligned}
\tilde{Q}_{mj}^{F} &= (-i)^{\delta_{mj}}k^2\Delta^{-1}(k)\left(\delta_1 P_1^F\Gamma_{mj}^{(1)} + iP_2^F\Gamma_{mj}^{(2)}\right),\\
\dot{u}_j^F &= i^{2-j}\mathrm{v}k^2\Delta^{-1}(k)\left(\delta_1 P_1^F\Gamma_2^{(1)} + iP_2^F\Gamma_2^{(2)}\right), \quad m,j = 1,2,\\
M^F &= k\Delta^{-1}(k)\left(i\delta_1 P_1^F\Gamma_\phi^{(1)} - P_2^F\Gamma_\phi^{(2)}\right),
\end{aligned}
\tag{6.22}
$$

where

$$
\begin{aligned}
\Gamma_{mm}^{(j)} =& k\left(\gamma_1\gamma_{mm}^{(11)}+\delta_{\eta_1\eta_2}\gamma_{mm}^{(21)}\right)U_j^{(1)}e^{k\gamma_1(y_2+h)} \\
&-\left\{\delta_{\eta_1\eta_2}\left[2\gamma_2\left(\gamma_2\alpha_{mm}^{(21)}+\alpha_{mm}^{(22)}\right)-\gamma_{mm}^{(12)}\right]\right. \\
&-k\left\{\delta_{\eta_1\eta_2}\left(y_2+h\right)\left(\gamma_2\gamma_{mm}^{(12)}+\gamma_{mm}^{(22)}\right)\right. \\
&\left.\left.+\left(1-\delta_{\eta_1\eta_2}\right)\gamma_2\gamma_{mm}^{(12)}\right\}\right\}U_j^{(2)}e^{k\gamma_2(y_2+h)}, \\
\Gamma_{mn}^{(j)} =& k\left(\gamma_{mn}^{(11)}-\delta_{\eta_1\eta_2}\gamma_1\gamma_{mn}^{(21)}\right)U_j^{(1)}e^{k\gamma_1(y_2+h)} \\
&+\left\{\delta_{\eta_1\eta_2}\left[2\gamma_2\left(\gamma_2\alpha_{mn}^{(22)}-\alpha_{21}^{(21)}\right)-\gamma_{mn}^{(22)}\right]\right. \\
&+k\left[\delta_{\eta_1\eta_2}\left(y_2+h\right)\left(\gamma_{mn}^{(12)}-\gamma_2\gamma_{mn}^{(22)}\right)\right. \\
&\left.\left.+\left(1-\delta_{\eta_1\eta_2}\right)\gamma_{mn}^{(12)}\right]\right\}U_j^{(2)}e^{k\gamma_2(y_2+h)}, \\
\Gamma_{\varphi}^{(j)} =& \frac{4}{3}\frac{G_1h^3U_j}{1-\nu_1}.
\end{aligned}
$$

Thus, the solution to the problem of the steady motion of a two-layer elastic half-space with initial stresses under the influence of a moving load in the region of Fourier images has the form (6.22).

The results of determining $\Delta(k)$ for compressible and incompressible half-space and various cases of conjugation of the plate and half-space are presented in Guz et al. (2015).

From (6.22), it follows that the value of the quantities characterizing the stress-strain state of a two-layer elastic half-space increases without limit at $\Delta(k) \to 0$. Under the condition that real positive multiple roots of the equation $\Delta(k) = 0$ exist, resonance is possible (Achenbach et al. 1967).

It follows from the results obtained in Guz et al. (2015) that the number of critical velocities of the load movement significantly depends on the initial stresses in the half-space, the mechanical characteristics of the plate and the half-space, and the conditions of their contact. The effect of initial stresses on critical velocities is more significant for relatively soft plates and non-rigid contacts. The value of the lowest critical speed for a non-rigid contact is always less than for a rigid one.

6.4 Numerical Study

In order to return to the original domain in (6.22), we use the inverse Fourier transform. It follows from the results obtained in Guz et al. (2015) that the calculation of the inversion integrals significantly depends on the speed of the load. Depending on the velocity v, the denominator $\Delta(k)$ in the inversion integrals may or may not have real positive roots. If no root exists on the real axis, then the inversion integrals have no singularities and can be calculated directly. In the presence of unequal real positive roots of the denominator $\Delta(k)$, the integrals along the integration contour from $-\infty + i\gamma$ to $+\infty + i\gamma$ can be replaced by the sum of the principal value of the integral and the sum of all residues multiplied by $(-i\pi)$ (Wright et al. 1970). In the case of a double positive root, the inversion integrals do not exist even in the Cauchy sense, i.e., resonance appears. Since it was assumed in the formulation of the problem that the perturbations caused by the moving load are very small, the resonant region was excluded from consideration.

In Guz et al. (2015), a study was made on the dependences of the bending moment in the plate on the initial stresses, the speed of the load, and the conditions of contact between the plate and the half-space.

Figures 6.1, 6.2, 6.3 and 6.4 show how the initial stresses in the base affect the characteristics of the stress-strain state in a two-layer half-space at different speeds of the load (subsonic, transonic, and supersonic) and the conditions of contact between the plate and the half-space.

Introduced notations mean the following: c_{11} and c_{12} are the spread in the direction of the y_1-axis properly to longitudinal and transverse polarized waves in unlimited compressible body with initial stress, c_1 and c_2 are the spread of transverse waves in the direction of axes Oy_1 and Oy_2 in the unlimited incompressible body with initial stress, c_s is the velocity of shear waves in a layer, v* is the critical speed of moving load (Guz et al. 2015).

As examples, a compressible half-space with an elastic potential of harmonic type and an incompressible half-space with the Bartenev–Khazanovich potential are considered (Guz 2004). It was assumed that the initial stress state is plane, and there is no surface load.

The calculation results are given for a concentrated linear load, the normal and tangential components of which are determined by the following formulas

$$P_1 = P\delta(y_1)\cos\alpha, \quad P_2 = P\delta(y_1)\sin\alpha, \quad P = G_1,$$

where α is the tilt angle to the axis load Oy_1.

6.5 Results

An analysis of the results shows that the presence of initial stresses has a significant effect on the distribution of stresses and displacement velocities in the half-space and the bending moment in the plate. This effect is different depending on the position of the considered point of the layered body relative to the point of application of the load.

The values of the parameters of the stress-strain state at a particular point of the layered body depend on the initial stresses, its coordinates, and contact conditions.

For subcritical speeds of movement of the load with rigid contact of stress, the speed of movement in the half-space is less than with non-rigid contact. At the same time, in the studied range of values λ_1, the growth rate of the amplitude of the studied

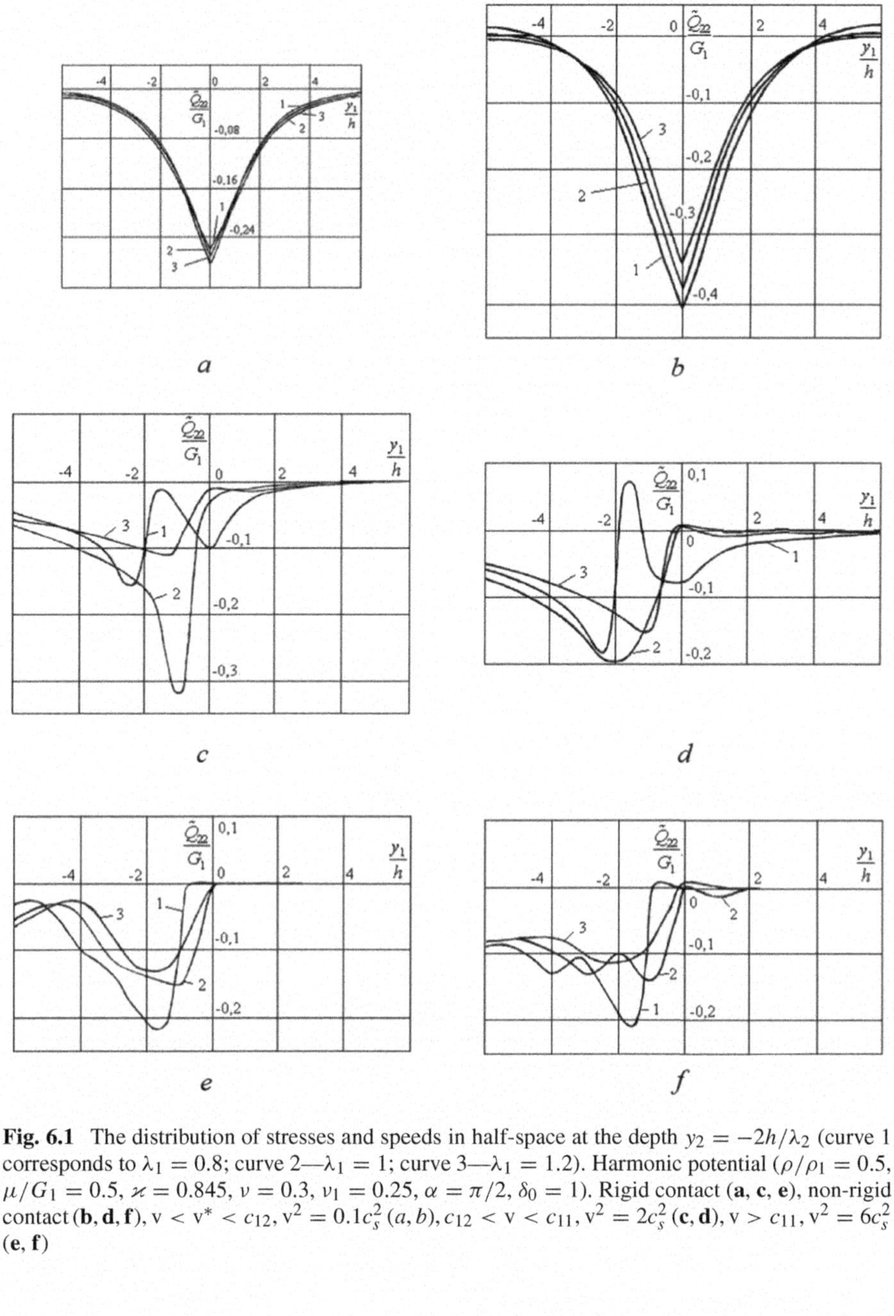

Fig. 6.1 The distribution of stresses and speeds in half-space at the depth $y_2 = -2h/\lambda_2$ (curve 1 corresponds to $\lambda_1 = 0.8$; curve 2—$\lambda_1 = 1$; curve 3—$\lambda_1 = 1.2$). Harmonic potential ($\rho/\rho_1 = 0.5$, $\mu/G_1 = 0.5$, $\varkappa = 0.845$, $\nu = 0.3$, $\nu_1 = 0.25$, $\alpha = \pi/2$, $\delta_0 = 1$). Rigid contact (**a**, **c**, **e**), non-rigid contact (**b**, **d**, **f**), $v < v^* < c_{12}$, $v^2 = 0.1c_s^2$ (a, b), $c_{12} < v < c_{11}$, $v^2 = 2c_s^2$ (**c**, **d**), $v > c_{11}$, $v^2 = 6c_s^2$ (**e**, **f**)

quantities during compression is greater than during tension. The attenuation with distance from the load application point in compression is slower than in tension.

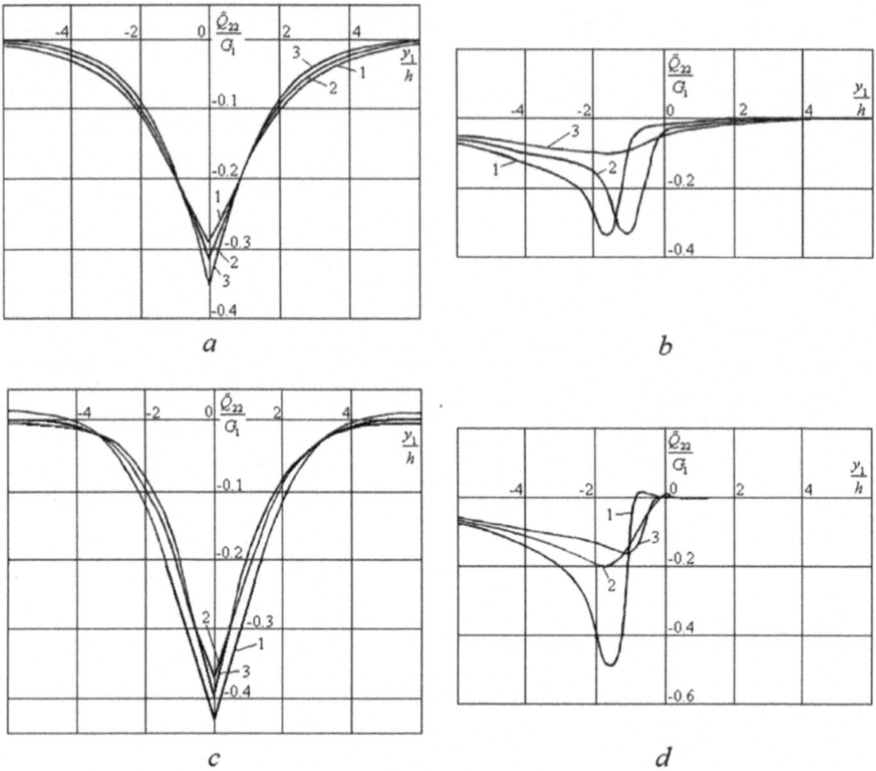

Fig. 6.2 The distribution of stresses and speeds in half-space at the depth $y_2 = -2h/\lambda_2$ (curve 1 corresponds to $\lambda_1 = 0.8$; curve 2—$\lambda_1 = 1$; curve 3—$\lambda_1 = 1.2$). Elastic potential of Bartenev–Khazanovich type ($\rho/\rho_1 = 0.5, \mu/G_1 = 0.5, \varkappa = 0.845, \nu_1 = 0.25, \alpha = \pi/2, \delta_0 = 1$). Rigid contact (**a**, **b**), non-rigid contact (**c**, **d**), $\mathrm{v} < \mathrm{v}^* < c_1$, $\mathrm{v}^2 = 0.1c_s^2$ (**a**, **c**), $\mathrm{v} > c_1$, $\mathrm{v}^2 > 2c_s^2$ (**b**, **d**)

The influence of the initial stresses increases significantly with the increase in the speed of the load. This is especially true during pre-compression. The influence of velocity and initial stresses is less significant with rigid contact than with non-rigid contact.

Accounting for the inertia of rotation within the considered speeds of movement of the surface load and values λ_1 in the case of rigid contact introduces a minor correction (less than 2.6 %). However, in the case of non-rigid contact, the difference in the results will be huge (up to 30 %). Therefore, it is essential to take into account the inertia of rotation at $\lambda_1 < 1$ and the high speeds of the load.

It can be seen that as the velocity increases, the symmetry is more and more violated, and the direct wave decays much faster and is practically absent in the supersonic case. However, it does not entirely disappear. The layering of the medium explains this.

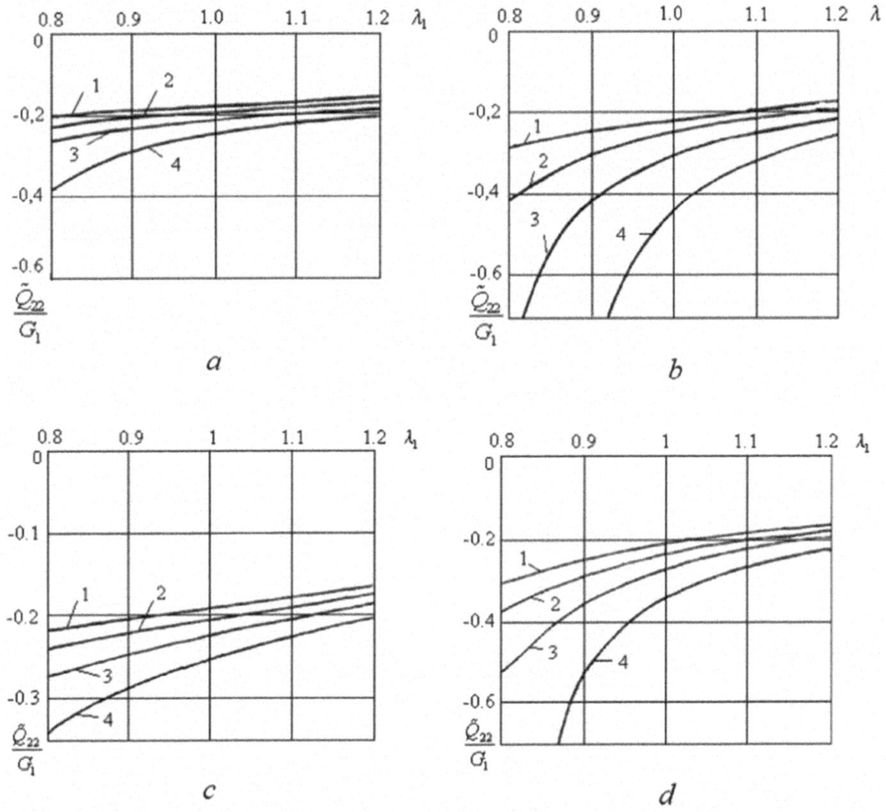

Fig. 6.3 Dependence of values that characterize the stress-strain state of underlying half-space, from the initial stresses at different subcritical speeds of loads at the point $y_1 = -\lambda_1 h$, $y_2 = -2h/\lambda_2$ (curve 1 corresponds to $v^2 = 0.1c_s^2$ curve 2—$v^2 = 0.2c_s^2$, curve 3—$v^2 = 0.3c_s^2$, curve 4—$v^2 = 0.4c_s^2$). Rigid contact (**a**, **c**), non-rigid contact (**b**, **d**), harmonic potential (**a**, **b**), elastic potential of Bartenev–Khazanovich type (**c**, **d**)

In the case of a rigid contact, the direct wave decays much faster than in a non-rigid contact.

6.6 Conclusions

Within the framework of the linearized theory of elasticity for bodies with initial stresses, the statement is considered, and the solution of the plane steady-state problem of the perturbation of a two-layer prestressed base, moving at a constant speed, consisting of a plate and an underlying half-space, is obtained. The equations of motion of the plate take into account shear and rotational inertia. Formulas are given

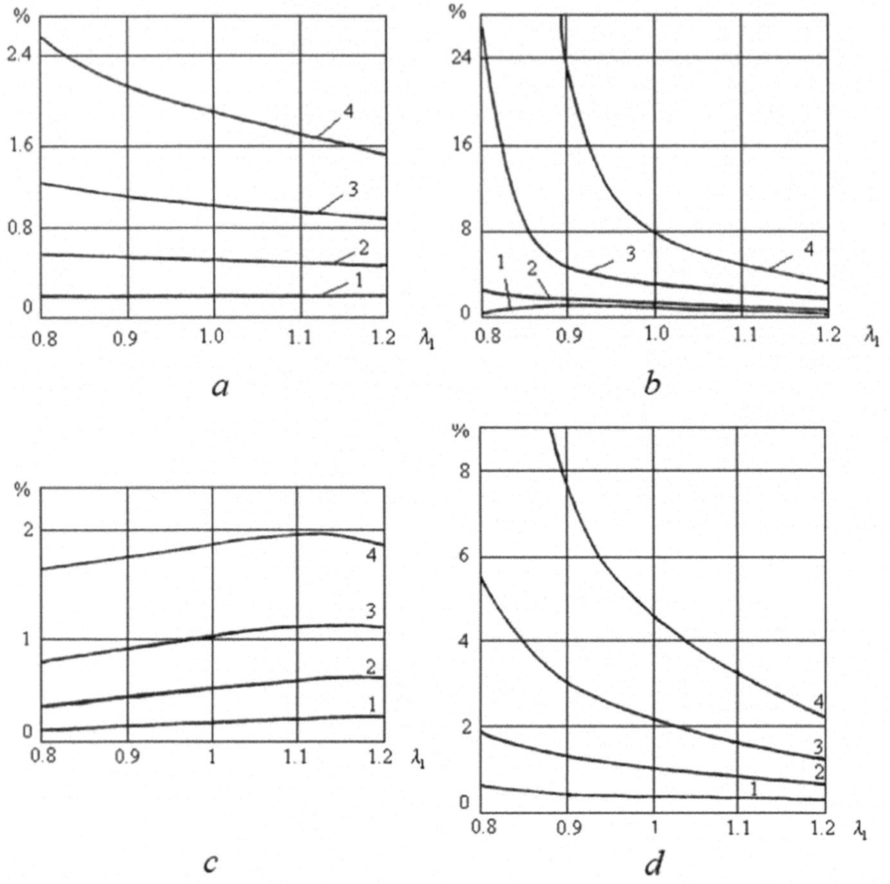

Fig. 6.4 The impact of the accounting rotational inertia at various speeds of loads and primary strains on the value of stress $\tilde{Q}_{22}$ at the point $y_1 = -\lambda_1 h$, $y_2 = -2h/\lambda_2$ (curve 1 corresponds to $v^2 = 0.1c_s^2$ curve 2—$v^2 = 0.2c_s^2$, curve 3—$v^2 = 0.3c_s^2$, curve 4—$v^2 = 0.4c_s^2$). Rigid contact (**a**, **c**), non-rigid contact (**b**, **d**), harmonic potential (**a**, **b**), elastic potential of Bartenev–Khazanovich type (**c**, **d**)

for the transformants of the characteristics of the stress-strain state of elements of a two-layered medium.

A study was made of the dependences of the parameters of the stress-strain state of the base on the initial stresses, the speed of the load, and the contact conditions for the elements of a two-layered medium. Initial (residual) stresses have a significant impact on the parameters of the stress-strain state of the base. This effect depends on the velocity of the surface load, the mechanical parameters of the elements of the layered medium, and the conditions of their conjugation.

References

Achenbach JD, Keshawa SP, Herrmann G (1967) Moving load on a plate resting on an elastic half space. Trans ASME Ser E J Appl Mech 34(4):183–189

Guz AN (2004) Elastic waves in bodies with initial (residual) stresses. A.S.K, Kyiv

Guz AN, Babich SYu, Glukhov YuP (2015) Mixed tasks of an elastic ground with initial stresses. Lap Lambert Academic Publishing, Saarbrucken

Wright JP, Baron ML (1970) Exponentially decaying pressure pulse moving with contact velocity on the surface of a layered elastic material (superseismic half space). Trans ASME Ser E J Appl Mech 37(1):148–159

Chapter 7
Numerical Modeling of Fatigue Cracks Growth in Thin Isotropic Plates Considering the Damage Accumulation History

Vladislav Golub and Alla Plashchynska

7.1 Introduction

Modeling and solving problems on the propagation of fatigue cracks in solids under cyclic elastic loading or so-called multi-cycle loading is supposed to be one of the most urgent tasks of brittle fracture mechanics (Bolotin 1990; Nott 1978). Fatigue failure occurs at stresses lower than the yield strength of the material and is not accompanied by the accumulation of macroscopic plastic deformations.

In this field of fracture mechanics, many experimental studies were performed, and several empirical and semiempirical relationships were formulated, establishing the dependence of fatigue crack growth rate on the parameters of cyclic loading. In particular, it is shown that at the stage of subcritical growth of fatigue cracks, the most reasonable is the Paris concept, according to which the rate of fatigue crack growth is considered a characteristic function of the amplitude of the stress intensity factor (Paris and Erdogan 1963; Paris et al. 1961). The results of most studies in this area are reduced to the refinement of the characteristic function by introducing supplementary parameters enabling consideration of peculiarities of cyclic loading, for example, the effect of asymmetry of a stress cycle and the maximum stress in a cycle.

According to the authors, the problem of constructing a theoretical model of fatigue crack growth was stated and solved for the first time in Cherepanov (1968). The analytical dependence for the growth rate was constructed based on dimensional analysis and validated experimentally for specimens made of aluminum alloys. It was used to estimate the effect of loading frequency and the external environment. The dimensional theory was also used in Andreikiv (1982) to construct a mathematical model of nucleation and propagation of fatigue cracks in three-dimensional

V. Golub (✉) · A. Plashchynska
S.P. Timoshenko Institute of Mechanics, National Academy of Science of Ukraine, Kyiv, Ukraine
e-mail: creep@inmech.kiev.ua

A. N. Guz et al. (eds.), *Advances in Mechanics*, Advanced Structured Materials 191,
https://doi.org/10.1007/978-3-031-37313-8_7

elastic-plastic bodies. According to the model, the tensile strain in the fracture process zone is used as an invariant characteristic of fatigue fracture.

An approach combining the energy concept of crack mechanics and the concept of continuous damage mechanics (Bolotin 1983, 1985, 1987, 1990) appears more promising for constructing theoretical models describing the growth of fatigue cracks in elastic bodies. Considering the accumulation of damages along the crack front as the main factor of fatigue crack growth was further continued in Golub et al. (1993); Golub and Panteleyev (2000); Golub and Plashchiskaya (1994, 2018); Nott (1978), where a two-stage model of fatigue crack growth was constructed for infinite thin isotropic plates under uniaxial tension-compression loading. In this case, the concept of continuous damage mechanics is combined with the force approach of crack mechanics. The model is validated experimentally by determining the kinetics of fatigue crack growth in thin plates made of aluminum alloys under uniaxial reversed tension-compression loading.

We use the approach combining the force concept of crack mechanics and the concept of damage mechanics to construct a two-stage model of propagation of fatigue cracks in thin isotropic plates of finite size, which makes it possible to take into account the amount of damage that precedes crack propagation.

7.2 Problem Statement. Initial Equations

Let us consider the two-stage process of growth of fatigue mode I crack in thin rectangular plates of finite dimensions made of isotropic linearly elastic materials, including the incubation stage and the stage of crack propagation.

The plate is under loading $\tilde{\sigma}$ varying in an asymmetric cycle, so that

$$\tilde{\sigma} = \sigma_m + \sigma_a \sin(2\pi n), \quad \sigma_{max} = (\sigma_m + \sigma_a) < \sigma_Y, \tag{7.1}$$

where σ_m is the mean stress of the cycle, σ_a is the amplitude of cyclic stress, σ_Y is the yield strength of the material, n is the number of loading cycles or discrete time ($n = ft$), t is physical time, and f is the loading frequency. The cyclic loading $\tilde{\sigma}$ is uniformly distributed along the edges of the plate perpendicular to the crack. The crack faces are free from loads.

It is also assumed that σ_a in (7.1) is independent of discrete time n (stationary mode), the frequency of loading $f \geqslant 10$ Hz, and the maximum cycle stress σ_{max} does not exceed the yield strength σ_Y (high-cycle fatigue). In this case, the plate deformation is mainly linearly elastic, and the fatigue fracture is quasi-brittle.

It is accepted further that repeated cyclic loading (7.1) can be reduced to an equivalent (in the number of cycles to fracture n_R) fully reversed cycle so that the amplitude of the equivalent stress, $(\sigma_a)_{eqv}$ is given by the relation (Golub and Plashchiskaya 1994)

$$(\sigma_a)_{eqv} = \left[\cos\left(\frac{\pi}{2}\frac{\sigma_m}{\sigma_B}\right)\right]^{-\eta} \sigma_a, \tag{7.2}$$

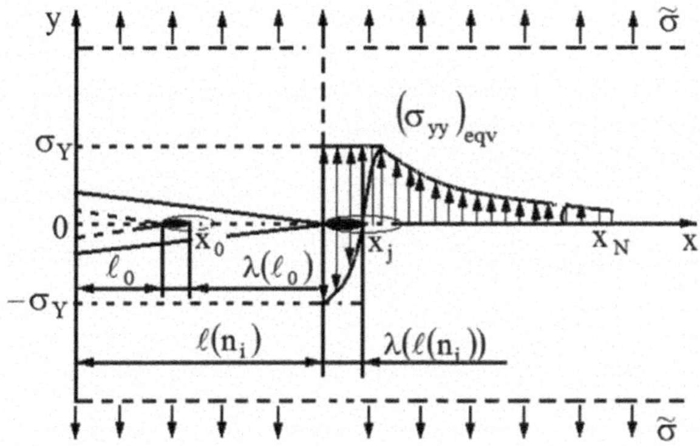

Fig. 7.1 Fatigue crack-tip model

where σ_B is the short-term strength limit of the plate material; η is the coefficient of sensitivity of the plate material to the asymmetry of the stress cycle.

We model the fatigue crack by a narrow sharp-pointed cut (Fig. 7.1) located along the x-axis with y-axis perpendicular to the crack. We denote the initial half-length of the fatigue crack by ℓ_0 and the current half-length of the crack, depending on the number of loading cycles n, by $\ell(n)$.

Two plastic zones (monotonic and cyclic) are formed in the vicinity of the fatigue crack tip (Rice 1967). The stresses along the monotonic plastic zone are bounded by the material's yield strength σ_Y. The initial $\lambda(\ell_0)$ and current $\lambda(\ell(n))$ lengths of the cyclic plastic zone are determined by the stress range $\Delta(\sigma_{yy})_{eqv}$ in the cycle, and the stress along the zone varies from $-\sigma_Y$ to $+\sigma_Y$ (Newman 1992; Rice 1967).

We consider the accumulation of fatigue damages along the front of a growing crack as the main mechanism controlling its propagation. Assuming that fatigue damage accumulates at each arbitrary point x_i of the crack front at time n, the kinetics of this process is described by the evolution equation

$$\frac{d\omega(x,n)}{dn} = D\left[\frac{\Delta\left(\sigma_{yy}\right)_{eqv}(x,n)}{1-\omega(x,n)}\right]^q \tag{7.3}$$

with initial and boundary conditions

$$\omega(x,n) = \begin{cases} 0, & \forall x, & n=0 & (a) \\ 1, & x=\ell_0+\lambda(\ell_0), & n=n_* & (b) \\ \omega(x_j,n_i), & x=\ell(n_i)+\lambda(\ell(n_i)), & n=n_i & (c) \\ 1, & x=\ell(n_i)+\lambda(\ell(n_i)), & n=n_{R_i} & (d) \end{cases} \tag{7.4}$$

where $i = 1, \ldots, N$, $j = i, \ldots, N$, $\omega(\cdot) = 1$ is the scalar damage parameter, D and q are material constants.

Conditions (*a*, *b*, *c*, and *d*) in (7.4) specify the initial conditions, duration of the incubation stage, the level of damage at the point x_j in time n_i, condition of local failure at the point x_j of the crack front, respectively. Condition (c) takes into account the history of damage accumulation along the growing crack front.

The stress distribution $\Delta\left(\sigma_{yy}\right)_{\mathrm{eqv}}(x, n)$ along the x-axis near the fatigue crack tip during the incubation period ($0 \leqslant n \leqslant n_*$) is given by the relation (Newman 1971)

$$\Delta\left(\sigma_{yy}\right)_{\mathrm{eqv}}(x, n) = \frac{(\sigma_{\mathrm{a}})_{\mathrm{eqv}}}{\sqrt{2}} \sqrt{\frac{\ell_0}{x - \ell_0}} f_0\left(\frac{h}{w}, \frac{\ell_0}{w}\right), \tag{7.5}$$

where h is the plate height, w is the plate width, and $\lambda(\ell_0)$ is given by the relation (Newman 1992; Rice 1967)

$$\lambda(\ell_0) = \tfrac{1}{8}\left[\frac{\pi}{2}\frac{(\sigma_{\mathrm{a}})_{\mathrm{eqv}}}{\sigma_{\mathrm{Y}}} f_0\left(h/w,\ \ell_0/w\right)\right]^2 \ell_0.$$

The stress distribution $\Delta\sigma_{\mathrm{eqv}}(x, n)$ along the x-axis near the fatigue crack tip at the stage of crack propagation ($n_* \leqslant n \leqslant n_R$) is given by the relation (Newman 1971)

$$\Delta\left(\sigma_{yy}\right)_{\mathrm{eqv}}(x, n) = \frac{(\sigma_{\mathrm{a}})_{\mathrm{eqv}}}{\sqrt{2}} \sqrt{\frac{\ell(n)}{x - \ell(n)}} f\left(\frac{h}{w}, \frac{\ell(n)}{w}\right), \tag{7.6}$$

where

$$\lambda\left(\ell(n)\right) = \tfrac{1}{8}\left[\frac{\pi}{2}\frac{(\sigma_{\mathrm{a}})_{\mathrm{eqv}}}{\sigma_{\mathrm{Y}}} f\left(h/w,\ \ell(n)/w\right)\right]^2 \ell(n).$$

The corrective function $f(\cdot)$ in (7.5) and (7.6) takes into account the influence of boundary conditions and represents an approximation of k numerical solutions for cracks of several lengths $\ell\,(n)$.

Further, we model the growth of mode I fatigue crack in thin isotropic finite plates under uniaxial asymmetric tension-compression loading by taking into account the history of damage accumulation along the front of the growing crack.

7.3 Simulation of the Fatigue Crack Growth Process

The two-stage model of fatigue crack growth is based on the joint solution of two problems: the problem of determining the local stress field in the vicinity of the tip of the growing crack and formulating the fatigue crack growth criterion based on the concepts of continuous damage mechanics. First, we divide the crack line (Fig. 7.1) into N segments with nodal points $x_j = \ell(n_i) + \lambda\left(\ell(n_i)\right)$ ($i = 0, \ldots, N$ and $j = i, \ldots, N$) to simulate the fatigue crack growth, taking into account the history of damage accumulation.

Then the integral equation of damage distribution along the fracture front based on (7.3)–(7.6) is defined in the form:

$$\sum_{i=0}^{N} \int_{\omega(n_i, x_j)}^{1} \left[1 - \omega\left(n, x_j\right)\right]^q d\omega$$
$$= D \sum_{i=0}^{N} \int_{n_i}^{n_{i+1}} \left((\sigma_a)_{eqv} \sqrt{\frac{\ell(n_i)}{2\left(x_j - \ell(n_i)\right)}} \cdot f\left(h/w, \ell(n_i)/w\right) \right)^q dn, \tag{7.7}$$

($i = 0, \ldots, N$ and $j = i, \ldots, N$), which represents a superposition for N discrete time intervals $n_i \leqslant n \leqslant n_{i+1}$ including the incubation stage and the crack growth stage.

7.3.1 Incubation Stage

The incubation stage corresponds to the loading period when the plate material is damaged but not destroyed. The duration of the incubation period is determined by Eq. (7.7) under the conditions $0 \leqslant \omega(n, x_0) \leqslant 1$ and $n_0 \leqslant n \leqslant n_1$ ($n_0 = 0$ and $n_1 = n_*$):

$$\int_0^1 [1 - \omega(n, x_0)]^q d\omega$$
$$= D \int_0^{n_*} \left[(\sigma_a)_{eqv} \sqrt{\frac{\ell_0}{2(x_0 - \ell_0)}} f\left(h/w, \ell_0/w\right) \right]^q dn. \tag{7.8}$$

Integrating in (7.8), we get

$$n_* = \left((1+q) D \left[(\sigma_a)_{eqv} \sqrt{\frac{\ell_0}{2(x_0 - \ell_0)}} \cdot f\left(h/w, \ell_0/w\right) \right]^q \right)^{-1}, \tag{7.9}$$

where n_* is the duration of the incubation stage.

7.3.2 Fatigue Crack Growth Stage

At the stage of fatigue crack growth, the stress state is determined by the front of each new crack length $\ell(n_i)$.

The moment of each subsequent crack jump at a point x_j is determined by solving Eq. (7.7) using the recursive formula

$$n_{i+1} = n_i + \frac{\left[1-\omega(n_i, x_j)\right]^{q+1}}{(1+q)D\left[(\sigma_\mathrm{a})_\mathrm{eqv}\sqrt{\dfrac{\ell(n_i)}{2(x_j-\ell(n_i))}}\cdot f\left(h/w,\ \ell(n_i)/w\right)\right]^q}, \tag{7.10}$$

under the conditions $\omega\left(n_i, x_j\right) \leqslant \omega(n, x_j) \leqslant 1$, $n_i \leqslant n \leqslant n_{i+1}$, $i = 1, \ldots, N$, and $j = i, \ldots, N$.

The level of accumulated fatigue damage at an arbitrary point x_j along the fracture front over time $n_i \leqslant n \leqslant n_{i+1}$ is determined by the equation

$$\int_{\omega(n_i, x_j)}^{\omega(n_{i+1}, x_j)} \left[1-\omega\left(n, x_j\right)\right]^q d\omega = D\int_{n_i}^{n_{i+1}} \left((\sigma_\mathrm{a})_\mathrm{eqv}\sqrt{\frac{\ell(n_i)}{2(x_j-\ell(n_i))}}\cdot f\left(h/w, \ell(n_i)/w\right)\right)^q dn, \tag{7.11}$$

integrating that, we get the relation

$$\omega(n_{i+1}, x_j) = 1 - \left(\left[1-\omega(n_i, x_j)\right]^{q+1} - (1+q)D \times \left[(\sigma_\mathrm{a})_\mathrm{eqv}\sqrt{\frac{\ell(n_i)}{2(x_j-\ell(n_i))}}\cdot f\left(h/w, \ell(n_i)/w\right)\right]^q (n_{i+1}-n_i)\right)^{\frac{1}{1+q}}, \tag{7.12}$$

which allows taking the history of damage accumulation along the front of growing fatigue crack into account.

7.3.3 Technique of Determining the Functions and Constants of the Model

The two-stage model of fatigue crack propagation in thin isotropic plates of finite size, presented in Sect. 7.2, contains two groups of material constants, subject to determination from the experiment, as well as the corrective function $f(\cdot)$, which allows for the finiteness of the plate size.

The first group of material constants includes yield strength σ_Y and ultimate strength σ_B. The values of σ_Y and σ_B are determined from the tensile stress-strain diagram $\sigma - \varepsilon$ plotted from test data for smooth cylindrical specimens made of a material for short-term ultimate strength.

The second group of material constants includes the coefficients D, q, and the exponent η. The coefficients D and q characterize the material resistance to fatigue failure and are determined from the fatigue tests on smooth cylindrical specimens under fully reversed cyclic loading. Experiments data are fitted by the equation

$$n_R = \left((1+q)D(\sigma_a)^q)\right)^{-1}, \tag{7.13}$$

derived by integrating the evolutionary damage Eq. (7.4) d, where $(\Delta\sigma_{yy})_{eqv} = \sigma_a$ and $n_* = n_R$. Finally, the coefficients D and q in (7.13) are determined by minimizing the functional

$$\Phi(D,q) = \sum_{i=1}^{s} \left\{ n_{Ri}(\sigma_{ai}) - \left[(1+q)D(\sigma_{ai})^q\right]^{-1} \right\}^2 \to \min, \tag{7.14}$$

where σ_{ai} and n_{Ri} are discrete values of amplitudes of cyclic stresses and associated numbers of cycles to failure.

The exponent η characterizes the sensitivity of the material to the asymmetry of the stress cycle and is determined by processing the data of fatigue tests on smooth cylindrical specimens for different values of the mean stress σ_m. The value η is determined by minimizing the following functional

$$\Phi\left[\psi\left[\sigma_m/\sigma_B\right], \eta\right] = \sum_{i=1}^{k} \{\psi\left[\sigma_{mi}/\sigma_B\right] - \left[\sigma_{ai}/\sigma_n\right]\}^2 \to \min, \tag{7.15}$$

which represents the best fit of experimental data to the linearized limit diagram. Here the function $\psi(\cdot)$ is selected in the form stated in (7.4).

To determine the exponent η, a single reference fatigue test on smooth cylindrical specimens can also be used in a zero-to-tension stress cycle ($\sigma_m = \sigma_a$). In this case, we determine η from the following relationship

$$\eta = \lg\left(\frac{\sigma_a^0}{\sigma_n^0}\right) \Big/ \lg\left(\cos\left(\frac{\pi}{2}\frac{\sigma_m^0}{\sigma_B}\right)\right), \tag{7.16}$$

where σ_a^0 and σ_m^0 are the amplitude of the cyclic stress and the mean stress of a zero-to-compression stress cycle; σ_n^0 is the fatigue limit for a reversed stress cycle corresponding to fatigue life n_R^0 under stresses σ_m^0 and σ_a^0.

The corrective functions $f_0(\cdot)$ and $f(\cdot)$, which allow for the effect of the finiteness of plate size on the growth kinetics of fatigue cracks, are specified by polynomials of the fourth order. In this case, we write the corrective function for the incubation stage as

$$f_0\left(h/w, \ell_0/w\right) = A_0 + A_1\ell_0/w + A_2\left(\ell_0/w\right)^2 + A_3\left(\ell_0/w\right)^3 + A_4\left(\ell_0/w\right)^4, \tag{7.17}$$

and for the stage of propagation of the fatigue crack as

$$f\left(h/w, \ell(n)/w\right) = A_0 + A_1\ell(n)/w + A_2\left(\ell(n)/w\right)^2 + A_3\left(\ell(n)/w\right)^3 + A_4\left(\ell(n)/w\right)^4. \quad (7.18)$$

Here A_0, A_1, A_2, A_3, and A_4 are polynomial coefficients dependent on the ratio between the height h and the width w of the plate, as well as on the ratio between the crack length ℓ and the plate width.

The values of the polynomial coefficients in (7.17) and (7.18) are determined by polynomial approximation of the numerical solutions of the problem on the stress state of a plate of finite size with a mode I crack under uniaxial tension. Numerical solutions are constructed by the method of boundary collocation (Tada 1971).

7.4 Calculation of the Kinetics of Fatigue Cracks Growth in Finite Plates

The problem of calculating the growth kinetics of mode I fatigue crack in thin plates of finite size under uniaxial asymmetric tension-compression $\tilde{\sigma}$ (7.1) is solved. Figure 7.2a shows the geometry and loading conditions of the plate with a crack. The geometric parameters of the plate are $h = 0.891$ m, $w = 0.305$ m, $2\ell_0 = 0.005$ m, $h/w = 2.92$.

The numerical solutions are approximated by polynomials of the fourth order (7.17) and (7.18). In this case, the corrective function $f(\cdot)$ is as follows

$$f\left(h/w, \ell/w\right) = 1.0106 - 0.1996(2\ell/w) + 1.829\left(2\ell/w\right)^2 - 3.068\left(2\ell/w\right)^3 + 3.2197\left(2\ell/w\right)^4. \quad (7.19)$$

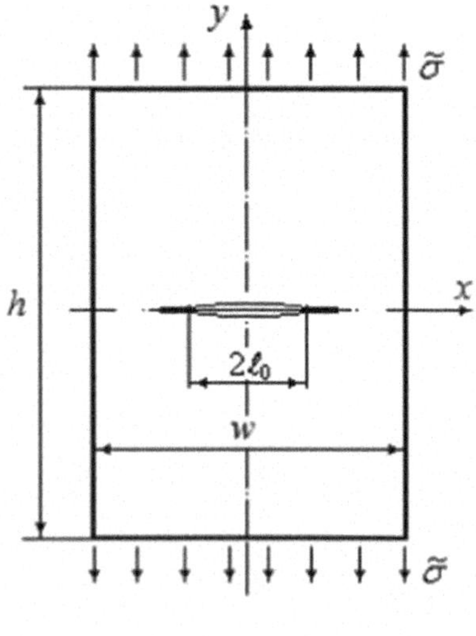

Fig. 7.2 Loading scheme of the finite plate with a crack

The computation results obtained using the numerical algorithm described in Sect. 7.3 are compared with the experimental data and with the computation results obtained using the analytical relations (Golub and Plashchiskaya 2018)

$$\begin{cases} n = n_* + \left(\dfrac{\pi}{4}\right)^{q-2} \dfrac{\left[\cos\left(\dfrac{\pi}{2}\dfrac{\sigma_\mathrm{m}}{\sigma_\mathrm{B}}\right)\right]^{2\eta}}{\left(1+q^{-1}\right) D(\sigma_\mathrm{Y})^{q-2}} \times \displaystyle\int_{\ell_0}^{\ell(n)} [\sigma_\mathrm{a} f\ (h/w, \ell(n)/w)]^{-2}\, \ell(n)^{-1} d\ell, \\ n_* = \left[(1+q)D\ [4\sigma_\mathrm{Y}/\pi]^q\right]^{-1}, \end{cases} \quad (7.20)$$

where all notations coincide with those adopted in (7.1)–(7.6).

The plate material is 7075-T6 aluminum alloy used in aviation and space technology. Table 7.1 shows the values of the material constants, which are determined from experimental data (Grover et al. 1953) by the method described in Sect. 7.3.3.

Figure 7.3 shows the results of calculating the kinetics of damage accumulation along the front of a fatigue crack growth in a plate made of 7075-T6 alloy at σ_m = 138 MPa and σ_a = 69 MPa, performed according to (7.9)–(7.12). Lines 1, 2, 3, and 4 indicate the level of accumulated damage $\omega(n_i, x_j)$ along the crack front during the time n_i ($i = 1, 2, 3, 4$), respectively. Line 5 corresponds to the damage $\omega(n_i, x_i)$ accumulated at the point x_i to each moment of time n_i. Fracture at a point x_j occurs when $\omega(n_{i+1}, x_j) = 1$ and $j = i$.

Table 7.1 The material constants of the model

Alloyσ_Y, MPa	σ_B, MPa	D, (MPa^q· cycle)$^{-1}$	q	η
7075-T6523	571	$3.33 \cdot 10^{-29}$	9.23	3.57

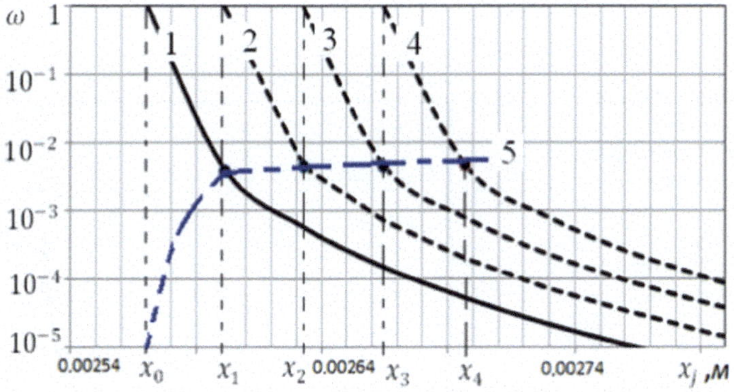

Fig. 7.3 The damage accumulation along the fatigue crack growth front in 7075 alloy plate

The kinetics of fatigue crack growth under uniaxial symmetric and asymmetric tension-compression is computed, including an asymmetric zero-to-compression cycle, an asymmetric sign-variable stress cycle, and an asymmetric constant-sign cycle. The results of numerical computations performed using (7.9) and (7.10) are compared with experimental data and with the results of analytical solutions performed using (7.20). Experimental data are taken from Hudson and Scardina (1967).

7.4.1 Fully Reversed Uniaxial Tension-Compression

The loading condition (7.1) is specified in the form

$$\tilde{\sigma} = \sigma_a \sin(2\pi n), \quad \sigma_m = 0, \tag{7.21}$$

where it is assumed that $\sigma_{min} = \sigma_m - \sigma_a = -\sigma_a$.

The results of computations performed by (7.9), (7.10), and (7.20) with (7.21) taken into account are compared in Fig. 7.4 ($\sigma_m = 0$, $\sigma_a = 207$ MPa, stress ratio $R = -1$) with experimental data. Here and below, the results of computations according to (7.9) and (7.10) are plotted by dash-dotted lines (lines 1), according to (7.20), by dashed lines (lines 2), and dots plot experimental data.

7.4.2 Asymmetric Sign-Variable Stress Cycle

The loading condition (7.1) is specified in the form

$$\tilde{\sigma} = \sigma_m + \sigma_a \sin(2\pi n), \quad \sigma_a > \sigma_m, \tag{7.22}$$

where it is assumed that $\sigma_{min} = \sigma_m - \sigma_a < 0$.

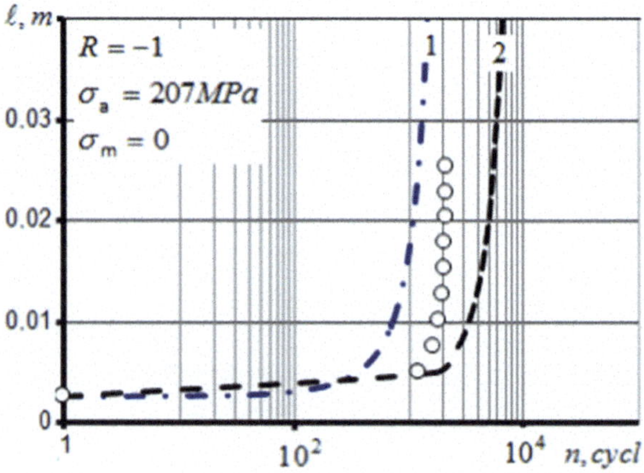

Fig. 7.4 Computed (1, 2) and experimental (o) dependences of the fatigue crack length ℓ versus the number of stress cycles n in aluminum alloy 7075-T6 thin plate under fully reversed uniaxial tension-compression

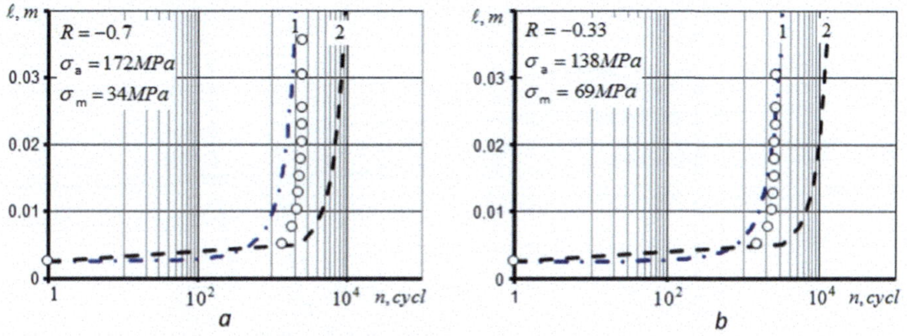

Fig. 7.5 Computed (1, 2) and experimental (o) dependences of the fatigue crack length ℓ versus the number of stress cycles n in aluminum alloy 7075-T6 thin plate under asymmetric sigh-variable cycle

The results of computations performed by (7.9), (7.10) and (7.20) taking into account (7.22) are compared in Fig. 7.5a ($\sigma_m = 34$ MPa, $\sigma_a = 172$ MPa, $R = -0.7$) and Fig. 7.5b ($\sigma_m = 69$ MPa, $\sigma_a = 138$ MPa, $R = -0.33$) with experimental data.

7.4.3 Repeated Zero-to-Compression Stress Cycle

The loading condition (7.1) is specified in this case in the form

$$\tilde{\sigma} = \sigma_m + \sigma_a \sin(2\pi n), \quad \sigma_m = \sigma_a, \tag{7.23}$$

where it is assumed that $\sigma_{min} = \sigma_m - \sigma_a = 0$.

The results of computations performed according to (7.9), (7.10), and (7.20) taking into account (7.23) are compared in Fig. 7.6 ($\sigma_m = \sigma_a = 69$ MPa and $R = 0$) with experimental data.

7.4.4 Asymmetric Constant-Sign Stress Cycle

The loading condition (7.1) is specified in this case in the form

$$\tilde{\sigma} = \sigma_m + \sigma_a \sin(2\pi n), \quad \sigma_m > \sigma_a, \tag{7.24}$$

where it is assumed that $\sigma_{min} = \sigma_m - \sigma_a > 0$.

The results of computations performed according to (7.9), (7.10), and (7.20), taking into account the loading conditions (7.24), are compared in Fig. 7.7a

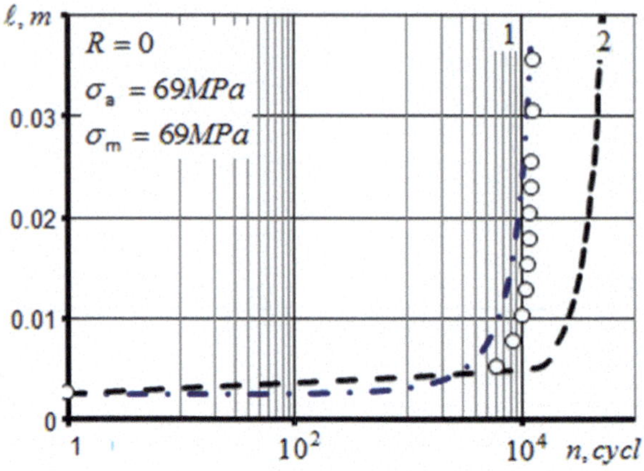

Fig. 7.6 Computed (1, 2) and experimental (○) dependences of the fatigue crack length ℓ versus the number of stress cycles *n* in aluminum alloy 7075-T6 thin plate under zero-to-compression stress cycle

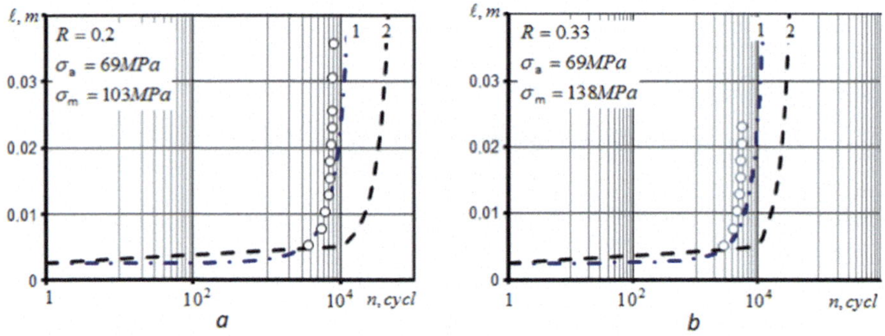

Fig. 7.7 Computed (1, 2) and experimental (○) dependences of the fatigue crack length ℓ versus the number of stress cycles *n* in aluminum alloy 7075-T6 thin plate under asymmetric constant-sign stress cycle

($\sigma_m = 103$ MPa, $\sigma_a = 69$ MPa, and $R = 0.2$) and Fig. 7.7b ($\sigma_m = 138$ MPa, $\sigma_a = 69$ MPa, and $R = 0.33$ MPa) with experimental data.

7.5 Analysis of the Obtained Results

In this work, the approach to modeling the process of fatigue crack growth, based on the use of the concepts of continuous damage mechanics, was further developed. A numerical algorithm of the model is constructed, which makes it possible to discretize the process of fatigue crack growth by the moments when the damage parameter at the growing crack tip reaches a value equal to 1. At each discretization moment, the amount of damage accumulated up to this point in time is taken into account, which made it possible to take into account the history of damage accumulation.

The numerical algorithm of the fatigue crack growth model has been experimentally tested on the problems of the kinetics of fatigue cracks growth in thin plates of aluminum alloy 7075-T6 under uniaxial symmetric, sign-variable, and sign-constant tension-compression. The results of comparing the computed and experimental data are shown in Figs. 7.4, 7.5, 7.6 and 7.7. For comparison, the results of analytical solutions are also given.

In general, as it follows from Figs. 7.4, 7.5, 7.6 and 7.7, the results of numerical computations (visualized by dashed-dotted lines), given that the process of fatigue failure is characterized by a significant (by 1–2 orders of magnitude) dispersion of fatigue properties, are in satisfactory agreement with the experimental results (visualized by points). The maximum error in the number of cycles for reaching the same length by a fatigue crack does not exceed 30%, and the maximum error between the lengths of fatigue cracks reached in the same number of loading cycles does not exceed 20%. The results of analytical solutions (visualized by dashed lines), which do not take into account the history of damage accumulation along the front of a growing fatigue crack, may differ from the experimental data by order of magnitude. The results of computations specify, in this case, slower growth of fatigue cracks.

7.6 Conclusion

The approach to modeling the process of fatigue crack growth in elastic bodies, based on the concepts of damage mechanics, seems to be the most effective way to solve this complex and very important problem of technical mechanics. The results of computations of the kinetics of growth of fatigue cracks in thin isotropic plates under symmetric and asymmetric tension-compression are in satisfactory agreement with the experimental data. The best agreement with the experiment is obtained when the solution is constructed using a numerical algorithm that allows one to take into account the history of damage accumulation along the growing crack front. An analytical solution that does not take into account the history of damage accumulation specifies a significantly slower process of fatigue crack growth.

References

Andreikiv A (1982) Spatial problems of the theory of cracks. Nauk, Dumka, Kyiv

Bolotin V (1983) Fatigue crack growth equations. Izv. Academy of sciences of the USSR. Mech Rigid Body 7:153–160

Bolotin V (1985) A unified approach to damage accumulation and fatigue crack growth. Eng Fract Mech 22(3):387–398

Bolotin V (1987) Model of a fatigue crack with a tip region. Sov Appl Mech 23:1159–1165

Bolotin V (1990) Resource of machines and constructions. Mashinostroenie, Moscow

Cherepanov G (1968) On the growth of cracks under cyclic loading. J Appl Mech Tech Phys 6:64–75
Golub V, Panteleyev E (1993) Fatigue damage and cyclic life-time of cracked isotropic plates considering two-stage fracture. Fatigue 93. Proc Intern Cong Fatigue EMAS 1:275–281
Golub V, Panteleyev E (2000) Subcritical growth of high-cycle fatigue cracks in finite thin isotropic plates. Int Appl Mech 36:938–947
Golub V, Plashchiskaya A (1994) Fatigue fracture model for thin isotropic plates with cracks in axial loading. Int Appl Mech 30(7):520–529
Golub V, Plashchiskaya A (2018) On the theory of growth of fatigue mode i cracks in thin isotropic plates of finite size under uniaxial tension-compression. Int Appl Mech 54(2):188–206
Golub V, Pelykh V, Pogrebnyak A (2010) Prediction of fatigue life of prismatic metal rods under asymmetric tension-compression by the method of equivalent stresses. Bull Nat Tech Univ Ukraine KPI 58:177–182
Grover H, Hyler W, Kuhn P, Landers C, Hawell F (1953) Axial-load fatigue properties of 24S-T and 75S-T aluminum alloys as determined in several laboratories. NACA-TN-2928. Washington
Hudson C, Scardina J (1967) Effect of stress ratio on fatigue crack growth in 7075-T6 aluminum-alloy sheet. NASA TMX-60125. Washington
Newman J (1971) An improved method of collocation for the stress analysis of cracked plate with various shaped boundaries. NASA TN D-6376. Washington
Newman J (1992) FASTRAN-II–A fatigue crack growth structural analysis program. NASA-TM-104159. Washington
Nott J (1978) Fundamentals of fracture mechanics. Metallurgy, Moscow
Paris P, Erdogan F (1963) A critical analysis of crack propagation laws. J Basic Eng 85(4):528–534
Paris PC, Gomez MP, Anderson WE (1961) A rational analytic theory of fatigue. Trend Eng 13(1):9–14
Rice J (1967) Mechanics of crack tip deformation and extension by fatigue. Fatigue crack propagation. ASTM STP 415. Am Soc Testing Mats 247–309
Tada H (1971) A note on the finite width correction to the stress intensity factor. Eng Fract Mech 3(3):345–347

Chapter 8
Solving Plates and Shells of Complex Shape Static Problems Using Coordinates Transformations

Alexander Grigorenko, Mykola Kryukov, Wolfgang H. Müller, and Serhii Yaremchenko

8.1 Introduction

The boundary value problems of the theory of plates and shells of canonical form, the geometry of which can be specified in orthogonal curvilinear coordinates (Cartesian, cylindrical, spherical), are sufficiently well studied, and there are many effective ways of solving them. However, with the need to consider non-orthogonal coordinates, more complications arise since it is often challenging to transform the general formulas of tensor analysis to a specific non-orthogonal system. It can be argued that at this stage of the development of the theory of plates and shells using the finite element method (FEM) or the finite difference method (FDM), it is possible to set any shape of the surface and choose any element. But we should consider that these methods should be used with caution since the careless application of FEM or FDM to any class of problems can lead to significant errors in calculations or even to physically unrealistic results (Blaauwendraad 2010).

Some studies that use numerical methods for structural analysis of complex shape plates in non-orthogonal coordinates are presented in Buragohain and Patodi (1978); Civalek and Gurses (2009); Malekzadeh and Fiouz (2007); Mohajerani (2015); Paimushin and Andreev (1983); Shahidi et al. (2007); Shufrin et al. (2010); Srinivasan and Ramachandran (1976). One of the methods of solving boundary value problems of the theory of plates and shells in non-orthogonal coordinates is described in this

A. Grigorenko (✉) · M. Kryukov · S. Yaremchenko
S.P. Timoshenko Institute of Mechanics, National Academy of Science of Ukraine, Kyiv, Ukraine
e-mail: ayagrigorenko1991@gmail.com

M. Kryukov
e-mail: mmkryukov@ukr.net

W. H. Müller
Institute of Mechanics, Technische Universität Berlin, Berlin, Germany
e-mail: wolfgang.h.mueller@tu-berlin.de

A. N. Guz et al. (eds.), *Advances in Mechanics*, Advanced Structured Materials 191,
https://doi.org/10.1007/978-3-031-37313-8_8

chapter. First, the original equations of the theory of plates and shells are written in orthogonal coordinates. After that, a specific substitution of variables is introduced, which makes it possible to transform the canonical orthogonal coordinate system to a non-orthogonal one. As a result, we get a two-dimensional boundary value problem on a rectangular domain that allows us to apply all available methods developed for solving boundary value problems for the mentioned domain. The basics of such transformations are outlined in Kornishin et al. (1989).

In this chapter, the spline-collocation method is used to reduce two-dimensional problems of the theory of plates and shells to one-dimensional ones, followed by their solution by the method of discrete orthogonalization (Bellman and Kalaba 1965; Godunov 1961).

8.2 Constitutive Equations

8.2.1 Bending of Oblique and Trapezoidal Plates

The bending of thin laminated orthotropic plates under a lateral load $q_z = q_0(x, y)$ is described in rectangular Cartesian coordinate system xOy by the following differential equation (Birman 2011; Kryukov 1997):

$$
\begin{aligned}
D_{11}\frac{\partial^4 w}{\partial x^4} + D_{22}\frac{\partial^4 w}{\partial y^4} + 2(D_{12} + 2D_{66})\frac{\partial^4 w}{\partial x^2 \partial y^2} + 2\frac{\partial D_{11}}{\partial x}\frac{\partial^3 w}{\partial x^3} \\
+ 2\frac{\partial D_{22}}{\partial y}\frac{\partial w}{\partial y^3} + 2\frac{\partial}{\partial y}(D_{12} + 2D_{66})\frac{\partial^3 w}{\partial x^2 \partial y} \\
+ 2\frac{\partial}{\partial x}(D_{12} + 2D_{66})\frac{\partial^3 w}{\partial x \partial y^2} + \left(\frac{\partial^2 D_{11}}{\partial x^2} + \frac{\partial^2 D_{12}}{\partial y^2}\right)\frac{\partial^2 w}{\partial x^2} \\
+ \left(\frac{\partial^2 D_{12}}{\partial x^2} + \frac{\partial^2 D_{22}}{\partial y^2}\right)\frac{\partial^2 w}{\partial x^2} + 4\frac{\partial^2 D_{66}}{\partial x \partial y}\frac{\partial^2 w}{\partial x \partial y} - q_z = 0, \qquad (8.1)
\end{aligned}
$$

where x and y are the Cartesian coordinates of the median surface of the plate, and w is the deflection.

The rigidity characteristics of the plate D_{ij} are defined as

$$
D_{ij} = \sum_{m=1}^{n} \int_{z_{m-1}}^{z_m} B_{ij}^m z^2 dz \quad (i, j = 1, 2, 6), \qquad (8.2)
$$

where B_{ij}^m are expressed in terms of the mechanical characteristics of the mth orthotropic layer:

$$
B_{11}^m = \frac{E_1^m}{1 - \nu_1^m \nu_2^m}, \quad B_{12}^m = \frac{\nu_2^m E_1^m}{1 - \nu_1^m \nu_2^m}, \quad B_{22}^m = \frac{E_2^m}{1 - \nu_1^m \nu_2^m}, \quad B_{66}^m = G_{12}^m,
$$

Fig. 8.1 Oblique plate

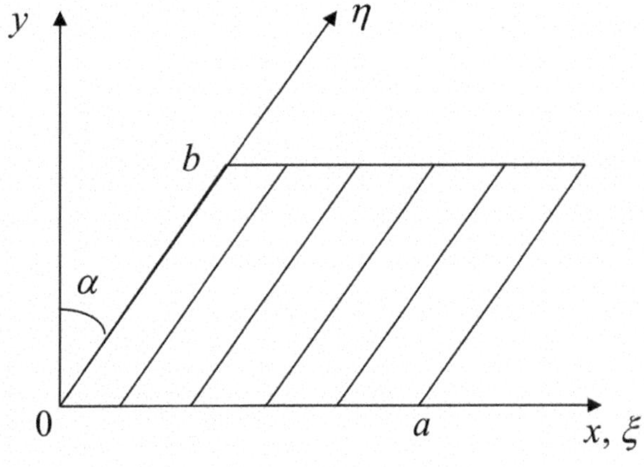

where E_1^m, E_2^m, G_{12}^m, ν_1^m and ν_2^m are the elastic and shear moduli and Poisson's ratios of the mth layer; $z = z_m$ are the interfaces between layers in a stack; $z = 0$ is the median surface of the plate; n is the number of layers in a stack, which is symmetric about the median surface.

We examine an oblique plate (Fig. 8.1) with the sides a and b. Let us parametrize it to transform the problem domain to canonical. We go to the distribution of bending of the plate in the oblique coordinate system $\xi O \eta$,

$$\xi = x - y \tan\alpha, \quad \eta = \frac{y}{\cos\alpha}. \tag{8.3}$$

After the transformation of (8.1), taking into account (8.3), the equation of bending of thin orthotropic plates in a new coordinate system can be obtained

$$a_0 \frac{\partial^4 w}{\partial \xi^4} + a_1 \frac{\partial^4 w}{\partial \eta^4} + a_2 \frac{\partial^4 w}{\partial \xi^3 \partial \eta} + a_3 \frac{\partial^4 w}{\partial \xi^2 \partial \eta^2} + a_4 \frac{\partial^4 w}{\partial \xi \partial \eta^3}$$
$$+ a_5 \frac{\partial^3 w}{\partial \xi^3} + a_6 \frac{\partial^3 w}{\partial \xi^2 \partial \eta} + a_7 \frac{\partial^3 w}{\partial \xi \partial \eta^2} + a_8 \frac{\partial^3 w}{\partial \eta^3} +$$
$$+ a_9 \frac{\partial^2 w}{\partial \xi^2} + a_{10} \frac{\partial^2 w}{\partial \xi \partial \eta} + a_{11} \frac{\partial^2 w}{\partial \eta^2} + a_{12} \frac{\partial^3 w}{\partial \eta^3} - q_z = 0. \tag{8.4}$$

Coefficients a_i are defined by the geometry and material of the plate (Kryukov, 1997).

The edge of the plate is clamped:

$$w = 0, \quad \frac{\partial w}{\partial \xi} = 0 \quad (\xi = 0,\ \xi = a); \quad w = 0, \quad \frac{\partial w}{\partial \eta} = 0 \quad (\eta = 0,\ \eta = b).$$

Fig. 8.2 Trapezoidal plate

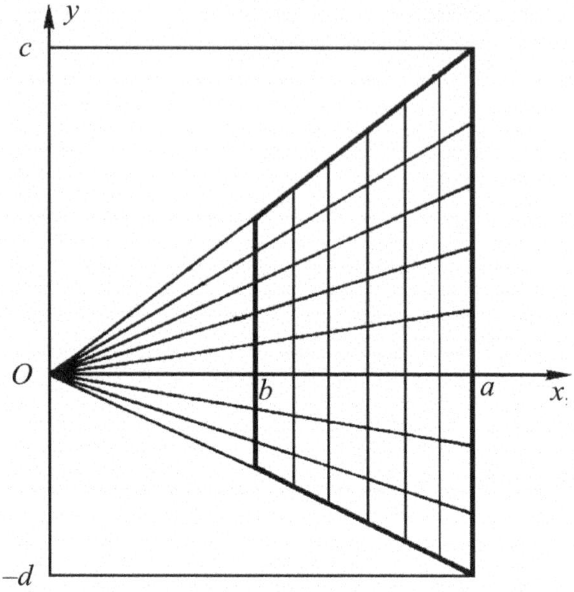

Consider a trapezoidal plate with base $c + d$ and height $a - b$ (Fig. 8.2). We describe the bending of the plate in a triangular (polar-oblique) coordinate system $\xi O \eta$ in which the plate boundaries coincide with the coordinate lines ξ = const and η = const.

In this case, we have

$$\xi = \frac{x}{a}, \quad \eta = k\frac{y}{x} \quad \left(k = \frac{a}{c}\right). \tag{8.5}$$

In the triangular coordinate system (ξ, η), (8.1) takes the form

$$\begin{aligned} \frac{\partial^4 w}{\partial \xi^4} &= a_{31}\frac{\partial^4 w}{\partial \xi^3 \partial \eta} + a_{22}\frac{\partial^4 w}{\partial \xi^2 \partial \eta^2} + a_{13}\frac{\partial^4 w}{\partial \xi \partial \eta^3} + a_{04}\frac{\partial^4 w}{\partial \eta^4} \\ &+ a_{30}\frac{\partial^3 w}{\partial \xi^3} + a_{21}\frac{\partial^3 w}{\partial \xi^2 \partial \eta} + a_{12}\frac{\partial^3 w}{\partial \xi \partial \eta^2} + a_{03}\frac{\partial^3 w}{\partial \eta^3} + a_{20}\frac{\partial^2 w}{\partial \xi^2} \\ &+ a_{11}\frac{\partial^2 w}{\partial \xi \partial \eta} + a_{02}\frac{\partial^2 w}{\partial \eta^2} + a_{01}\frac{\partial w}{\partial \eta} + a_0 q_z. \end{aligned} \tag{8.6}$$

Variable coefficients a_{ij} are defined by the geometry and material of the plate (Kryukov 1997). The edge of the plate is clamped:

$$w = 0, \quad \frac{\partial w}{\partial \xi} = 0 \quad \left(\xi = \frac{b}{a}, \ \xi = 1\right); \quad w = 0, \quad \frac{\partial w}{\partial \eta} = 0 \quad \left(\eta = -\frac{d}{c}, \ \eta = 1\right).$$

8.2.2 Quadrilateral Plates in Refined Formulation

Resolving equations in refined formulation (for total rotation angles ψ_x, ψ_y and deflection of the plate w, under normal uniform load q) take the form (Grigorenko et al. 2018):

$$
\begin{aligned}
&\frac{\partial \psi_x}{\partial x} + \frac{\partial^2 w}{\partial x^2} + \frac{\partial \psi_y}{\partial y} + \frac{\partial^2 w}{\partial y^2} = -\frac{6q}{5Gh},\\
&D\left(\frac{\partial^2 \psi_x}{\partial x^2} + \nu \frac{\partial^2 \psi_y}{\partial x \partial y}\right) + \frac{Gh}{6}\left(\frac{h^2}{2}\left[\frac{\partial^2 \psi_x}{\partial y^2} + \frac{\partial \psi_y}{\partial x \partial y}\right] - 5\left[\psi_x + \frac{\partial w}{\partial x}\right]\right) = 0,\\
&D\left(\frac{\partial^2 \psi_y}{\partial y^2} + \nu \frac{\partial^2 \psi_x}{\partial x \partial y}\right) + \frac{Gh}{6}\left(\frac{h^2}{2}\left[\frac{\partial^2 \psi_y}{\partial x^2} + \frac{\partial \psi_x}{\partial x \partial y}\right] - 5\left[\psi_y + \frac{\partial w}{\partial y}\right]\right) = 0,
\end{aligned}
\tag{8.7}
$$

where $D = Eh^3/[12(1-\nu^2)]$, $G = E/[2(1+\nu)]$, ν is the Poisson's ratio, E is the elastic modulus, h is the plate thickness.

The potential strain energy of the plate reads

$$
\begin{aligned}
\Pi = \frac{1}{2}\iint_S \Bigg\{ D\left[\left(\frac{\partial \psi_x}{\partial x}\right)^2 + \left(\frac{\partial \psi_y}{\partial y}\right)^2 + 2\nu \frac{\partial \psi_x}{\partial x}\frac{\partial \psi_y}{\partial y}\right] + \frac{Gh^3}{12}\left(\frac{\partial \psi_x}{\partial y}\right.\\
\left. + \frac{\partial \psi_y}{\partial x}\right)^2 + \frac{5Gh}{6}\left[\left(\psi_x + \frac{\partial w}{\partial x}\right)^2 + \left(\psi_y + \frac{\partial w}{\partial y}\right)^2\right] - 2qw \Bigg\} dxdy.
\end{aligned}
\tag{8.8}
$$

To transform a quadrilateral domain into a rectangular one, we use new coordinates ξ, η connected with x, y by the following relations:

$$
\begin{aligned}
x &= a_1 + a_2\xi + a_3\eta + a_4\xi\eta,\\
y &= b_1 + b_2\xi + b_3\eta + b_4\xi\eta.
\end{aligned}
\tag{8.9}
$$

The coefficients a_i and b_i can be determined from a system of eight linear equations by substituting into (8.8) of four points (x_i, y_i) in the previous system and four points (ξ_i, η_i) in the new one.

If, for example, $(\xi_i, \eta_i) = (\pm 1, \pm 1)$, transformation (8.9) becomes

$$
x = \sum_{i=1}^{4} x_i N_i, \quad y = \sum_{i=1}^{4} y_i N_i,
\tag{8.10}
$$

where $N_i = (1 + \xi\xi_i)(1 + \eta\eta_i)/4$ are the shape functions of the quadrangular finite element of the first order.

In what follows, we will express all the derivatives with respect to x and y in terms of the derivatives with respect to ξ and η. The transformation formulas for the first derivatives of an arbitrary function f reads

$$
\begin{aligned}
\frac{\partial f}{\partial x} &= \frac{1}{J}\left(\frac{\partial f}{\partial \xi}\frac{\partial y}{\partial \eta} - \frac{\partial f}{\partial \eta}\frac{\partial y}{\partial \xi}\right),\\
\frac{\partial f}{\partial y} &= \frac{1}{J}\left(\frac{\partial f}{\partial \eta}\frac{\partial x}{\partial \xi} - \frac{\partial f}{\partial \xi}\frac{\partial x}{\partial \eta}\right),
\end{aligned}
\tag{8.11}
$$

where J is the Jacobian of transformation (8.9). Using (8.11), we can get explicit expressions for the second derivatives.

Let us map the original quadrangle onto the square $[0; 1] \times [0; 1]$ and substitute into (8.7) the expressions for the derivatives with respect to x, y in terms of the derivatives with respect to ξ, η. As a result, we get the following equation:

$$
L\bar{u} = 0, \tag{8.12}
$$

where L is a linear differential operator of the second order within the domain ξ, η, $\bar{u} = \{w, \psi_x, \psi_y\}$ is an unknown vector. The clamped boundary condition is $\bar{u} = \bar{0}$.

8.2.3 Oblique Cylindrical Shells in Classical Formulation

Equations describing the bending of shells under uniform normal load q in a cylindrical system of coordinates take the form

$$
\begin{aligned}
\frac{\partial^2 u}{\partial z^2} + \frac{1-\nu}{2R^2}\frac{\partial^2 u}{\partial \theta^2} + \frac{1+\nu}{2R}\frac{\partial^2 v}{\partial z \partial \theta} - \frac{\nu}{R}\frac{\partial w}{\partial z} &= 0,\\
\frac{1+\nu}{2}\frac{\partial^2 u}{\partial z \partial \theta} + R\frac{1-\nu}{2}\frac{\partial^2 v}{\partial z^2} + \frac{1}{R}\frac{\partial^2 v}{\partial \theta^2} - \frac{1}{R}\frac{\partial w}{\partial \theta} &= 0,\\
\nu\frac{\partial u}{\partial z} + \frac{1}{R}\frac{\partial v}{\partial \theta} - \frac{1}{R}w - \frac{h^2}{12}\left(R\frac{\partial^4 w}{\partial z^4} + \frac{2}{R}\frac{\partial^4 w}{\partial z^2 \partial \theta^2} + \frac{1}{R^3}\frac{\partial^4 w}{\partial \theta^4}\right) &= -\frac{R(1-\nu^2)}{Eh}q,
\end{aligned}
\tag{8.13}
$$

$(0 \leqslant z \leqslant L, 0 \leqslant \theta \leqslant 2\pi)$, u, v and w are the axial, circumferential, and normal displacements, respectively, h is the shell thickness, R is the mid-surface radius, E is the elastic modulus, and ν is Poisson's ratio.

The mid-surface of the shell with oblique ends is given in the rectangular Cartesian coordinate system xyz using parameters ξ, η as follows

$$
\begin{aligned}
x = R\cos\eta, \quad y = R\sin\eta,\\
z = \xi + R\,[\tan\beta_1 - \xi(\tan\beta_1 + \tan\beta_2)/L]\,(1 - \cos\eta),
\end{aligned}
\tag{8.14}
$$

where L is the cylinder generatrix in the section $\eta = 0$; β_1 and β_2 are the angles of the lower and upper cuts, respectively; $0 \leqslant \xi \leqslant L$; if the cylinder is closed, then $0 \leqslant \eta \leqslant 2\pi$. If the angles β_1 and β_2 are positive, then the projection of the cylinder onto the plane xOz has the form shown in Fig. 8.3. When one of the angles is negative

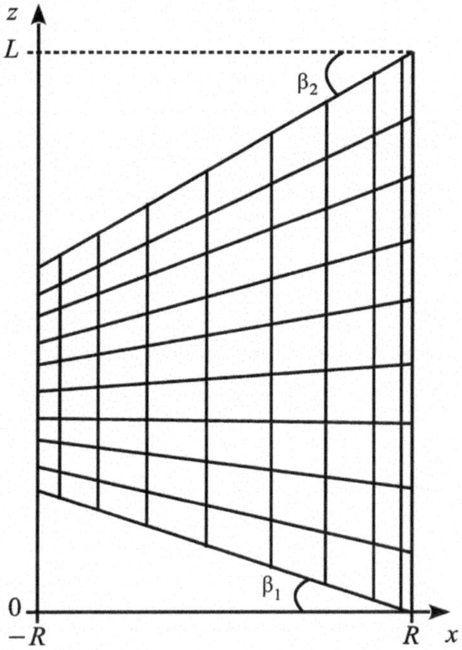

Fig. 8.3 Cylinder projection at both positive angles β_1 and β_2

($\beta_1 < 0$), the shell projection is shown in Fig. 8.4. It is obvious that if both angles are negative, they can be made positive by the appropriate selection of the length L.

The relationship between the coordinate systems $\xi\eta$ and $z\theta$ can be found from (8.14):

$$\xi = \frac{az + b(\cos\theta - 1)}{d\cos\theta + e}, \quad \eta = \theta, \tag{8.15}$$

where $a = L, b = LR\tan\beta_1, d = R(\tan\beta_1 + \tan\beta_2), e = a - d$.

To transform in (8.19) from $z\theta$ to $\xi\eta$, we can use the following formulas for arbitrary function f:

$$\frac{\partial f}{\partial z} = \frac{\partial f}{\partial \xi}\frac{\partial \xi}{\partial z} + \frac{\partial f}{\partial \eta}\frac{\partial \eta}{\partial z}, \quad \frac{\partial f}{\partial \theta} = \frac{\partial f}{\partial \xi}\frac{\partial \xi}{\partial \theta} + \frac{\partial f}{\partial \eta}\frac{\partial \eta}{\partial \theta}. \tag{8.16}$$

Using of (8.15) and (8.16) yields

$$\begin{aligned} &\frac{\partial f}{\partial z} = \lambda_1\frac{\partial f}{\partial \xi}, \quad \frac{\partial^2 f}{\partial z^2} = \lambda_1^2\frac{\partial^2 f}{\partial \xi^2},\\ &\frac{\partial^2 f}{\partial \theta^2} = \lambda_3\frac{\partial f}{\partial \xi} + \lambda_2^2\frac{\partial^2 f}{\partial \xi^2} + \frac{\partial^2 f}{\partial \eta^2} + 2\lambda_2\frac{\partial^2 f}{\partial \xi\eta},\\ &\frac{\partial^2 f}{\partial z\partial\theta} = \lambda_4\frac{\partial f}{\partial \xi} + \lambda_1\lambda_2\frac{\partial^2 f}{\partial \xi^2} + \lambda_1\frac{\partial^2 f}{\partial \xi\partial\eta}, \end{aligned} \tag{8.17}$$

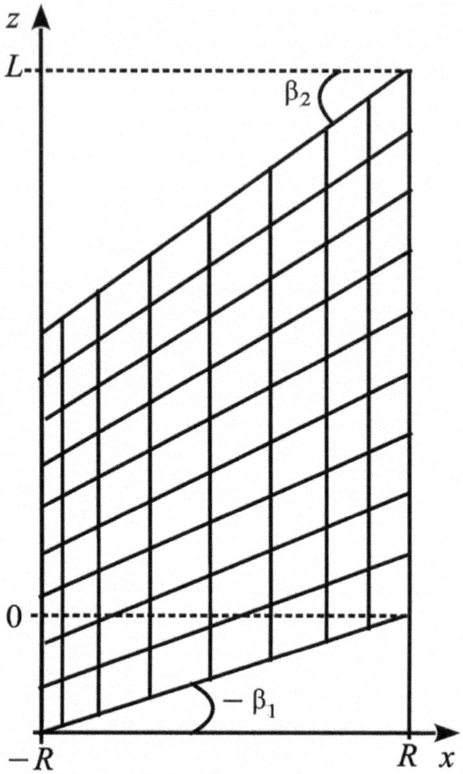

Fig. 8.4 Cylinder projection if β_1 is negative

etc., where

$$\lambda_1 = a/(d\cos\eta + e), \quad \lambda_2 = \lambda_1\,(d\xi - b)\sin\eta/a,$$
$$\lambda_3 = d\lambda_1[2d\lambda_2\sin\eta + (d\xi - b)\cos\eta]/a, \quad \lambda_4 = d\lambda_1^2\sin\eta/a.$$

With (8.17) for the functions u, v, w, ψ_z and ψ_θ, we transform expressions (8.13) to the form

$$\begin{aligned}\frac{\partial^2 u}{\partial\xi^2} &= C_1\frac{\partial^2 u}{\partial\xi\partial\eta} + C_2\frac{\partial^2 u}{\partial\eta^2} + C_3\frac{\partial u}{\partial\xi} + C_4\frac{\partial^2 v}{\partial\xi\partial\eta}\\ &\quad + C_5\frac{\partial^2 v}{\partial\eta^2} + C_6\frac{\partial v}{\partial\eta} + C_7\frac{\partial w}{\partial\xi} + C_8\frac{\partial w}{\partial\eta},\\ \frac{\partial^2 v}{\partial\xi^2} &= D_1\frac{\partial^2 u}{\partial\xi\partial\eta} + D_2\frac{\partial^2 u}{\partial\eta^2} + D_3\frac{\partial u}{\partial\xi} + D_4\frac{\partial^2 v}{\partial\xi\partial\eta}\\ &\quad + D_5\frac{\partial^2 v}{\partial\eta^2} + D_6\frac{\partial v}{\partial\eta} + D_7\frac{\partial w}{\partial\xi} + D_8\frac{\partial w}{\partial\eta},\end{aligned} \tag{8.18}$$

$$
\begin{aligned}
\frac{\partial^2 w}{\partial \xi^2} &= F_1\frac{\partial u}{\partial \xi} + F_2\frac{\partial v}{\partial \xi} + F_3\frac{\partial v}{\partial \eta} + F_4\frac{\partial^4 w}{\partial \xi^3 \partial \eta} + F_5\frac{\partial^4 w}{\partial \xi^2 \partial \eta^2} \\
&\quad + F_6\frac{\partial^4 w}{\partial \xi \partial \eta^3} + F_7\frac{\partial^4 w}{\partial \eta^4} + F_8\frac{\partial^3 w}{\partial \xi^3} + F_9\frac{\partial^3 w}{\partial \xi^2 \partial \eta} + F_{10}\frac{\partial^3 w}{\partial \xi \partial \eta^2} \\
&\quad + F_{11}\frac{\partial^2 w}{\partial \xi^2} + F_{12}\frac{\partial^2 w}{\partial \xi \partial \eta} + F_{13}\frac{\partial w}{\partial \xi} + F_{14}w + F_{15}q.
\end{aligned}
$$

Coefficients C_i, D_i and F_i depend on variables ξ and η. We have rigidly clamped oblique ends at $\eta = \text{const}$ and symmetry conditions at $\xi = 0$ and $\xi = \pi$ (Grigorenko et al. 2009).

8.2.4 *Circular Oblique Cylinders in Refined Formulation*

Consider a cylindrical shell in an orthogonal coordinate system z, θ, where z is a longitudinal coordinate directed along the axis of revolution, θ is the central angle in cross section. Then, using the equations of the refined shell theory based on the straight-line hypothesis, we can describe the stress-strain state of the shell acted upon by a surface normal load q by the following system of partial differential equations (Grigorenko et al. 2020):

$$
\begin{aligned}
\frac{\partial^2 u}{\partial \theta^2} &= a_{11}\frac{\partial^2 u}{\partial z^2} + a_{12}\frac{\partial^2 v}{\partial z \partial \theta} + a_{13}\frac{\partial w}{\partial z}, \\
\frac{\partial^2 v}{\partial \theta^2} &= a_{21}\frac{\partial^2 u}{\partial z \partial \theta} + a_{22}v + a_{23}\frac{\partial^2 v}{\partial z^2} + a_{24}\frac{\partial w}{\partial \theta} \\
&\quad + a_{25}\frac{\partial^2 \psi_1}{\partial z \partial \theta} + a_{26}\psi_2 + a_{27}\frac{\partial^2 \psi_2}{\partial z^2}, \\
\frac{\partial^2 w}{\partial \theta^2} &= a_{31}\frac{\partial u}{\partial z} + a_{32}\frac{\partial v}{\partial \theta} + a_{33}w + a_{34}\frac{\partial^2 w}{\partial z^2} + a_{35}\frac{\partial \psi_1}{\partial z} + a_{36}\frac{\partial \psi_2}{\partial \theta} + a_{37}q, \\
\frac{\partial^2 \psi_1}{\partial \theta^2} &= a_{41}\frac{\partial^2 u}{\partial z^2} + a_{42}\frac{\partial^2 v}{\partial z \partial \theta} + a_{43}\frac{\partial w}{\partial z} + a_{44}\psi_1 + a_{45}\frac{\partial^2 \psi_1}{\partial z^2} + a_{46}\frac{\partial^2 \psi_2}{\partial z \partial \theta}, \\
\frac{\partial^2 \psi_2}{\partial \theta^2} &= a_{51}\frac{\partial^2 u}{\partial z \partial \theta} + a_{52}v + a_{53}\frac{\partial^2 v}{\partial z^2} + a_{54}\frac{\partial w}{\partial \theta} \\
&\quad + a_{55}\frac{\partial^2 \psi_1}{\partial z \partial \theta} + a_{56}\psi_2 + a_{57}\frac{\partial^2 \psi_2}{\partial z^2},
\end{aligned}
\tag{8.19}
$$

where a_{ij} are variable coefficients that depend on the geometrical and mechanical parameters of the shell.

Using (8.17) for u, v, w, ψ_1, ψ_2, from (8.19) we obtain for oblique shell:

$$\begin{aligned}
\frac{\partial^2 u}{\partial \eta^2} &= -\lambda_3 \frac{\partial u}{\partial \xi} + \left(a_{11}\lambda_1^2 - \lambda_2^2\right) \frac{\partial^2 u}{\partial \xi^2} - 2\lambda_2 \frac{\partial^2 u}{\partial \xi \partial \eta} + a_{12}\lambda_4 \frac{\partial v}{\partial \xi} \\
&\quad + a_{12}\lambda_1\lambda_2 \frac{\partial^2 v}{\partial \xi^2} + a_{12}\lambda_1 \frac{\partial^2 v}{\partial \xi \partial \eta} + a_{13}\lambda_1 \frac{\partial w}{\partial \xi}, \\
\frac{\partial^2 v}{\partial \eta^2} &= a_{21}\lambda_4 \frac{\partial u}{\partial \xi} + a_{21}\lambda_1\lambda_2 \frac{\partial^2 u}{\partial \xi^2} + a_{21}\lambda_1 \frac{\partial^2 u}{\partial \xi \partial \eta} + a_{22} v - \lambda_3 \frac{\partial v}{\partial \xi} \\
&\quad + \left(a_{23}\lambda_1^2 - \lambda_2^2\right) \frac{\partial^2 v}{\partial \xi^2} - 2\lambda_2 \frac{\partial^2 v}{\partial \xi \partial \eta} + a_{24}\lambda_2 \frac{\partial w}{\partial \xi} + a_{24} \frac{\partial w}{\partial \eta} \\
&\quad + a_{25}\lambda_4 \frac{\partial \psi_1}{\partial \xi} + a_{25}\lambda_1\lambda_2 \frac{\partial^2 \psi_1}{\partial \xi^2} + a_{25}\lambda_1 \frac{\partial^2 \psi_1}{\partial \eta \partial \xi} + a_{26}\psi_2 + a_{27}\lambda_1^2 \frac{\partial^2 \psi_2}{\partial \xi^2} \\
\frac{\partial^2 w}{\partial \eta^2} &= a_{31}\lambda_1 \frac{\partial u}{\partial \xi} + a_{32}\lambda_2 \frac{\partial v}{\partial \xi} + a_{32} \frac{\partial v}{\partial \eta} + a_{33} w - \lambda_3 \frac{\partial w}{\partial \xi} \\
&\quad + \left(a_{34}\lambda_1^2 - \lambda_2^2\right) \frac{\partial^2 w}{\partial \xi^2} - 2\lambda_2 \frac{\partial^2 w}{\partial \xi \partial \eta} + a_{35}\lambda_1 \frac{\partial \psi_1}{\partial \xi} \\
&\quad + a_{36}\lambda_2 \frac{\partial \psi_2}{\partial \xi} + a_{36} \frac{\partial \psi_2}{\partial \eta} + a_{37} q, \\
\frac{\partial^2 \psi_1}{\partial \eta^2} &= a_{41}\lambda_1^2 \frac{\partial^2 u}{\partial \xi^2} + a_{42}\lambda_4 \frac{\partial v}{\partial \xi} + a_{42}\lambda_1\lambda_2 \frac{\partial^2 v}{\partial \xi^2} + a_{42}\lambda_1 \frac{\partial^2 v}{\partial \eta \partial \xi} \\
&\quad + a_{43}\lambda_1 \frac{\partial w}{\partial \xi} + a_{44}\psi_1 - \lambda_3 \frac{\partial \psi_1}{\partial \xi} + \left(a_{45}\lambda_1^2 - \lambda_2^2\right) \frac{\partial^2 \psi_1}{\partial \xi^2} - 2\lambda_2 \frac{\partial^2 \psi_1}{\partial \xi \partial \eta} \\
&\quad + a_{46}\lambda_4 \frac{\partial \psi_2}{\partial \xi} + a_{46}\lambda_1\lambda_2 \frac{\partial^2 \psi_2}{\partial \xi^2} + a_{46}\lambda_1 \frac{\partial^2 \psi_2}{\partial \xi \partial \eta}, \\
\frac{\partial^2 \psi_2}{\partial \eta^2} &= a_{51}\lambda_4 \frac{\partial u}{\partial \xi} + a_{51}\lambda_1\lambda_2 \frac{\partial^2 u}{\partial \xi^2} + a_{51}\lambda_1 \frac{\partial^2 u}{\partial \xi \partial \eta} + a_{52} v + a_{53}\lambda_1^2 \frac{\partial^2 v}{\partial \xi^2} \\
&\quad + a_{54}\lambda_2 \frac{\partial w}{\partial \xi} + a_{54} \frac{\partial w}{\partial \eta} + a_{55}\lambda_4 \frac{\partial \psi_1}{\partial \xi} + a_{55}\lambda_1\lambda_2 \frac{\partial^2 \psi_1}{\partial \xi^2} + a_{55}\lambda_1 \frac{\partial^2 \psi_1}{\partial \xi \partial \eta} \\
&\quad + a_{56}\psi_2 - \lambda_3 \frac{\partial \psi_2}{\partial \xi} + \left(a_{57}\lambda_1^2 - \lambda_2^2\right) \frac{\partial^2 \psi_2}{\partial \xi^2} - 2\lambda_2 \frac{\partial^2 \psi_2}{\partial \xi \partial \eta}.
\end{aligned} \tag{8.20}$$

The information on the boundary form is taken into account in expressions for λ_i. Oblique cuts are rigidly clamped. Symmetry conditions are used on straight contours.

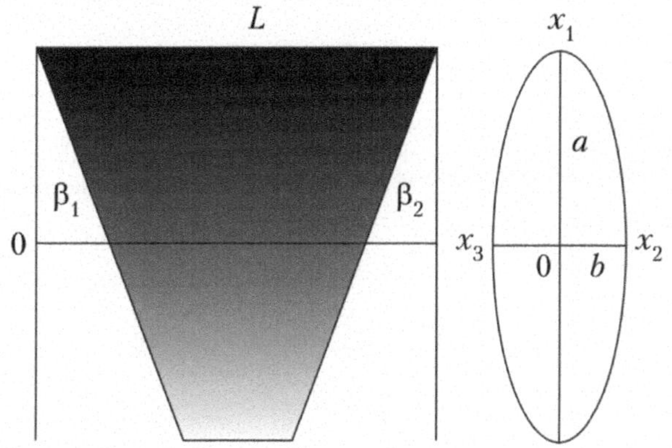

Fig. 8.5 Midsurface projections of elliptical cross-sectional shell with oblique cuts

8.2.5 *Cylindrical Shells with Elliptic Cross Section with Oblique Cuts*

By analogy with the approach used in previous sections, it is possible to obtain a resolving equation for the elliptical cross-sectional cylindrical shell with oblique cuts (Grigorenko et al. 2020). The projections of this shell are presented in Fig. 8.5.

The equations for non-circular cylindrical shells under uniform normal load q in the orthogonal curvilinear coordinate system $O\alpha_1\alpha_2$ take the form:

$$a_{11}\frac{\partial^2 u}{\partial\alpha_1^2}+a_{12}\frac{\partial u}{\partial\alpha_2}+a_{13}\frac{\partial^2 u}{\partial\alpha_2^2}+a_{14}\frac{\partial^2 v}{\partial\alpha_1\partial\alpha_2}+a_{15}\frac{\partial w}{\partial\alpha_1}=0,$$

$$a_{21}\frac{\partial^2 u}{\partial\alpha_2\partial\alpha_1}+a_{22}v+a_{23}\frac{\partial^2 v}{\partial\alpha_1^2}+a_{24}\frac{\partial v}{\partial\alpha_2}+a_{25}\frac{\partial^2 v}{\partial\alpha_2^2}+a_{26}w$$
$$+a_{27}\frac{\partial w}{\partial\alpha_2}+a_{28}\frac{\partial^2\psi_1}{\partial\alpha_2\partial\alpha_1}+a_{29}\psi_2+a_{2,10}\frac{\partial^2\psi_2}{\partial\alpha_1^2}=0,$$

$$a_{31}\frac{\partial u}{\partial\alpha_1}+a_{32}v+a_{33}\frac{\partial v}{\partial\alpha_2}+a_{34}w+a_{35}\frac{\partial^2 w}{\partial\alpha_1^2}+a_{36}\frac{\partial w}{\partial\alpha_2}$$
$$+a_{37}\frac{\partial^2 w}{\partial{\alpha_2}^2}+a_{38}\frac{\partial\psi_1}{\partial\alpha_1}+a_{39}\frac{\partial\psi_2}{\partial\alpha_2}+a_{3,10}q=0,$$

$$a_{41}\frac{\partial^2 u}{\partial\alpha_1^2}+a_{42}\frac{\partial u}{\partial\alpha_2}+a_{43}\frac{\partial^2 v}{\partial\alpha_2\partial\alpha_1}+a_{44}\frac{\partial w}{\partial\alpha_1}+a_{45}\psi_1$$
$$+a_{46}\frac{\partial^2\psi_1}{\partial\alpha_1^2}+a_{47}\frac{\partial\psi_1}{\partial\alpha_2}+a_{48}\frac{\partial^2\psi_1}{\partial{\alpha_2}^2}+a_{49}\frac{\partial^2\psi_2}{\partial\alpha_2\partial\alpha_1}=0,$$

$$a_{51}\frac{\partial^2 u}{\partial\alpha_2\partial\alpha_1}+a_{52}v+a_{53}\frac{\partial^2 v}{\partial\alpha_1^2}+a_{54}\frac{\partial v}{\partial\alpha_2}+a_{55}w+a_{56}\frac{\partial w}{\partial\alpha_2}$$

$$+ a_{57}\frac{\partial^2\psi_1}{\partial\alpha_1\partial\alpha_2} + a_{58}\psi_2 + a_{59}\frac{\partial^2\psi_2}{\partial\alpha_1^2} + a_{5,10}\frac{\partial\psi_2}{\partial\alpha_2} + a_{5,11}\frac{\partial^2\psi_2}{\partial\alpha_2^2} = 0, \quad (8.21)$$

where variable coefficients a_{ij} depend on shape and material of shell.

The mid-surface of the shell with oblique cuts in Cartesian coordinates x, y, z is defined as follows

$$\begin{gathered} x = f_1(\eta), \quad y = f_2(\eta), \\ z = \frac{\xi}{L}\big(L + [f_1(\eta) - f_1(0)][\tan\beta_1 + \tan\beta_2]\big) + [f_1(0) - f_1(\eta)]\tan\beta_1, \end{gathered} \quad (8.22)$$

$0 \leqslant \xi \leqslant L, 0 \leqslant \eta \leqslant 2\pi$; β_1 and β_2 are angles of oblique cuts.

Let us change variables in (8.21). The relationship between ξ, η and α_1, α_2 reads

$$\begin{aligned} \alpha_1 &= \frac{\xi}{L}\big(L + [f_1(\eta) - f_1(0)][\tan\beta_1 + \tan\beta_2]\big) + (f_1(0) - f_1(\eta))\tan\beta_1, \\ \alpha_2 &= \eta. \end{aligned} \quad (8.23)$$

We can use the following formulas for an arbitrary function f:

$$\begin{gathered} \frac{\partial f}{\partial\alpha_1} = \lambda_1\frac{\partial f}{\partial\xi}, \quad \frac{\partial^2 f}{\partial\alpha_1^2} = \lambda_1^2\frac{\partial^2 f}{\partial\xi^2}, \quad \frac{\partial f}{\partial\alpha_2} = \lambda_2\frac{\partial f}{\partial\xi} + \frac{\partial f}{\partial\eta}, \\ \frac{\partial^2 f}{\partial\alpha_2^2} = \lambda_3\frac{\partial f}{\partial\xi} + \lambda_2^2\frac{\partial^2 f}{\partial\xi^2} + \frac{\partial^2 f}{\partial\eta^2} + 2\lambda_2\frac{\partial^2 f}{\partial\xi\partial\eta}, \\ \frac{\partial^2 f}{\partial\alpha_1\partial\alpha_2} = \lambda_4\frac{\partial f}{\partial\xi} + \lambda_1\lambda_2\frac{\partial^2 f}{\partial\xi^2} + \lambda_1\frac{\partial^2 f}{\partial\xi\partial\eta}, \end{gathered} \quad (8.24)$$

where

$$\begin{gathered} \lambda_1 = \frac{L}{\Delta_1}, \quad \lambda_3 = \frac{\Delta_2}{\Delta_1^2}\left[f_1''\Delta_1 - 2(f_1')^2(\tan\beta_1 + \tan\beta_2)\right], \\ \lambda_2 = \frac{\Delta_2}{\Delta_1}f_1', \quad \lambda_4 = -\frac{L}{\Delta_1^2}f_1'(\tan\beta_1 + \tan\beta_2), \\ \Delta_1 = [f_1 - f_1(0)](\tan\beta_1 + \tan\beta_2) + L, \\ \Delta_2 = L\tan\beta_1 - \xi(\tan\beta_1 + \xi\tan\beta_2). \end{gathered} \quad (8.25)$$

Resolving equations in a new non-orthogonal coordinate system $O\xi\eta$ are the following

$$\begin{aligned} \frac{\partial^2 u}{\partial\eta^2} = &-\frac{a_{12}\lambda_2 + a_{13}\lambda_3}{a_{13}}\frac{\partial u}{\partial\xi} - \frac{a_{11}\lambda_1^2 + a_{13}\lambda_2^2}{a_{13}}\frac{\partial^2 u}{\partial\xi^2} - \frac{a_{12}}{a_{13}}\frac{\partial u}{\partial\eta} - 2\lambda_2\frac{\partial^2 u}{\partial\eta\partial\xi} \\ &- \frac{a_{14}\lambda_4}{a_{13}}\frac{\partial v}{\partial\xi} - \frac{a_{14}\lambda_1\lambda_2}{a_{13}}\frac{\partial^2 v}{\partial\xi^2} - \frac{a_{14}\lambda_1}{a_{13}}\frac{\partial^2 v}{\partial\eta\partial\xi} - \frac{a_{15}\lambda_1}{a_{13}}\frac{\partial w}{\partial\xi}, \\ \frac{\partial^2 v}{\partial\eta^2} = &-\frac{a_{21}\lambda_4}{a_{25}}\frac{\partial u}{\partial\xi} - \frac{a_{21}\lambda_1\lambda_2}{a_{25}}\frac{\partial^2 u}{\partial\xi^2} - \frac{a_{21}\lambda_1}{a_{25}}\frac{\partial^2 u}{\partial\eta\partial\xi} \end{aligned}$$

$$
\begin{aligned}
& -\frac{a_{22}\,v}{a_{25}} - \frac{a_{24}\,\lambda_2 + a_{25}\,\lambda_3}{a_{25}}\frac{\partial v}{\partial \xi} - \frac{a_{23}\,\lambda_1^2 + a_{25}\,\lambda_2^2}{a_{25}}\frac{\partial^2 v}{\partial \xi^2} - \frac{a_{24}}{a_{25}}\frac{\partial v}{\partial \eta} \\
& - 2\,\lambda_2\,\frac{\partial^2 v}{\partial \eta \partial \xi} - \frac{a_{26}\,w}{a_{25}} - \frac{a_{27}\,\lambda_2}{a_{25}}\frac{\partial w}{\partial \xi} - \frac{a_{27}}{a_{25}}\frac{\partial w}{\partial \eta} - \frac{a_{28}\,\lambda_4}{a_{25}}\frac{\partial \psi_1}{\partial \xi} \\
& - \frac{a_{28}\,\lambda_1\,\lambda_2}{a_{25}}\frac{\partial^2 \psi_1}{\partial \xi^2} - \frac{a_{28}\,\lambda_1}{a_{25}}\frac{\partial^2 \psi_1}{\partial \eta \partial \xi} - \frac{a_{29}\,\psi_2}{a_{25}} - \frac{a_{2,10}\,\lambda_1^2}{a_{25}}\frac{\partial^2 \psi_2}{\partial \xi^2},
\end{aligned}
$$

$$
\begin{aligned}
\frac{\partial^2 w}{\partial \eta^2} = & -\frac{a_{31}\,\lambda_1}{a_{37}}\frac{\partial u}{\partial \xi} - \frac{a_{32}\,v}{a_{37}} - \frac{a_{33}\,\lambda_2}{a_{37}}\frac{\partial v}{\partial \xi} - \frac{a_{33}}{a_{37}}\frac{\partial v}{\partial \eta} - \frac{a_{34}\,w}{a_{37}} \\
& - \frac{a_{36}\,\lambda_2 + a_{37}\,\lambda_3}{a_{37}}\frac{\partial w}{\partial \xi} - \frac{a_{35}\,\lambda_1^2 + a_{37}\,\lambda_2^2}{a_{37}}\frac{\partial^2 w}{\partial \xi^2} - \frac{a_{36}}{a_{37}}\frac{\partial w}{\partial \eta} \\
& - 2\,\lambda_2\,\frac{\partial^2 w}{\partial \eta \partial \xi} - \frac{a_{38}\,\lambda_1}{a_{37}}\frac{\partial \psi_1}{\partial \xi} - \frac{a_{39}\,\lambda_2}{a_{37}}\frac{\partial \psi_2}{\partial \xi} - \frac{a_{39}}{a_{37}}\frac{\partial \psi_2}{\partial \eta} - \frac{a_{3,10}\,q}{a_{37}},
\end{aligned}
$$

$$
\begin{aligned}
\frac{\partial^2 \psi_1}{\partial \eta^2} = & -\frac{a_{42}\,\lambda_2}{a_{48}}\frac{\partial u}{\partial \xi} - \frac{a_{41}\,{\lambda_1}^2}{a_{48}}\frac{\partial^2 u}{\partial \xi^2} - \frac{a_{42}}{a_{48}}\frac{\partial u}{\partial \eta} - \frac{a_{43}\,\lambda_4}{a_{48}}\frac{\partial v}{\partial \xi} \\
& - \frac{a_{43}\,\lambda_1\,\lambda_2}{a_{48}}\frac{\partial^2 v}{\partial \xi^2} - \frac{a_{43}\,\lambda_1}{a_{48}}\frac{\partial^2 v}{\partial \eta \partial \xi} - \frac{a_{44}\,\lambda_1}{a_{48}}\frac{\partial w}{\partial \xi} - \frac{a_{45}\,\psi_1}{a_{48}} \\
& - \frac{a_{47}\,\lambda_2 + a_{48}\,\lambda_3}{a_{48}}\frac{\partial \psi_1}{\partial \xi} - \frac{a_{46}\,{\lambda_1}^2 + a_{48}\,{\lambda_2}^2}{a_{48}}\frac{\partial^2 \psi_1}{\partial \xi^2} - \frac{a_{47}}{a_{48}}\frac{\partial \psi_1}{\partial \eta} \\
& - 2\,\lambda_2\,\frac{\partial^2 \psi_1}{\partial \xi \partial \eta} - \frac{a_{49}\,\lambda_4}{a_{48}}\frac{\partial \psi_2}{\partial \xi} - \frac{a_{49}\,\lambda_1\,\lambda_2}{a_{48}}\frac{\partial^2 \psi_2}{\partial \xi^2} - \frac{a_{49}\,\lambda_1}{a_{48}}\frac{\partial^2 \psi_2}{\partial \xi \partial \eta},
\end{aligned}
$$

$$
\begin{aligned}
\frac{\partial^2 \psi_2}{\partial \eta^2} = & -\frac{a_{51}\,\lambda_4}{a_{5,11}}\frac{\partial u}{\partial \xi} - \frac{a_{51}\,\lambda_1\,\lambda_2}{a_{5,11}}\frac{\partial^2 u}{\partial \xi^2} - \frac{a_{51}\,\lambda_1}{a_{5,11}}\frac{\partial^2 u}{\partial \eta \partial \xi} - \frac{a_{52}\,v}{a_{5,11}} \\
& - \frac{a_{54}\,\lambda_2}{a_{5,11}}\frac{\partial v}{\partial \xi} - \frac{a_{53}\,{\lambda_1}^2}{a_{5,11}}\frac{\partial^2 v}{\partial \xi^2} - \frac{a_{54}}{a_{5,11}}\frac{\partial v}{\partial \eta} - \frac{a_{55}\,w}{a_{5,11}} - \frac{a_{56}\,\lambda_2}{a_{5,11}}\frac{\partial w}{\partial \xi} \\
& - \frac{a_{56}}{a_{5,11}}\frac{\partial w}{\partial \eta} - \frac{a_{57}\,\lambda_4}{a_{5,11}}\frac{\partial \psi_1}{\partial \xi} - \frac{a_{57}\,\lambda_1\,\lambda_2}{a_{5,11}}\frac{\partial^2 \psi_1}{\partial \xi^2} - \frac{a_{57}\,\lambda_1}{a_{5,11}}\frac{\partial^2 \psi_1}{\partial \eta \partial \xi} \\
& - \frac{a_{58}\,\psi_2}{a_{5,11}} - \frac{a_{5,10}\,\lambda_2 + a_{5,11}\,\lambda_3}{a_{5,11}}\frac{\partial \psi_2}{\partial \xi} \\
& - \frac{a_{5,11}\,{\lambda_2}^2 + a_{59}\,{\lambda_1}^2}{a_{5,11}}\frac{\partial^2 \psi_2}{\partial \xi^2} - \frac{a_{5,10}}{a_{5,11}}\frac{\partial \psi_2}{\partial \eta} - 2\,\lambda_2\,\frac{\partial^2 \psi_2}{\partial \eta \partial \xi}.
\end{aligned}
\tag{8.26}
$$

Boundary conditions for elliptical shells are the same as boundary conditions for circular shells with oblique cuts.

8.3 Technique for Solving the Boundary Value Problem

Most problems in this chapter were solved by reducing two-dimensional problems of plates and shell theory to one-dimensional ones by the spline-collocation method (Bellman and Kalaba 1965; Godunov 1961).

For instance, for the problem of bending the trapezoidal plate, the solution to the boundary value problem is sought in the form:

$$w = \sum_{i=0}^{N} w_i(\xi)\varphi_i(\eta), \tag{8.27}$$

where w_i are unknown functions, and φ_i are linear combinations of quintic B-splines that satisfy the boundary conditions on contours $\eta = \text{const}$.

Substituting (8.27) into resolving equations and boundary conditions at contours $\xi = \text{const}$ and demanding to satisfy them in $N + 1$ collocation points, we obtain one-dimensional boundary value problem in Cauchy form:

$$\frac{d\bar{Y}}{d\xi} = A(\xi)\bar{Y} + \bar{f}, \tag{8.28}$$

$$B_1\bar{Y} = \bar{b}_1, \quad B_2\bar{Y} = \bar{b}_2, \tag{8.29}$$

where, in this case,

$$\bar{Y} = \{w_0, w_1, \ldots, w_N, w_0', w_1', \ldots, w_N', w_0'', \ldots, w_N'', w_0''', \ldots, w_N'''\}^T.$$

For shells, in classical formulation, solution takes the form

$$\bar{Y} = \{u_0, u, \ldots, u_N, u_0', \ldots, u_N', w_0, v_1, \ldots, v_N, v_0', v_1', \ldots, v_N', w_0, \ldots, w_N, \\ w_0', \ldots, w_N', w_0'', \ldots, w_N'', w_0''', \ldots, w_N'''\}^T.$$

For quadrilateral plates and shells, in refined formulation, we obtain similar expressions:

- for plates

$$\bar{Y} = \{w_0, w_1, \ldots, w_N, w_0', w_1', \ldots, w_N', \psi_{x0}, \ldots, \psi_{xN}, \\ \psi_{x0}', \ldots, \psi_{xN}', \psi_{y0}, \ldots, \psi_{yN}, \psi_{y0}', \ldots, \psi_{yN}', \}^T,$$

- for shells

$$\bar{Y} = \{u_0, u, \ldots, u_N, u_0', \ldots, u_N', w_0, v_1, \ldots, v_N, v_0', v_1', \ldots, v_N',$$
$$w_0, \ldots, w_N, w_0', \ldots, w_N',$$
$$\psi_{10}, \ldots, \psi_{1N}, \psi_{10}', \ldots, \psi_{1N}', \psi_{20}, \ldots, \psi_{2N}, \psi_{20}', \ldots, \psi_{2N}'\}^T.$$

Since, for problems that were solved using the refined formulation, spline-approximation was carried out along generatrix, (8.28) take the form

$$\frac{d\bar{Y}}{d\eta} = A(\eta)\bar{Y} + \bar{f}, \tag{8.30}$$

Resolving equations for oblique plates have the same form.

One of the specified dimensional boundary value problems was solved by the discrete-orthogonalization method (Bellman and Kalaba 1965; Godunov 1961).

8.4 Numerical Results

8.4.1 Oblique Plate

Let us consider the bending of the oblique plate (of thickness h) with rigidly clamped contours under the normal uniform load $q_z = q_0$. The input data are the following: $a = b = 10$, $h = 0.1$, $E_y = E_0$, $E_x = E_\mu E_0$, $G_{xy} = \lambda E_0$ and $\nu_y = \nu_0$. Five cases of elastic constants of the plate material are analyzed (Table 8.1). The third case corresponds to isotropic material.

Table 8.2 summarizes the peak values of the deflection for five cases of orthotropy and six values of angle $\alpha = \{0°, 15°, 30°, 45°, 60°, 75°\}$. Thirty-two collocation points along axis were used for solving the system of 128 ordinary differential equations. For example, for $\alpha = 15°$ with E_x decreasing, the deflection in plate center ($\xi = 6, \eta = 5$) increases by factors of 1.44, 1.55, 2.52, and 3.7 compared with the first case of orthotropy. As the angle α increases, the deflection decreases. For example, for the second case of orthotropy, the deflection decreases by factors 1.12, 1.62, 3.34, 12.18, and 157.9 compared with the rectangular plate.

Table 8.1 Five cases of elastic constants of the plate material

	μ	λ	ν_0
i	2	0.3	0.075
ii	1.35	0.215	0.122
iii	1	0.385	0.3
iv	0.741	0.159	0.165
v	0.5	0.125	0.15

Table 8.2 The peak values of the deflection for five cases of orthotropy and six values of angle α

α	i	ii	iii	iv	v
0°	$0.8818 \cdot 10^5$	$0.1272 \cdot 10^5$	$0.1381 \cdot 10^6$	$0.2224 \cdot 10^6$	$0.3276 \cdot 10^5$
15°	$0.7878 \cdot 10^5$	$0.1136 \cdot 10^5$	$0.1225 \cdot 10^6$	$0.1982 \cdot 10^6$	$0.2917 \cdot 10^6$
30°	$0.5448 \cdot 10^5$	$0.7845 \cdot 10^5$	$0.8378 \cdot 10^5$	$0.1366 \cdot 10^6$	$0.2008 \cdot 10^6$
45°	$0.2639 \cdot 10^5$	$0.3805 \cdot 10^5$	$0.4095 \cdot 10^5$	$0.6639 \cdot 10^5$	$0.9796 \cdot 10^5$
60°	$0.7204 \cdot 10^4$	$0.1044 \cdot 10^5$	$0.1168 \cdot 10^5$	$0.1842 \cdot 10^5$	$0.2722 \cdot 10^5$
75°	$0.5519 \cdot 10^3$	$0.8055 \cdot 10^3$	$0.9599 \cdot 10^3$	$0.1447 \cdot 10^4$	$0.2153 \cdot 10^4$

Table 8.3 The peak values of the deflection in the section $y = 0$

β	i	ii	iii	iv	v
5°	1527	1766	1755	2079	2259
10°	1121	1319	1330	1591	1750
15°	729.1	927.7	937.5	1118	1231
20°	536.3	627.1	631.3	755.1	832.0
25°	376.1	427.0	421.1	469.2	537.6

8.4.2 Trapezoidal Plate

We have solved the bending problem for a three-layer trapezoidal plate with constant thickness h_0 and all sides rigidly restrained under a uniform lateral load q_0. The input data are the following: $a - b = 10$, $c = d = 5$, $\beta_1 = \beta_2 = \beta$, $E_x = E_0$, $E_y = E_\mu E_0$, $G_{xy} = \lambda E_0$ and $\nu_x = \nu_0$. The thickness of the middle orthotropic layer $h_m = 0.8h_0$, and the thickness of the face layers $h_f = 0.1h_0$. Five values of the angle $\beta = \{5°, 10°, 15°, 20°, 25°\}$ and five cases (i)–(v) of elastic constants of the middle layer have been analyzed (see Sect. 8.4.1).

Table 8.3 summarizes the peak values of the deflection in the section $y = 0$. We deduce that the deflection increases with decreasing E_y. For example, for $\beta = 15°$, the deflection increases by factors of 1.17, 1.18, and 1.55 compared with the first case of orthotropy. As the angle β increases, the deflection decreases. For example, for the fourth case of orthotropy, the deflection decreases by factors of 1.31, 1.86, 2.75, and 4.43 compared with $\beta = 5°$.

8.4.3 Quadrilateral Plate in Refined Formulation

Using the above approaches, we have analyzed the stress-strain state of plates whose outline and deflection distributions wE/q are shown in Figs. 8.6, 8.7, 8.8 and 8.9. One plate is square with a side length equal to 2, and the other two plates are isosceles trapeziums with such apexes that the quadrangle area remains constant. The

Fig. 8.6 Square plate (case 1)

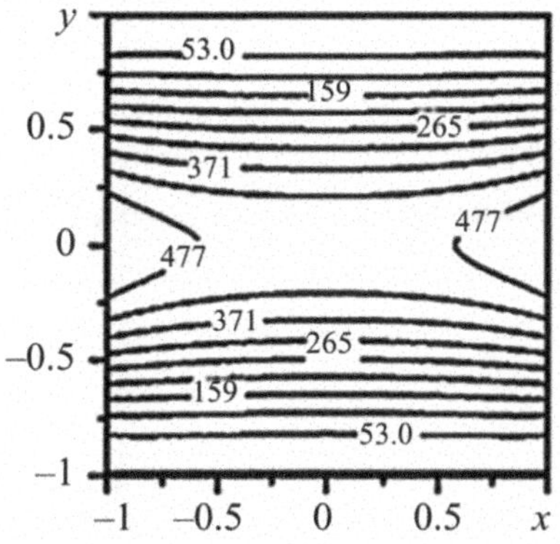

Fig. 8.7 Trapezoidal plate (case 2)

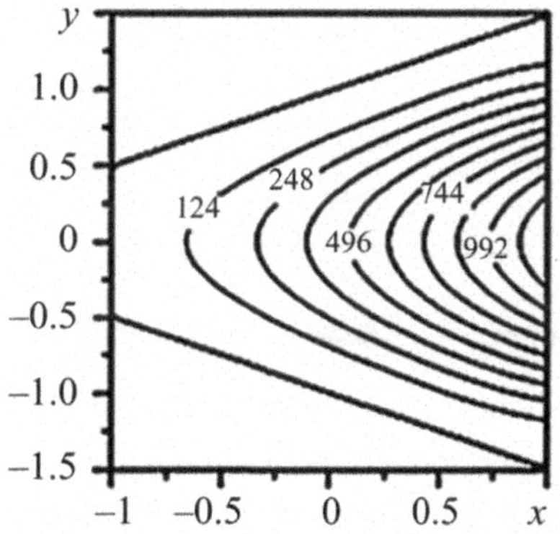

trapezium bases are equal to 1 and 3 in Fig. 8.7 and 0.2 and 3.8 in Fig. 8.8. The coordinates of the apexes can be found in Table 8.2.

The following parameter values were used: $h = 0.1$ and $\nu = 0.3$. The trapezium bases are free, and the sides are clamped. The figures show the results obtained with the spline-collocation and discrete-orthogonalization methods using 60 collocation points and 1500 integration points.

When using FEM, the square $[0; 1] \times [0; 1]$ is divided into elements with sizes $[0; 1/200] \times [0; 1/200]$. Then, (8.9) is used to determine the coordinates of the nodes of the finite elements in the coordinates xy. The resulting matrix $3 \times 201 \times 201$, while the band width is 3×203. Such an approach made it possible to obtain results (Table 8.4) in agreement with those obtained with the above two methods. We use the Gaussian method to solve the systems of linear algebraic equations. Table 8.4 compares the maximum deflections wE/q achieved at the point $x = 1$, $y = 0$. We failed to calculate the deflection of the trapezium close to a triangle (with bases 0.02 and 3.98). FEM solved this problem with the maximum deflection ($wE/q = 1617.7$) differing from that for the trapezium (case 3) insignificantly.

The stress-strain state of the triangle (case 4) was determined using both approaches. To this end, three points of the four were placed on the same line for the

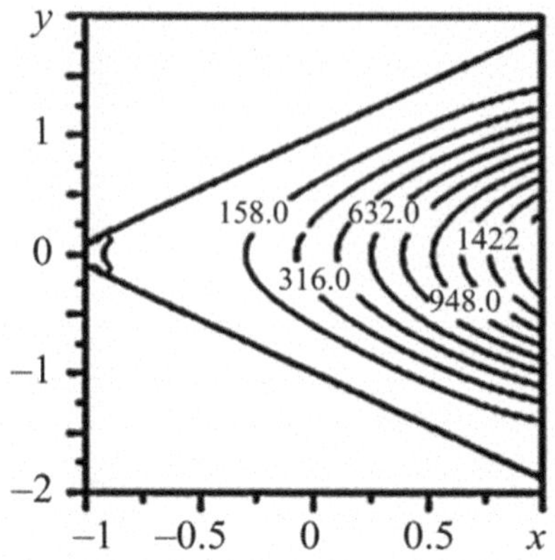

Fig. 8.8 Trapezoidal plate (case 3)

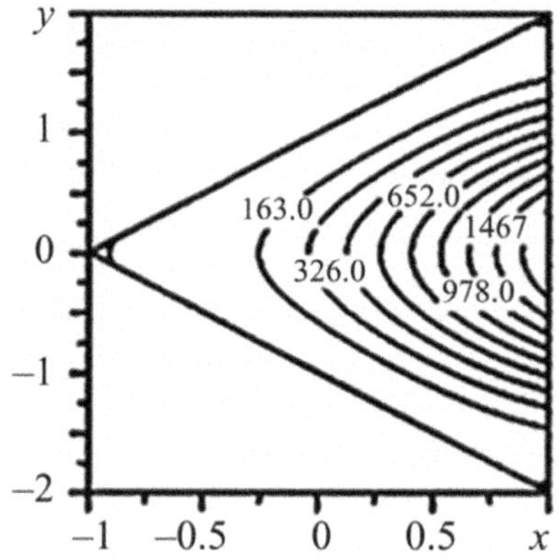

Fig. 8.9 Triangular plate (case 4)

Table 8.4 Results obtained by two methods

	Figure 8.6	Figure 8.7	Figure 8.8	Figure 8.9
SCM	529.34	1235.6	1576.7	1628.5
FEM	527.75	1231.6	1570.8	1622.2

quadrangle to degenerate it into a triangle. If the first apex had coordinates (–1; 0), the other apexes of the "quadrangle" (anticlockwise) were (1; –2), (1; 2), and, for example, (0; 1). The reliability and accuracy of this calculation can be evaluated by comparing the maximums obtained for case 3 and the trapezium with bases 0.02 and 3.98.

Table 8.5 Deflections wE/q in section $\xi = L/2$

η	$\beta = 6°$			$\beta = 18°$			$\beta = 30°$		
	I	II(20)	II(30)	I	II(20)	II(30)	I	II(20)	II(30)
0	383.4	383.8	383.1	411.2	413.4	410.7	437.1	440.5	436.3
$\pi/10$	382.7	383.0	382.4	409.3	410.9	408.7	433.2	435.5	432.4
$2\pi/10$	380.6	380.7	380.3	403.6	403.7	402.8	422.2	422.0	421.2
$3\pi/10$	377.3	377.1	377.0	394.4	393.0	393.4	406.8	404.1	405.3
$4\pi/10$	373.1	372.5	372.8	382.6	380.2	381.6	391.3	387.7	389.3
$5\pi/10$	368.3	367.4	368.0	369.2	366.8	368.5	379.4	376.6	377.7
$6\pi/10$	363.2	362.2	363.0	355.5	354.1	355.5	369.7	368.4	368.9
$7\pi/10$	358.6	357.5	358.4	343.0	342.8	343.5	352.2	352.7	352.8
$8\pi/10$	354.7	353.7	354.7	332.8	333.5	333.6	317.6	319.8	319.8
$9\pi/10$	352.2	351.2	352.2	326.0	327.3	326.9	275.3	279.5	279.5
π	351.3	350.4	351.3	323.5	325.1	324.5	255.5	260.6	260.6

8.4.4 Beveled Cylindrical Shells in Classical and Refined Formulations

Results obtained using classical theory equations (Grigorenko et al. 2009) have been compared with results obtained using refined theory equations. The input data are the following: $L = 60$, $R = 20$, $h = 1$, $\beta_1 = \beta_2 = \beta$ and $\nu = 0.3$. The projection of shell is shown in Fig. 8.3. Table 8.5 contains data for deflections wE/q in section $\xi = L/2$ at $\beta = 6°; 18°; 30°$. Column I contains data from Grigorenko et al. (2009), columns II (20) and II (30) contain the data obtained by the approach presented in Grigorenko et al. (2020) with the number of collocation points $N + 1 = 20$ and $N + 1 = 30$, respectively. The interval of integration is divided into 400 subintervals. As can be seen, the data obtained using both approaches are in good agreement, the difference being no more than 0.5%.

8.4.5 Noncircular Beveled Shells

Next, we find the stress-strain state of the elliptic cylindrical cross section $x_1 = a\cos\eta$, $x_2 = b\sin\eta$ shell with semiaxises $a = 25$, $b = 16$ and $a = 16$, $b = 25$; $L = 60$, $h = 1$, $\nu = 0.3$. Angles β_1 and β_2 are equal.

Figure 8.10 shows distribution of the displacements for $b = 16$ and $a = 25$ at $\beta_1 = \beta_2 = 0°; 15°; 25°$. Graphics show that elliptical cross-sectional peak displacements are much greater than for the circular cylinder. In cross section $\eta = 0$, displacements are almost unchanged depending on cutting angles, but their absolute values decrease when approaching $\eta = \pi$. There are almost not deformation at

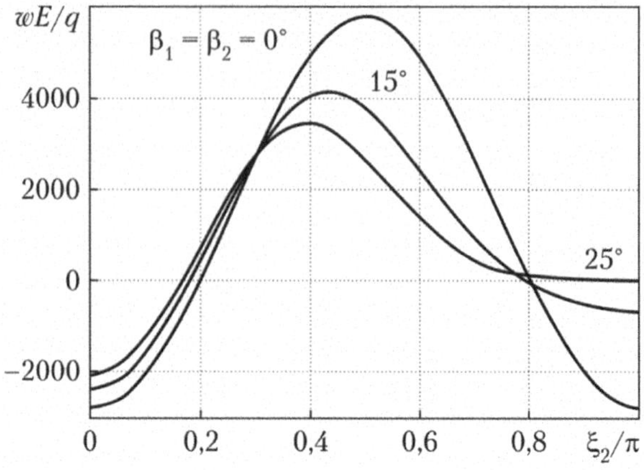

Fig. 8.10 Displacements distribution in elliptical cross-sectional shell with semiaxes $a = 25$, $b = 16$ and $\xi = L/2$

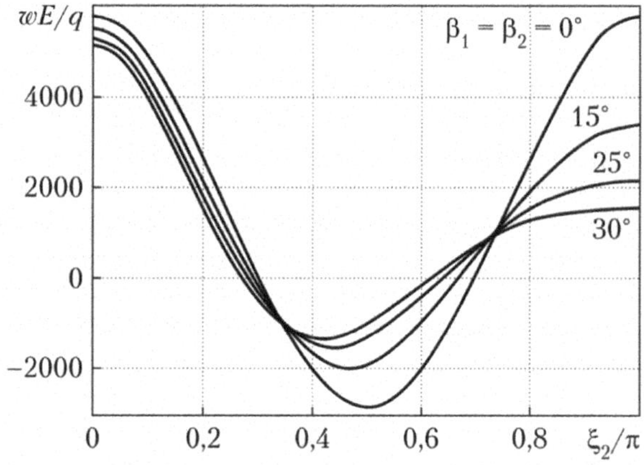

Fig. 8.11 Displacements distribution in elliptical cross-sectional shell with semiaxes $a = 16$, $b = 25$ and $\xi = L/2$

$\beta_1 = \beta_2 = 25°$ near $\eta = \pi$. For $a = 16$ and $b = 25$ (Fig. 8.11), peak displacements wE/q are at $\eta = 0$ and, as in the previous case, are almost independent of cutting angles. However, displacements decrease by factor 3 in cross section $\eta = \pi$ at $\beta_1 = \beta_2 = 30°$ compared with the shell without cuts.

8.5 Conclusions

Using the classical and refined theory of shells of the Timoshenko type, the equations of the stress-strain state of quadrilateral plates and cylindrical shells of circular and non-circular cross sections with beveled sections at the ends in a new non-orthogonal coordinate system were obtained.

The two-dimensional boundary values problems obtained are reduced to one-dimensional ones using the spline-collocation method when approximating one of the coordinate directions. The latter are solved by the discrete-orthogonalization method.

The solution of the bending problems of trapezoidal plates in the classical formulation and arbitrary quadrilateral plates in the refined formulation of circular and elliptical shells with oblique sections are given.

A comparison of the results obtained using the spline-collocation method and the finite element method was carried out for quadrilateral plates in the refined formulation.

In the case of cylindrical shells of circular cross sections with beveled cuts, the solutions to the problems are obtained using the refined theory compared to the results obtained using the classical theory.

References

Alwar RS, Ramachandra Rao N (1973) Nonlinear analysis of orthotropic skew plates. AIAA J 11(4):495–498

Bellman R, Kalaba R (1965) Quasilinearization and nonlinear boundary-value problems. American Elsevier Publishing Company, New York

Birman V (2011) Plate structures. Springer, New York

Blaauwendraad J (2010) Plates and FEM surprises and pitfalls. Springer, New York

Buragohain DN, Patodi SC (1978) Large deflection analysis of skew plates by lumped triangular element formulation. Comput Struct 9:183–189

Civalek O, Gurses M (2009) Frequency analysis of trapezoidal plates and membrane using discrete singular convolution. Asian J Civil Eng (Build Housing) 9(6):593–605

Godunov SK (1961) Numerical solution of boundary-value problems for systems of linear ordinary differential equations. Usp Mat Nauk 16(3):171–174 [in Russian]

Grigorenko YM, Kryukov NN, Kholkina NS (2009) Spline-approximation solution of stress-strain problems for beveled cylindrical shells. Int Appl Mech 45(12):413–420

Grigorenko AY, Pankrat'ev SA, Yaremchenko SN (2018) Analysis of the stress-strain state of complex-shaped plates. Int Appl Mech 54(6):695–701

Grigorenko YM, Grigorenko AY, Kryukov NN et al (2020) Design of cylindrical shells with oblique cuts in refined statement using spline-Approximation. Int Appl Mech 56(3):326–333

Grigorenko YM, Grigorenko AY, Kryukov NN et al (2020) Stress-strain state of elliptical cross-section cylindrical shells with beveled cuts. Dopov Nac akad nauk Ukr 6:21–29 [in Ukrainian]

Kornishin MS, Paimushin VN, Snigirev VF (1989) Computational geometry in problems of shell mechanics. Science, Moscow [in Russian]

Kryukov NN (1997) Design of oblique and trapezoidal plates with the use of spline functions. Int Appl Mech 33(5):414–417

Malekzadeh P, Fiouz AR (2007) Large deformation analysis of orthotropic skew plates with non-linear rotationally restrained edges using DQM. Compos Struct 80(2):196–206

Malekzadeh P, Karami G (2006) Differential quadrature nonlinear analysis of skew composite plates based on FSDT. Eng Struct 28:1307–1318

Mohajerani P (2015) The thick orthotropic plates analysis method, Part 1: A review. IOSR J Mech Civil Eng 12(2):69–77

Paimushin VN, Andreev SV (1983) Numerical study of the stress-strain state of one- and three-layer plates and shells of complex geometry. Sov Appl Mech 19(7):583–588

Shahidi AR et al (2007) Nonlinear static analysis of arbitrary quadrilateral plates in very large deflections. Commun Nonlin Sci Numer Simul 12:832–848

Shufrin I et al (2010) A semi-analytical approach for the geometrically nonlinear analysis of trapezoidal plates. Int J Mech Sci 52:1588–1596

Srinivasan RS, Ramachandran SV (1976) Large deflection of clamped skew plates. Comp Meth Appl Mech Eng 7:219–233

Chapter 9
Wave Scattering on Finite Wedge-Shaped Objects

Viktor Grinchenko, Ihor Vovk, Vitalii Husak, and Volodymyr Matsypura

9.1 Introduction

Problems about wave scattering on bodies of different configurations have important theoretical and applied significance since the results of their solutions can serve as a basis for the construction of location systems of object detection and classification.

The most important problem in the design of aircraft and surface vehicles is to reduce their radar visibility. When an object (e.g., an aircraft) has a complex shape, and its dimensions are several wavelengths or more, its one-position (backscattering) or two-position scattering cross section is characterized by significant fluctuations in values with a small change in the angles of incidence of the electromagnetic wave. Such a complex situation arises due to the diffraction and interference phenomena of the electromagnetic wave reflected from different parts of the aircraft.

The problem of stealth technology is to reduce the positional scattering cross section of various objects as much as possible. Many stealth technology techniques and methods are classified information, but the basic principles of reducing radar visibility are studied and described in the literature (see, for example, Lagarkov and Pogosyan (2003); Lvova (2003); Alekseev et al. (2007); Knott et al. (2004); Suharevskiy et al. (2009); Vozhdaev and Teperin (2018)). Stealth technology includes

V. Grinchenko · I. Vovk
Institute of Hydromechanics, National Academy of Science of Ukraine, Kyiv, Ukraine
e-mail: grinchenko@nas.gov.ua

I. Vovk
e-mail: vovkiv@nas.gov.ua

V. Husak (✉) · V. Matsypura
Faculty of Mechanics and Mathematics, Taras Shevchenko National University of Kyiv, Kyiv, Ukraine
e-mail: gusakw7@gmail.com

A. N. Guz et al. (eds.), *Advances in Mechanics*, Advanced Structured Materials 191,
https://doi.org/10.1007/978-3-031-37313-8_9

the following main directions: the theory of diffraction on complex bodies, the development and research of radio-absorbing materials, coating technology, and, finally, the radiophysical experiment used for control in each listed direction (Lagarkov and Pogosyan 2003).

Currently, stealth technology begins with mathematical modeling of electromagnetic wave scattering on the object, radar visibility of which is to be reduced. This step is fundamental for preliminary estimating the achievable result and allows for optimizing the shape and electrophysical characteristics of the object. Mathematical and computational models are based on the solution of boundary problems of electromagnetic wave diffraction on bodies of complex shape, having special materials and coatings in their composition. The capabilities of modern computing technology make it possible to create software for modeling electromagnetic wave scattering even on such complex objects as aircraft and ships.

Several software packages suitable for such modeling are available on the world market. They all use "facet" models of objects and the method of physical diffraction theory (Ufimtsev 2007; Chatzigeorgiadis 2004) in their development. The application of asymptotic and hybrid algorithms (Shenderov 1989) presupposes a clear idea of the physical phenomena that lead to the formation of the scattered field in each particular case: specular reflections, diffraction on edges, and others. When calculating the positional scattering cross section of large (in wavelengths) complex bodies, it is essential that at relatively high frequencies, it is possible to distinguish the individual fragments of the object structure, which at a given angle make the main contribution to the formation of the scattered field. This makes it possible to carry out calculations for such scattering centers independently of each other if necessary, taking into account their mutual influence at later stages. Therefore, much attention is paid to the development and improvement of appropriate electrodynamic models.

A wedge is considered to be a classical model of this kind. In this model, infinite boundaries are taken into account using the results of physical diffraction theory, and the fields in a finite domain of space near the edge are determined by numerically solving the electromagnetic field equations.

The study of wave scattering on finite wedge-shaped bodies is of interest since the shape of such bodies is close to the shapes of some parts of aircraft, surface, and space objects. In this chapter, based on the method of partial domains (Grinchenko et al. 2018), a rigorous solution to the problem of wave scattering on models of finite wedge-shaped objects in the presence of surfaces of large wave dimensions is constructed.

9.2 Formulation and Construction of the Analytical Solution to a Scattering Problem of a Plane Wave on a Finite Wedge-Shaped Object

Let us distinguish two models of a finite wedge for analysis: a sharp and a rounded.

9.2.1 Sharp Wedge

Consider the plane problem of plane wave scattering on a wedge-shaped object (Fig. 9.1); let us assume that the object has an infinite extent along the axis perpendicular to the plane of the figure. To describe the object's geometry and construct the solution to the problem, let us introduce a polar coordinate system $rO\theta$ centered at the corner point of the wedge. The object is a finite wedge with an outer angle of 2ϕ, the sides of which are closed by an arc of a circle of radius R. The surfaces of the object are considered acoustically rigid. The wedge-shaped object is in an ideal medium with density ρ and sound velocity c. The angle θ_0 gives the direction of the incident plane wave.

According to the idea of the partial domain method (Grinchenko et al. 2018), the entire sound field existence space is divided into two domains: 1 is the outer circle of radius R, i.e., $r \geqslant R$, $-\pi \leqslant \theta \leqslant \pi$; 2 is the sector of a circle of radius R, i.e., $0 \leqslant r \leqslant R$, $-\phi \leqslant \theta \leqslant \phi$.

Let us divide the incident wave field $p_0 = \exp(-ikr\cos(\theta - \theta_0))$ into even and odd components:

$$p_0 = p_0^+ + p_0^-, \tag{9.1}$$

$$\begin{aligned} p_0^+ &= \tfrac{1}{2}\left[\exp(-ikr\cos(\theta-\theta_0)) + \exp(-ikr\cos(\theta+\theta_0))\right], \\ p_0^- &= \tfrac{1}{2}\left[\exp(-ikr\cos(\theta-\theta_0)) - \exp(-ikr\cos(\theta+\theta_0))\right]. \end{aligned}$$

Then the solution to the problem is found as a sum of solutions to even and odd problems. Dividing the problem into even and odd problems will allow us to consider the range of angle change $0 \leqslant \theta \leqslant \pi$. This makes it possible to halve the size of the matrix of the system of equations.

Let us write the pressure field in partial domain 1 in the form:

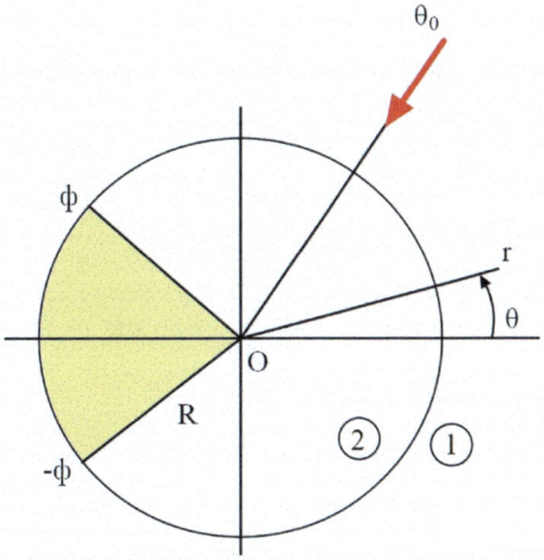

Fig. 9.1 The geometry of the finite sharp wedge model

$$p_1 = p_1^+ + p_1^-, \tag{9.2}$$

$$p_1^+ = \sum_{n=0}^{\infty} A_n \cos(n\theta) \frac{H_n^{(1)}(kr)}{H_n^{(1)\prime}(kR)} + p_0^+,$$
$$p_1^- = \sum_{n=1}^{\infty} B_n \sin(n\theta) \frac{H_n^{(1)}(kr)}{H_n^{(1)\prime}(kR)} + p_0^-.$$

The pressure field in partial domain 2:

$$p_2 = p_2^+ + p_2^-, \tag{9.3}$$

$$p_2^+ = \sum_{n=0}^{\infty} C_n \cos(\alpha_n \theta) \frac{J_{\alpha_n}(kr)}{J'_{\alpha_n}(kR)},$$
$$p_2^- = \sum_{n=1}^{\infty} D_n \sin(\beta_n \theta) \frac{J_{\beta_n}(kr)}{J'_{\beta_n}(kR)},$$

where $\alpha_n = n\pi/\phi$ and $\beta_n = (2n-1)\pi/2\phi$.

The conjugation conditions of the fields on the boundary of the partial domains look like this:

$$\frac{\partial p_1^\pm}{\partial r} = \begin{cases} \dfrac{\partial p_2^\pm}{\partial r}, & r = R,\ \theta = [0, \phi] \\ 0, & r = R,\ \theta = [\phi, \pi], \end{cases} \tag{9.4}$$

$$p_1^\pm = p_2^\pm, \quad r = R,\ \theta = [0, \phi]. \tag{9.5}$$

When solving the even problem, sets of coefficients A_n and C_n are defined, and during the solution of the odd problem, sets of coefficients B_n and D_n are defined.

9.2.2 Rounded Wedge

Let us construct a flat model of a wedge-shaped object in which the sharp edge of the wedge (Fig. 9.2) is rounded by a cylindrical surface. Figure 9.2 shows the geometry of such a wedge and all the necessary notations needed when solving the diffraction problem.

To construct the solution to the problem, let us introduce two polar coordinate systems: $rO\theta$ and $r_1O_1\theta_1$. The radius of the rounded wedge $a = O_1A = O_1B$ is given. The angle of the rounded wedge 2ϕ is defined as the angle of the corresponding wedge without rounding. From the rectangular triangle O_1AO, we determine the following geometric parameters of the model:

$$\gamma = \pi - \phi, \quad \gamma_1 = \pi/2 - \gamma, \quad b = \frac{\sin\gamma_1}{\sin\gamma} a, \quad d = OO_1 = \sqrt{a^2 + b^2}. \tag{9.6}$$

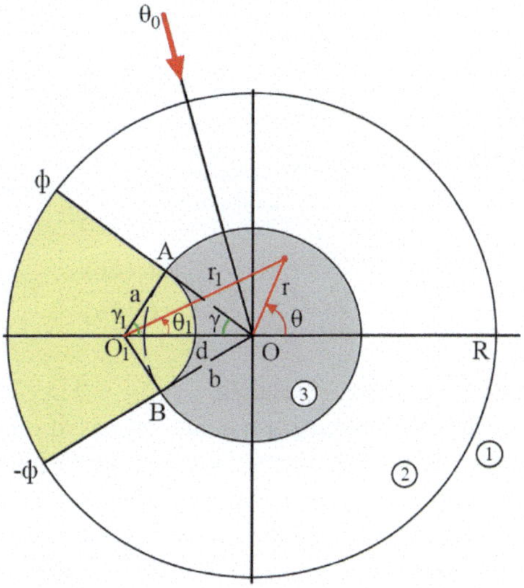

Fig. 9.2 The geometry of the finite rounded wedge model

According to the partial domain method, the entire sound field existence space is divided into three domains:

$$1 - r \geqslant R, \qquad -\pi \leqslant \theta \leqslant \pi;$$
$$2 - b \leqslant r \leqslant R, \quad -\phi \leqslant \theta \leqslant \phi;$$

3—the interior of the circle of radius b minus the intersection zone with the circle of radius a centered at point O_1.

The surface of the object, minus the rounded part ($r_1 = a, \ -\gamma_1 \leqslant \theta_1 \leqslant \gamma_1$), is assumed to be acoustically rigid. The rounding surface is characterized by the normal conductivity $Y = (1/\rho c)(1 - V)/(1 + V)$, which is defined through a given reflection coefficient at pressure V. The relative normal conductivity $\overline{Y} = \rho c Y = (1 - V)/(1 + V)$.

Let us divide the incident wave field $p_0 = \exp(-ikr\cos(\theta - \theta_0))$ into even and odd components, as written in (9.1). The pressure field in partial domain 1 is defined by (9.2).

The pressure field in partial domain 2:

$$p_2 = p_2^+ + p_2^-, \tag{9.7}$$

$$p_2^+ = \sum_{n=0}^{\infty} C_n \cos(\alpha_n\theta)\frac{J_{\alpha_n}(kr)}{J'_{\alpha_n}(kR)} + \sum_{n=0}^{\infty} D_n \cos(\alpha_n\theta)\frac{N_{\alpha_n}(kr)}{N'_{\alpha_n}(kb)},$$
$$p_2^- = \sum_{n=1}^{\infty} C_n^s \sin(\beta_n\theta)\frac{J_{\beta_n}(kr)}{J'_{\beta_n}(kR)} + \sum_{n=1}^{\infty} D_n^s \sin(\beta_n\theta)\frac{N_{\beta_n}(kr)}{N'_{\beta_n}(kb)},$$

where $\alpha_n = n\pi/\phi$ and $\beta_n = (2n - 1)\pi/2\phi$.

The pressure field in partial domain 3:

$$p_3 = p_3^+ + p_3^-, \quad p_3^\pm = p_3^{(1)\pm} + p_3^{(2)\pm}, \tag{9.8}$$

$$p_3^{(1)+} = \sum_{n=0}^{\infty} E_n \cos(n\theta)\frac{J_n(kr)}{J_n'(kb)}, \quad p_3^{(2)+} = \sum_{n=0}^{\infty} F_n \cos(\eta_n\theta_1)\frac{H_{\eta_n}^{(1)}(kr_1)}{H_{\eta_n}^{(1)\prime}(ka)},$$
$$p_3^{(1)-} = \sum_{n=1}^{\infty} E_n^s \sin(n\theta)\frac{J_n(kr)}{J_n'(kb)}, \quad p_3^{(2)-} = \sum_{n=1}^{\infty} F_n^s \sin(\mu_n\theta_1)\frac{H_{\mu_n}^{(1)}(kr_1)}{H_{\mu_n}^{(1)\prime}(ka)},$$

where $\eta_n = n\pi/\gamma_1$ and $\mu_n = (2n-1)\pi/2\gamma_1$.

The main results are achieved based on the partial domain method for the cases when partial domains do not overlap (Grinchenko et al. 2018). In Grinchenko (1996), an approach is described that can serve as a basis for considering a special class of problems admitting the construction of a general solution within the partial domain method in the absence of physically determined boundary conditions over the whole coordinate surface to which the complete system of functions is bound. According to this approach, physical boundaries are complemented by "non-physical" sections. Here it should be emphasized that on the "non-physical" sections of the boundary, the boundary conditions can be continued by arbitrary functions. In this case, quantitative characteristics of the wave field should not depend on the type of these functions. However, since in practical use of the approach, it is a question of solving infinite systems to which the reduction procedure is applied, a rational choice of conditions on "non-physical" boundaries can significantly improve the quality of the execution of the energy conservation law. Taking this into account, let us write the boundary conditions in the form:

$$\frac{\partial p_1^\pm}{\partial r} = \begin{cases} \dfrac{\partial p_2^\pm}{\partial r}, & r = R,\ \theta = [0, \phi], \\ 0, & r = R,\ \theta = [\phi, \pi]. \end{cases} \tag{9.9}$$

$$p_1^\pm = p_2^\pm,\ r = R,\ \theta = [0, \phi] \tag{9.10}$$

$$p_2^\pm = p_3^\pm,\ r = b,\ \theta = [0, \phi]. \tag{9.11}$$

$$\begin{cases} \dfrac{\partial p_3^\pm}{\partial r} = \dfrac{\partial p_2^\pm}{\partial r}, & r = b,\ \theta = [0, \phi], \\ \dfrac{\partial p_3^{(1)\pm}}{\partial r} = g(\theta), & r = b,\ \theta = [\phi, \pi]. \end{cases} \tag{9.12}$$

$$\frac{1}{i\omega\rho}\frac{\partial p_3^\pm}{\partial r_1} = -Yp_3^\pm,\ r_1 = a,\ \theta_1 = [0, \gamma_1]. \tag{9.13}$$

Since the function sets $\cos(n\theta)$ and $\sin(n\theta)$ have the property of orthogonality on the interval $[0, \pi]$, the boundary condition for the field ${p_3}^{(1)\pm}$ in (9.12) continues on the non-physical region $r = b,\ \theta = [\phi, \pi]$ as function $g(\theta)$. Let us assume that $g(\theta) = 0$ in the calculations.

The minus sign in (9.12) is due to the fact that the direction of the external normal (it coincides with the direction of the radial coordinate r_1) is opposite to the direction of wave propagation.

During the solution of the even problem, the sets of coefficients A_n, C_n, D_n, E_n, and F_n are defined, and during the solution of the odd problem, the sets of coefficients B_n, C_n^s, D_n^s, E_n^s, and F_n^s are defined. Passing from infinite series in (9.2), (9.7) and (9.8) to finite series, leave the number of coefficients A_n, B_n equal to N_1, coefficients C_n, C_n^s, D_n, D_n^s—N_2, E_n, E_n^s—N_3, F_n, F_n^s—N_4.

9.3 Scattering Cross Section

In the problems of wave scattering by objects, the most important characteristics are the scattering cross sections. These are the energy characteristics, which are determined in the far-field of scattering of the object. Let us recall the definitions of these characteristics, taking into account the fact that plane problems are considered:

- the total scattering cross section $\sigma_s(\theta_0)$ is equal to the ratio of the power P_s scattered by the object to the intensity of the plane harmonic wave I_0, incident on the object. The total scattering cross section $\sigma_s(\theta_0)$ is a function of the angle of incidence of the plane wave θ_0.

$$\sigma_s(\theta_0) = \frac{P_s}{I_0} = \frac{1}{I_0}\int_{-\pi}^{\pi} I_s(\theta) r d\theta, \tag{9.14}$$

where I_s is the intensity of the scattered wave;

- the positional scattering cross section $\sigma(\theta, \theta_0)$ is equal to the ratio of the power of a non-directional source with an intensity equal to the intensity of the scattered wave in a given direction θ to the intensity of a plane harmonic wave impinging on the object at an angle θ_0.

$$\sigma(\theta, \theta_0) = \frac{4\pi r^2 I_s(\theta)}{I_0}. \tag{9.15}$$

- if in (9.15) $\theta = \theta_0$, we obtain the backscattering cross section:

$$\sigma_L(\theta_0) = \sigma(\theta = \theta_0, \theta_0). \tag{9.16}$$

From (9.2), the expression for the scattering field follows:

$$p_s = \sum_{n=0}^{\infty} A_n \cos(n\theta) \frac{H_n^{(1)}(kr)}{H_n^{(1)\prime}(kR)} + \sum_{n=1}^{\infty} B_n \sin(n\theta) \frac{H_n^{(1)}(kr)}{H_n^{(1)\prime}(kR)}. \tag{9.17}$$

In the far-field ($kr \gg 1$), the asymptotic of the Hankel function is as follows:

$$H_n^{(1)}(kr) = \sqrt{\frac{2}{\pi kr}} \exp\left(ikr - in\frac{\pi}{2} - i\frac{\pi}{4}\right).$$

Given this asymptotic, (9.17) can be written in the form:

$$p_s = \sqrt{\frac{2}{\pi kr}} \exp\left(ikr - i\frac{\pi}{4}\right) L(\theta), \tag{9.18}$$

where

$$L(\theta) = \sum_{n=0}^{\infty} \frac{A_n}{H_n^{(1)\prime}(kR)} \cos(n\theta) \exp\left(-in\frac{\pi}{2}\right) + \sum_{n=1}^{\infty} \frac{B_n}{H_n^{(1)\prime}(kR)} \sin(n\theta) \exp\left(-in\frac{\pi}{2}\right). \tag{9.19}$$

Then the formula for the intensity of the scattering field in the far zone:

$$I_s = \frac{|p_s|^2}{2\rho c} = \frac{2I_0}{\pi kr} L(\theta)L^*(\theta), \tag{9.20}$$

where the intensity of the incoming plane wave $I_0 =| p_0 |^2 /(2\rho c) = 1/(2\rho c)$.

Substituting (9.20) into (9.14)–(9.16), we obtain the following expressions for the scattering cross sections:

$$\sigma_s(\theta_0) = \frac{2}{\pi k} \int_{-\pi}^{\pi} L(\theta)L^*(\theta)d\theta, \tag{9.21}$$

$$\sigma(\theta, \theta_0) = \frac{4}{k} L(\theta)L^*(\theta), \tag{9.22}$$

$$\sigma_L(\theta_0) = \frac{4}{k} L(\theta_0)L^*(\theta_0). \tag{9.23}$$

The unit of measure for all three scattering cross sections is a meter.

To make formulas (9.21)–(9.23) dimensionless, these expressions are divided by the midsection of the object. The midsection $M(\theta_0)$ is the cross-sectional area of the geometric shadow zone of the object. The midsection depends on the angle of incidence of the wave θ_0 on the object under study. The total scattering cross section cannot exceed the doubled value of the midsection (Shenderov 1989; Grinchenko et al. 2018), i.e., $\sigma_s(\theta_0) \leqslant 2M(\theta_0)$.

9.4 Construction of the Numerical Algorithm and Verification of the Fulfillment of Boundary Conditions

9.4.1 Sharp Wedge

Let us first consider a sharp wedge (Fig. 9.1). Substituting the expressions for the fields in the partial domains (9.1)–(9.3) into the conjugation conditions for the fields (9.4) and (9.5), we arrive at the following equations (we describe only the even problem):

$$k\sum_{n=0}^{\infty} A_n \cos(n\theta) + \left.\frac{\partial p_0^+(r,\theta)}{\partial r}\right|_{r=R} = \begin{cases} k\sum\limits_{n=0}^{\infty} C_n \cos(\alpha_n\theta),\ \theta = [0,\phi], \\ 0,\ \theta = [\phi,\pi]. \end{cases} \tag{9.24}$$

$$\sum_{n=0}^{\infty} A_n \frac{H_n^{(1)}(kR)}{H_n^{(1)'}(kR)} \cos(n\theta) + p_0^+(r=R,\theta) = \sum_{n=0}^{\infty} C_n \frac{J_{\alpha_n}(kR)}{J'_{\alpha_n}(kR)} \cos(\alpha_n\theta), \quad \theta = [0,\phi]. \tag{9.25}$$

The functions included in (9.24) and (9.25) depend on the spatial coordinate θ with the specified limits of its variation.

Then the transition from (9.24) and (9.25) to the infinite system of linear algebraic equations of the second kind follows (Grinchenko et al. 2018). It should be noted that the analytical representations of the sound field constructed within the partial domain method always exactly satisfy the wave equation for any number of terms held in the series. Therefore, the estimation of the accuracy of meeting conjugation conditions on the boundaries of partial domains should be taken as the basis of estimations of the accuracy of the problem's solution. It is evident that in order to increase the accuracy of sound field estimation, it is necessary to increase the number of unknown complex coefficients when solving the algebraic system of equations by the reduction method. This is especially important for the high-frequency domain of the noise spectrum.

Note that the efficiency of algorithms for solving infinite systems of algebraic equations can be ensured by taking into account the known singularities in the vicinity of corner points (in our case, it is the wedge edge). This makes it possible to obtain quantitative estimates of the characteristics of sound fields in domains as close as possible to the corner points. If, on the other hand, the field characteristics in the points distant from the corner points are of primary interest, then sufficient accuracy of the results can be ensured by using the method of simple reduction, keeping a certain number of equations in the system. A numerical experiment allows us to estimate the degree of fulfillment of the sound field conjugation conditions for the selected model parameters. Such a procedure is standard when using the partial domain method and was tested by the authors more than once (Grinchenko et al. 2018).

The transition from (9.24) and (9.25) to an infinite system of linear algebraic equations can be performed using the conjugation by points of the fields at the interface of partial domains (let us call these points nodal points) (Grinchenko et al. 2021).

So, let us restrict the infinite series in (9.24) and (9.25) by setting the number of coefficients A_n and C_n to N_1 and N_2, respectively. The starting point in calculations at conjugation by points of fields is normalized to the wavelength $\lambda = c/f$ distance between the nodal points Δ along the interface between partial domains 1 and 2, that is, the value Δ/λ. Based on the value Δ/λ, the number of nodal points $N_1 = [(\pi R)/\Delta]$ and $N_2 = [(\phi/\pi)N_1]$ is determined.

Let us denote the coordinates of nodal points at the interface between partial domains 1 and 2 by $(r_m = R, \theta_m),\ m = 0, 1, 2, \ldots, N_1$. We choose angular coordinates of nodal points $0 < \theta_m < \pi,\ m = 0, 1, 2, \ldots, N_1$ and write (9.24) and (9.25) as a finite system of linear algebraic equations of order $N_1 + N_2$:

$$k\sum_{n=0}^{N_1} A_n \cos(n\theta_m) + \left.\frac{\partial p_0^+(r, \theta_m)}{\partial r}\right|_{r=R} = \begin{cases} k\sum\limits_{n=0}^{N_2} C_n \cos(\alpha_n\theta_m), & \theta_m = [0, \phi], \\ 0, & \theta_m = [\phi, \pi],\ m = 0, 1, \ldots, N_1. \end{cases} \tag{9.26}$$

$$\sum_{n=0}^{N_1} A_n \frac{H_n^{(1)}(kR)}{H_n^{(1)\prime}(kR)} \cos(n\theta_m) + p_0^+(r = R, \theta_m) = \sum_{n=0}^{N_2} C_n \frac{J_{\alpha_n}(kR)}{J'_{\alpha_n}(kR)} \cos(\alpha_n\theta_m),\quad \theta_m = [0, \phi],\ m = 0, 1, \ldots, N_2. \tag{9.27}$$

To control the possibility of using the conjugation by points of the fields in the scattering problems under consideration, let us algebraize (9.24) and (9.25) using the classical approach of the root mean square approximation of the fields on the boundary of partial domains (Grinchenko et al. 2018, 2021). Wherein we will not distinguish between even and odd problems.

In order to correlate the calculations performed using the conjugation by points of the fields and the root mean square approximation, it is necessary, when using the root mean square approximation, to put the number of modes in domain 1 equal to $2N_1$, and the number of modes in domain 2 is $2N_2$ (doubling the number of modes is due to the absence of dividing the problem into even and odd ones). In this case, the order of the system of linear algebraic equations is $2(N_1 + N_2)$. In this approach, we multiply (9.24) and (9.25) by $\cos(m\theta)$ ($m = 0, 1, \ldots, 2N_1$) and $\cos(\alpha_n\theta)$ ($m = 0, 1, 2, \ldots, 2N_2$), respectively. Let us use the property of orthogonality of the given sets of trigonometric functions on the intervals stated in the conjugation conditions for fields (9.24) and (9.25). As a result, we have a system of linear algebraic equations of the second kind (Grinchenko et al. 2018), the order of which is determined by the number of coefficients A_n and C_n taken into account, namely, $2(N_1 + N_2)$. The explicit form of the system of linear algebraic equations, because of its obvious structure, will not be written out.

Since we are interested in the sharp wedge model with the presence of surfaces of large wave sizes, we choose the following model parameters: $R/\lambda = 10.2, \phi = 165°$.

Figure 9.3 illustrates the quality of fulfillment of the conjugation conditions on the boundary of partial domains 1 and 2 ($r = R, \ -\phi \leqslant \theta \leqslant \phi$) and the boundary condition on the rigid surface of the wedge-shaped object ($r = R, \ -\pi \leqslant \theta \leqslant -\phi \bigcup \phi \leqslant \theta \leqslant \pi$). As can be seen, the boundary conditions are well fulfilled, and the solutions obtained by the conjugation by points of the fields and using the root mean square approximation of the fields are well matched.

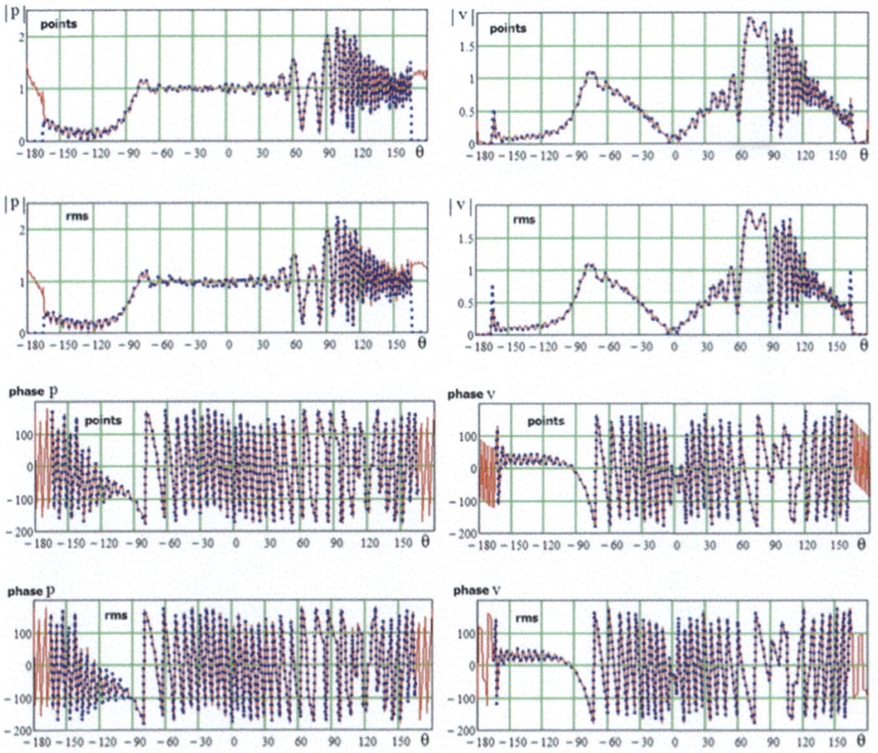

Fig. 9.3 Sharp wedge: illustration of conjugation conditions for the fields of pressure p and normal velocity υ in module and phase at the boundary of partial domains 1 and 2 ($r = R, \ -\phi \leqslant \theta \leqslant \phi$) and the boundary condition on the rigid surface $r = R, \ -\pi \leqslant \theta \leqslant -\phi \bigcup \phi \leqslant \theta \leqslant \pi, \ R/\lambda = 10.2, \ \phi = 165°, \ \theta_0 = 90°$: "points"—the conjugation by points, $\Delta/\lambda = 0.2, N_1 = 160, N_2 = 147$, "rms"—the root mean square approximation, $N_1 = 320, N_2 = 294$; lines and points correspond to the fields in domain 1 and 2, respectively

9.4.2 Rounded Wedge

Now let us see how the boundary conditions are fulfilled in the problem for the rounded wedge (Figs. 9.4, 9.5, and 9.6). All subsequent calculations were carried out using the conjugation by points of the fields with separation of the even and odd problems. Let us leave the parameters of the model $R/\lambda = 10.2,\ \phi = 165°$ for the sharp wedge. The rounding surface of the wedge (Fig. 9.2) is determined by the coordinates of $r_1 = a, -\gamma_1 \leqslant \theta_1 \leqslant \gamma_1$; at angle $\phi = 165°$ the value of $\gamma_1 = 75°$. Let us choose the radius of the wedge $a/\lambda = 0.8$. Let the rounding surface is characterized by the reflection coefficient $V = 0.1$, then its normalized conductivity $\bar{Y} = 0.818$.

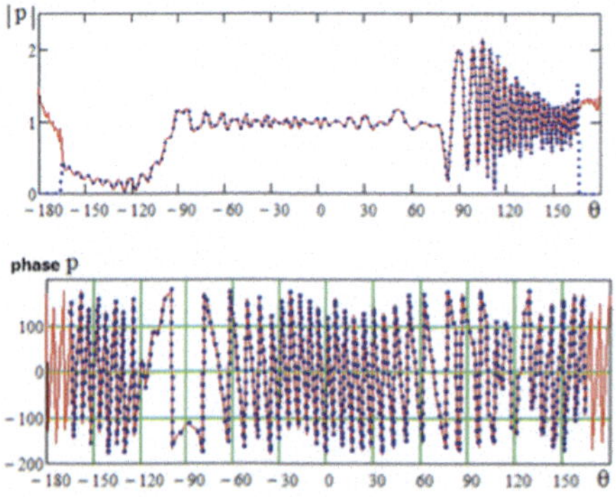

Fig. 9.4 Rounded wedge: illustration of conjugation conditions for the pressure p and normal velocity υ in module and phase at the boundary of partial domains 1 and 2 ($r = R,\ -\phi \leqslant \theta \leqslant \phi$) and the boundary condition on the rigid surface $r = R,\ -\pi \leqslant \theta \leqslant -\phi \bigcup \phi \leqslant \theta \leqslant \pi,\ R/\lambda = 10.2,\ \phi = 165°,\ \gamma_1 = 75°,\ a/\lambda = 0.8,\ V = 0.1,\ \theta_0 = 90°,\ \Delta/\lambda = 0.2,\ N_1 = 160,\ N_2 = 147,\ N_3 = 94,\ N_4 = 22$: lines and points correspond to the fields in domain 1 and 2 respectively

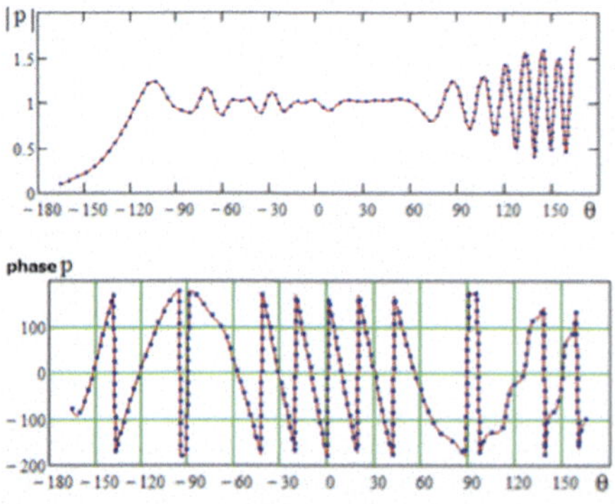

Fig. 9.5 Rounded wedge: illustration of conjugation conditions for the pressure p and normal velocity υ in module and phase at the boundary of partial domains 2 and 3 ($r = b, -\phi \leqslant \theta \leqslant \phi$), $\phi = 165°$: lines and points correspond to the fields in domain 2 and 3, respectively

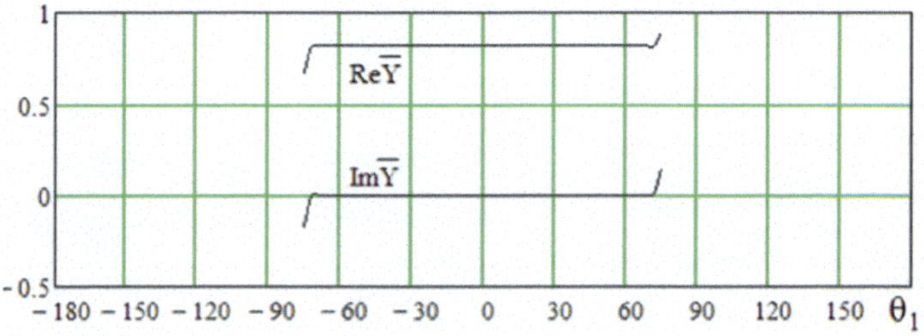

Fig. 9.6 Rounded wedge: real Re $\bar{Y}$ and imaginary Im $\bar{Y}$ parts of the normalized conductance $\bar{Y} = \rho c Y$ of the rounded wedge surface ($r_1 = b$, $-\gamma_1 \leqslant \theta_1 \leqslant \gamma_1$), $\gamma_1 = 75°$, $V = 0.1$, $\bar{Y} = 0.818$

All graphs in Figs. 9.4, 9.5 and 9.6 illustrate the good quality of fulfillment of the boundary conditions on the boundaries of the partial domains (1 and 2, 2 and 3), the boundary condition on the rigid surface of the wedge-shaped object ($r = R$, $-\pi \leqslant \theta \leqslant -\phi \bigcup \phi \leqslant \theta \leqslant \pi$) and the boundary condition on the rounded surface of the wedge.

9.5 Calculation of the Scattering Cross Section

Figure 9.7 shows plots of the total scattering cross section $\sigma_s(\theta_0)$ (curve 1) and inverse scattering cross section $\sigma_L(\theta_0)$ (curve 2) for three variants of the wedge-shaped object model. The scattering cross sections are normalized to the value of R. As can be seen, the value of σ_s (curve 1) over a wide range of angle changes θ_0 is practically equal to two for a sharp wedge and close to two for a rounded wedge. This is due to the large value of $R/\lambda = 10.2$.

For the backscattering cross section σ_L (curve 2), there are two highlights due to the normal incidence of the wave on the flat surface of an object of significant wave size ($R/\lambda = 10.2$). First, comparing Fig. 9.7a, b, it can be noted that the rounding of the wedge by a rigid surface significantly increases the backscattering cross section in the range of angles $-60° \leqslant \theta_0 \leqslant 60°$. At the same time, covering the rounded part of the wedge with absorbing material with an absorption coefficient $V = 0.1$ returns the situation close to a sharp wedge Fig. 9.7c.

Figure 9.8 shows plots, normalized to the value of R, of the two-position cross section of the scattering $\sigma(\theta, \theta_0)$ of the object from the angle θ, determining the direction of the scattered wave. The angle of incidence of the plane wave $\theta_0 = 0$. Two angle zones can be distinguished on the graphs: an illuminated zone in the range of angles approximately $-100° < \theta < 100°$ and a shadow zone $-180° < \theta < -100°$, $100° < \theta < 180°$. As can be seen, the illuminated zones for the sharp wedge (Fig. 9.8a) and the rounded with the absorbing surface rounded (Fig. 9.8c) are close in terms of scattering cross-sectional levels. This result is consistent with the backscattering cross-sectional plots in Fig. 9.7a, c. The rigid rounding of the

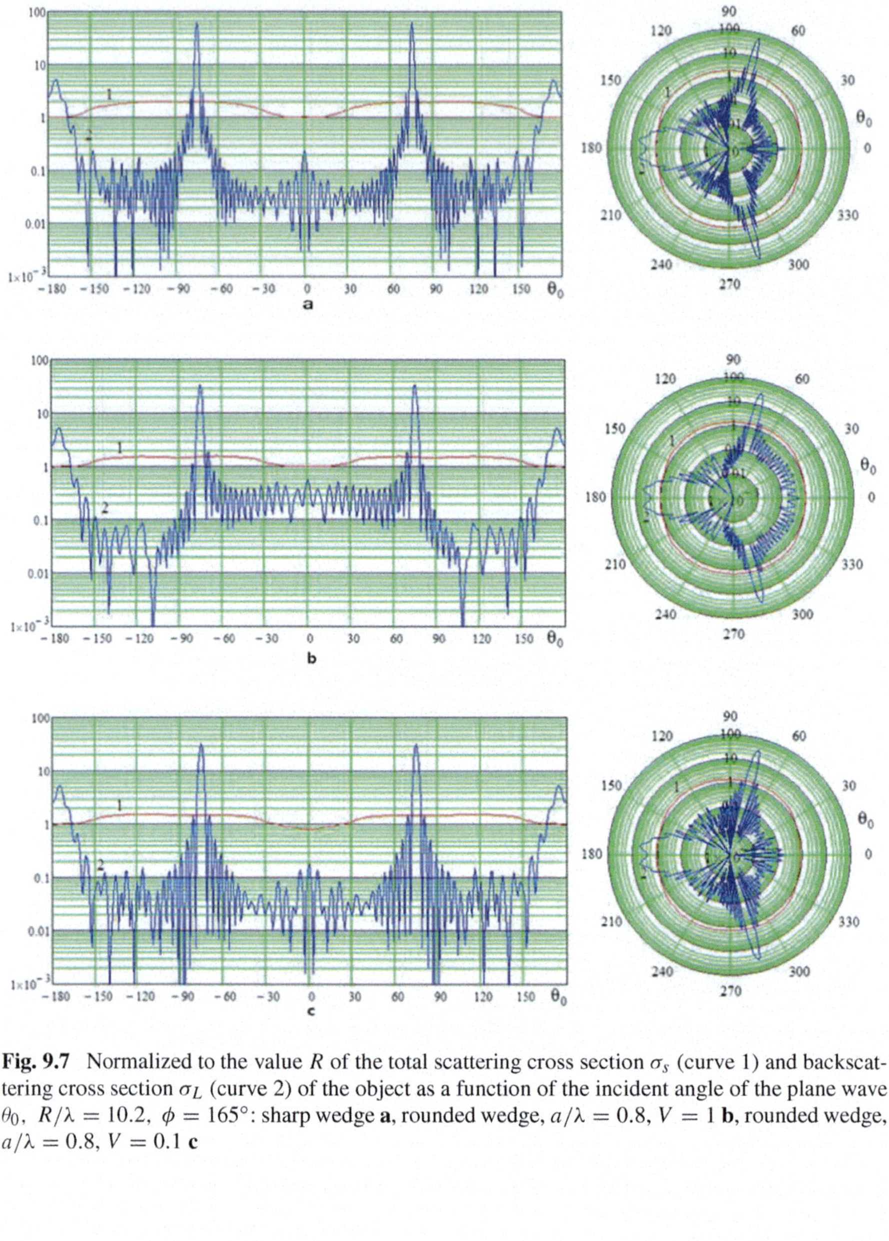

Fig. 9.7 Normalized to the value R of the total scattering cross section σ_s (curve 1) and backscattering cross section σ_L (curve 2) of the object as a function of the incident angle of the plane wave θ_0, $R/\lambda = 10.2$, $\phi = 165°$: sharp wedge **a**, rounded wedge, $a/\lambda = 0.8$, $V = 1$ **b**, rounded wedge, $a/\lambda = 0.8$, $V = 0.1$ **c**

wedge surface (Fig. 9.8b) noticeably increases the scattering cross section in the angle range $-100° < \theta < 100°$. As for the shadow zone, all three variants have significant shadow lobes due to the large wave size $R/\lambda = 10.2$ of the wedge-shaped object.

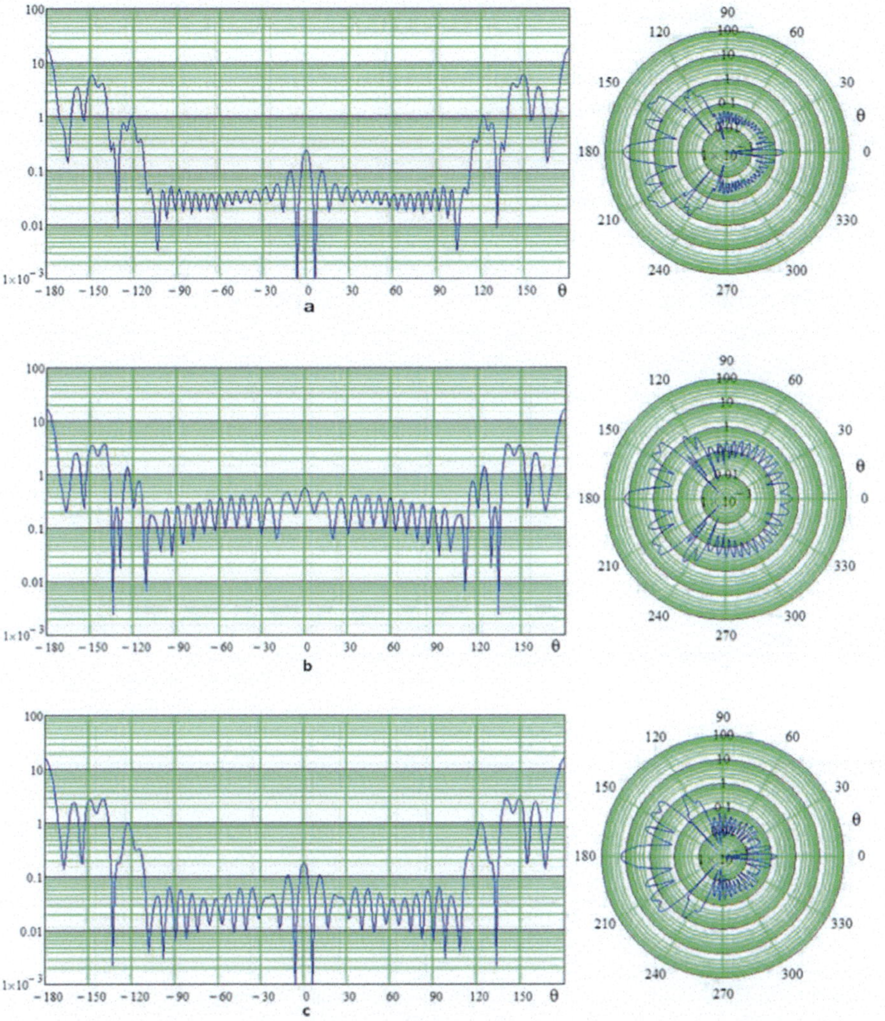

Fig. 9.8 Normalized to the value of R the two-position cross section of the scattering $\sigma(\theta, \theta_0)$ of the object from the angle θ, determining the direction of the scattered wave; the angle of incidence of the wave $\theta_0 = 0$, $R/\lambda = 10.2$, $\phi = 165°$: sharp wedge **a**, rounded wedge, $a/\lambda = 0.8$, $V = 1$ **b**, rounded wedge, $a/\lambda = 0.8$, $V = 0.1$ **c**

9.6 Conclusions

Based on the method of partial domains, the rigorous solution to the problem of plane wave scattering on finite wedge-shaped objects (a sharp and a rounded wedge) in the presence of surfaces of large wave size is constructed. The obtained rigorous solutions to the problems of wave scattering on finite wedge-shaped bodies are of interest because the shape of such bodies is close to the shape of individual parts of aircraft, surface, and space objects.

The reliability of the solution is conditioned by the quality control of conjugation conditions on the boundaries of partial domains and the boundary conditions on the corresponding surfaces of the model. Two variants of field conjugation were calculated for a sharp wedge: the root mean square approximation and the conjugation by points of the fields. A high degree of coincidence of the calculation results in the two variants is shown.

Calculations of the total, backscattering, and two-position scattering cross sections have been carried out. It is shown that the value of the total scattering cross section in a wide range of changes in the angle of incidence of a plane wave is practically equal to two for a sharp wedge and close to two for a rounded wedge. This is due to the presence of large wave size surfaces in wedge-shaped objects.

It is shown that the backscattering cross section depends significantly on the radius of the rounded edge and the normal conductivity of the rounded surface. At the same time, covering the rounded part of the wedge with an absorbing material significantly reduces the reflectivity of the object at the corresponding angles of incidence of the wave.

The use of the proposed method for solving scattering problems makes it possible to evaluate the potential possibilities of the method for monitoring the radar cross section by changing the geometry of the scatterer (Zohuri 2020).

References

Alekseev A, Shtager E, Kozyirev S (2007) Physical basis of stealth technology. VVM, SPb

Chatzigeorgiadis F (2004) Development of code for a physical optics radar cross section prediction and analysis application. Naval Postgraduate School, Monterey

Grinchenko V (1996) Development of the method for solving problems of sound radiation and scattering in noncanonical regions. Hydromechanics 70:27–40

Grinchenko V, Vovk I, Matsypura V (2018) Acoustic wave problems. Begell House, New York

Grinchenko V, Husak V, Matsypura V (2021) Using two variants of conjugation of the wave fields in the method of partial domains. J Numer Appl Math 136(2):5–16

Vozhdaev V, Teperin L (2018) Characteristics of the radar visibility of aircraft. Fizmatlit

Lagarkov A, Pogosyan M (2003) Fundamental and applied problems of stealth technologies. Bull RAS 73(9):779–787

Lvova L (2003) Radar visibility of aircraft. RFYaTs-VNIITF, Snezhinsk

Shenderov E (1989) Sound radiation and scattering. Sudostroenie

Suharevskiy O, Vasilets V, Kukobko S (2009) Scattering of electromagnetic waves by air and ground radar objects. Ivan Kozhedub Kharkiv National Air Force University, Kharkiv

Ufimtsev P (2007) Fundamentals of the physical theory of diffraction. Hoboken, New Jersey

Grinchenko V, Husak V, Matsypura V (2021) Using two variants of conjugation of the wave fields in the method of partial domains. J Numer Appl Math 136(2):5–16

Zohuri B (2020) Radar energy warfare and the challenges of stealth technology. Springer

Chapter 10
Heat and Stress State of Coated Bodies Under Dry Friction Taking into Account Radiation Heat Exchange

Oleksandr Hachkevych, Rostyslav Terlets'kyi, and Orest Gumenchuk

10.1 Introduction

The constantly increasing level of load-speed modes of operation of machinery and the need to increase its resource requires an increase in the physical-mechanical, tribotechnical, and other properties of materials of machine parts. Since the strengthening or modification of the surface layer allows to prevent the origin of deformation defects in them and subsequent destruction (wear), the formation of protective coatings on the surface is a promising way to create high-strength and wear-resistant materials with increased physical and mechanical properties and resistance to destruction in the process of frictional interaction and working in difficult operating conditions.

For the creation of such materials, it is important to develop theoretical studies aimed at studying the thermomechanical behavior of layered structural elements (in particular, with coatings) under conditions of intense thermal load at friction, the action of electromagnetic radiation of the light range (in particular, thermal) and heat transfer at high temperatures. For such studies, it is necessary to develop model representations of the process of deformation of layered bodies, taking into account the physical processes of radiation and heat propagation, as well as the development of specific models of thermomechanics.

O. Hachkevych · R. Terlets'kyi (✉) · O. Gumenchuk
Pidstryhach Institute for Applied Problems of Mechanics and Mathematics, Lviv, Ukraine
e-mail: teroxa55@gmail.com

O. Hachkevych
e-mail: dept13@iapmm.lviv.ua

O. Gumenchuk
e-mail: or1961@ukr.net

A. N. Guz et al. (eds.), *Advances in Mechanics*, Advanced Structured Materials 191,
https://doi.org/10.1007/978-3-031-37313-8_10

10.2 Statement of the Problem on the Study of Thermal and Stress States

Due to the roughness of the surfaces of real solids, rubbing bodies contact each other through individual actual spots of contact, that is, the frictional contact is generally not continuous, but discrete. Dry friction heat generation occurs in very thin surface layers of individual roughnesses in contact. From the actual contact spots, heat propagates deep into both contacting bodies. The actual contact spots have different shapes and sizes. If all this is taken into account, then even the statement of the thermal problem of heating bodies under the action of friction heat will be very difficult. Therefore, in the heat dynamics of dry friction, simplified computational models are used to study the thermal regime of friction (Amosov 2011).

One of them is a one-dimensional model of continuous frictional contact. Here, the friction heat dissipation is considered to be uniformly distributed over the entire area of the nominal contact. The surface temperature of rubbing bodies calculated under this assumption is called the average surface temperature. This model is realized at large pressing forces of rubbing bodies, when the area of actual contact becomes a significant part of the nominal friction contact area, or at significant friction times, when the temperature of all friction areas becomes close to the temperature of the actual contact spots. In simplified models, the contacting bodies are considered to be half-spaces. Their use is permissible only in the case of short-term processes, when the friction time is less than the warm-up time of each of the friction bodies. In this case, heat exchange with the environment can also be ignored.

10.2.1 Problem Statements on the Study of Thermal State

We consider a friction pair (tribological system) formed by two bodies. Body 1 is homogeneous. Body 2 is layered, consisting of the main body (substrate) and the coating applied to it to increase wear resistance and thermal stability. The system of bodies is subjected to a load that ensures their pressing (contact) and causes a certain load on the interface (contact). One of the bodies moves (slides) on the surface of the other. The system is in conditions of heat exchange with the external environment, in particular thermal radiation heat transfer with surrounding bodies.

In order to study the thermal state in coated bodies during operation (at dry friction) or at high-temperature process heating, consider the layered system shown in Fig. 10.1. It consists of layer 1 of thickness h_1 and layer 2 of thickness h_2 on which a coating (layer 0) of thickness h_0 is applied. When modeling friction processes and thermal processes, we will proceed from the calculation model described above (one-dimensional model of continuous friction contact, in which friction heat is considered to be uniformly distributed over the entire surface of the nominal contact, which is limited by the contour of the friction surface).

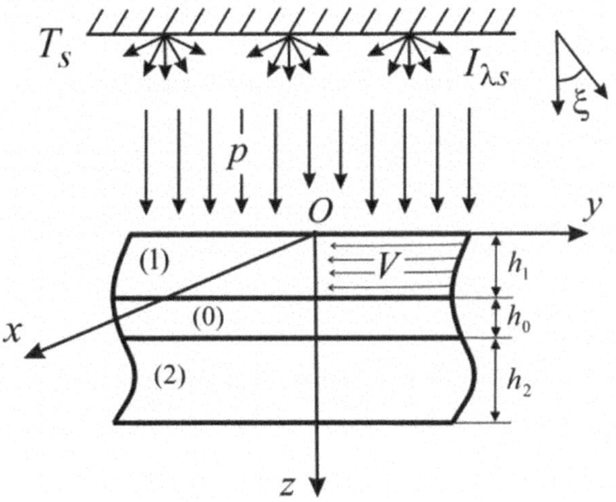

Fig. 10.1 Tribological system in conditions of thermal radiation heat transfer. Layer 2 (substrate) with a coating of 0 and layer 1 slides with velocity V under the action of a distributed pressing load p. $I_{\lambda s}$ is the intensity of thermal radiation from a parallel to the layers of the heated isothermal surface whose temperature T_s

A pressing distributed load p is applied to the surface $z = 0$, which causes a distributed load p_1 on the surface of the nominal contact. Layer 1 slides on the surface $z = h_1$ of the nominal contact with a speed of V. Then on this surface, there is heat generation with specific power $q_{fr} = \tau_{fr} V$, where the friction force will be $\tau_{fr} V = f p_1$ (Amosov 2011). The temperature dependence of the friction coefficient f can be assumed linear, that is $f = f_0 [1 - A(T - T_0)]$, where T_0 is the initial temperature in the system. If we do not take into account the temperature dependence, then $A = 0$. Here f_0 is the friction coefficient at the initial temperature T_0.

Layer 1 can be exposed to thermal radiation of intensity $I_{\lambda s}(\nu)$ from parallel to the layers of the heated isothermal surface, the temperature of which T_s is assumed to be given Hachkevych et al. (2017, 2022). Here $\nu = \cos\xi$, and ξ are acute angles that form the directions of incidence of rays from the surface with a positive direction of the axis z directed in the direction of the outer normal to layer 2, and λ is the wavelength of the radiation. For diffuse radiation (Modest 2003; Siegel and Howell 1972), such intensity can be given in the form Hachkevych et al. (2022); Terlets'kyi and Brukhal' (2017)

$$I_{\lambda s} = k \frac{2\pi c_1}{\lambda^5 \exp(c_2/\lambda T_s - 1)},$$

where $k = \text{const}$ is the coefficient by which the characteristics of the real radiation source and its location relative to the body can be taken into account, c_1 and c_2 are known constants. The system as a whole is in conditions of convective heat exchange with the external environment (Hachkevych et al. 2022). From the point of view of absorption properties, all layers are considered to be opaque to thermal radiation (different variants of layer transparency can be considered (Hachkevych et al. 2017; Modest 2003; Siegel and Howell 1972)).

According to the used friction model, we assume the conditions of ideal thermal contact (Terlets'kyi and Turii 2013) on the surface $z = h_1$ of the coating 0 and layer 1. We also assume perfect thermal contact at the interface $z = h_1 + h_0$ between the coating and the substrate (layer 2).

Then the thermal state in the system is described by the equations of thermal conductivity for the composite layers, which at constant densities ρ_j and thermophysical characteristics of the materials of the layers (coefficients of thermal conductivity κ_j, heat capacity $c_{\varepsilon j}$, $j = 0, 1, 2$) to determine the deviations $\theta_j = T_j - T_0$ of the temperature in the plate from the initial one have the form Terlets'kyi and Brukhal' (2017):

$$\frac{\partial^2 \theta_j (z, t)}{\partial z^2} - \frac{1}{\varpi_j} \frac{\partial \theta_j (z, t)}{\partial t} = 0, \quad j = 1, 0, 2, \tag{10.1}$$

where $\varpi_j = \kappa_j / \rho_j c_{\varepsilon j}$ are coefficients of temperature conductivity.

The system of Eq. (10.1) is supplemented with boundary conditions that define the convective heat exchange of the system with the external environment on the surfaces $z = 0$, $h = h_1 + h_0 + h_2$ (taking into account the absorption and emission of heat energy) and contact conditions that express the thermal equilibrium at the interface $z = h_1$, $z = h_1 + h_0$ between the layer 1 and coating and coating and layer 2, as well as heat generation due to friction.

The boundary and contact conditions for the system are as follows:

$$\begin{aligned} \kappa_1 \frac{\partial \theta_1(0, t)}{\partial z} &- \alpha_{s1} \left[\theta_1(0, t) + T_0 - T_1^{\text{ext}}(t) \right] \\ &= -k \left(1 - R^{(1)}\right) \sigma T_s^4 + \varepsilon^{(1)} \sigma \left[\theta_1(0, t) + T_0 \right]^4, \end{aligned} \tag{10.2}$$

$$\kappa_1 \frac{\partial \theta_1(h_1, t)}{\partial z} + q_{fr} = \kappa_0 \frac{\partial \theta_0(h_1, t)}{\partial z}, \quad \theta_1(h_1, t) = \theta_0(h_1, t), \tag{10.3}$$

$$\begin{aligned} \kappa_0 \frac{\partial \theta_0(h_1 + h_0, t)}{\partial z} &= \kappa_2 \frac{\partial \theta_2(h_1 + h_0, t)}{\partial z}, \\ \theta_0(h_1 + h_0, t) &= \theta_2(h_1 + h_0, t), \end{aligned} \tag{10.4}$$

$$\begin{aligned} \kappa_2 \frac{\partial \theta_2(h, t)}{\partial z} &+ \alpha_{s2} \left[\theta_2(h, t) + T^{(0)} - T_2^{\text{ext}}(t) \right] \\ &+ \varepsilon^{(2)} \sigma \left[\theta_2(h, t) + T_0 \right]^4 = 0, \end{aligned} \tag{10.5}$$

where α_{s1} and α_{s2} are heat transfer coefficients from the surfaces of layers 1 and 2 to the external environment (air), $T_1^{\text{ext}}(t)$ and $T_2^{\text{ext}}(t)$ are air temperatures in the regions $z < 0$ and $z > h$ the external environment (can be assumed equal to the initial temperature T_0). Besides, $\varepsilon^{(1)}$ and $\varepsilon^{(2)}$ are hemispherical total emissivity of the surfaces of layer 1 and layer 2, $R^{(1)}$ is the reflection coefficient of the surface of layer 2 ($1 - R^{(2)} = \varepsilon^{(2)}$), and σ is the Stefan Boltzmann constant (Modest 2003; Siegel and Howell 1972).

Note that if external irradiation is absent in relation (10.2) it is necessary to put $T_s = 0$.

On the basis of the obtained relations (10.1)–(10.5), we can write down the dependences describing the thermal state of layer 2 with coating 0 for the problem of technological heat treatment using thermal radiation. In this case, layer 1 is omitted, and the origin of coordinates is transferred to the surface of the coating. They include Eq. (10.1), where $j = 0, 2$ and the following boundary and contact conditions

$$\begin{aligned}
&\kappa_0 \frac{\partial \theta_0(0,t)}{\partial z} - \alpha_{s0}\,\theta_0(0,t) = -k\left(1 - R^{(0)}\right)\sigma T_s^4 + \varepsilon^{(0)}\sigma\left[\theta_0(0,t) + T_0\right]^4,\\
&\kappa_0 \frac{\partial \theta_0(h_0,t)}{\partial z} = \kappa_2 \frac{\partial \theta_2(h_0,t)}{\partial z}, \quad \theta_0(h_0,t) = \theta_2(h_0,t),\\
&\kappa_2 \frac{\partial \theta_2(h,t)}{\partial z} + \alpha_{s2}\left[\theta_2(h,t) + T^{(0)} - T_2^{\text{ext}}(t)\right] + \varepsilon^{(2)}\sigma\left[\theta_2(h,t) + T_0\right]^4 = 0,
\end{aligned} \tag{10.6}$$

where $h = h_0 + h_2$, $\varepsilon^{(0)}$ is the hemispherical total emissivity of the coating surface, and $R^{(0)}$ is the reflection coefficient.

10.2.2 Generalized Conditions of Heat Transfer Through a Thin Layer

Applying the approach described in Shevchuk (1996) (using certain assumptions about the distribution of temperature or heat fluxes through the thickness of a thin intermediate layer or the formal operator method (Podstrigach and Shvets 1978; Terlets'kyi and Turii 2013)), the study of the thermal behavior of a three-layer system can be reduced to the study of a two-layer system.

We present the generalized conditions of heat transfer through a thin interlayer, which we use to model the coating. In the considered case when all layers are opaque generalized contact thermal conditions will have the form:

$$2\left(\kappa_2 \frac{\partial \theta_2}{\partial z} - \kappa_1 \frac{\partial \theta_1}{\partial z} - q_{fr}\right) = C \frac{\partial\left(\theta_1 + \theta_2\right)}{\partial t}, \tag{10.7}$$

$$6\left(\kappa_2 \frac{\partial \theta_2}{\partial z} + \kappa_1 \frac{\partial \theta_1}{\partial z} + q_{fr}\right) - 12D\left(\theta_2 - \theta_1\right) = C \frac{\partial\left(\theta_1 - \theta_2\right)}{\partial t}. \tag{10.8}$$

Here $C = h_0 \rho_0 c_\varepsilon$ is the reduced heat capacity of the intermediate layer, $D = \kappa_0 / h_0$ ($1/D$ is thermal resistance).

Thus, the thermal state in the considered system can be described approximately by the system of two equations of thermal conductivity (10.1) ($j = 1, 2$) under boundary ((10.2) and (10.5)) and generalized ((10.7) and (10.8)) contact conditions.

It is also possible to obtain generalized boundary conditions (conditions of heat transfer through the coating) in the case of layer 2 with coating 0 for the problem of technological heat treatment. Then the thermal state in the layer is described by Eq. (10.1) ($j = 2$) with the following generalized boundary condition (Podstrigach and Shvets 1978)

$$\bar{\lambda}\frac{\partial\theta_2(0,t)}{\partial t} - \left[\kappa\frac{\partial\theta_2(0,t)}{\partial z} - k\left(1 - R^{(0)}\right)\sigma T_s^4\right]$$
$$-\alpha_{s0}\theta_2(0,t) + \sigma\varepsilon^{(0)}\left[\theta_2(0,t) + T_0\right]^4 = 0$$

and condition (10.6). Here $\bar{\lambda} = \kappa_0 h_0$ is the reduced thermal conductivity, and $R^{(0)}$ is the radiation characteristics of the coating surface.

10.2.3 Formulation of Problems on the Study of the Stress State

In the study of the stress state in the considered three-layer system to evaluate the effect of frictional heating (q_{fr}) and distributed pressing load p on the stress state, the components ${\sigma_{ik}}^{(j)}$ ($j = 1, 0, 2$; $i, k = x, y, z$) of the stress tensor in the layers are presented as the sum of two components ${\sigma_{ik}}^{(j)q}$ and ${\sigma_{ik}}^{(j)p}$due to heating and power load, respectively.

The stress state corresponding to the temperature distributions in three-layer infinite plates is described by the relations of the temperature problem of the elasticity theory formulated with respect to the components of the stress tensor (Hachkevych et al. 2017, 2022; Terlets'kyi and Brukhal' 2017). It is obtained (under conditions of ideal mechanical contact of layers and absence of force loads on their surfaces $z = 0, h$)

$$\sigma_{xx}^{(j)q} = -\frac{E_j}{1-\nu_j}\varphi_j + C_1^{(j)}z + C_2^{(j)},$$
$$\sigma_{yy}^{(j)q} = -\frac{E_j}{1-\nu_j}\varphi_j + C_3^{(j)}z + C_4^{(j)},$$
$$\sigma_{zz}^{(j)} = 0,$$

where $\varphi_j = \alpha_t^{(j)}\theta_j$; $E_j, \nu_j, \alpha_t^{(j)}$($j = 0,\ 1,\ 2$) are Young's modulus, Poisson's ratio and linear coefficient of thermal expansion, respectively. In this case, if the edges are equally fixed ${C_1}^{(j)} = {C_3}^{(j)}$, ${C_2}^{(j)} = {C_4}^{(j)}$ and

$$\sigma_{xx}^{(j)q} = \sigma_{yy}^{(j)q} = -\frac{E_j}{1-\nu_j}\varphi_j + C_1^{(j)}z + C_2^{(j)}, \quad \sigma_{zz}^{(j)q} = 0. \tag{10.9}$$

The integration constants $C_k^{(j)}$ $(k = 1, 2, 3, 4)$ are determined from the conditions of fixing the edges $x, y \to \pm\infty$ of the plate (Hachkevych et al. 2022; Terlets'kyi and Brukhal' 2017) and the conditions of mechanical contact at the interface. If both edges are free from external forces and moments, then we have a system of equations to find $C_k^{(j)}(k = 1, 2)$:

$$C_1^{(1)}\left(-\frac{h_1}{2}(h_1 + h_0) + \frac{h_2(h_2 + h_0)}{K_{12}}\right) + C_2^{(1)}\left(h_1 - \frac{h_0}{K_{10}} + \frac{h_2}{K_{12}}\right) = S,$$

$$\begin{aligned} C_1^{(1)}\left(\frac{h_1^3}{3} + \frac{h_1 h_0}{2}\left(h_1 + \frac{h_0}{2}\right) + \frac{h_0^3}{12K_{10}} + \frac{1}{K_{12}}\left(\frac{h_1 h_0}{2}\left(h_1 + \frac{h_0}{2}\right) + \frac{h_2^3}{3}\right)\right) \\ + C_2^{(1)}\left(-\frac{h_1}{2}(h_1 + h_0) + \frac{h_2(h_2 + h_0)}{2K_{12}}\right) \end{aligned} = P,$$

$$C_1^{(0)} = \frac{C_1^{(1)}}{K_{10}}, \quad C_2^{(0)} = \frac{C_2^{(1)}}{K_{10}}, \quad C_1^{(2)} = \frac{C_1^{(1)}}{K_{12}}, \quad C_2^{(2)} = \frac{C_2^{(1)}}{K_{12}}.$$

Here

$$K_{10} = \frac{E_1}{(1 - \nu_1)}\frac{(1 - \nu_0)}{E_0}, \quad K_{02} = \frac{E_0}{(1 - \nu_0)}\frac{(1 - \nu_2)}{E_2}, \quad K_{12} = \frac{E_1}{(1 - \nu_1)}\frac{(1 - \nu_2)}{E_2},$$

$$S = \frac{E_1}{1 - \nu_1}\int_0^{h_1}\varphi_1 dz + \frac{E_0}{1 - \nu_0}\int_{h_1}^{h_1 + h_0}\varphi_0 dz + \frac{E_2}{1 - \nu_2}\int_{h_1 + h_0}^{h}\varphi_2 dz,$$

$$\begin{aligned} P = \frac{E_1}{1 - \nu_1}\int_0^{h_1}\left(z - h_1 - \frac{h_0}{2}\right)\varphi_1 dz \\ + \frac{E_0}{1 - \nu_0}\int_{h_1}^{h_1 + h_0}\left(z - h_1 - \tfrac{h_0}{2}\right)\varphi_0 dz \\ + \tfrac{E_2}{1 - \nu_2}\int_{h_1 + h_0}^{h}\left(z - h_1 - \tfrac{h_0}{2}\right)\varphi_2 dz. \end{aligned}$$

If the edges are rigidly pinched

$$C_2^{(0)} = \frac{S}{K_{10}h_1 + K_{02}h_2 + h_0}, \quad C_2^{(1)} = K_{10}C_2^{(0)}, \quad C_2^{(2)} = K_{02}C_2^{(0)}. \tag{10.10}$$

With rigid fixing of edges, $C_k^{(j)} = 0$ $(k = 1, 2, 3, 4)$ and

$$\sigma_{xx}^{(j)q} = \sigma_{yy}^{(j)q} = -\frac{E_j}{1 - \nu_j}\varphi_j.$$

The stresses due to the force load are found from the equation

$$\frac{\partial^2 \sigma_{zz}^{(j)p}}{\partial z^2} = 0. \tag{10.11}$$

Under the following boundary conditions

$$\sigma_{zz}^{(1)p}(0,\,t) = -p,\quad \sigma_{zz}^{(2)p}(h,\,t) = 0. \tag{10.12}$$

Contact conditions that ensure equality of displacement are as follows:

$$\begin{gathered}
\sigma_{zz}^{(1)p}(h_1,t) = \sigma_{zz}^{(0)p}(h_1,t),\\
\frac{\partial\sigma_{zz}^{(1)p}(h_1,t)}{\partial z} = \frac{\partial\sigma_{zz}^{(0)p}(h_1,t)}{\partial z},\\
\sigma_{zz}^{(1)p}(h_1+h_0,t) = \sigma_{zz}^{(0)}(h_1+h_0,t),\\
\frac{\partial\sigma_{zz}^{(1)p}(h_1+h_0,t)}{\partial z} = \frac{\partial\sigma_{zz}^{(0)p}(h_1+h_0,t)}{\partial z}.
\end{gathered} \tag{10.13}$$

The solution of Eq. (10.11) under boundary and contact conditions (10.12), (10.13) has the form

$$\sigma_{zz}^{(j)p} = \frac{p}{h}z - p,\quad \sigma_{xx}^{(j)p} = \sigma_{yy}^{(j)p} = \frac{\nu_j}{1-\nu_j}\sigma_{zz}^{(j)p},\quad j = 0,1,2. \tag{10.14}$$

From expression (10.14) we find the distributed load $p_1$on the surface $z = h_1$ of the nominal contact $p_1 = p(1 - h_1/h)$.

Similarly, we obtain expressions for the thermal stresses for the considered case of irradiated layer 2 with a coating 0. In this case, the stresses are determined by the relations (10.9) at $j = 0, 2$, where the expressions for the integration constants are:

- both edges are free from external forces and moments:

$$\begin{aligned}
C_1^{(0)} &= 6K_{02}\left[\left(-K_{02}h_0^2 + h_2^2\right)S - \left(K_{02}h_0 + h_2\right)P\right]/L,\\
C_2^{(0)} &= 2K_{02}\Big[\left(3K_{02}h_0^2 + h_2\left(2h_0 + h_2\right)\right)P - \left(-K_{02}h_0^3 + h_2^2\left(3h_0 + 2h_2\right)\right)S\Big]/L,\\
C_1^{(2)} &= C_1^{(0)}/K_{02},\quad C_2^{(2)} = C_2^{(0)}/K_{02}.
\end{aligned}$$

$$\begin{aligned}
L &= 3\left[K_{02}h_0^2 + h_2\left(2h_0 + h_2\right)\right]\left(-K_{02}h_0^2 + h_2^2\right)\\
&\qquad -2\left[-K_{02}h_0^3 + h_2^2\left(3h_0 + 2h_2\right)\right]\left(K_{02}h_0 + h_2\right),\\
P &= \frac{E_0}{1-\nu_0}\int_0^{h_0}(z-h_0)\theta_0 dz + \frac{E_2}{1-\nu_2}\int_{h_0}^{h}(z-h_0)\theta_2 dz,\\
S &= \frac{E_0}{1-\nu_0}\int_0^{h_0}\varphi_0 dz + \frac{E_2}{1-\nu_2}\int_{h_0}^{h}\varphi_2 dz,\quad h = h_0 + h_2.
\end{aligned}$$

- both edges are rigidly pinched:

$$C_1^{(j)} = 0, \quad C_2^{(0)} = \frac{S}{h_0 + h_2/K_{02}},$$
$$C_2^{(2)} = C_2^{(0)}/K_{02}, \quad K_{02} = \frac{E_0}{(1-\nu_0)} \frac{(1-\nu_2)}{E_2}.$$

10.2.4 Approximate Expressions for Thermal Stresses

As for a two-layer plate (Hachkevych et al. 2017; Terlets'kyi and Turii 2013), for the considered three-layer plate (tribological system), it is also possible to obtain approximate expressions for thermal stresses in layers 1 and 2 under different conditions of fixing the edges at infinity. To do this, let us pass in the found corresponding stress expressions to the boundary while $h_0/2 \to 0$ maintaining the reduced stiffness characteristics for bending $g_0^* = K_{02}h_0^3/2$ and tension $g_0 = 2\,K_{02}h_0$ of the intermediate layer (coating). Then in the case of a plate with edges free from forces and moments, the constants $C_1{}^{(j)}$ and $C_2{}^{(j)}$ $(j = 1, 2)$ included in the expressions for stresses will be:

$$C_1^{(1)} = 6K_{12}\left[\left(h_2^2 - K_{12}h_1^2\right)S - \left(2K_{12}h_1 - g_0 + h_2\right)P\right]/M,$$
$$C_2^{(1)} = 2K_{12}\left[3\left(h_2^2 - K_{12}h_1^2\right)P - \left(2K_{12}h_1^3 + g_0^* + 2h_2^3\right)S\right]/M,$$
$$C_1^{(2)} = C_1^{(1)}/K_{12}, \quad C_2^{(2)} = C_2^{(1)}/K_{12},$$
$$M = 3\left(h_2^2 - K_{12}h_1^2\right)^2 - \left(2K_{12}h_1^3 + g_0^* + 2h_2^3\right)\left(2K_{12}h_1 - g_0 + h_2\right).$$

In case of fixed support, constants $C_1{}^{(j)}$ and $C_2{}^{(j)}$ $(j = 1, 2)$ have the form:

$$C_2^{(1)} = K_{10}C, \quad C_2^{(2)} = K_{02}C, \quad C = \frac{S}{K_{10}h_1 + K_{02}h_2}.$$

We also give approximate expressions for thermal stresses for the case of a coated plate. If the edges are free from forces and moments, then

$$C_2^{(1)} = K_{10}C, \quad C_2^{(2)} = K_{02}C, \quad C = \frac{S}{K_{10}h_0 + K_{02}h_2}.$$

In case of fixed support,

$$C_1^{(2)} = 0, \quad C_2^{(2)} = \frac{S}{g_0/2 + h}, \quad S = \frac{E_2}{1-\nu_2}\int_0^h \varphi_2 dz, \quad h = h_0 + h_2.$$

The obtained relations for stresses are complementary to the relations for temperature given in Section 1.3. Together they make it possible to use the model of a two-

or single-layer plate under complicated thermal contact or boundary conditions to describe the heat transfer and thermal stress state in a system containing a coating.

10.3 Investigation of Temperature and Stress State at Dry Friction

Numerical studies were carried out for the case when external thermal irradiation is absent $T_s = 0$. In this case, the thermal state of the system was determined on the basis of relations (10.1)–(10.5), (10.9) numerically by the finite difference method using an implicit difference scheme and iterative methods. The convergence of the iterative process is provided by the "high" (≈ 0.01) compression factor of the constructed corresponding operator. The running technique (for a system of equations with a three-diagonal matrix when finding the temperature at each time step) and the Gaussian method with the choice of the main element (when finding the heat fluxes at the bases of the layers) were used.

The thermal stresses in the system caused by friction heating, at the numerically found temperature distribution, were determined for the case of rigid fixing of the edges of the layers, that is, according to the relations (10.10) using numerical integration (Simpson's method).

For the materials of layers 1 and 2, Steel Ca8 was chosen, and the coating was Steel 12Ch18Ni10Ti (as in the experiments on the wear of coated bodies (Kukareko et al. 2013), the thermophysical and mechanical characteristics of which (average integral over the heating interval) are as follows (Livshits et al. 1980):

Steel Ca8:

$$\begin{aligned} &\kappa = 43\,\mathrm{W/(K\cdot m)}, \quad \rho = 7920\,\mathrm{kg/m^3}, \quad C_\varepsilon = 528\,\mathrm{J/(kg\cdot K)},\\ &\varpi = 1.038\cdot 10^{-5}\,\mathrm{m^2/s}, \quad \alpha_s = 50\,\mathrm{W/m^2\cdot K}, \quad \varepsilon = 0.5,\\ &E = 18\,\mathrm{GPa}, \quad \nu = 0.28, \quad \alpha_t = 1.35\cdot 10^{-5}\,\mathrm{K^{-1}}. \end{aligned}$$

Steel 12Ch18Ni10Ti:

$$\begin{aligned} &\kappa = 20\,\mathrm{W/(K\cdot m)}, \quad \rho = 7839\,\mathrm{kg/m^3}, \quad C_\varepsilon = 500\,\mathrm{J/(kg\cdot K)},\\ &\varpi = 0.505\cdot 10^{-5}\,\mathrm{m^2/s}, \quad E = 174\,\mathrm{GPa}, \quad \nu = 0.28, \quad \alpha_t = 1.75\cdot 10^{-5}\,\mathrm{K^{-1}}. \end{aligned}$$

The compressive distributed load is chosen equal to 1 MPa, according to the experiments in Kukareko et al. (2013). The results of numerical studies of temperature T and thermal stresses σ^q are shown in Figs. 10.2, 10.3, 10.4, 10.5 and 10.6. They are carried out for the thicknesses of the moving layer 1 and substrate 2 equal to $h_1 = h_2 = 0.01$. The dashed curves are calculated without taking into account the heat transfer by radiation from the surfaces of layers 1 and 2.

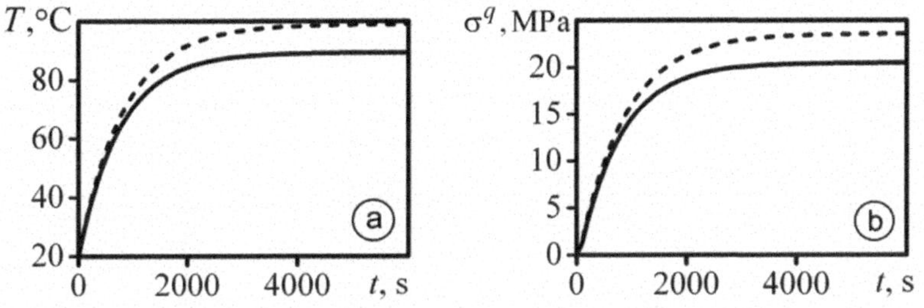

Fig. 10.2 Variation with time of temperature **a** and thermal stresses **b** on the friction surface at coating thickness $h_0 = 0.001$ m and friction speed $V = 0.1$ m/s

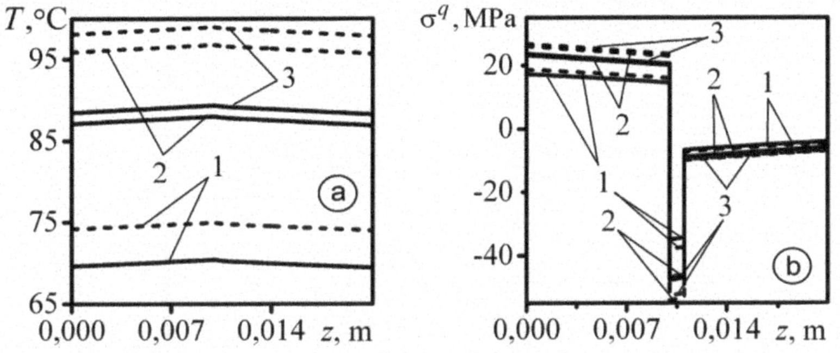

Fig. 10.3 Coordinate distribution of temperature **a** and thermal stresses **b** at different time points $t = 1000, 3000, 6000$ s (curves 1–3, respectively) at coating thickness $h_0 = 0.001$ m and friction velocity $V = 0.1\,\text{m/s}$

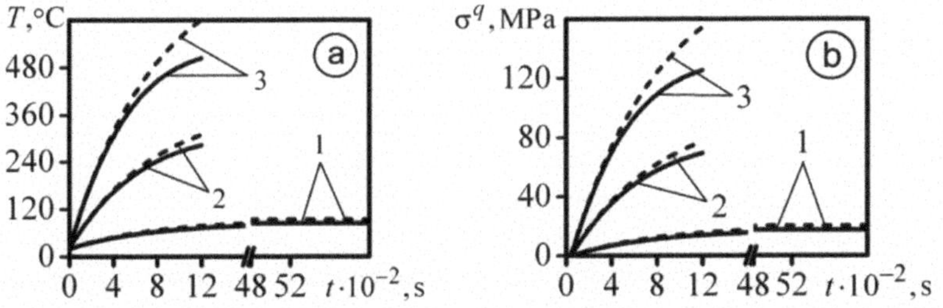

Fig. 10.4 Variation with time of temperature **a** and thermal stresses **b** on the friction surface with coating thickness $h_0 = 0.0002$ m at different friction speeds $V = 0.1; 0.5; 1\,\text{m/s}$ (curves 1–3, respectively)

The figures show that the consideration of thermal radiation heat transfer by is essential in the thermal and stress states assessment, and neglecting its influence leads to overestimation of temperature and thermal stress levels. The zone of compressive stresses is located in the substrate and coating, and tensile stresses are in the moving layer 1. On the contact surfaces there are stress jumps. At the considered layer thicknesses and fixing conditions, the temperature and stress levels increase with

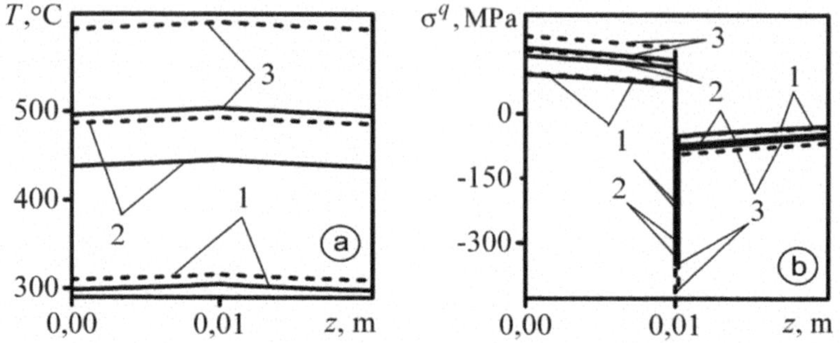

Fig. 10.5 Coordinate distributions of temperature **a** and thermal stresses **b** (at coating thickness $h_0 = 0.0002\,\text{m}$ and friction velocity $V = 1.0\,\text{m/s}$) at different moments of time $t =$ 400, 800, 1200 s (curves 1–3)

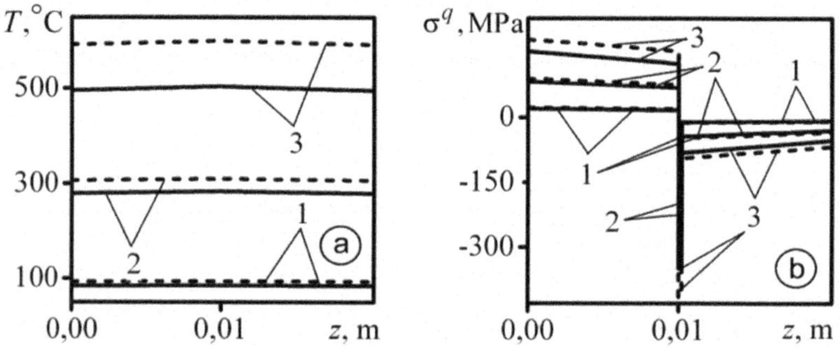

Fig. 10.6 Coordinate distribution of temperature **a** and thermal stress **b** at different friction velocities $V = 0.1, 0, 5, 1$ m/s (curves 1–3) at coating thickness $h_0 = 0.0002\,\text{m}$. Curve 1 corresponds to $t = 6000\,\text{s}$, and curves 2 and 3 – $t = 1200\,\text{s}$

increasing friction speed. From the data shown in the figures and the calculation of the force stresses (10.14), it follows that the determining factors in the assessment of the stressed state are thermal stresses.

The obtained graphical results can be used to evaluate the thermal and stress states in the surface layers of materials under dry friction depending on the test conditions (pressure and friction speed).

10.4 Conclusions

The proposed new mathematical models of the thermomechanics of coated plates allow us to study thermal and stress states during high-temperature technological heating in heat treatment by thermal radiation or operation under dry friction, taking into account radiation heat exchange. The obtained approximate relations for determining the temperature and stresses make it possible to use a model of a two- or single-layer plate under complicated thermal contact or boundary conditions to describe heat transfer and thermal stress state in a three- or two-layer plate containing a coating. Taking into account the radiation heat exchange during dry friction is essential in assessing thermal and stress states, and neglecting its influence leads to an overestimation of temperature and thermal stresses. Thermal stresses are the primary parameters in the assessment of the stress state.

References

Amosov AP (2011) Thermophysical models of friction of inert and explosive materials. Mashinostroenie, Moscow [in Russian]

Hachkevych O, Terlets'kyi R, Turii O (2017) The formulation and development of methods of solving thermomechanics problems for irradiated layered solids. Math Model Comput 4(1):21–36

Hachkevych OR, Kushnir RM, Terletskii RF (2022) Mathematical problems of thermomechanics for deformable bodies subjected to thermal Irradiation. Ukr Math J 73(10):1522–1536

Kukareko VA, Byeli AV, Belotserkovskii MA, Grigorchik AN (2013) Effect of ion-beam nitriding on structure, phase state, and tribological behavior of efficient thermal spray coatings deposited from various classes of rod steels. J Frict Wear 34(6):475–480

Livshits BG, Kraposhin VS, Lipetskiy YL (1980) Physical properties of metals and alloys. Metallurgy, Moscow [in Russian]

Modest MF (2003) Radiative heat transfer. Press, San Diego, Acad

Podstrigach YS, Shvets RN (1978) Thermoelasticity of thin shells. Nauk, Dumka, Kyiv [in Russian]

Shevchuk VA (1996) Generalized boundary conditions for heat transfer between a body and the surrounding medium through a multilayer thin covering. J Math Sci 81(6):3099–3102

Siegel R, Howell JR (1972) Thermal radiation heat transfer. McGraw Hill, New York

Terlets'kyi R, Brukhal' M (2017) Nonstationary heat exchange and stressed state in the irradiated system of layers with different transparencies. Mater Sci 52(4):533–541

Terlets'kyi R, Turii O (2013) Modeling and investigation of heat transfer in plates with thin coatings with regard for the influence of radiation. J Math Sci 192(6):703–722

Chapter 11
Problems of Thermomechanics of Multilayered Electroconductive Bodies Under the Action of the Pulsed Electromagnetic Fields with Modulation of Amplitude

Oleksandr Hachkevych, Roman Musii, and Nataliya Melnyk

11.1 Introduction

The structural elements of many modern machines and mechanisms are layered conductive bodies (LCB), which, during the operation of these devices, are affected by various physical and mechanical fields, e.g., pulsed electromagnetic fields (PEMFs) (Herlah 1999; Knoepfel 1970; Mikhailov 1979; Shneerson et al. 2014).

Mathematical models of homogeneous and lump-homogeneous bodies under non-stationary power and thermal loads are considered in the literature (Gribanov and Panichkin 1984; Ionov and Ogibalov 1975; Lurie 2005; Moon 1984; Podstrigach and Kolyano 1976; Timoshenko and Goodier 1970). Mathematical models of the thermomechanics of homogeneous electrically conductive bodies under the action of electromagnetic fields (EMFs) of various types are known (Burak et al. 2006; Hachkevich 1992; Hachkevych and Musii 2019; Hachkevych et al. 2011, 2019; Ionov and Ogibalov 1975). These models consider the adiabaticity of the processes of heating and deformation of the electroconductive body of the PEMF (Herlah 1999; Knoepfel 1970; Mikhailov 1979; Moon 1984; Mozhen 1991; Shneerson et al. 2014), which is experimentally observed under the action of different types of PEMFs, in particular with amplitude modulation in the mode with a pulse modulated signal (MPMS), in the mode with a decaying sine wave (MDSW) and a single electromagnetic pulse (SEMP). Such PEMFs are widely used in the technologies of pulsed electromagnetic processing of materials (Asai 2012; Batygin et al. 2003). Based on these models,

O. Hachkevych
Pidstryhach Institute for Applied Problems of Mechanics and Mathematics, Lviv, Ukraine
e-mail: dept13@iapmm.lviv.ua

R. Musii · N. Melnyk (✉)
Lviv Polytechnic National University, Lviv, Ukraine
e-mail: nataliia.b.melnyk@lpnu.ua

R. Musii
e-mail: roman.s.musii@lpnu.ua

A. N. Guz et al. (eds.), *Advances in Mechanics*, Advanced Structured Materials 191,
https://doi.org/10.1007/978-3-031-37313-8_11

the thermomechanical behavior and performance of uniform electrically conductive bodies of canonical form under the action of the considered characteristic types of PEMFs were investigated (Hachkevych et al. 2011; Musii et al. 2021; Musii 2010). In addition to the temperature caused by Joule's heat (Hachkevich 1992; Podstrigach et al. 1977), an additional component of temperature created by the heating of an electroconductive body because of its volumetric deformation due to the action of volumetric ponderomotive force was investigated (Burak et al. 2006; Hachkevych et al. 2011; Musii 2010). It was found that this temperature component is negligible for the effects of PEMF and the effects in the MPMS and MDSW when using frequencies of carrier electromagnetic oscillations outside the resonant frequencies of the electromagnetic field and is 10–20% of the value of the temperature component caused by heat at the first resonant frequency Joule (Hachkevych et al. 2011).

The analysis of the obtained qualitative and quantitative results of the deformation of homogeneous electroconductive bodies of PEMF of characteristic types based on the considered model confirms the rationality of its use to determine the thermostressed state of homogeneous electroconductive bodies under the action of PEMF with modulation of the amplitude of the radio frequency range.

When considering LCB, the question arises of considering the electromagnetic, thermal, and mechanical contact between the constituent layers and the influence of this contact on the distribution of physical and mechanical fields in each constituent layer. Therefore, it is necessary:

1. Build an appropriate physical-mathematical model of LCB under the action of PEMF.
2. Based on this model, formulate a new class of problems of thermomechanics of electrically conductive bodies.
3. Develop a methodology for solving these problems to determine the electromagnetic, temperature, and mechanical fields in each component layer of LCB.
4. Formulate the conditions under which the performance of the LCB as a structural element as a whole and the mechanical properties of the contact connection of the constituent layers of the LCB are preserved.

11.2 The Calculated Physicomathematical Model

Considering the above-mentioned problems, let us formulate a physicomathematical model for determining the thermostressed state of LCB under the action of PEMF. This model includes four stages (Musii et al. 2018).

At the first stage, in each nth ($n = 1, \ldots, N$) component layer of the LCB, the non-stationary EMF is determined from Maxwell's relations, which is described by the magnetic field intensity vector $\mathbf{H}^{(n)}$, and the Joule heat $Q^{(n)}$ and ponderomotive forces $\mathbf{F}^{(n)}$ caused by it taking into account the given boundary conditions on the outer surfaces of the first and last layers and the conditions of ideal electromagnetic contact on all surfaces of the connection of the component layers.

At the second stage, from the thermal conductivity equation, in which Joule's heat $Q^{(n)}$ is a volume source of heat, the non-stationary temperature field $T^{(n)}$ in each nth component layer of LCB is determined under the conditions of convective heat exchange with the environment on the outer surfaces of the first and last layers and ideal thermal contact on all surfaces with the connection of constituent layers.

At the third stage, the displacement vector $\mathbf{u}^{(n)}$ and the dynamic stress tensor $\hat{\sigma}^{(n)}$ in each nth component layer of LCB from the equations of dynamic thermoelasticity based on the found temperature $T^{(n)}$ and ponderomotive force $\mathbf{F}^{(n)}$ were determined.

At the fourth stage, to evaluate the efficiency of the LCB, we determine the intensities $\sigma_i{}^{(n)}$ of the total stresses $\hat{\sigma}^{(n)} = \hat{\sigma}^{(n)Q} + \hat{\sigma}^{(n)F}$ (due to both Joule heat and ponderomotive forces) (Hachkevych et al. 2011; Musii 2010) in each nth constituent layer, which are calculated according to the formula (Gribanov and Panichkin 1984; Ionov and Ogibalov 1975; Lurie 2005)

$$\sigma_i^{(n)} = \sqrt{\left[3I_2(\hat{\sigma}^{(n)}) - I_1^2(\hat{\sigma}^{(n)})\right]/2} \tag{11.1}$$

and compare their value with the elastic limit deformation of the material $\sigma_d{}^{(n)}$ of the nth layer. Here $I_j\left(\hat{\sigma}^{(n)}\right)$ is the jth invariant of the stress tensor.
The LCB loses its performance as a whole, as a structural element, if at least one of the nth component layers loses it, i.e., in the component layer, the condition

$$\max \sigma_i^{(n)} \leqslant \sigma_d{}^{(n)} \tag{11.2}$$

is fulfilled (Musii et al. 2018).

This condition corresponds to the well-known Huber–Mises condition (Lurie 2005; Timoshenko and Goodier 1970) for an isotropic layer, according to which plastic deformations occur when the stress intensity σ_i reaches the yield point of the material σ_d during its stretching.

In technical devices, the structural elements of which are LCB, there is an important problem of preserving the properties of the contact connection of their component layers. To assess the limit of preservation of these properties, the values of the stress intensities $\sigma_i{}^{(n)*}$ in the component layers on the contact surface of the nth and $(n+1)$th component layers are compared with the strength limit $\sigma_M{}^{(n)}$ of the nth contact joint. If the condition (Musii et al. 2018)

$$\sigma_i^{(n)*} \leqslant \sigma_M^{(n)} \tag{11.3}$$

is fulfilled, then the properties of the nth contact connection are preserved. Note that the values of $\sigma_d{}^{(n)}$ and $\sigma_M{}^{(n)}$ for the materials used in the engineering of the component layers of LCB are known from experiments.

11.3 Initial Relations of the Thermomechanics of LCB Under the Action of PEMF

Let us formulate the initial physical and mathematical conditions and write down the initial relations for the quantitative description of the parameters characterizing electromagnetic, thermal, and mechanical processes in LCB under the action of PEMF with amplitude modulation. Such PEMFs are given by the values of the magnetic field intensity vector $\mathbf{H}^{(n)}(\mathbf{r}, t)$ on the outer surfaces of the first and last component layers of the LCB in the form (Hachkevych et al. 2011)

$$\mathbf{H}\left(\mathbf{r}_s^{(n)}, t\right) = \mathbf{H}_0\left(\mathbf{r}_s^{(n)}, t\right) \equiv \mathbf{H}^0\left(\mathbf{r}_s^{(n)}, t\right) \sin\left(\omega t + \phi_n\right). \tag{11.4}$$

Here $\mathbf{H}^0\left(\mathbf{r}_s{}^{(n)}, t\right) = H_*(t)\mathbf{H}_0^*\left(\mathbf{r}_s{}^{(n)}\right)$ is modulated amplitude; $\mathbf{H}_0^*\left(\mathbf{r}_s{}^{(n)}\right)$ is the amplitude of the carrier signal; $H_*(t)$ is a dimensionless function of the pulse character of the change in time t of the signal, which modulates the amplitude of electromagnetic oscillations (the amplitude of the carrier signal), and satisfies the conditions $H_*(t) \leqslant 1$ at $t \in [0, t_i]$, $H_*(0) = 0$, $H_*(t_i) = 0$; ω is the frequency of the carrier signal; t_i is the duration of the electromagnetic action; ϕ_n is the initial phase, $\mathbf{r}$ and $\mathbf{r}_s{}^{(n)}$ are radius vectors of volume points V and outer surfaces S_n of the first and last component layers of LCB.

Assume that the parameters ω, t_i, $\mathbf{H}_0^*\left(\mathbf{r}_s{}^{(n)}\right)$ and $H_*(t)$ are such that the active PEMF belongs to the class of pulsed "non-destructive" PEMFs, the action of which does not yet lead to the occurrence of shock waves (with a pulse length of less than a fraction of a second ($t_i < 0.01$ s) and with the highest value of the magnetic field strength on the outer surfaces the first and last constituent layers $H_{\max} \leqslant 10^7$ A/m.

The materials of each nth component layer of LCB are considered homogeneous, isotropic, non-dielectric, and non-ferromagnetic, for which electromechanical and thermoelectric effects are insignificant, induction vectors $\mathbf{D}^{(n)}$ and $\mathbf{B}^{(n)}$ are parallel to vectors of electric $\mathbf{E}^{(n)}$and magnetic $\mathbf{H}^{(n)}$ field strengths, and conduction currents $\mathbf{j}^{(n)}$ are parallel to $\mathbf{E}^{(n)}$. Assume that the material equations of electrodynamics in each nth layer have the form (Mozhen 1991; Podstrigach et al. 1977)

$$\mathbf{D}^{(n)} = \varepsilon_n \mathbf{E}^{(n)}, \quad \mathbf{B}^{(n)} = \mu_n \mathbf{H}^{(n)}, \quad \mathbf{j}^{(n)} = \sigma_n \mathbf{E}^{(n)}, \tag{11.5}$$

where $\varepsilon_n = \varepsilon_0 \varepsilon_{*n}$, $\mu_n = \mu_0 \mu_{*n}$, ε_{*n} and μ_{*n} are relative electric and magnetic permeability, σ_n is coefficient of electroconductivity of the material of the nth constituent layer, ε_0 and μ_0 are electric and magnetic steel, respectively.

Under such assumptions, the effect of PEMF on the processes of thermal conductivity and deformation in the conductive layer, as in the models for quasi-steady EMF known in the literature (Hachkevich 1992), is taken into account due to the physical factors caused by this field, i.e., Joule heat and ponderomotive forces (Hachkevych et al. 2011; Podstrigach et al. 1977)

$$Q^{(n)} = \sigma_n \mathbf{E}^{(n)} \mathbf{E}^{(n)}, \quad \mathbf{F}^{(n)} = \sigma_n \mu_n \mathbf{E}^{(n)} \times \mathbf{H}^{(n)}. \tag{11.6}$$

These factors lead to the emergence of non-stationary temperature and mechanical fields. Ponderomotive moments

$$\mathbf{M}^{*(n)} = \mathbf{M}_E^{*(n)} + \mathbf{M}_M^{*(n)},$$

due to the parallelism of the induction vectors, the electric and magnetic field strengths are equal to zero. The following notations are introduced:

$$\mathbf{M}_E^{*(n)} = \mathbf{P}^{(n)} \times \mathbf{E}^{(n)}$$

are electrical moments,

$$\mathbf{M}_M^{*(n)} = \mathbf{M}^{(n)} \times \mathbf{B}^{(n)}$$

are magnetic moments,

$$\mathbf{P}^{(n)} = \mathbf{D}^{(n)} - \varepsilon \mathbf{E}^{(n)}, \quad \mathbf{M}^{(n)} = \mathbf{B}^{(n)}/\mu_{0n} - \mathbf{H}^{(n)}$$

are the polarization and magnetization vectors, respectively. In this approximation and for constant characteristics of the material (in particular, equal to the average value in the considered interval of temperature changes), the initial relations for the quantitative description of the parameters characterizing the electromagnetic, thermal, and mechanical processes in each nth constituent layer of the LCB under pulsed electromagnetic action are formulated for the first three stages, according to the selected physical and mathematical model.

At the first stage, we determine the EMF parameters in the nth constituent layer of the LCB. We choose the magnetic field intensity vector $\mathbf{H}^{(n)}(\mathbf{r}, t)$ as the key function. Disregarding the bias currents for durations of the fronts of pulsed electromagnetic action greater than 10^{-6} s and frequencies of electromagnetic oscillations less than 10^8 Hz, based on Maxwell's relations and Ohm's law under accepted assumptions (Hachkevich 1992; Podstrigach et al. 1977)

$$\operatorname{rot} \mathbf{H}^{(n)} = \mathbf{j}^{(n)}, \quad \operatorname{rot} \mathbf{E}^{(n)} = -\mu_n \frac{\partial \mathbf{H}}{\partial t}, \quad \mathbf{j}^{(n)} = \sigma_n E^{(n)} \tag{11.7}$$

we get the equation for determining $\mathbf{H}^{(n)}(\mathbf{r}, t)$ in the nth constituent layer

$$\Delta \mathbf{H}^{(n)} - \sigma_n \mu_n \frac{\partial \mathbf{H}^{(n)}}{\partial t} = 0, \quad \operatorname{div} \mathbf{H}^{(n)} = 0 \tag{11.8}$$

under boundary conditions

$$\mathbf{H}^{(n)}(\mathbf{r}_s, t) = \mathbf{H}_*^{(n)}(t)\, H_0^*\left(\mathbf{r}_s^{(n)}\right) \tag{11.9}$$

under the boundary conditions on the outer surfaces of the first and last layers of the LCB and the conditions of ideal electromagnetic contact, which on the surface

$\mathbf{r} = \mathbf{r}_n$ of the junction of the nth and $(n + 1)$th component layers of the LCB are written in the form (Musii et al. 2018; Podstrigach et al. 1977)

$$\mathbf{H}^{(n)} = \mathbf{H}^{(n+1)}, \quad \text{rot}\,\mathbf{H}^{(n)} = k_\sigma^n \text{rot}\,\mathbf{H}^{(n+1)} \tag{11.10}$$

and zero initial conditions

$$\mathbf{H}^{(n)}(\mathbf{r}, 0) = 0. \tag{11.11}$$

We use the notations: Δ is the Laplace operator, $k_\sigma^n = \sigma_n / \sigma_{n+1}$.

The specific densities of Joule heat $Q^{(n)}(\mathbf{r}, t)$ and ponderomotive forces $\mathbf{F}^{(n)}(\mathbf{r}, t)$ in the nth component layer are as follows

$$Q^{(n)}(\mathbf{r}, t) = \frac{1}{\sigma_n}\left(\text{rot}\, H^{(n)}(\mathbf{r}, t)\right)^2, \tag{11.12}$$

$$F^{(n)}(\mathbf{r}, t) = \mu_n \cdot \text{rot}\, H^{(n)}(\mathbf{r}, t) \times H^{(n)}(\mathbf{r}, t). \tag{11.13}$$

At the second stage, the temperature $T^{(n)}(\mathbf{r}, t)$ in nth layer of the LCB is determined based on the found Joule heat distribution $Q^{(n)}(\mathbf{r}, t)$ from the thermal conductivity equation

$$\Delta T^{(n)} - \frac{1}{\kappa_n}\frac{\partial T^{(n)}}{\partial t} = -\frac{Q^{(n)}}{\lambda_n}, \tag{11.14}$$

considering the conditions of convective heat exchange with the environment

$$\frac{\partial T^{(n)}}{\partial n} + H\left(T^{(n)} - T_s^{(n)}\right) = 0 \tag{11.15}$$

on the outer surfaces of the first and last layers of LCB and conditions

$$T^{(n)} = T^{(n+1)}, \quad \text{grad}\, T^{(n)} = k_\lambda^n \,\text{grad}\, T^{(n+1)} \tag{11.16}$$

of ideal thermal contact on the surface of the connection of the n th and $(n + 1)$th constituent layers of LCB and initial conditions

$$T^{(n)}(\mathbf{r}, 0) = 0. \tag{11.17}$$

Here $T^{(n)}$ is temperature deviation from the initial temperature $T_0{}^{(n)}$; $\partial T^{(n)}/\partial n = \mathbf{n}\,\text{grad}\, T^{(n)}$; $\mathbf{n}$ is the normal vector to the outer surfaces of the first and last layers of LCB; $T_s{}^{(n)}$ is given constant temperature of the environment outside the first and last layers of LCB; $H_n = H_{*n}/\lambda_n$ is the relative heat transfer coefficient; H_{*n} is the coefficient of heat transfer from the surfaces of the first and last layers of LCB; $k_\lambda{}^n = \lambda_{n+1}/\lambda_n$; κ_n and λ_n are coefficients of temperature and thermoconductivity of materials of the nth layer.

At the third stage, using the known expressions of temperature $T^{(n)}(\mathbf{r}, t)$ and ponderomotive force $\mathbf{F}^{(n)}(\mathbf{r}, t)$, the thermostressed state of LCB is determined from the equations of the uncoupled dynamic problem of thermoelasticity written for the nth layer (Musii et al. 2018; Podstrigach et al. 1977)

$$
\begin{gathered}
\operatorname{div}\hat{\sigma}^{(n)} + \mathbf{F}^{(n)} = \rho_n \frac{\partial \mathbf{u}^{(n)}}{\partial t}, \\
\hat{\sigma}^{(n)} = 2G_n \left\{ \hat{e}^{(n)} + \frac{\nu_n}{1-2\nu_n} \left[e_{jj}^{(n)} - \alpha_n \frac{1+\nu_n}{1-2\nu_n} T^{(n)} \right] \hat{I} \right\}, \\
\hat{e}^{(n)} = \operatorname{Def}\mathbf{u}^{(n)} \equiv \tfrac{1}{2}\left(\nabla \mathbf{u}^{(n)} + \mathbf{u}^{(n)} \nabla \right).
\end{gathered}
\tag{11.18}
$$

Here $\hat{e}^{(n)} = \left\{ e_{ij}{}^{(n)} \right\}$ is the deformation tensor; Def is the deformer; ∇ is the Hamilton operator; $\hat{I} = \left\{ \delta_{ij} \right\}$ is unit tensor; δ_{ij} is Kronecker's symbol; $\nabla \mathbf{u}^{(n)}$ and $\mathbf{u}^{(n)} \nabla$ is the dyadic product of vectors ∇ and $\mathbf{u}^{(n)}$; α_n and ν_n are coefficients of linear thermal expansion and Poisson; $G_n = E_n/[2(1+\nu_n)]$ is shear modulus; E_n is Young's modulus; ρ_n is the material density of the n th layer. All thermophysical and physicomechanical characteristics of the materials of the constituent layers of LCB are considered to be constant.

We solve (11.18) under the appropriate boundary (thermal and mechanical) and initial conditions. Under the initial conditions, the values of the displacement vector $\mathbf{u}^{(n)}$, velocity $\partial \mathbf{u}^{(n)}/\partial t$, and temperature $T^{(n)}$ in the entire region of each n th component layer at the initial moment $t = 0$ of time are preferably set. If these functions are equal to zero, then the initial conditions have the form

$$
u_i^{(n)}(\mathbf{r},0) = 0, \quad \frac{\partial u_i^{(n)}(\mathbf{r},0)}{\partial t} = 0, \quad T^{(n)}(\mathbf{r},0) = 0, \quad i = 1,2,3. \tag{11.19}
$$

Depending on the type of boundary conditions, the system of Eq. (11.18) is solved in displacements or stresses. If the defining functions of the thermoelasticity problem are the temperature $T^{(n)}(\mathbf{r},t)$ and the stress tensor $\hat{\sigma}^{(n)}(\mathbf{r},t)$, then we get a system of equations

$$
\operatorname{Def}\left(\operatorname{div}\hat{\sigma}^{(n)} + \mathbf{F}^{(n)}\right) = \rho_n \frac{\partial^2}{\partial t^2}\left[\frac{1+\nu_n}{E_n}\hat{\sigma}^{(n)} + \left(\alpha_n T^{(n)} - \frac{\nu_n}{E_n}\sigma_*^{(n)} \right)\hat{I} \right], \tag{11.20}
$$

and, in the case of determining functions $T^{(n)}(\mathbf{r},t)$ and $\mathbf{u}^{(n)}(\mathbf{r},t)$, we get a system of equations

$$
\begin{aligned}
\left(\Delta + \frac{1}{1-2\nu_n}\operatorname{grad}\operatorname{div} \right)\mathbf{u}^{(n)} - \frac{2(1+\nu_n)\rho_n}{E_n}\frac{\partial^2 \mathbf{u}^{(n)}}{\partial t^2} \\
= \frac{\alpha_n E_n}{1-2\nu_n}\operatorname{grad} T^{(n)} - \frac{2(1+\nu_n)}{E_n}\mathbf{F}^{(n)}.
\end{aligned}
\tag{11.21}
$$

Here $\sigma_*^{(n)} \equiv I_1\left(\hat{\sigma}^{(n)}\right)$. We solve the system of Eq. (11.20) under the boundary conditions

$$\hat{\sigma}^{(n)}\mathbf{n} = 0, \tag{11.22}$$

which correspond to the absence of a force load on the outer surfaces of the first and last component layers of LCB, as well as the conditions of ideal mechanical contact

$$\hat{\sigma}^{(n)} = \hat{\sigma}^{(n+1)}, \quad \operatorname{div}\hat{\sigma}^{(n)} + \mathbf{F}^{(n)} = k_\rho^n\left(\operatorname{div}\hat{\sigma}^{(n+1)} + \mathbf{F}^{(n+1)}\right) \tag{11.23}$$

on the surface $\mathbf{r} = \mathbf{r}_n$ of the junction of the nth and $(n+1)$ th constituent layers of LCB, as well as under the initial $t = 0$ conditions

$$\begin{gathered}\hat{\sigma}^{(n)}(\mathbf{r}, 0) = 0,\\ \frac{1}{2G_n}\frac{\partial\hat{\sigma}^{(n)}(\mathbf{r}, 0)}{\partial t} + \left(\alpha_n\frac{\partial T^{(n)}(\mathbf{r}, 0)}{\partial r} - \frac{\nu_n}{E_n}\frac{\partial\hat{\sigma}_*^{(n)}(\mathbf{r}, 0)}{\partial t}\right)\hat{I} = 0.\end{gathered} \tag{11.24}$$

Here $k_\rho^n = \rho_n/\rho_{n+1}$.

Equation (11.21) written in displacements is solved under boundary conditions

$$\left\{\operatorname{Def}\mathbf{u}^{(n)} + \frac{\nu_n}{1-2\nu_n}\left[\operatorname{div}\mathbf{u}^{(n)} - \frac{\alpha_n(1+\nu_n)}{1-2\nu_n}T^{(n)}\right]\hat{I}\right\}\mathbf{n} = 0, \tag{11.25}$$

corresponding to conditions (11.22), written in terms of the displacement $\mathbf{u}^{(n)}$ on the external surfaces of the first and last constituent layers of LCB, free from force load, and under conditions of ideal mechanical contact (Musii et al. 2018; Podstrigach et al. 1977)

$$\begin{aligned}&\mathbf{u}^{(n)} = \mathbf{u}^{(n+1)},\\ &\operatorname{Def}\mathbf{u}^{(n)} + \frac{\nu_n}{1-2\nu_n}\left[\operatorname{div}\mathbf{u}^{(n)} - \frac{\alpha_n(1+\nu_n)}{1-2\nu_n}T^{(n)}\right]\hat{I}\\ &= k_E^n\left\{\operatorname{Def}\mathbf{u}^{(n+1)} + \frac{\nu_{n+1}}{1-2\nu_{n+1}}\left[\operatorname{div}\mathbf{u}^{(n+1)} - \frac{\alpha_{n+1}(1+\nu_{n+1})}{1-2\nu_{n+1}}T^{(n+1)}\right]\hat{I}\right\}\end{aligned} \tag{11.26}$$

on the surface $\mathbf{r} = \mathbf{r}_n$ of the junction of the n th and $(n+1)$ th constituent layers of LCB, as well as initial at $t = 0$ conditions

$$\mathbf{u}^{(n)}(\mathbf{r}, 0) = 0, \quad \frac{\partial\mathbf{u}^{(n)}(\mathbf{r}, 0)}{\partial t} = 0. \tag{11.27}$$

Here $k_E^n = E_{n+1}(1+\nu_n)/E_n(1+\nu_{n+1})$.

Based on the found displacement vector $\mathbf{u}^{(n)}$, we determine the stress tensor $\hat{\sigma}^{(n)}$ in the n th component layer of the LCB from the relation (Podstrigach et al. 1977)

$$\hat{\sigma}^{(n)} = 2G_n \left\{ \text{Def}\,\mathbf{u}^{(n)} + \frac{\nu_n}{1-2\nu_n}\left[\text{div}\,\mathbf{u}^{(n)} - \frac{\alpha_n (1+\nu_n)}{1-2\nu_n} T^{(n)}\right]\hat{I}\right\}. \quad (11.28)$$

To estimate the contribution of Joule's heat $Q^{(n)}(\mathbf{r}, t)$ and ponderomotive forces $\mathbf{F}^{(n)}(\mathbf{r}, t)$ to the thermally stressed state of the LCB, the stress tensor $\hat{\sigma}^{(n)}$ and vector of displacements $\mathbf{u}^{(n)}$ in the n th component layer of the LCB are presented as the sum of two components

$$\hat{\sigma}^{(n)} = \hat{\sigma}^{Q(n)} + \hat{\sigma}^{F(n)}, \quad \mathbf{u}^{(n)} = \mathbf{u}^{Q(n)} + \mathbf{u}^{F(n)}, \quad (11.29)$$

caused by Joule's heat and ponderomotive force $\mathbf{F}^{(n)}$, respectively. The components $\hat{\sigma}^{Q(n)}$ and $\mathbf{u}^{Q\,(n)}$ were determined from Eqs. (11.20)–(11.21) at $\mathbf{F}^{(n)} = 0$, and the components $\hat{\sigma}^{F\,(n)}$ and $\mathbf{u}^{F\,(n)}$ were determined from these equations at $T^{(n)} = 0$.

Based on the found total displacement vectors $\mathbf{u}^{(n)}$ and stress tensors $\hat{\sigma}^{(n)}$, taking into account formula (11.28) for the found displacements $\mathbf{u}^{(n)}$, we determine the stress intensities in the n th component layer of LCB using (11.1). After that, based on the fulfillment of inequalities (11.2) and (11.3), we investigate the operability and preservation of the properties of the contact connection of the constituent layers of LCB.

11.4 Formulation of Initial-Boundary Problems of Thermomechanics of Multilayer Electroconductive Bodies of Canonical Form

Let us formulate the initial-boundary problems of thermomechanics for the LCB of the canonical form—one-dimensional for the LCB with plane-parallel boundaries, plane axisymmetric for a hollow cylinder, and centrally symmetric for a hollow sphere under the action of a homogeneous non-stationary EMF, which is given by the values of the corresponding tangent component $H_j^{(n)}(\gamma, t)$ of the vector $\mathbf{H}^{(n)}$ on the outer surfaces of the first and last layers of the considered LCB

$$H_j^{(1)}\left(r_s^{(1)}, t\right) = H_j^{-}(t), \quad H_j^{(n)}\left(r_s^{(n)}, t\right) = H_j^{+}(t), \quad (11.30)$$

where $H_j^{\pm}(t)$ are given function of time for the LCB with plane-parallel boundaries; $j = z, \gamma = r$ for a cylinder; $j = \varphi, \gamma = r$ for a sphere. The function $H_j^{(n)}(\gamma, t)$ in nth layer is described by the equation

$$\frac{\partial^2 H_j^{(n)}}{\partial \gamma^2} + \frac{k}{\gamma}\frac{\partial H_j^{(n)}}{\partial \gamma} - \sigma_n \mu_n \frac{\partial H_j^{(n)}}{\partial t} = 0 \quad (11.31)$$

under the appropriate boundary conditions (11.30) and conditions of ideal electromagnetic contact on the surface $\gamma = \gamma_1$ of the connection of the constituent layers, as well as zero initial conditions at $t = 0$. Here $k = 0$, $\gamma_1 = 0$ and $\gamma_n = h_n$ for LCB with plane-parallel boundaries, where h_k $(k = 1, \dots, N)$ are the thicknesses of its component layers, $\gamma_1 = r_0, \gamma_p = r_p$ $(p = 1, \dots, N-1)$, $\gamma_N = r_N$ for a cylinder and a sphere, $k = 1$ for a cylinder, $k = 2$ for a sphere.

The temperature field $T^{(n)}(\gamma, t)$ in the LCB layer is determined from the thermal conductivity equation

$$\frac{\partial^2 T^{(n)}}{\partial \gamma^2} + \frac{k}{\gamma}\frac{\partial T^{(n)}}{\partial \gamma} - \frac{1}{\kappa_{\text{n}}}\frac{\partial T^{(n)}}{\partial t} = -\frac{Q^{(n)}}{\lambda_n} \tag{11.32}$$

under the boundary conditions of convective heat exchange on the outer surfaces of the first and last component layers of the LCB and the conditions of ideal thermal contact on the surfaces $\gamma_p = r_p$ $(p = 1, \dots, N-1)$ of the connection of the component layers, as well as zero initial conditions at $t = 0$. Here, κ_{n} and λ_n are the coefficients of temperature and thermoconductivity of materials of the n th layer.

The components $\sigma_{jj}{}^{(n)}$ $(j = x, y, z)$ of the dynamic stress tensor in LCB with plane-parallel boundaries are found from the equations

$$\begin{aligned} &\frac{\partial^2 \sigma_{zz}^{(n)}}{\partial z^2} - \frac{1}{c_{1n}^2}\frac{\partial^2 \sigma_{zz}^{(n)}}{\partial t^2} = \alpha_n \rho_n \frac{1+\nu_n}{1-\nu_n}\frac{\partial^2 T^{(n)}}{\partial t^2} - \frac{\partial F_z^{(n)}}{\partial z}, \\ &\sigma_{xx}^{(n)} = \sigma_{yy}^{(n)} = \frac{\nu_n}{1-\nu_n}\sigma_{zz}^{(n)} - \frac{\alpha E_n T^{(n)}}{1-2\nu_n}, \end{aligned} \tag{11.33}$$

where $A_{1n} = \left(E_n\left(1-\nu_n\right) / \left(\rho_n\left(1+\nu_n\right)\left(1-2\nu_n\right)\right)\right)^{-1/2}$ is velocity of elastic expansion waves in the nth component layer of LCB. The radial component $u_r{}^{(n)}$ of the vector of displacements $\mathbf{u}^{(n)}$ for a hollow cylinder and sphere is determined from the equation

$$\begin{aligned} &\frac{\partial^2 u_r^{(n)}}{\partial r^2} + \frac{m}{r}\frac{\partial u_r^{(n)}}{\partial r} - \frac{m}{r^2}u_r^{(n)} - \frac{1}{c_{1n}^2}\frac{\partial^2 u_r^{(n)}}{\partial t^2} \\ &\qquad = \alpha_n \frac{1+\nu_n}{1-\nu_n}\frac{\partial T^{(n)}}{\partial r} - \frac{\left(1+\nu_n\right)\left(1-2\nu_n\right)}{E_n\left(1-\nu_n\right)}F_r^{(n)} \end{aligned} \tag{11.34}$$

under specified boundary conditions on the outer surfaces of the first and last layers, conditions of ideal mechanical contact on the surfaces of the connection of all constituent layers, and corresponding initial conditions. Here $m = 1$ for the cylinder, and $m = 2$ for the sphere. We calculate the components $\sigma_{jj}(n)$ $(j = r, \varphi, z, \theta)$ of the stress tensor from the ratios

$$\begin{aligned} \sigma_{rr}^{(n)} &= \frac{\text{E}_{\text{n}}}{\left(1+\nu_n\right)\left(1-2\nu_n\right)} \\ &\quad \times \left[\left(1-\nu_n\right)\frac{\partial u_r^{(n)}}{\partial r} + \nu_n \frac{m}{r}u_r^{(n)} - \alpha_n\left(1+\nu_n\right)T^{(n)}\right] \end{aligned} \tag{11.35}$$

for both LCB (the cylinder and the sphere) from ratios

$$\sigma_{\varphi\varphi}^{(n)} = \frac{\mathrm{E_n}}{(1+\nu_n)(1-2\nu_n)} \times \left[(1-\nu_n)\frac{u_r^{(n)}}{r} + \nu_n \frac{\partial u_r^{(n)}}{\partial r} - \alpha_n (1+\nu_n) T^{(n)} \right], \tag{11.36}$$

$$\sigma_{zz}^{(n)} = \nu_n \left(\sigma_{rr}^{(n)} + \sigma_{\varphi\varphi}^{(n)} \right) - \alpha_n E_n T^{(n)}$$

for the cylinder, as well as from ratios

$$\sigma_{\varphi\varphi}^{(n)} = \sigma_{\theta\theta}^{(n)} = \frac{\mathrm{E_n}}{(1+\nu_n)(1-2\nu_n)} \left[\frac{u_r^{(n)}}{r} + \nu_n \frac{\partial u_r^{(n)}}{\partial r} - \alpha_n (1+\nu_n) T^{(n)} \right] \tag{11.37}$$

for the sphere.

11.5 Methodology for Constructing Solutions of Formulated Initial-Boundary Problems

Let us develop a methodology for constructing solutions to the above-formulated initial-boundary problems of thermomechanics, which describe the EMF, temperature, and stress state of the canonical LCB.

The technique is based on the approximation with respect to the corresponding spatial variable γ of the key functions

$$\Phi^{(n)}(\gamma,t) = \left\{ H_j^{(n)}(\gamma,t), T^{(n)}(\gamma,t), \sigma_{zz}^{(n)}(\gamma,t), u_j^{(n)}(\gamma,t) \right\}$$

in the nth component layer of the considered LCB by polynomials of the corresponding p degree (Hachkevych et al. 2011; Musii et al. 2021; Musii 2010)

$$\Phi^{(n)}(\gamma,t) = \sum_{i=0}^{p} a_i^{\Phi(n)}(t)\, \gamma^i. \tag{11.38}$$

The degree p of the approximation polynomials (11.38) in each nth component layer is chosen in such a way as to satisfy the boundary conditions on the external surfaces and the surfaces of the connection of these layers in this LCB. The coefficients $a_i{}^{\Phi(n)}(t)$ of the approximation polynomials (11.38) are determined through the integral (total over the $n = 1, \ldots, N$ layer package) characteristics $\Phi_s(t)$ of the key functions $\Phi^{(n)}(\gamma,t)$

$$\Phi_s(t) = \sum_{n=1}^{N} \int_{\gamma_{n-1}}^{\gamma_n} \Phi^{(n)}(\gamma,t)\gamma^{s+l} d\gamma, \quad s = 1,2 \quad \left(\gamma_0 = r_s^{(1)}, \gamma_n = r_s^{(N)} \right) \tag{11.39}$$

and given boundary conditions on the outer surfaces of the first and last component layers of the considered LCB and on the surfaces of the connection of each nth and $(n+1)$th component layers. Here $l=-1$ for the LCB with plane-parallel boundaries; $l=0$ for a cylinder and a sphere when integrating Eqs. (11.31) and (11.32); $l=1$ for a cylinder and a sphere—for Eq. (11.34).

To find the integral characteristics $\Phi_s(t)$ of the sought functions $\Phi^{(n)}(\gamma,t)$, the original Eqs. (11.31)–(11.34) are integrated according to (11.39) considering expressions (11.38). As a result, we reduce the original initial-boundary value problems for the key functions to the corresponding Cauchy problems for the integral characteristics of these functions described by systems of equations

$$\begin{cases} \dfrac{dH_1(t)}{dt} - d_1H_1(t) - d_2H_2(t) = d_3H_j^-(t) + d_4H_j^+(t) \\ \dfrac{dH_2(t)}{dt} - d_5H_1(t) - d_6H_2(t) = d_7H_j^-(t) + d_8H_j^+(t), \end{cases} \tag{11.40}$$

$$\begin{cases} \dfrac{dT_1}{dt} + d_1^TT_1 + d_2^TT_2 = \displaystyle\sum_{n=1}^{N} \frac{\kappa_n}{\lambda_n} \int_{\gamma_{n-1}}^{\gamma_n} Q^{(n)}(\gamma,t)\gamma^{l+1}d\gamma \\ \dfrac{dT_2}{dt} + d_3^TT_1 + d_4^TT_2 = \displaystyle\sum_{n=1}^{N} \frac{\kappa_n}{\lambda_n} \int_{\gamma_{n-1}}^{\gamma_n} Q^{(n)}(\gamma,t)\gamma^{l+2}d\gamma, \end{cases} \tag{11.41}$$

$$\begin{cases} \dfrac{d^2\sigma_{zz1}}{dt^2} - d_1^*\sigma_{zz1} - d_2^*\sigma_{zz2} = W_1^*(t) \\ \dfrac{d^2\sigma_{zz2}^Q}{dt^2} - d_3^*\sigma_{zz1} - d_4^*\sigma_{zz2}^Q = W_2^*(t), \end{cases}$$

$$\begin{cases} \dfrac{d^2u_{r1}}{dt^2} - d_1^{**}u_{r1} - d_2^{**}u_{r2} = W_1^{**}(t) \\ \dfrac{d^2u_{r2}}{dt^2} - d_6^{**}u_{r1} - d_7^{**}u_{r2} = W_2^{**}(t) \end{cases} \tag{11.42}$$

and solve under appropriate initial conditions using the Laplace transform. The coefficients $d_{1\div8}$, ${d_{1\div4}}^T$, $d_{1\div4}^*$, and $d_{1\div4}^{**}$ are determined through the geometric parameters and physical and mechanical characteristics of the constituent layers of the LCB under consideration, $W_s{}^*(t)$, $W_s{}^{**}(t)$ $(s=1,2)$ are the integrated right-hand sides of (11.33) and (11.34) in accordance with (11.39) using $T^{(n)}$, $F_z^{(n)}$, and $F_r^{(n)}$ on the surfaces $\gamma=\gamma_{n-1}$ $(n=1,\ldots,N)$ of layers connection.

We write the solutions of Cauchy problems (11.40)–(11.42) in the form of convolutions of functions describing given boundary conditions and homogeneous solutions. Expressions of the corresponding tangent component $H_j{}^{(n)}(\gamma,t)$ of the vector $\mathbf{H}^{(n)}$ are obtained

$$H_j^{(n)}(\gamma,t)=\sum_{i=0}^{p}\left\{\sum_{s=1}^{2}a_{is}^{(n)}\sum_{k=1}^{2}\int_0^t\left[A_{s1}(k)H_j^-(\tau)\right.\right.$$
$$\left.\left.+A_{s2}(k)H_j^+(\tau)\right]e^{p_k(t-\tau)}d\tau+a_{i3}^{(n)}H_j^-(t)+a_{i4}^{(n)}H_j^+(t)\right\}\gamma^i,\quad(11.43)$$

the temperatures

$$T^{(n)}(\gamma,t)=\sum_{k=0}^{p}\sum_{s=1}^{2}\left(b_{ks}^{(n)}\sum_{m=1}^{2}\int_0^t\left[B_{s1}(m)\,W_1^Q(\tau)\right.\right.$$
$$\left.\left.+B_{s2}(m)\,W_2^Q(\tau)\right]e^{p_m(t-\tau)}d\tau\right)\gamma^k.\quad(11.44)$$

The components $\sigma_{zz}^{(n)}(z,t)$ of stress tensor in LCB with plane-parallel boundaries

$$\sigma_{zz}^{(n)}(z,t)=\sum_{i=0}^{p}\left(\sum_{s=1}^{2}c_{is}^{(n)}\sum_{\alpha=1}^{4}\int_0^t\left[R_{s1}^*(\alpha)W_1^*(\tau)\right.\right.$$
$$\left.\left.+R_{s2}^*(\alpha)\,W_2^*(\tau)\right]e^{p_\alpha(t-\tau)}d\tau\right)z^i,\quad(11.45)$$

radial component $u_r^{(n)}(r,t)$ of displacement vector $\mathbf{u}^{(n)}$ for a cylinder and a sphere

$$u_r^{(n)}(r,t)=\sum_{i=0}^{p}\left\{\sum_{s=1}^{2}c_s^{(n)}\sum_{\beta=1}^{4}\int_0^t\left[R_{s1}^{**}(\beta)W_1^{**}(\tau)\right.\right.$$
$$\left.\left.+R_{s2}^{**}(\beta)\,W_2^{**}(\tau)\right]e^{p_\beta(t-\tau)}d\tau\right\}r^i.\quad(11.46)$$

Here $A_{sj}(k)$, $B_{sj}(m)$, $R_{sj}{}^*(\alpha)$, and $R_{sj}{}^{**}(\beta)$ $(s,\,j=1,2)$ are expressions depending on roots p_k, p_m, p_α, and p_β of the corresponding characteristic equations of problems for determining functions $H_j{}^{(n)}$, $T^{(n)}$, $\sigma_{zz}{}^{(n)}$, and $u_r{}^{(n)}$. Based on the known expressions (11.46) of the functions $u_r{}^{(n)}(r,t)$ from relations (11.35)–(11.37), we determine the components $\sigma_{rr}{}^{(n)}$ and $\sigma_{\varphi\varphi}{}^{(n)}$ of the stress tensor and the corresponding axial stresses $\sigma_{zz}{}^{(n)}$ in the cylinder and meridian stresses $\sigma_{\theta\theta}{}^{(n)}$ in the sphere.

11.6 Solutions of Problems for Bimetallic Bodies of Canonical Form Under Electromagnetic Action in MPMS

Let us present a mathematical description of uniform EMFs, which are given by the values $H_j^{\mp}(t)$ of the corresponding tangent component $H_j^{(n)}(\gamma, t)$ of the vector $\mathbf{H}^{(n)}$ on the outer surfaces of the first and last constituent layers of the LCB in the form (Hachkevych et al. 2011)

$$H_j^{\mp}(t) = H_0 H_*(t) \cos \omega t. \tag{11.47}$$

For action in MPMS, the function $H_*(t)$ has the form $H_*(t) = k\left(\mathrm{e}^{-\beta_1 t} - \mathrm{e}^{-\beta_2 t}\right)$. Here k is the normalization factor, H_0 is the maximum value of the magnetic field strength on the external surfaces of the LCB, β_1 and β_2 are the parameters characterizing, respectively, the times of the rising t_{incr} and falling t_{decr} fronts of the modulating pulse.

For the convenience of transformations, expression (11.47) is given in complex form (Ionov and Ogibalov 1975)

$$H_j^{\mp}(t) = \frac{kH_0}{2} \sum_{k=1}^{4} \left(\mathrm{e}^{-\beta_1 - i\omega} + \mathrm{e}^{-\beta_1 + i\omega} - \mathrm{e}^{-\beta_2 - i\omega} - \mathrm{e}^{-\beta_2 + i\omega}\right), \tag{11.48}$$

where $i = \sqrt{-1}$.

Based on the obtained general solutions (11.43)–(11.46) of the problem of thermomechanics for the considered LCB under uniform non-stationary electromagnetic action, we write the solutions of this problem under electromagnetic action in MPMS. By substituting expression (11.47) into (11.43)–(11.46) and considering the expressions of the corresponding approximation polynomials (11.38), the expressions of the corresponding tangent component $H_j^{(n)}(\gamma, t)$ of the vector $\mathbf{H}^{(n)}$ are obtained in the following form

$$H_j^{(n)}(\gamma, t) = \frac{k_0 H_0}{2} \sum_{i=0}^{2} \left(\mathrm{e}^{-(\beta_1 - i\omega)t} B_{i1}^{(n)} + \mathrm{e}^{-(\beta_2 - i\omega)t} B_{i2}^{(n)} \right.$$
$$\left. + \mathrm{e}^{-(\beta_1 + i\omega)t} B_{i3}^{(n)} + \mathrm{e}^{-(\beta_2 + i\omega)t} B_{i4}^{(n)} + \mathrm{e}^{p_1 t} B_{i5}^{(n)} + \mathrm{e}^{p_2 t} B_{i6}^{(n)}\right)\gamma^i,$$

specific densities of Joule heat and ponderomotive forces

$$Q^{(n)}(\gamma, t) = \frac{1}{\sigma_n} \frac{k_0^2 H_0^2}{4} \sum_{i=1}^{2} \sum_{j=1}^{2} ij \sum_{l=1}^{20} C_{ijl}^{(n)} \mathrm{e}^{\alpha_l t} \gamma^{i+j-2},$$

$$F_r^{(n)}(\gamma, t) = -\mu_n \frac{k_0^2 H_0^2}{4} \sum_{i=1}^{2} \sum_{j=1}^{2} i \sum_{l=1}^{20} C_{ijl}^{(n)} \mathrm{e}^{\alpha_l t} \gamma^{i+j-1},$$

temperatures

$$T^{(n)}(\gamma,t)=\frac{k_0^2H_0^2}{4}\sum_{p=0}^{2}\sum_{m=1}^{2}\sum_{i=1}^{2}\sum_{j=1}^{2}ij\sum_{l=1}^{20}M_{ijlm}^{(n)}\frac{e^{p_m t}-e^{\alpha_l t}}{p_m-\alpha_l}\gamma^p.$$

Here

$$M_{ijlm}^{(n)}=\sum_{s=1}^{2}b_{ps}^{(n)}\left(N_{1ijl}B_{s1m}+N_{2ijl}B_{s2m}\right),$$

$$N_{sijl}=\sum_{n=1}^{2}\frac{\kappa_n}{\sigma_n\lambda_n}\frac{\gamma_n^{i+j+s-1}-\gamma_{n-1}^{i+j+s-1}}{i+j+s-1}C_{ijl}^{(n)},\quad s=1,2.$$

The expressions for $B_{ij}{}^{(n)}$, $C_{ijl}{}^{(n)}$, $M_{ijlm}{}^{(n)}$, N_{sijl} and B_{sjm} are of the same type for the considered LCBs and differ only in the values of the corresponding coefficients and roots of the characteristic equations depending on the geometric parameters of the layers.

The components $\sigma_{zz}{}^{(n)Q}$ and $\sigma_{zz}{}^{(n)F}$ are the components of stress tensor $\sigma_{zz}{}^{(n)Q}(z,t)$ for LCB with plane-parallel boundaries have the form

$$\sigma_{zz}^{(n)Q}(z,t)=\frac{k_0^2H_0^2}{4}\sum_{f=1}^{2}\left(c_{0f}^{(n)}+c_{1f}^{(n)}z+c_{2f}^{(n)}z^2\right)\sum_{\alpha=1}^{4}\sum_{q=1}^{2}R_{f\alpha q}$$

$$\times\sum_{x=1}^{2}\left\{c_{11}^2\alpha_1\rho_1\frac{1+\nu_1}{1-\nu_1}\left(b_{0x}^{(1)}\frac{(-h_1)^q}{q}+b_{1x}^{(1)}\frac{(-h_1)^{q+1}}{q+1}+b_{2x}^{(1)}\frac{(-h_1)^{q+2}}{q+2}\right)\right.$$

$$\left.-c_{12}^2\alpha_2\rho_2\frac{1+\nu_2}{1-\nu_2}\left(b_{0x}^{(2)}\frac{h_2^q}{q}+b_{1x}^{(2)}\frac{h_2^{q+1}}{q+1}+b_{2s}^{(2)}\frac{h_2^{q+2}}{q+2}\right)\right\}$$

$$\times\left\{\sum_{m=1}^{2}\sum_{k=1}^{2}\sum_{i=1}^{2}\sum_{j=1}^{2}\frac{ijB_{xmk}}{i+j+k-2}\left[\frac{\kappa_2h_2^{i+j+k-2}}{\lambda_2\sigma_2}\right.\right.$$

$$\times\sum_{l=1}^{20}C_{ijl}^{(2)}\left(\frac{p_m^2e^{p_\alpha t}-p_m^2e^{p_m t}}{(p_m-\alpha_l)(p_\alpha-p_m)}-\frac{\alpha_l^2e^{\alpha_l t}-\alpha_l^2e^{p_\alpha t}}{(p_m-\alpha_l)(p_\alpha-\alpha_l)}\right)$$

$$-\frac{\kappa_1(-h_1)^{i+j+k-2}}{\lambda_1\sigma_1}$$

$$\left.\left.\times\sum_{l=1}^{20}C_{ijl}^{(1)}\left(\frac{p_m^2e^{p_\alpha t}-p_m^2e^{p_m t}}{(p_m-\alpha_l)(p_\alpha-p_m)}-\frac{\alpha_l^2e^{\alpha_l t}-\alpha_l^2e^{p_\alpha t}}{(p_m-\alpha_l)(p_\alpha-\alpha_l)}\right)\right]\right\},$$

$$\sigma_{zz}^{(n)F}(z,t)=\frac{k_0^2H_0^2}{4}\sum_{m=1}^{2}z^m\Bigg(\sum_{s=1}^{2}c_{ms}^{(n)F}\Bigg\{\sum_{\alpha=1}^{4}\sum_{q=1}^{2}P_{s\alpha q}\Bigg[\sum_{i=1}^{2}\sum_{j=0}^{2}\frac{i}{i+j+q}$$
$$\times\left(C_{11}^2\mu_1(-h_1)^{i+j+q}\sum_{l=1}^{20}C_{ijl}^{(1)}\frac{e^{\alpha_l t}-e^{p_\alpha t}}{p_\alpha-\alpha_l}-C_{12}^2\mu_2h_2^{i+j+q}\sum_{l=1}^{20}C_{ijl}^{(2)}\frac{e^{\alpha_l t}-e^{p_\alpha t}}{p_\alpha-\alpha_l}\right)$$
$$+d_5^F\frac{k_0^2H_0^2}{4}\left(\mu_1\sum_{l=1}^{20}C_{10l}^{(1)}\frac{e^{\alpha_l t}-e^{p_\alpha t}}{p_\alpha-\alpha_l}-\mu_2\sum_{l=1}^{20}C_{10l}^{(1)}\frac{e^{\alpha_l t}-e^{p_\alpha t}}{p_\alpha-\alpha_l}\right)\Bigg]\Bigg\}$$
$$+c_{m3}^{(n)F}F_0(t)\Bigg).$$

Accordingly, the components $u_r^{(n)Q}$ and $u_r^{(n)F}$ of the radial displacements $u_r^{(n)}$ for the cylinder and sphere can be written in the form

$$u_r^{(n)Q}(\gamma,t)=\sum_{\alpha=0}^{2}\big[c_{\alpha1}^{(n)Q}u_{r1}^Q(t)+c_{\alpha2}^{(n)Q}u_{r2}^Q(t)+c_{\alpha3}^{(n)Q}T^{(1)}(r_0,t)$$
$$+c_{\alpha4}^{(n)Q}T^{(2)}(r_2,t)+c_{\alpha5}^{(n)Q}N_0^Q(r_1,t)\big]\gamma^\alpha,$$
$$u_r^{(n)F}(r,t)=\sum_{\alpha=0}^{2}\left[c_{\alpha1}^{(n)F}u_{r1}^F(t)+c_{\alpha2}^{(n)F}u_{r2}^F(t)\right]r^\alpha,$$

where

$$u_{rs}^Q(t)=\sum_{\beta=1}^{4}\sum_{p=0}^{2}\sum_{m=1}^{2}\sum_{i=1}^{2}\sum_{j=1}^{2}ij\sum_{l=1}^{20}\frac{M_{ijlm}^{(n)}}{p_m-\alpha_l}\left(\frac{e^{p_m t}-e^{p_\beta t}}{p_m-p_\beta}-\frac{e^{\alpha_l t}-e^{p_\beta t}}{\alpha_l-p_\beta}\right)$$
$$\times\left\{R_{s\beta1}^Q\left[d_3^Qr_0^p+d_4^Qr_2^p+d_5^QK_{\alpha E}r_1^p-pg_{p1}\right]\right.$$
$$\left.+R_{s\beta2}^Q\left[d_8^Qr_0^p+d_9^Qr_2^p+d_{10}^QK_{\alpha E}r_1^p-pg_{p2}\right]\right\},$$
$$u_{rs}^F(t)=-\mu_n\frac{k_0^2H_0^2}{4}\sum_{\beta=1}^{4}\sum_{n=1}^{2}d_n\sum_{i=1}^{2}\sum_{j=1}^{2}i$$
$$\times\sum_{l=1}^{20}C_{ijl}^{(n)}\left[R_{s\beta1}^Ff_{1ij}+R_{s\beta2}^Ff_{2ij}\right]\frac{e^{\alpha_l t}-e^{p_\beta t}}{\alpha_l-p_\beta},\quad s=1,2.$$

Here $F_0(t)$ is the combination of values ${F_j}^{(n)}$ $(j=z,r)$ on the surface of the connection of layers; expressions $P_{s\alpha q}$, ${R_{s\beta k}}^Q$, ${R_{s\beta k}}^F$, g_{ps}, f_{sij}, $K_{\alpha E}$, ${c_{0f}}^{(n)}$, ${b_{ij}}^{(n)}$, and ${c_{ms}}^{(n)F}$ are given in terms of the geometric, physical and mechanical parameters of LCB.

11.7 Study of Thermomechanical Behavior and Properties of the Contact Joint of a Bimetallic Cylinder Under Electromagnetic Action in MPMS

Numerical studies were carried out for a bimetallic cylinder, the constituent layers of which have a thickness of 1 mm and are made of stainless steel and copper (Livshits et al. 1980). The durations t_i of the modulation pulse of the electromagnetic action in the MPMS were assumed to be equal to $t_i = 10^{-4}$ s, 10^{-3} s and 10^{-2} s.

Figures 11.1, 11.2, 11.3, 11.4, 11.5 and 11.6 show the results of studies of the thermal stress state and performance of a hollow bimetallic cylinder with inner $r_0 =$ 8 mm and outer $r_2 = 10$ mm radii and the surface of the connection $r_1 = 9$ mm under action in MPMS of duration $t_i = 10^{-3}$ s. Figures 11.1 and 11.2 show the change in time of the components $\sigma_{rr}{}^{(n)j}$, $\sigma_{\varphi\varphi}{}^{(n)j}$ $(j = Q, F)$ of radial $\sigma_{rr}{}^{(n)}$ and circular $\sigma_{\varphi\varphi}^{(n)}$ components of the stress tensor at frequencies $\omega = 6.28 \cdot 10^4$ 1/s of carrier electromagnetic oscillations (outside the range of resonance frequencies ω_{rs} $(s = 1, 2)$ of the electromagnetic field (Hachkevich 1992; Hachkevych et al. 2011; Podstrigach et al. 1977)).

The change in time of the component $\sigma_{\varphi\varphi}{}^{(n)F}$ and $\sigma_{\varphi\varphi}{}^{(n)Q}$ of circular stresses $\sigma_{\varphi\varphi}{}^{(n)}$ with frequency $\omega = \omega_{r1} = 2.32 \cdot 10^5$ 1/s is shown in Figs. 11.3, 11.4, 11.5 and 11.6. All values in Figs. 11.1, 11.2, 11.3, 11.4, 11.5 and 11.6 are calculated on the surface $r = r_1$ of the connection of the constituent layers of the cylinder.

Figure 11.7 shows the dependence of the maximum values of stress intensity $\sigma_{i\ \max}^{(n)}$ on the magnitude H_0 of the action in the MPMS ($H_{0\,\max} = 10^5$ A/m) and the

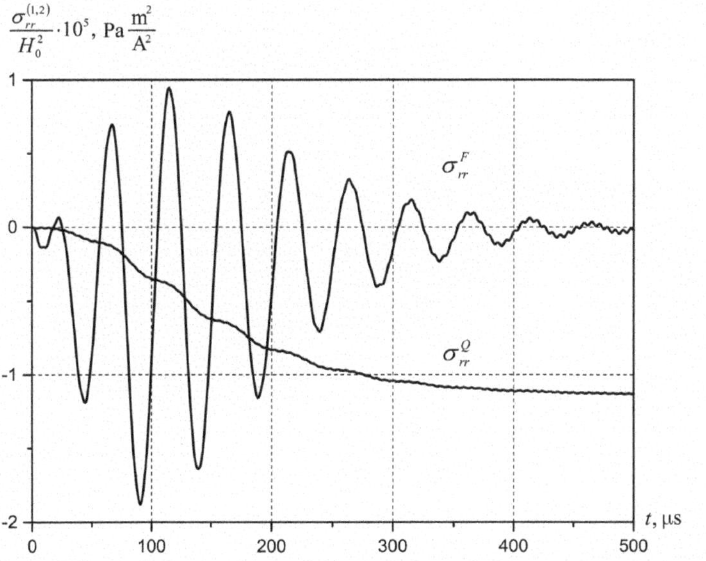

Fig. 11.1 The change in time of the components $\sigma_{rr}^{(n)Q}$ and $\sigma_{rr}^{(n)F}$ of radial components $\sigma_{rr}^{(n)}$ of the stress tensor at frequencies $\omega = 6.28 \cdot 10^4$ 1/s on the connection surface $r = r_1$ of the cylinder's constituent layers

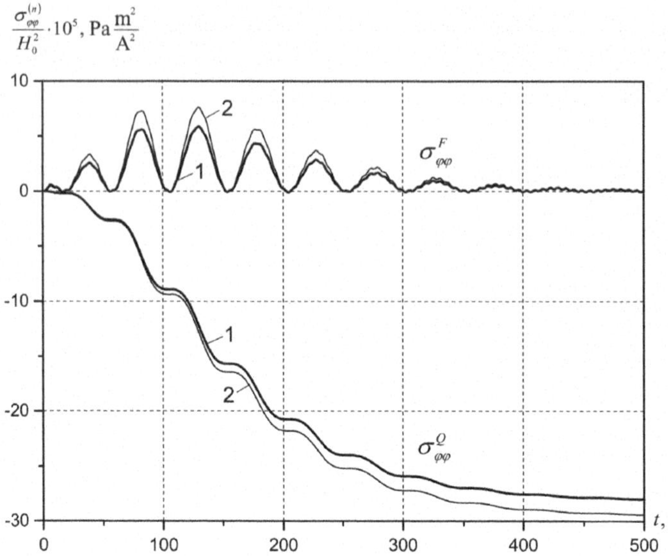

Fig. 11.2 The change in time of the components $\sigma_{\varphi\varphi}^{(n)Q}$ and $\sigma_{\varphi\varphi}^{(n)F}$ of circular components $\sigma_{\varphi\varphi}^{(n)}$ of the stress tensor at frequencies $\omega = 6.28 \cdot 10^4$ 1/s on the connection surface $r = r_1$ of the cylinder's constituent layers

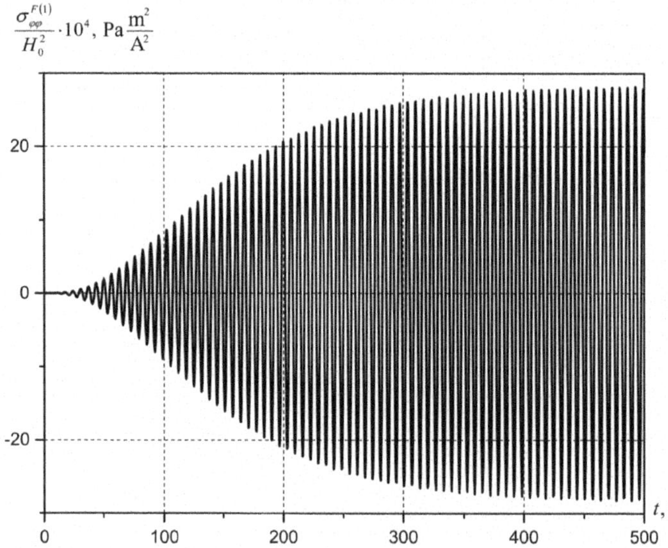

Fig. 11.3 The change in time of the component $\sigma_{\varphi\varphi}^{(1)F}$ of circular stresses $\sigma_{\varphi\varphi}^{(1)}$ in the first layer at the resonant frequency $\omega = \omega_{r1} = 2.32 \cdot 10^5$ 1/s on the connection surface $r = r_1$ of the cylinder's constituent layers

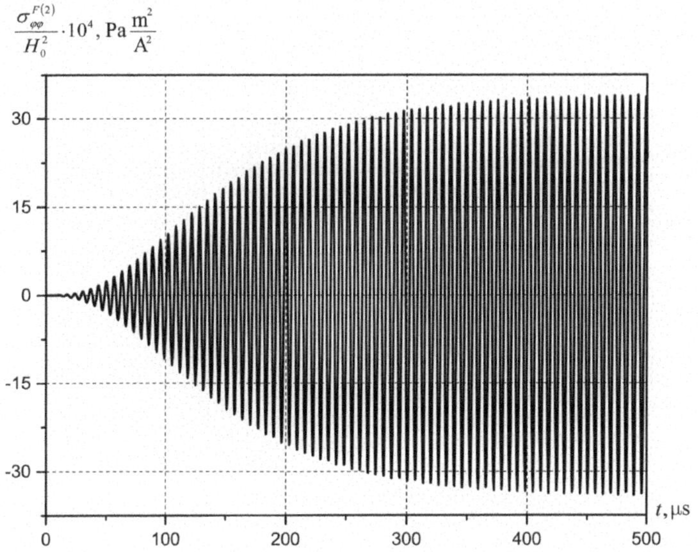

Fig. 11.4 The change in time of the component $\sigma_{\varphi\varphi}^{(2)F}$ of circular stresses $\sigma_{\varphi\varphi}^{(2)}$ in the second layer at the resonant frequency $\omega = \omega_{r1} = 2.32 \cdot 10^5$ 1/s on the connection surface $r = r_1$ of the cylinder's constituent layers

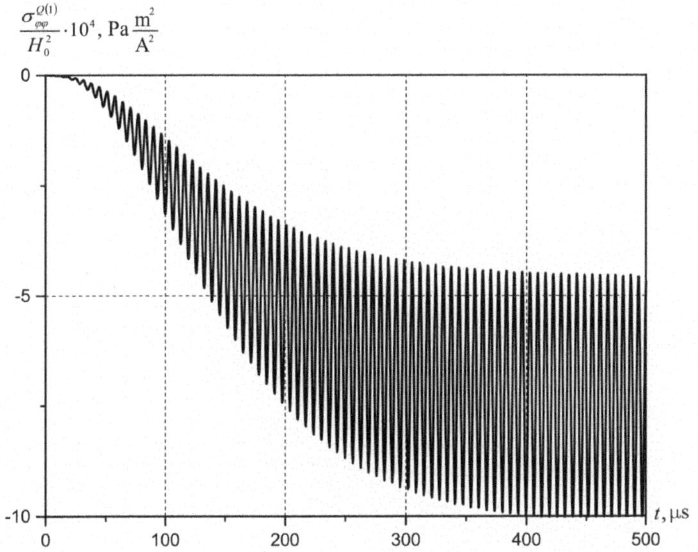

Fig. 11.5 The change in time of the component $\sigma_{\varphi\varphi}^{(1)Q}$ of circular stresses $\sigma_{\varphi\varphi}^{(1)}$ in the first layer at the resonant frequency $\omega = \omega_{r1} = 2.32 \cdot 10^5$ 1/s on the connection surface $r = r_1$ of the cylinder's constituent layers

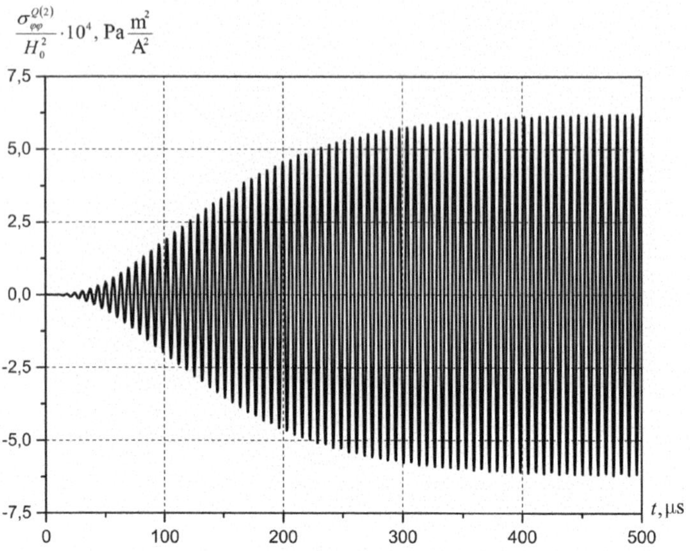

Fig. 11.6 The change in time of the component $\sigma_{\varphi\varphi}^{(2)Q}$ of circular stresses $\sigma_{\varphi\varphi}^{(2)}$ in the second layer at the resonant frequency $\omega = \omega_{r1} = 2.32 \cdot 10^5$ 1/s on the connection surface $r = r_1$ of the cylinder's constituent layers

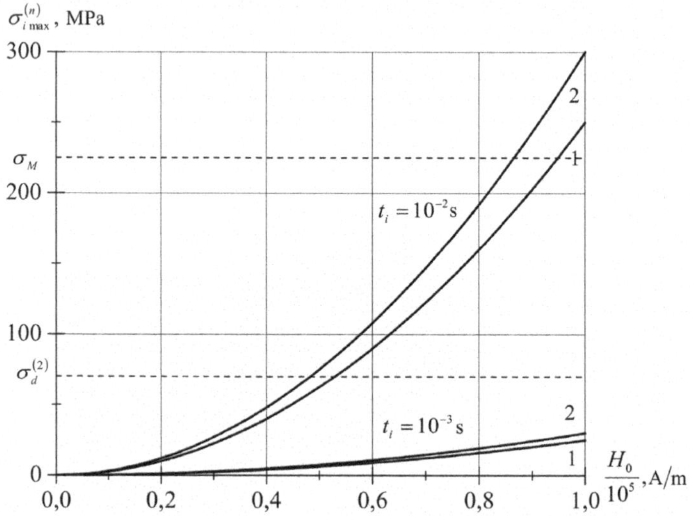

Fig. 11.7 The dependence of the maximum values of stress intensity $\sigma_{i\,\max}^{(n)}$ on the magnitude H_0 at the resonant frequency $\omega = \omega_{r1} = 2.32 \cdot 10^5$ 1/s in the cylinder's constituent layers on the surface of their connection for durations $t_i = 10^{-3}$ s, 10^{-2} s, (1 and 2 are layer's numbers)

frequency $\omega = \omega_{r1}$ in the constituent layers of the cylinder on the surface of their connection for durations $t_i = 10^{-3}$ s and 10^{-2} s. Lines 1 and 2 correspond to the steel and copper layers. It was found that the performance and properties of the contact connection are preserved at $t_i = 10^{-3}$ s and are lost at $t_i = 10^{-2}$ s.

11.8 Conclusions

Physicomathematical models are proposed for researching the thermal stress state, performance, and properties of the contact connection of multilayer electrically conductive bodies of canonical form under the action of PEMF with amplitude modulation. New classes of problems of thermomechanics of multilayer electrically conductive bodies of canonical form under the action of considered PEMFs are formulated. A methodology for constructing solutions of formulated initial-boundary value problems for determining the magnetic field strength, temperature, and components of the dynamic stress tensor using a polynomial approximation of the determining functions for the thickness of each constituent layer of LCB has been developed. This technique makes it possible to accurately satisfy all given boundary conditions on the outer surfaces of the first and last layers and on the surfaces of the contact connection of all constituent layers. This made it possible to reduce the original initial-boundary value problems to the defining functions to the corresponding Cauchy problems to the integral characteristics of these functions, summed over the package of all layers. Using the integral Laplace transform for the given initial conditions on the defining functions, general solutions of the Cauchy problems on the integral characteristics of these functions are written. They are obtained in the form of convolutions of functions that describe the given expressions of the determining functions on the outer surfaces of the first and last component layers of the LCB and homogeneous solutions of the corresponding Cauchy problems.

The thermomechanical behavior, performance, and preservation of the properties of the contact joint of a bimetallic hollow cylinder under electromagnetic action in MPMS, depending on its time and amplitude-frequency characteristics, were studied.

The revealed regularities of the thermomechanical behavior of LCB under electromagnetic action in MPMS can be a theoretical basis for choosing the optimal modes of technologies for pulsed electromagnetic processing of layered conductive elements of structures.

References

Asai S (2012) Electromagnetic processing of materials. Springer, Netherlands

Batygin YV, Lavinsky VI, Himenko LT (2003) Impulse magnetic fields for advanced technologies. MOST-Tornado Publ, Harkov [in Russian]

Burak YI, Hachkevych OR, Musii RS (2006) Thermoelasticity of electroconductive bodies under conditions of action of pulsed electromagnetic fields. Math Meth Phys Mech Fields 49(1):75–84 [in Ukranian]

Gribanov VF, Panichkin NG (1984) Coupled and dynamic problems of thermoelasticity. Mashinostroenie, Moscow [in Russian]
Hachkevich AR (1992) Thermomechanics of electroconductive bodies under the influence of quasi-established electromagnetic fields. Nauk, Dumka, Kyiv [in Russian]
Hachkevych O, Musii R (2019) Mathematical modeling in thermomechanics of electroconductive bodies under the action of the pulsed electromagnetic fields with modulation of amplitude. Math Model Comput 6(1):30–36
Hachkevych OR, Musii RS, Tarlakovskyi DV (2011) The thermomechanics of non-ferromagnetic conductive bodies for the actions of pulsed electromagnetic fields with amplitude modulation. SPOLOM, Lviv [in Ukranian]
Hachkevych OR, Musii RS, Melnyk NB, Dmytruk VA (2019) Dynamic thermoelastic processes in a conductive plate under the action of electromagnetic pulses of microsecond and nanosecond durations. J Therm Stresses 42(9):1110–1122
Herlah F (1999) Pulsed magnets. Rep Prog Phys 62(6):859–920
Ionov VN, Ogibalov PM (1975) Stresses in bodies under impulsive loading. Vyshayia Shkola, Moscow [in Russian]
Ionov VN, Ogibalov PM (1975) Stresses in bodies under impulsive loading. Vyshayia Shkola, Moscow [in Russian]
Knoepfel H (1970) Pulsed high magnetic fields. North-Holland Publ, Amsterdam, London
Livshits BG, Kraposhin VS, Lipetsky YL (1980) Physical properties of metals and alloys. Metallurgy, Moscow [in Russian]
Lurie AI (2005) Theory of elasticity. Springer-Verlag, Berlin, Heidelberg
Mikhailov VM (1979) Pulse electromagnetic fields. Publishing House at Kharkov University, Kharkov, Vishcha shkola [in Russian]
Moon FC (1984) Magnetosolid mechanics. J. Wileg, New York
Mozhen J (1991) Mechanics of electromagnetic continuum. Mir, Moscow [in Russian]
Musii R, Melnyk N, Dmytruk V (2018) Thermoelastic processes analyzer for piecewise homogeneous conductive structures subjected to pulsed electromagnetic action. J Therm Stresses 41(9):1125–1135
Musii RS, Melnyk NB, Drohomyretska KT, Zakaulova Y (2021) Thermomechanical behavior of an electrically conductive cylindrical implant under the action of external unstable electromagnetic fields. Math Model Comput 8(2):184–191
Musii RS (2010) Dynamic problem of thermomechanics for conductive bodies of canonical form. Rastr–7 Publ., Lviv [in Ukranian]
Podstrigach YS, Kolyano YM (1976) Generalized thermomechanics. Nauk, Dumka, Kyiv [in Russian]
Podstrigach YS, Burak YI, Gachkevich AR, Chernyavskaya LV (1977) Thermoelasticity of conductive bodies. Nauk, Dumka, Kyiv [in Russian]
Shneerson GA, Dolotenko MI, Kryvosheev SI (2014) Strong and superstrong pulsed magnetic fields generation. De Gruyter, Berlin
Timoshenko SP, Goodier JN (1970) Theory of elasticity. McGraw-Hill Kogakusha Ltd, Tokyo

Chapter 12
Substantiation of Reliability of Calculation of Strength of Rocket and Space Technology Structures Without Destructive Tests

Anatoly Dzyuba and Volodymyr Sirenko

12.1 Introduction

The process of designing and creating load-bearing structures of new equipment is inevitably associated with the need to develop adequate design schemes, reliable mathematical models, and effective numerical algorithms for their implementation and further conducting appropriate computer modeling and numerical analysis using effective computing tools, the use of experimental research methods, non-destructive testing, and modern manufacturing technologies (Dashchenko et al. 2011; Degtyarev 2014; Drobenko et al. 2020; Gudramovich et al. 2017; Hudramovich and Dzyuba 2009; Mossakovsky et al. 1990; Usyukin 1988).

Such a process has a multi-stage interactive nature, at the final stage of which, despite the stringent requirements for the quality of work and means of their control, often forced, due to existing doubts about the reliability of the obtained design, calculation, and technological results, experimental field tests of the created fragments and (or) the entire structure are carried out (Degtyarev 2014; Drobenko et al. 2020; Marchuk et al. 2020; Tkachuk et al. 2017; Veretelnik et al. 2017).

Conducting such (usually destructive and often repeated) tests significantly increases the cost of developing new complex engineering structures. At the same time, the costs of destructive tests are especially significant for the development and creation of non-serial (one-time) products.

With the growing competition in the market of services, in particular, rocket and space technology (launch vehicles), destructive tests of individual components, and

A. Dzyuba (✉)
Oles Honchar Dnipro National University, Dnipro, Ukraine
e-mail: dzb@ua.fm

V. Sirenko
Yuzhnoye State Design Office, Dnipro, Ukraine
e-mail: v.n.sirenko@i.ua

A. N. Guz et al. (eds.), *Advances in Mechanics*, Advanced Structured Materials 191,
https://doi.org/10.1007/978-3-031-37313-8_12

in many cases the structure as a whole, significantly increase the cost of creating such products.

The solution to the problem of increasing the safety guarantees of such structures by overestimating the safety factor obviously worsens the weight performance of products, which in turn also leads to a decrease in their competitiveness. Solving this problematic conflict is one of the priority tasks of this industry (Degtyarev 2014).

12.2 Statement of the Problem

The present work is aimed at developing a concept and methodology for more reliable prediction of the bearing capacity and determination of the limiting (destructive) parameters of the stress–strain state of heterogeneous shell elements of modern engineering structures, in particular, rocket and space technology, tanks, dry compartments of launch vehicles, etc., without conducting (or with a significant reduction in the volume) of their full-scale destructive tests.

Structural elements of rocket and space technology are quite costly, highly stressed, heterogeneous, often interconnected, composite, wafer-type, or metal shells of rotation, supported, as a rule, by longitudinal and transverse structural frames. Dry compartments, in addition, usually have several irregularly spaced technological-edged hatch holes of different sizes (Fig. 12.1). The stress–strain state of such elements is quite heterogeneous.

Despite the fact that the calculations of such structures are carried out using modern certified, sophisticated, and sufficiently accurate computing tools (Dashchenko et al. 2011; Degtyarev 2014; Hudramovich and Dzyuba 2009; Marchuk et al. 2020), the results of the calculated parameters of the strength reliability of such power structures, in particular, the destructive loads, have deviations from their physically real

Fig. 12.1 Fragment of the compartment

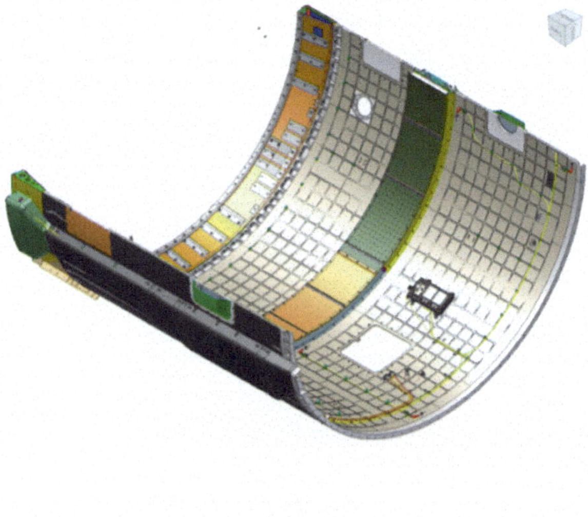

values, which requires their additional verification, usually by means of rather costly experimental tests.

In the study, it is assumed that the input data of such calculations: geometric dimensions, physical and mechanical characteristics, boundary conditions (fixation), parameters of external loads, etc., for reasons of different nature are approximate values, which leads to the accumulation of calculation errors in the process of implementing the corresponding computational algorithms. The causes and consequences of such possible deviations in solving complex engineering problems are analyzed.

One of the main tasks for solving this problem is to develop a methodology for estimating the error of the calculation result due to possible deviations of such input calculation data, as well as to establish their values that would correspond to the results of the finite element calculation of such a structure within a given error, which would allow making appropriate adjustments to the design and calculation schemes.

12.3 Analysis of Errors in Strength Calculation

12.3.1 Errors of Numerical Operations

Numerical modeling of the behavior of complex structures consists of a basic finite element program for calculating their main design parameters, engineering analysis tools, and a set of computational operations related, in particular, to the preparation of input parameters.

The main causes of errors in the numerical implementation of algorithms for solving applied problems can be classified as follows (Berezin and Zhidkov 1966):

1. Errors of the mathematical model.
   Such errors are associated with the accepted physical assumptions and certain idealizations in the construction of the mathematical model, the peculiarities of taking into account the inhomogeneities of the geometry of the object, the physical and mechanical characteristics of the material, external loads, etc., and can be reduced only by more accurate implementation of the fundamental laws of mechanics and description of the physical (mechanical) aspects of the problem.
2. Errors of the numerical method of implementation.
   In modern conditions, one of the main means of numerical research in mechanics is the use of finite element analysis methods. Errors in the use of this approach are directly related to the way of taking into account the inhomogeneities of geometry and material, the choice of the type and number of elements, nodes, and methods of their thickening in places of rapidly changing parameters, etc. Such errors also arise in the case of replacing infinite series with a limited number of terms, functions with an approximate polynomial, stopping the iteration process with a finite number of steps, etc.

3. Technical computational errors.
   To date, the problem of such errors associated with the limited capabilities of computers for processing and storing the appropriate number of numerical digits has ceased to be relevant due to the wide possibilities of using modern powerful computers for the calculation of critical engineering structures.

In the future, it is believed that the correct use of modern FEM software packages makes it possible to achieve a sufficiently high accuracy of the results (at the level of estimates of the error in solving systems of linear algebraic equations of high dimensionality) (Berezin and Zhidkov 1966; Dashchenko et al. 2011). Errors of operations in modern computers are also close to zero. Therefore, it is assumed that the accuracy of the calculation results, which is associated with overcoming the errors in 1–3, is ensured by the professional use of modern certified finite element analysis tools.

Possible errors here can only be explained by the incompetence (or human factor) of the performers of the relevant calculations. The causes of such errors are not investigated separately in this work. At the same time, they can be identified in the general scheme of the developed concept.

12.3.2 Errors Caused by Deviations of Input Data

When solving applied problems of mechanics, it is necessary to take into account that the values of many input calculation parameters included in the mathematical description or algorithm for solving the problem are determined experimentally with some error. This applies, in particular, to the physical and mechanical characteristics of the material, as well as the reliability of determining the parameters of external loads (aero-, hydrodynamic, thermal, etc.), forces of interaction with other substructures in the form of boundary conditions, etc.

In the process of manufacturing the structure, geometric dimensions such as sections of power elements, wall thicknesses of shells, etc., may receive technological inaccuracies. Deviations of some of the input data are random and, therefore, poorly controlled. This applies in particular to structural elements made of composite and powder materials since, in this case, the material is manufactured together with the structure itself and, therefore, its physical and mechanical characteristics are predictive (approximate) values (Dzyuba and Sirenko 2022; Gomeniuk et al. 2009; Guz 1993, 2003; Marchuk et al. 2018).

Obviously, the error of the results of calculations with approximate numbers is a consequence of the errors of the components. Thus, in particular, it is known that the largest absolute error of an algebraic sum is equal to the sum of the absolute errors of the terms. Similar conclusions can be drawn for other operations (Berezin and Zhidkov 1966).

The possibility of accumulating the error of the result of calculating the deviations of input data is obvious from the simplest case of calculating the stresses in the rod under axial load or bending.

Let the longitudinal force $N_{\max}$, moment $M_{\max}$, area F, and moment of resistance W of the cross section of the rod are predefined (or specified) within the technical theories with an accuracy of 5 %. Then, under adverse circumstances, the deviation of these components in the same direction (increase or decrease), the calculation result $\sigma_{\max} = N_{\max}/F$, $\sigma = M_{\max}/W$ can actually double the error. Obviously, as a result of the recursion of such operations, the calculation error can hypothetically be significantly larger. On the other hand, in the presence of errors of individual input data in different directions, the error of the calculation result can be reduced or completely leveled.

The task of assessing the accuracy of the result of the computational process includes a number of other problematic aspects. Thus, if the solution of a certain problem continuously depends on the input data, and a small change in the input data corresponds to a small change in the solution, the problem is called input-stable.

The simplest example of instability to errors in input parameters can be a system of linear algebraic equations:

$$\begin{aligned} 10.1x + 9.9y &= 20.0 \\ 9.9x + 10.1y &= 20.0, \end{aligned}$$

which has an obvious solution (1; 1). If we consider a system

$$\begin{aligned} 10.1x + 9.9y &= 20.1 \\ 9.9x + 10.1y &= 19.9, \end{aligned}$$

whose right-hand sides are changed only by 0.5%, then its solution will be (1.5; 0.5), which differs from the solution of the original system by 50%.

Thus, the determination of the accuracy of the result of even the simplest computational process in the conditions of the approximate nature of the input data is a rather difficult task, the analysis of which cannot be ignored in the process of performing a sequence of a large number of computational operations, which are algorithms for solving problems of mechanics.

12.3.3 Influence of Strength Criteria on the Result of Critical Loads Calculation

The source of deviations in the results of prediction of the critical (destructive) load on the structure (loss of its bearing capacity) is also the choice of the material failure criterion (Lou et al. 2012; Matvienko 2006; Obraztsov et al. 1977; Wierzbicki et al. 2005).

Thus, in particular, even the results of formal application of well-known classical strength criteria:

$$\sigma_1 \leqslant [\sigma_+], \quad |\sigma_3| \leqslant [\sigma_-], \tag{12.1a}$$

$$\sigma_{\text{equiv}} = \sigma_1 - \mu(\sigma_2 + \sigma_3), \tag{12.1b}$$

$$\sigma_{\text{equiv}} = \sigma_1 - \sigma_3, \tag{12.1c}$$

$$\sigma_{\text{equiv}} = \frac{\sqrt{2}}{2}\sqrt{(\sigma_1 - \sigma_2)^2 + (\sigma_2 - \sigma_3)^2 + (\sigma_3 - \sigma_1)^2}, \tag{12.1d}$$

which are sufficiently reliable for certain materials, types of deformation, and applications, have discrepancies (for example, at $\sigma_1 \neq 0$, $\sigma_2 = 0$, $\sigma_3 \neq 0$, and $\mu \neq 0$).

For a power structure, which consists of a certain number of components of different shapes, materials, connecting elements, purpose, etc., which is under extreme spatial loading, the complexity of this problem increases significantly.

As an example, the destruction of the composite material of the shell of rotation is determined by many factors: the rupture of fibers in the tensile zones, the loss of their stability in compression, delamination between the binder and reinforcing material, as well as between individual layers, cracking of the binder, the formation and development of cracks, as well as the appearance of other damages of different nature and in different sequences (Guz 1993, 2003; Obraztsov et al. 1977).

Therefore, when assessing the strength of heterogeneous composite structures, their material is often forced to be endowed with some mechanical characteristics of the orthotropic environment averaged in certain directions. Different variants of the approach to averaging the mechanical characteristics of such an anisotropic material are based, as a rule, on computational and experimental methods (Dzyuba and Sirenko 2022; Gomeniuk et al. 2009; Marchuk et al. 2018).

One of such criterion is, in particular, Hill's energy criterion (Obraztsov et al. 1977):

$$\left(\frac{\sigma_1}{\sigma_1^*}\right)^2 - \frac{\sigma_1\sigma_2}{\sigma_1^{*2}} + \left(\frac{\sigma_2}{\sigma_2^*}\right)^2 + \left(\frac{\tau_{12}}{\tau_{12}^*}\right)^2 \leqslant 1, \tag{12.2}$$

Here, σ_1, σ_2, and τ_{12} are the calculated values of stresses, σ_1^*, σ_2^*, and τ_{12}^* are the boundary values of the strength of the material in tension–compression along and across the reinforcing elements and in shear tests, respectively. In this case, the strength of the material in tension and compression is considered to be the same.

Obviously, with such "averaging", an error in the assessment of the destructive stresses is also inevitable.

In some cases, to assess the strength of composite material can be used and simpler first theory of strength $|\sigma_1| \leqslant [\sigma]$, $|\sigma_2| \leqslant [\sigma]$, which determines the strength of the reinforcing material along the paths of maximum principal stresses.

In general, today there is a significant number of fracture criteria based on the principles of achieving the values of stresses, deformations, absorbed energy, their combinations, and other mechanical factors of certain critical values. At the same time, the assessment of critical (destructive) load with their help often gives significant (up to 30% or more) discrepancies.

In addition, a number of criteria are unacceptable for determining critical loads in the field of aerospace engineering. These are (for obvious reasons) the criteria based on cracking models, including Griffith's energy criterion, the Irwin force criterion, the Rice–Cherepanov integral (Matvienko 2006), and criteria that use statistical data.

Among the strength criteria that are used for aluminum alloys in numerical FEM analysis, it should be noted the criterion (Dashchenko et al. 2011; Lou et al. 2012), which is based on calculation:

- equivalent strains:

$$\varepsilon_{\text{equiv}} = \sqrt{\frac{2}{3}\left(\varepsilon_1^2 + \varepsilon_2^2 + \varepsilon_3^2\right)}, \tag{12.3}$$

  where ε_i $(i = 1, 2, 3)$ are the principal deformations;
- maximal shear stresses:

$$\tau_{\max} = \max\left(\frac{\sigma_1 - \sigma_2}{2}, \frac{\sigma_2 - \sigma_3}{2}, \frac{\sigma_3 - \sigma_1}{2}\right), \tag{12.4}$$

  where σ_i $(i = 1, 2, 3)$ are the principal stresses;
- the maximal equivalent stress criterion by Mises:

$$\sigma_{\text{equiv}} = \frac{1}{\sqrt{2}} \times \sqrt{(\sigma_{11} - \sigma_{22})^2 + (\sigma_{22} - \sigma_{33})^2 + (\sigma_{33} - \sigma_{11})^2 + 6\left(\sigma_{12}^2 + \sigma_{23}^2 + \sigma_{13}^2\right)}, \tag{12.5}$$

  which is one of the main packages of FEM analysis applications.

In these criteria, the yield strength of the material σ_T is considered to be the ultimate stress value.

Among more general criteria that take into account deformation beyond the elastic limit, the following should be noted:

- Johnson–Cook criterion, which defines the strain values at which failure occurs, in the form of

$$\varepsilon_{\text{equiv}} = C_1 + C_2 \exp(C_3 \eta), \tag{12.6}$$

  where

$$\eta = \frac{\sigma_m}{\sigma_{\text{equiv}}} = \frac{(\sigma_1 + \sigma_2 + \sigma_3)/3}{\sqrt{(\sigma_1 - \sigma_2)^2 + (\sigma_2 - \sigma_3)^2 + (\sigma_3 - \sigma_1)^2 / \sqrt{2}}}.$$

  Here, σ_m is average (hydrostatic) stress; σ_{equiv} is equivalent Mises stress; constants C_1, C_2 and C_3 are determined experimentally.
- The Xue–Wierzbicki criterion (Wierzbicki et al. 2005) in known publications is considered the most used in finite element analysis for elastic–plastic deformation. According to this criterion, the limit values of deformations are set in the form:

$$\sigma_{\text{equiv}} = C_1 \exp(-C_2\eta) - \left[C_1 \exp(-C_2\eta) - C_3 \exp(-C_4\eta)\right] \left(1 - \zeta^{1/n}\right)^n , \tag{12.7}$$

where

$$\eta = \frac{\sigma_{\text{m}}}{\sigma_{\text{equiv}}}, \quad \zeta = \frac{27\sigma_1\sigma_2\sigma_3}{2\sigma_{\text{equiv}}^3}.$$

Here, n is the hardening index, which can take values on the interval [0, 1]: 0 and 1 correspond to perfectly plastic and elastic materials, respectively. For most metals, $n \in (0.1,\ 0.5)$. In particular, for aluminum alloys $n \simeq 0.16$. Constants C_i, $(i = 1, \ldots, 4)$ are determined experimentally.

The disadvantage of using criteria (12.6) and (12.7) is the use of special physical and mathematical characteristics of the material, which are determined (usually with a certain error) experimentally.

Thus, the choice of the failure criterion is also a source of uncertainty in determining the value of the critical load on the structure.

12.3.4 Numerical Studies

Modern application packages of finite element analysis programs provide ample opportunities for using the full range of tools of this effective approach (Dashchenko et al. 2011). This includes, in particular, the use of different numbers and types of finite elements, their shape, size, ways to take into account the conditions of interaction with other substructures, external influences, etc.

Further, the results of numerical calculations of structural elements with specially selected geometric, physical, mechanical, and computational parameters are presented in order to more clearly demonstrate certain aspects of the problem under discussion.

The calculation schemes (Fig. 12.2) of a rectangular flat panel (plate) clamped along one of the edges, which has a rectangular hole symmetrically and asymmetrically located relative to the axes of symmetry, for cases of its deformation under the action of a uniformly distributed load q (Fig. 12.2a) and displacement Δ = const (Fig. 12.2b) are considered.

Such a problem may arise as a result of the fragmentation of a complex structure into simpler components. As an example, a flat panel or cylindrical shell, the individual belts of which have one or more rectangular holes. Under compression, these can be compartments or elements of the rocket body (Fig. 12.1) or the shaft, loaded with the mass of the above fragments, etc.

The peculiarity of the problem lies in the choice of boundary conditions as conditions for the interaction of individual substructures. In this case, a decision can be made about the uniformity of q distribution or constancy Δ (Fig. 12.2), although in the presence of heterogeneity (of different nature) of the substructure, these parameters may deviate from uniform values.

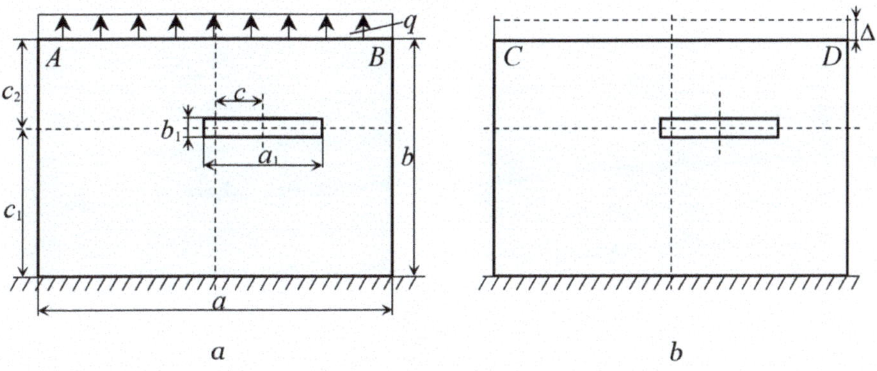

Fig. 12.2 Calculation schemes of plates

The numerical experiment was carried out for the cases of asymmetric arrangement of a rectangular hole at $c = 0, 0.06, 0.12, 0.18, and 0.24$ m of the plate with the following parameters: $a = 0.9$ m; $b = 0.6$ m; plate thickness $d = 0.003$; $a_1 = 0.3$ m; $b_1 = 0.05$ m; $c_1 = 0.225$ m; $c_2 = 0.375$ m; modulus of elasticity of the material $E = 200$ GPa; Poisson's ratio $\mu = 0.3$; the grid is square; and the number of nodes is 8440. For the convenience of comparisons, q and Δ are taken from the condition of equality of the equal acting force $P = 8 \cdot 10^4$ N at the clamping of both design schemes.

The stress distribution is shown in Fig. 12.3 for the design schemes of Fig. 12.2 (at $c = 0$ in Fig. 12.2a, b at $c = 0.24$ m in Fig. 12.3c, d).

From the calculation results, it follows that for the case of asymmetrically located cutout (Fig. 12.3b) the maximum stresses $\sigma_{\max} = 177.65$ MPa, and for the scheme (Fig. 12.3a) $\sigma_{\max} = 146.86$ MPa. The difference is about 21 %.

This is due to the fact that for the design scheme with asymmetrically arranged holes, an eccentric tensile scheme is implemented with the emergence of an additional moment, as well as the presence of a bend of the plate fragment over the hole under the action of a longitudinal load.

During the numerical experiment, the expectedly high sensitivity of the calculation result in places with a high gradient of change (concentration) of stresses from the number of finite elements was also found. In addition, the coordinates of the maximum stress point location and the general picture of stress distribution in the plate and displacements of its loaded contour have changed.

For the plate loaded by displacement, the limitation of the degrees of freedom $\Delta = \text{const}$ along the CD contour (Fig. 12.2b) reduces the stress level compared to the schemes (Fig. 12.3a, c), and the influence of the asymmetric location of the cutout on the stress value is significantly less (about 1 %) and changes from $\sigma_{\max} = 101.49$ MPa (Fig. 12.3b) to $\sigma_{\max} = 106.8$ MPa (Fig. 12.3d). The qualitative picture of stress distribution also practically does not change.

The case of weak sensitivity of the calculation results to changes in the input data is demonstrated by the example of numerical studies of the effect of changes in the

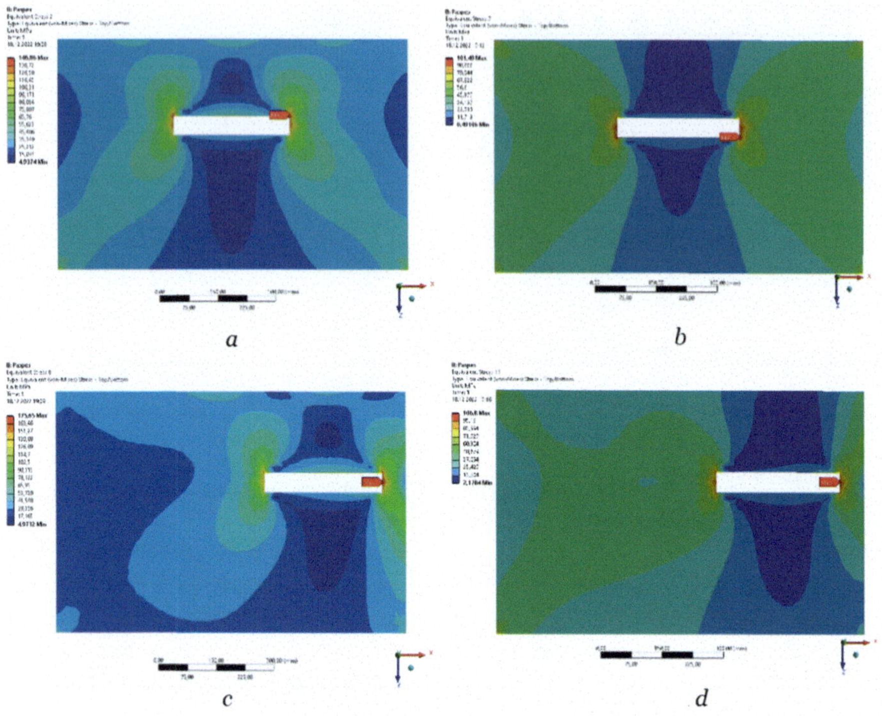

Fig. 12.3 Results of calculation of the plate with a hole

Fig. 12.4 Calculation scheme of the composite shell

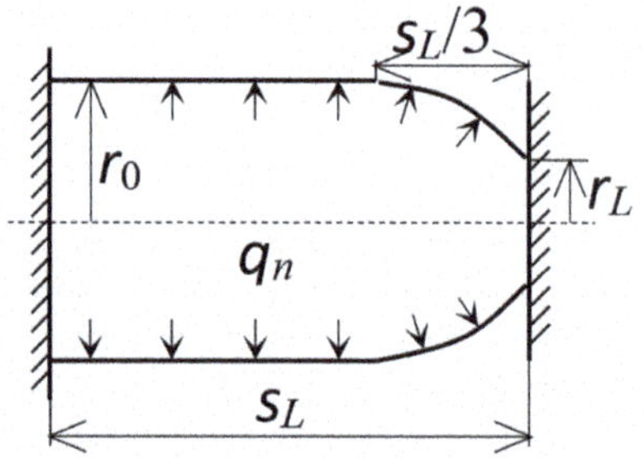

values of physical and mechanical characteristics of the material E_α and E_β on the maximum stresses in a clamped fragment of a two-layer fiberglass shell of rotation under the action of internal pressure q_n (Fig. 12.4). The parameters of the shell are as follows: $s_0 = 0$ m; $s_L = 1.08$ m; $r_0 = 0.788$ m; $r_L = 0.57$ m; $h = 0.4 \cdot 10^{-2}$ m; $q_n = 0.1$ MPa; Poisson's ratios, $\mu_\alpha = 0.1$ and $\mu_\beta = 0.1$; shear modulus, $G_{\alpha\beta} = E_{\alpha(\beta)}/(2[1 + \mu_{\alpha(\beta)}])$; h is the shell wall thickness.

From the results of the calculations shown in Fig. 12.5, it follows that when changing the value of the longitudinal modulus of elasticity from $E_\alpha = 5$ GPa (Fig. 12.5a) to $E_\alpha = 10$ GPa (Fig. 12.5b) at $E_\beta = 15$ GPa the maximum stress by Mises did

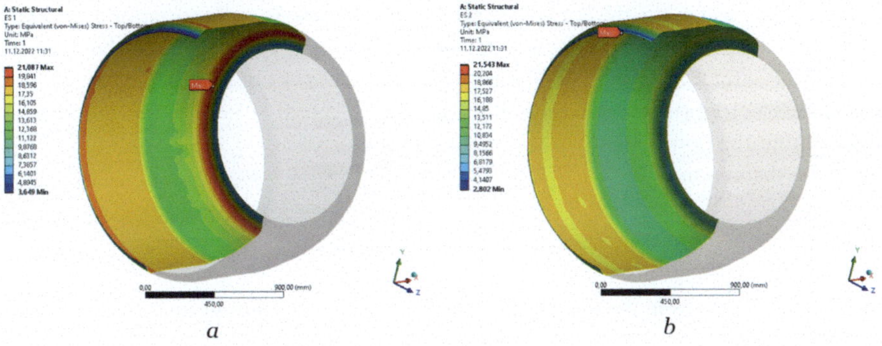

Fig. 12.5 Results of composite shell calculation

not actually change (from $\sigma_{\max} = 21.1$ MPa to $\sigma_{\max} = 21.5$ MPa). There was only a slight change in the picture of the stress–strain state in the composite shell as a whole. Thus, for the parameters adopted in the calculation, the calculation scheme turned out to be weakly sensitive to the input data parameter E_α.

At the same time, the coordinates of the location point $\sigma_{\max}$ have changed (from $s = 1.085$ m to $s = 0.04$ m), which is important for determining the points of possible structural failure and installing appropriate measuring devices for experimental destructive tests.

From the above results of the numerical experiment, it follows that the choice of input data and boundary conditions can have a significant impact on the calculation results, which indicates the need for additional computer analysis to substantiate the reliability of the calculation results.

12.4 Methods of Forecasting the Destructive Load

12.4.1 Estimation of the Error of the Strength Calculation Result From Deviations of Input Data

It is proposed to solve the problem of increasing the reliability of predicting the value of the critical (destructive) load of the structure without conducting costly destructive field tests by replacing them with the process of systematic, although rather laborious, computer modeling of the relationship between the results of calculations and the input parameters of the problem using elements of the theory of sensitivity (Diskovskiy et al. 2012; Nazarenko et al. 2019).

Provided that a perfect (accurate) finite element analysis tool is used, the results of its calculation will be a function $P^s(\bar{q})$ of the vector of input parameters $\bar{q} = (q_1, q_2, \ldots, q_n)$.

Taking $P^s_{\exp}$ as known the data of measurements of stresses, deformations, or other parameters of the strength criterion in the *s-th* (arbitrary or the most dangerous) point of the structure with coordinates $\bar{X}^s = (x^s, y^s, z^s)$ obtained from the results of non-destructive (in the elastic region) testing of a real structure, the problem is to determine the components of the input parameter vector $\bar{q} = (q_1, q_2, \ldots, q_n)$ at which the discrepancy

$$\Delta P = \left| P^s(\bar{q}) - P^s_{\exp} \right| \tag{12.8}$$

between these values in the most dangerous for destruction s th point of the structure will be minimal.

If it is necessary to determine the discrepancy between the calculated and experimental data simultaneously for s measurement points, the criterion (12.8) is adopted by the least squares scheme (Berezin and Zhidkov 1966)

$$\Delta P = \sqrt{\sum_s \left(P^s(\bar{q}) - P^s_{\exp} \right)^2}$$

To study the influence of possible deviation of input data from their nominal values on the deviation of certain parameters of the final result of calculations, that is, the sensitivity of the n-dimensional function $P^s(\bar{q})$, the method of coordinate descent (the method of alternate change of varied Gauss–Seidel parameters q_i $(i = 1, \ldots, n)$ is used (Boyarinov and Kafarov 1969). To construct the response function, the basic calculation of the structure $P^s_{0i} = P^s\left(q_1, q_2, \ldots, q_i^0, \ldots, q_n\right)$ and the sequence of calculations $P^s_{\pm i} = P^s(q_1, q_2, \ldots, q_i \pm \Delta q_i, \ldots, q_n)$ for each of the arguments q_i $(i = 1, \ldots, n)$ is performed.

By performing such a computational process for different node points Δq_i by, in particular, the scanning method (Boyarinov and Kafarov 1969; Reklaitis et al. 1983), it is possible to find the value of each of the input parameters $q_i^{\min}$ $(i = 1, \ldots, n)$ corresponding to the smallest deviation of the calculation results and experimental data.

To reduce the number of such direct calculations in the construction of the response function of the calculated values $P^s_{\pm i}$ for each of the parameters q_i $(i = 1, \ldots, n)$, the method of quadratic approximation (Powell's method) is used (Reklaitis et al. 1983).

After introducing the notation $q_i^- = q_i^0 - \Delta q_i$, $q_i^+ = q_i^0 + \Delta q_i$ $(i = 1, \ldots, n)$ and calculating the values P^s_{-i}, P^s_{0i}, and P^s_{+i} for a certain point s, the quadratic polynomial of the response function on a uniform grid $(\pm \Delta q_i)$ is given as

$$f_i(\bar{q}) = a_0^i + a_1^i\left(q_i - q_i^-\right) + a_2^i\left(q_i - q_i^-\right)\left(q_i - q_i^0\right). \tag{12.9}$$

Assuming the equality of the function $f_i(\bar{q})$ and P^s_{-i}, P^s_{0i}, P^s_{+i} at the interpolation nodes q_i^-, q_i^0, q_i^+, we can obtain:

$$a_0^i = P_{-i}^s, \quad a_1^i = \frac{P_{0i}^s - P_{-i}^s}{q_i^0 - q_i^-}, \quad a_2^i = \frac{1}{q_i^+ - q_i^0}\left(\frac{P_{+i}^s - P_{-i}^s}{q_i^+ - q_i^-} - \frac{P_{0i}^s - P_{-i}^s}{q_i^0 - q_i^-}\right). \tag{12.10}$$

Thus, according to the three calculations at $q_i = q_i^-$, $q_i = q_i^0$, $q_i = q_i^+$ and the corresponding values of the result function P_{-i}^s, P_{0i}^s P_{+i}^s, we can calculate the values of the coefficients a_0^i , a_1^i, and a_2^i of the approximating quadratic polynomial for all q_i $(i = 1, \ldots, n)$.

It should be noted that to improve the accuracy of approximation of the result function $P^s(\bar{q})$, if necessary, an approximating polynomial of a higher order can be used, in particular, Newton's polynomial on an uneven grid. At the same time, the construction of the approximating polynomial $f_i(q_1^*, q_2^*, \ldots, q_i, \ldots, q_n^*)$ of the n th order requires more design calculations.

Having approximating polynomials in the form of one-dimensional functions $f_i(q_1^*, q_2^*, \ldots, q_i, \ldots, q_n^*)$, we find the values $q_i^{\min}$ that correspond to the smallest deviation of the function f_i (which approximates the calculation results $P^s(\bar{q})$) and the data $P_{\exp}^s$ of non-destructive (in the elastic region) test in the form

$$\min_{q_i} \left| P_{\exp}^s - f_i(q_i) \right| .$$

Given the simplicity of calculating the approximating quadratic polynomial, this problem is easily solved by the scanning method with a uniform or (to reduce the number of calculations) nonuniform step (Boyarinov and Kafarov 1969).

The obtained calculation results $\delta_i = q_i^{\min} - q_i^0$ allow us to determine the direction and value of the change in the input parameters q_i $(i = 1, \ldots, n)$ necessary to achieve compliance of the calculation results $P^s\left(\bar{q}^{\min}\right)$ with the non-destructive test data $P_{\exp}^s$. The value of this deviation is calculated as

$$\Delta P_i^s = P^s\left(q_1^*,\ q_2^*, \ldots, q_i^{\min}, \ldots, q_n^*\right) - P_{\exp}^s.$$

The obtained results make it possible to identify the causes of such deviations, in particular, by additional measurements of the relevant parameters of the real structure, to make appropriate adjustments and to build a process of its more correct calculation and creation.

If it is necessary to further reduce this discrepancy, taking into account the mutual influence of changes in input data on the calculation result, an algorithm for the method of successive approximations can be built by calculating the values of the input parameters of the next $k + 1$ search step in the form:

$$q_i^{k+1} = q_i^k + \gamma\left(q_i^{\min} - q_i^k\right), \tag{12.11}$$

where $0 < \gamma \leqslant 1$ is the relaxation coefficient.

In this case, the components of the input data q_i $(i = 1, \ldots, n)$, the influence of which on the result of the calculation is insignificant $|\delta_i| \leqslant \varepsilon$, as well as the parameters determined by the results of expert assessments, are excluded from consideration.

Algorithms for accelerating the convergence of iterative processes (12.11) are presented in Dzyuba et al. (2020).

12.4.2 Application of the Approach

The control of errors in the calculation of the stress–strain state of the structure is carried out by conducting a comparative analysis of the obtained numerical results with the data of parallel non-destructive (under elastic deformation) testing of a real object using strain and stress measuring instruments (Tkachuk et al. 2017; Wierzbicki et al. 2005) or holographic interferometry methods (Dzyuba and Selivanov 2021; Gudramovich et al. 2017).

This approach allows us to establish the reliability of the calculated characteristics by identifying their possible deviations and making appropriate adjustments to reduce the errors of numerical analysis caused by a combination of the above factors: the introduction of initial simplifying hypotheses and assumptions related, in particular, to the choice of the calculation scheme, mathematical model, as well as the error of software for FEM analysis and the influence of input data errors. The result of this stage will be the algorithm of structural analysis in the elastic domain.

Experimental tests of the real structure are carried out under the conditions

$$\sigma_{5q} \leqslant \lambda\sigma_{kr}, \tag{12.12}$$

where $0 < \lambda \leqslant 1$ is the coefficient of "confidence" in the preliminary calculation results; σ_{5q} and σ_{kr} are determined at the points of installation of measuring devices (the most dangerous for destruction) simultaneously according to several selected strength criteria (Sect. 12.3.3), according to the results of which the criterion with the closest calculated values of critical stresses to the data of experimental tests in the elastic region is selected.

When the limit values (12.12) in the process of such a parallel numerical–experimental study are reached, the increase in the load on the real structure stops. In the presence of unacceptable deviations in the calculation results and experimental data, computer simulation of the influence of deviations in the values of input parameters on the results of strength calculation is carried out (Sect. 12.4.1).

After identifying the causes of inaccuracy in the calculation and making appropriate adjustments, the numerical study of the stress–strain state of structures continues beyond the limits of elasticity using the failure criterion determined at the first stage and further reaching the critical state and virtual failure.

The results of this stage allow us to establish more reliably the value of the critical (destructive) load and the general picture of the loss of bearing capacity beyond the elastic limit.

Additional substantiation (if necessary) of the results of control of design errors and, most importantly, the choice of the failure criterion can be carried out by comparative analysis of the results of numerical calculations and destructive test data of analogues of the structure in the form of simplified small-sized models, fragments, or samples made of the material of the structure.

12.5 Conclusions

The analysis of the causes of errors in calculations of the strength of complex structures of new technology and ways to identify and correct them is presented.

A methodology for studying the influence of deviations of input data on the calculation result has been developed, the essence of which is to systematically evaluate and control the error accumulated in the process of creating a structure in order to replace rather costly destructive tests of full-scale structures with less costly, although rather laborious, additional comparative computer modeling using data from non-destructive (in the elastic region of deformation) testing of the structure.

The approach allows:

- to obtain data for controlling the magnitude of the accumulated error and to build and justify the methodology for more reliable prediction of the limiting (destructive) values of the parameters of the stress–strain state of complex structures;
- to establish the quantitative influence of the scatter of the main input parameters on the result of calculations and to determine the maximum tolerances for input data errors, including those that are introduced into the structure during its manufacture, and, as a result, are not taken into account in the calculations;
- to justify the choice of a rational value of the coefficient k of safety margin, which in this case will be really consistent with the results of numerical calculations, errors of calculated mathematical models, the influence of deviations of input data, manufacturing technology, and the results of non-destructive tests.

The results of the research can be used to make adjustments to the design and calculation data in order to increase the reliability of the bearing capacity calculation results and reduce the volume or complete rejection of destructive tests.

References

Berezin IS, Zhidkov NP (1966) Methods of calculations. Nauka, Moscow

Boyarinov LI, Kafarov BV (1969) Optimization methods in chemical technology. Chemistry, Moscow

Dashchenko AF, Lazareva DV, Suryaninov NG (2011) Ansys in problems of engineering mechanics. Burun and K, Kharkov

Degtyarev AV (2014) Problems and prospects. ART-PRESS, Dnipro

Diskovskiy AA, Prudko EI, Khoroshmanenko PG (2012) Sensitivity analysis in designing structures from functionally graded materials. Probl Comput Mech Strength Struct 18:57–70

Drobenko BD, Klimenko DV, Kushnir RM, et al (2020) Methodology of investigation of structural integrity of rocket engineering. Space Technol Rocket Armament 2

Dzyuba AP, Selivanov YuM (2021) Research of strength characteristics and optimization of parameters of case structures using holographic interferometry. J Phys: Conf Ser 1741:1–7

Dzyuba AP, Sirenko VN (2022) Algorithmization of the determination of physicomechanical characteristics of the material of multilayer composite shells of revolution varying along the meridian. J Math Sci 2224:93–103

Dzyuba AP, Safronova IA, Levitina LD (2020) Algorithm for computational costs reducing in problems of calculation of asymmetrically loaded shells of rotation. Strength Mater Theory Struct 105:107–117

Gomeniuk SI, Grebeniuk SN, Olshansky VE, Lavrenko AC (2009) Application of various theories of determination of elastic characteristics of composite materials for structural calculation. Vestnik Engin 2:139–142

Gudramovich VS, Dzyuba AP, Selivanov YM (2017) Methods of holographic interferometry in mechanics of inhomogeneous thin-walled structures. Lyra, Dnipro

Gudramovich VS, Dzyuba AP, Selivanov YM (2017) Methods of holographic interferometry in mechanics of inhomogeneous thin-walled structures. Lyra, Dnipro

Hudramovich VS, Dzyuba AP (2009) Contact interaction and optimization of locally loaded shell structures. J Math Sci 162(2):231–245

Lou Y, Huh H, Lim S, Pack K (2012) New ductile fracture criterion for prediction of fracture forming limit diagrams of sheet metals. Int J Solids Struct 49:3605–3615

Marchuk MV, Kharchenko VM, Hom'yak MM (2018) Mathematical model for determining the effective physical and mechanical characteristics of a rechristened reinforced composite ball. Appl Probl Mech Math 16:64–73

Marchuk MV, Sirenko VM, Drobenko BD (2020) Methodology of determining ruinous stresses on large-sized thin-walled structures taking into account the results of non-ruinous tests. Appl Probl Mech Math 18:133–138

Matvienko YG (2006) Models and criteria of fracture mechanics. Fizmatlit, Moscow

Mossakovsky VI, Makarenkov AG, Nikitin PI et al (1990) Rocket structure integrity. Vysshaya Shkola, Moscow

Obraztsov IF, Vasiliev VV, Bunakov VA (1977) Optimal reinforcement of shells of rotation from composite materials. Mashinostroenie, Moscow

Obraztsov IF, Vasiliev VV, Bunakov VA (1977) Optimal reinforcement of shells of rotation from composite materials. Mashinostroenie, Moscow

Tkachuk NA, Hlan AV, Sheiko AI et al (2017) Development of mathematical apparatus for solving problems of the calculation and experimental investigation of elements of mechanical systems. J NTU "KPI" Ser Mech Eng CAD 12:110–131

Wierzbicki T, Bao Y, Lee YW, Bai Y (2005) Colibration and evaluation of seven fracture models. Int J Mech Sci 47(4–5):719–743

Usyukin VI (1988) Structural mechanics of space engineering structures. Mashinostroenie, Moscow

Veretelnik YV, Tkachuk AV, Kokhanovskaya OV et al (2017) Computer modeling of processes and states of complex systems: Substantiation of model parameters by experimental and computational method. J NTU "KPI" Ser Mech Eng CAD 12:14–25

Wierzbicki T, Bao Y, Lee YW, Bai Y (2005) Colibration and evaluation of seven fracture models. Int J Mech Sci 47(4–5):719–743

Chapter 13
Strength Assessment of Critical Elements of Nuclear Power Plant Equipment: State-of-the-Art Calculation Approaches

Valery Kharchenko and Alexander Chirkov

13.1 Introduction

Extension of operation of nuclear power plants (NPPs) is a global trend, the strategy of which is to gradually extend their service life. In general, a significant part of NPP power units in the world is already operated beyond the design life. Highly qualified scientific and technical support is required to ensure their safe service conditions, in particular, the current task is to calculate the structural strength and predict the residual service life of NPP equipment components.

In the G.S. Pisarenko Institute of Problems of Strength of the National Academy of Sciences of Ukraine (IPS NASU), theoretical and applied results on the development and application of refined models and methods for solving actual problems of mechanics related to the justification of the strength and service life of the equipment of the first circuit of the reactor plant with WWER water-water power reactor were obtained (Chirkov 2020; Kharchenko et al. 2018).

Scientific and applied developments of the IPS NASU take into account modern trends in world practice in the calculation justification of the strength of NPP equipment elements, as well as contain new conceptual approaches to solving nonlinear boundary value problems of mechanics of deformed structures. The results of fundamental and applied research are the methodological basis of the refined design analysis for solving urgent problems of modern nuclear power engineering-ensuring the conditions for safe operation of NPP power units and justification of their service life extension.

It should be noted that the development of models and methods of strength calculation and their application to the analysis of structural mechanics problems is one of the priority directions in the activities of IPS NASU for the last 40 years. Scientific and applied research aimed at improving the methodology for the refined calculation

V. Kharchenko · A. Chirkov (✉)
G.S. Pisarenko Institute of Problems of Strength, National Academy of Science of Ukraine, Kyiv, Ukraine
e-mail: chirkale82@gmail.com

A. N. Guz et al. (eds.), *Advances in Mechanics*, Advanced Structured Materials 191,
https://doi.org/10.1007/978-3-031-37313-8_13

of the stress-strain state and fracture resistance of NPP equipment and the development of the relevant regulatory framework have been implemented in practice within the justification of the extended service life of many Ukrainian NPP power units.

The authors did not seek to fully cover all issues related to the methodological aspects of the strength calculation justification, consideration, and analysis of all factors affecting the calculated assessment of NPP structures and fracture resistance. Below, some important results of scientific and applied research are presented, which, in our opinion, complement the known data of other authors and may be of interest to specialists in the field of calculation methods for strength analysis of NPP equipment. Practical results demonstrate the capabilities and advantages of the developed methods of the refined calculation of stress-strain state for assessing the strength of critical elements of the equipment of the first circuit of the WWER-1000 reactor unit.

13.2 Mathematical Models of Material Deformation Processes

The refined analysis of the NPP primary circuit equipment stress-strain state is based on the solution of boundary value problems of thermo-visco-plasticity taking into account the combined influence of the following factors: spatial nature of the stress-strain state of structures; mutual influence of the elements of the connected reactor equipment; inhomogeneity and non-stationarity of heating and cooling; inelastic deformation and loading history; residual technological stresses and deformations; subcritical metal damage; radiation effects, swelling and creep; contact interaction; dependence of metal properties on temperature and irradiation dosage.

Thus, the modeling of the kinetics of stress-strain state and the analysis of fracture resistance of NPP equipment elements belong to the most complex problems of mechanics of materials and structures, mathematical and computational methods of analysis.

The equations of state are formulated taking into account the specifics of physical and mechanical processes that occur in the material during the loading of the studied structural element, using modern and developed mathematical models. In particular, models that allow describing the processes of radiation swelling and radiation creep of irradiated metal, taking into account the accumulated damaging dose, irradiation temperature, and the effect of stresses and accumulated irreversible deformation on the swelling and creep of the material, were used to assess the progressive shape change of the reactor core lining (Chirkov 2020; Kharchenko et al. 2018).

The nucleation and growth of the volume concentration of ductile fracture pores in the material are taken into account by the Rice–Tracey–Huang models (Huang 1991; Rice and Tracey 1969), and for the conditions of prolonged neutron irradiation—by the proposed equation from the Kachanov solution for a spherical cavity in an elastic-plastic medium (Kachanov 1969). Irreversible deformations take into account instantaneous plastic deformations, radiation swelling, radiation creep, and structural volumetric deformations describing the growth of pore concentration in the material.

Thus, the analysis of deformation processes in the general case is based on the definition of the total strain e in the form

$$e = e_e + e_T + R/3 + e_p + e_R + K/3, \tag{13.1}$$

where e_e, e_T are the elastic and thermal strain components, respectively, R is radiation swelling of irradiated metal taking into account the influence of stresses, e_p is instantaneous plastic strain, e_R is radiation creep in case of prolonged irradiation; K is volumetric concentration of ductile fracture micro-pores (Chirkov 2020; Kharchenko et al. 2018).

Dependence of the free swelling value R_0 on the damaging dose Z and irradiation temperature T is taken on the basis of experimental studies of radiation porosity of irradiated austenitic steels (Chirkov 2022):

$$R_0 = c_0 Z^{\beta} \exp\left(-r(T - T_{max})^2\right), \tag{13.2}$$

where $T_{\max}$ is the temperature of maximum swelling; c_0, β and r are the material constants.

The equation describing the effect of stresses and accumulated irreversible deformation on swelling R was proposed in Chirkov (2022) based on the analysis of experimental data on the determination of radiation swelling of austenitic steels for different stress states. The dependence of the swelling value R is taken in the form

$$R = R_0 R_1 e^{-\lambda Q}, \quad R_1 = 1 + C_R\left(\omega\sigma_m + (1-\omega)\bar{\sigma}\right), \tag{13.3}$$

where σ_m is the average normal stress; $\bar{\sigma}$ is stress intensity; C_R is material constant, which generally depends on the damaging dose and irradiation temperature; ω is a weighting factor that determines the degree of stress influence of σ_m and $\bar{\sigma}$ on swelling R; λ is the material constant, which is determined by experimental data; Q is a parameter that takes into account the accumulated irreversible deformation (Chirkov 2020).

The coefficient ω depends on the level of damaging dose and irradiation temperature, i.e., in general, on the swelling value, and the choice of ω is carried out taking into account the condition that the swelling R is positive. If $R < 1\%$, the main influence on swelling is the stress intensity $\bar{s}$ for $R > 1\%$, the average normal stress dominates σ_m. Hence, in the case of $R < 1\%$ the coefficient ω is close to zero, and for $R > 1\%$ it increases significantly and approaches the value $\omega_0 = 0.85$ (Chirkov 2022). Using the coefficient $0 < \omega \leq \omega_0$, which depends on swelling, it is possible to describe the material behavior in more detail compared to the choice of $\omega = \omega_0$, especially at the beginning of irradiation. However, it is necessary to ensure that the conditions of the correctness of the equations of state of the material are not violated due to small values of the coefficient ω.

According to Eq. (13.3), the swelling R is affected by both the value of σ_m and $\bar{\sigma}$ which makes it possible to take into account the asymmetry of the influence of stresses σ_m on the process of material swelling for a more accurate approximation of experimental data (MT-D.0.03.391-06 2009). In addition, the swelling R depends on the value of the accumulated irreversible deformation Q, which includes instantaneous plastic strain and radiation creep strain.

On the basis of Eq. (13.3), we obtain the relationship for the increments

$$\frac{dR}{R} = \frac{dR_0}{R_0} + \frac{C_R}{R_1}\left(\omega d\sigma_m + (1-\omega)\, d\bar{\sigma}\right) - \lambda dQ, \tag{13.4}$$

which suggests that the average tensile stress σ_m speeds up or slows down the swelling R depending on the sign of the stress σ_m, whereas the stress intensity $\bar{\sigma}$ accelerates the swelling process, in contrast to the accumulated irreversible deformation Q, which slows down the swelling of the material.

Since swelling R characterizes the volume change, the structural deformations of swelling are the components of the ball tensor, which according to (13.3) depend on both the ball component and the deviatoric components of the stress tensor. This concept of volume change is not generally accepted in the mechanics of materials, but it is consistent with experimental results on radiation swelling.

In general, equation (MT-D.0.03.391-06 2009) is used to describe the constant rate of radiation creep:

$$\frac{\overline{d\varepsilon^c}}{dt} = \left(B_0\frac{dZ}{dt} + C_0\frac{d\tilde{R}}{dt}\right)\bar{\sigma}, \tag{13.5}$$

where $\overline{d\varepsilon^c}$ is the intensity of creep strain growth; t is time or any other parameter that characterizes the load change; B_0 and C_0 are material constants that are weakly dependent on temperature; dZ/dt is the rate of the damaging dose; $d\tilde{R}/dt$ is swelling rate; $\tilde{R}$ is swelling, which is determined by the formula of full or free swelling.

Substituting parameter $\tilde{R}$ in Eq. (13.5) by the compressed swelling R allows one to describe radiation creep more adequately than using free swelling R_0 (Chirkov 2022). In this case, the creep rate depends not only on the stress intensity $\bar{\sigma}$ but also on the average normal stress σ_m because the swelling R also depends on σ_m. Thus, according to the adopted swelling model, the deviatoric components of radiation creep depend on both deviatoric stress components and the ball component of the stress tensor. Such a creep model is not generally accepted in classical theories of plasticity and creep, but it is consistent with experimental data in case of their approximation by Eq. (13.5).

To take into account the porosity of the irradiated material, the equation obtained on the basis of Kachanov's solution for the problem of comprehensive stretching of a spherical cavity in unlimited space is used (Kachanov 1969). It is assumed that the increase in the volume concentration of pores K in the material is due to inelastic deformation, which includes deformations of instantaneous plasticity and radiation creep. Then the increase in pore concentration dK can be described as follows:

$$dK = 3h_1\phi\,(g)\exp\,(h_2 g)\,\overline{d\varepsilon^n}, \tag{13.6}$$

where $g = \sigma_m/\bar{\sigma}$ is the stiffness of the stressed state; $\overline{d\varepsilon^n}$ is the intensity of increments of accumulated irreversible deformations; coefficients h_1, h_2, and the continuous function $\phi\ (g)$ are defined as follows: under the condition of elastic deformation we have $h_1 = 0.75$, $h_{12} = 0$, $\phi(g) = g$; in the presence of plastic deformation: $h_1 = 0.184$, $h_2 = 1.5$, $\phi(g) = 1$. The increment of pore concentration dK is a continuous function with respect to the parameter g and its first derivative also satisfies this condition. These properties of the function dK follow from the Kachanov solution of the deformation of a spherical cavity, because this solution ensures the continuity of radial displacements, deformations, and stresses at the interface between the elastic and plastic parts of the body.

The application of the Kachanov solution to modeling the process of growth of the volume concentration of pores allows to take into account the radiation creep in the elastic region of the deformation diagram of the irradiated material, in contrast to the classical Rice–Tracey–Huang equations, in which the elastic region is not considered. Taking into account this factor affects the results of the analysis of porous material behavior, since radiation hardening occurs with an increase in the dose, which leads to a decrease in the plasticity of the material, and therefore, in the case of prolonged neutron irradiation, the role of radiation creep in the elastic region of the deformation diagram increases. Under such conditions, the main component in the deviatoric components of irreversible deformations is the radiation creep of the irradiated material, and a significant contribution of plastic deformations is local in the zones of stress concentrators. In particular, this peculiarity of material behavior concerns the analysis of irreversible shape change of the WWER reactor core lining during operation, as well as elements of internals subjected to prolonged neutron irradiation.

13.3 Validity of Governing Equations via Drucker's Postulate

Mathematical models that allow describing the processes of inelastic deformation taking into account the influence of radiation effects and accumulated damage lead to non-classical equations of state of the material, and therefore it is necessary to determine the conditions that ensure their correctness. It is worth noting that the analysis of the conditions for the correctness of the equations of state, which take into account the porosity of the irradiated material, is quite relevant, since it allows to prove an unambiguous solution of the nonlinear boundary value problem corresponding to the adopted model of inelastic deformation, as well as to justify the convergence and accuracy of approximate methods for its solution.

On the basis of developed energy approaches based on general principles of nonlinear mechanics of continuous media, the following condition was established (Chirkov 2020). If in the process of loading the elements of elastic-plastic medium are exposed to radiation swelling and radiation creep and there is a growth of ductile fracture pores, the dissipation power of additional stresses $d\boldsymbol{\sigma}$ on the irreversible deformations caused by them $d\boldsymbol{\varepsilon}^n$ will be non-negative only in the case of the inequality:

$$\left(d\boldsymbol{\sigma}, d\boldsymbol{\varepsilon}^n\right) \geqslant 0. \tag{13.7}$$

This condition ensures the correctness of the equations of state and generalizes Drucker's postulate for irradiated porous material, because the inequality in increments is formulated for the total stress and strain components in contrast to the classical theories of plasticity and creep, in which this constraint is defined only for deviatoric components.

13.4 Boundary Problem Statement

The generalized boundary value problem, which allows to describe non-isothermal processes of inelastic deformation taking into account radiation effects and metal damage, is formulated in a quasi-static formulation in the form of integral identities for displacements, strains, and stresses (Chirkov 2020; Kharchenko et al. 2018):

$$\begin{gathered}(\boldsymbol{\varepsilon}(t), \delta\boldsymbol{\sigma})_L = (B\boldsymbol{u}(t), \delta\boldsymbol{\sigma})_L\,; \\ (\boldsymbol{\sigma}(t), \delta\boldsymbol{\varepsilon})_L = \left(D\left(\boldsymbol{\varepsilon}(t), \boldsymbol{\xi}(t)\right)\left(\boldsymbol{\varepsilon}(t) - \boldsymbol{\xi}(t)\right), \delta\boldsymbol{\varepsilon}\right)_L\,; \\ (\boldsymbol{\sigma}(t),\ B\delta\boldsymbol{u})_L = \langle \boldsymbol{f}(t), \delta\boldsymbol{u}\rangle\,,\end{gathered} \tag{13.8}$$

where $\delta\boldsymbol{\sigma}$, $\delta\boldsymbol{\varepsilon}$, and $\delta\boldsymbol{u}$ are arbitrary continuous functions that can be interpreted as variations of stresses, strains and displacements, respectively; t is time or any other parameter that characterizes the loading process; B is linear differential operator for calculating small deformations $\boldsymbol{\varepsilon}$ by the given displacements $\boldsymbol{u}$; D is nonlinear operator that establishes the relationship between stresses and strains; $\boldsymbol{\xi}$ is the initial strain, including thermal and irreversible components accumulated before the beginning of loading; $\langle \boldsymbol{f}, \delta\boldsymbol{u}\rangle$ is linear form, which is identical to the work of surface loads on possible displacements $\delta\boldsymbol{u}$; $(\cdot, \cdot)_L$ is scalar product in space L tensor-functions of stresses and strains integrated with the square on Ω, where Ω is the area occupied by the body. These notations are described in more detail in Kharchenko et al. (2018), Chirkov (2020).

To substantiate the correctness of the boundary value problem, represented by the system of Eq. (13.8), it is transformed into a nonlinear operator equation in the Hilbert space with respect to displacements. On the basis of the formulated principle of non-negativity of dissipation under the condition of loading of the irradiated porous material, as well as with the involvement of the functional analysis apparatus, the properties of strong monotonicity and Lipschitz continuity of the operator corresponding to Eq. (13.8) are established. Therefore, in accordance with the general results of the theory of nonlinear operators, there is a single solution of Eq. (13.8) and its continuous dependence on perturbations of the initial data (Chirkov 2020; Kharchenko et al. 2018).

Generalized and modified iterative methods of elastic solutions and variable elasticity parameters were developed for solving nonlinear boundary value problems describing inelastic deformation processes with account of radiation effects and accumulated metal damage, and their convergence was proved taking into account the deformation history of loading (Chirkov 2020; Kharchenko et al. 2018).

13.5 Main Provisions of the Calculation Technique

It is noteworthy that the solution of most practical problems of structural mechanics by analytical methods is impossible. Therefore, approximate calculation methods are used. In particular, the most universal of them is the finite element method (FEM).

At the same time, it is necessary to take into account the disadvantages of the classical FEM schemes, which consist in discontinuous distribution and lower order of convergence of the stress-displacement approximation. However, stresses are usually the main desired functions in the problems of deformable body mechanics, and therefore must be determined with a high degree of accuracy.

Traditional approaches to improve the accuracy by increasing the density of the finite element partitioning or transition to more complex finite elements are not always effective even in the case of linear problems. For evolutionary and nonlinear three-dimensional problems, they are practically unacceptable, because the large dimensionality of the discrete model, significant nonlinearity of metal properties in the vicinity of the crack front, and modeling of conditions with increased dose and temperature of irradiation can lead to loss of stability or violation of the convergence of computational processes.

Thus, it is promising in the numerical analysis of problems of mechanics to apply the concept of the so-called mixed formulations of the FEM, in which stresses are included in the solving equations along with displacements as equal unknowns, and not determined by the solution of the problem in displacements. This allows stresses and displacements to be approximated by a different set of basis functions, which improves the accuracy of stress and strain determination.

The main advantages of the developed mixed schemes are in the continuous approximation of stresses and strains, as well as in solving the boundary value problem taking into account the exact fulfillment of static conditions on the body surface (Chirkov 2020; Kharchenko et al. 2018).

At the same time, it should be taken into account that the approaches used to analyze classical schemes in displacements are not suitable for studying the conditions of correctness and convergence of solutions based on the mixed method. The condition, which is key in substantiating the stability and convergence of the mixed method, is determined due to the developed alternative approaches to the analysis of projection-grid schemes with the involvement of constructed projection operators. It is on the basis of the formulated stability condition that the convergence of the mixed approximation for stresses and displacements is proved. The obtained a priori estimates establish not only the convergence of the mixed method in the problems of nonlinear deformation mechanics, but also indicate a more accurate distribution of stresses and strains compared to the conventional finite element approximation (Chirkov 2020; Kharchenko et al. 2018).

One of the main provisions of the revised stress-strain state calculation of NPP equipment is a universal approach, which consists in fragmentation of the structure model. According to this approach, the calculation of its individual node is performed in two stages. At the first stage, the global stress-strain state is calculated for the complete structural model, and at the second stage, the refined calculation of the local state of the studied node is performed using the fragmentation procedure and finer finite element partitioning. The fracture resistance is estimated on the basis of additional calculation of the local state with the inclusion of the postulated defect in the model of a fragment of the structure, which is fully consistent with the IAEA recommendations.

The developed methods are implemented in a specialized software complex created in IPS NASU and approved for use in the nuclear industry of Ukraine (Software 2004). Methodological recommendations and methods for calculation of NPP equipment are implemented in the developed standard of the Ukrainian organization - the standard of IPS NASU (SOU 2020). To assess the strength and service life of WWER reactor pressure vessels (RPVs), an industry normative document was developed (MT-D.0.03.391-06 2009), which was used to justify the service life extension of many Ukrainian NPPs.

13.6 Calculations for Critical Elements of the WWER-1000 Nuclear Reactor Pressure Vessel

The following are practical results that demonstrate the capabilities and advantages of the developed calculation methods for assessing the strength of critical elements of the WWER-1000 reactor primary circuit equipment.

Figure 13.1 shows the equipment elements for which the results of the design analysis were obtained: WWER-1000 nuclear RPV including the cylindrical part and

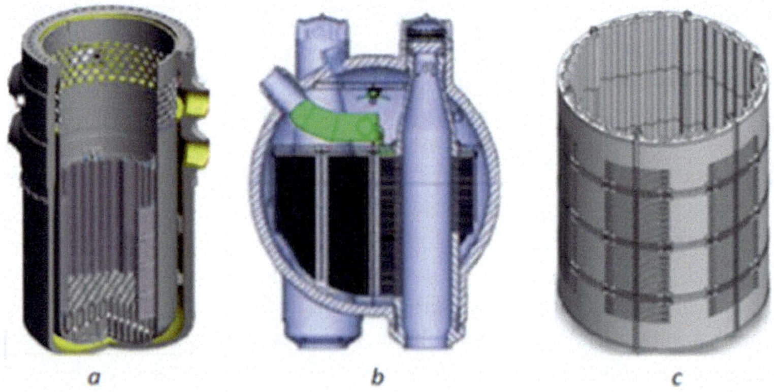

Fig. 13.1 The studied elements of the equipment of the first circuit of the WWER-1000 reactor unit: nuclear reactor vessel **a**; coolant collector-steam generator vessel welding unit **b**; reactor core containment **c**

the area of nozzles as the most critical element of the reactor installation; the unit of welding of the coolant collector to the nozzle of the PGV-1000M steam generator vessel as the most problematic element of the steam generator design; WWER-1000 reactor core lining as the most irradiated element of the reactor internals.

13.6.1 Reactor Pressure Vessel

Under reactor operating modes caused by emergencies, in particular thermal shock, the main criterion for the strength and integrity of the vessel is its ability to resist destruction during thermal shock.

Three-dimensional finite element models of the WWER-1000 reactor pressure vessel (RPV), including the elliptical bottom, spherical lid, nozzle area, and cylindrical part with the support shoulder and welded butt welds of the shells with postulated semi-elliptical cracks built into the computational model, were created to simulate the kinetics of the NDC. For the cylindrical part of the RPV, in particular, an underfloat circumferential semi-elliptical crack with a depth of 15 mm and a semi-axis ratio of 0.3, located in the metal of weld No. 4 on the axis of the cold water tongue, was modeled. Calculated finite element models of WWER-1000 RPV are shown in Fig. 13.2.

It is shown that the computational modeling of nonlinear effects of the RPV metal behavior during thermal shock is one of the determining factors for substantiating its safe operation and predicting the service life, as evidenced by the international

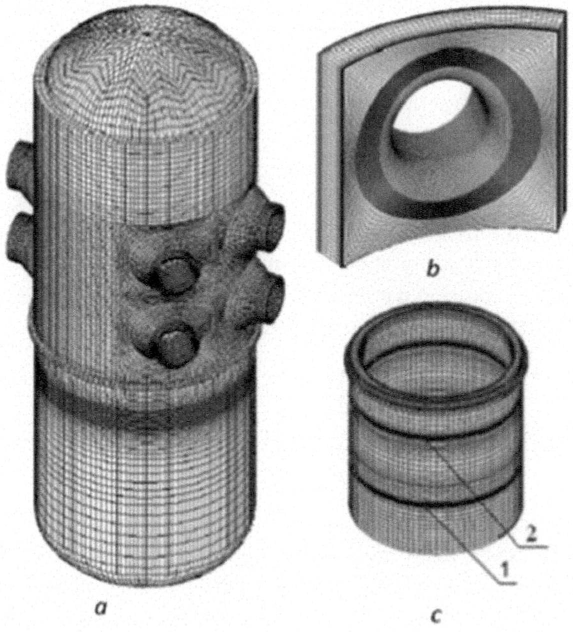

Fig. 13.2 Finite element models of WWER-1000 reactor vessel: full model **a**; fragment of the nozzle area **b**; model of the cylindrical part with welds **c**: 1 is seam No. 3 and 2 is seam No. 4

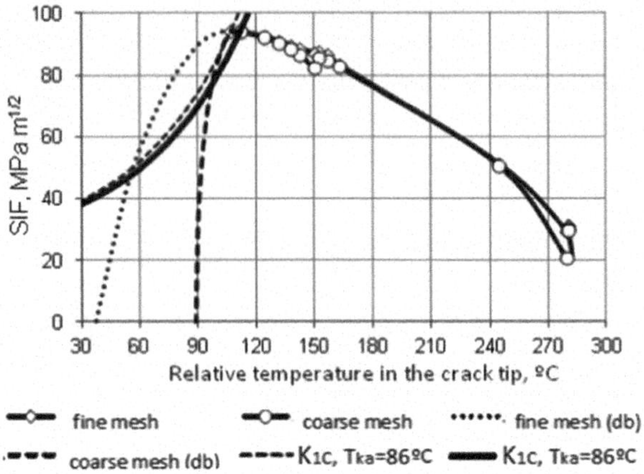

Fig. 13.3 Determination of the maximum allowable critical brittle temperature T_{ka} of reactor vessel metal based on the proposed approach using liquid and dense finite element meshes; db is descending branch

experience in solving practical problems of extending the service life of NPP power units.

An important feature of modeling the nonlinear effects of the RPV metal behavior is to take into account the history of inelastic deformation of the metal in the vicinity of the postulated crack front. The fracture resistance was evaluated taking into account the data of kinetics calculation and the nature of metal loading in front of the crack front to determine the change of stress intensity factor (SIF) with temperature. The so-called "descending branch", which is typical for many modes of thermal shock, is caused by the formation of a local compressive stress zone during metal unloading in front of the crack front during emergency cooling of the reactor core (Kharchenko et al. 2018).

Figure 13.3 depicts the results on the updated elastic-plastic calculations for estimation of fracture resistance of the cylindrical part of the WWER-1000 RPV using modern concepts for prediction of ductile fracture of RPV steels are presented.

The proposed approach to assessing the fracture resistance of the RPV takes into account the descending branch in the calculated dependence of the RPV on temperature for modes with thermal shock. According to this approach, the processes of active loading and local metal unloading in front of the crack front are analyzed, and an additional condition is formulated, which allows to justify additional strength reserves and service life of the RPV.

13.6.2 Welded Joint of the Coolant Collector and the Steam Generator Vessel

The problematic element of the primary circuit of the WWER-1000 reactor unit equipment is the welded joint of the coolant collector and the steam generator vessel, which has damages detected within the design service life. Damage initiation is caused by the influence of corrosive environment and high level of tensile stresses arising during operation of the steam generator.

To determine the residual stresses and the global state of the steam generator vessel, three-dimensional finite element models of the equipment of the loop of the first circuit of the reactor unit were created, consisting of the following structural elements of the equipment: steam generator vessel, "hot" and "cold" coolant collectors, hot and cold collector welding units, hot and cold branches of the main circulation pipeline, main circulation pump, RPV with coolant inlet and outlet pipes. The constructed models take into account the mutual influence of the steam generator and elements of the connected equipment on the determination of the stress-strain state of the steam generator as a whole. The calculated finite element models for determining the global and local stress-strain state of the PGV-1000M steam generator are shown in Fig. 13.4.

It should be emphasized that the first results of calculating the stress-strain state of the above weld assembly using three-dimensional models, which take into account the influence of the attached equipment, were obtained in the IPS NASU, showing a high level of tensile stresses under operating modes of loading exactly at the place of detected damage on the inner surface of the nozzle (Kharchenko et al. 2018).

The calculated determination of residual stresses and strains includes modeling of the operations of filling the processing of weld No. 111 and local heat treatment of the nozzle area under the high tempering mode during the manufacture or repair of the steam generator. The results of the assessment of residual technological stresses and strains in the zone of weld No. 111 are given in Kharchenko et al. (2018).

Below are the results of determining the calculated values of the stress intensity factor for a surface defect in the form of a groove extending along the circumferential coordinate of the nozzle in combination with a semi-elliptical crack that has the same semi-axes as the crack postulated in the regulatory documents. The length of

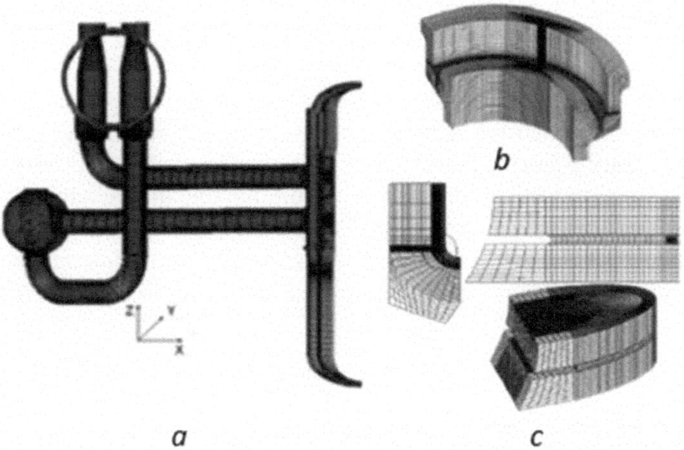

Fig. 13.4 Finite element models: reactor plant **a**; computational model of the welding unit **b**; finite element meshes in the filet area with a surface defect in the form of a long groove combined with a semi-elliptical crack **c**

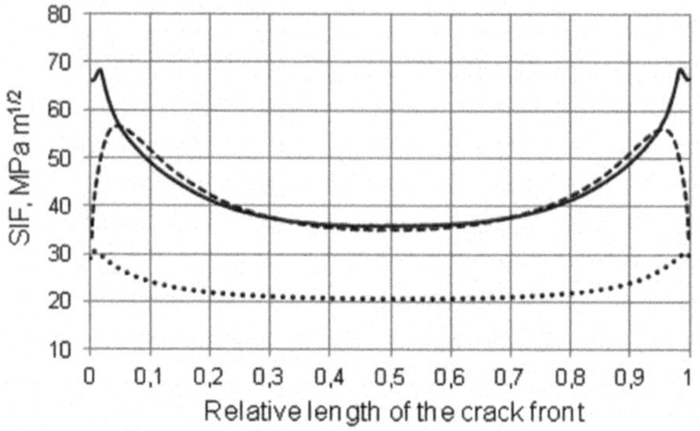

Fig. 13.5 Stress intensity factor distribution along the crack front for normal operating conditions. Here solid line corresponds to the elastic-plastic calculation taking into account the extended groove; dashed line is elastic-plastic calculation taking into account the postulated semi-elliptical crack; dotted line is linear-elastic calculation with the postulated crack

the groove along the circumferential coordinate of the nozzle was determined to be 500 mm, the height of the groove (the distance between its banks) is 1.25 mm, the total depth of the groove and the crack was set to 18 mm, which corresponds to a quarter of the nozzle thickness. The geometric dimensions of the groove were determined on the basis of test calculations to obtain the most conservative estimate of the resistance to fracture of the welded joint.

The proposed model of the calculated defect is a more adequate schematization of the surface damage of the welded joint compared to the traditional modeling of the defect in the form of a semi-elliptical crack. The physical model of the corrosion defect nucleation and propagation was not considered, since the expediency of using the proposed calculated schematization of the surface defect to take into account the corrosion damage of the inner surface of the nozzle of the welding assembly was proved.

Figure 13.5 shows that the history of loading, residual technological inheritance, and consideration of extensive damage of corrosion origin on the inner surface of the nozzle significantly affect the resistance to fracture of the welded joint. For the mode of hydraulic strength tests on the second circuit, the maximum value of the stress intensity factor increases by almost 60%, and under normal operating conditions - more than twice (Kharchenko et al. 2018). Consequently, the traditional linear-elastic calculation of the weld assembly's fracture resistance leads to an overestimation of its strength.

13.6.3 Baffle of the Reactor's Active Zone

At high damaging doses and irradiation temperatures, austenitic steels of 08Kh18N10T type undergo radiation swelling and radiation creep, which can lead

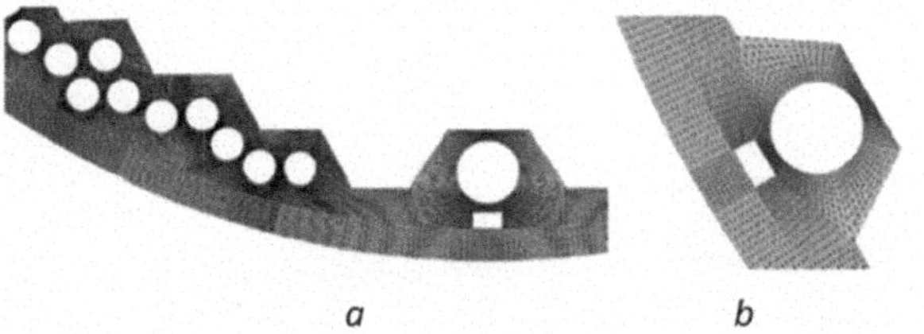

Fig. 13.6 Fragments of the finite element model of the cross section of the partition and WWER-1000 reactor internals: meshing in the cooling zone channels **a**; meshing in the channel area with the largest diameter and the groove channel **b**

to irreversible shape change of such element of reactor internals as the lining. Prediction of radiation swelling and radiation creep is necessary to justify the possibility of reloading fuel assemblies from the reactor core, as well as to control the gap between the containment and the internals during operation. Closure of the lining with the shaft as a result of radiation swelling and radiation creep of the metal can lead to contact stresses, as well as redistribution of primary coolant flow rate in the reactor and, as a result, to changes in the reactor operating temperature.

Therefore, the calculated assessment of the progressive shape change of the containment during operation is one of the main conditions for justifying the normal operation of the reactor during the design and long-term operation of the NPP power unit.

Figure 13.6 shows fragments of a two-dimensional finite element model of the cross section of the WWER-1000 reactor shell and internals, which were used to perform stress-strain state calculations.

The calculated data were obtained on the basis of solving the coupled contact problem of thermal conductivity and radiation creep. The stress-strain state calculations were performed in two-dimensional formulation for a cross section with the maximum damaging dose and irradiation temperature at the maximum height of the shielding under the condition of generalized plane deformation.

Figures 13.7 and 13.8 depict the data on gaps after 50 years of operation during the reactor operation at full power, as well as after the planned reactor shutdown, calculated with and without radiation creep. These data were obtained taking into account nominal gaps of 2.5 and 4 mm, respectively, provided that the core dimensions after reactor shutdown correspond to the initial values (Chirkov 2020). The distance on the graphs is counted along the circumferential coordinate from the largest opening of the cooling channel counterclockwise by an angle of 180°.

According to the results obtained, which are shown in the graphs, the exhaustion of the nominal gap between the lining and the shaft is not observed within the design service life, although some values of the gap are quite close to zero. The minimum values of the gap are local in the area of the longitudinal cooling channel opening with the largest diameter and the groove channel for the coolant flow between the partition and the shaft. According to the results of calculations, it should be expected

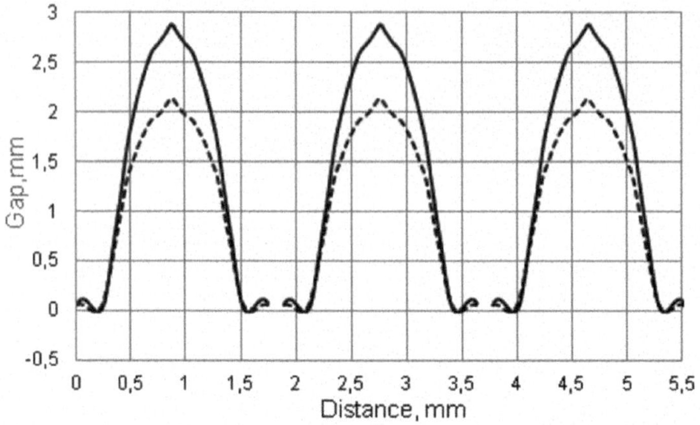

Fig. 13.7 Comparison of the radial gap between the lining and the shaft for the conditions of the reactor operation at full capacity after 50 years of operation. Here and in Fig. 13.8, the solid line corresponds to the calculation with an account of radiation creep, the dashed line—with no account of radiation creep

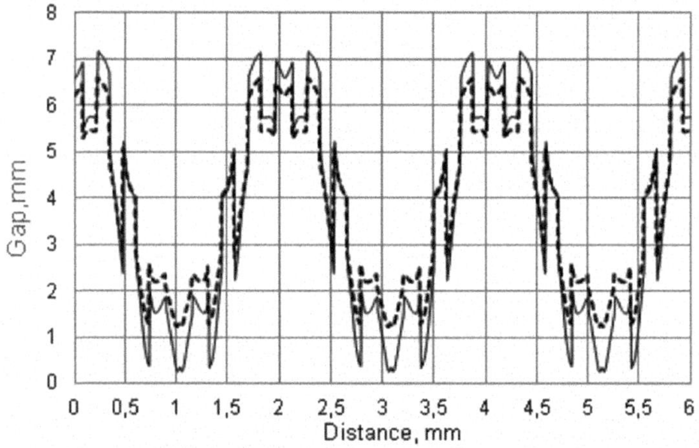

Fig. 13.8 Comparison of the residual gap between the edges of the shielding and spacer grids of peripheral fuel assemblies after 50 years of operation

that during reactor operation beyond the design basis there may be local contact of the containment with the shaft in the area of the channel with the largest hole and the groove channel.

After 50 years of operation and shutdown of the reactor, the residual gap between the edges of the baffle and the fuel assembly lattices is not exhausted, resulting in no jamming after the fuel campaign termination.

13.7 Conclusions

Summarizing the results of the updated calculations on strength assessment and service life prediction of NPP equipment elements, the following conclusions are obtained:

- the proposed approach to the analysis of fracture resistance of the RPV takes into account the processes of active loading and local unloading of the metal in front of the postulated crack front, which allows to justify additional strength and service life reserves;
- the history of elastic-plastic deformation, residual technological inheritance, and consideration of extensive damage of corrosion origin on the inner surface of the nozzle significantly affect the fracture resistance of the welded joint, and therefore the traditional linear-elastic calculation does not have a sufficient degree of conservatism, which leads to strength overestimation;
- failure to take into account the processes of radiation swelling and radiation creep of irradiated metal leads to an incorrect assessment of the reactor core shell shape change even within the design lifetime.

The developed complex of calculation studies has been widely implemented in the practice of calculations for justification of the beyond-design service life of Ukrainian NPP power units, in particular: for the state examination of works on justification of strength and fracture resistance of WWER-1000 RPVs at Zaporizhzhya, Rivne, and Pivdennoukrainsk NPPs; for justification of the strength of the base metal and welds of steam generators of Rivne NPP power unit No. 3; for strength calculation of the vessel (tank) and its components for the research nuclear power reactor WWER-M of the Institute of Nuclear Research of the National Academy of Sciences of Ukraine.

References

Chirkov OY (2020) Radiation creep in problems of mechanics of inelastic deformation of materials and structural elements. G.S. Pisarenko Institute of Problems of Strength of NAS of Ukraine, Kyiv

Chirkov OY (2022) Modified elastic solution processes and interchangeable elasticity parameters in the problems of radiation creep. Strength Mater 54:167–177

Huang Y (1991) Accurate dilatation rates for spherical voids in triaxial stress fields. J Appl Mech 58:1084–1086

Kachanov LM (1969) Fundamentals of the theory of plasticity. Nauka, Moscow

Kharchenko VV, Chirkov AYu, Kobelsky SV, Kravchenko VI (2018) Methods of computational analysis in problems of strength of equipment elements of WWER reactor plants. G.S.Pisarenko Institute of Problems of Strength of NAS of Ukraine, Kyiv

MT-D.0.03.391-06 (2009) Methodology of Strength and Service Life Assessment of WWER Reactor Pressure Vessels in Operation, Kyiv

Rice JR, Tracey DM (1969) On the ductile enlargement of voids in triaxial stress fields. J Mech Phys Solids 17:201–217

Software (2004) Three-dimensional finite element modeling of thermal and thermal stress state of elements of engineering structures (SPACE). UkrSERO Certification System. Certificate of conformity No. UA1.017.0054634-04

SOU 56-47-2020 (2020) Methods for calculation of stress state and fracture resistance of reactor vessels and primary circuit equipment elements of NPPs with WWER. G.S. Pisarenko Institute of Problems of Strength of NAS of Ukraine, Kyiv

Chapter 14
Mathematical Modeling of the Stress State for the Displacement and Rotation of the Rigid Penny-Shaped Inclusion Embedded in a Piezoelectric Space

Vitaly Kirilyuk and Olga Levchuk

14.1 Introduction

The widespread use of piezoelectric materials in manufacturing energy converters and measuring elements for various industries causes interest in analyzing force and electric fields in electroelastic (piezoelectric) bodies near stress concentrators. A study of stress distribution based on solutions to spatial problems of electroelasticity for anisotropic piezoelectric solids in a mathematically rigorous formulation takes into account coupled force and electric fields. Thus, the study is significantly complicated because there is a need to solve the boundary value problem for a complex system of coupled differential equations.

Spatial problems of the theory of elasticity concerning the determination of the stress state in isotropic and transversely isotropic materials with rigid inclusions of a circular or elliptical shape were considered in Kassir and Sih (1968), Keer (1965), Kotousov et al. (2014), Nategh et al. (2018), Rahman (2001), Selvadurai (1979, 1980, 1982). For electroelastic materials, such studies were carried out in Kirilyuk (2008), Kyrylyuk and Levchuk (2021). The problems of the displacement and rotation of a circular rigid disk located in the isotropy plane of elastic transversely isotropic space were considered in Selvadurai (1979, 1980).

In this chapter, exact solutions to the problems of displacement and rotation of a rigid disk-shaped inclusion in a piezoelectric space are found. Moreover, closed expressions for stress under a circular disk are obtained. Finally, the influence of the coupled elastic and electric fields on the stress state in a piezoelectric material is investigated.

When stating the problem, it is assumed that the circular rigid disk is in the plane of isotropy of an electroelastic transversely isotropic material, and the two-sided surface of the disk is non-electrode coated. By representing the solution of the

V. Kirilyuk (✉) · O. Levchuk
S.P. Timoshenko Institute of Mechanics, National Academy of Science of Ukraine, Kyiv, Ukraine
e-mail: kirilyuk_v@ukr.net

A. N. Guz et al. (eds.), *Advances in Mechanics*, Advanced Structured Materials 191,
https://doi.org/10.1007/978-3-031-37313-8_14

equations of statics of electroelasticity for a transversely isotropic solid in terms of harmonic functions and reducing the problems to the consideration of dual integral equations, exact solutions of the problems are obtained, expressions for stresses under a circular disk are found in analytic form. In a particular case, the obtained results yield expressions for determining the stress state corresponding to displacement and rotation of a rigid circular disk in a purely elastic transversely isotropic material.

14.2 Formulation of the Problem

Problem 1. First, we consider a circular disk occupying an area Ω in the plane $z = 0$ (Fig. 14.1) perpendicular to the axis of symmetry of the electroelastic transversely isotropic material (z-axis). We also assume that the double-sided surface of the disk has no electrode coating. Under force P, the disc-shaped inclusion moves along the z-axis by distance δ.

By taking into account the asymmetry of the normal components of the stress tensor and electrical induction vector relative to the symmetry plane $z = 0$, the problem can be reduced to the case of purely elastic transversely isotropic half-space $z \geqslant 0$ (Selvadurai 1979) (Fig. 14.1). Boundary conditions of the problem of electroelasticity are the following

$$u_z = \delta, (x, y) \in \Omega, \quad z = 0^+; \quad \sigma_{zz} = 0, \quad (x, y) \notin \Omega, \quad z = 0^+,$$

$$u_x = u_y = 0, \quad z = 0^+, \tag{14.1}$$

$$D_z(x, y) = 0, \quad z = 0^+, \tag{14.2}$$

where $\Omega = \{(x, y): \ x^2/a^2 + y^2/a^2 \leqslant 1\}$, D_z is the normal component of the vector of electrical induction (electrical displacements). According to the electrical state, condition (14.2) corresponds to the case when the disk has no electrode coating and the absence of free charges in the plane $z = 0$. The conditions (14.1) are the usual boundary conditions for a purely elastic problem.

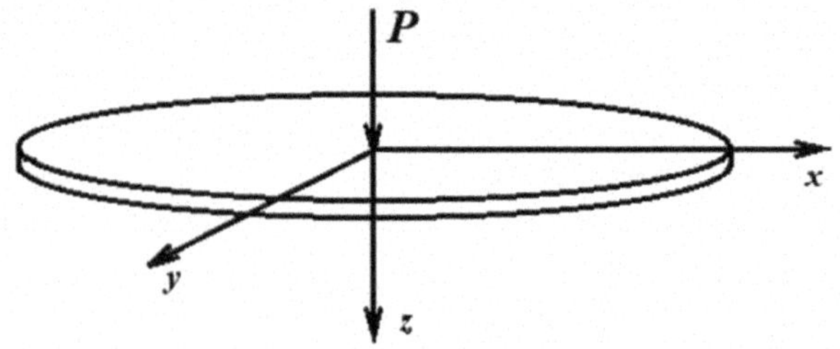

Fig. 14.1 Displacement of the rigid circular disk along the axis of symmetry of the piezoelectric material

14.3 Basic Equations and Relations

The equations of statics of electroelasticity for a piezoelectric transversely isotropic body can be written in terms of the components of the displacement vector and the electric potential Ψ as follows (Podil'chuk 1998)

$$
\begin{aligned}
&c_{11}^E u_{x,xx} + \tfrac{1}{2}\left(c_{11}^E - c_{12}^E\right)u_{x,yy} + c_{44}^E u_{x,zz} \\
&\qquad + \tfrac{1}{2}\left(c_{11}^E + c_{12}^E\right)u_{y,xy} + \left(c_{13}^E + c_{44}^E\right)u_{z,xz} + (e_{31} + e_{15})\Psi_{,xz} = 0, \\
&c_{11}^E u_{y,yy} + \tfrac{1}{2}\left(c_{11}^E - c_{12}^E\right)u_{y,xx} + c_{44}^E u_{y,zz} \\
&\qquad + \tfrac{1}{2}\left(c_{11}^E + c_{12}^E\right)u_{x,xy} + \left(c_{13}^E + c_{44}^E\right)u_{z,yz} + (e_{31} + e_{15})\Psi_{,yz} = 0, \\
&\left(c_{13}^E + c_{44}^E\right)(u_{x,xz} + u_{y,yz}) + c_{44}^E(u_{z,xx} + u_{z,yy}) \\
&\qquad\qquad + c_{33}^E u_{z,zz} + e_{15}(\Psi_{,xx} + \Psi_{,yy}) + e_{33}\Psi_{,zz} = 0, \\
&(e_{31} + e_{15})(u_{x,xz} + u_{y,yz}) + e_{15}(u_{z,xx} + u_{z,yy}) \\
&\qquad\qquad + e_{33}u_{z,zz} - \varepsilon_{11}^S(\Psi_{,xx} + \Psi_{,yy}) - \varepsilon_{33}^S\Psi_{,zz} = 0.
\end{aligned}
\tag{14.3}
$$

In (14.3), c_{11}^E, c_{12}^E, c_{13}^E, c_{33}^E, and c_{44}^E are independent modulus of elasticity, e_{31}, e_{15}, and e_{33} are piezomodules, ε_{11}^S and ε_{33}^S are dielectric constants.

The solution of (14.3), according to Podil'chuk (1998), can be obtained using four potential functions Φ_j $(j = 1, \ldots, 4)$:

$$
\begin{aligned}
&u_x = \sum_{j=1}^{3}\Phi_{j,x} + \Phi_{4,y}, \quad u_y = \sum_{j=1}^{3}\Phi_{j,y} - \Phi_{4,x}, \quad u_z = \sum_{j=1}^{3}k_j\Phi_{j,z}, \\
&\Psi = \sum_{j=1}^{3} l_j\Phi_{j,z},
\end{aligned}
\tag{14.4}
$$

where k_j and l_j are some constants, and functions Φ_j satisfy the equation

$$
\Phi_{j,xx} + \Phi_{j,yy} + \nu_j\Phi_{j,zz} = 0 \quad (j = 1, 2, 3), \tag{14.5}
$$

$\nu_4 = 2c_{44}^E / \left(c_{11}^E - c_{12}^E\right)$ and ν_i $(i = 1, 2, 3)$ are the roots of third-order algebraic equation

$$
\begin{aligned}
&\nu^3(A_1B_2 - C_1D_2) + \nu^2(A_1B_3 + A_2B_2 - C_1D_3 - C_2D_2) \\
&\qquad + \nu(A_2B_3 + A_3B_2 - C_2D_3 - C_3D_2) + A_3B_3 - C_3D_3 = 0.
\end{aligned}
\tag{14.6}
$$

The constants k_j and l_j $(j = 1, 2, 3)$ in (14.4) are related to ν_j introduced in (14.5) and can be determined from algebraic equation (14.6) by the following relations

$$\frac{a_j + c_{13}^E k_j + e_{31} l_j}{c_{11}^E} = \frac{c_{33}^E k_j + e_{33} l_j}{c_{13}^E + a_j} = \frac{c_{33}^E k_j - \varepsilon_{33}^S l_j}{e_{31} + d_j} = \nu_j \quad (j = 1, 2, 3), \quad (14.7)$$

$$a_j = c_{44}^E (1 + k_j) + e_{15} l_j, \quad d_j = e_{15}(1 + k_j) - \varepsilon_{11}^S l_j \quad (j = 1, 2, 3, 4), \quad (14.8)$$

where expressions (14.7) and (14.8) are used in subsequent calculations.

Note that, according to (14.5), the functions $\Phi_1(x, y, z_1)$, $\Phi_2(x, y, z_2)$, $\Phi_3(x, y, z_3)$ and $\Phi_4(x, y, z_4)$ $(z_j = z\nu_j{}^{-1/2}, j = 1, \ldots, 4)$, are harmonic in the corresponding coordinate systems.

14.4 Solution Method

Since the problem is axisymmetric with respect to the z-axis, we use the cylindrical coordinate system (r, θ, z), in which the components of displacement and the electric potential have the following form

$$\begin{aligned}
u_r &= \frac{\partial}{\partial r}\left(\sum_{j=1}^{3} \Phi_j(r, \theta, z_j)\right) + \frac{1}{r}\frac{\partial}{\partial \theta}\Phi_4(r, \theta, z_4), \\
u_\theta &= \frac{1}{r}\frac{\partial}{\partial \theta}\left(\sum_{j=1}^{3} \Phi_j(r, \theta, z_j) - \frac{\partial}{\partial r}\Phi_4(r, \theta, z_4)\right), \\
u_z &= \frac{\partial}{\partial z}\left(\sum_{j=1}^{3} k_j \Phi_j(r, \theta, z_j)\right), \\
\Psi &= \frac{\partial}{\partial z}\left(\sum_{j=1}^{3} l_j \Phi_j(r, \theta, z_j)\right).
\end{aligned} \quad (14.9)$$

A solution to the boundary value problem of electroelasticity for a piezoelectric material can be found based on (14.4)–(14.9). Let us represent the functions $\Phi_i(r, \theta, z_i)$ in the form

$$\Phi_i(r, \theta, z_i) = \alpha_i f(r, \theta, z_i) \quad (i = 1, 2, 3), \quad \Phi_4 \equiv 0, \quad (14.10)$$

where α_i are unknown constants, $f(r, \theta, z)$ is a harmonic function that satisfies the Laplace equation in cylindrical coordinates.

The boundary conditions for the problem in cylindrical coordinates have the form

$$\begin{aligned}
&u_z = \delta, \quad 0 \leqslant r \leqslant a, \quad z = 0^+, \\
&\sigma_{zz} = 0, \quad r > a, \quad z = 0^+, \\
&u_r = u_\theta = 0, \quad z = 0^+,
\end{aligned} \quad (14.11)$$

$$D_z = 0, \quad z = 0^+. \tag{14.12}$$

Using (14.7)–(14.9), we find the displacement u_z, stress σ_{zz} and electrical induction D_z in the plane $z = 0$:

$$u_z|_{z=0} = \sum_{j=1}^{3} \alpha_j \frac{k_j}{\sqrt{v_j}} f_{,z}|_{z=0},$$
$$\sigma_{zz}|_{z=0} = \sum_{j=1}^{3} \alpha_j \left[c_{44}^E(1+k_j) + e_{15}l_j\right] f_{,zz}|_{z=0},$$
$$D_z|_{z=0} = \sum_{j=1}^{3} \alpha_j \left[e_{15}(1+k_j) - \varepsilon_{11}^S l_j\right] f_{,zz}|_{z=0}.$$

We use the following form for the function $f(r, \theta, z_i)$

$$f(r,\theta,z) = F(r,z) = \frac{1}{a^2}\int_0^\infty \xi A(\xi) \mathrm{e}^{-(\xi/a)z} J_0(\xi r/a) d\xi,$$

where J_0 is a zero-order Bessel function. Unknown values of constants α_i in the expressions of potential functions (14.10) are taken from the system of linear equations

$$\sum_{j=1}^{3} \alpha_j = 0,$$
$$\sum_{j=1}^{3} \alpha_j \left[e_{15}(1+k_j) - \varepsilon_{11}^S l_j\right] = 0, \tag{14.13}$$
$$\sum_{j=1}^{3} \alpha_j \frac{k_j}{\sqrt{v_j}} = 1.$$

Using the representations (14.4) and (14.9), the expressions of the components of stress and electrical induction, and the system (14.13), it follows that the boundary conditions for displacements u_r, u_θ and electrical induction D_z in (14.11) and (14.12) are satisfied. The remaining boundary conditions lead to the system of dual integral equations for determining the unknown function $A(\xi)$

$$\sum_{j=1}^{3} \alpha_j \frac{k_j}{\sqrt{v_j}} \frac{1}{a^3} \int_0^\infty \xi^2 A(\xi) J_0(\xi r/a) d\xi = \delta, \quad 0 \leqslant r \leqslant a,$$
$$\int_0^\infty \xi^3 A(\xi) J_0(\xi r/a) d\xi = 0, \quad r > a. \tag{14.14}$$

Introducing new variables

$$w_0 = \frac{\delta a^3}{\sum_{j=1}^{3} \alpha_j \frac{k_j}{\sqrt{v_j}}}, \quad \rho = \frac{r}{a}, \quad C(\xi) = \xi^2 A(\xi),$$

we reduce (14.14) to the following form

$$\int_0^\infty C(\xi) J_0(\xi\rho) d\xi = w_0, \quad 0 < \rho \leqslant 1,$$

$$\int_0^\infty \xi C(\xi) J_0(\xi\rho) d\xi = 0, \quad \rho \geq 1.$$

According to the results of Selvadurai (1979), we find

$$C(\xi) = \xi^2 A(\xi) = \frac{2w_0 \sin \xi}{\pi \xi}.$$

Next, we have (Erdelyi et al. 1954; Sneddon 1966)

$$\sigma_{zz}|_{z=0^\pm} = \begin{cases} \mp \dfrac{2\sum_{j=1}^{3} \alpha_j [c_{44}^E(1+k_j) + e_{15} l_j]\delta}{\pi\sqrt{a^2 - r^2} \sum_{j=1}^{3} \alpha_j \frac{k_j}{\sqrt{v_j}}}, & r \leqslant a \\ 0, & r > a. \end{cases} \tag{14.15}$$

The force P acting on the disk-shaped inclusion to move it along the z-axis by the value δ is calculated based on (14.15)

$$\begin{aligned} P &= 2\pi \int_0^a r \left[(\sigma_{zz})_{z=0^-} - (\sigma_{zz})_{z=0^+} \right] dr \\ &= 8\delta a \frac{\sum_{j=1}^{3} \alpha_j \left[c_{44}^E(1+k_j) + e_{15} l_j \right]}{\sum_{j=1}^{3} \frac{\alpha_j k_j}{\sqrt{v_j}}}. \end{aligned} \tag{14.16}$$

Note that from the obtained expressions (14.15) and (14.16) for an electroelastic material, the results for a purely elastic transversely isotropic material in Selvadurai (1979) follow as a special case

Problem 2. Let us consider the problem of rotating a hard circular disk located in a piezoelectric material (Fig. 14.2). As in the previous problem, we assume that the inclusion is in the isotropy plane of the material $z = 0$ (perpendicular to the axis of symmetry of the piezoelectric material)

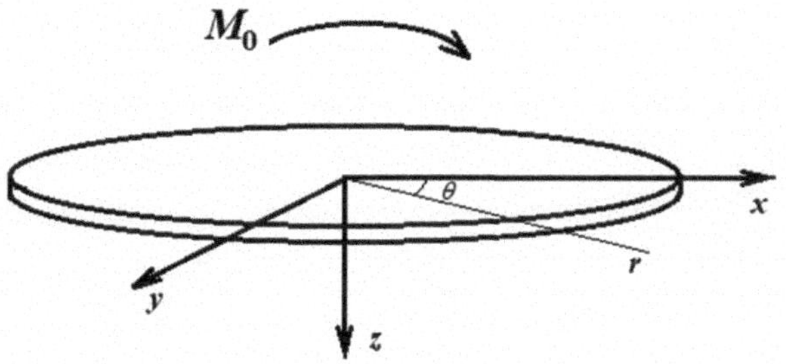

Fig. 14.2 Rotation of the rigid circular disk in the electroelastic material

In the case of rotation of the disk-shaped inclusion, we obtain the following boundary conditions

$$\begin{aligned} &u_z = \omega r \cos\theta, \quad 0 \leqslant r \leqslant a, \quad z = 0^+, \\ &\sigma_{zz} = 0, \quad r > a, \quad z = 0, \\ &u_r = u_\theta = 0, \quad z = 0^+, \\ &D_z = 0, \quad z = 0^+. \end{aligned} \tag{14.17}$$

Let us take the harmonic functions $\Phi_i(r, \theta, z_i)$ for this case in the form

$$\Phi_i(r, \theta, z_i) = \alpha_i g(r, \theta, z_i), \, (i = 1, 2, 3)\, ; \, \Phi_4 \equiv 0, \tag{14.18}$$

where the function $g(r, \theta, z_i)$ is selected as follows

$$g(r, \theta, z) = \frac{1}{a^2} \int_0^\infty \xi B(\xi) \mathrm{e}^{-(\xi/a)z} J_1(\xi r/a) d\xi \cos\theta, \tag{14.19}$$

where J_1 is the Bessel function of the first order. Unknown values of constants α_i in the expressions of potential functions (14.18) and (14.19) are determined from the same system of linear equations (14.13) as for the first problem. When determining unknown values α_i from this system, it follows that the boundary conditions in displacements and with respect to electrical induction in the plane $z = 0$ are satisfied, and from the remaining boundary conditions, we obtain a system of dual integral equations

$$\begin{aligned} &\sum_{j=1}^{3} \alpha_j \frac{k_j}{\sqrt{v_j}} \frac{1}{a^3} \int_0^\infty \xi^2 B(\xi) J_1(\xi r/a) d\xi \cos\theta = \omega r \cos\theta, \quad 0 \leqslant r \leqslant a, \\ &\int_0^\infty \xi^3 B(\xi) J_1(\xi r/a) d\xi \cos\theta = 0, \quad r > a. \end{aligned} \tag{14.20}$$

The system (14.20) is like the system of pair integral equations for a purely elastic problem (the systems differ only in the factors before the integral in the first equation). Similarly to Selvadurai (1980), Sneddon (1966), reducing the system of dual integral equations to the solution of the Abel integral equation, we find

$$B(\xi) = 4\omega a^4 \sum_{j=1}^{3} \alpha_j \frac{k_j}{\sqrt{\upsilon_j}} \frac{\sin\xi - \xi\cos\xi}{\pi\xi^4}.$$

Next, like it was proposed in Selvadurai (1980) using results of Erdelyi et al. (1954), we obtain the analytical expression for stress distribution under a rigid circular disk when it rotates in a piezoelectric material

$$\sigma_{zz}|_{z=0^{\pm}} = \begin{cases} \mp \dfrac{4\omega\left(\sum\limits_{j=1}^{3}\alpha_j[c_{44}^E(1+k_j)+e_{15}l_j]\right) r\cos\theta}{\left(\sum\limits_{j=1}^{3}\alpha_j \dfrac{k_j}{\sqrt{\upsilon_j}}\right)\pi\sqrt{a^2-r^2}}, & r \leqslant a \\ 0, & r > a. \end{cases} \tag{14.21}$$

The moment acting on the disk-shaped inclusion, using formula (14.21) is found as follows

$$\begin{aligned} M_0 &= \int_0^a \int_{-\pi}^{\pi} \left[(\sigma_{zz})_{z=0^-} - (\sigma_{zz})_{z=0^+}\right] r^2 \cos\theta dr d\theta \\ &= \frac{16}{3}\omega a^3 \left(\sum_{j=1}^{3}\alpha_j \frac{k_j}{\sqrt{\upsilon_j}}\right)^{-1} \sum_{j=1}^{3}\alpha_j[c_{44}^E(1+k_j)+e_{15}l_j]. \end{aligned} \tag{14.22}$$

Note that from (14.21) and (14.22), we obtain the results for elastic transversely isotropic materials (Selvadurai 1980) as a particular case (see Appendix A).

Analysis of the results of numerical research. Using the obtained analytical expressions, we study the stress distribution under a rigid penny-shaped disk, as well as the effect of the connection of force and electric fields on the displacement of the disk in the material. We use the data of Kirilyuk (2006), which shows the properties of several piezoelectric materials (see the Table 14.1 in Appendix A).

After calculations, we conclude that to move the circular rigid disk to the same value δ (along the axis of symmetry of the material) for different piezoelectric materials, we need to apply different values of P. So, force P for materials PZT-7A, $BaTiO_3$, PZT-5H exceeds its corresponding value for the same displacement in the material PZT-5 by 26.02, 46.37, 36.81%.

We will also study the effect of coupled force and electric fields on the displacement of a circular disk in piezoelectric space. To do this, we compare the displacement of the disk in the electroelastic material and pure elastic transversely isotropic material, which has the same elastic properties as the electroelastic material. As a result of the comparison, we find that a smaller value of the force must be applied to move the circular disk in a purely elastic transverse-isotropic space. Thus, for materials PZT-5, PZT-7A, $BaTiO_3$ and PZT-5H, we obtained the ratios $P^{\text{Elast}}/P^{\text{Piezo}} = 61.39$, 64.99,

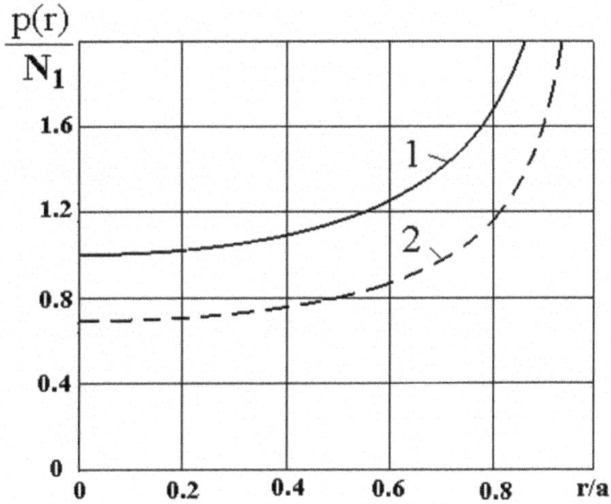

Fig. 14.3 Stress distribution for the displacement of a circular disk, the influence of the connection of force and electric fields

80.03, 69.21%. That is, for the given piezoelectric materials, the least influence of the connection of force and electric fields was observed in the material $BaTiO_3$.

Figure 14.3 shows the influence of the relationship between force and electric fields on the distribution of stresses under a penny-shaped disk for its displacement in an electroelastic space. The case of identical disk displacement in electroelastic and purely elastic materials is considered. Curve 1 in Fig. 14.3 corresponds to the displacement of the disk in the piezoelectric material PZT-5H, and line 2 corresponds to displacement in a pure elastic transverse-isotropic material, the elastic properties of which coincide with the properties of the electroelastic material PZT-5H.

In Fig. 14.3, the following notation is used

$$N_1 = 2N^*\delta/(\pi a),$$

where

$$N^* = \left(\sum_{j=1}^{3}\alpha_j\frac{k_j}{\sqrt{v_j}}\right)^{-1}\sum_{j=1}^{3}\alpha_j\left[c_{44}^E(1+k_j)+e_{15}l_j\right].$$

Figure 14.4 shows the effect of the connection of force and electric fields on the distribution of stresses under the circular disk when it rotates in the elastic space (material PZT-5H) at $\theta = 0$. The case of rotation of the disk through the same angle in electroelastic and purely elastic materials is considered. Notation $N_2 = 4\omega N^*/\pi$ is used in this figure. Curve 1 in Fig. 14.4 corresponds to the case of rotation of the disk in the piezoelectric material PZT-5H, line 2 corresponds to rotation of the circular disk in a purely elastic transversely isotropic material, the elastic properties of which coincide with the properties of the electroelastic material PZT-5H. The connection of force and electric fields significantly affects the distribution of stresses under the rigid disk as it moves and rotates in the piezoelectric material.

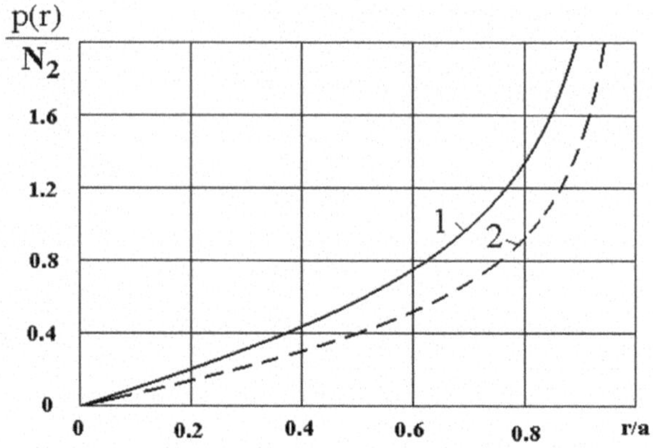

Fig. 14.4 Distribution of stresses under the circular disk during its rotation (cross section $\theta = 0$)

14.5 Conclusion

In this paper, using a rigorous mathematical model that considers the coupled force and electric fields, the exact solutions of the problems of displacement and rotation of a circular rigid disk in piezoelectric material were obtained, and the stress distribution under a circular rigid disk was researched. Numerical calculations have been performed, and the influence of the connection of force and electric fields on the distribution of stresses during displacement and rotation of a circular rigid disk in an electroelastic material has been studied.

Appendix A. The Transition from an Electroelastic Problem to the Elasticity Problem for a Transversely Isotropic Elastic Material (Particular Case)

We show that the results of Selvadurai (1979, 1980) for elastic transversely isotropic material are the partial cases of both obtained solutions to the problems for the electroelastic material. In the case of purely elastic transversely isotropic material, according to Elliot (1948), for the values of the roots of the quadratic equation

$$c_{11}c_{44}\nu^2 - \left[c_{44}^2 + c_{33}c_{11} - (c_{13} + c_{44})^2\right]\nu + c_{33}c_{44} = 0$$

there are relations

$$\frac{k_i(c_{13} + c_{44}) + c_{44}}{c_{11}} = \frac{k_i c_{33}}{k_i c_{44} + (c_{13} + c_{44})} = \nu_i. \tag{14.23}$$

By means of the relations (14.23), we have

$$
\begin{gathered}
k_i c_{33} = \left[k_i c_{44} + (c_{13} + c_{44})\right] \nu_i, \\
\left(c_{33}\frac{k_i}{\nu_i} - c_{13}\right) \nu_i = c_{44}\left(1 + k_i\right) \nu_i .
\end{gathered}
\tag{14.24}
$$

Next, from (14.24), we obtain

$$
c_{33}\left(\frac{k_1}{v_1} - \frac{k_2}{v_2}\right) = c_{44}\left(k_1 - k_2\right). \tag{14.25}
$$

In the transition to a purely elastic transversely isotropic material, the sum of two harmonic functions ($\Phi_3 \equiv 0$) must be calculated in representation (14.4) instead of three (according to Elliot (1948)), assuming in the expressions for the electroelastic material $l_1 = l_2 = l_3 = 0$ and $k_3 = 0$.

Then in (14.15), (14.16), (14.21) and (14.22) instead of the factor

$$
\left(\sum_{j=1}^{3} \alpha_j \frac{k_j}{\sqrt{v_j}}\right)^{-1} \sum_{j=1}^{3} \alpha_j \left[c_{44}^E (1 + k_j) + e_{15} l_j\right]
$$

for the case of pure elastic problems, we must use the expression

$$
\left(\sum_{j=1}^{2} \alpha_j \frac{k_j}{\sqrt{v_j}}\right)^{-1} \sum_{j=1}^{2} \alpha_j \left[c_{44}(1 + k_j)\right].
$$

If the boundary conditions (14.11) and (14.17) are satisfied in Problems 1 and 2 for the pure elastic material, the condition on the electrical state must be rejected (the second equation for system (13)), since the material is already purely elastic, and from the first condition for both systems (absence of displacements $u_r = u_\theta = 0$ in the plane $z = 0$) it follows that $\alpha_2 = -\alpha_1$ for both problems. Considering $\alpha_2 = -\alpha_1$ and the relation (14.25), we have

$$
\begin{aligned}
&\left(\sum_{j=1}^{2} \alpha_j \frac{k_j}{\sqrt{v_j}}\right)^{-1} \sum_{j=1}^{2} \alpha_j \left[c_{44}(1 + k_j)\right] \\
&\qquad = A_{33}\left(\frac{k_1}{v_1} - \frac{k_2}{v_2}\right)\left(\frac{k_1}{\sqrt{v_1}} - \frac{k_2}{\sqrt{v_2}}\right)^{-1} \\
&\qquad = A_{33}\left(k_1 v_2 - k_2 v_1\right)\left[\sqrt{v_1 v_2}\left(k_1\sqrt{v_2} - k_2\sqrt{v_1}\right)\right]^{-1}.
\end{aligned}
\tag{14.26}
$$

Substituting (14.26) in (14.15), (14.16), (14.21), and (14.22), we obtain the results of Selvadurai (1979, 1980) for the displacement and rotation of the rigid disk-shaped inclusion in the elastic transversely isotropic material.

Table 14.1 Properties of piezoelectric materials PZT-5, PZT-7A, $BaTiO_3$, PZT-5H

Value	Dimension	PZT-5	PZT-7A	$BaTiO_3$	PZT-5H
c_{11}^E	GPa	121	148	150	126
c_{12}^E	GPa	75.4	76.2	66	55
c_{13}^E	GPa	75.2	74.2	66	53
c_{33}^E	GPa	111	131	146	117
c_{44}^E	GPa	21.1	25.4	44	35.3
$\varepsilon_{11}^S/\varepsilon_0$	$\varepsilon_0 = 8.85 \times 10^{-12}$ F/m	916	460	1115	1706.2
$\varepsilon_{33}^S/\varepsilon_0$	$\varepsilon_0 = 8.85 \times 10^{-12}$ F/m	830	235	1260	1468.9
e_{31}	C/m^2	−5.4	−2.1	−4.35	−6.5
e_{33}	C/m^2	15.8	9.5	17.5	23.3
e_{15}	C/m^2	12.3	9.7	11.4	17.0

References

Elliot HA (1948) Three-dimensional stress distribution in hexagonal aeolotropic crystals. Proc Camber Phil Soc 44:522–533

Erdelyi A, Magnus W, Oberhettinger F, Tricomi FG (1954) Tables of integral transforms. McGraw-Hill, New York

Kassir MK, Sih GC (1968) Some three-dimensional inclusion problems in elasticity. Int J Solids Struct 4:225–241

Keer LM (1965) A note on the solutions for two asymmetric boundary-value problems. Int J Solids Struct 1:257–263

Kirilyuk VS (2008) Stress state of a piezoceramic body with a plane crack opened by a rigid inclusion. Int Appl Mech 44:757–768

Kirilyuk VS (2006) On the relationship between the solutions of static contact problems of elasticity and electroelasticity for a half-space. Int Appl Mech 42:1256–1269

Kotousov A, Neto LB, Khanna A (2014) On a rigid inclusion pressed between two elastic half spaces. Mech Mater 68:38–44

Kyrylyuk VS, Levchuk OI (2021) Stress state of an orthotropic electroelastic medium with an arbitrarily oriented elliptic inclusion. Int Appl Mech 57:320–326

Nategh S, Khojasteh A, Rahimian M (2018) Bonded contact of a rigid disk inclusion with a penny-shaped crack in a transversely isotropic solid. J Eng Math 110:123–146

Podil'chuk YuN (1998) Representation of the general solution of statics equations of the electroelasticity of a transversally isotropic piezoceramic body in terms of harmonic function. Int Appl Mech 34:624–628

Rahman M (2001) The normal shift of a rigid elliptical disc in a transversely isotropic solid. Int J Solids Struct 38:3965–3977

Selvadurai APS (1979) On the displacement of a penny-shaped rigid inclusion embedded in a transversely isotropic elastic medium. Solid Mech Arch 4:163–172

Selvadurai APS (1980) Asymmetric displacements of a rigid disc inclusion embedded in a transversely isotropic elastic medium of infinite extent. Int J Eng Sci 18:979–986

Selvadurai APS (1982) Axial displacement of a rigid elliptical disc inclusion embedded in a transversely isotropic elastic solid. Mech Res Comm 9:39–45

Sneddon IN (1966) Mixed boundary value problem in potential theory. North-Holland Publishing Company, Amsterdam

Chapter 15
State-of-the-Art Methods of Material's Current State Assessment and Life-Time Prediction for NPP Primary Circuit Equipment

Ievgen Kondryakov

15.1 Introduction

The global practice of justifying the structural integrity of critical nuclear power plant (NPP) components shows that the regulatory framework in Ukraine requires improvement for the correct justification of the strength and lifetime of the first-circuit NPP's equipment components.

The most critical element of the first-circuit NPP's equipment components is the reactor pressure vessel (RPV), so special attention is paid to the development of methods for assessing its strength and lifetime prediction. At the same time, world experience shows that the RPV service life is limited by the results of brittle fracture resistance calculations. When evaluating the RPV brittle strength, on the one hand, numerical calculations of the stress intensity factor (SIF) of RPV elements with a postulated crack under thermal shock conditions are required. On the other hand, it is necessary to know the current state of the RPV metal to build a fracture toughness–temperature dependence. Precisely, the use of modern improved calculation methods and methods of predicting the degree of RPV metal degradation level during its operation makes it possible to assess RPV durability and lifetime accurately.

Monitoring the condition of the RPV metal during the entire service life is one of the main conditions for ensuring the reliable and safe operation of the RPV and the entire power plant. The RPV is the main safety barrier of the reactor plant and cannot be replaced due to technical and economic expediency, and in practice, its condition determines the lifetime of the entire power unit.

A necessary condition for the safe operation of the VVER-1000 reactor plant is the preservation of the fracture toughness temperature margin of the RPV materials. Over time, as a result of radiation embrittlement, this margin, that is, the difference between the maximum allowable critical brittleness temperature (T_k^a) and the actual one (T_k),

I. Kondryakov (✉)
G.S. Pisarenko Institute of Problems of Strength, National Academy of Science of Ukraine, Kyiv, Ukraine
e-mail: kondryakov@ipp.kiev.ua

A. N. Guz et al. (eds.), *Advances in Mechanics*, Advanced Structured Materials 191,
https://doi.org/10.1007/978-3-031-37313-8_15

decreases, increasing the probability of brittle fracture of the reactor. According to the PNAE G-7-002-86 (1989) standard, the critical brittleness temperature T_K is considered one of the key parameters characterizing the fracture toughness of RPV materials. The critical brittleness temperature is used in determining safety limits for thermohydraulic parameters during RPV routine hydraulic tests. Calculations on the RPV brittle strength for normal operation conditions, violations of normal operation conditions, and emergency situations are also carried out using the results of mechanical tests of surveillance specimens (SS). In addition, comparing the T_K value with the maximum permissible brittleness temperature T_k^a allows us to confirm the RPV design lifetime and assess the possibility of extending its service life. Thus, safe operation of the RPV VVER-1000 is ensured only in the case when the T_K value remains below T_k^a at any operation regime.

The program for controlling the properties of the RPV metal using SS, made from the material of this vessel, is the basis for establishing changes in the RPV metal mechanical characteristics under operating conditions. The SS test results are used to predict the RPV lifetime from the point of view of resistance to brittle failure.

Cylindrical specimens for uniaxial tension tests, Charpy specimens for impact bending tests, and compact CT specimens (or COD) for fracture toughness tests are used as test specimens.

Many studies show that radiation embrittlement makes the most significant contribution to the shift of the critical brittleness temperature of the metal (Kevorkyan et al. 2004; Margolyn et al. 2010; Nikolaev 2007; Wallin 1991; Wallin et al. 1984). When developing methods for predicting the RPV lifetime, great attention is paid to the radiation effect on the change of mechanical characteristics, primarily the critical brittleness temperature. Unfortunately, the limited number of SSs does not allow for developing a unified methodology for predicting metal radiation embrittlement. Therefore, today several such methodologies are used in various regulatory documents (VVER-1000 2009; IAEA-TECDOC-1442 2005; PNAE G-7-002-86 1989; MRKR CHR 2004; VERLIFE 2008). The issue of the use of certain methods is still debatable, and now in Ukraine, there is still no unified state standard for assessing the RPV strength and lifetime.

With the development of computer technology and calculation methods, the task of evaluating the resistance to brittle fracture of the first-circuit NPP's equipment components can be solved with quite a high accuracy. However, the accuracy of the obtained solution is affected by many factors that require their analysis. In modern regulatory documents, the use of various approaches is allowed, which involve the use of both engineering methods (Marie and Chapuliot 2008; MR 108.7-86 1980) and finite element (FE) models (two and three dimensional) with a built-in crack to determine the SIF (Choi et al. 2019; González-Albuixech et al. 2016; Kharchenko et al. 2018; Kharchenko and Kondryakov 2007; Qian and Niffenegger 2013; VERLIFE 2008).

Recently, along with the classical FE method (FEM), other methods of calculation have been actively developing. This is due to the shortcomings of FEM, which are caused by the significant dependence of calculation results on the FE mesh density. In particular, mixed FEM schemes are being developed (Kharchenko et al. 2018). Also, one of the promising methods is the extended finite element method (XFEM), which makes it possible to obtain satisfactory results in calculating the resistance to brittle fracture without modeling a built-in crack (Belytschko and Black 1999; González-Albuixech et al. 2014; Mora et al. 2019). Using XFEM allows us to calculate the resistance to brittle fracture for different shapes, sizes, and locations of cracks in complex units, for example, the nozzle area, where generating a high-quality FE mesh is difficult when modeling a pre-existing crack (Chapuliot 2016; Liu 2018). At the same time, to obtain more accurate results, it is advisable to use the submodeling technique (Kim 2005; Marenic et al. 2010), which has proven itself well in assessing the strength of complex units of equipment elements.

This chapter considers several factors that can significantly influence the determination of RPV structural integrity and lifetime. A detailed analysis of the well-known methods for estimating RE and predicting the fracture toughness–temperature dependences was carried out using the SS test results of weld metal of the representative RPV with a high Ni content. Existing methods are compared, and their advantages and disadvantages are shown. Also, the results of numerical modeling of the RPV cylindrical part and the nozzle zone using classical and extended FE methods are compared.

15.2 Methods of Radiation and Thermal Embrittlement Assessment of NPP Reactor Pressure Vessel Materials

The use of different RE assessment methods can significantly affect the estimation of the first-circuit NPP's equipment components lifetime, particularly the RPV. At the same time, there is currently no universal method. Their development and improvement continue, especially to take into account new experimental data from SS tests. Therefore, using the most conservative approach with enough justification is advisable for such assessments.

The assessment of RPV metal RE involves the construction of trend curves Tk(F) based on the SS tests results. Next, the critical value T_k^{crit} is determined based on the known design fluence value. If the condition $T_k^{\text{crit}} = T_{ka}$ is fulfilled, the RPV lifetime can be extended. Different methods are used to construct trend curves, which differ in processing SS tests experimental data (Charpy specimens).

Different methods of approximation are used to determine ΔT_F according to the SS test results. In this case, ΔT_F is usually written in the following form:

$$\Delta T_F = A_F \cdot \left(\frac{F}{F_0}\right)^n \tag{15.1}$$

To define A_F and n parameters, different methods of analysis are used:

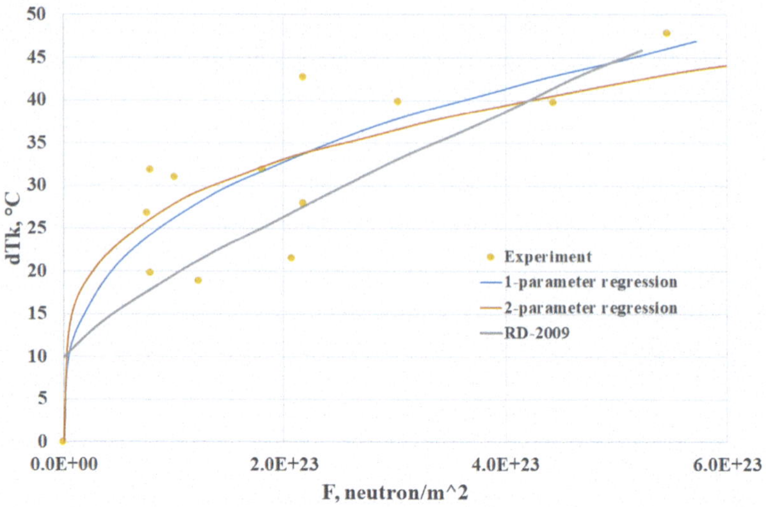

Fig. 15.1 Radiation embrittlement curves constructed by various methods of regression analysis

1. One-parameter regression method (regression parameter is only coefficient A_F, $n = 1/3$). This method is used in the regulatory document PNAE G-7-002-86 (1989).
2. Two-parameter regression method (regression parameters are both A_F and n coefficients). This method is used in the regulatory document Verlife-2008 (VERLIFE 2008).
3. RD-2009 technique. The method for evaluating material properties in the radiation and thermal embrittlement process, which is set out in the regulatory document (VVER-1000 2009), allows estimating the shift of critical brittleness temperature based both on the standard and side-grooved Charpy specimen test results and constructing a theoretical curve based on empirical coefficients using chemical composition data.

Figure 15.1 shows RE curves constructed by different methods for the weld metal of one of the representative RPVs. As can be seen, the choice of the method of processing experimental data and evaluating the metal RE affects the obtained results significantly. Currently, there is no universal technique for a large scatter of experimental data and their insufficient number, especially for high fluence values.

15.3 Methods of Construction of Fracture Toughness–Temperature Dependences and RPV Lifetime Assessment

To estimate the residual RPV lifetime, a comparison is made between the calculated SIF temperature curves, obtained from the results of calculations on the resistance to brittle fracture of the RPV element with a crack, and the fracture toughness–temperature dependences obtained from the results of the SS tests and constructed for a given fluence level.

The limited amount of SSs and experimental data, especially at high fluence values, does not make it possible to accurately predict the state of the metal for overdesign service life. Therefore, today there are several methods of constructing fracture toughness–temperature dependences.

Direct and indirect methods can be used to construct fracture toughness–temperature dependences. With direct methods, fracture toughness tests are carried out, and temperature dependences are determined by the single-temperature or multi-temperature method. In this case, the fracture toughness characteristics can be determined directly for the degraded material without data on its initial properties. Direct methods include the methods of Master Curve (MC) (Wallin 1999), Basic Curve (BC) (MRKR CHR 2004), and Unified Curve (UC) (VVER-1000 2009). Indirect methods use the Charpy impact bending test data and a correlation formula relating the critical brittleness temperature and the fracture toughness–temperature dependences. In this case, it is necessary to know the initial critical brittleness temperature T_{k0}, and SS tests are needed only to determine $\Delta T k_0$.

According to the MC method, the fracture toughness–temperature dependences are described by (15.2) for CT-specimen with a thickness of $B = 150$ mm and a probability of failure $P_f = 5\,\%$, where T_0 is the reference temperature, which is determined according to ASTM E 1921-13a (2015)

$$K_{JC}^{5\%}(T) = 23.32 + 23.385\exp\{[0.019\,(T - T_0)\}. \tag{15.2}$$

The BC method is a modification of the MC method and allows the use of test results for both fracture toughness and impact strength to predict $K_{JC}(T)$ curves (MRKR CHR 2004). The dependence of $K_{JC}(T)$ according to the BC method for the reference thickness of the specimen $B = 150$ mm with the probability of failure $P_f = 5\%$ is written as

$$K_{Jc}(0.05) = 23 + 48\exp\{0.019\,(T - T_k)\}. \tag{15.3}$$

The UC method was developed at the Prometheus Research and Development Center based on the probabilistic "Prometheus model" for predicting fracture toughness, using a new local brittle fracture criterion (VVER-1000 2009; Margolyn et al.

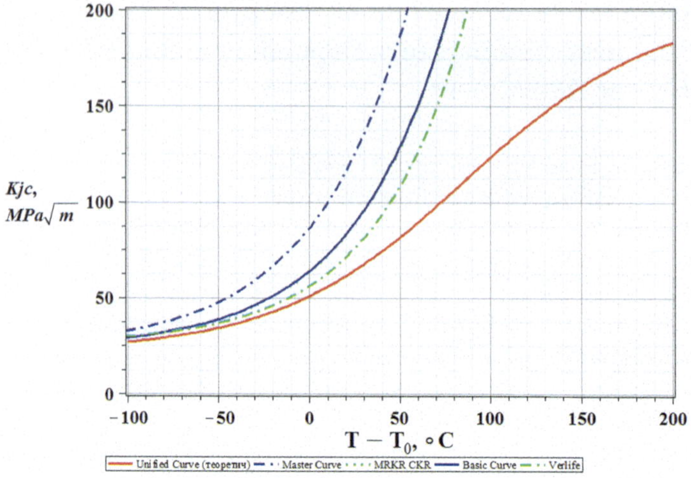

Fig. 15.2 Fracture toughness–temperature dependences for the weld metal of one of the representative RPVs

2010). According to the UC method, the fracture toughness temperature of specimens with a thickness of $B = 25$ mm at $P_f = 50\%$ of RPV steels at any degree of embrittlement can be described by the dependence

$$K_{JC}^{\text{calc}}(T) = k\left[K_{\text{JC}}^{\text{shelf}} - K_{\min}\Omega_{\text{calc}}\left(1 + \tanh\left(\frac{T - \delta T_{\text{type}} - 130}{105}\right)\right)\right] + K_{\min}. \tag{15.4}$$

The parameter Ω can be determined based on the results of tests on fracture toughness or impact strength (VVER-1000 2009).

Within the framework of the international project TAREG 2.01/00 (2008), an analysis of all described methods was carried out. Figure 15.2 shows fracture toughness–temperature dependences obtained by various methods for the weld metal of one of the representative RPVs. As can be seen, the differences between fracture toughness–temperature dependences are quite significant. It is primarily due to insufficient experimental data, especially at high fluence values corresponding to the design ones. Since there is currently no unified regulatory framework in Ukraine for assessing the RPV strength and lifetime, it is necessary to analyze the SS test results using different methods and justify their choice using the principle of conservatism of results.

Further, various methods of determining the SIF are considered, and the results obtained with these methods are compared.

15.4 Engineering Methods of the SIF Assessment

For any load conditions, the SIF for surface cracks can be calculated using engineering methods for determining the stress state of a solid without cracks (VERLIFE 2008). Two such methods are considered in this work.

According to the first method (MR 108.7-86 1980), the SIF is calculated according to the formula

$$K_I = \sigma_k Y a^{1/2}, \tag{15.5}$$

where Y is the shape factor, a is the crack length, and σ_k is the equivalent stresses.

The following equations are used for the surface crack (Fig. 15.3):

- at point A (deep point) $Y = Y_A$ and $\sigma_k = \sigma_{kA}$, where Y_A is defined as follows:

$$Y_A = \frac{2 - 0.82\,a/c}{\left\{1 - \left[0.89 - 0.57(a/c)^{1/2}\right]^3 (a/s)^{1.5}\right\}^{3.25}}; \tag{15.6}$$

- at point B (point on the surface) $Y = Y_B$, $\sigma_k = \sigma_{kB}$ and

$$Y_B = \left[1.1 + 0.35(a/s)^2\right](a/c)^{1/2} Y_A.$$

The equivalent stress σ_K required for calculating the SIF is determined by the components of σ_A, σ_B and σ_C, which are normal to the crack plane at points A, B, and C, respectively.

If the stresses do not change along the thickness of the RPV wall, then $\sigma_{kA} = \sigma_{kB} = \sigma_A = \sigma_B$. If the stresses change linearly across the wall thickness, then

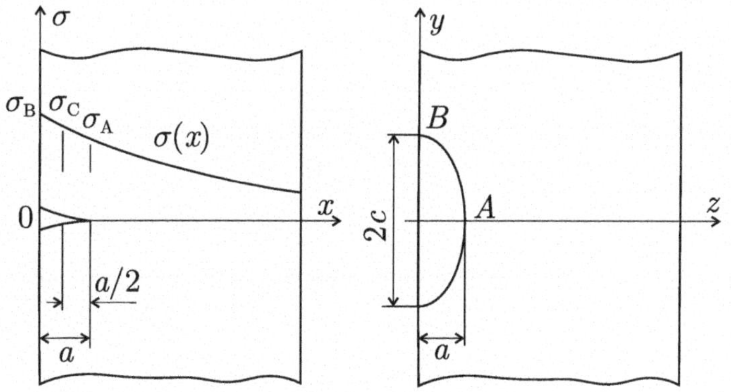

Fig. 15.3 Geometry of a surface crack (VERLIFE 2008)

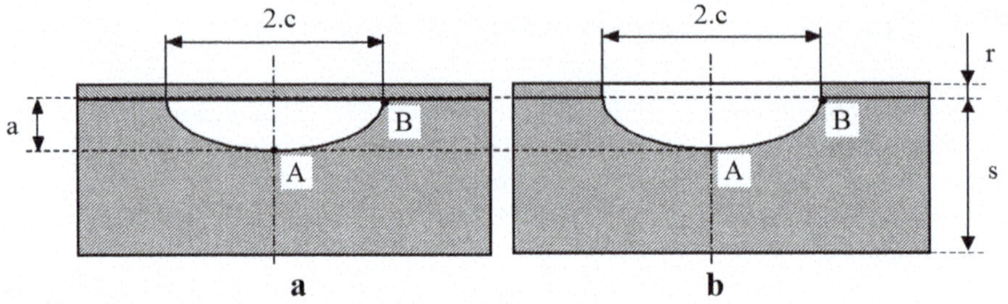

Fig. 15.4 Geometry of the surface (**b**) and undercladding (**a**) crack (VERLIFE 2008)

$$\sigma_{KA} = 0.61\sigma_A + 0.39\sigma_B + \Big\{0.11\,a/c - 0.28\,a/s\left[1 - (a/c)^{1/2}\right]\Big\}(\sigma_A - \sigma_B), \tag{15.7}$$

$$\sigma_{KB} = 0.18\,\sigma_A + 0.82\,\sigma_B. \tag{15.8}$$

If the stresses along the wall thickness change according to the parabolic law, then

$$\sigma_{KA} = 0.111\,(3\;\;\sigma_A + \sigma_B + 5\,\sigma_C) + 0.4a/c\big(0.38\sigma_A - 0.17\sigma_B - 0.21\,\sigma_C\big) - 0.28a/s\left[1 - (a/c)^{1/2}\right](\sigma_A - \sigma_B), \tag{15.9}$$

$$\sigma_{KB} = 0.64\,\sigma_B + 0.36\,\sigma_C. \tag{15.10}$$

The second method (Marie and Chapuliot 2008; VERLIFE 2008) makes it possible to estimate the SIF taking into account the cladding (Fig. 15.4). According to this technique, the stress distribution along the thickness of the RPV wall is described by a polynomial of the third or fourth degree (Fig. 15.5):

$$\sigma = \sigma_0 + \sigma_1\frac{x}{s+r} + \sigma_2\left(\frac{x}{s+r}\right)^2 + \sigma_3\left(\frac{x}{s+r}\right)^3 + \sigma_4\left(\frac{x}{s+r}\right)^4. \tag{15.11}$$

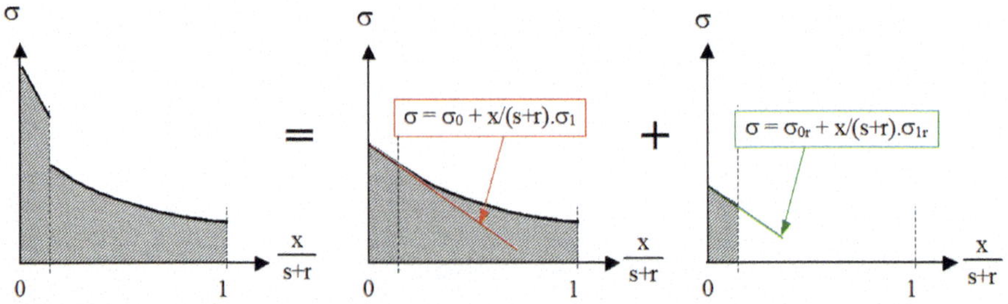

Fig. 15.5 Description of the stress distribution along the RPW wall thickness (VERLIFE 2008)

Then the stresses in the cladding are added. They are approximated by the following linear law:

$$\sigma = \sigma_{0r} + \sigma_{r1} \frac{x}{s+r}. \tag{15.12}$$

Then the stress coefficients $\sigma_0, \ldots, \sigma_4$, σ_{0r}, and σ_{1r} are calculated. After determining such coefficients, the SIF can be calculated using influence functions. The values of such functions are given in tabular form (VERLIFE 2008) for different values of crack sizes, cladding thickness, and the ratio of Young's moduli of the cladding and base metal. With the help of table values, the SIF values for the crack deep point can be calculated:

$$K_I = \left(\sigma_0 i_0 + \sigma_1 i_1 \frac{a+r}{s+r} + \sigma_2 i_2 \left(\frac{a+r}{s+r} \right)^2 + \sigma_3 i_3 \left(\frac{a+r}{s+r} \right)^3 + \sigma_4 i_4 \left(\frac{a+r}{s+r} \right)^4 + \sigma_{0r} i_{0r} + \sigma_{1r} i_{1r} \frac{a+r}{s+r} \right) \sqrt{\pi (a+r)}, \tag{15.13}$$

and for the point on the border of the base metal–cladding:

$$K_I = \left(\sigma_0 i_0 + \sigma_1 i_1 \frac{a+r}{s+r} + \sigma_2 i_2 \left(\frac{a+r}{s+r} \right)^2 + \sigma_3 i_3 \left(\frac{a+r}{s+r} \right)^3 + \sigma_4 i_4 \left(\frac{a+r}{s+r} \right)^4 \right) \sqrt{\pi a}, \tag{15.14}$$

where a is the crack depth in the base metal and r is the cladding thickness.

15.5 Calculation of SIF of a Pre-Existing Crack

Most of the modern regulatory documents (IAEA-TECDOC-1442 2005; VERLIFE 2008) recommend calculating the SIF using FEM and a postulated defect of a given size built into the model. This method requires using a mesh with sufficient density around the crack tip, which leads to an increase in calculation time. However, it makes it possible to obtain more accurate results and take into account elastic–plastic effects.

Regulatory documents recommend calculating semi-elliptical cracks with semi-axis ratios $a/c = 0.3$ and 0.7 in the axial and circumferential directions. An example of an FE mesh of RPV cylindrical part with built-in postulated semi-elliptical axial crack, with a ratio of semi-axes $a/c = 0.3$ and a depth $a = 17$ mm, is shown in Fig. 15.6a.

Numerical modeling of the RPV cylindrical part with cracks of different sizes and orientations was carried out using the above-described SIF calculation methods.

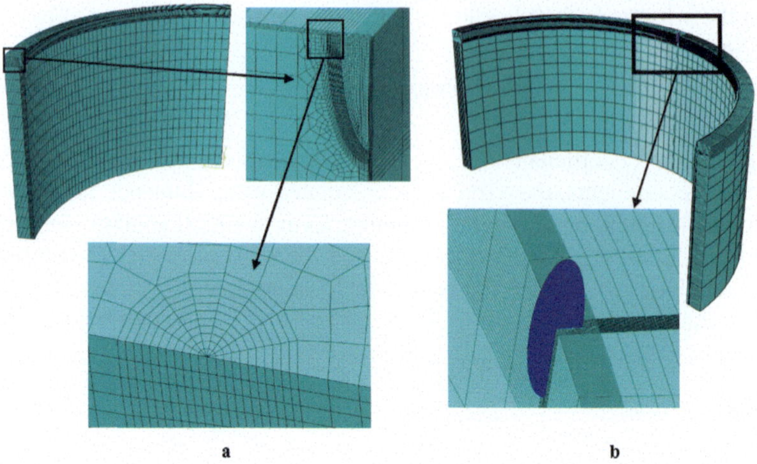

Fig. 15.6 Finite element model (**a**) and model using the XFEM approach (**b**) for the RPV cylindrical part with pre-existing crack

Calculations were performed for simplified 2D models and three-dimensional models with a pre-existing crack in an elastic formulation with the material's mechanical properties according to PNAE G-7-002-86 (1989) for one of the characteristic thermal shock emergency regimes. Figure 15.7a shows the SIF–temperature dependences for an axial semi-elliptical crack with a ratio of semi-axes $a/c = 0.3$. The curves are obtained using two engineering methods and a numerical method for a model with the crack in a two-dimensional formulation. It can be seen that the engineering methods give similar results, while the numerical calculation for the pre-existing crack model in 2D formulation gives overestimated results. Therefore, three-dimensional finite element models are usually used to determine brittle fracture resistance. It can be seen in Fig. 15.7b that using 3D models reduces the conservatism of the results. Furthermore, the crack orientation affects the results. Figure 15.7c shows that an axial crack, in this case, is more dangerous than a circumferential one with the same ratio $a/c = 0.3$.

15.6 Application of the Extended Finite Element Method

The XFEM approach, developed based on FEM by T. Belychko and T. Black (1999), is a numerical method for solving differential equations with discontinuous functions. The XFEM approach was created to overcome difficulties in solving problems with localized singularities. Solving these problems by the classical FEM is relatively inefficient compared with XFEM. The main idea of XFEM is to use a special approximation of the displacement field, which can model any irregular crack surfaces and the asymptotic stress field near the crack tip. At the same time, there is

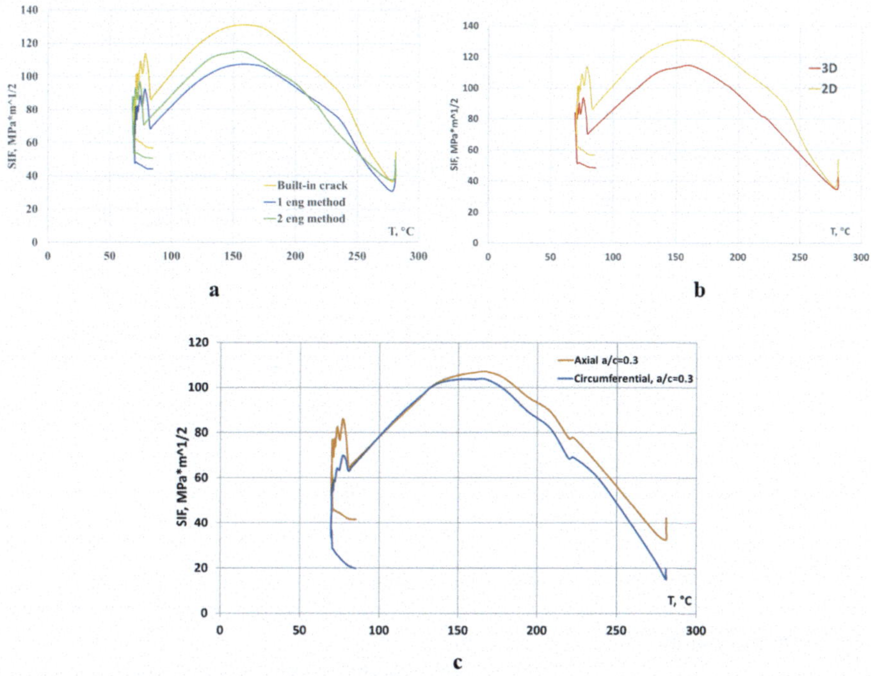

Fig. 15.7 SIF–temperature dependences: for an axial crack in a two-dimensional formulation, calculated according to the first and second engineering methods and built-in crack method (**a**); for built-in axial crack, using 2D and 3D models (**b**); for circumferential and axial cracks, calculated according to the first engineering method in 3D formulation (**c**)

no need to change the finite element mesh and embed the crack in the calculation model. It is desirable to have a sufficiently dense FE mesh around the crack tip to obtain more accurate results.

The need to study brittle fracture resistance for different sizes, orientations, and locations (undercladding, with penetration into the cladding) of cracks makes it necessary to build separate FE meshes when using classical FEM (Choi et al. 2019) or use approximate engineering and analytical methods (Li et al. 2020). Therefore, the XFEM approach is often used to study the brittle fracture resistance of first-circuit NPP's equipment components.

To verify the accuracy of XFEM, the RPV cylindrical part with an axial crack with the ratio of semi-axes $a/c = 0.3$ was analyzed. The FE model for using XFEM is shown in Fig. 15.6b. The obtained SIF–temperature dependences for the characteristic thermal shock regime agree well for both methods (Fig. 15.8).

Many studies have been developed to evaluate the brittle fracture resistance of cylindrical parts of the first-circuit NPP's equipment components. These methods are implemented in some international regulatory documents. At the same time, evaluating brittle fracture resistance for complex units of the first-circuit NPP's equipment

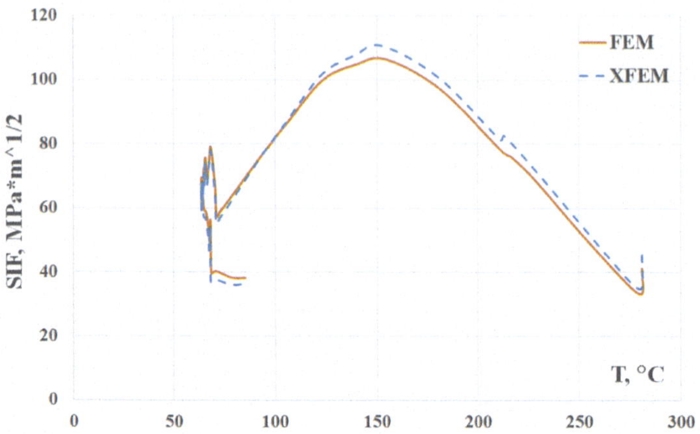

Fig. 15.8 SIF–temperature dependences for an axial crack with $a/c = 0.3$ in the RPV cylindrical part

components is a rather difficult task. First, it applies to the RPV nozzle zone, and there is currently no universal approach for estimating the brittle fracture resistance of such complex units. Developing FE models of the nozzle zone with pre-existing cracks, especially considering the cladding, is highly complicated. Hence, the use of XFEM helps solve such problems.

The "submodeling" technique can also be used to obtain more accurate results and save calculation time. This technique consists in "cutting out" the necessary part (constructive unit) from the global model with the addition to the cut planes of the boundary conditions (displacements) obtained during the calculation of the global model. In this case, the global model can have a sufficiently sparse mesh, significantly reducing the calculation time. At the same time, the cut part (submodel) can have a denser mesh and be supplemented with other geometric elements and conditions. The "submodeling" technique uses the Saint-Venant principle. If the valid boundary conditions at the cut plane are replaced by equivalent boundary conditions for a static formulation, then sufficiently accurate results can be obtained for complex geometric units with stress concentrators. Recently, the "submodeling" technique has been widely used in assessing the strength and durability of first-circuit NPP's equipment components (Kim 2005; Marenic et al. 2010), which significantly reduces the calculation time and increases the accuracy of the obtained results.

In this chapter, calculations of the RPV nozzle zone were carried out using the "submodeling" technique and XFEM. The calculations were performed for four cracks with a depth of 29.4 mm (0.1 of the RPV wall thickness) with the ratio of semi-axes $a/c = 0.3$ and 0.7 in the axial and circumferential directions. At the same time, the axial cracks are located in the inlet nozzles in such a way that the plane of the crack is formed by the axis of the nozzle and the RPV axis, and the main axis of the semi-ellipse forms an angle of 45° with the horizontal axis (the axis of the nozzle). The circumferential cracks are turned 90° relative to the axial ones (Fig. 15.9).

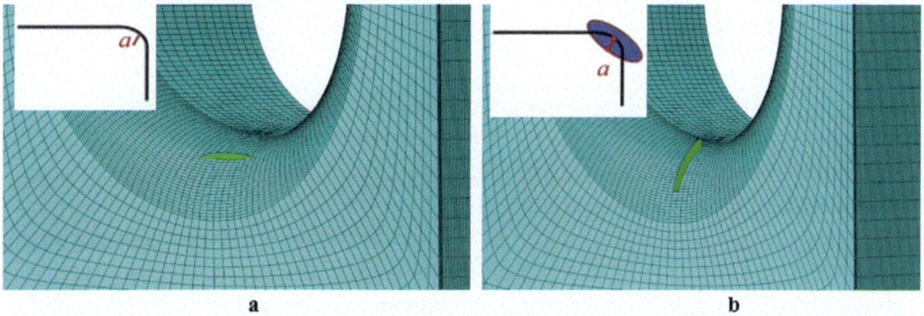

Fig. 15.9 The location of the circumferential (**a**) and axial (**b**) cracks in the RPV nozzle zone

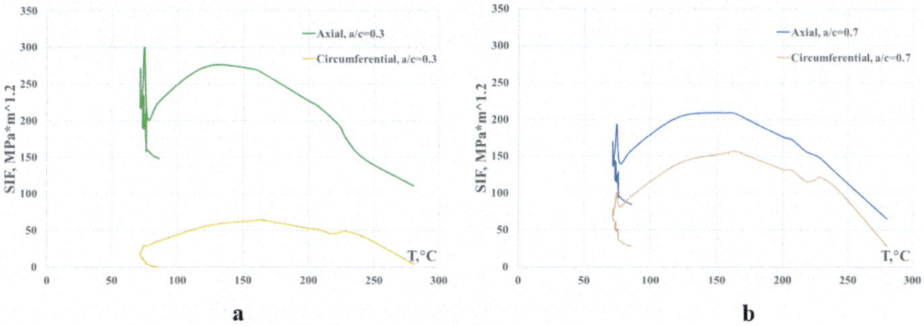

Fig. 15.10 SIF–temperature dependences in the RPV nozzle zone for the characteristic regime of thermal shock for an axial and circumferential crack with $a/c = 0.3$ (**a**) and $a/c = 0.7$ (**b**)

Figure 15.10 compares the SIF–temperature dependences for all cracks. The dimensions being equal, axial cracks are more dangerous than circumferential ones. At the same time, cracks with a ratio of semi-axes $a/c = 0.3$ (Fig. 15.10a) are more sensitive to the location direction than cracks with $a/c = 0.7$ (Fig. 15.10b).

Thus, applying the extended finite element method makes it possible to carry out an express analysis of the influence of the size, shape, and location direction of cracks in the RPV nozzle zone and determine the most dangerous cracks in terms of brittle fracture resistance.

Based on XFEM, an express technique of resistance to brittle fracture assessment with the possibility of varying the shape, size, and location of the crack has been developed. The effectiveness of this technique was illustrated by numerical modeling of the RPV cylindrical and nozzle part for one of the dangerous regimes of thermal shock.

15.7 Conclusions

- Based on a detailed analysis of known methods of radiation embrittlement assessment and prediction of fracture toughness–temperature dependences, as well as their comparison, were carried out using the surveillance specimen's test results of the weld metal of a representative RPV with a high Ni content.
- According to various regression analysis methods, the RE trend curves for the weld metal of one of the representative RPVs were constructed. In addition, it was shown that the methods for experimental data procession and RPV metal radiation embrittlement evaluation significantly affect the results.
- A significant difference was shown in fracture toughness temperature dependences obtained using different methods for the weld metal of one of the characteristic RPVs, primarily due to insufficient experimental data, especially at high fluence values that correspond to the design values.
- With the use of various methods, the evaluation of brittle fracture resistance for the RPV cylindrical part under the characteristic regime of thermal shock was fulfilled.
- Based on the extended finite element method, an express technique of resistance to brittle fracture assessment with the possibility of varying the shape, size, and location of the crack has been developed, which will allow us to effectively determine the crack's critical size and the most dangerous place in the structural element. The effectiveness of using this technique was illustrated by the example of numerical modeling of the RPV cylindrical and nozzle part for one of the dangerous regimes of thermal shock.

The research results presented in this chapter were used within the framework of the international project TAREG 2.01/00, during the performance of the state examination of works on justification of the strength and fracture resistance of the RPV VVER-1000 reactors at the power units of the Zaporizhzhya, Rivne, and South-Ukrainian NPPs, as well as in works on the lifetime extension and safe operation of power units No. 1,2,3 of the VP "RAES".

References

ASTM E 1921-13a Standard Test Method for Determination of Reference Temperature, To, for Ferritic Steels in the Transition Range (2015)

Belytschko T, Black T (1999) Elastic crack growth in finite elements with minimal remeshing. Int J Num Meth Eng 45:601–620

Chapuliot S (2016) Stress Intensity Factor calculation in sharp and beveled edge nozzle corners. Int J Press Vessel Pip 141:11–18

Choi S, Surh H-B, Kim J-W (2019) Effect of postulated crack location on the pressure-temperature limit curve of reactor pressure vessel. Nucl Eng Technol 51:1681–1688

Determination method viscosity destruction according to test results sample witnesses for calculation strength and resource corps VVER-1000 reactors. ROSATOMENERGO, Moscow (2009)

González-Albuixech VF, Qian G, Niffenegger M (2014) Integrity analysi sof reactor pressure vessels subjected to pressurized thermal shocks by XFEM. Nuclear Eng Des 275:336–343
González-Albuixech VF, Qian G, Sharabi M, Niffenegger M, Niceno B, Lafferty N (2016) Integrity analysis of a reactor pressure vessel subjected to a realistic pressurized thermal shock considering the cooling plume and constraint effects. Eng Fract Mech 162:201–217
Guidelines for prediction of irradiation embrittlement of operating WWER-440 reactor pressure vessels. IAEA-TECDOC-1442. IAEA, Wien (2005)
Kevorkyan YR, Erak DYu, Kryukov AM et al (2004) Development physical and empirical models of radiation embrittlement materials corps VVER reactors. Vopr Atom Sci Technol Ser Radiat Mater Sci New Mater 62(1):116–130
Kharchenko VV, Chirkov AYu, Kobelsky SV, Kravchenko VI (2018) Methods of computational analysis in problems of strength of equipment elements of WWER reactor plants. G.S. Pisarenko Institute of Problems of Strength of NAS of Ukraine, Kyiv
Kharchenko VV, Kondryakov EA (2007) Numerical modeling spread and stop cracks in the reactor casing. Sat. Reliability and durability of machines and construction, vol 29, pp 163–168
Kim T-W et al (2005) Application of submodel method for the structural analysis of integral reactor internals. Transactions of the Korean nuclear society autumn meeting. Busan, Korea, pp 27–28
Li Y, Jin T, Wang Z, Wang D (2020) Engineering critical assessment of RPV with nozzle corner cracks under pressurized thermal shocks. Nuclear Eng Technol 52:2638–2651
Liu R et al (2018) Analysis for crack growth regularities in the nozzle-cylinder intersection area of Reactor Pressure Vessel. Ann Nuclear Energy 112:779–793
Marenic E, Skozrit I, Tonkovic Z (2010) On the calculation of stress intensity factor and J-integrals using the submodeling technique. J Press Vessel Technol Trans ASME 132(4):041203
Margolyn BZ, Nikolaev VA, Yurchenko EV et al (2010) New approach to description embrittlement materials corps VVER -1000 reactors in progress operation. Problem Strength 1:7–26
Marie S, Chapuliot S (2008) Improvement of the calculation of the stress intensity factors for underclad and through-clad defects in a reactor pressure vessel subjected to a pressurized thermal shock. Int J Press Vessel Pip 85:517–531
Mora DF, Niffenegger M, Qian G, Jaros M, Niceno B (2019) Modelling of reactor pressure vessel subjected to pressurized thermal shock using 3D-XFEM. Nuclear Eng Des 353:110237
Nikolaev Y (2007) Radiation Embrittlement of Cr – Ni – Mo and Cr – Mo RPV Steels. J ASTM Int 4:8–17
PNAE G-7-002-86 (1989) Norms strength calculation equipment and pipelines atomic power plants. Energoatomizdat, Moscow
Procedure for calculation on brittle fracture resistance for WWER RPV in service. MRKR CHR 2004, RD EO 0606-2005. St-Petersburg, Moscow (2004)
Project TAREG 2.01/00 Validation of neutron embrittlement for VVER 1000 & 440/213 RPVs with emphasis on integrity assessment. INR of NAS of Ukraine Report, Kyiv (2008)
Qian G, Niffenegger M (2013) Integrity analysis of a reactor pressure vessel subjected to pressurized thermal shocks by considering constraint effect. Eng Fract Mech 112–113:14–25
State scientific center of the Russian Federation, Methodical recommendations. MR 108.7-86. Power Components. Calculations and strength experiments. Calculation of stress intensity factors, CNIITMASH, Moscow (1980)
Unified Procedure for Lifetime Evaluation of Components and Piping in WWER NPPs ("VERLIFE"). European Commission (2008)
Wallin K (1991) Fracture toughness transition curve shape for ferritic structural steels. Fracture of engineering materials and structures. Elsevier Applied Science, London, pp 83–88
Wallin K (1999) The master curve method: a new concept for brittle fracture. Int J Mater Prod Technol 14:342–354
Wallin K, Saario T, Torronen K (1984) Statistical model for carbide induced brittle fracture in steel. Met Sci 18:13–16

Chapter 16
Research Methods on Strength and Reliability of Composite Structural Elements of Rocket and Space Technology under Extreme Thermomechanical Loading Conditions

Leonid Kravchuk, Kostiantyn Buiskykh, and Mykola Kucher

16.1 Introduction

The development of modern rocket and aerospace technology is marked by the practical ability to implement operational processes with extreme thermodynamic parameters with an appropriate level of the material base, capable of ensuring high reliability and performance of the structure under challenging conditions of thermomechanical loading and corrosion–erosion effects of high-speed gas flows. The effectiveness of this technique is primarily associated with the development of new classes of materials, taking into account their specifics in creating a structure concerning actual operating conditions.

Taking into account the high requirements for safe operation with simultaneous minimization of weight indicators of such objects, it is of fundamental importance to have correct information on the physical and mechanical characteristics of such materials in the operating temperature and load ranges, as well as the degree of validity of the criteria for assessing the performance of structural elements, taking into account the specifics of operational and technological factors, to quantitatively determine the possibility of establishing the terms of guaranteed safe operation. The availability of such information about the behavior of materials is crucial at all stages of the design and production of rocket and space technology structures, forecasting their functionality and reliability for the entire period of operation. To solve these issues promptly, it is necessary to develop appropriate methods that allow

L. Kravchuk · K. Buiskykh (✉) · M. Kucher
G.S. Pisarenko Institute of Problems of Strength, National Academy of Science of Ukraine, Kyiv, Ukraine
e-mail: bkp@ipp.kiev.ua

L. Kravchuk
e-mail: lvk@ipp.kiev.ua

M. Kucher
e-mail: kucher@ipp.kiev.ua

A. N. Guz et al. (eds.), *Advances in Mechanics*, Advanced Structured Materials 191,
https://doi.org/10.1007/978-3-031-37313-8_16

us to obtain data on structural elements' strength and bearing capacity, assess the effect of technological and structural factors, and study the kinetics of damage under simulated conditions.

The analysis of the specifics of the class of composite materials, technologies for their production, and further manufacturing of structural elements indicate the originality of approaches, methodology, and algorithms for solving these problems. At the same time, it is fundamental to achieve the highest functional and economic efficiency of the created products of new technology, which is possible only as a result of comprehensive research and scientific support of design developments in the process of bringing them to operational status. It was for the development of this promising area of mechanics of materials and structural strength that the Institute of Problems of Strength of the National Academy of Sciences of Ukraine (IPS NASU) was established in 1966; in 1988, it was named after G.S. Pisarenko.

Regarding fundamental research and solving a wide range of applied problems in various engineering fields, the IPS NASU has developed original methods and an appropriate experimental base for research in the field of extreme temperatures and loads. A specific feature of the empirical base of the IPS NASU is the availability of a complex of unique test benches, which allows implementation of a complete closed cycle of research, which provides consistent scientific support of all stages of creating structures for operation in extreme conditions according to the criteria of strength and reliability: development of new original material with the corresponding study of its properties in a wide range of temperatures from cryogenic ones to 3000 °C and stresses up to 200 MPa, optimal design of products with their testing. It should be noted that the main ideas and their practical implementation primarily concerned the fields of technology, where Ukraine did not occupy the last place in the world community. Taking into account that real progress, in whatever field it may be, is always based on new materials and the latest technologies that can ensure the operation of the structure in increasingly harsh operating conditions, the relevant methodological and experimental base of the institute was modified and adapted to current requirements. Examples of such developments briefly described below usually have no analogs.

16.2 Methods of Studying the Behavior and Properties of Carbon–Carbon Composite Materials

The class of composite materials is distinguished by the peculiarity of having specific thermomechanical characteristics, which none of the elements of the structure separately does not manifest, as well as the vast possibilities of forming specific compositions with programmable effects at the structural level. Among them, a special place is occupied by carbon–carbon composite materials (CCCMs), both in their properties and their application. First is their ability to maintain high strength at

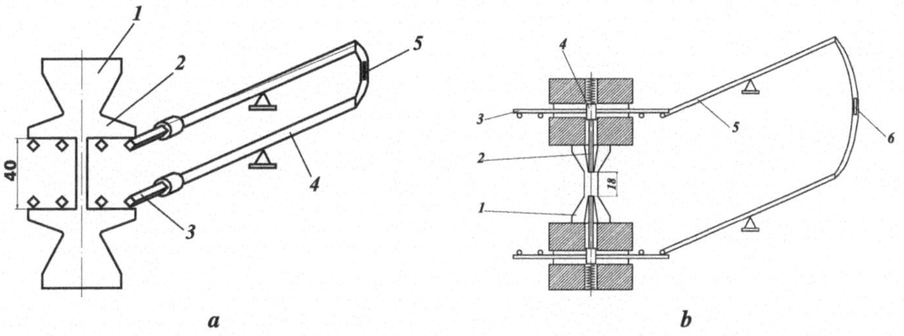

Fig. 16.1 Scheme of measuring the deformation of the working part of the specimen in tensile (**a**) and compression (**b**) tests

temperatures up to 3000 °C with a significant dependence on the heating rate. At the same time, high strength is well combined with favorable thermophysical properties.

Among the developed original experimental test facilities and methods for studying the strength of composite materials (Pisarenko 1980), the most original is a multifunctional complex for studying carbon–carbon composite materials (Dzyuba et al. 2018; Dzyuba and Oksiyuk 2005). Created based on the 1958U10-1 test bench, the complex is equipped with original modern systems of mechanical loading, heating of specimens, creation of a gas environment in the test chamber, and an information-measuring system with the means of experimental control for testing in automatic mode. Using the capabilities of these systems allows the tensile, compression, torsion, bending, shear, short-term creep, relaxation, and low-cycle fatigue of metals and structural polymers to be carried out over a wide range of loads and strain rates. It is possible to work out almost any independent program for loading and heating, which allows reproduction close to the operating conditions of the material.

For example, the following methodological details regarding high-temperature tensile and compression tests illustrate the possibility of increasing the results' reliability, particularly the proposed original specimens and methods of measuring deformations (Fig. 16.1), for which five patents have been obtained. During tension (Fig. 16.1a), the movement of the working part of the specimen (*1*) is determined by the mutual movement of its protrusions (*2*), on which four pairs of levers (*4*) with prismatic tips (*3*) made of heat-resistant ceramics were installed. The other ends of the levers, placed in the cooled compartment of the test chamber, are connected in pairs by elastic brackets (*5*) with glued strain gauges.

At each moment, the computer program determines the shape of the protrusion lines into which the prisms rest and calculates the displacement of the point on the axis of symmetry of the specimen.

During compression (Fig. 16.1b), the movement of the working part of the specimen (*1*) is tracked by thin rods (*2*) made of the specimen material, installed in blind conical holes drilled from the ends of the specimen along its axis to the boundaries of the working part. The opposite ends of the rods rest against the bottom of the spring-

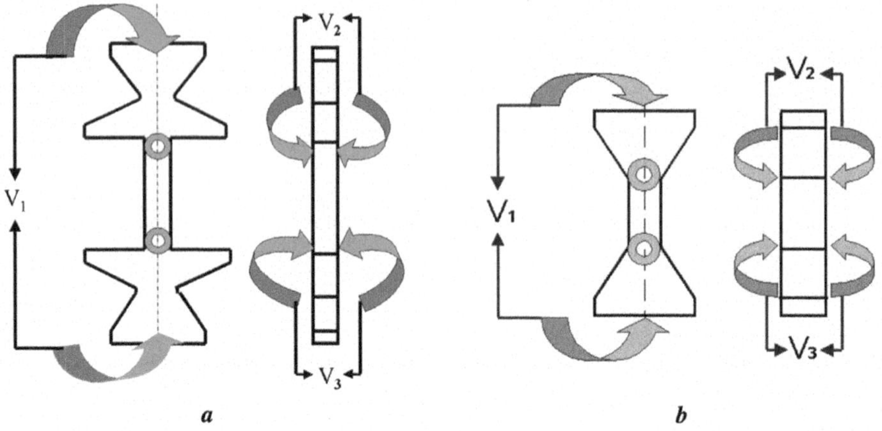

Fig. 16.2 Scheme of implementing the method of heating specimens in tensile (**a**) and compression (**b**) tests by passing an electric current

loaded plungers (*4*), rigidly connected to the rockers (*3*). The mutual movement of the latter through levers (*5*) is transformed into the deformation of strain gauges (*6*). Such techniques made it possible to measure the displacement only within the working part of the specimen, which provides a highly accurate study of the deformation characteristics, both thermal during heating and mechanical, from the action of the load after temperature stabilization.

The installation uses heating the specimen by passing an electric current (Fig. 16.2).

A uniform temperature field along the length of the working part of the specimen is provided by additional heating of its edges (V2) and longitudinal heating (V1). Control over the uniformity of heating is carried out using thermocouples installed along the working part of the specimen or by scanning with a pyrometer "INFRATHERM IGA-100".

An information and measuring system provides the control of tests under mechanical and/or thermal loading and determination of their respective strength parameters and deformed state. TDF (Drozdov et al. 2016) was developed based on the PMX measuring system of the well-known company HBM (Germany). Adapted to this test complex (Drozdov et al. 2016), the 16-channel PMX system fulfills the conditions for the implementation of multichannel measurement of deformations with the connection of strain gauge, inductive and other transducers, as well as load values (forces, pressures),

The information and measuring system is supplemented with a temperature measurement unit created based on a programmable indicator of technological parameters Mikra I3 (Ukraine), with the ability to connect various sensors, particularly the pyrometer "INFRATHERM IGA-100".

The test complex software provides for data visualization in real time, both in digital and graphical form, accumulation, and storage of experimental data on a personal computer.

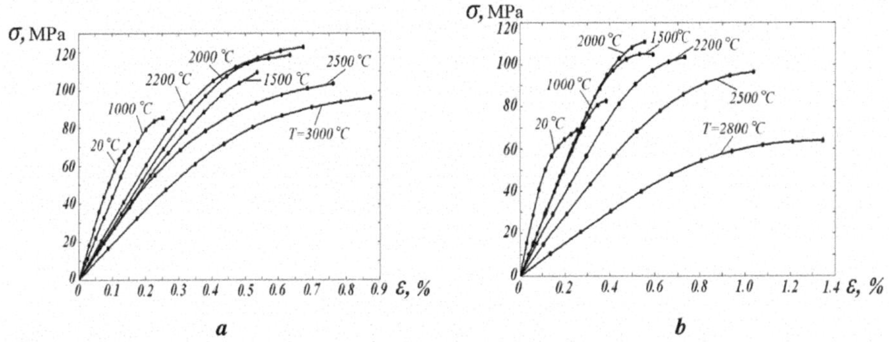

Fig. 16.3 Temperature effect on the type of stress–strain diagram of CCCM in tension (**a**) and compression (**b**)

Since the creation of the complex and its further modification, a large volume of tests of materials (and structures) has been performed on the tasks of rocket and aerospace technology developers.

As an example, the results of studies of the thermomechanical characteristics of one of the carbon-carbon composite materials at temperatures of 20–3000 °C (heating rate of 1000 °C/s) in tension and compression in an inert environment (argon) are shown in Fig. 16.3, indicating specific properties of the studied compositions of the CCCM.

The results of tensile tests (Fig. 16.3a) indicate that in the range of 20–1000 °C, the strength characteristics of the CCCM change insignificantly. In the range of 1000–2300 °C, material hardening is observed, and in comparison with room temperature, the deformation at fracture increases by 6–8 times. When the test temperature is increased to 2500–3000 °C, the value of the fracture stress decreases with a significant rise in deformability.

Analysis of the results of compression tests (Fig. 16.3b) shows that the general trend of the temperature effect on the type of stress-strain diagram has a character similar to tensile. At temperatures above 2700 °C, there is a change in the form of fracture of the specimen, which is associated with a loss of stability.

When solving the problems of the high-temperature strength of structural elements, it is necessary to consider the combined effect of different natural mechanical, thermal, physical, and chemical factors. Each of them contributes to the overall damage of the material. Indirectly, this is reflected in the change in the deformation diagram.

16.3 Analytical and Experimental Methods for Assessing the Effect of Ablation Processes on the Characteristics of Composite Materials

At high temperatures, physical and mechanical destructive processes occur in the matrices and fibers of polymer composite materials, which proceed at different speeds and determine the internal mechanisms of ablation—a self-regulating process of heat and mass transfer, which due to chemical reactions, leads to irreversible changes in the material in combination with heat absorption and removal of part of the mass from the surface of the solid, creating a heat-shielding layer—a barrier to the action of the incoming flow during flights in the dense layers of the Earth's atmosphere. Ablative materials can absorb a significant amount of thermal energy. Therefore, they have been used and improved for more than 50 years for the thermal protection of structural elements of rocket technology, especially disposable components.

Depending on the conditions under which these processes occur, thermal and thermo-oxidative degradations of polymers are distinguished. Oxidation of polymers by molecular oxygen is one of the most common chemical reactions and is accelerated by chemical reagents and various physical factors, such as thermal effects. To a large extent, this applies to carbon fiber composites, for which there are steady trends toward their use in various industries.

The main advantages of carbon plastics are high stiffness and specific strength characteristics of elasticity, resistance to aggressive environments, etc. The reasons that limit their application in practice include the lack of reliable experimental methods for determining the thermomechanical characteristics of reinforced structures, the difficulties of mathematical modeling of deformation processes, and strength assessment of layered composite structures, in particular, taking into account degradation processes.

For example, the calculations of elasticity characteristics and strength estimation of a unidirectional layered composite based on carbon tape are given in Kucher and Samusenko (2018), Kucher and Samusenko (2017). Such composites belong to the class of transversally isotropic materials. Symmetric laying patterns relative to their median surface are reduced to orthotropic materials.

To predict the thermomechanical behavior of such materials, the relations of the mechanics of multiphase media can be used (Dimitrienko 2016; Grinfeld 1990).

Model of ablative composite material. A unidirectional composite at room temperature is believed to have two phases: the polymer phase of the matrix and the reinforcing fibers. At high temperatures, ablation processes occur in the polymer matrix according to the pyrolysis mechanism, and two new phases appear, namely pyrolytic and gas. The first is solid pyrolysis products (coke), and the second is gas products that fill the pores in the composite. Reinforcing fibers are also affected by ablation. The initial phase of the fiber will be called the amorphous phase. At high temperatures, another phase, called "crystalline", appears in the reinforcement.

When heated, the initial and new phases of the matrix and fibers interact with each other, resulting in the composite transitioning into a new phase state. The adopted structure formation scheme is based on the hypothesis that all phases exist simultaneously at any given time in each element of the composite volume. It is believed that heating changes only the ratio of the volume phases of the composite.

The solving system of thermomechanics equations is derived based on the hypothesis of the regularity of their internal structure and the idea of asymptotic expansion for partial differential equations with rapidly oscillating coefficients. It consists of equations that describe the change in the mass of fibers and polymer matrix during deformation, equations of filtration of the gas phase in the pores, relations that describe heat and mass transfer, equilibrium equations, and Cauchy's relation between the tensors of averaged deformation and averaged displacement. In addition, it is necessary to add initial and boundary conditions.

Note that the problems of heat and mass transfer and thermal conductivity are coupled and cannot be solved separately. At the same time, the problem of heat and mass transfer can be solved independently from the problem of deformation since mechanical stresses do not affect the parameters of internal heat and mass transfer.

The considered model and the calculated dependences are based on the following hypotheses. Based on the continuum hypothesis, it is assumed that the layered composite is a homogeneous orthotropic material. The unreinforced epoxy matrix and carbon fiber bundle are supposed to be isotropic linearly elastic materials.

Under other conditions and assumptions given in Kucher and Samusenko (2018), Kucher and Samusenko (2017), the calculation dependencies and examples of computational–experimental studies were obtained: changes in the specific density and elastic characteristics of the epoxy matrix and carbon fibers at high-temperature heating.

Thus, the equation describing the change in the density of the ablative matrix during heating is obtained:

$$\frac{\partial \rho_m}{\partial t} = -\frac{J_m^0}{\rho_b} \Gamma_m \left(\rho_b - \rho_b^\infty\right) \exp\left(\frac{E_{am}}{R\theta}\right). \tag{16.1}$$

The following expression describes the solution of Eq. (16.1):

$$\frac{\rho_m}{\rho_m^0} = 1 - \Gamma_m + \Gamma_m \exp\left(\frac{-J_m^0}{\rho_b} \Gamma_m \int_0^t \exp\left\{\frac{-E_{am}}{R\theta(\tau)} d\tau\right\}\right). \tag{16.2}$$

It follows from (16.2) that the density of the ablating matrix depends not only on the temperature θ but also on the rate of its change. Three parameters determine the change in the relative density of the matrix during heating: activation energy E_{am}, gasification coefficient Γ_m, and the factor before the exponent $J_m^0 G_m / \rho_b$. The values of these parameters depend on the environment in which the heating takes place (inert or oxidizing). In an inert environment, the process of thermal destruction occurs, and in the case of an oxidizing environment, the process of OD occurs. The constants mentioned above for different polymeric matrices are given in Dimitrienko (2016), Lebedev (1963).

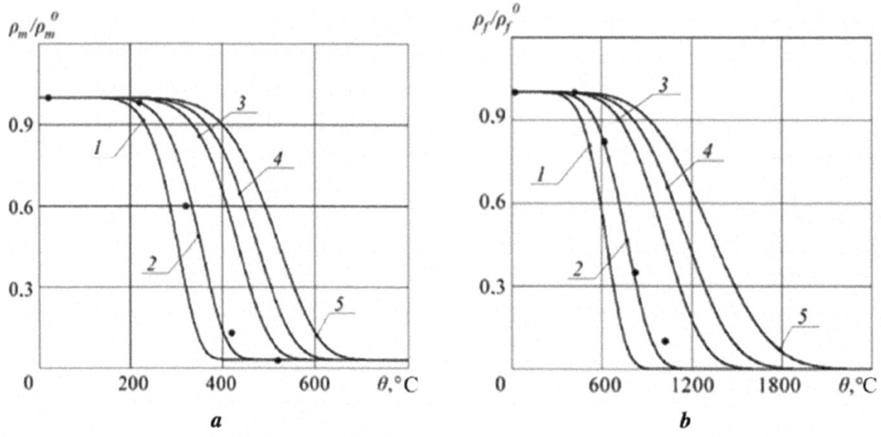

Fig. 16.4 Change in the relative density of the epoxy matrix (**a**) and carbon fiber bundle (**b**) in the process of thermal oxidative degradation: 1—$\theta_0 = 0.03$ K/s, 2—$\theta_0 = 0.1$ K/s, 3—$\theta_0 = 0.5$ K/s, 4—$\theta_0 = 1.0$ K/s, 5—$\theta_0 = 2.0$. Solid lines are calculated using (16.2), and dots correspond to experimental results

Figure 16.4a shows the calculated dependences that determine the change in the density of the epoxy matrix at different heating rates in air (the process of thermal oxidative degradation). Calculations were performed for the following values of parameters: $\varphi_g^0 = 0.06$, $\rho_b = 1200\ \text{kg/m}^{-3}$, $\rho_p = 2000\ \text{kg/m}^{-3}$, $E_{am}/R = 7500$ K, $\Gamma_m = 0.97$, $J_m^0 = 3.6 \cdot 10^{-5}\ \text{kg/(m}^3\ \text{s)}$.

As shown in Fig. 16.4, there is a good agreement between the calculated and experimental results for the unreinforced epoxy matrix and the carbon fiber bundle. This correspondence also indicates the correct choice of the thermophysical characteristics of the epoxy matrix.

Even though the ablation of carbon fibers occurs at higher temperatures than for the epoxy matrix, it is necessary to consider both factors for polymer composites.

The change in elastic characteristics, in particular, the elastic modulus of an isotropic matrix E_m and a bundle of isotropic fibers E_f under high-temperature heating, is considered from the standpoint of the mechanics of multiphase media depending on the temperature and heating rate (Kucher and Samusenko 2017).

As a result of averaging the cubic cell of periodicity, we obtain

$$E_m = E_b^0 \cdot \tilde{a}_m, \quad E_f = E_f^0 \cdot \tilde{a}_f, \tag{16.3}$$

where E_m^0 is the initial elastic modulus of the polymer phase, $\tilde{a}_m$ is a function describing the change in the elastic characteristics of the ablative matrix, E_f^0 is the value of the elastic modulus of the fibers at $t = 0$, $\tilde{a}_f$ is a function relating the change in the elastic characteristics of the fiber bundle during heating.

At the same time,

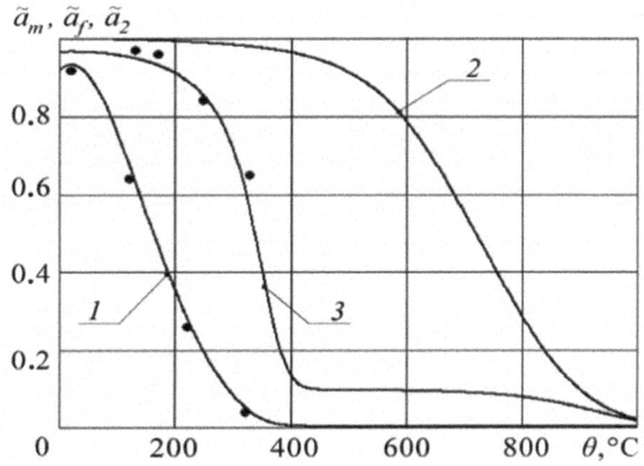

Fig. 16.5 Distributions of the functions describing the change in Young's moduli of the epoxy matrix (16.1), carbon fiber bundle (16.2), and carbon fiber (16.3) at the heating rate $\theta_0 = 0.1$ K/s (points correspond to the experimental results)

$$\tilde{a}_m = \left[\frac{S_p}{a_0(S_p^2 - S_b^2) + m_p(1 - S_p^2)} + \frac{S_p - S_b}{a_0 S_p^2 + m_p(1 - S_p^2)} + \frac{1 - S_p}{m_p}\right]^{-1},$$
$$S_b = \varphi_g^{1/3}, \quad S_p = (1 - \varphi_b)^{1/3}, \quad m_p = E_p/E_b^0, \quad a_0 = \exp\left\{-\alpha_m \Delta\hat{\theta}\right\},$$
$$\Delta\hat{\theta} = \Delta\theta - P\int_0^t \exp\{-Q(t-\tau)\}\Delta\theta(\tau)\,d\tau, \quad \Delta\theta = \theta - \theta_0. \tag{16.4}$$

The constants α_m, P, and Q are determined based on the best fit between the calculated and experimental data describing the change in the elastic characteristics of the epoxy matrix at constant heating rates. As noted in Kucher and Samusenko (2017), the parameters P and Q can be considered the same for all ablative materials. In this case, we take $P = 1.56\ \text{s}^{-1}$ and $Q = 2.5 \cdot 10^{-4}\ \text{s}^{-1}$ (Kucher and Samusenko 2017). Our calculations show that we can take $\alpha_m = 4.395 \cdot 10^{-6}$ K for the epoxy matrix.

Note that at low temperatures, when the ablation process of the epoxy matrix has not yet begun, the change in the function a_m is determined by the value a_0, which describes the change in the elastic characteristics of the polymer phase. Then, $S_p = \text{const}$ and $S_b = 1$.

Note that the ablated epoxy matrix and its solid phases are isotropic materials, which follows from the accepted hypothesis of the central symmetry of the pore shape.

Figure 16.5 shows the dependences of $\tilde{a}_m$ and $\hat{a}_f$ on the temperature θ at a fixed strain rate $\dot{\theta}_0 = 0.1$ K/s. Note that the temperature is listed for convenience and is given in °C. Calculations are performed at the following values of parameters $\varphi_f^0 = 0.5$, $\alpha_f = 2.93 \cdot 10^{-8}$ K and $m_c = 1.3$.

A good agreement between the calculated and experimental data for the composite indirectly confirms the correctness of the choice of the function $\tilde{a}_f$ in the calculation of the effective characteristics of the ablating unidirectional composite at a constant heating rate. The effective characteristics of elastic moduli E_1, E_2, Poisson's ratios

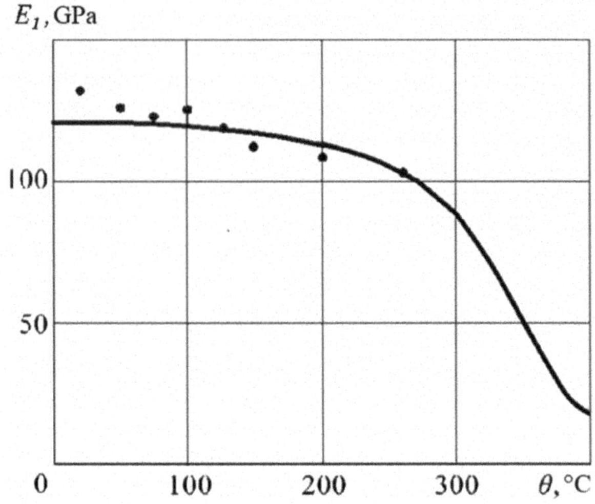

Fig. 16.6 The effective modulus of elasticity E_1 of a unidirectional composite at a fixed strain rate as a function of temperature (solid line corresponds to calculation, dots—our experiment)

ν_{12}, ν_{23} and inplane shear moduli G_{12} are determined in the first approximation based on the characteristics of the ablating matrix and fiber:

$$E_1 = E_f^0 \tilde{a}_2 \varphi_f, \quad E_2 = E_b^0 \tilde{a}_5, \quad \nu_{12} = \nu_m, \quad \nu_{21} = \nu_m \tilde{a}_4, \quad G_{12} = G_b^0 \tilde{a}_3, \quad (16.5)$$

the modulus in the plane of transverse isotropy, $G_{23} = E_2/(2(1 + \nu_{12}))$.

The functions $\tilde{a}_2$, $\tilde{a}_3$, $\tilde{a}_4$, and $\tilde{a}_5$ depend on the structural parameters φ_f, Φ, the ratios of elastic characteristics of the phases E_f^0/E_b^0, ν_f/ν_m and the functions $\tilde{a}_f$, $\tilde{a}_m$ that describe the change in elastic moduli of the unreinforced matrix and fiber depending on temperature.

Figure 16.6 shows the results of calculating the elasticity characteristics of the composite E_1. It is observed that there is a reasonably good fit for the elastic modulus E_1 as a function of temperature for quasistatic heating conditions. When calculating other elastic characteristics, we can only discuss a qualitative description in their definition. When describing the dependence of the shear modulus on temperature, it is necessary to consider that G_{12} is not equal to G_{21}. For accuracy, it is necessary to use their average value.

The possibility of estimating the ultimate stresses of a transversely isotropic composite at elevated temperatures is based on the mechanics of multiphase media, taking into account the thermomechanical characteristics of the ablating polymer matrix and carbon fiber bundle (Kucher and Samusenko 2018), provided that the mass transfer problem can be solved separately from the coupled problem of thermomechanics of ablating materials since mechanical stresses do not affect its parameters. It is established that as a result of the calculation of the ultimate stresses based on the hypothesis of complete tightness of surfaces between phases or under the assumption of low pressure in the pores, the lower and upper limits of the change in the strength

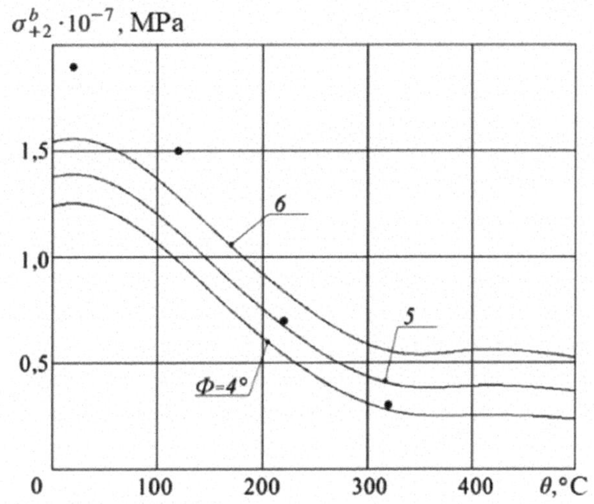

Fig. 16.7 Distribution of the tensile strength of epoxy carbon fiber reinforced plastic in transverse tension (solid lines correspond to calculations, dots—experimental results)

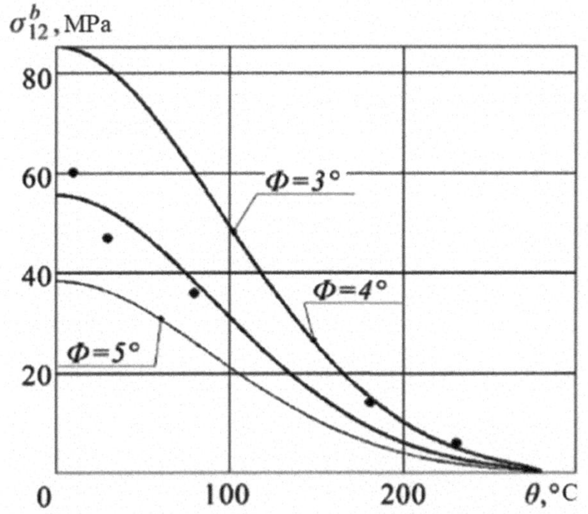

Fig. 16.8 Shear strength distribution of epoxy carbon fiber reinforced plastic along the fibers (solid lines correspond to calculations, dots—experimental results)

of the composite under different types of stress states are determined. The results shown in Figs. 16.7 and 16.8 are obtained for carbon plastics with varying angles of fiber orientation (Φ) under heating in air at a speed of 0.1 K/s. They indicate that the results of calculations of the strength of a unidirectional composite under tension in the transverse direction and under shear along and across the fibers correlate well with the experimental data obtained by the authors Danil'chuk et al. (2015).

In summary, the obtained results confirmed the effectiveness of the developed methods for predicting the strength characteristics of ablating polymeric unidirectional carbon plastics at elevated temperatures in thermal oxidative degradation under different stress states.

16.4 Bench Tests of Composite Structural Elements

The most unique and valuable components of the experimental bench complex are gas-dynamic high-temperature benches, which were created in the period 1955–1960 as an adequate experimental base for the operative solution of questions concerning new special heat-resistant alloys, refractory metal and polymer composite materials, structural ceramics, etc., optimization of the composition and technologies of their production and subsequent manufacture of structural elements from them at different stages of development and implementation of new technical solutions in the field of engineering. The effectiveness of this technique is determined, first, by the practical ability to implement thermal processes with extreme thermodynamic parameters and an appropriate level of scientific base, which allows for ensuring high reliability and performance of the structure under challenging conditions of thermomechanical loading, high temperatures, corrosion and erosion effects of high-speed gas flows.

Gas-dynamic stands are a complex of systems and units for creating and forming the flow of gases—products of combustion of the air-fuel mixture, program regulation of its parameters, and control of the state of the objects of study during the tests. Several variants of installations based on standard power equipment were created depending on the level of thermal and mechanical loads and peculiarities of the solved scientific and applied problems. The equipment of gas-dynamic stands with different variants of systems makes it possible to carry out modes of thermal loading in the temperature range of 300–2500 °C of arbitrary duration with varying rates of temperature change and gas flow rate, including trans- and supersonic ranges, simulating the state of natural structural elements in almost any mode of their operation. At certain stages of equipment development, the stands were modified and equipped with additional systems, and the necessary methodological base was developed. In recent years, a significant modernization of stand systems has been carried out. It mainly concerns the systems of control, registration, and processing of received information, for which the modern element base and computer capabilities are used. Among the most significant stages of development of this experimental base and the conducted studies, it is expedient to note the following.

From the early 1960s to the late 1980s, the fundamental problem solved on gas-dynamic stands was the study of composite materials to create optimal designs for the thermal protection of spacecraft and rocket engine elements. At different stages of development, urgent problems concerning the thermal protection structures of spacecraft and rocket engines were solved. Several other problems were also solved, including choosing optimal classes and combinations of materials and methods of designing heat shielding packages and determining the dimensions of heat shielding structures for different heat-loaded elements of spacecraft surfaces. The conditions of bench tests were modeled by the intensity, duration, and frequency of thermal effects of actual spacecraft trajectories in the dense layers of the atmosphere (Tretyachenko 1979).

Analysis of loading conditions and patterns of damage to structural elements of rocket and space technology allows us to identify three main groups of factors that determine the process of thermal damage to structures. Common to them is that all of them are, to some extent, determined by the thermal state of the object of study and the kinetics of this state in space–time coordinates.

The absolute values of the material temperatures determine the first group of factors. They define actual values of physical and mechanical characteristics of materials and also values of thermal stresses of the second kind in structurally heterogeneous materials.

Non-stationary processes of temperature change in time determine the second group of factors. It reflects the specifics of the effect of thermal load on an element of complex geometric shape, determining its thermal stress state and the kinetics of this state in the thermal cycle and from cycle to cycle. The main elements of this group are amplitudes of change and absolute values of extreme thermal stresses in the cycle, the law of change of stresses in time, and gradients of thermal stresses.

The third group of factors reflects the impact on the object of study of the external environment. They are responsible for the deterioration of the mechanical properties of the surface material, changes in the chemical composition of the material, and, as a consequence, in its physical properties, the emergence of surface thermal stresses associated with the anisotropy of the properties of the basic, and changes near the surface of the material.

The methodological basis of research on gas-dynamic stands is a set of specialized methods for modeling the thermal stress states of the material and the intensity of the external effect of the environment, which, when implemented, ensure the equivalence of the processes of material damage and the boundary state of the structural element under study in model and full-scale conditions (Tretyachenko 1979).

16.4.1 Experimental and Calculation Methods of Ablative Thermal Protection Research

The model of the behavior of the thermal protective coating and the mechanisms of its destruction in high-temperature and high-speed gas flow are determined by the conditions of actual operation. The most effective method of thermal protection of disposable structural elements of rocket technology is ablative sacrificial coatings that are destroyed. Thermal protection of structures with the help of a sacrificial layer is more reliable and efficient in terms of weight than other methods. At the same time, practically no restrictions are imposed on the maximum heat flux. All this leads to the fact that destroyed ablative thermal protective coatings have no competition at high thermal loads and limited service life. That is why the most significant progress in the field of creation and functioning of rocket and space technology objects has been achieved in the development of disposable (sacrificial) heat-protective coatings, for which ablation is the primary and necessary mechanism of material destruction under the thermal action of an external heating source (Polezhaev and Frolov 2005).

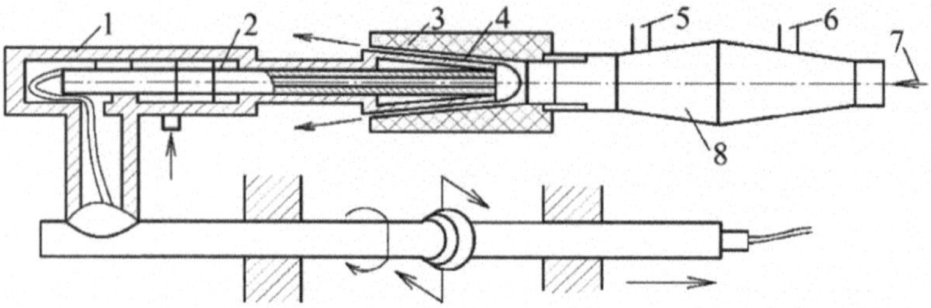

Fig. 16.9 Scheme of the manipulator: 1—hydraulic cylinder, 2—water-cooled casing, 3—graphite chamber, 4—the model under study, 5—oxygen supply, 6—fuel supply, 7—air supply, 8—combustion chamber

When developing the methodology for modeling the conditions of thermal protection tests on models of rocket and space technology in a high-temperature gas flow on a gas-dynamic stand, some approaches are used to provide modeling of external effects on the structural element and the equivalence of material damage processes in model and full-scale conditions. The fundamental basis of these approaches is the classical theories of similarity and dimensionality, the main results of which are transformed and adapted to studying the strengths of materials and damage to structural elements under aerodynamic heating to extremely high temperatures.

As the essential equipment, the test complex of gas-dynamic stands was used, the principal design features of which and methodological solutions for testing thermal protection are presented in the scheme of Fig. 16.9. A specific feature of the bench for such tests is a three-stage combustion chamber designed to create a gas flow of a given temperature, speed, and chemical composition by varying the ratio of components of the kerosene–air–oxygen mixture.

The heated gas flow generator consists of an air supply system (7), fuel storage, supply, and flow control system (6), and an oxygen supply system (5). A specific feature of the test bench for such tests is a three-stage combustion chamber designed to create a gas flow of a given temperature, velocity, and chemical composition by varying the ratio of components of the fuel–air mixture. The first stage is a standard uncooled combustion chamber. The second water-cooled stage is designed to support the combustion process and the formation of the mixture with an additional supply of fuel (6) and oxygen (5). In the third stage—a variable test chamber—an uncooled graphite subsonic nozzle is designed to form a gas flow of specified parameters. The profiling of the graphite chamber can be carried out along the model's contour, providing programming of the flow velocity in different sections of the channel, which allows us to simulate the effect of convective and radiation heat flows. The radiation component is modeled by the temperature level of the preheated walls of the graphite nozzle, which has a high emissivity ($\varepsilon = 0.9 \ldots 1$).

Since the gas flow along the axis of the chamber has a different temperature, the model of the heat shielding coating (Fig. 16.10) is moved to the position in which the specified thermal test conditions are achieved. Simultaneous movement along

Fig. 16.10 Models of heat protective coating during tests and before the start-up of the bench

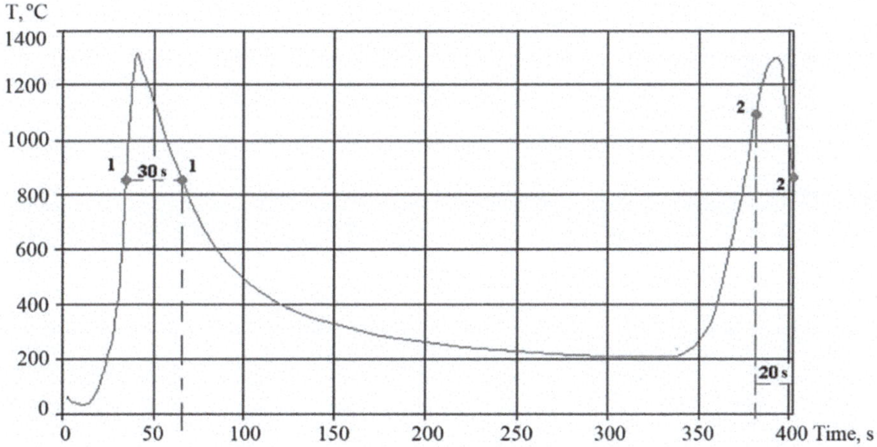

Fig. 16.11 Mode of thermal loading of structural elements with a thermal protective coating

the axis of the chamber and adjustment of the fuel components during the test cycle provides programmed heating of the surface.

For example, we consider the thermal load mode of the model with the TZP presented in Fig. 16.11. For the correct reproduction of the specified heating mode, computational and experimental studies were carried out, which consisted of thermometry of the gas flow in the modes of both a gradual increase in temperature on the simulator of the model with a heat-protective coating and the introduction of the latter into the gas flow with the maximum peak temperature. These data served as basic information in solving the problem of one-sided heating of the wall of the heat shielding coating with the corresponding thermophysical characteristics by the gas flow.

In Figs. 16.12 and 16.13, two variants of the modes of change in the temperature of the gas flow and the coating in the first half-cycle of heating after the introduction

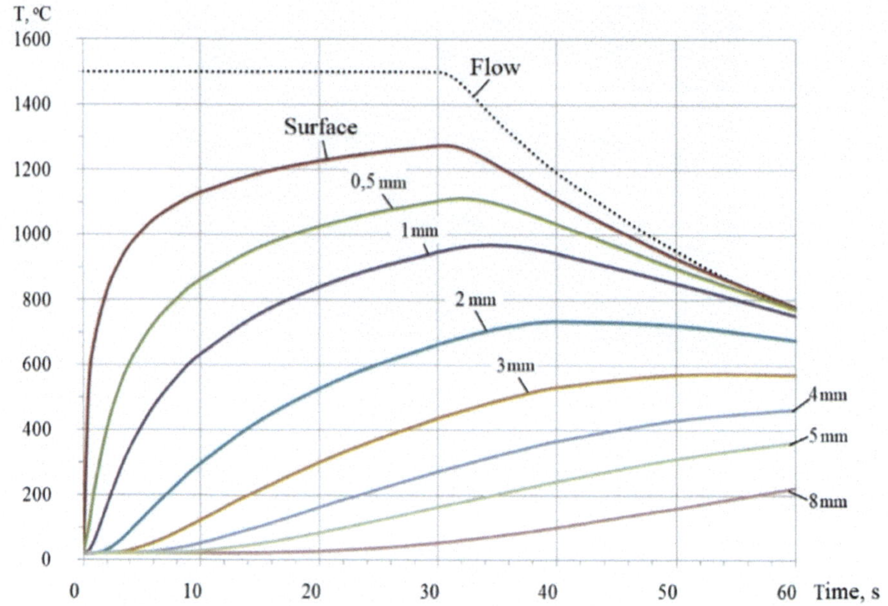

Fig. 16.12 Calculated temperature distribution over the thickness of the thermal protective coating after entering the model into the flow of extreme parameters

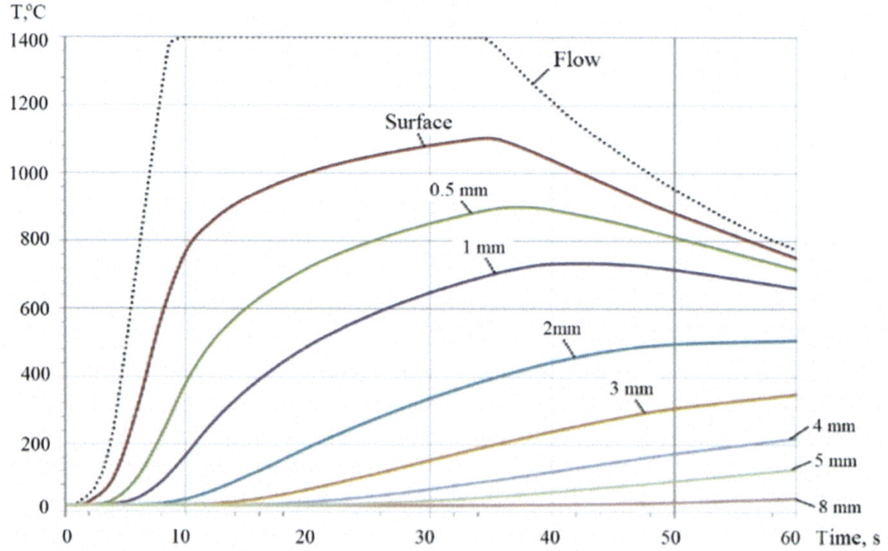

Fig. 16.13 Calculated temperature distribution over the coating thickness after entering the model into the flow with the subsequent programmatic change of its parameters

of the model imitator under the nozzle of the test chamber are shown. The analysis of the results shows that the mode of thermal loading of the thermal protective coating can be realized by adjusting the rate of change of the gas flow temperature directly during the experiment according to the thermometric data.

16.4.2 Methods of Simulation of Aerodynamic Heating Conditions of Reusable Spacecraft Structural Elements

An essential problem in developing and operating reusable rocket and space technology products is ensuring their integrity and safe return to Earth. For other types of descending space objects, solving problems associated with the aerodynamic heating of individual structural elements and aerospace vehicles remains fundamental. At the same time, the two directions differ in the approaches to the design of reusable space system elements.

The first direction solves the problem of ensuring the reliable functioning of active and passive external flight control systems; first of all, the most thermally stressed rotating elements of the airframe, which are structures with large angles of attack, such as the toe and the leading edge of the fuselage, wing toe, elevator rudder, and air intake edges (Gusarova 2017). At significant thermal force loads, determined by the vehicle's speed in the atmosphere, their surface temperature can reach 1100–1200 °C.

The main task of the second direction is to maintain the permissible temperature level of the spacecraft power shell at all parts of the flight trajectory, especially during re-entry into the Earth's atmosphere. For reusable spacecraft, the concept of ablative heat protection is unacceptable, providing adequate heat protection for disposable spacecraft parts. Therefore, one of the tasks in creating reusable space stations is to develop a reliable, reusable heat shield with acceptable dimensions, mass parameters, and cost.

Regarding the wide range of these tasks, the IPS NASU's developments are considered on the example of bench test methods for solving a set of tasks related to the assessment of the performance of the air intake edges of the direct-flow air-jet engine of the airframe of the reusable transport space station and the multilayer heat shielding structure with an external metal three-layer panel, implemented on a complex of gas-dynamic stands. The methodological basis of bench studies is a set of specialized methods for modeling the thermal stress states of the material and the intensity of the external effect of the environment, which, when implemented, ensure the equivalence of the processes of material damage and the limit state of the structural element under study in model and full-scale conditions.

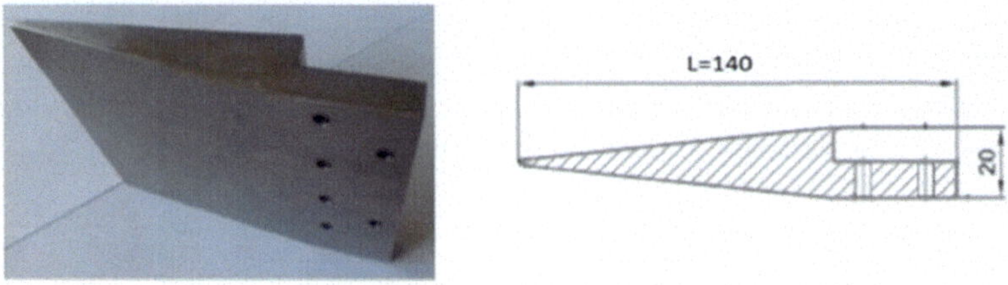

Fig. 16.14 Model of the air intake edge

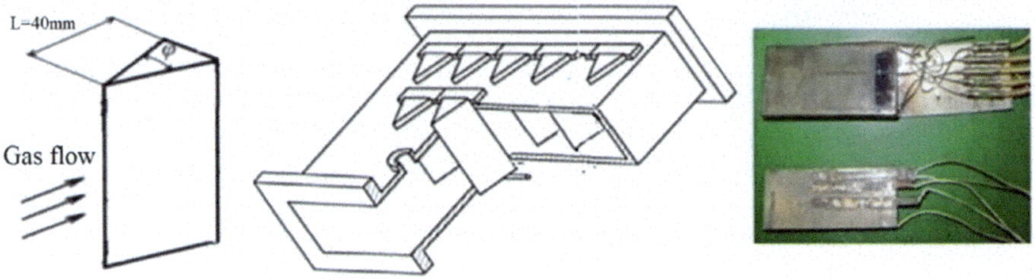

a *b*

Fig. 16.15 Test objects on the stand: model (**a**) and test chamber of the gas-dynamic test bench (**b**)

16.4.2.1 Methods of Investigation of Air Intake Edges

The edges of the air intake have the shape of a wedge-shaped prism (Fig. 16.14), which allows solving this problem quite effectively using the approaches described in regulatory documents (DSTU 2367-94 1994). The correctness of the results obtained during bench tests according to these standards is ensured by a reasonable choice of the model's geometry (shape, dimensions) and the mode of changing the parameters of the gas flow. The proposed methods specify the principles of choosing the geometric dimensions of the model. Under the given conditions of external effect, the mode of thermal loading (temperature, heating rate, and temperature gradient at the specimen edge) is regulated by choice of the deflection angle φ and the radius of curvature of the specimen edge R. The level of thermal stresses, ceteris paribus, is determined by the size of the specimen chord L.

An experimental study of the models of air intake edges in a high-temperature gas flow was carried out at one of the installations of the gas-dynamic stand using test chambers (Fig. 16.15). Methodologically, the test program is assigned based on the results of computational–experimental modeling on the stand of natural heat flows (Troshchenko 2005).

The practical implementation of this condition is reduced to the experimental determination of the law of change in the temperature of the gas flow, which ensures the identity of the kinetics of the temperature state of the edges of the models in time on the bench and the specified programs (Fig. 16.3), estimated by the flight conditions by calculation. When choosing the geometry of the model for bench tests following

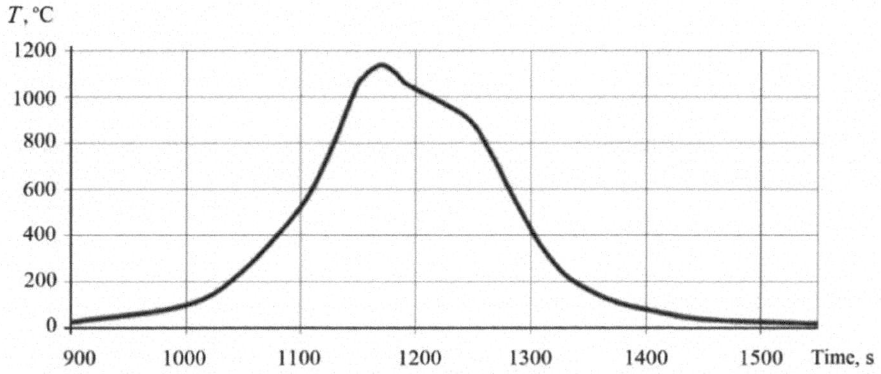

Fig. 16.16 Mode of thermal loading of edge models

the previously developed methods for this task, it is advisable to keep the shape and dimensions (R and φ) of the edge zone identical (Fig. 16.14). The procedure for the rational choice of the size L, which determines the bearing capacity of the structure according to the strength criteria, involves several calculation estimates (DSTU 2367-94 1994), taking into account the actual physical and mechanical properties of the materials under study and the parameters of the gas flow.

To determine the mode of thermal loading of the air intake edge models, as well as to determine the boundary conditions of the first kind, which are necessary for calculating the thermal stress state of the models, thermometry of wedge-shaped specimens was carried out. For thermometry, it is necessary to make thermocouple-reinforced witness specimens (Fig. 16.15c) from an alloy similar in chemical composition and thermophysical characteristics to the material from which the air intake edges were supposed to be made. The number of thermocouples is chosen in such a way as to obtain sufficiently complete information about the surface temperature without significantly distorting the temperature fields due to thinning of the specimen body during the milling of the grooves and without changing the aerodynamic characteristics of the profile. Thermocouples were placed, taking into account the inhomogeneity of the temperature distribution on the side face of the specimen. In the area of maximum temperature gradients, thermocouples were installed with a minimum distance, considering the adopted technology's capabilities.

As a result of such experiments, the task of obtaining initial data for controlling the parameters of the gas-dynamic bench, which ensure the implementation of the mode of change in the temperature of the gas flow and the thermal state of the edge, was solved (Fig. 16.16). The information obtained about the kinetics of the thermal state of the witness specimen (specimen with thermocouples) in the cycle during tests under the selected mode of thermal load is the basis for the justification and optimal choice of the chord size L of the specimen, which has certain features, for example, when solving material science problems regarding the selection of material and the assessment of the bearing capacity of a structure made of a particular material. This information is also essential for quantitative analysis of bench test results in general. The significance of such procedures is illustrated by the results obtained during the cycle of bench tests of specimen models of air intake edges in the mode considered in Fig. 16.16.

The problem of determining the thermal stress-strain state of the studied specimens was solved in a spatial formulation using the SPACE application package (Software 2002). Considering the spatial and temporal heterogeneity of thermal processes, the thermal state of the material of the wedge-shaped specimen was determined by solving the nonlinear spatial problem of unsteady heat conduction under mixed boundary conditions of heat transfer.

The need to calculate models of different materials and different sizes to determine the effect of coatings, material degradation, including ablation and its damage on the tubular fluidized bed based on the results of tests of specimens of the same size determines the practicality of applying boundary conditions of the third kind in the calculations of unsteady thermal conductivity problems. The latter were determined by solving the direct and inverse thermal conductivity problems using numerical modeling methods (Kravchuk et al. 2018).

When solving the inverse problem, the initial conditions are the boundary conditions of the first kind, which are determined by the corresponding thermometry. To determine the boundary conditions of the third kind by solving the gas-dynamic (direct) problem, the experiment records information about the change in time of the gas flow temperature and pressure at the inlet to the test chamber of a specific geometry, on which the thermophysical properties of the gas and its gas-dynamic characteristics depend (Kravchuk et al. 2018). The conformity of the obtained results is checked by comparing the calculated distribution of the specimen surface temperature with the experimental data (thermometry). Figure 16.17 shows the results of the thermometry of the witness specimen.

Comparing the results of determining the boundary conditions by the first and second methods with experimental data indicates the chosen calculation model's validity and reliability. The analysis of the results of the calculation of the TNDS suggests several methodological features of solving material science problems. In particular, when selecting materials according to the criteria of their performance, the size of the chord of the specimen can be limited to approximately 40–50 mm. To solve the strength, the size of the models (specimens) must be selected following the standard (DSTU 2367-94 1994). The above methodological solutions were implemented during a series of tests, the results of which are partially considered (Pisarenko 1980).

The developed methods have certain specific features due to the study of the performance of modern thermal protection of reusable space systems under extreme temperatures and power loads. The main methodological approaches and experimental bench base for solving this class of problems were created and modernized following the requirements of a particular stage of development of the rocket and space industries.

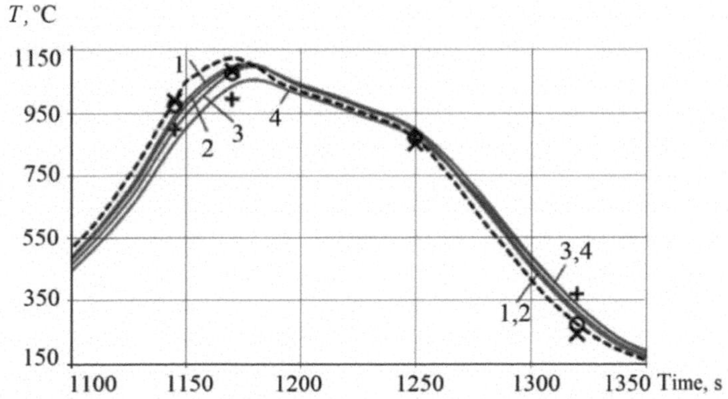

Fig. 16.17 Comparison of the calculated temperature values (1—at the edge; 2,3, and 4—at distances of 3, 10, and 20 mm from the edge, respectively; dashed line—flow temperature) with the experimental values (×—at the edge, o, +—at distances of 3 and 20 mm from the edge) for individual points in time

16.4.2.2 Research Methods of Multiple Thermal Protection

The main principal approaches of testing were considered concerning the research of thermal protection of disposable spacecraft, for the operation mode of which critically high levels of thermal loads and the use of destructive composites—sacrificial thermal protective coatings are characteristic. The second direction—the development of thermal protective coatings for reusable spacecraft—provided for the development of technology for the production of composite materials and the creation of optimal designs for the thermal protection of spacecraft, taking into account their design features. A range of tasks was solved, including the choice of optimal classes and combination of materials and methods of designing heat shielding packages, determining their sizes for different heat-loaded elements of spacecraft surfaces, choosing the best ways to fasten blocks of heat shielding materials to each other and the entire structure on the surface of the vehicle, including the protection of hatches, portholes, etc. The created specialized methodological and experimental base ensured comprehensive studies of the behavior of heat-shielding materials and structures in almost all parameters that determine their functional specificity, the maximum reflection of the effect of operational loads—thermal and mechanical.

The intensity, duration, and frequency of thermal effects of actual spacecraft trajectories in the dense layers of the atmosphere modeled the conditions of the bench tests. Methodologically, the test program is assigned according to the results of the computational–experimental modeling of natural heat flows on the stand. For different tasks, such programs are created based on the results of tests with calibration of special metrological (calorimetric) models, in the process of which the dependence of the flow parameters on the flow rate of the main components and their distribution

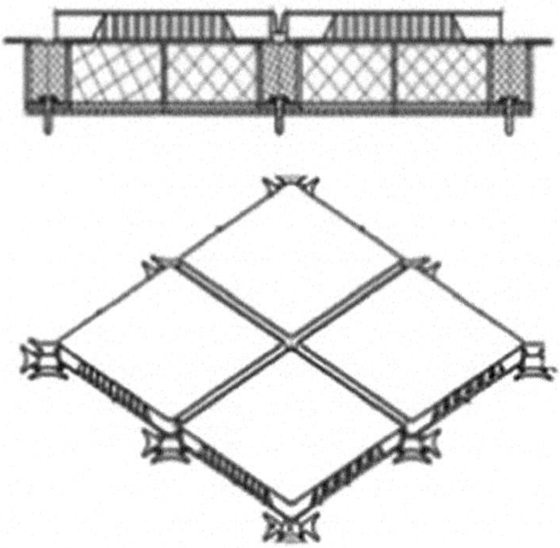

Fig. 16.18 Construction of multilayer heat-protective coating

over the volume of the gas flow are established, followed by the implementation of the bench control system elements (Kravchuk et al. 2018).

Currently, for the designed and partially functioning reusable space stations, the most promising and reliable are heat-shielded tile structures that combine heat protection of the spacecraft hull and load-bearing function. Certain technologies for manufacturing external panels from heat-resistant alloys can ensure their reliability during repeated use (up to 100 flights) with re-entry into the Earth's atmosphere.

Satisfactory characteristics at temperatures up to 1100 °C have been obtained for alloys based on nichrome with a dispersed oxide strengthening phase (Solntsev et al. 2017; Tikhiy et al. 2015). To evaluate the use of materials, development of design, and technological solutions in the creation of panel heat-shielding coatings of this class, the previously used methods (Danil'chuk et al. 2015; Pisarenko 1980; Troshchenko 2005) were modified, or original methods of modeling the loading conditions of coatings were developed, as well as the modernization of individual systems of the gas-dynamic test bench, the principal features of which are discussed below. The methodological issues arising in assessing the performance of the structure of the studied type are considered (Fig. 16.18) under the given conditions of single and multiple (cyclic) temperature exposure.

For all cases, the thermal loading mode, as in the previous tasks, is represented by a program for changing the temperature of the coating surface, which is identical to full-scale and bench conditions, and contact temperature sensors and an infrared radiation pyrometer monitor its progress. Their information is used as control signals for working out the thermal load program. An example of such a program is shown in Fig. 16.19 (Kravchuk et al. 2018).

In bench tests of ablative coatings, high-temperature loading conditions are realized by fixing the model in the gas flow zone, usually using manipulators (Figs. 16.9 and 16.10). Panel-type structures (Fig. 16.18) can be investigated according to the

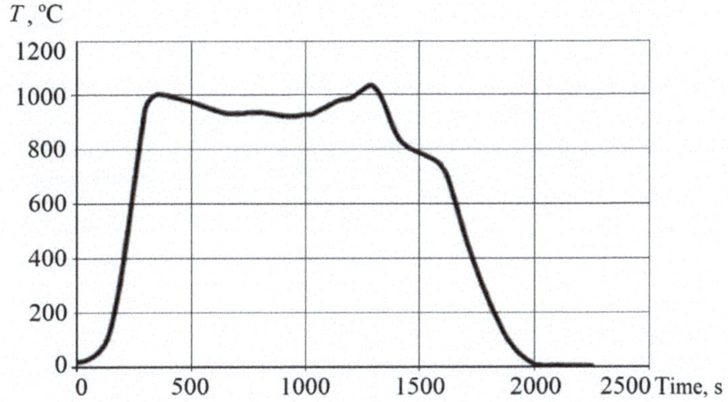

Fig. 16.19 Mode of thermal loading of the heat-protective metal three-layer structure

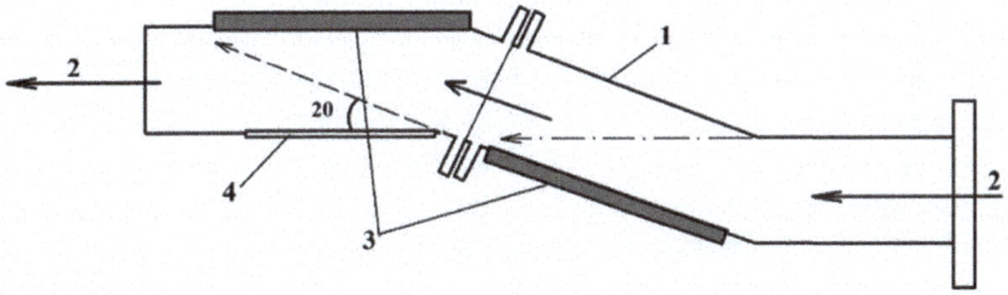

Fig. 16.20 Scheme of panel arrangement in test chambers: 1—test chamber, 2—gas flow, 3—heat protection panels, 4—window for video recording

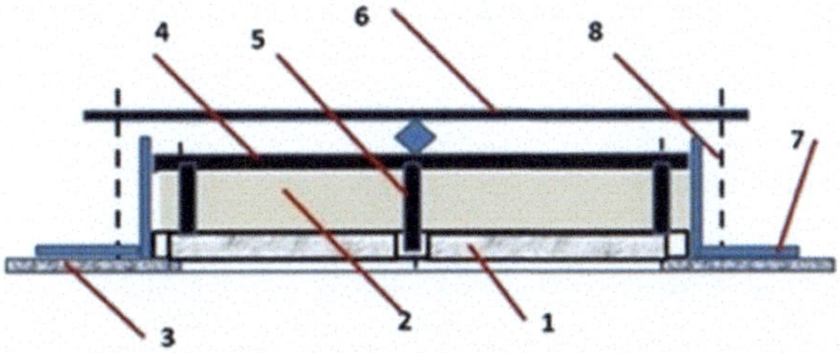

Fig. 16.21 Scheme of fastening the heat shield panel at eight points: 1—heat shielding panel; 2—thermal insulation; 3—chamber wall; 4—rigid "cold" plate; 5—racks; 6—panel fastening to the chamber; 7—frame; 8—fastening bolts

scheme shown in Fig. 16.20. In this variant, the tested panel is a fragment of a specially designed test chamber, which allows testing at different angles of flow attack.

The chambers of the gas-dynamic test bench for testing a metal three-layer structure are designed with a minimum live section at the inlet to maintain the same angle of attack at all points of the blown models. The need to solve the above-mentioned problems led to the implementation of two schemes of fastening the heat shielding package to the test chamber. The first scheme allows relatively free thermal deformation of the panel. Figure 16.21 shows the method of its fastening.

The mounting system of the heat shielding model on the test chamber reflected the mounting conditions in actual structures. The thermal protection panel was attached at eight points utilizing eight racks to a rigid plate that simulated the inner wall of the protected ("cold") structure in actual conditions. The mounting plate had neither thermal nor mechanical connection with the walls of the test chamber. The rate of heating of the thermal insulation determines its temperature deformations.

Using modern measuring and original information systems during the experiments ensures the receipt of data sufficient in volume and accuracy for the calculated estimates of the TNDS and the analysis of test results.

16.5 Conclusions

Based on the basic provisions of the classical theory of similarity and dimensionality, which are transformed and adapted to the problems of studying the strength of materials and damage to structural elements under thermocyclic loading in corrosive environments, the methodology for modeling the conditions of aerodynamic heating of structural elements of single and reusable spacecraft is presented.

A set of specialized methods for modeling the thermostress state of the material and the intensity of the external effect of the environment has been developed, which, being the methodological basis of bench studies, allowed us to ensure the equivalence of the processes of material damage and the limit state of the studied structural element in model and full-scale conditions.

It is shown that the developed methods and experimental means provided the possibility of assessing the functional characteristics, obtaining a set of properties, and working out the technology of forming elements of aerospace structures of aircraft operating under conditions of aerodynamic heating to extremely high temperatures.

References

Composite Materials Handbook. Vol. 2: Polymer Matrix Composites Materials Properties. MIL-HDBK- 17-2F. US Department of Defense (2002)

Danil'chuk EL, Kucher NK, Kushnarev AP et al (2015) Deformation and strength of unidirectional carbon-fiber-reinforced plastics at elevated temperatures. Strength Mater 47(4):573–578

Dimitrienko YuI (2016) Thermomechanics of composite structures structures under high temperatures. Springer, Dordrecht

Drozdov AV, Kharchenko VV, Dziuba VS et al (2016) Multifunctional data acquisition and control system rigging the beds and units for strength tests of models and structure elements from composite materials. Strength Mater 48(5):632–638

DSTU 2367-94 (1994) Metals, alloys, heat-resistant coatings. Method of high-temperature corrosion and thermal fatigue testing in the fuel combustion product stream [in Ukrainian], Valid since Feb 25, 1994

Dzyuba VS, Potapov AM, Gusarova IA et al (2018) Strength of carbon-carbon composite materials at high temperatures (up to 3300K). Technol Syst (3) 49–59

Dzyuba VS, Oksiyuk SV (2005) Investigation of strength of carbon-carbon composite materials at temperatures from 293 to 3300 K under high-rate heating. Strength Mater 37(1):99–104

Grinfeld MA (1990) Methods of continuum mechanics in the theory of phase transformations. Nauka, Moscow
Gusarova IA (2017) Selection of materials for external structures of the reusable transport-space system. Technol Syst 1:63–69
Kravchuk LV, Buiskikh KP, Gusarova IA et al (2018) Methods for the simulation of the aerodynamic heating conditions of the structural elements of space shuttles. Strength Mater 50(4):565–574
Kucher MK, Samusenko AA (2017) Assessment of elasticity characteristics and strength parameters of unidirectional polymer composites at high temperatures. Strength Mater 49(3):453–463
Kucher NK, Samusenko AA (2018) Mathematical prediction of strength of ablating polymer unidirectional composites in transverse direction and in shear along and across fibers. Strength Mater 50(4):575–583
Lebedev NN (1963) Special functions and their applications. GIFML, Moscow [In Russian]
Pisarenko GS (1980) Strength of materials and structural elements under extreme conditions (in 2 volumes), 2. Naukova Dumka, Kyiv [in Russian]
Pisarenko GS (1992) Actual problems of strength in modern mechanical engineering. Naukova Dumka, Kyiv [In Russian]
Polezhaev YuV, Frolov GA (2005) Thermal destruction of material. Izvod IPM NASU, Kyiv [In Russian]
Proncevich EV, Gusarova IA, Karpikova OA et al (2016) Multifunctional coatings with special properties in rocket engineering. Technol Syst 1:30–33
Software "Three-Dimensional Finite Element Simulation of the Thermal and Thermally Stressed State of the Elements of Engineering Structures (SPACE)", UkrSEPRO Sertification System, Compliance Certificate No. UA1.017.0084261-02.-2002. [in Ukrainian]
Solntsev V, Frolov G, Kravchuk L et al (2017) Development of rolling modes for samples made from nichrome powder alloy and their testing at operating temperatures. Machines Technologies Materials 11(5):254–258
Tikhiy VG, Gusev VV, Potapov AM et al (2015) Tile heat shield structures of space shuttles with different outer load-bearing members. Avtomat Svarka 3–4:66–71
Tretyachenko GN (1979) Modeling in the study of structural strength of structures. Naukova Dumka, Kyiv [In Russian]
Troshchenko VT (2005) Strength of materials and structures. Akademperiodika, Kyiv [in Russian]

Chapter 17
Scattering of *SH*-Waves by an Elastic Fiber of Non-canonical Shape with a Thin Interphase Layer

Yaroslav Kunets, Roman Kushnir, and Valery Matus

17.1 Introduction

The development of technologies for obtaining composite materials, improvement of nondestructive testing and defectoscopy, modern problems of medicine, geophysics, and seismology have significantly expanded the scope of studying the contact interaction of thin deformable bodies with the surrounding environment. Thin elastic inhomogeneities are not only stress concentrators but are also widely used as composite fillers, resonators, waveguides, sound diffusers, and nodes of measuring equipment systems. In these cases, in addition to solving the traditional problems of assessing the strength and deformability of structures, the study of the spectral characteristics of scattered fields, and the sensitivity of elastic systems to dynamic disturbances, including non-stationary ones, becomes important. The modes of operation of structures with thin-walled elements are also diversified, and the contact conditions of the phases of the heterogeneous elastic system are modified.

The effect of thin inhomogeneities on the physicomechanical fields in media has been studied in many publications (see, for example, Pasternak et al. (2021), Emets and Rogowski (2013), Benveniste and Miloh (2001) and their literature reviews). In the vast majority of works, static problems were considered. Their solution was based on some additional assumptions (hypotheses) of a physical or mathematical nature, which leads to a decrease in the dimensionality of the original problem in the vicinity of fine inhomogeneity.In this regard, there is a great variety of models of thin elastic inclusions, which are proposed for inhomogeneities with different material

Y. Kunets · R. Kushnir · V. Matus (✉)
Pidstryhach Institute for Applied Problems of Mechanics and Mathematics, Lviv, Ukraine
e-mail: matus@iapmm.lviv.ua

Y. Kunets
e-mail: kunets@iapmm.lviv.ua

R. Kushnir
e-mail: dyrector@iapmm.lviv.ua

A. N. Guz et al. (eds.), *Advances in Mechanics*, Advanced Structured Materials 191,
https://doi.org/10.1007/978-3-031-37313-8_17

properties. The inclusion of low or high stiffness was considered separately, and models have been created that cover a wide range of properties.

The methods of singular perturbation theory were also applied to the problems of modeling the interaction of thin inhomogeneities with the environment. Its basic provisions are outlined in Friedrichs (1955), Fraenkel (1969). L. Prandtl formulated the basic idea of these methods during the study of the phenomenon of the boundary layer, which is mathematically determined by the presence of a small parameter (inverse Reynolds number) in the higher derivatives of the Navier–Stokes equations, which occurs in the case of flow bodies by hydrodynamic flows. It consists in introducing internal local coordinates (stretching of coordinates) near the object's surface, on which the boundary conditions are set and near which a rapid change of solution in the direction of these coordinates is observed. Since a similar structure of the solution was found in other problems with a small parameter, an approach was formed according to which these solutions are presented in the form of asymptotic (outer) expansions with further additional study (construction of boundary layers, inner asymptotic expansions) in the regions of rapid change of the considered processes (Fraenkel 1969; Friedrichs 1955). Within the framework of the asymptotic approach, a number of problems for regions containing thin inhomogeneities have been investigated. Note that the structure of the asymptotic expansions of the solutions of such problems depends significantly on the ratio of parameters characterizing the material properties of the components of the composite. Static problems for bodies with inhomogeneities of large and small stiffness were studied in the case of inclusions (Kanaun 1984) and in the case of interphase layers (Benveniste and Miloh 2001). In Kunets (1987), the asymptotic approach is applied to problems of axisymmetric torsion, and the case of inhomogeneity of arbitrary stiffness is considered.

Problems with thin inhomogeneities within the framework of the asymptotic approach were studied in several ways. According to one of them, a system of integral equations equivalent to the initially posed problem is first written down, after which, using the theory of asymptotic expansions of integrals for a small parameter, the theory of generalized functions, etc., an integral equation given on the middle surface (line) of inhomogeneity, which asymptotically approximately determines the sought integrand functions (Kanaun 1984; Poddubnyak 1986). In another approach, the asymptotic representations of the solution are substituted into the differential equation and the boundary conditions of the problem, followed by decoupling of the corresponding operators by a small parameter (Benveniste and Miloh 2001; Mishuris et al. 2006; VanDyke 1964). Sometimes, instead of a strong formulation of the problem (differential equations, constitutive relations), a variational (weak) formulation is used (Rizzoni et al. 2014).

In this chapter, an asymptotic analysis of the equation of motion of the thin interphase separating the matrix from the fiber is carried out. The case of antiplane deformation of an elastic system is considered. Before, similar elastic structures under wave loading were studied for the case of circular fibers and interfacial layers of constant thickness. At the same time, the interphase was considered an elastic body in ideal contact with the matrix and the fiber (Cai 2004), or its interaction with the external environment was modeled under the conditions of non-ideal spring-type contact between the fiber and the matrix (Butrak et al. 2012; Wang and Sudak 2007). Here, the case of interphase of variable thickness and fiber of non-canonical cross section is investigated. As a result of the application of asymptotic analysis, a simplified variant of the problem of SH-wave scattering in a matrix with fiber under effective conditions of their contact is formulated.

The most popular research tool in the far wave zone of waves scattered by objects of complex geometry is the null-field method. With its help, the interaction of elastic waves with volume inclusions was considered mainly under conditions of ideal contact between the heterogeneity and the surrounding medium (Martin 2006; Waterman 2009). Below, this method is transferred to the case of SH-wave scattering by inclusion of a non-canonical form in the presence of a thin interphase of variable thickness and low stiffness.

17.2 Formulation of the Problem

Consider the unbounded elastic matrix W_1 containing fiber W_2 in the presence of thin-walled interphase inhomogeneity W_0. The wave load and the geometry of the scatterer satisfy the conditions of the anti-planet strain state. The cross section of the fiber has a non-canonical shape bounded by a closed curve S. The area of location of the inhomogeneity is formed by removing from the areas W_1 and W_2 some two-sided neighborhood around the line S so that $W_0 = W_0^+ \cup W_0^-$, $W_0^+ = W_1 \cap W_0$, $W_0^- = W_2 \cap W_0$ (see Fig. 17.1).

The matrix, fiber, and inhomogeneity are isotropic elastic bodies with densities ρ_1, ρ_2, ρ_0 and shear moduli μ_1, μ_2, μ_0, respectively. The system's motion varies harmonically in time t with the circular frequency ω; therefore, the time factor $\exp(-i\omega t)$ is common to all values and is omitted to simplify the notation.

Under antiplane deformation, the displacements in the composite satisfy the two-dimensional Helmholtz equations (Emets and Rogowski 2013; Martin 2006)

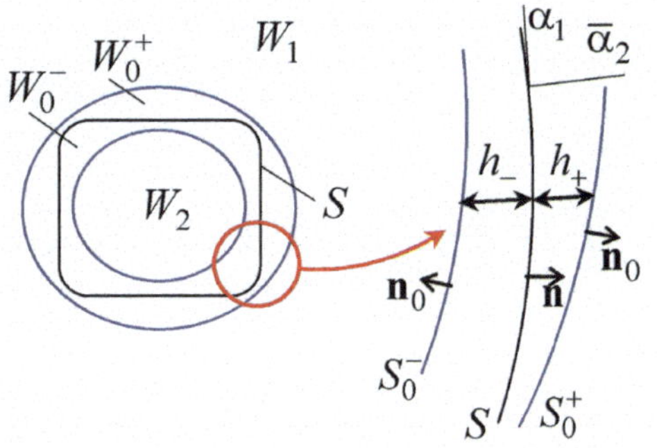

Fig. 17.1 The geometry of the composite body and "microscopic" image of part of the interphase inhomogeneity

$$\left(\nabla^2 + k_{S1}^2\right) u_1(\mathbf{x}) = 0, \quad u_1(\mathbf{x}) = u^{sc}(\mathbf{x}) + u^{in}(\mathbf{x}), \quad \mathbf{x} \in W_1 \backslash W_0^+, \tag{17.1}$$

$$\left(\nabla^2 + k_{S2}^2\right) u_2(\mathbf{x}) = 0, \quad \mathbf{x} \in W_2 \backslash W_0^-, \tag{17.2}$$

$$\left(\nabla^2 + k_{S0}^2\right) u_0(\mathbf{x}) = 0, \quad \mathbf{x} \in W_0, \tag{17.3}$$

conditions of ideal mechanical contact between the components of the composite

$$\begin{aligned} u_1(\mathbf{x}) = u_0(\mathbf{x}), \quad \mu_1 \frac{\partial u_1(\mathbf{x})}{\partial n_0} = \mu_0 \frac{\partial u_0(\mathbf{x})}{\partial n_0}, \quad \mathbf{x} \in S_0^+, \\ u_2(\mathbf{x}) = u_0(\mathbf{x}), \quad \mu_2 \frac{\partial u_2(\mathbf{x})}{\partial n_0} = \mu_0 \frac{\partial u_0(\mathbf{x})}{\partial n_0}, \quad \mathbf{x} \in S_0^-, \end{aligned} \tag{17.4}$$

and radiation conditions at infinity

$$u^{sc}(\mathbf{x}) = \frac{\exp i\,(k_{S1} r + \pi/4)}{\sqrt{8\pi k_{S1} r}} f(\theta) + O\left(r^{-1}\right), \quad r \to \infty, \tag{17.5}$$

where $\mathbf{x} = (x_1, x_2)$ are Cartesian coordinates of points,

$$x_1 = r\cos\theta, \quad x_2 = r\sin\theta, \quad 0 \leqslant \theta \leqslant 2\pi;$$

$u_j(\mathbf{x})$, $k_{Sj} = \omega / c_{Sj}$, and $c_{Sj} = (\mu_j/\rho_j)^{1/2}$ are displacements, wave numbers and speeds of shear waves in the composite, respectively; n_0 is the external unit normal to $S_0 = S_0^+ \cup S_0^-$; $f(\theta)$ is the complex amplitude of SH-wave scattering; (r, θ) are polar coordinates. The problem is to determine the unknown scattered field $u^{sc}(\mathbf{x})$ by a given value of the incident field $u^{in}(\mathbf{x})$ ($u_1(\mathbf{x}) = u^{in}(\mathbf{x}) + u^{sc}(\mathbf{x})$).

The solutions of equations (17.1), (17.2), and (17.3) can be given in integral form (Emets and Rogowski 2013; Martin 2006)

$$\int_{S_0^+} \left[u_1\left(\mathbf{x}'\right) \frac{\partial}{\partial n_0} G_1\left(\mathbf{x}', \mathbf{x}\right) - G_1\left(\mathbf{x}', \mathbf{x}\right) \frac{\partial}{\partial n_0} u_1\left(\mathbf{x}'\right)\right] dS_0' = \begin{cases} u^{sc}(\mathbf{x}), & \mathbf{x} \in W_1 \backslash \overline{W}_0^+ \\ -u^{in}(\mathbf{x}), & \mathbf{x} \in W_2 \cup \overline{W}_0^+ \end{cases}, \tag{17.6}$$

$$\int_{S_0^-} \left[u_2\left(\mathbf{x}'\right) \frac{\partial}{\partial n_0} G_2\left(\mathbf{x}', \mathbf{x}\right) - G_2\left(\mathbf{x}', \mathbf{x}\right) \frac{\partial}{\partial n_0} u_2\left(\mathbf{x}'\right)\right] dS_0' = \begin{cases} 0, & \mathbf{x} \in W_1 \cup \overline{W}_0^- \\ u_2(\mathbf{x}), & \mathbf{x} \in W_2 \backslash \overline{W}_0^- \end{cases}, \tag{17.7}$$

$$\int_{S_0} \left[u_0\left(\mathbf{x}'\right) \frac{\partial}{\partial n_0} G_0\left(\mathbf{x}', \mathbf{x}\right) - G_0\left(\mathbf{x}', \mathbf{x}\right) \frac{\partial}{\partial n_0} u_0\left(\mathbf{x}'\right) \right] dS_0'$$

$$= \begin{cases} -u_0(\mathbf{x}), & \mathbf{x} \in W_0 \\ 0, & \mathbf{x} \in W_1 \backslash \overline{W_0^+} \\ 0, & \mathbf{x} \in W_2 \backslash \overline{W_0^-} \end{cases}, \quad (17.8)$$

where $dS_0' = dS_0\left(\mathbf{x}'\right)$ and $G_j\left(\mathbf{x}', \mathbf{x}\right)$ are the fundamental solutions of the Helmholtz equation

$$G_j\left(\mathbf{x}', \mathbf{x}\right) = \frac{i}{4} H_0^{(1)}\left(k_{Sj}\left|\mathbf{x} - \mathbf{x}'\right|\right).$$

Integral representations (17.6)–(17.8) are valid for an arbitrary thickness of the interfacial inhomogeneity W_0 and can be used to solve the problem of scattering of elastic waves under contact conditions (17.4). However, this approach is significantly complicated by the need to satisfy boundary conditions on close surfaces S_0^+ and S_0^-. The problem can be simplified if the contact conditions (17.4), using the thinness of the region W_0, are transferred to the surface S of the bonding of the regions W_1 and W_2.

17.3 Asymptotic Model of Thin Interphase Inhomogeneity

To construct a model of thin interphase inhomogeneity in an elastic medium, it is convenient to use an orthogonal coordinate system connected to the contour of the scatterer. If contour S is described by the parametric equation $\mathbf{x} = \mathbf{r}(\alpha_1)$, then the curvilinear coordinates (see Fig. 17.1) are defined as follows

$$\mathbf{x} = \mathbf{r}(\alpha_1) + \alpha_2 \mathbf{n}(\alpha_1),$$

where $\mathbf{n}(\alpha_1)$ is the external unit normal to S. The following relations give the Lame metric coefficients of this coordinate system

$$H_1(\boldsymbol{\alpha}) = H(\boldsymbol{\alpha}) = A(\alpha_1)\left[1 + \alpha_2/R(\alpha_1)\right], \quad H_2(\boldsymbol{\alpha}) = 1,$$

where $A(\alpha_1) = \left|d\mathbf{r}(\alpha_1)/d\alpha_1\right|$, $R(\alpha_1)$ is the radius of curvature of the curve S.

The region $W_0 = W_0^+ \cup W_0^-$ and its border $S_0 = S_0^+ \cup S_0^-$ in coordinates (α_1, α_2) are set as follows (see Fig. 17.1):

$$W_0^{\pm} = \{(\alpha_1, \alpha_2) : \alpha_1 \in S,\ 0 \leqslant \pm\alpha_2 < \varepsilon g_{\pm}(\alpha_1)\},$$
$$S_0^{\pm} = \{(\alpha_1, \alpha_2) : \alpha_1 \in S,\ \alpha_2 = \pm\varepsilon g_{\pm}(\alpha_1) = \pm h_{\pm}(\alpha_1)\},$$
$$\varepsilon = \max_{\mathbf{x}\in S} h(\mathbf{x})/a \ll 1.$$

Here $h(\mathbf{x}) = h_+(\mathbf{x}) + h_-(\mathbf{x})$ is a variable thickness of the interphase inhomogeneity, $g_{\pm}(\alpha_1)$ are sufficiently smooth positive definite functions, a is the characteristic size of the region W_2, ε is a small dimensionless parameter that characterizes a small (relative to a) thickness of the inhomogeneity.

We believe that the wavelengths propagating in the composite body are much longer than the thickness of the interphase inhomogeneity, i.e.,

$$k_{Sj} \max_{\mathbf{x}\in S} |h(\mathbf{x})| \ll 1, \quad j = 0, 1, 2. \tag{17.9}$$

Under such conditions, we use asymptotic expansions on a small parameter ε for the displacement fields in the matrix, fiber, and inhomogeneity

$$u^{sc}(\mathbf{x}) = \sum_n \varepsilon^n u^{(n)}(\mathbf{x}), \quad u_2(\mathbf{x}) = \sum_n \varepsilon^n u_2^{(n)}(\mathbf{x}), \quad \varepsilon \to 0, \tag{17.10}$$

$$u_0(\boldsymbol{\alpha}) = \sum_n \varepsilon^n u_0^{(n)}(\alpha_1, \bar{\alpha}_2), \quad \alpha_2 = \varepsilon\bar{\alpha}_2, \quad \varepsilon \to 0, \quad n = 0, 1, \ldots. \tag{17.11}$$

The normal derivative and components (n_{01} and n_{02}) of the normal $\mathbf{n}_0$ on S_0 are given as follows

$$\frac{\partial}{\partial n_0} = \frac{n_{01}}{H(\boldsymbol{\alpha})}\frac{\partial}{\partial\alpha_1} + n_{02}\frac{\partial}{\partial\alpha_2}, \quad \boldsymbol{\alpha} \in S_0,$$
$$n_{01} = -\varepsilon A^{-1}(\alpha_1)\frac{dg_{\pm}(\alpha_1)}{d\alpha_1} + O(\varepsilon^2), \quad n_{02} = \pm 1 + O(\varepsilon^2), \quad \varepsilon \to 0.$$

The equation of motion of inhomogeneity is written in variables $(\alpha_1, \bar{\alpha}_2)$. As a result, we obtain a singularly perturbed differential equation

$$\frac{\partial}{\partial\bar{\alpha}_2} H(\boldsymbol{\alpha})\frac{\partial u_0(\boldsymbol{\alpha})}{\partial\bar{\alpha}_2} + \varepsilon^2\left[\frac{\partial}{\partial\alpha_1}\frac{\partial u_0(\boldsymbol{\alpha})}{H(\boldsymbol{\alpha})\partial\alpha_1} + H(\boldsymbol{\alpha})k_0^2 u_0(\boldsymbol{\alpha})\right] = 0, \quad \boldsymbol{\alpha} \in W_0.$$

Let us substitute (17.11) into this equation. Equating the coefficients of the same powers of ε, we obtain a recurrent sequence of ordinary differential equations in the variable $\bar{\alpha}_2$ with regard to the sought terms of the expansions (17.11). Satisfying the contact conditions (17.4), we take into account the solutions of these equations, and the terms of the asymptotic series (17.10) are given by Taylor series in the neighborhood of $\alpha_2 = 0$, which are convergent under the conditions (17.9). We obtain the contact conditions of the matrix and the scatterer by equating the terms with the same powers of the parameter ε. The type of these conditions is determined by the order of smallness of the mechanical contrast parameters $\bar{\gamma}_j = \mu_0/\mu_j$ ($j = 1, 2$) of thin-walled inhomogeneity in comparison with the value of ε. Two ranges of changes in values $\bar{\gamma}_j$ are considered: (i) $0 \leqslant \bar{\gamma}_j \leqslant \varepsilon$, (ii) $\varepsilon \leqslant \bar{\gamma}_j \leqslant 1/\varepsilon$.

Soft inhomogeneities. Range 1 corresponds to interphase thin-walled inhomogeneity, the stiffness of which is less than the stiffness of the surrounding medium ($\bar{\gamma}_j/\varepsilon = \text{const}$, $\varepsilon \to 0$). Then, with accuracy up to the main terms of the asymptotic expansion (17.10), we obtain the conditions for the dynamic interaction of such compliant inhomogeneity with the external environment

$$\begin{aligned} u_1(\mathbf{x}) - u_2(\mathbf{x}) &= \frac{h(\mathbf{x})}{\bar{\gamma}_1}\frac{\partial u_1(\mathbf{x})}{\partial \alpha_2}, \\ \mu_1\frac{\partial u_1(\mathbf{x})}{\partial \alpha_2} &= \mu_2\frac{\partial u_2(\mathbf{x})}{\partial \alpha_2}, \quad \mathbf{x} \in S, \end{aligned} \tag{17.12}$$

where $u_1 = u^{in} + u^{(0)} + O(\varepsilon)$, $u_2 = u_2^{(0)} + O(\varepsilon)$.

Non-contrast inhomogeneities. Range 2 describes non-contrast inhomogeneities. It is assumed that $\bar{\gamma}_j = \text{const} \neq 0$ when $\varepsilon \to 0$. In this case, the main term of the asymptotic expansions (zero terms in (17.10)) corresponds to the problem of scattering under conditions of ideal contact of the scatterer with the matrix, and the presence of thin-walled inhomogeneity is described by the next term (the first term of the expansions in (17.10)) of the expansion in the form

$$\begin{aligned} u_1(\boldsymbol{\alpha}, \omega) - u_2(\boldsymbol{\alpha}, \omega) =& h_\mu \mu_1 \frac{\partial u_1^{(0)}(\boldsymbol{\alpha}, \omega)}{\partial \alpha_2}, \\ \mu_1 \frac{\partial u_1(\boldsymbol{\alpha}, \omega)}{\partial \alpha_2} - \mu_2 \frac{\partial u_2(\boldsymbol{\alpha}, \omega)}{\partial \alpha_2} & \\ = \frac{\partial}{A\partial \alpha_1}\mu_h \frac{\partial u_1^{(0)}(\boldsymbol{\alpha}, \omega)}{A\partial \alpha_1} &+ \omega^2 \rho_h u_1^{(0)}(\boldsymbol{\alpha}, \omega), \quad \boldsymbol{\alpha} \in S, \end{aligned} \tag{17.13}$$

where

$$\begin{gathered} \mu_1 \frac{\partial u_1^{(0)}(\boldsymbol{\alpha}, \omega)}{\partial \alpha_2} = \mu_2 \frac{\partial u_2^{(0)}(\boldsymbol{\alpha}, \omega)}{\partial \alpha_2}, \\ u_1^{(0)}(\boldsymbol{\alpha}, \omega) = u_2^{(0)}(\boldsymbol{\alpha}, \omega), \quad \boldsymbol{\alpha} \in S, \\ u_1^{(0)} = u^{(0)} + u^{in}, \quad u_1 = u_1^{(0)} + \varepsilon u^{(1)} + O(\varepsilon^2), \quad u_2 = u_2^{(0)} + \varepsilon u_2^{(1)} + O(\varepsilon^2), \\ \mu_h = h_+\mu_1 + h_-\mu_2 - h\mu_0, \quad h_\mu = h\mu_0^{-1} - h_+\mu_1^{-1} - h_-\mu_2^{-1}, \\ \rho_h = h_+\rho_1 + h_-\rho_2 - h\rho_0. \end{gathered}$$

Equations (17.12) and (17.13) are the effective contact conditions, which model the interaction of the matrix, fiber, and the thin layer between them with an accuracy of order ε and ε^2 in expansions (17.10), respectively. Conditions with continuous stresses and displacement jump proportional to them are known in the literature as conditions of the spring contact type (Benveniste and Miloh 2001). Here they are obtained under the assumption of the layer thickness variability, the effect of which on the effective contact conditions depends on the type of inhomogeneity. The reaction of compliant interfacial layers is determined by the total thickness and non-contrast one by the thickness fraction in the matrix and the fiber. Conditions of type (17.13) were previously obtained for thin inclusions in an infinite elastic matrix in the case of the torsion problem (Kunets 1987) and the dynamic antiplane problem (Kit et al. 1999). Analogous conditions under static loading of two elastic bodies connected by a thin layer were studied in Rizzoni et al. (2014).

17.4 Null-Field Method in Scattering Problems with Effective Contact Conditions

The contact conditions (17.12) and (17.13) make it possible to simplify the integral representations (17.6)–(17.8), which are used to solve the problem (17.1)–(17.4) (Kushnir et al. 2019). Under effective contact conditions, the scattering problem is formulated for the fiber W_2 contained in the matrix W_1. We assume that displacements and stresses are discontinuous on the surface S of the fiber. Then the integral representations are as follows

$$\begin{aligned} &u^{in}(\mathbf{x},\omega) \\ &+\int_S\left[u_2(\mathbf{x}',\omega)\frac{\partial}{\partial n}G_1(\mathbf{x}',\mathbf{x})-\mu_{21}G_1(\mathbf{x}',\mathbf{x})\frac{\partial}{\partial n}u_2(\mathbf{x}',\omega)\right]dS \\ &\qquad+\int_S\left\{[u]_S\frac{\partial}{\partial n}G_1(\mathbf{x}',\mathbf{x})-\mu_1^{-1}[t_n]_S G_1(\mathbf{x}',\mathbf{x})\right\}dS \\ &\qquad\qquad=\begin{cases}u_1(\mathbf{x},\omega), & \mathbf{x}\in W_1\\ 0, & \mathbf{x}\in W_2\end{cases}, \end{aligned} \tag{17.14}$$

$$\begin{aligned} &\int_S\left[u_2(\mathbf{x}')\frac{\partial}{\partial n}G_2(\mathbf{x}',\mathbf{x})-G_2(\mathbf{x}',\mathbf{x})\frac{\partial}{\partial n}u_2(\mathbf{x}')\right]dS \\ &\qquad\qquad=\begin{cases}0, & \mathbf{x}\in W_1\\ -u_2(\mathbf{x}), & \mathbf{x}\in W_2\end{cases}, \end{aligned} \tag{17.15}$$

where $\mu_{21}=\mu_2/\mu_1$,

$$[u]_S=u_1(\mathbf{x},\omega)-u_2(\mathbf{x},\omega),\quad [t_n]_S=\frac{\partial u_1(\mathbf{x},\omega)}{\partial n}-\mu_{21}\frac{\partial u_2(\mathbf{x},\omega)}{\partial n},\quad \mathbf{x}\in S.$$

We expand the incident and scattered field in terms of cylindrical wave functions (Kushnir et al. 2019)

$$u^{in}(\mathbf{x}) = u_0 \sum_{\sigma,m} \varepsilon_m b^{in}_{\sigma m} \mathrm{Re}\, \psi^1_{\sigma m}(\mathbf{x}), \quad |\mathbf{x}| < r_0, \tag{17.16}$$

$$u^{sc}(\mathbf{x}) = u_0 \sum_{\sigma,m} \varepsilon_m f_{\sigma m}\, \psi^1_{\sigma m}(\mathbf{x}), \quad |\mathbf{x}| > r_1, \tag{17.17}$$

$$\begin{aligned} \psi^j_{\sigma m}(\mathbf{x}) = H^{(1)}_m\left(k_{Sj} r\right) C_{\sigma m}(\theta), \quad \mathrm{Re}\, \psi^j_{\sigma m}(\mathbf{x}) = J_m\left(k_{Sj} r\right) C_{\sigma m}(\theta), \\ C_{1m}(\theta) = \cos m\theta, \quad C_{2m}(\theta) = \sin m\theta, \end{aligned} \tag{17.18}$$

where $H^{(1)}_m(\mathbf{x})$ are the Hankel functions of the first kind, $J_m(\mathbf{x})$ are the Bessel functions of the first kind, $\varepsilon_m = 2 - \delta_{m0}$ (δ_{m0} is the Kronecker symbol), r_0 and r_1 are the radii of the circles inscribed in S and circumscribed around S . We also present the fundamental solutions as expansions by the system of wave functions (17.18)

$$G_j\left(\mathbf{x}', \mathbf{x}\right) = \frac{i}{4} \sum_{\sigma=1}^{2} \sum_{m=0}^{\infty} \varepsilon_m \psi^j_{\sigma m}(\mathbf{x}') \mathrm{Re}\, \psi^j_{\sigma m}(\mathbf{x}), \quad \left|\mathbf{x}'\right| > |\mathbf{x}|. \tag{17.19}$$

By substituting (17.16), (17.17) and (17.19) into the integral representations (17.14) and (17.15), taking into account the orthogonality of the functions (17.18) on the circle, we obtain the equations of the moments of the null-field method

$$\begin{aligned} &\int_S \left(u_2 \frac{\partial \psi^1_{S\sigma m}}{\partial n} - \mu_{21} \psi^1_{S\sigma m} \frac{\partial u_2}{\partial n} \right) dS \\ &+ \int_S \left\{ [u]_S \frac{\partial \psi^1_{S\sigma m}}{\partial n} - \mu_1^{-1} [t_n]_S\, \psi^1_{S\sigma m} \right\} dS = 4 i u_0 b^{in}_{\sigma m}, \end{aligned} \tag{17.20}$$

$$\int_S \left[u_2 \frac{\partial \mathrm{Re}\, \psi^2_{S\sigma m}}{\partial n} - \left(\mathrm{Re}\, \psi^2_{S\sigma m}\right) \frac{\partial u_2}{\partial n} \right] dS = 0, \tag{17.21}$$

$$\begin{aligned} &\frac{i}{4} \int_S \left\{ [u]_S \frac{\partial}{\partial n} \left(\mathrm{Re}\, \psi^1_{S\sigma m}\right) - \mu_1^{-1} [t_n]_S \,\mathrm{Re}\, \psi^1_{S\sigma m} \right\} dS \\ &+ \frac{i}{4} \int_S \left[u_2 \frac{\partial Re \psi^1_{S\sigma m}}{\partial n} - \mu_{21} \left(\mathrm{Re}\, \psi^1_{S\sigma m}\right) \frac{\partial u_2}{\partial n} \right] dS = u_0 f_{\sigma m}, \end{aligned} \tag{17.22}$$

where $\sigma = 1, 2$, $m = 0, 1, \ldots$.

In the case of a non-contrast layer, the numerical solution to the scattering problem is found in two stages. At the first stage there are zero terms $u^{(0)}(\mathbf{x})$ and ${u_2}^{(0)}(\mathbf{x})$ of the asymptotic series (17.10). These terms satisfy the ideal contact condition on the contour S, therefore $[u]_S = [t_n]_S = 0$. The unknown quantities ${u_2}^{(0)}$ and $\partial {u_2}^{(0)}/\partial n$ on the contour of the scatterer are found in the form of trigonometric series. Substituting them into (17.20) and (17.21), we get a system of linear algebraic equations of infinite order with respect to the coefficients of these series. We find the solution of this system numerically and proceed to the second stage, at which, according to the boundary conditions (17.13), jumps of displacements and stresses are known functions. Again, the unknown values $u_2^{(1)}$and $\partial u_2^{(1)}/\partial n$ on the scatterer contour are found in the form of trigonometric series. From (17.20) and (17.21), we obtain a system of linear algebraic equations, the left part of which is the same as in the first stage. Having the solution of this system obtained from (17.22), we calculate the coefficients of series in (17.17) and the scattering amplitude in the far field according to (17.5).

In more detail, we examine the problem of wave scattering by fiber in the presence of a soft layer and the peculiarities of using the null-field method for its solution.

17.5 The Effect of a Thin Compliant Interphase Layer on *SH*-Wave Scattering Amplitudes

Let a plane *SH*-wave falls on the fiber in the elastic matrix in the presence of an interphase layer (see Fig. 17.2)

$$u^{in}(\mathbf{x}) = u_0 \exp\left[-ik_{S1}\left(x_1 \cos\theta_{in} + x_2 \sin\theta_{in}\right)\right]. \tag{17.23}$$

The elastic characteristics and thickness of the layer are such that its interaction with the matrix and the fiber can be modeled by effective conditions (17.12). The scattered field that satisfies the radiation condition (17.5) is determined using the null-field method. In the integral expressions (17.20) and (17.21) of this method, the unknown values on the fiber contour are specified in a series of trigonometric functions:

$$\begin{aligned} u_1(\mathbf{x}) - u_2(\mathbf{x}) &= u_0 \sum_{\sigma,m} x_{1\sigma m} C_{\sigma m}(\theta), \\ u_2(\mathbf{x}) &= u_0 \sum_{\sigma,m} x_{2\sigma m} C_{\sigma m}(\theta), \quad \mathbf{x} \in S. \end{aligned} \tag{17.24}$$

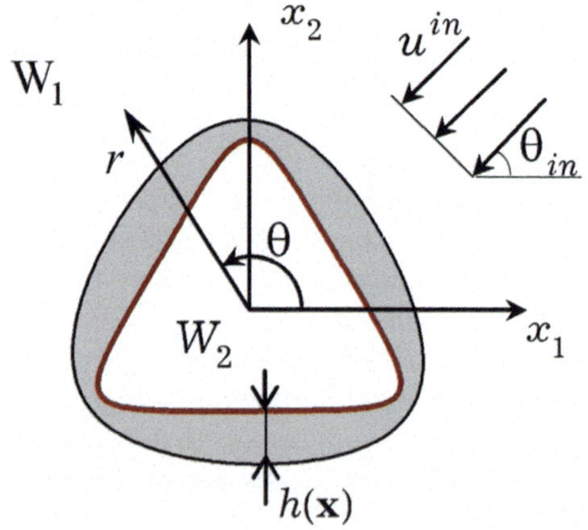

Fig. 17.2 Two-dimensional geometry of an elastic matrix with a fiber and an interphase layer (θ_n is the angle of incidence of the plane wave)

Taking into account conditions (17.12), from (17.20) and (17.21), we obtain a system of linear algebraic equations of infinite order for determining unknown coefficients $x_{1\sigma m}$ and $x_{2\sigma m}$:

$$\sum_{\tau',\sigma',m'} a_{1\sigma m,\tau'\sigma'm'}\ x_{\tau'\sigma'm'} = 4ib^{in}_{H\sigma m}, \qquad \sum_{\tau',\sigma',m'} a_{2\sigma m,\tau'\sigma'm'}\ x_{\tau'\sigma'm'} = 0, \tag{17.25}$$

$(m, m' = 0, 1, \ldots, \sigma, \sigma', \tau' = 1, 2)$,

$$a_{\tau\sigma m,\tau'\sigma'm'} = \int_S \bar{a}_{\tau\sigma m,\tau'}(r) C_{\sigma'm'}(\theta)\, dS(\theta),$$

$$\bar{a}_{1\sigma m,1} = \frac{\partial \psi^1_{\sigma m}}{\partial n} - \frac{\mu_{01}}{h(\theta)}\psi^1_{\sigma m}, \quad \bar{a}_{1\sigma m,2} = \frac{\partial \mathrm{Re}\,\psi^1_{\sigma m}}{\partial n},$$

$$\bar{a}_{2\sigma m,1} = \bar{\gamma}_1\mu_{21}^{-1}\frac{1}{h(\theta)}\mathrm{Re}\,(\psi^2_{\sigma m}), \quad \bar{a}_{2\sigma m,2} = \frac{\partial \mathrm{Re}\,\psi^2_{\sigma m}}{\partial n},$$

$$b^{in}_{H\sigma m} = i^{-m} C_{\sigma m}(\theta_{in}).$$

From (17.17), (17.22), (17.5), and (17.24), we have

$$f(\theta) = u\_0 \sum_{\sigma,m} \varepsilon_m i^{-m} A_{\sigma m} C_{\sigma m}(\theta), \qquad A_{\sigma m} = \sum_{\sigma',m'} \left(x_{1\sigma'm'}\mathrm{Re}\, a_{1\sigma m,1\sigma'm'} + x_{2\sigma'm'}\ \mathrm{Re}\, a_{1\sigma m,2\sigma'm'}\right). \tag{17.26}$$

Knowing the solution of (17.25), which is obtained numerically by the reduction method, we calculate the *SH*-wave scattering amplitude $f(\theta)$ from (17.26). Another quantity of practical interest is the scattering cross section

$$\sigma^H = \frac{1}{k_{S1}} u_0^{-1} \mathrm{Im}\, f(\pi + \theta_{in}).$$

For example, consider the scattering of plane *SH*-waves (17.23) by fibers, the cross section of which has an area πa^2 and a contour given by the parametric equation

$$r(\beta) = a\sqrt{\frac{1 + \varepsilon_0^2 + 2\varepsilon_0 \cos[(N+1)\beta]}{1 - \varepsilon_0^2 N}}, \qquad \theta(\beta) = \arctan\frac{\sin\beta + \varepsilon_0 \sin(N\beta)}{\cos\beta - \varepsilon_0\cos(N\beta)}, \quad 0 \leqslant \beta \leqslant 2\pi, \tag{17.27}$$

where a is the characteristic size of the cross section of the fiber, which can be in the form of ellipses ($N = 1$), equilateral triangles ($N = 2$, $\varepsilon_0 = 0.25$) or squares ($N = 3$, $\varepsilon_0 = 1/9$) with rounded corners.

The variable thickness of the layer is taken as follows

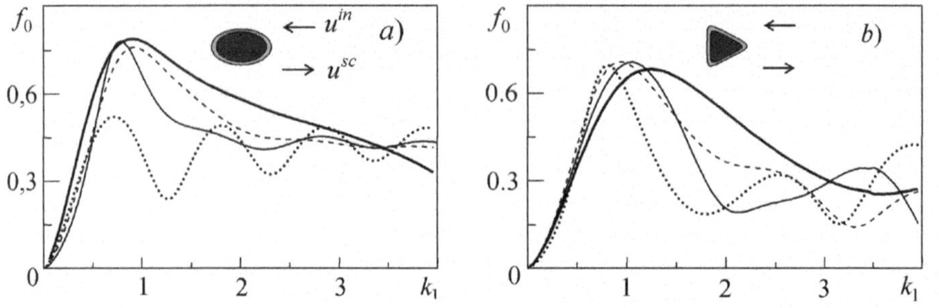

Fig. 17.3 Frequency dependences of normalized scattering amplitudes for different forms of fiber during monostatic probing $\theta_{in} = \theta_{sc} = 0°$: solid, thin, and dashed curves correspond to the values of $\gamma_1 = 0.5$ and $\gamma_1 = 2$; solid, bold, and dotted curves correspond to the cases of ideal matrix-fiber contact and a hole in the matrix

$$h(\mathbf{x}) = \varepsilon a \left[1 + L_1 \sin(L_2\theta + L_3)\right], \quad 0 \leqslant \theta \leqslant 2\pi, \quad \mathbf{x} \in S, \tag{17.28}$$

where L_1, L_2, and L_3 are the given parameters that characterize the shape of the layer.

Figure 17.3 shows the spectral dependences of the normalized scattering amplitudes

$$f_0 = \frac{u_0^{-1} \left| f\left(\theta_{sc}\right) \right|}{\sqrt{k_1}}, \tag{17.29}$$

where $k_1 = k_{S1}a$ and θ_{sc} is the angle of observation of SH-waves scattered in the Fraunhofer zone. Glass fiber in an epoxy matrix ($\rho_1 = 1250$ kg/m^3, $\mu_1 = 1.28$ GPa, $\rho_2 = 2550$ kg/m^3, $\mu_2 = 29.9$ GPa) was studied. It is assumed that the interphase layer has a constant thickness $h(\mathbf{x}) = \varepsilon a$ and contrast parameter $\gamma_1 = \mu_0/(\varepsilon\mu_1) = O(1)$, $\varepsilon \to 0$. Elliptical fibers with the ratio of semi-axes $b/a = 0.6$ (in (17.27), $N = 1$ and $\varepsilon_0 = 0.25$) and triangular fibers are shown in Fig. 17.3a, b, respectively. For not too small values of the parameter γ_1 in the low-frequency range, the scattering amplitude spectra increase monotonically and have an oscillatory character in the medium-frequency range. Moreover, in this frequency range, an increase in the stiffness of the interphase layer leads to an increase in the level of normalized scattering amplitudes, which is usually between the corresponding levels for the cases of ideal matrix-fiber contact and a free hole in the matrix. As the wavelength of the scatterer k_1 (dimensionless frequency) increases, the difference between these levels is leveled off.

Figures 17.4, 17.5, 17.6 and 17.7 illustrate the characteristics of SH-wave scattering by inclusions in the matrix for the presence of very soft interfacial layers.

Figures 17.4 and 17.5 show the normalized scattering cross sections $\bar{\sigma} = \sigma^H/(2a)$ as a function of the wavelength k_1 for different values of the interphase layer stiffness γ_1. In all cases, the wave's incidence angle is $\theta_{in} = 0$. Matrix/fiber combinations were analyzed: (i) composite is epoxy resin/glass; (ii) composite is stainless

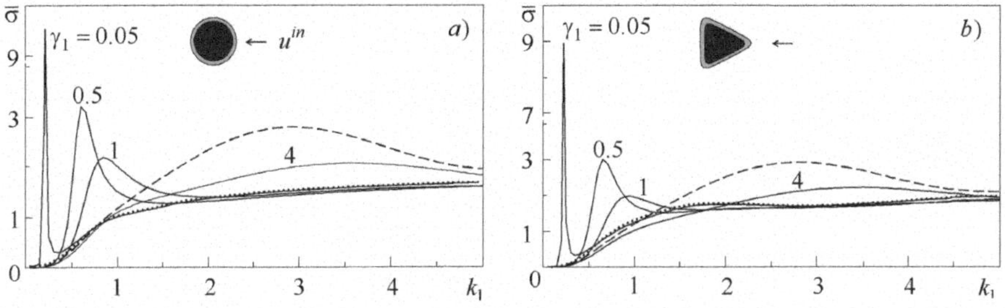

Fig. 17.4 Spectra of normalized *SH*-wave scattering cross sections by glass circular (**a**) and triangular (**b**) fibers in an epoxy matrix for different interlayer stiffnesses γ_1

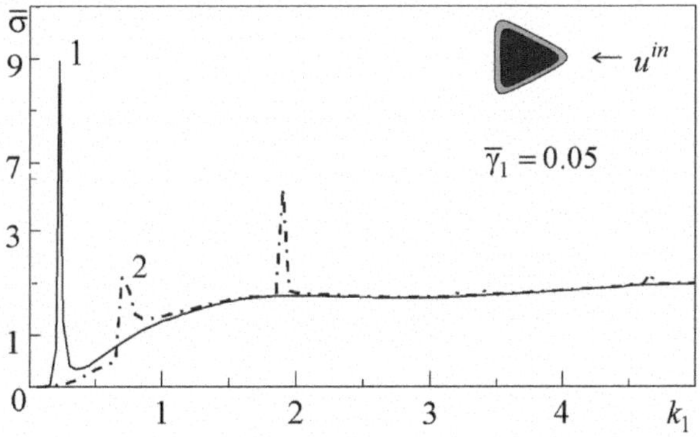

Fig. 17.5 Frequency dependences of the normalized *SH*-wave scattering cross sections by a triangular fiber: 1 is a glass fiber in an epoxy matrix; 2 is magnesium fiber in stainless steel

steel/magnesium ($\nu_1 = 0.3$, $\nu_2 = 0.3$, $\mu_2/\mu_1 = 0.22$, $\rho_2/\rho_1 = 0.21$); (iii) composite is aluminum/stainless steel ($\mu_1/\mu_2 = 0.478$, $c_{S1}/c_{S2} = 1.171$). Numerical calculations were performed for fibers in the presence of a constant thickness layer ($L_1 = L_2 = L_3 = 0$ in (17.28)).

Dashed curves in Fig. 17.4 illustrate the cross sections of *SH*-wave scattering by fibers under the conditions of their ideal contact with the matrix, and the dotted curves are the cases of cavities of the corresponding shapes. Comparing the effect of the stiffness of the interlayer on the wave properties of fibers, we see that it is qualitatively the same for both circular and triangular shapes. The scattering cross-sectional spectra have a resonant character at very small values of the interlayer stiffness. Fibers stiffer than the matrix in the given frequency range have only low-frequency resonance. In the case of fibers, the stiffness of which is less than the stiffness of the matrix (dashed-dotted curve in Fig. 17.5), the low-frequency resonance shifts toward higher frequencies, and resonances of other forms of oscillation of the cross section of the fiber appear in the spectrum. Outside of resonant frequencies, the scattering cross sections of the considered forms of fibers are almost identical to the scattering cross sections of holes of the corresponding form. As the stiffness of the fibers increases, the width of the low-frequency resonance increases and disappears at $\gamma_1 > 1$.

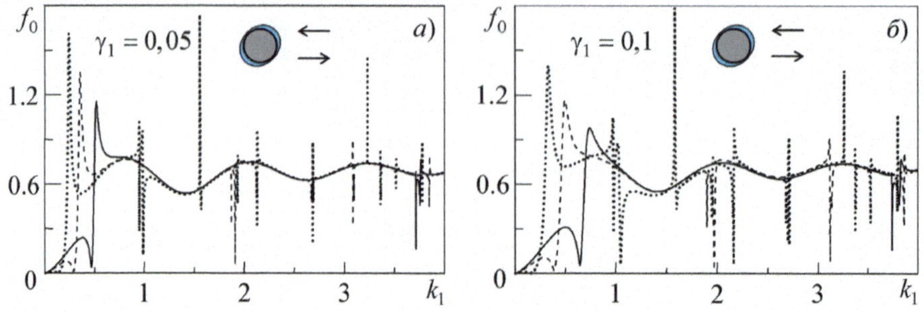

Fig. 17.6 Frequency dependences of the normalized *SH*-wave scattering amplitudes for low stiffnesses $\gamma_1 \ll 1$ of interphase layers and monostatic probing $\theta_{in} = \theta_{sc} = 0°$: solid, dashed, and dotted curves correspond to $\mu_1/\mu_2 = k_2/k_1 = 0.5, 1, 2$, respectively

Figure 17.6 shows the spectral dependences of the normalized scattering amplitudes f_0 at low stiffnesses of the interfacial layers $\gamma_1 \ll 1$ and various mechanical properties of the matrix and fiber (the cross section of the fiber is a circle of radius a). At the same time, it was assumed that $k_2/k_1 = \mu_1/\mu_2 = \gamma_2$ ($k_2 = k_{S2}a$). For the thickness of the inclusion in (17.28) $L_1 = 0.7$, $L_2 = 2$, and $L_3 = 0$ (corrugated layer) was taken.

Curves in Fig. 17.6 have a clearly expressed resonant character in the entire considered frequency range. Their background for $\gamma_1 \ll 1$ is practically independent of the mechanical parameters of the composite and coincides with the corresponding curves for the case of a circular hole of radius a in the matrix. The location of the resonances in Fig. 17.6, except for the first low-frequency one, approximately coincides with the natural frequencies for the regions defining the cross section of the fibers. In the considered case, the solutions of the equation $J_n'(\gamma_2 k_1) = 0$ ($J_n'(\mathrm{x})$ is derivative by argument from the Bessel function $J_n(\mathrm{x})$ of the nth order) exists. Moreover, in the case of an increase in the contrast parameter γ_1, there is an increase in the width of the resonances and their shift to the region of higher frequencies.

The splitting of some resonance frequencies of the amplitude spectra shown in Fig. 17.6 occurs. The reason for this effect is revealed by the analysis of the curves in Fig. 17.7. Spectral dependences of normalized backscatter amplitudes f_0 ($\theta_{in} = \theta_{sc} = 0°$ in (17.29)) are shown here. The stiffness parameter of the interphase layer $\gamma_1 = 0.1$. Figure 17.7a compares the amplitudes in the case of a circular fiber in the presence of a constant thickness layer (dotted curve) and a corrugated layer (solid curve, $L_1 = 0.3$, $L_2 = 2$ and $L_3 = 0$ in (17.28)). The corrugation of the layer causes the growth of the first low-frequency resonance (shift to the right from the resonance line in the case of a layer of constant thickness) and the splitting of the next resonance. Figure 17.7b examines the effect of fiber shape on scattering amplitudes. A triangular fiber and two cases of backscattering $\theta_{in} = \theta_{sc} = 0°$ (solid curve) and $\theta_{in} = \theta_{sc} = 60°$ (dashed curve) are considered. For comparison, the backscattering amplitude in the case of a circular fiber is given by a dotted curve. In both cases, it was assumed that the interfacial layer has a constant thickness. We see that the

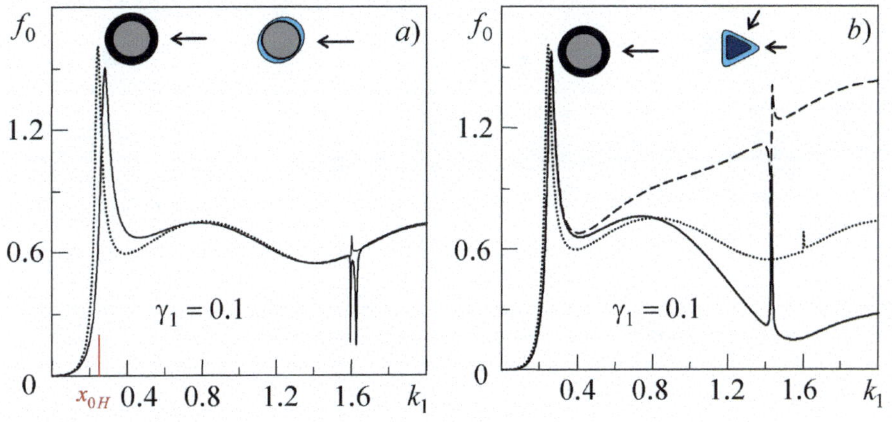

Fig. 17.7 Frequency dependences of the normalized *SH*-wave scattering amplitudes in an aluminum matrix with a steel fiber for different forms of interlayer (**a**) and fiber (**b**)

shape of the fiber determines the shift of the corresponding resonant frequencies. The reliability of numerical calculations is confirmed by the fact that the values of resonant frequencies do not depend on the direction of wave incidence.

17.6 Pulses of Echo Signals in a Matrix with Fiber and a Thin Compliant Interphase Layer

The influence of resonant frequencies on the time dependence of waves scattered by a steel fiber in an aluminum matrix in the presence of a soft interfacial layer is illustrated in Fig. 17.8. The reduced displacement in the far field was calculated:

$$\begin{aligned}\bar{u}(\theta_{sc}, \tau_1) &= \lim_{r\to\infty} \sqrt{r/a}\, u_0^{-1} u^{sc}(\mathbf{x}, \tau_1)\\ &= \frac{u_0^{-1}}{(2\pi)^{3/2}} \int_0^{k_m} \mathrm{Re}\left[\varphi(k_1) f(\theta_{sc}) \exp(-ik_1\tau_1 + i\pi/4)\right] \frac{dk_1}{\sqrt{k_1}}.\end{aligned}$$

Here, $\tau_1 = (c_{S1}t - r)/a$, $\varphi(k_1)$ is the frequency modulation of the incident plane *SH*-wave

$$\varphi(k_1) = \frac{4\sqrt{\pi}k_1^2}{k_0^3} \exp\left(-\frac{k_1^2}{k_0^2}\right),$$

which is the spectrum of Ricker wavelet (Murai 2007)

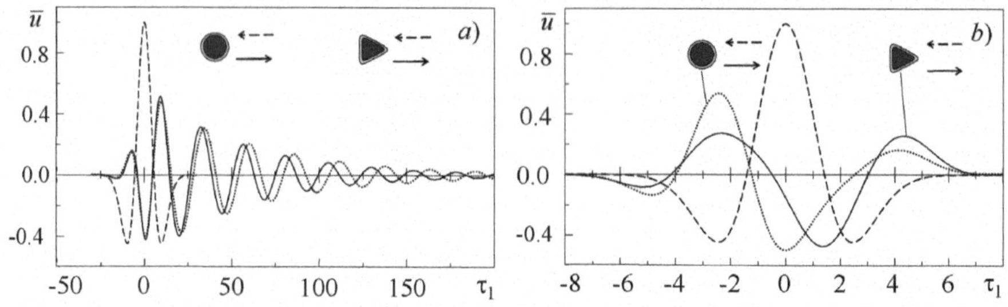

Fig. 17.8 Time dependences of normalized movements in the far zone, caused by non-stationary *SH*-waves with a carrier resonant (**a**) and non-resonant (**b**) frequencies

$$\varphi(\tau_1) = \pi^{-1} \int_0^{\infty} \varphi(k_1)\cos(k_1\tau_1)dk_1$$
$$= \left[1 - 2(k_0\tau_1/2)^2\right]\exp\left[-(k_0\tau_1/2)^2\right],$$

k_0 is the characteristic wave size (when $k_1 = k_0$, the function $\varphi(k_1)$ takes the maximum value). Dashed curves in Fig. 17.8 correspond to the incident pulse, the dotted and solid curves are echo signals from circular and triangular fibers with layers of constant thickness.

The curves in Fig. 17.8a are obtained for k_0 equal to the low-frequency resonances of the scattering amplitude: $k_0 = 0.247$ for a circular fiber, and $k_0 = 0.262$ for a triangular fiber. Time is counted from the moment of arrival of the maximum value of the incident signal at the beginning of the coordinate system. Figure 17.8b is obtained at non-resonant values of the characteristic wave size $k_0 = 1$. The duration of echo signals excited by pulses, the spectrum of which is formed around resonant frequencies, significantly exceeds the duration of the incident pulse and is equal to this duration in the case of non-resonant frequencies. This property of echo pulses is important for analyzing the nature of fiber-matrix connections under nondestructive testing of composite materials. Since the resonant frequencies change with the shape of the fibers and the stiffness of the layer, the information about their location on the frequency scale simplifies the diagnosis of composite materials.

We determine the value of resonance frequencies using the resonance theory (Uberal 1973) when the scatterer is a circular fiber. In this case, for displacements (17.17) in a scattered field, the following expression is valid

$$u^{sc}(\mathbf{x}) = \frac{1}{2}\sum_{m=0}^{\infty} \varepsilon_m i^{-m}\,(S_m - 1)\,H_m^{(1)}(k_1\bar{r})\cos m\theta,$$

where $\bar{r} = r/a$,

$$
\begin{gathered}
S_m = S_m^{(s)} \frac{F_m - z_m^{(2)}}{F_m - z_m^{(1)}}, \\
F_m = \mu_{12} k_1 \frac{J_m(k_2)}{J_m'(k_2)} + \frac{h}{\bar{\gamma}_1} k_1 k_2, \quad z_m^{(i)} = k_2 \frac{H_m^{(i)}(k_1)}{H_m^{(i)\prime}(k_1)}, \\
S_m^{(s)} = -\frac{H_m^{(2)\prime}(k_1)}{H_m^{(1)\prime}(k_1)}, \\
J_m'(k_2) = \frac{dJ_m(k_2)}{dk_2}, \quad H_m^{(i)\prime}(k_1) = \frac{dH_m^{(i)}(k_1)}{dk_1}.
\end{gathered}
\tag{17.30}
$$

According to the resonance theory, the resonance of the system occurs when the real part of the denominator in (17.30) is zero, i.e.,

$$
F_m(k_1) = \operatorname{Re} z_m^{(1)}(k_1). \tag{17.31}
$$

Analysis of the scattering amplitude by modes showed that in the vicinity of the low-frequency resonance, sharp changes are characteristic of the zero mode ($m = 0$). Therefore, the equation for determining low-frequency resonance is the condition (17.31) when $m = 0$. We take into account that in this frequency range $k_1 \to 0$, $k_2 \to 0$, and the stiffness of the interlayer was small: $\bar{\gamma}_1 / h \to 0$. We apply asymptotic expressions for Bessel functions for small values of the argument. As a result, we obtain the value x_{0H} of the resonance in the region of low frequencies (wavelengths) in the case of *SH*-wave scattering by a circular fiber in the presence of a soft layer

$$
x_{0H} = \frac{c_{S2}}{c_{S1}} \sqrt{2 \frac{\mu_0}{h} \frac{a}{\mu_2}}. \tag{17.32}
$$

The paper Cai (2004) analyzes the spectral characteristics of *SH*-wave scattering by a circular inclusion in the presence of an interlayer of a certain thickness. It is assumed that in the case of soft layers, the cause of low-frequency resonances is the oscillation of the inclusion as a rigid whole, and their value can be determined from the condition that the forced oscillation frequency of the system is equal to the natural frequency of the simplest oscillator "weightless mass on a spring". Using this hypothesis, we will determine the value x_H of such resonances in cases where the scatterer is a fiber of a non-canonical shape given by the parametric equation (17.27).

The square of the natural circular frequency ω_{0H} of oscillations of the "weightless mass on the spring" oscillator is proportional to the coefficient of elasticity of the spring K and inversely proportional to the mass M. In this case, we have

$$
K = \frac{\mu_0}{h} L, \quad M = \pi a^2 \rho_2,
$$

where L is the length of the inclusion circuit

$$
L = 2a\pi \left(1 + \varepsilon_0^2 N(4N - 1)/16\right) + O\left(\varepsilon_0^4\right).
$$

Let us first consider a circular fiber. For the value of the resonant frequency (resonant wave size) x_{0H}, in this case, we have an expression

$$x_{0H} = \frac{\omega_{0H}}{c_{S1}} a = \frac{a}{c_{S1}} \sqrt{\frac{2a\pi\mu_0}{h} \frac{1}{\pi a^2 \rho_2}} = \frac{c_{S2}}{c_{S1}} \sqrt{2\frac{\mu_0}{h}\frac{a}{\mu_2}}. \quad (17.33)$$

If the fiber has a non-canonical shape, then the resonance frequency x_H is as follows

$$x_H = x_{0H}\left(1 + \varepsilon_0^2 N(4N - 1)/16\right).$$

Comparing (17.33) and (17.32), we see that both the resonance theory and the oscillator hypothesis give the same resonance frequency values. Therefore, the application of this hypothesis to non-circular fibers is justified, which is also confirmed by numerical calculations.

17.7 Conclusion

An approach for studying wave phenomena in an elastic medium containing an elastic fiber of a non-canonical shape surrounded by a thin interfacial layer of variable thickness is proposed. The composite is in the conditions of a longitudinal shear fixed in time. The technique is based on an asymptotically accurate model of the dynamic interaction of thin inhomogeneity with the surrounding elastic medium and the null-field method. Mathematical models of the interlayer characterize jumps of displacements and stresses on the fiber contour. Their appearance depends on the ratio of the elastic parameters of the composite components. For the model of non-contrast thin-walled interfacial inhomogeneity, when the ratio of the shear modulus of the layer to the shear modulus of the matrix and fiber is of the order of unity compared to the thickness of the inhomogeneity, the corresponding displacement and stress jumps are known quantities and are determined from the solution of the problem for complete ideal contact of the matrix and the fiber. Unlike the static versions of such conditions obtained earlier in Rizzoni et al. (2014), dynamic conditions contain inertial components. The model of interphase thin-walled inhomogeneity of low stiffness is described by the connection of dynamic stresses with displacement jumps on the surface of the non-ideal contact of the matrix and the fiber and assumes the ratio of the elastic parameters of the thin-walled element to the elastic parameters of the adjacent components of the same order of magnitude as the relative thickness of the inhomogeneity. At the same time, the stress on the fiber contour is a continuous function. Such conditions are known in the literature as spring-type conditions. With the application of the obtained contact conditions of the matrix and the fiber, a simplified version of the problem of the scattering of plane *SH*-waves is formulated.

The numerical analysis of the wave field in the far scattering zone was carried out on the basis of the ratios of the null-field method. It was found that at low stiffnesses of interphase inhomogeneity, the spectra of scattered waves have a pronounced resonant character, and their background practically does not depend on the ratio of the shear moduli of the matrix and the fiber. For a family of inclusions, the contour of which is given by a parametric equation and layers of constant thickness, analytical values of low-frequency resonances were obtained depending on the mechanical parameters of the layer, its thickness, and parameters characterizing the shape of the inclusion. For the values of the contrast parameter of the material of a thin interphase layer of the order of its thickness in the low- and medium-frequency ranges, the scattering amplitude spectra do not have a resonant character; with an increase in the wave size of the scatterer, the difference between the levels of these amplitudes is equalized.

References

Benveniste Y, Miloh T (2001) Imperfect soft and stiff interphases in two-dimensional elasticity. Mech Mater 33:309–323

Butrak IO, Kil'nitskaya TI, Mykhas'kiv VV (2012) The scattering of an harmonic elastic wave by a volume inclusion with a thin interlayer. J Appl Math Mech 76:342–347

Cai L-V (2004) Scattering of antiplane shear waves by layered circular elastic cylinder. J Acoust Soc Am 115:515–522

Emets VF, Rogowski J (2013) Mathematical-numerical modelling of ultrasonic scattering data from a closed obstacles and inverse analysis. Academic Publishers House EXIT, Warsaw

Fraenkel LE (1969) On the method of matched asymptotic expansions. Part I: a matching principle. Math Proc Camb Phil Soc 65:209–231

Friedrichs KO (1955) Asymptotic phenomena in mathematical physics. Bull Am Math Soc 61:485–504

Kanaun SK (1984) On singular models of a thin inclusion in a homogeneous elastic medium. J Appl Math Mech 48:50–58

Kit GS, Emets' VF, Kunets' YaI (1999) A model of the elastodynamic interaction of a thin-walled inclusion with a matrix under antiplanar shear. J Math Sci 97:3810–3816

Kunets YaI (1987) Axisymmetric torsion of an elastic space with a thin elastic inclusion. J Appl Math Mech 51:497–503

Kushnir R, Kunets Y, Matus V, Trofymchuk O (2019) Plane scattering problem for an inclusion of non-classical shape with a thin interphase layer. In: Gdoutos E (ed) Structural integrity, vol 8. Proceedings of the 2nd international conference on theoretical, applied and experimental mechanics, pp 209–214. Springer

Martin PA (2006) Multiple scattering interaction of time-harmonic waves with N obstacles. Cambridge University Press, Cambridge

Mishuris GS, Movchan NV, Movchan AB (2006) Steady-state motion of a mode-III crack on imperfect interfaces. Q J Mech Appl Math 59:487–516

Murai Y (2007) Scattering attenuation, dispersion and reflection of SH waves in two-dimensional elastic media with densely distributed cracks. Geophys J Int 168:211–223

Pasternak IaI, Sulym HO, Ilchuk N (2021) Interaction of physicomechanical fields in bodies with thin structural inhomogeneities: a survey. J Math Sci 253:63–83

Poddubnyak AP (1986) Integral equations of the problem of the torsion of an elastic body with a thin disc-like inclusion. J Appl Math Mech 50:492–497

Rizzoni R, Dumont S, Lebon F, Sacco E (2014) Higher order model for soft and hard elastic interfaces equation. Int J Solids Struct 51:4137–4148

Uberal H (1973) Surface wave in acoustic. In: Physical acoustics: principles and methods, vol 10, pp 1–60. Academic Press, N.Y., London

VanDyke MD (1964) Perturbation methods in fluid mechanics. Academic Press, New York
Wang X, Sudak LJ (2007) Scattering of elastic waves by multiple elastic circular cylinders with imperfect interface. Waves Random Complex Media 17:159–187
Waterman PC (2009) T-matrix methods in acoustic scattering. J Acoust Soc Am 125:42–51

Chapter 18
3D Time-Harmonic Elastic Waves Scattering on Shell-Like Rigid Movable Inclusions

Roman Kushnir, Iaroslav Pasternak, and Heorhiy Sulym

18.1 Introduction

Inclusion and inhomogeneity problems are widely studied since they have many applications in modeling the behavior of composite materials (Kachanov and Sevostianov 2018), as well as analysis of fracture of such inhomogeneous solids (Murakami 2002). In modeling material inhomogeneities (including cracks and rigid inclusions), two main approaches are used (Pasternak and Sulym 2021): (1) modeling, which considers the real geometry of the inhomogeneity, or (2) those using simplified geometry. The second approach is very important in the analysis of the influence of thin inhomogeneities because it efficiently couples with the boundary layer effect that significantly affect the accuracy of the first approach in this case.

However, due to the mathematical complexity, there are few studies of the elastodynamic problems for solids containing thin inhomogeneities. In the case of the 3D problems, penny-shaped cracks (Kit et al. 1996) or plane disk-shaped inclusions (Mykhas'kiv and Khay 2009; Mykhas'kiv et al. 2008) are mainly considered. The authors did not find the studies of the elastodynamics of solids containing thin shell-like inhomogeneities of arbitrary shape, particularly perfectly rigid ones.

Therefore, this paper presents a general boundary element approach, which allows analysis of time-harmonic elastodynamic problems for solids containing rigid movable arbitrarily shaped shell-like inhomogeneities (or even fixed inclusions if their mass tends to infinity). Boundary integral equations and the mathematical models

R. Kushnir · H. Sulym
Pidstryhach Institute for Applied Problems of Mechanics and Mathematics, Lviv, Ukraine
e-mail: dyrector@iapmm.lviv.ua

H. Sulym
e-mail: h.sulym@pb.edu.pl

I. Pasternak (✉)
Lesya Ukrainka Volyn National University, Lutsk, Ukraine
e-mail: iaroslav.m.pasternak@gmail.com

A. N. Guz et al. (eds.), *Advances in Mechanics*, Advanced Structured Materials 191,
https://doi.org/10.1007/978-3-031-37313-8_18

of rigid shell-like inclusions are derived, and the issues on the problem's numerical solution are studied in detail.

18.2 Problem Statement. Boundary Integral Equations

Consider the elastodynamic problem for a linear elastic anisotropic solid. The equations of its motion in a fixed rectangular coordinate system $Ox_1x_2x_3$ are well known and can be written as (Achenbach 1973)

$$\sigma_{ij,j} - \rho \ddot{u}_i = -f_i, \tag{18.1}$$

where $\sigma_{ij}(\mathbf{x}, t)$, $u_i(\mathbf{x}, t)$, and $f_i(\mathbf{x}, t)$ are the components of stress tensor, displacement vector, and body force vector, respectively; ρ is the mass density. Einstein summation and differentiation (with respect to spatial coordinate) conventions are assumed. Newton's (dot) notation is used for differentiation with respect to time variables.

The stress–strain relation (Hook's law) for a linear elastic anisotropic solid can be written as (Achenbach 1973)

$$\sigma_{ij} = C_{ijkm} u_{k,m}, \tag{18.2}$$

where C_{ijkm} is the fourth-order tensor of elastic moduli. This tensor possesses several symmetry properties, one of which is

$$C_{ijkm} = C_{kmij}. \tag{18.3}$$

Consider the steady-state time-harmonic motion of the solid when the displacement field can be presented as

$$u_i(\mathbf{x}, t) = \tilde{u}_i(\mathbf{x}) \exp(-i\omega t). \tag{18.4}$$

Then (18.1) can be written as

$$\tilde{\sigma}_{ij,j} + \rho\omega^2 \tilde{u}_i = -\tilde{f}_i, \tag{18.5}$$

which defines the steady-state time-harmonic motion of the solid.

In general, boundary integral dependencies for elastodynamic problems are derived using the Maxwell–Betti reciprocal work theorem (Dominguez 1993). However, this approach has less extensibility, particularly in the case of quasicrystal materials.

In contrast, this chapter uses the symmetry property (18.3) and the governing Eqs. (18.2) and (18.5) to obtain the boundary integral formula and equations. For this purpose, first consider two arbitrary C_2-differentiable vector functions ϕ_i, ψ_i and a symmetric fourth-order tensor $A_{ijkm} = A_{kmij}$ resembling the symmetry property (18.3). With the help of direct differentiation, it can be easily shown that the following differential identity holds

$$\frac{\partial}{\partial x_j}\left(\phi_i A_{ijkm}\psi_{k,m} - \psi_i A_{ijkm}\phi_{k,m}\right) = \phi_{i,j} A_{ijkm}\psi_{k,m} + \phi_i A_{ijkm}\psi_{k,jm} - \psi_{i,j} A_{ijkm}\phi_{k,m} - \psi_i A_{ijkm}\phi_{k,jm}. \quad (18.6)$$

Since $A_{ijkm} = A_{kmij}$, we can write that

$$\phi_{i,j} A_{ijkm}\psi_{k,m} = \phi_{i,j} A_{kmij}\psi_{k,m} = \psi_{i,j} A_{ijkm}\phi_{k,m},$$

and therefore (18.6) can be simplified:

$$\frac{\partial}{\partial x_j}\left(\phi_i A_{ijkm}\psi_{k,m} - \psi_i A_{ijkm}\phi_{k,m}\right) = \phi_i A_{ijkm}\psi_{k,jm} - \psi_i A_{ijkm}\phi_{k,jm}. \quad (18.7)$$

Integrating of (18.7) over the 3D domain $\mathfrak{B}$ and utilizing Gauss divergence theorem, the following integral identity is obtained:

$$\iint_{\partial\mathfrak{B}}\left(\phi_i A_{ijkm}\psi_{k,m} - \psi_i A_{ijkm}\phi_{k,m}\right) n_j dS = \iiint_{\mathfrak{B}}\left(\phi_i A_{ijkm}\psi_{k,jm} - \psi_i A_{ijkm}\phi_{k,jm}\right) dV, \quad (18.8)$$

where $\partial\mathfrak{B}$ is the boundary of the domain $\mathfrak{B}$ and n_j are the components of a unit outward normal vector to the surface $\partial\mathfrak{B}$.

Now, we can assume that $\mathfrak{B}$ is the domain occupied by the solid, $\phi_i \equiv \tilde{u}_i$, $A_{ijkm} = C_{ijkm}$ and $\psi_i \equiv \tilde{U}_{pi}$, where $\tilde{U}_{pi}$ is the fundamental solution of (18.5) for $f_i = \delta_{ip}\delta(x - x_0)$. Here, δ_{ij} is the Kronecker symbol, and $\delta(x)$ is the Dirac delta function. Applying these substitutions to (18.8), we obtain

$$\iint_{\partial\mathfrak{B}}\left(\tilde{u}_i C_{ijkm}\tilde{U}_{pk,m} - \tilde{U}_{pi} C_{ijkm}\tilde{u}_{k,m}\right) n_j dS = \iiint_{\mathfrak{B}}\left(\tilde{u}_i C_{ijkm}\tilde{U}_{pk,jm} - \tilde{U}_{pi} C_{ijkm}\tilde{u}_{k,jm}\right) dV. \quad (18.9)$$

According to (18.2) and (18.5),

$$\begin{aligned} C_{ijkm}\tilde{u}_{k,jm} + \rho\omega^2\tilde{u}_i &= -\tilde{f}_i, \\ C_{ijkm}\tilde{U}_{pk,jm} + \rho\omega^2\tilde{U}_{pi} &= -\delta_{ip}\delta(\mathbf{x} - \mathbf{x}_0); \end{aligned} \quad (18.10)$$

using the standard notation for the traction vector

$$\tilde{t}_i = \tilde{\sigma}_{ij} n_j = C_{ijkm} \tilde{u}_{k,m} n_j, \quad \tilde{T}_{pi} = C_{ijkm} \tilde{U}_{pk,m} n_j, \tag{18.11}$$

Equation (18.9) can be rewritten as

$$\iint_{\partial \mathfrak{B}} \left(\tilde{T}_{pi} \tilde{u}_i - \tilde{U}_{pi} \tilde{t}_i \right) dS = \iiint_{\mathfrak{B}} \tilde{u}_i \left(-\rho \omega^2 \tilde{U}_{pi} - \delta_{ip} \delta \left(\mathbf{x} - \mathbf{x}_0 \right) \right) dV \\ - \iiint_{\mathfrak{B}} \tilde{U}_{pi} \left(-\rho \omega^2 \tilde{u}_i - \tilde{f}_i \right) dV. \tag{18.12}$$

The terms with $\rho\omega^2$ cancel each other, and (18.12) can be rearranged as

$$\tilde{u}_p \left(\mathbf{x}_0, \omega \right) = \iint_{\partial \mathfrak{B}} \tilde{U}_{pi} \left(\mathbf{x} - \mathbf{x}_0, \omega \right) \tilde{t}_i \left(\mathbf{x}, \omega \right) dS \left(\mathbf{x} \right) \\ - \iint_{\partial \mathfrak{B}} \tilde{T}_{pi} \left(x - x_0, \omega \right) \tilde{u}_i \left(x, \omega \right) dS \left(x \right) \\ + \iiint_{\mathfrak{B}} \tilde{U}_{pi} \left(x - x_0, \omega \right) \tilde{f}_i \left(x, \omega \right) dV \left(x \right). \tag{18.13}$$

Equation (18.13) presents Somigliana integral identity for time-harmonic elastodynamics of linear elastic anisotropic solids. It should be emphasized that (18.13) is obtained accounting only for the partial symmetry property (18.3) of the elasticity tensor. Thus, the proposed approach for obtaining such integral formulae can be naturally extended to the cases of other materials, which possess symmetry of their material properties similar to (18.3). One of the possible extensions of these results can be obtained for quasicrystal materials, in which material constants can be arranged in the fourth-order array, which satisfies (18.3) (Fan et al. 2022).

Approaching internal point $\mathbf{x}_0 \in \mathfrak{B}$ to some boundary point $\mathbf{x}_B \in \partial\mathfrak{B}$, (18.13) results in the boundary integral equations of elastodynamics, which naturally accounts for the Dirichlet $\left(\tilde{u}_i\left(\mathbf{x},\omega\right) = \tilde{u}_i^g\left(\mathbf{x},\omega\right)\right)$ or the Neumann $\left(\tilde{t}_i\left(\mathbf{x},\omega\right) = \tilde{t}_i^g\left(\mathbf{x},\omega\right)\right)$ boundary conditions at $\mathbf{x} \in \partial\mathfrak{B}$. Besides, mixed boundary conditions can also be easily considered. Due to the properties of the time-harmonic elastodymanic Green's function (Wang and Achenbach 1995) for a smooth boundary at the point $\mathbf{x}_B \in \partial\mathfrak{B}$, we obtain

$$\frac{1}{2} \tilde{u}_p \left(\mathbf{x}_B, \omega \right) = \iint_{\partial \mathfrak{B}} \tilde{U}_{pi} \left(\mathbf{x} - \mathbf{x}_B, \omega \right) \tilde{t}_i \left(\mathbf{x}, \omega \right) dS \left(\mathbf{x} \right) \\ - \iint_{\partial \mathfrak{B}} \tilde{T}_{pi} \left(\mathbf{x} - \mathbf{x}_B, \omega \right) \tilde{u}_i \left(\mathbf{x}, \omega \right) dS \left(\mathbf{x} \right) \\ + \iiint_{\mathfrak{B}} \tilde{U}_{pi} \left(\mathbf{x} - \mathbf{x}_B, \omega \right) \tilde{f}_i \left(\mathbf{x}, \omega \right) dV \left(\mathbf{x} \right). \tag{18.14}$$

If there is a smooth surface S of field discontinuity, (18.14) can be rewritten as (Pasternak et al. 2017)

$$\frac{1}{2}\Sigma\tilde{u}_p(\mathbf{x}_B,\omega) = \iint_S \tilde{U}_{pi}(\mathbf{x}-\mathbf{x}_B,\omega)\,\Sigma\tilde{t}_i(\mathbf{x},\omega)\,dS(\mathbf{x}) - \iint_S \tilde{T}_{pi}(\mathbf{x}-\mathbf{x}_B,\omega)\,\Delta\tilde{u}_i(\mathbf{x},\omega)\,dS(\mathbf{x}) + \iiint_{\mathfrak{B}} \tilde{U}_{pi}(\mathbf{x}-\mathbf{x}_B,\omega)\,\tilde{f}_i(\mathbf{x},\omega)\,dV(\mathbf{x}), \tag{18.15}$$

where $\Sigma f = f^+ + f^-$, $\Delta f = f^+ - f^-$, and superscripts "+" and "–" stand for field values on the faces S^+ and S^- of the cut obtained with the discontinuity surface S.

Differentiating (18.15) with respect to spatial coordinates and applying Hook's law (18.2), the hypersingular integral equations can be obtained, which along with (18.15) allows determination of field values on each of the faces S^+ and S^-,

$$\frac{1}{2}\Delta\tilde{t}_i(\mathbf{x}_B,\omega) = \iint_S \tilde{D}_{ijk}(\mathbf{x}-\mathbf{x}_B,\omega)\,n_j(\mathbf{x})\,\Sigma\tilde{t}_k(\mathbf{x},\omega)\,dS(\mathbf{x}) - \iint_S \tilde{S}_{ijk}(\mathbf{x}-\mathbf{x}_B,\omega)\,n_j(\mathbf{x})\,\Delta\tilde{u}_k(\mathbf{x},\omega)\,dS(\mathbf{x}) + \iiint_{\mathfrak{B}} \tilde{D}_{ijk}(\mathbf{x}-\mathbf{x}_B,\omega)\,n_j(\mathbf{x})\,\tilde{f}_k(\mathbf{x},\omega)\,dV(\mathbf{x}). \tag{18.16}$$

Here,

$$\tilde{D}_{ijk} = -C_{ijmp}\tilde{U}_{mk,p}, \quad \tilde{S}_{ijk} = -C_{ijmp}C_{kqrs}n_q\tilde{U}_{mr,ps}. \tag{18.17}$$

18.3 A Mathematical Model of a Rigid Shell-Like Inclusion

In this study, a rigid shell-like inclusion is proposed to model with a surface of traction discontinuity, simplifying its modeling and excluding the influence of the boundary layer effect that is significant in the numerical analysis of problems involving thin shapes.

Assume that a solid contains n rigid shell-like inhomogeneities with finite mass density ρ^{i}. Accounting for kinematics of a rigid body motion, the mathematical model of the rigid shell-like inclusion can be written as

$$\frac{1}{2}\Sigma\tilde{u}_i\left(x^k\right) = \tilde{u}_i^k + \varepsilon_{ijm}\tilde{\omega}_j^k x_m^k, \quad \Delta\tilde{u}_i = 0 \quad \forall x^k \in S_k \quad (k = 1, \ldots, n), \tag{18.18}$$

where $\tilde{u}_i^k$ is the amplitude of the rigid translation of the kth inclusion and $\tilde{\omega}_j^k$ is its rotation; ε_{ijm} is a permutation tensor.

Accounting for its small thickness $2h_k$, the equations of steady-state time-harmonic motion of the kth inclusion can be written as

$$\iint_{S_k} \Sigma\tilde{t}_i\,(x)\,dS\,(\mathbf{x}) - h_k\rho^{\mathrm{i}}\omega^2 \iint_{S_k} \Sigma\tilde{u}_i\,(\mathbf{x})\,dS\,(\mathbf{x}) - \tilde{P}_i^k = 0, \qquad (18.19)$$

$$\varepsilon_{ijm}\left(\iint_{S_k} x_j\Sigma\tilde{t}_m\,(\mathbf{x})\,dS - h_k\rho^{\mathrm{i}}\omega^2 \iint_{S_k} x_j\Sigma\tilde{u}_m\,(\mathbf{x})\,dS\right) - M_i^k = 0, \qquad (18.20)$$

where $\tilde{P}_i^k$ and M_i^k are the resultant amplitudes of the external forces and couples applied to the kth inclusion, respectively.

Substituting (18.18) into (18.19) and (18.20), accounting for the triple product expansion, we obtain

$$\iint_{S_k} \Sigma\tilde{t}_i\,(\mathbf{x})\,dS\,(\mathbf{x}) - 2h_kS_k\rho^{\mathrm{i}}\omega^2\left[\tilde{u}_i^k + \varepsilon_{ijm}\tilde{\omega}_j^k x_m^{Ck}\right] - \tilde{P}_i^k = 0, \qquad (18.21)$$

$$\varepsilon_{ijm}\iint_{S_k} x_j\Sigma\tilde{t}_m\,(\mathbf{x})\,dS\,(\mathbf{x}) - M_i^k$$
$$- 2h_k\rho^{\mathrm{i}}\omega^2\left[\varepsilon_{ijm}\tilde{u}_m^k x_j^{Ck}S_k + \tilde{\omega}_i^k J_{pp}^{(k)} - \tilde{\omega}_p^k J_{ip}^{(k)}\right] = 0, \qquad (18.22)$$

where

$$x_m^{Ck} = S_k^{-1}\iint_{S_k} x_m dS\,(\mathbf{x})$$

is the center of mass of the kth inclusion, and

$$J_{ip}^{(k)} = \iint_{S_k} x_i x_p dS\,(\mathbf{x})$$

are its moments of inertia about the coordinate axes.

When $\rho^{\mathrm{i}} \to \infty$, (18.18), (18.21), and (18.22) result in the model of a rigid fixed (unmovable) shell-like inclusion, i.e., $\Sigma u_i^k = 0$.

Equations (18.15), (18.18), (18.21), and (18.22) allow us to solve the time-harmonic elastodynamics problems for anisotropic solid with rigid shell-like inclusions.

18.4 Boundary Element Solution of the Problem

18.4.1 General Issues of the Boundary Element Solution

For the boundary element solution of derived boundary integral equations for a particular problem, for the surface $\partial\mathfrak{B}$ of the solid along with inclusion surfaces S_k, mesh of quadrilateral quadratic discontinuous boundary elements is generated. The local curvilinear coordinate system $O\xi\eta$ is associated with each boundary element, moreover, $-1 \leqslant \xi \leqslant 1$, $-1 \leqslant \eta \leqslant 1$. The collocation points are placed at nodes $\xi = (-2/3; 0; 2/3)$; $\eta = (-2/3; 0; 2/3)$. Therefore, nine collocation points are associated with each boundary element (Pasternak et al. 2017).

Boundary conditions along with unknown boundary and discontinuity functions are interpolated within the collocation points at each boundary element Γ_N as

$$\mathbf{b}_N(\xi, \eta) = \sum_{i=1}^{3}\sum_{j=1}^{3} \mathbf{b}_N^{i,j} \phi_i(\xi)\, \phi_j(\eta), \tag{18.23}$$

where $\mathbf{b} = \left(\Delta\tilde{u}_i, \Sigma\tilde{t}_i\right)^{\mathrm{T}}$, and the discontinuous shape functions $\phi_i(\xi)$ are selected the same as in Pasternak et al. (2019).

Substituting (18.23) into the boundary integral Eqs. (18.15), (18.18), (18.21), and (18.22), we obtain the system of linear algebraic equations for unknown nodal values of sought discontinuity functions and translations and rotations of inclusions.

Due to the $1/r$ singularity of the kernel $\tilde{U}_{ij}(\mathbf{x}, \omega)$ (Wang and Achenbach 1995), (18.15) results in weakly and strongly singular integrals. Techniques for evaluating arising regular, weakly, strongly, and hypersingular integrals are explicitly described in Pasternak et al. (2017, 2019). Weakly singular integrals are evaluated in polar coordinates; thus, the Jacobian cancels the $1/r$ singularity. Singular integrals are evaluated in polar coordinates using special modified Kutt quadratures with Chebyshev nodes (Pasternak et al. 2017), which allow accurate determination of Cauchy principal value and Hadamard finite part of the arising integrals. These techniques allow accurate analysis of solids containing discontinuity surfaces of arbitrary smooth shape.

18.4.2 Evaluation of Kernels for Anisotropic Materials

Wang and Achenbach (1995) used the Radon transform technique to obtain the solution of (18.10) in the form

$$\tilde{U}_{ij}(\mathbf{x} - \mathbf{x}_0, \omega) = \tilde{U}_{ij}^{S}(\mathbf{x} - \mathbf{x}_0) + \tilde{U}_{ij}^{R}(\mathbf{x} - \mathbf{x}_0, \omega), \tag{18.24}$$

where $\tilde{U}_{ij}^{S}$ is the static Green's function for infinite space,

$$\tilde{U}_{ij}^{S}(\mathbf{x},\mathbf{y}) = \frac{1}{8\pi^2|\mathbf{x}|} \oint\limits_{\substack{|\mathbf{n}|=1 \\ \mathbf{n}\cdot\mathbf{x}=0}} \Gamma_{ij}^{-1}(\mathbf{n})\, dl(\mathbf{n})$$

and

$$\tilde{U}_{ij}^{R}(\mathbf{x},\omega) = \frac{i}{4\pi^2} \int\limits_{\substack{|\boldsymbol{\xi}|=1 \\ \boldsymbol{\xi}\cdot\mathbf{x}>0}} \sum_{m=1}^{3} \frac{k_m E_{im} E_{jm}}{2\rho c_m^2} \mathrm{e}^{ik_m|\boldsymbol{\xi}\cdot\mathbf{x}|} dS(\boldsymbol{\xi}). \tag{18.25}$$

Here, E_{ij} are eigenvectors corresponding to eigenvalues λ_j of the symmetric and positive defined matrix $\Gamma_{ik}(\boldsymbol{\xi}) = C_{ijkm}\xi_j\xi_m$, and

$$c_m = \sqrt{\lambda_m/\rho}, \quad k_m = \omega/c_m. \tag{18.26}$$

The static Green's function $\tilde{U}_{ij}^{S}$ for anisotropic solids and its derivatives are reduced to the contour integrals, which are efficiently evaluated using the trapezoid rule (Pasternak et al. 2017) due to its exponential convergence for integrals over a periodic interval.

However, the numerical evaluation of $\tilde{U}_{ij}^{R}(\mathbf{x}-\mathbf{x}_0,\omega)$ is a challenging problem, since the term $\exp\{ik_m|\boldsymbol{\xi}\cdot\mathbf{x}|\}$ is highly oscillating with increase in ω and ρ. Therefore, special techniques should be used for its evaluation (e.g., see (Evans and Webster 1997; Iserles and Norsett 2005)).

To proceed with the derivatives of Green's function, first, recall that

$$\frac{\partial \exp(ik_m|\boldsymbol{\xi}\cdot\mathbf{x}|)}{\partial x_p} = ik_m\xi_p \operatorname{sign}(\boldsymbol{\xi}\cdot\mathbf{x}) \exp(ik_m|\boldsymbol{\xi}\cdot\mathbf{x}|),$$

and since $d\operatorname{sign}(x)/dx = 2\delta(x)$, and $\exp(0) = 1$,

$$\frac{\partial^2 \exp(ik_m|\boldsymbol{\xi}\cdot\mathbf{x}|)}{\partial x_q \partial x_p} = ik_m\xi_q\xi_p\left[2\delta(\boldsymbol{\xi}\cdot\mathbf{x}) + ik_m \exp(ik_m|\boldsymbol{\xi}\cdot\mathbf{x}|)\right].$$

Therefore,

$$\tilde{U}_{ij,p}^{R} = -\frac{1}{4\pi^2} \int\limits_{\substack{|\boldsymbol{\xi}|=1 \\ \boldsymbol{\xi}\cdot\mathbf{x}>0}} \sum_{m=1}^{3} \frac{k_m^2 E_{im} E_{jm}\xi_p}{2\rho c_m^2} \mathrm{e}^{ik_m|\boldsymbol{\xi}\cdot\mathbf{x}|} dS(\boldsymbol{\xi}), \tag{18.27}$$

$$\tilde{U}_{ij,pq}^{R} = -\frac{i}{4\pi^2} \int\limits_{\substack{|\boldsymbol{\xi}|=1 \\ \boldsymbol{\xi}\cdot\mathbf{x}>0}} \sum_{m=1}^{3} \frac{k_m^3 E_{im} E_{jm}\xi_p\xi_q}{2\rho c_m^2} \mathrm{e}^{ik_m|\boldsymbol{\xi}\cdot\mathbf{x}|} dS(\boldsymbol{\xi})$$

$$-\frac{\rho\omega^2}{4\pi^2}\int\limits_{\substack{|\boldsymbol{\xi}|=1\\ \boldsymbol{\xi}\cdot\mathbf{x}>0}}\sum_{m=1}^{3}\frac{E_{im}E_{jm}\xi_p\xi_q}{\lambda_m^2}\delta\left(\boldsymbol{\xi}\cdot\mathbf{x}\right)dS\left(\boldsymbol{\xi}\right). \tag{18.28}$$

Since $E_{im}E_{jm}\lambda_m^{-2} = \Gamma_{ij}^{-2}(\boldsymbol{\xi})$, (18.28) is reduced to

$$\tilde{U}_{ij,pq}^{R} = -\frac{i}{4\pi^2}\int\limits_{\substack{|\boldsymbol{\xi}|=1\\ \boldsymbol{\xi}\cdot\mathbf{x}>0}}\sum_{m=1}^{3}\frac{k_m^3E_{im}E_{jm}\xi_p\xi_q}{2\rho c_m^2}e^{ik_m|\boldsymbol{\xi}\cdot\mathbf{x}|}dS\left(\boldsymbol{\xi}\right)$$

$$-\frac{\rho\omega^2}{8\pi^2|\mathbf{x}|}\int\limits_{\substack{|\boldsymbol{\xi}|=1\\ \boldsymbol{\xi}\cdot\mathbf{x}=0}}\Gamma_{ij}^{-2}\left(\boldsymbol{\xi}\right)\xi_p\xi_q dl\left(\boldsymbol{\xi}\right). \tag{18.29}$$

Note that the second term in (18.29) is real-valued and weakly singular. Also, the second integral in (18.29) is a contour integral over a unit circle evaluated in the plane $\boldsymbol{\xi}\cdot\mathbf{x} = 0$.

Thus, the techniques proposed in Pasternak et al. (2017) to find kernel function integrals can also be applied in the case of time-harmonic elastodynamic problems.

18.4.3 Evaluation of the Generalized Stress Intensity Factors

Since the kernel functions possess the same singular behavior as in the static problems, dynamic stresses also possess square root singularity at the front line of the shell-like inhomogeneity (Pasternak et al. 2019). According to Pasternak et al. (2019), the stress field in front of the inclusion in a local coordinate system $A\boldsymbol{\tau}\mathbf{nm} \sim Ax_1x_2x_3$ (where $\boldsymbol{\tau}$ is a unit vector normal to inclusion front line at the point A; $\mathbf{n}$ is a unit vector normal to inclusion surface at the point A, and $\mathbf{m} = \boldsymbol{\tau}\times\mathbf{n}$ is a unit vector tangential to inclusion front line) is defined as

$$\begin{aligned}\tilde{\sigma}_1 = \left[\tilde{\sigma}_{i1}\right] = \frac{2}{\sqrt{2\pi}}\operatorname{Im}\left\{\mathbf{B}\left\langle p_* Z_*^{-1/2}\right\rangle\mathbf{A}^{\mathrm{T}}\tilde{\mathbf{k}}^{(2)}\right\},\\ \tilde{\sigma}_2 = \left[\tilde{\sigma}_{i2}\right] = \frac{-2}{\sqrt{2\pi}}\operatorname{Im}\left\{\mathbf{B}\left\langle Z_*^{-1/2}\right\rangle\mathbf{A}^{\mathrm{T}}\tilde{\mathbf{k}}^{(2)}\right\},\end{aligned} \tag{18.30}$$

where

$$\left\langle Z_*^{-1/2}\right\rangle = \operatorname{diag}\left[\left(x_1+p_1x_2\right)^{-1/2},\ldots,\left(x_1+p_3x_2\right)^{-1/2}\right];$$

the matrices $\mathbf{A}$, $\mathbf{B}$ and constants p_α ($\alpha = 1,\ldots,3$) are determined from the Stroh eigenvalue problem, and the generalized stress intensity factor vector $\tilde{\mathbf{k}}^{(2)} = [K_{12}, K_{22}, K_{32}]^{\mathrm{T}}$ is defined as

$$\tilde{\mathbf{k}}^{(2)} = -\lim_{\mathbf{x}\to\mathbf{x}(A)} \sqrt{\frac{\pi s(\mathbf{x})}{2}} \Sigma\tilde{\mathbf{t}}(\mathbf{x}). \tag{18.31}$$

Extended stress discontinuity $\Sigma\tilde{\mathbf{t}}$ in (18.31) is computed in a local coordinate system $A\boldsymbol{\tau}\mathbf{nm} \sim Ax_1x_2x_3$. Here, $s(\mathbf{x})$ is an arc length evaluated from $\mathbf{x}$ to A along the cross section of the inclusion with the plane $(\mathbf{n}, \boldsymbol{\tau})$.

Special shape functions (Pasternak et al. 2019) are used on the front boundary elements to account for a square root singularity (18.31) of stress discontinuity.

18.5 Numerical Example

Without loss in generality, consider an isotropic elastic medium with Poisson's ratio $\nu = 0.3$. Such a choice of material is made to show the effect of a shell-like inhomogeneity on time-harmonic wave scattering. Then we have a test solution, which can be applied to the study of anisotropy influence on wave scattering.

Assume that a plane longitudinal wave is acting on the medium, which equation writes as

$$u_3^{\text{in}}(\mathbf{x}) = -ia\omega^{-1}\omega^* \exp\left(i\omega x_3/\left(a\omega^*\right)\right), \quad u_1^{\text{in}}(\mathbf{x}) = 0, \quad u_2^{\text{in}}(\mathbf{x}) = 0.$$

Here, $\omega^* = c_2/a$, $c_2^2 = \rho^{-1}\mu$, a is a linear distance and μ is the shear modulus of the material. The corresponding stress components in the incident wave are as follows:

$$\sigma_{ij}^{in}(\mathbf{x}) = C_{ij33} \exp\left(\frac{i\omega x_3}{a\omega^*}\right).$$

Assume that the considered medium contains a shell-like inclusion, which surface is given by the following equation of elliptic paraboloid of revolution:

$$x_3 = \frac{\varepsilon}{a}\left(x_1^2 + x_2^2\right), \quad x_1^2 + x_2^2 \leq a^2,$$

where ε is the dimensionless parameter. If $\varepsilon = 0$, we obtain a penny-shaped inclusion of a radius a.

Assuming $2h = 0.02a$ (the thickness of the shell-like inclusion) and $\rho^i = 10^3\rho$ (the mass density of the inclusion), consider the influence of $\tilde{\omega} = \omega/\omega^*$ and ε on inclusion vibration $\tilde{u}_i^0 = u_i^0/a$ and generalized stress intensity $\tilde{K}_i = K_{i2}/K_0$ on its front line. Here $K_0 = C_{3333}\sqrt{\pi a}$.

The numerical solution is performed with the developed boundary element approach. Only 12 quadrilateral boundary elements are used following (Pasternak et al. 2019), where results were obtained for static problems.

Figures 18.1, 18.2, and 18.3 depict the magnitude of $\tilde{K}_i$ over the inclusion front line for different values of ε, which changes from 0 to 0.8 with a step 0.2. Solid lines

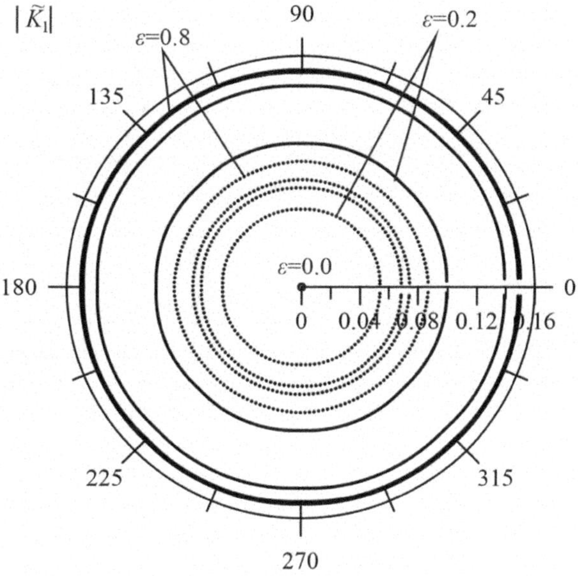

Fig. 18.1 Magnitude of $\tilde{K}_1$ at the front line of paraboloid rigid inclusion

correspond to $\omega/\omega^* = 1$, and dotted ones to $\omega/\omega^* = 2$. We observe the significant influence of vibration frequency ω on the stress intensity at the inclusion front line. Also, the obtained values demonstrate high consistency with the underlying physical processes. Since the problem is geometrically symmetric, the symmetry in generalized stress intensity factors is also observed. For rigid disk inclusions, obtained values of generalized stress intensity factors agree with the analytic solution given in Mykhas'kiv and Khay (2009), Achenbach (1973).

It is interesting that in contrast to static problems (Pasternak et al. 2019), generalized stress intensity factor $\tilde{K}_3$ also possesses nonzero value for $\varepsilon > 0$ due to wave scattering and interference, which can also be referred to modeling of the geometry of the paraboloid of revolution with a discrete number of quadratic quadrilateral boundary elements. Nevertheless, this does not sufficiently influence the solution since $\tilde{K}_3$ is much smaller than other generalized stress intensity factors. Moreover, if the number of boundary elements is doubled, the value of $\tilde{K}_3$ decreases approximately ten times, which also proves that this effect is due to the boundary element modeling of the shape of inhomogeneity.

Figure 18.4 depicts the dependence of displacement amplitude of rigid inclusion $\tilde{u}_3^0$ on ε. Solid lines correspond to $\omega/\omega^* = 1$, and dotted ones to $\omega/\omega^* = 2$. Other components of u_i^0 and their rotation amplitudes $\tilde{\omega}_i^0$ are zeros due to the symmetry of the problem.

According to Fig. 18.4, with the elongation of the paraboloid, its displacement amplitude possesses a significant phase shift. With the increase in ω, displacement amplitude decreases because the selected inclusion's mass density is much greater than the density of the medium. Nevertheless, there can be some resonance effects that definitely should be studied further.

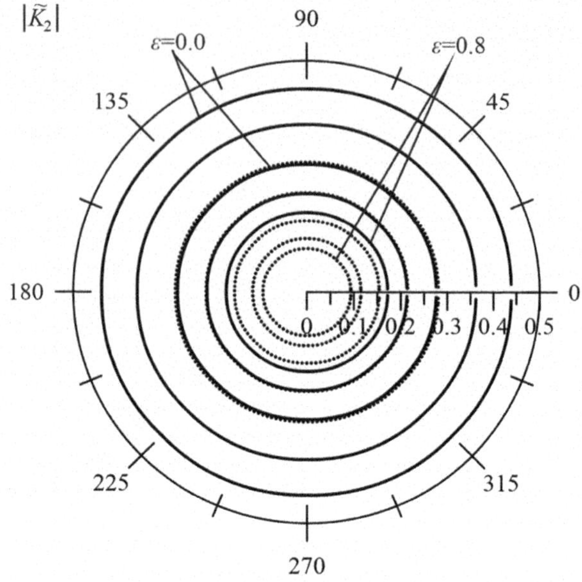

Fig. 18.2 Magnitude of $\tilde{K}_2$ at the front line of paraboloid rigid inclusion

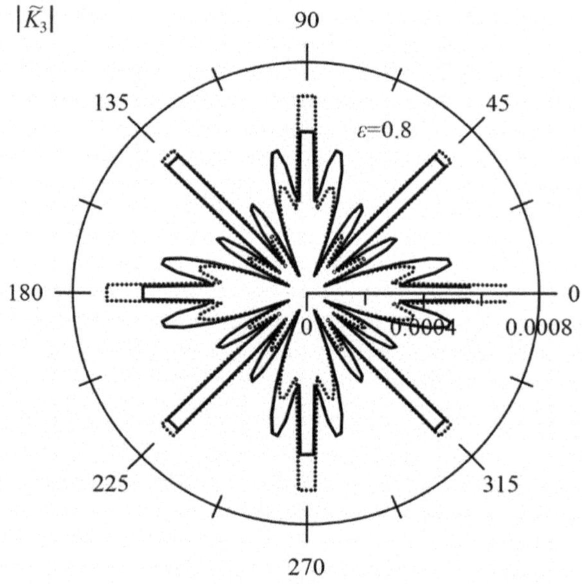

Fig. 18.3 Magnitude of $\tilde{K}_3$ at the front line of paraboloid rigid inclusion

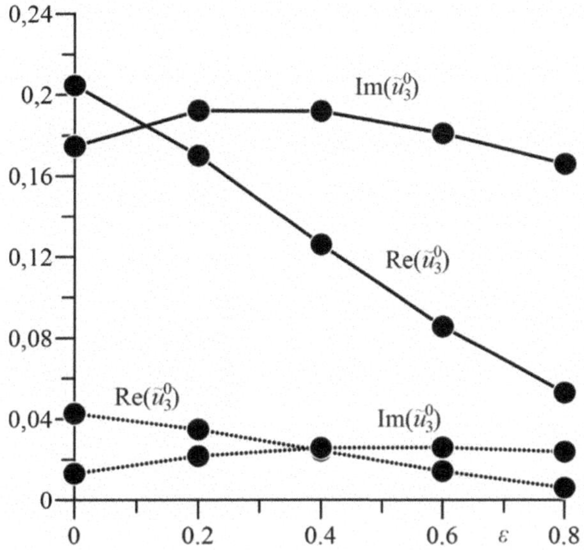

Fig. 18.4 Dependence of inclusion rigid displacement amplitude $\tilde{u}_3^0$ on ε

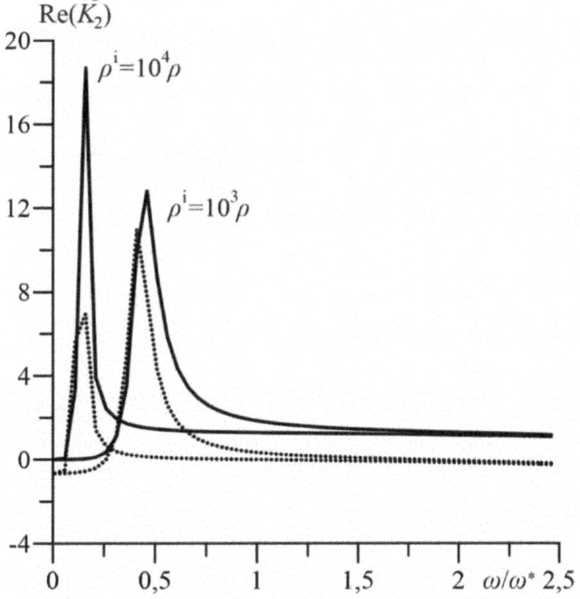

Fig. 18.5 Dependence of $\mathrm{Re}\left(\tilde{K}_2\right)$ on ω

Let us consider the influence of vibration frequency ω and the mass density ρ^i of inhomogeneity on the stress intensity and rigid displacement of the inclusion. The dependence of $\mathrm{Re}(\tilde{K}_2)$ on ω, ρ^i and ε is presented in Fig. 18.5, and corresponding plots for $\mathrm{Re}\left(\tilde{u}_3^0\right)$ are given in Fig. 18.6. Solid lines correspond to $\varepsilon = 0$ and dotted ones to $\varepsilon = 0.4$. Figures 18.5 and 18.6 show some resonance phenomena exist. For bigger

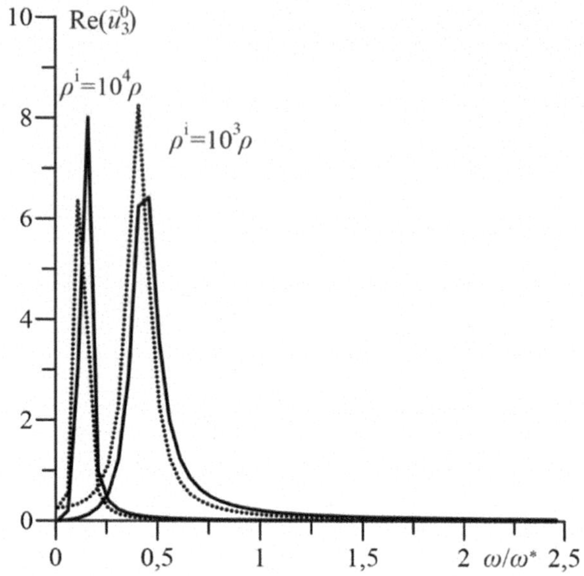

Fig. 18.6 Dependence of $\mathrm{Re}(\tilde{u}_3^0)$ on ω

values of ρ^i, the resonant frequency is lower, which also fits the physical background of the problem since the frequency of mass vibrating on a spring is lower for bigger mass $(\omega^2 = k/m)$. We can also observe that the maximum values of $\mathrm{Re}\,(K_2)$ a slightly lower for $\varepsilon = 0.4$ compared to $\varepsilon = 0$ since the paraboloid ($\varepsilon = 0.4$) has a more streamlined shape than a rigid disk ($\varepsilon = 0$). The resonant frequencies for $\tilde{u}_3^0$ are the same that those for $\tilde{K}_2$.

18.6 Conclusions

The paper presents a thorough approach to the analysis of isotropic and anisotropic solids containing rigid movable shell-like inclusions. Obtained boundary integral equations and inclusion models, along with the resulting boundary element approach, are presented so that they can be extended to more general problems of time-harmonic magnetoelectroelasticity of anisotropic solids.

Presented numerical examples show good agreement with the physical nature of the considered wave processes, which verifies obtained results and approves the efficiency of the proposed approach.

References

Achenbach JD (1973) Wave propagation in elastic solids. North-Holland, Amsterdam

Dominguez J (1993) Boundary elements in dynamics. Elsevier, London

Evans GA, Webster JR (1997) A high order, progressive method for the evaluation of irregular oscillatory integrals. Appl Numer Math 23:205–218

Fan TY, Yang W, Cheng H, Sun XH (2022) Generalized dynamics of soft-matter quasicrystals: Mathematical models. Solutions and applications, 2nd edn. Springer, Singapore

Iserles A, Norsett SP (2005) Efficient quadrature of highly oscillatory integrals using derivatives. Proc R Soc A 461:1383–1399

Kachanov M, Sevostianov I (2018) Micromechanics of materials, with applications. Springer, Cham

Kit HS, Khaj MV, Mykhas'kiv VV (1996) Analysis of dynamic stress concentration in an infinite body with parallel penny-shaped cracks by BIEM. Eng Fract Mech 55(2):191–207

Murakami Y (2002) Metal fatigue: Effects of small defects and nonmetallic inclusions. Elsevier

Mykhas'kiv VV, Khay OM (2009) Interaction between rigid-disc inclusion and penny-shaped crack under elastic time-harmonic wave incidence. Int J Solids Struct 46:602–616

Mykhas'kiv V, Khay O, Sladek J, Sladek V, Zhang C (2008) Micromechanical dynamic influence of rigid disk-shaped inclusion on neighboring crack in 3D elastic matrix. Mater Sci Forum 133–136

Pasternak IM, Sulym H (2021) Thermoelasticity of solids containing thread-like inhomogeneities. I. Nondeformable thread-like inclusions. Int J Solids Struct 232(1):111176

Pasternak I, Pasternak R, Pasternak V, Sulym H (2017) Boundary element analysis of 3D cracks in anisotropic thermomagnetoelectroelastic solids. Eng Anal Boundary Elem 74:70–78

Pasternak I, Sulym H, Ilchuk N (2019) Boundary element analysis of 3D shell-like rigid electrically conducting inclusions in anisotropic thermomagnetoelectroelastic solids. Z Angew Math Mech e201800319

Wang CY, Achenbach JD (1995) Three-dimensional time-harmonic elastodynamic green's functions for anisotropic solids. Proc R Soc Lond A 449:441–458

Chapter 19
Dynamics of an Asymmetrical Three-Layer Spherical Dome With an Inhomogeneous Filler Under a Concentrated Impact

Peter Lugovoi and Serhii Orlenko

19.1 Introduction

The effective load-bearing capacity of three-layer shell structures makes them very useful in various engineering applications. The continuous development of new structures leads to the creation of materials for complex structures with predefined properties, which requires careful analysis. Many works in the literature study the dynamics of layered shells with reinforcements and technical features by various methods (Librescu and Hause 2000; Lugovoi et al. 2020; Lugovyy et al. 2019; Surianinov et al. 2019). However, the creation of advanced technologies and special-purpose objects often leads to the need to develop structural three-layer shell elements with an aggregate of a complicated geometric structure. These elements are often subjected to various dynamic loads, including non-stationary loads. The dynamic behavior of such shells has not been sufficiently studied. This chapter considers asymmetric three-layer spherical shells with a discrete-symmetric light filler reinforced with ribs. The reinforcing ribs are located along the lines of the main curvatures and connect the load-bearing layers. The distances between the reinforcing ribs are much larger than the size of their cross sections. The load-bearing layers of such shells have different thicknesses and can be made of different materials. In such structures, there is a stress–strain state with significant displacement and stress gradients when dynamic loads are applied. In this case, it is advisable to apply the theory of layered shells using independent hypotheses for each layer (Li et al. 2008; Piskunov and Rasskazov 2002). It increases the general order of the system of equations but allows for a more detailed study of the dynamic behavior of the three-layer structure under dynamic loads. The problem's solution is based on the theory of shells and rods,

P. Lugovoi (✉) · S. Orlenko
S.P. Timoshenko Institute of Mechanics, National Academy of Science of Ukraine, Kyiv, Ukraine
e-mail: plugovyy@inmech.kiev.ua

S. Orlenko
e-mail: orlenko_sergey@ukr.net

A. N. Guz et al. (eds.), *Advances in Mechanics*, Advanced Structured Materials 191,
https://doi.org/10.1007/978-3-031-37313-8_19

based on Timoshenko's shear model using independent static and kinematic hypotheses for each layer. The variational principle of Hamilton–Ostrogradsky stationarity is used to derive the equations of oscillations of a three-layer non-homogeneous structure. Many works (Hause and Librescu 1998, 2007; Hohe and Librescu 2003; Malekzadeh et al. 2014; Senitsky et al. 1997; Timoshenko and Woinowsky-Krieger 1959) are devoted to studying the dynamics of layered spherical shells. The study of the dynamics of three-layer spherical shells with a discrete heterogeneous aggregate under non-stationary loads was initiated in Orlenko (2020), Lee and Lee (1997), Lugovoi et al. (2020a), Lugovoi et al. (2020b), Meish and Shtantsel (2002), Shul'ga et al. (2003), where the related problems were solved by the finite difference method. However, the light aggregate between the load-bearing layers and the reinforcing ribs significantly complicates the problem. Therefore, the finite element method was used to study the dynamics of three-layer asymmetric spherical shells with a discrete-symmetric light, reinforcing rib aggregate (Lugovoi et al. 2020; Lugovyi and Orlenko 2021; Lugovyi et al. 2021).

Some new numerical results of solving specific problems are given in this chapter, and new mechanical effects are revealed.

19.2 Formulation of the Problem—Basic Equations

A three-layer asymmetric spherical shell with light aggregate reinforced with discrete-symmetric ribs is an elastic structure consisting of internal (index 1), external (index 2) load-bearing layers, light aggregate (index t), and a set of discrete ribs (index j), rigidly connected to the specified load-bearing layers. The shell has a constant total thickness h with a smooth median surface in the orthogonal coordinate system α, z. The coordinate line $R\alpha$ on the shell's middle surface at $z = 0$ coincides with the generating line; the coordinate line z is a straight line orthogonal to the median surface. We consider the value of z to be positive if the point is on the side of the convexity of the middle surface. Discrete ribs and light aggregate rigidly connect the shells. The type of deformed state of the inner and outer load-bearing layers can be determined by components of the generalized displacement vector $\bar{U}_1 = (u_s^1, u_3^1, \varphi_1^1)^T$ and $\bar{U}_2 = (u_s^2, u_3^2, \varphi_1^2)^T$ (Timoshenko and Woinowsky-Krieger 1959). The displacement fields for a light filler are determined by the generalized displacement vector $\bar{U}_t = (u_s^t, u_3^t, \varphi_1^t)^T$ according to the model proposed in Frostig and Thomsen (2004). The deformed state of the reinforcing rib directed along the circumferential coordinate will be determined by the generalized displacement vector $\bar{U}_j = (u_s^j, u_3^j, \varphi_1^j)^T$ (Meish and Shtantsel 2002).

The coefficients of the first quadratic form and the curvature of the coordinate surface are written as follows: $A_1 = R_i$, $A_2 = R_i$, $k_1 = 1/R_i$, and $k_2 = 1/R_i$.

According to the theory of shear deformation in shells (Timoshenko and Woinowsky-Krieger 1959), the displacements u_1^i and u_3^i in the load-bearing layers with small linear displacements in the direction α (longitudinal), z (thickness), and t (time) are expressed as

$$
\begin{aligned}
u_1^i(s,z,t) &= u_0^i(s,t) + z_i\varphi_1^i(s,t),\\
u_3^i(s,z,t) &= u_{03}^i(s,t) \quad (i=1,2),
\end{aligned}
\tag{19.1}
$$

where φ_1^i is the angle determining the middle surface normal to load-bearing layers, $s = R\alpha$.

Based on the assumptions of a rigid connection of reinforcing ribs with spherical load-bearing layers, the contact conditions of mass centers of the ribs with the load-bearing layers are written as (Meish and Shtantsel 2002)

$$
u_1^j = u_1^{jk}(s_j) \mp \frac{H_j}{2}\varphi_1^{jk}(s_j), \quad u_3^j = u_3^j k(s_j), \quad \varphi_i^{jk}(s_j) \quad (k=1,2), \tag{19.2}
$$

where $s_j = R\alpha_j$ is the coordinate of the line that contains projections of mass centers of jth rib cross sections onto the corresponding median surface of a load-bearing layer:

$$
h_j^i = 0.5h_i + \frac{H_i}{2},
$$

where h_i $(i = 1, 2)$ are the thicknesses of spherical load-bearing layers; $H_j/2$ is the distance from the axis of jth edge to the surface of the smooth shells; and $h_t = H_j$ is the thickness of light aggregate.

The deformation ratios for the load-bearing layers and the jth edge are taken as follows:

$$
\begin{aligned}
\varepsilon_{11}^i &= \frac{\partial U_0^i}{\partial S_i} + \frac{u_{03}^i}{r_i}, \quad \varepsilon_{22}^i = \frac{u_0^i}{r_i}\cot\alpha + \frac{u_{03}^i}{r_i},\\
\varepsilon_{13}^i = \varphi_1^i, \quad \kappa_{11}^i &= \frac{\partial\varphi_1^i}{\partial S_i}, \quad \kappa_{22}^i = \frac{\varphi_1^i}{r_i}\cot\alpha, \quad \varepsilon_{22j} = \frac{u_{3j}}{r_j}.
\end{aligned}
\tag{19.3}
$$

Expressions for displacements of light filler are written according to the model (Frostig and Thomsen 2004):

$$
u_1^t(s,z,t) = \left(1 + \frac{z_t}{R_t}\right)u_0^t(s,t) + z_t u_1^t(s,t), \quad u_3^t(s,z,t) = u_{03}^t(s,t). \tag{19.4}
$$

Kinematic dependences for the aggregate are taken based on small deformations:

$$
\begin{aligned}
\varepsilon_{11}^t &= \frac{1}{1+(z_t/R_t)}\left(\frac{\partial u_0^t}{\partial s} + \frac{u_{03}^t}{R_t}\right), \quad \varepsilon_{22}^t = \frac{u_3^t}{R_t + z_t},\\
2\varepsilon_{13}^t &= \frac{1}{1+(z_t/R_t)}\left(\frac{\partial u_{03}^t}{\partial s} - \frac{u_0^t}{R_t}\right) + u_1^t.
\end{aligned}
\tag{19.5}
$$

The compatibility conditions, which provide for an ideal connection between the aggregate and the load-bearing layers without detachment and slipping, can be presented in the following form (Kheirikhah et al. 2011):

$$u_1^t(z = z_t^i) = u_0^i + \tfrac{1}{2}(-1)^k h_i \varphi_1^i$$
$$u_{03}^t = u_{03}^i, \tag{19.6}$$

where

$$k = \begin{cases} 0, & i = 1 \\ 1, & i = 2 \end{cases}, \quad z_t^i = \frac{h_t}{2}\begin{cases} -1, & i = 1 \\ 1, & i = 2 \end{cases}.$$

Using the expressions for the field of displacements for the load-bearing layers (19.1) and the light aggregate (19.4) as well as the conditions of continuity of displacements for interlayer (19.6), we derive the simplified compatibility conditions:

$$\begin{aligned} u_0^t &= \frac{u_0^1 + u_0^2}{2} - \tfrac{1}{4}\left(h_2\varphi_1^2 - h_1\varphi_1^1\right), \\ u_1^t &= \frac{u_0^1 - u_0^2}{h_t} - \frac{1}{2h_t}\left(h_2\varphi_1^2 + h_1\varphi_1^1\right), \\ u_{03}^t &= \tfrac{1}{2}\left(u_{03}^1 + u_{03}^2\right). \end{aligned} \tag{19.7}$$

The equations of motion for load-bearing layers and light aggregate are derived using the Hamilton–Ostrogradsky variational principle of stationarity, according to which

$$\delta \int_{t_1}^{t_2} (K - P + A)dt = 0, \tag{19.8}$$

where P is the total potential energy of the elastic system, K is the total kinetic energy of the elastic system, A is the work of external forces, and t_1 and t_2 are fixed moments of time. When deriving the equations of oscillations of three-layer shells with a light aggregate, the components of the movements of the load-bearing layers, reinforcing ribs, and aggregate made of light material are subject to independent variation.

Expressions for variations of the total potential and kinetic energy of the indicated components are written in the form:

$$\begin{aligned} \delta P &= \delta \sum_{i=1}^{2} P^i + \delta \sum_{j=1}^{J} P^j + \delta \sum_{S_t} P^t, \\ \delta K &= \delta \sum_{i=1}^{2} K^i + \delta \sum_{j=1}^{J} K^j + \delta \sum_{S_t} K^t, \end{aligned} \tag{19.9}$$

where

$$\delta P^i = \int_{S_i} \left[\int_{-h_i/2}^{h_i/2} (T_{11}^i \delta\varepsilon_{11}^i + T_{22}^i \delta\varepsilon_{22}^i + T_{13}^i \delta\varepsilon_{13}^i \right. \tag{19.10}$$
$$\left. + M_{11}^i \delta\kappa_{11}^i + M_{22}^i \delta\kappa_{22}^i) dz_i \right] dS_i$$

$$\delta P^t = \int_{S_t}\left[\int_{-h_t/2}^{h_t/2}(T_{11}^t\delta\varepsilon_{11}^t + T_{22}^t\delta\varepsilon_{22}^t + T_{13}^t\delta\varepsilon_{13}^t + M_{11}^t\delta\kappa_{11}^t + M_{22}^t\delta\kappa_{22}^t)dz_t\right]dS_t \tag{19.11}$$

$$\delta P^j = \int_{L_j}(T_{11j}\delta\varepsilon_{11j} + T_{22j}\delta\varepsilon_{22j} + T_{13j}\delta\varepsilon_{13j} + M_{11j}\delta\kappa_{11j})dL_j \tag{19.12}$$

$$\delta K^i = \int_{S_i}\left\{\int_{-h_i/2}^{h_i/2}\left[\rho_i h_i\left(\frac{\partial^2 u_1^i}{\partial t^2}\partial u_1^i + \frac{\partial^2 u_{03}^i}{\partial t^2}\partial u_{03}^i\right) + \rho_i\frac{h_i^3}{12}\left(\frac{\partial^2\varphi_1^i}{\partial t^2}\partial\varphi_1^i\right)\right]dz_i\right\}dS_i \tag{19.13}$$

$$\delta K^t = \int_{S_t}\left[\int_{-h_t/2}^{h_t/2}\rho_t h_t\left(\frac{\partial^2 u_0^t}{\partial t^2}\partial u_0^t + \frac{\partial^2 u_1^t}{\partial t^2}\partial u_1^t + \frac{h_t^2}{12}\frac{\partial^2 u_{03}^t}{\partial t^2}\partial u_{03}^t\right)dz_t\right]dS_t \tag{19.14}$$

$$\delta K^j = \int_{L_j}\left[\rho_j F^j\left(\frac{\partial^2 u_1^j}{\partial t^2}\partial u_1^j + \frac{\partial^2 u_3^j}{\partial t^2}\partial u_3^j\right) + \rho_j\left(I_{kr}^j\frac{\partial^2\varphi_1^j}{\partial t^2}\partial\varphi_1^j\right)\right]dL_j. \tag{19.15}$$

In (19.10)–(19.15), quantities F^j and I_{kr}^j correspond to geometric characteristics of the transverse reinforcing ribs, ρ_j is the specific weight of reinforcing rib material, and ρ_j $(j = 1, 2)$ and ρ_t are the specific weights of the load-bearing layer and light aggregate materials, respectively. An internal axisymmetrically distributed non-stationary normal load $P_1(s, t)$ (s and t are spatial and temporal coordinates) is applied to an asymmetric three-layer spherical structure. It should be noted that when calculating potential and kinetic energy for light aggregate in the expressions δP^t and δK^t, integrals are evaluated over a volume that is increased by the volume of reinforcing ribs. However, it does not practically affect the general error of the shells theory since the volume of reinforcing ribs in the volume of light aggregate for three-layer shells of rotation is less than 5 %.

For a three-layer asymmetric spherical shell with a light aggregate reinforced with discrete ribs under an axisymmetric impulse load, after the standard transformations in the variational Eq. (19.8), taking into account (19.10)–(19.15), we obtain a ninth-order system of hyperbolic equations of motion and the corresponding boundary and initial conditions:

$$\frac{1}{\sin\alpha}\frac{\partial}{\partial s_i}(\sin\alpha T_{11}^i) - \frac{1}{R_i}(\cot\alpha T_{22}^i - T_{13}^i) - \frac{1}{R_t}T_{13}^i = \rho_i h_i\frac{\partial^2 u_1^i}{\partial t^2},$$

$$\frac{1}{\sin\alpha}\frac{\partial}{\partial s_i}(\sin\alpha T_{13}^i) - \frac{1}{R_i}(T_{11}^i + T_{22}^i) - \frac{1}{R_t h_t}(M_{11}^t + M_{22}^t) + P_2 = \rho_i h_i\frac{\partial^2 u_3^i}{\partial t^2},$$

$$
\begin{aligned}
\frac{1}{\sin\alpha}\frac{\partial}{\partial s_i}\left(\sin\alpha M_{11}^i\right)-\frac{\cot\alpha}{R_i}M_{22}^i-T_{13}^i&=\frac{\rho_i h_i^3}{12}\frac{\partial^2\varphi_1^i}{\partial t^2},\\
\frac{1}{\sin\alpha}\frac{\partial}{\partial s_t}\left(\sin\alpha T_{11}^t\right)+\frac{1}{R_t}\left(T_{11}^t-T_{13}^t\right)+\frac{8}{h_t^2}M_{13}^t&=\rho_t h_t\frac{\partial^2 u_0^t}{\partial t^2},\\
\frac{1}{\sin\alpha}\frac{\partial}{\partial s_t}\left(\sin\alpha M_{11}^t\right)-T_{13}^t+\frac{1}{R_t}M_{13}^t&=\rho_t h_t\frac{\partial^2 u_1^t}{\partial t^2},\\
\frac{1}{\sin\alpha}\frac{\partial}{\partial s_t}\left(\sin\alpha T_{13}^t\right)-\frac{1}{R_t}\left(T_{11}^t+T_{22}^t\right)&=\rho_t h_t\frac{\partial^2 u_{03}^t}{\partial t^2},\\
\left[T_{11}^{i\pm}\right]_j=\rho_j F_j\frac{\partial^2 u_{1j}}{\partial t^2},\quad \left[T_{13}^{i\pm}\right]_j&=\rho_j F_j\frac{\partial^2 u_{3j}}{\partial t^2},\\
\left[M_{11}^{i\pm}\right]_j&=\rho_j I_{krj}\frac{\partial^2\varphi_{1j}}{\partial t^2},
\end{aligned}
\tag{19.16}
$$

$(i=1,2)$. The oscillations of asymmetric three-layer elastic structures are described by two ninth-order systems of hyperbolic equations, which take into account the discontinuity coefficients of the load-bearing layers-reinforcing elements and the difference of material of the load-bearing layers. On the lines of breaks in the oscillation Eq. (19.16), $[T_{11}^{i\pm}]_j$, $[T_{13}^{i\pm}]_j$ and $[M_{11}^{i\pm}]_j$ correspond to the forces–moments acting on the jth discrete element from the load-bearing layers.

The relations between the magnitudes of forces–moments and the corresponding magnitudes of deformations for load-bearing layers and reinforcing ribs have the form:

$$
\begin{aligned}
&T_{11}^i=\frac{E_i h_i}{1-\nu_i^2}\left(\varepsilon_{11}^i+\nu_i\varepsilon_{22}^i\right),\quad T_{22}^i=\frac{E_i h_i}{1-\nu_i^2}\left(\varepsilon_{22}^i+\nu_i\varepsilon_{11}^i\right),\\
&T_{13}^i=k^2G_{13}^i\varepsilon_{13}^i,\\
&M_{11}^i=\frac{E_i h_i^3}{12(1-\nu_i^2)}\left(\kappa_{11}^i+\nu_i\kappa_{22}^i\right),\quad M_{22}^i=\frac{E_i h_i^3}{12(1-\nu_i^2)}\left(\kappa_{22}^i+\nu_i\kappa_{11}^i\right),\\
&T_{22j}=E_jF_j\varepsilon_{22j},
\end{aligned}
\tag{19.17}
$$

$(i=1,2)$; E_j, G_{13}^i, and ν_i are mechanical parameters of the material of the load-bearing layers; k^2 is the integral transverse shear coefficient of the theory of plates and shells; and F_j are the cross-sectional area of the jth rib.

Forces and moments for light aggregate can be determined as follows:

$$
\begin{aligned}
&T_{11}^t=\int_{-h_t/2}^{h_t/2}\left(1+\frac{z_t}{R_t}\right)\sigma_{11}^t dz_t,\quad T_{13}^t=\int_{-h_t/2}^{h_t/2}\sigma_{13}^t dz_t,\\
&M_{11}^t=\int_{-h_t/2}^{h_t/2}z_t\left(1+\frac{z_t}{R_t}\right)\sigma_{11}^t dz_t.
\end{aligned}
\tag{19.18}
$$

The equations of oscillations (19.16) are supplemented by the corresponding boundary and initial conditions.

19.3 Numerical Results

The problem of dynamic deformation of an asymmetric three-layer spherical dome is considered. We solve the problem on the spatial interval $[-\pi/2, \pi/2]$. Due to the symmetry about the top of the sphere at $\alpha = 0$, the interval $[0, \pi/2]$ is considered. To reveal some effects, we write down the boundary conditions at $\alpha = 0$ in the following form:

$$u_1^k = \varphi_1^k = 0, \quad 2\frac{\partial T_{13}^k}{\partial s} - P_2(s_0, t)\delta_{2k} = \rho h \frac{\partial^2 u_3^k}{\partial t^2}, \tag{19.19}$$

and besides, at $\alpha = \pi/2$, we have rigid clamping:

$$u_1^k = u_3^k = \varphi_1^k = 0, \quad k = 1, 2. \tag{19.20}$$

In (19.19), δ_{2k} is the Kronecker symbol. The external load is as follows:

$$P_2(s, t) = \begin{cases} A(1 - t/T), & 0 \leqslant t \leqslant T \\ 0, & t > T, \end{cases} \tag{19.21}$$

$A = 10^6$ Pa, $T = R_1/c = 5 \cdot 10^{-5}$ s, $c^2 = E_1/[\rho(1 - \mu_1^2)]$. The initial conditions are zero for the load-bearing layers at $t = 0$:

$$u_1^i = u_3^i = \varphi_1^i = 0, \quad \frac{\partial u_1^i}{\partial t} = \frac{\partial u_3^i}{\partial t} = \frac{\partial \varphi_1^i}{\partial t} = 0 \quad (i = 1, 2). \tag{19.22}$$

The discretely symmetric reinforcing edges (parallels) are located at the points $\alpha_k = [41 + 40(k - 1)]\Delta\alpha_1, k = 1, \ldots, 5$.

The corresponding initial boundary value problem (19.16), (19.19)–(19.22) is solved using the finite element method. We use the dependence between the potential energy of deformations in a solid and the potential energy of tractions:

$$P = E - W, \tag{19.23}$$

where E is the potential energy of deformations and W is the potential energy of applied forces. After dividing the entire region into elements, (19.23) takes the form

$$P = \sum_{e=1}^{E} \left(E^{(e)} - W^{(e)}\right) = \sum_{e=1}^{E} \pi^{(e)}. \tag{19.24}$$

Global stiffness matrix and global residual vector in the matrix equation

$$[K]\{U\} = \{F\} \tag{19.25}$$

can be defined in the following way:

$$[K] = \sum_{e=1}^{E} [k^{(e)}], \quad F = -\sum_{e=1}^{E} \{f^{(e)}\}. \tag{19.26}$$

For the numerical solutions, we use the following values for geometric, physical, and mechanical parameters: $E_1^1 = E_j = 7 \cdot 10^{10}$ Pa, $\nu_j = \nu_1^1 = 0.3$, $\rho_1 = \rho_j = 2.1 \cdot 10^3$ kg/m^3, $h_1 = h_2 = 0.005$ m, $h_j = 2h_2$, $R_1 = 0.3$ m, $R_2 = 0.302$ m, $F_j = 4 \cdot 10^{-5}$ m^2, $E_2^2 = 2 \cdot 10^{11}$ Pa, $\rho_2 = 7.8 \cdot 10^3$ kg/m^3, $\nu_2^2 = 0.3$, $(E_1^1 + E_2^2)/2E_t = 50$, $\rho_t = 25$ kg/m^3, $\nu_t = 0.33$, and $h/R = 0.075$. The outer and inner load-bearing layers are steel and AMG6 alloy, respectively.

To study the dynamics of the asymmetric three-layer spherical shell with a discrete-symmetric light aggregate reinforced with ribs, we created the finite-element model (Fig. 19.1), which included 43 200 finite elements and 52 680 nodes (A hemisphere with five edges).

In Fig. 19.1a, the XZ cross-sectional view of the finite element model for the foam is shown. Next, in Fig. 19.1b, c, d, the models of the internal load-bearing layer, the five ribs, and the outer load-bearing layer, respectively, are presented. Finally, in Fig. 19.1e, a finite-element model of the three-layer shell with five ribs and foam is shown. All the figures present the XZ section. Following (Rychkov 2013), three-dimensional finite elements are used to guarantee the accuracy and reliability of the obtained results.

We can obtain the stress state of a three-layer elastic structure of a spherical type at any moment of time (we found the solution for $0 \leqslant t \leqslant 40T$). In Fig. 19.2, the dependences of normal deflections in the middle surfaces of the load-bearing layers u_3 on the angular coordinate α are given in the reduced coordinates. The curves with indexes 1 and 2 correspond to the internal and external load-bearing layers, respectively, and are obtained at the moment of time $t = 5.85T$ (time of reaching the maximum of (1) and (2)). There is no light aggregate. Based on the presented results, it is possible to evaluate the influence of the sphericity of the structure on the antisymmetry of (1) and (2) along the spatial coordinate. The first natural frequency of the design is 1440 Hz.

In Fig. 19.3, the dependence of normal stresses in the middle surfaces of the load-bearing layers σ_{22}^1 (1) and σ_{22}^2 (2) on the coordinate α are shown. The light aggregate is absent. Curve 1 corresponds to the stress σ_{22}^1 in the inner load-bearing layer of the spherical shell and 2—σ_{22}^2 in the outer load-bearing layer (both curves are obtained for $t = 5.85T$). The first natural frequency of the design is 1440 Hz.

In Fig. 19.4, the dependence of normal deflections u_3^1 (1) and u_3^2 (2) in the middle surfaces of the load-bearing layers on the angular coordinate α are given. Curve 1 corresponds to the deflection u_3^1 in the internal load-bearing layer of the spherical shell, and 2—the deflection u_3^2 of the external load-bearing layer (both curves correspond to the moment of time $t = 9.55T$). The ratio of the modulus of elasticity of the load-bearing layers to the modulus of elasticity of the light aggregate $(E_1^1 + E_2^2)/2E_t = 50$. The results show the influence of the sphericity of the structure and the light aggregate on the antisymmetry of the distribution of u_3^1 (1) and u_3^2 (2) along the angular coordinate. The first natural frequency of the design is 1576 Hz.

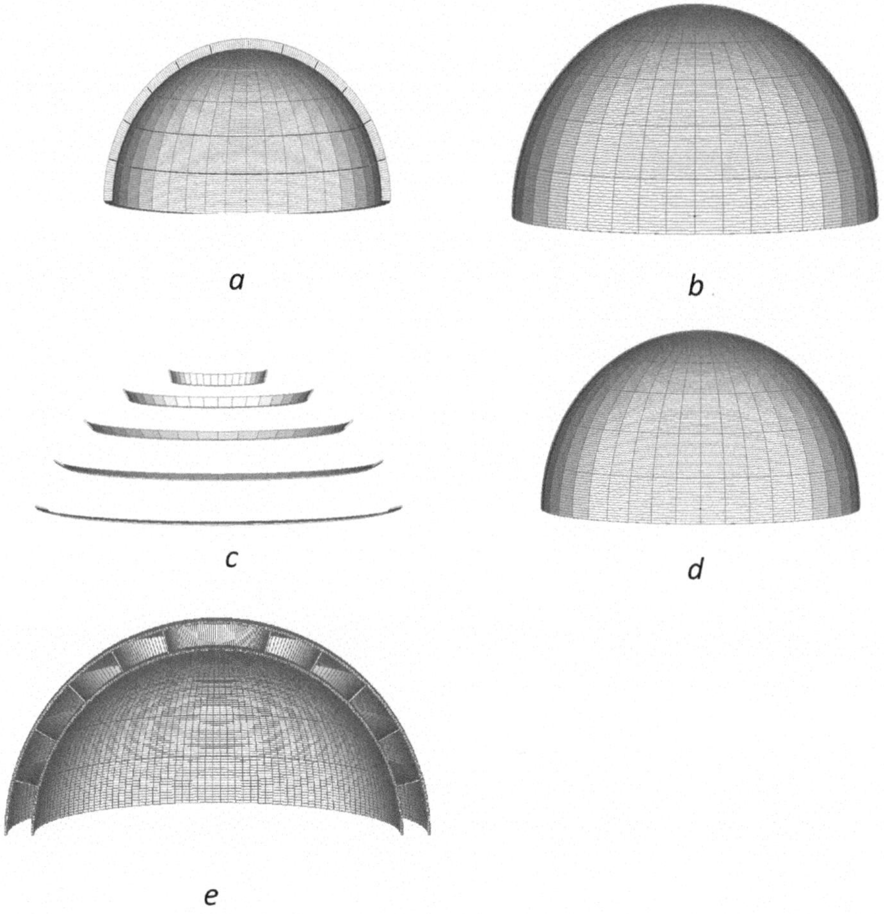

Fig. 19.1 Components of the finite-element model of a three-layer spherical shell with five ribs and foam

In Fig. 19.5, the dependence of the normal stresses in the middle surfaces of the load-bearing layers σ_{22}^1 (1) and σ_{22}^2 (2) on the coordinate are shown. The ratio of the modulus of elasticity of the load-bearing layers to the modulus of elasticity of the light aggregate $(E_1^1 + E_2^2)/2E_t = 50$. Curve 1 corresponds to the stress σ_{22}^1 in the internal load-bearing layer of the spherical shell, and 2—the stress σ_{22}^2 in the external load-bearing layer (both curves are obtained for the moment of time $t = 9.55T$). The first natural frequency of the design is 1576 Hz.

Comparing the graphs in Figs. 19.2 and 19.5, we can conclude that light aggregate significantly affects the nature of vibrations of the internal load-bearing layer. Its frontal point changes the value of the normal deflection to the opposite, and the first natural frequency of the structure changes by 9.5 %. In addition, both graphs demon-

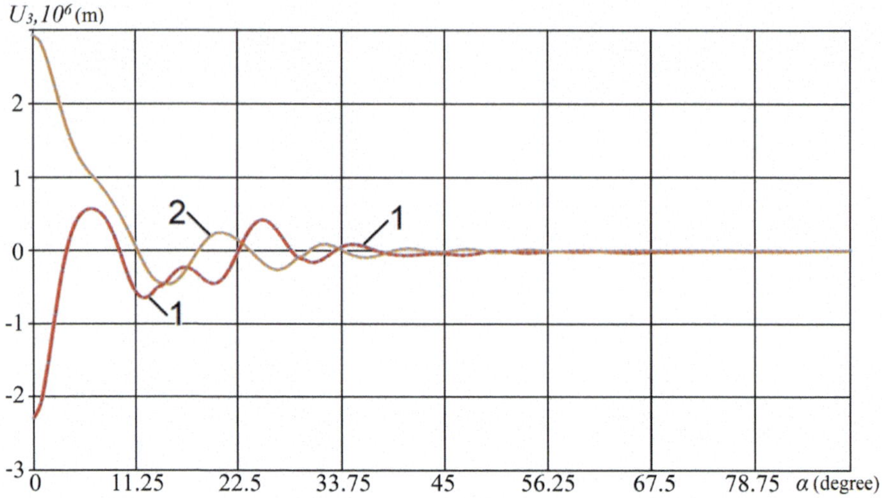

Fig. 19.2 Dependencies of the maximum deflections u_3 on the angle coordinates α at $t = 5.85T$ without light aggregate

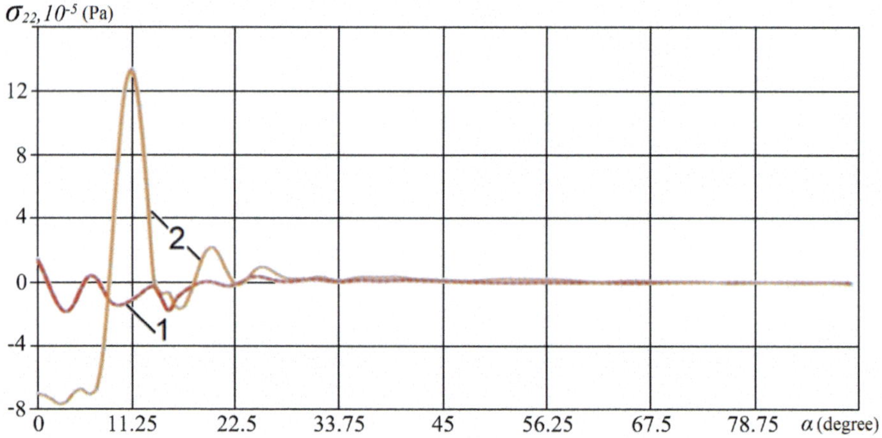

Fig. 19.3 Dependencies of the maximum stresses σ_{22} on the angle coordinates α at $t = 5.85T$ without light filler

strate the ability of such asymmetric three-layer structures to dampen disturbances from local shock loads.

It should also be noted that the compatibility conditions (see (19.6) and (19.7)), which provide an ideal connection between the aggregate and the load-bearing layers without detachment and slippage, cause the phenomenon of structural damping. It can be explained by the fact that inertial terms of the light aggregate are added to (19.16) when deriving the equations of oscillations with accounting for the continuity of movements (19.16). We can also conclude that a decrease in the thickness of the

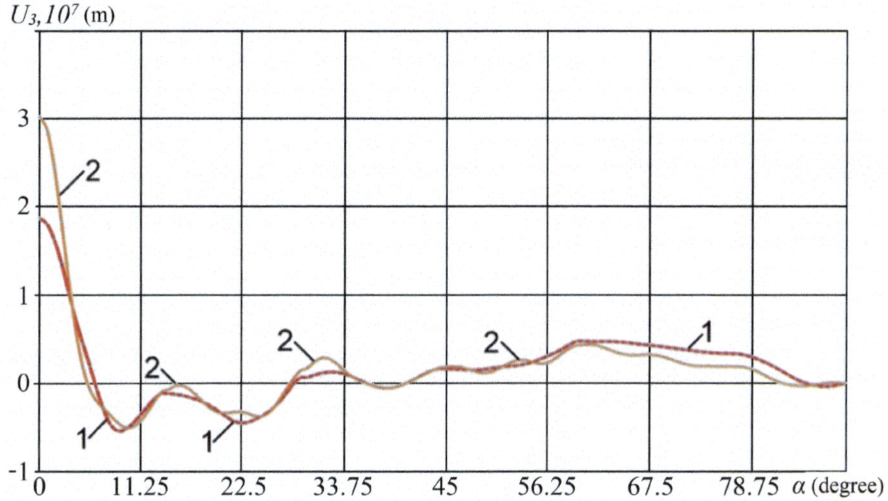

Fig. 19.4 Dependencies of the maximum u_3 deflections on the angle coordinates α at $t = 9.55T$ at $(E_1^1 + E_2^2)/2E_t = 50$

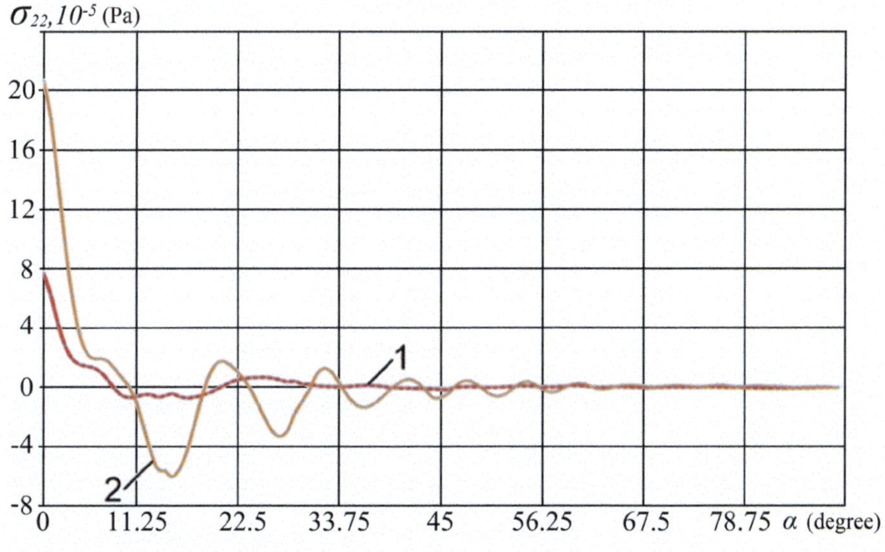

Fig. 19.5 Dependencies of the maximum stresses σ_{22} on the angle coordinates α at $t = 9.55T$ and $(E_1^1 + E_2^2)/2E_t = 50$

load-bearing layers and an increase in the thickness of the light aggregate leads to a significant increase in the spectrum of natural frequencies of the structure.

References

Frostig Y, Thomsen OT (2004) Higher-order free vibration of sandwich panels with a flexible core. Int J Solids Struct 41:1697–1724
Hause T, Librescu L (1998) Postbucling of anisotropic flat and doubly-curved sandwich panels under complex loading conditions. Int J Solids Struct 35(23):3007–3027
Hause T, Librescu L (2007) Dynamic response of doubly-curved anisotropic sandwich panels impacted by blast loadings. Int J Solids Struct 44(20):6678–6700
Hohe J, Librescu L (2003) Nonlinear theory for double-curved anisotropic sandwich shells with cross-compressed cores. Int J Solids Struct 40(5):1059–1088
Kheirikhah MM, Khalili SMR, Malekzadeh Fard K (2011) Biaxial buckling analysis of soft-core composite sandwich plates using improved high-order theory. Eur J Mech A/Solids 31(1):54–66
Lee YS, Lee KD (1997) Dynamic response of laminated circular cylindrical shells under impulse loads. Comput Struct 63(1):149–157
Li R, Kardomateas GA, Simitses GJ (2008) Nonlinear response of a shallow sandwich shell with compressible core to blast loading. J Appl Mech 75:061023–1
Librescu L, Hause T (2000) Recent developments in the modeling and behavior of advanced sandwich constructions: a survey. Compos Struct 48:1–7
Lugovoi PZ, Meish VF, Meish YA, Orlenko SP (2020) Dynamic design of compound shell structures of revolution under non-stationary loads. Int Appl Mech 56:22–32
Lugovoi PZ, Shugailo OP, Orlenko SP, Diemienkov VM (2020) Oscillation of thermal insulation three-layer cylindrical pipes under operating loads. Yaderna ta Radyiatsyijna Bezpeka 3(87):55–61
Lugovoi PZ, Meish VF, Orlenko SP (2020) Numerical simulation of the dynamics of spherical sandwich shells reinforced with discrete ribs under a shockwave. Int Appl Mech 56:590–598
Lugovoi PZ, Shugailo OP, Orlenko SP, Diemienkov VM (2020) Oscillation of thermal insulation three-layer cylindrical pipes under operating loads. Yaderna ta Radyiatsyijna Bezpeka 3(87):55–61
Lugovyi PZ, Orlenko SP (2021) Effect of the asymmetry of cylindrical sandwich shells on their stress-strain state under non-stationary loading. Int Appl Mech 57:543–553
Lugovyi PZ, Gaidaichuk VV, Skosarenko YV, Kotenko KE (2021) Stress-strain state of three-layer cylindrical shells with reinforced light core under non-stationary loading. Int Appl Mech 57:395–404
Lugovyy PZ, Skosarenko YV, Orlenko SP (2019) Application of the method of spline-collocations to the solution of dynamic and static problems for structurally inhomogeneous cylindrical shells. Dopov Nac akad nauk Ukr 8:25–33
Malekzadeh FK, Livani M, Veisi A, Gholami M (2014) Improved high-order bending analysis of double curved sandwich panels subjected to multiple loading conditions. Lat Am J Solids Struct 11:1591–1614
Meish VF, Shtantsel SE (2002) Dynamic problems in the theory of sandwich shells of revolution with a discrete core under non-stationary loads. Int Appl Mech 38:1501–1507
Meish VF, Shtantsel SE (2002) Dynamic problems in the theory of sandwich shells of revolution with a discrete core under non-stationary loads. Int Appl Mech 38:1501–1507
Orlenko SP (2020) Numerical simulation of the dynamics of a three-layer spherical shell with a discretely inhomogeneous filler. Dopov Nac akad nauk Ukr 3:19–27 [In Ukrainian]
Piskunov VG, Rasskazov AO (2002) Evolution of the theory of laminated plates and shells. Int Appl Mech 38:135–166

Rychkov SP (2013) Modeling of structures in the Femap with NX Nastran environment. DMK Press, Moscow

Shul'ga NA, Meish VF (2003) Forced vibration of three-layered spherical and ellipsoidal shells under axisymmetric loads. Mech Comp Mater 39:439–446

Surianinov M, Yemelianova T, Lazarieva D (2019) Investigation of free vibrations of threelayered cylindrical shell supported by transverse ribs. Int J Eng Technol 11(1):61–66

Timoshenko S, Woinowsky-Krieger S (1959) Theory of plates and shells, 2nd edn. McGraw-Hill

Chapter 20
Analytical and Numerical Solution of Static Problems of Non-Circular Cylindrical Shells

Volodymyr Maksymyuk, Evgen Storozhuk, and Ivan Chernyshenko

20.1 Introduction

Cylindrical shells of non-circular cross-sections (elliptical, oval, supercircular, and parabolic) are widely used in modern engineering. In some cases, they are stronger, more stable, and lighter (Soldatos 1999). For example, elliptical hollow profiles combine in construction the advantages of round and rectangular profiles (Chan et al. 2010). In aviation technology, the fuselage (Boulle et al. 2015) and hermetic cabin (Zheleznov et al. 2006) of modern aircraft often have a non-circular cross-section for reasons of aerodynamics, the layout of technological connections, and the efficiency of using the internal volume.

In short shells, the ends are fixed in a certain way, so their stress–strain state (SSS) is two-dimensional. On the other hand, in long shells, the SSS in any cross-section can be assumed to be the same, so the boundary value problem is one-dimensional. At first glance, two-dimensional problems should be more complicated than one-dimensional ones. However, this is true for analytical methods, and for numerical methods, it depends on the characteristics of the deformation of the shells and the effectiveness of the methods.

Analytical (exact) solutions of problems for long cylindrical shells of non-circular cross-sections have a double significance: they are important both from an engineering point of view and are tests for numerical methods. Moreover, they are significant for mechanics problems, in which locking phenomena are considered (Maksimyuk et al. 2012). More precisely, it is a problem of the deformation of a long cylindrical

V. Maksymyuk (✉) · E. Storozhuk · I. Chernyshenko
S.P. Timoshenko Institute of Mechanics, National Academy of Science of Ukraine, Kyiv, Ukraine
e-mail: desc@inmech.kiev.ua

E. Storozhuk
e-mail: stevan@ukr.net

I. Chernyshenko
e-mail: desc@inmech.kiev.ua

A. N. Guz et al. (eds.), *Advances in Mechanics*, Advanced Structured Materials 191,
https://doi.org/10.1007/978-3-031-37313-8_20

shell with an elliptical cross-section under internal pressure due to large bends under small tension.

This one-dimensional problem has a 150-year history (Bresse 1866, p. 326–338). However, the first analytical solutions were oversimplified. For a shell (actually, a ring) under internal pressure, the solution mentioned above (Bresse 1866) gave moments and displacements in two sections of symmetry planes that have different signs (which is correct for high ellipticity) but equal in absolute value (which is a simplification).

Correct results, apparently for the first time (1930), were obtained by Timoshenko and published in the first edition of the monograph (Timoshenko 1941) (second edition). The moments were calculated using a table of some coefficients for several values of the ellipse aspect ratio. However, the procedure for obtaining these coefficients was not specified and remained unknown. Further, in 1965, Roark (Young and Budynas 2002) presented extended similar tables and gave formulas for calculating corrections in the case of a non-uniform moment of inertia of the ring (of variable thickness). The subsequent history of the development of analytical and graphical methods is briefly described in Holland (1976). In fact, this problem and Timoshenko's table (Timoshenko 1941) in particular became a reliable test for further research.

In the publications on this problem, which appeared after the works mentioned above, SSS (Grigorenkoc et al. 1999; Kiseleva et al. 2015; Romano and Ramlet 1967), stability (Chen and Kempner 1976; Sheinman and Firer 1994; Tennyson et al. 1971), and oscillations (Meish and Kepenach 2014; Tornabene et al. 2015; Yamada et al. 1984) of oval and elliptical cylindrical shells under the action of various types of loads were investigated with the help of numerical, analytical (approximate) and experimental methods. However, only in Grigorenko and Kharitonova (2007), using the Kirchhoff–Love hypotheses and relations of the theory of shallow shells (in a simplified classical formulation), exact analytical solutions of boundary value problems were obtained for an open long non-circular cylindrical shell, the curvature of which varies according to the quadratic law.

Therefore, in view of the above, numerical methods are used in this chapter to study the SSS of physically nonlinear thin non-circular cylindrical shells made of composite materials. The methods are aimed at overcoming possible membrane locking.

Below, the formulation of linear and physically nonlinear problems of statics for infinitely long non-circular cylindrical shells is given, analytical (exact), numerical, and analytical–numerical methods for their solution are developed, and specific numerical results are obtained for shells of oval, elliptical, and supercircular cross-sections under the action of uniform normal pressure.

20.2 Boundary Value Problem Statement, Basic Linear Relations, and General Solution for a Long Non-Circular Cylindrical Shell

Let us consider an infinitely long cylindrical shell of arbitrary cross-section, made of transversely isotropic material and loaded with surface and linear forces. Assume that the directrix of the shell cross-section changes smoothly. The thickness h of the shell along the generatrix is constant and continuously changes along the directrix. The load acting on the shell is distributed smoothly along the generatrix. It follows from the above that the displacements, strains, and stresses in each cross-section of the shell are the same, and all the searched values will change only along the directrix.

The geometry of the middle surface of the shell is set in the global Cartesian coordinate system (X, Y, Z), the OX-axis of which is directed along the axis of the cylinder (Fig. 20.1).

We refer the shell to the curvilinear orthogonal coordinate system (x, φ, ζ) aligned with the lines of principal curvatures of the shell. Here, x and ζ are the lengths of the generatrix and the normal to the mid-surface, and φ is the angle between the normal to the mid-surface and the vertical (OZ) axis.

To study the SSS of this class of shells, we use the equations of the refined theory of deep shells, which take into account transverse shear deformations (Grigorenko et al. 2010). In this case, we have the following expressions for geometric relations:

$$\varepsilon = \frac{du}{rd\varphi} + \frac{w}{r}, \quad \gamma = \vartheta + \frac{dw}{rd\varphi} - \frac{u}{r}, \quad \mu = \frac{d\vartheta}{rd\varphi}, \tag{20.1}$$

where ε, γ, and μ are the strain components of the shell; u and w are the tangential displacement and bending of the shell mid-surface, respectively; ϑ is the rotation angle of the normal; r is the cross-sectional radius curvature.

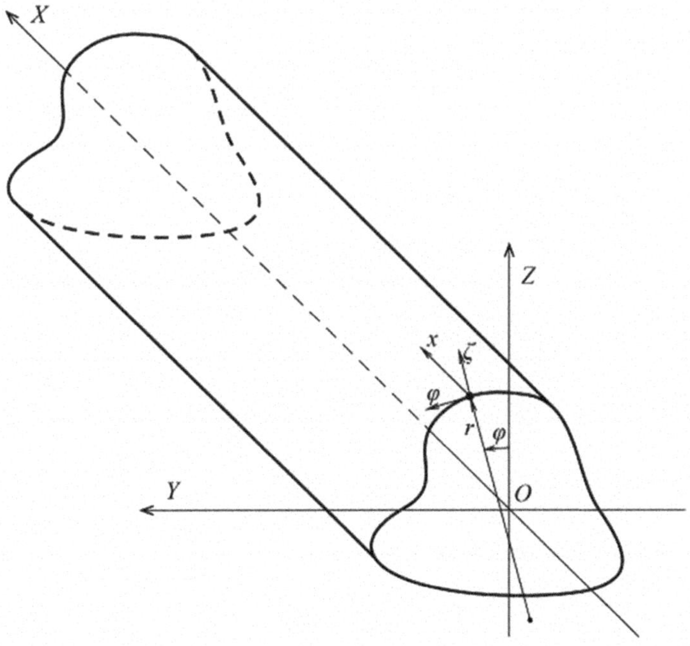

Fig. 20.1 Long cylindrical shell of non-circular section

The relationship between internal forces/moment and strain components is written based on Hooke's law:

$$N = D_N \varepsilon, \quad D_N = \frac{Eh}{1-\nu^2}, \quad Q = D_Q \gamma, \quad D_Q = kG_{\varphi\zeta} h,$$
$$M = D_M \mu, \quad D_M = \frac{Eh^3}{12(1-\nu^2)}. \tag{20.2}$$

In (20.2), N and Q are the tangential and shearing forces, respectively; M is the bending moment; D_N, D_M, and D_Q are the stiffness characteristics of the shell; E and ν are the elastic modulus and Poisson's ratio in the isotropy plane, respectively; $G_{\varphi\zeta}$ is the shear modulus in the cross-sectional plane; k is a coefficient depending on how the shear is distributed across the thickness.

The equilibrium equations read

$$\frac{dN}{rd\varphi} + \frac{1}{r} Q + q_\varphi = 0, \quad \frac{dQ}{rd\varphi} - \frac{1}{r} N + q_\zeta = 0, \quad \frac{dM}{rd\varphi} - Q = 0, \tag{20.3}$$

where q_φ and q_ζ are the components of the surface load.

In solving specific problems, the equilibrium equations (20.3), geometric and physical relations (20.1) and (20.2) must be accompanied by appropriate boundary conditions.

Performing successive integration, we obtain a general solution to the given problem

$$Q = C_1 \cos\varphi + C_2 \sin\varphi + \cos\varphi \int_0^\varphi q_1 dx - \sin\varphi \int_0^\varphi q_2 dx + (rq_\zeta)\,|_0 \sin\varphi,$$

$$N = C_1 \sin\varphi + C_2 \cos\varphi - \sin\varphi \int_0^\varphi q_1 dx - \cos\varphi \int_0^\varphi q_2 dx + (rq_\zeta)\,|_0 \cos\varphi, \tag{20.4}$$

$$M = \int_0^\varphi rQdx + C_3 = M^* + C_3,$$

$$\vartheta = \int_0^\varphi r\mu dx + C_4 = \int_0^\varphi \frac{rM^* dx}{D_M} + C_3 \int_0^\varphi \frac{rdx}{D_M} + C_4,$$

$$u = C_5 \cos\varphi + C_6 \sin\varphi + \cos\varphi \int_0^\varphi f_1 dx + \sin\varphi \int_0^\varphi f_2 dx - r\varepsilon|_0 \sin\varphi,$$

$$w = C_5 \sin\varphi - C_6 \cos\varphi + \sin\varphi \int_0^\varphi f_1 dx - \cos\varphi \int_0^\varphi f_2 dx + r\varepsilon|_0 \cos\varphi.$$

Here, we introduced

$$q_1 = r(q_\varphi \sin x - q_\zeta \cos x), \quad q_2 = r(q_\varphi \cos x + q_\zeta \sin x),$$
$$\varepsilon = \frac{N}{D_N}, \quad \mu = \frac{M}{D_M}, \quad \gamma = \frac{Q}{D_Q}, \quad M^* = \int_0^\varphi rQdx,$$
$$f_1 = r\left[\varepsilon \cos x - (\vartheta - \gamma)\sin x\right], \quad f_2 = r\left[\varepsilon \sin x + (\vartheta - \gamma)\cos x\right].$$

Solving particular problems, the integration constants $C_1, C_2, \ldots, C_6$ are determined from the boundary conditions.

20.3 Analytical (Exact) Solutions of Statics Problems for Non-Circular Cylindrical Shells

For two types (closed and open) of a long oval cylindrical shell of constant thickness h, we have managed to express the primitive integral functions in terms of elementary functions in formulas (20.4) and obtain analytical (exact) expressions for internal force factors and generalized displacements of the shell under the action of a static load.

20.3.1 Exact Solutions of Boundary Value Problems for a Closed Cylindrical Shell of Oval Cross-Section

A closed long cylindrical shell of the oval cross-section with a continuous radius of curvature. Let us find the exact solution to the boundary value problem for an infinitely long non-circular cylindrical shell (i.e., a pipe) closed along the directrix $(-\pi \leqslant \varphi \leqslant \pi)$ under uniform internal pressure of intensity q (Fig. 20.2).

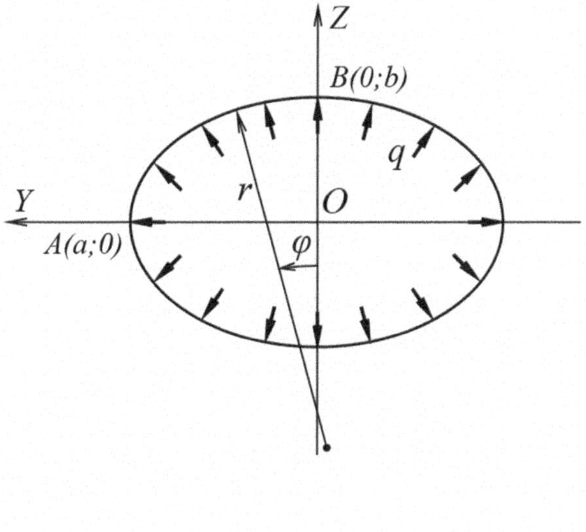

Fig. 20.2 Cross-section of a non-circular cylindrical shell closed along the directrix

Let the cross-section of the shell have the form of an oval with two mutually perpendicular axes of symmetry. The radius of curvature of the cross-section is a continuous function of the angle φ:

$$r = r_0\,(1 + \xi\cos 2\varphi)\,. \tag{20.5}$$

Here, a and b are the major and minor semi-axes of the cross-section;

$$r_0 = \frac{a+b}{2}, \quad \xi = 3\frac{a-b}{a+b}, \quad -\pi \leqslant \varphi \leqslant \pi.$$

Considering geometrical and mechanical symmetries, we solve the problem for one-fourth of the shell ($0 \leqslant \varphi \leqslant \pi/2$) as a design model and specify symmetry conditions ($u = \vartheta = Q = 0$) at the points $\varphi = 0$ and $\varphi = \pi/2$.

The exact solution of the boundary value problem (20.1)–(20.3) for a closed infinitely long oval cylindrical shell under the action of uniform internal pressure, which satisfies the symmetry conditions at the points $\varphi = 0$ and $\varphi = \pi/2$, has the following form:

$$\begin{aligned}
Q &= -\frac{2qr_0\xi}{3}\sin 2\varphi, \quad N = r_0 q - \frac{qr_0\xi}{3}\cos 2\varphi,\\
M &= \frac{qr_0^2\xi}{3}\left(-\frac{\xi}{2} + \cos 2\varphi + \frac{\xi}{4}\cos 4\varphi\right),\\
\vartheta &= \frac{qr_0^3\xi}{6D_M}\left[\left(1 - \frac{3\xi^2}{8}\right)\sin 2\varphi + \frac{3\xi}{8}\sin 4\varphi + \frac{\xi^2}{24}\sin 6\varphi\right],\\
u &= A_2\sin 2\varphi + A_4\sin 4\varphi + A_6\sin 6\varphi + A_8\sin 8\varphi,\\
w &= \frac{r_0^2 q}{D_N}\left(1 - \frac{\xi^2}{6} + \frac{2\xi}{3}\cos 2\varphi - \frac{\xi^2}{6}\cos 4\varphi\right)\\
&\quad - 2A_2\cos 2\varphi - 4A_4\cos 4\varphi - 6A_6\cos 6\varphi - 8A_8\cos 8\varphi,
\end{aligned} \tag{20.6}$$

where

$$\begin{aligned}
A_2 &= \frac{4qr_0^2\xi}{9D_N} - \frac{qr_0^4\xi}{18D_M}\left(1 - \frac{\xi^2}{16}\right) - \frac{2qr_0^2\xi}{9D_Q}, \quad A_6 = -\frac{11qr_0^4\xi^3}{10080D_M},\\
A_4 &= -\frac{2qr_0^2\xi^2}{45D_N} - \frac{qr_0^4\xi^2}{60D_M}\left(\frac{7}{4} - \frac{\xi^2}{3}\right) - \frac{qr_0^2\xi^2}{45D_Q}, \quad A_8 = -\frac{qr_0^4\xi^4}{18144D_M}.
\end{aligned}$$

Closed oval cylindrical shell with a stepped radius of curvature. Consider an infinitely long cylindrical shell whose cross-section is formed by two pairs of smoothly joined circular arcs of radii r_1 (arcs I) and r_2 (arcs II) (Fig. 20.3), where α is the angle that defines the point of junction of the circular arcs.

Let us determine the SSS of the shell subject to uniform normal pressure q.

The shell consists of four circular cylindrical plates, two of which (plates I) deform symmetrically about the plane XOZ ($\varphi_1 = 0$), and the other two (plates II) about

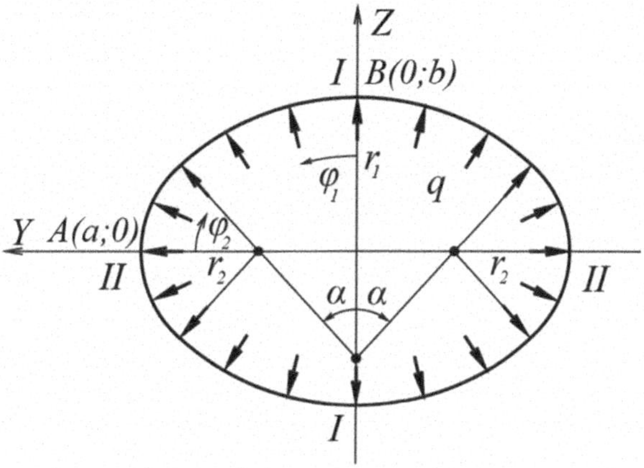

Fig. 20.3 Cross-section of an oval cylindrical shell with stepped radius of curvature

the plane XOY ($\varphi_2 = 0$). Therefore, it is sufficient to analyze only halves of two cylindrical plates (for example, plates I and II) owing to the symmetry conditions $u_1 = \vartheta_1 = Q_1 = 0$ at the point $\varphi_1 = 0$ and $u_2 = \vartheta_2 = Q_2 = 0$ at the point $\varphi_2 = 0$ and the following interface conditions at the point $\varphi_1 = \alpha$ ($\varphi_2 = \pi/2 - \alpha$):

$$\begin{aligned} &u_1(\alpha) = -u_2\left(\tfrac{\pi}{2} - \alpha\right), \quad w_1(\alpha) = w_2\left(\tfrac{\pi}{2} - \alpha\right), \quad \vartheta_1(\alpha) = -\vartheta_2\left(\tfrac{\pi}{2} - \alpha\right), \\ &N_1(\alpha) = N_2\left(\tfrac{\pi}{2} - \alpha\right), \quad Q_1(\alpha) = -Q_2\left(\tfrac{\pi}{2} - \alpha\right), \quad M_1(\alpha) = M_2\left(\tfrac{\pi}{2} - \alpha\right). \end{aligned} \tag{20.7}$$

Here and below, indices 1 and 2 refer to plates I and II, respectively.

The general solution of the boundary value problem (20.1)–(20.3) for a closed long cylindrical shell of oval cross-section with a stepped radius of curvature under uniform internal pressure has the following form:

$$\begin{aligned} &Q_1 = C_1 \cos\varphi_1 + C_2 \sin\varphi_1, \quad Q_2 = C_3 \cos\varphi_2 + C_4 \sin\varphi_2, \\ &N_1 = -C_1 \sin\varphi_1 + C_2 \cos\varphi_1 + r_1 q, \quad N_2 = -C_3 \sin\varphi_2 + C_4 \cos\varphi_2 + r_2 q, \\ &M_1 = -r_1 C_2 \cos\varphi_1 + C_5, \quad M_2 = -r_2 C_4 \cos\varphi_2 + C_6, \\ &\vartheta_1 = (r_1 C_5 \varphi_1 - r_1^2 C_2 \sin\varphi_1)/D_M + C_7, \\ &\vartheta_2 = (r_2 C_6 \varphi_2 - r_2^2 C_4 \sin\varphi_2)/D_M + C_8, \\ &u_1 = C_9 \cos\varphi_1 + C_{10} \sin\varphi_1 + L_1 \varphi_1 + B_1 \varphi_1 \cos\varphi_1, \\ &u_2 = C_{11} \cos\varphi_2 + C_{12} \sin\varphi_2 + L_2 \varphi_2 + B_2 \varphi_2 \cos\varphi_2, \\ &w_1 = r_1 N_1/D_N - C_{10} \cos\varphi_1 - L_1 - B_1(\cos\varphi_1 - \varphi_1 \sin\varphi_1), \\ &w_2 = r_2 N_2/D_N - C_{12} \cos\varphi_2 - L_2 - B_2(\cos\varphi_2 - \varphi_2 \sin\varphi_2). \end{aligned} \tag{20.8}$$

Here,

$$\begin{aligned} &L_1 = r_1^2 C_5/D_M, \; B_1 = r_1 C_2/2D_N + r_1 C_2/2D_Q + r_1^3 C_2/2D_M, \\ &L_2 = r_2^2 C_6/D_M, \; B_2 = r_2 C_4/2D_N + r_2 C_4/2D_Q + r_2^3 C_4/2D_M. \end{aligned}$$

Constants of integration $C_1, C_2, \ldots, C_{12}$ are determined from the symmetry conditions at the points $\varphi_1 = 0$ and $\varphi_2 = 0$, as well as the conjugation conditions at the point $\varphi_1 = \alpha$.

20.3.2 *Exact Solutions of Statics Problems for an Open Cylindrical Shell of Oval Cross-Section*

Let us study the deformation of the circumferentially ($-\alpha \leqslant \varphi \leqslant \alpha;\ 0 < \alpha \leqslant \pi$) open infinitely long cylindrical shell with oval cross-section described by (20.5) and clamped longitudinal edges ($\varphi = \pm\alpha$). The shell is loaded by a uniformly longitudinally distributed transverse force P applied at the vertex (Fig. 20.4).

The cross-section and the load are symmetric about the vertical plane XOZ. Therefore, we consider half of the shell ($0 \leqslant \varphi \leqslant \alpha$). The edge $\varphi = \alpha$ is clamped $u = w = \vartheta = 0$, and $u = \vartheta = 0$, $Q = P/2$ at $\varphi = 0$.

After integration in (20.4) and satisfying part of the boundary conditions ($\vartheta(0) = 0$, $Q(0) = P/2$, and $\vartheta(\alpha) = 0$), we obtain the following expressions for internal force factors and generalized displacements:

$$
\begin{aligned}
Q =& \frac{P}{2}\cos\varphi + C_2\sin\varphi, \quad N = -\frac{P}{2}\sin\varphi + C_2\cos\varphi,\\
M =& \frac{r_0P}{2}\left[\delta_1 + \left(1+\frac{\xi}{2}\right)\sin\varphi + \frac{\xi}{6}\sin 3\varphi\right]\\
&+ C_2r_0\left[\delta_2 + \left(\frac{\xi}{2}-1\right)\cos\varphi - \frac{\xi}{6}\cos 3\varphi\right],\\
\vartheta =& \frac{r_0^2P}{2D_M}\left[\delta + \delta_1\varphi + \left(\frac{\xi^2}{6}-1\right)\cos\varphi + \frac{\xi\delta_1}{2}\sin 2\varphi\right.\\
&\left. - \left(\frac{2\xi}{9}+\frac{\xi^2}{12}\right)\cos 3\varphi - \frac{\xi^2}{60}\cos 5\varphi\right] + \frac{C_2r_0^2}{D_M}\left[\delta_2\varphi + \left(\frac{\xi^2}{6}-1\right)\sin\varphi\right.\\
&\left. + \frac{\xi\delta_2}{2}\sin 2\varphi + \left(\frac{\xi^2}{12}-\frac{2\xi}{9}\right)\sin 3\varphi - \frac{\xi^2}{60}\sin 5\varphi\right],
\end{aligned}
$$

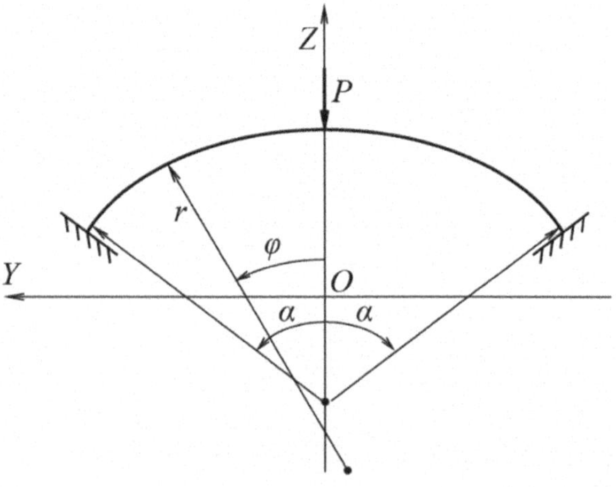

Fig. 20.4 A long circumferentially open cylindrical shell of an oval cross-section with clamped longitudinal edges

$$
\begin{aligned}
u =& C_5 \cos\varphi + C_6 \sin\varphi + A_0 + B_0\varphi + A_{11}\varphi \sin\varphi + B_{11}\varphi\cos\varphi \\
& + A_2 \sin 2\varphi + A_3 \sin 3\varphi + (B_2 + B_{21}\varphi)\cos 2\varphi + B_3 \cos 3\varphi \\
& + A_4 \sin 4\varphi + A_5 \sin 5\varphi + B_5 \cos 5\varphi + A_7 \sin 7\varphi + B_7 \cos 7\varphi, \\
w =& C_5 \sin\varphi - C_6 \cos\varphi + rN/D_N - B_0 - (A_{11} - B_{11}\varphi)\sin\varphi \\
& - (B_{11} + A_{11}\varphi)\cos\varphi - (2A_2 + B_{21})\cos 2\varphi + 2(B_2 + B_{21}\varphi)\sin 2\varphi \\
& - 3A_3 \cos 3\varphi + 3B_3 \sin 3\varphi - 4A_4 \cos 4\varphi - 5A_5 \cos 5\varphi \\
& + 5B_5 \sin 5\varphi - 7A_7 \cos 7\varphi + 7B_7 \sin 7\varphi,
\end{aligned}
\tag{20.9}
$$

where the coefficients δ, δ_1, δ_2, A_i, and B_j depend on ξ, r_0, α, D_Q, D_M, P, and C_2.

Note that the expressions for the coefficients A_i and B_j consist of two parts, one of which contains the transverse force P, and the other terms contain the unknown constant C_2.

Therefore, we first express the integration constants C_5 and C_6 in terms of the coefficients A_i and B_j using the boundary conditions $u(\alpha) = 0$ and $w(\alpha) = 0$.

After that, using (20.9), we calculate the internal forces, moment, rotation angle of the normal, tangential movement, and displacement of the shell.

20.3.3 Numerical Results and Their Analysis

We present the results of studying the influence of the aspect ratio of cross-section a/b and transverse shear deformations on the SSS of infinitely long closed oval cylindrical shells with a stepped radius of curvature under uniform internal pressure q (Storozhuk and Yatsura 2016). The shells have the following characteristics: $r_0/h = 10, k = 5/6$, $E = 38.4$ GPa, $G_{\varphi\zeta} = 0.2$ GPa, and $\nu = 0.1933$.

Tables 20.1 and 20.2 summarize the dimensionless deflection $\tilde{w} = wE_0/hq$ ($E_0 = 10$ MPa) and dimensionless stress $\tilde{\sigma} = \sigma/q$ ($\sigma = N/h + M\zeta/h^3$) on the outside ($\tilde{\zeta} = \zeta/h = 0.5$) and inside ($\tilde{\zeta} = -0.5$) surfaces of the shells at two points

Table 20.1 Values of deflections and stresses at the vertices of a closed oval cylindrical shell with a continuous radius of curvature

K	SSS	Point	$\tilde{\zeta}$	a/b					
				1.0	1.1	1.2	1.3	1.4	1.5
1,2	$\tilde{\sigma}$	B	0.5	10	37.075	59.917	79.301	95.833	110
			−0.5	10	−18.027	−41.736	−61.909	−79.167	−94
		A	0.5	10	−19.116	−47.355	−74.613	−100.833	−126
			−0.5	10	40.068	69.174	97.221	124.167	150
1	$\tilde{w}$	B	0.0	0.025	0.916	1.779	2.604	3.385	4.119
		A	0.0	0.025	−0.797	−1.478	−2.043	−2.511	−2.900
2	$\tilde{w}$	B	0.0	0.025	0.524	1.012	1.479	1.918	2.327
		A	0.0	0.025	−0.427	−0.791	−1.081	−1.311	−1.492

Table 20.2 Values of deflections and stresses at the vertices of a closed oval cylindrical shell with a stepped radius of curvature

K	SSS	Point	$\tilde{\xi}$	a/b					
				1.0	1.1	1.2	1.3	1.4	1.5
1,2	$\tilde{\sigma}$	B	0.5	10	37.046	59.814	79.087	95.485	109.5
			−0.5	10	−17.999	−41.632	−61.696	−78.818	−93.5
		A	0.5	10	−19.144	−47.459	−74.826	−101.182	−126.5
			−0.5	10	40.097	69.277	97.435	124.515	150.5
1	$\tilde{w}$	B	0.0	0.025	0.931	1.810	2.652	3.450	4.199
		A	0.0	0.025	−0.807	−1.494	−2.061	−2.528	−2.914
2	$\tilde{w}$	B	0.0	0.025	0.536	1.035	1.513	1.962	2.378
		A	0.0	0.025	−0.436	−0.807	−1.102	−1.335	−1.517

of the cross-section (vertex and covertex). These results are obtained for (a/b =1.0, 1.1, 1.2, 1.3, 1.4, 1.5) using the Timoshenko ($K = 1$) and Kirchhoff–Love ($K = 2$) models.

The results analysis reveals that the deflection is maximum at the covertex (point B). Allowing for the transverse-shear strains ($K = 1$) leads to an increase in the maximum deflection compared with that predicted by the Kirchhoff–Love model ($K = 2$). For $a/b = 1.0, 1.1, 1.2, 1.3, 1.4, 1.5$, the increase in the maximum deflection is 0, 75, 76, 76, 76, 77 (%), respectively, for the closed shell with a continuous radius of curvature, and 0, 74, 75, 75, 76, 77 (%) for the closed shell with a stepped radius of curvature.

The stresses are maximum at the vertex A (the end of the major axis corresponding to the inside surface of the closed shells). The solution of the boundary value problems shows that the stresses obtained with the involvement of both models ($K = 1, 2$) for closed shells are equal.

The presented results show that the deflections and stresses in the closed shells with a constant and stepped radius of curvature differ insignificantly. For example, for both models ($K = 1, 2$), the difference between the maximum deflections does not exceed 2.3%, and the difference between the maximum stresses is less than 0.33%.

The deflections at the vertex and covertex (points B and A) of the closed oval induced by the internal pressure are of opposite signs. This conclusion agrees with the results reported in Bresse (1866); Novozhilov (1964); Timoshenko (1941) for an ellipse and an oval with a continuous radius of curvature and qualitatively contradicts the statement made in Novozhilov (1964) on an oval with a stepped radius of curvature.

When the cross-section shape deviates from circular (the ratio a/b increases), the maximum values of all components of the SSS of closed shells increase significantly. For example, taking into account the transverse shear strains in the closed shell with $a/b = 1.5$ (Table 20.2) increases the maximum stress by a factor of 15 and the maximum deflection by a factor of 164.

20.4 Analytical–Numerical Approach to Solving Boundary-Value Problems for Cylindrical Shells of Non-Circular Cross-Section Under the Action of Static Loads

Since the integrands in expressions for internal forces, bending moment, rotation angle of the normal, tangential displacement, and shell deflection (20.4) are continuous, the integrals in these formulas exist, but the primitive functions for most of the cross-sections of the cylindrical shell cannot be expressed in terms of elementary functions. For this reason, the authors developed a method (Storozhuk and Yatsura 2017) in which the marked integrals are calculated numerically using the trapezoidal rule.

20.4.1 Closed Long Cylindrical Shell of Non-Circular Cross-Section

Assume that the cross-section of a long cylindrical shell closed along the directrix $(-\pi \leqslant \varphi \leqslant \pi)$ has two mutually perpendicular axes of symmetry (OY and OZ). The shell is loaded by two pairs of antipodal transverse forces $P_y = \text{const}$ and $P_z = \text{const}$, which are uniformly distributed along the generatrices, and normal surface forces $q_\zeta = q_\zeta(\varphi)$, which are applied symmetrically with respect to the XOY- and XOZ-planes (Fig. 20.5).

Since the cross-sectional geometry and force distribution is symmetrical, we consider a quarter $(0 \leqslant \varphi \leqslant \pi/2)$ of the shell cross-section as a design model and specify the following boundary conditions at the points $\varphi = 0$ and $\varphi = \pi/2$:

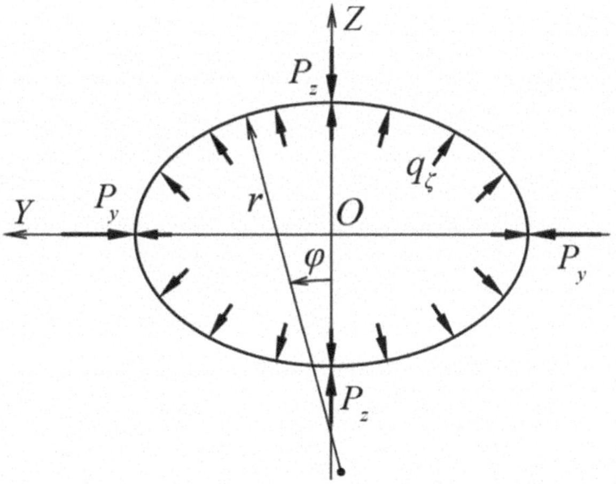

Fig. 20.5 A closed cylindrical shell of non-circular cross-section

$$u(0) = \vartheta(0) = 0, \quad Q(0) = \frac{P_z}{2}, \quad u\left(\frac{\pi}{2}\right) = \vartheta\left(\frac{\pi}{2}\right) = 0, \quad Q\left(\frac{\pi}{2}\right) = -\frac{P_y}{2}. \tag{20.10}$$

The SSS of the shell is described by the system of Eqs. (20.1)–(20.3), whose general solution is presented above. The integration constants are determined using the boundary conditions (20.10). We obtain the following solution to the boundary value problem for this shell:

$$Q = \frac{P_z}{2}\cos\varphi - \frac{P_y}{2}\sin\varphi - \cos\varphi\int_0^{\varphi} rq_\zeta \cos x dx + \sin\varphi\int_{\varphi}^{\pi/2} rq_\zeta \sin x\, dx,$$

$$N = -\frac{P_z}{2}\sin\varphi - \frac{P_y}{2}\cos\varphi + \sin\varphi\int_0^{\varphi} rq_\zeta \cos x dx + \cos\varphi\int_{\varphi}^{\pi/2} rq_\zeta \sin x dx, \tag{20.11}$$

$$M = M^* - \frac{1}{L}\int_0^{\pi/2}\frac{rM^*}{D_M}dx, \quad L = \int_0^{\pi/2}\frac{rdx}{D_M}, \quad \vartheta = \int_0^{\varphi}\frac{rM}{D_M}dx,$$

$$u = \cos\varphi\int_0^{\varphi} f_1 dx - \sin\varphi\int_{\varphi}^{\pi/2} f_2 dx,$$

$$w = \sin\varphi\int_0^{\varphi} f_1 dx + \cos\varphi\int_{\varphi}^{\pi/2} f_2 dx.$$

20.4.2 Open Long Cylindrical Shell With Non-Circular Cross-Section and Hinged Longitudinal Edges

Let us study the strain state of an open (along a directrix $(-\alpha \leqslant \varphi \leqslant \alpha)$) long cylindrical shell with arbitrary cross-section and hinged longitudinal edges ($\varphi = \pm\alpha$). The shell is loaded by normal $q_\zeta = q_\zeta(\varphi)$ and tangential $q_\varphi = q_\varphi(\varphi)$ surface forces as well as a transverse force $P = \text{const}$ (Fig. 20.6) uniformly distributed along the generatrix containing the vertexes of the cross-sections.

The cross-section and load are symmetrical with respect to the vertical plane XOZ. It makes it possible to consider only half ($0 \leqslant \varphi \leqslant \alpha$) of the shell. Let $u = w = M = 0$ at the edge $\varphi = \alpha$ and $u = \vartheta = 0$ and $Q = P/2$ at the edge $\varphi = 0$.

The solution to this boundary value problem, which satisfies the above boundary conditions, has the following form:

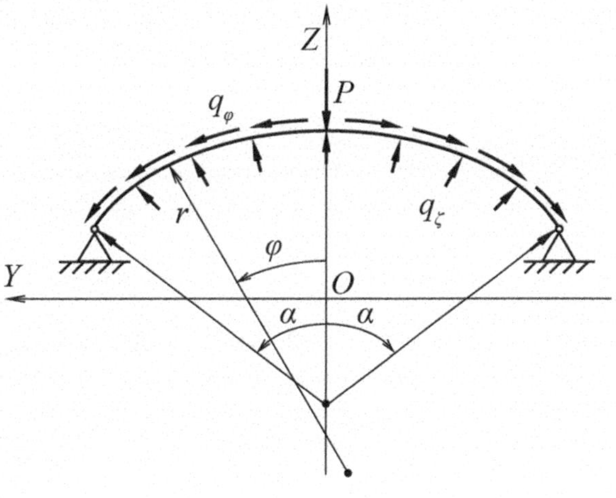

Fig. 20.6 Open cylindrical shell with non-circular cross-section and hinged longitudinal edges

$$
\begin{aligned}
Q &= C_2 \sin\varphi + Q_1, \quad N = C_2 \cos\varphi + N_1, \\
M &= C_2 l_1 + M_1, \quad \vartheta = C_2 l_2 + \vartheta_1, \\
u &= \cos\varphi \int_0^{\varphi} f_1 dx - \sin\varphi \int_{\varphi}^{\alpha} f_2 dx, \\
w &= \sin\varphi \int_0^{\varphi} f_1 dx + \cos\varphi \int_{\varphi}^{\alpha} f_2 dx,
\end{aligned}
\tag{20.12}
$$

where the integration constant C_2 is determined as follows:

$$
C_2 = -\frac{\displaystyle\int_0^{\alpha} r\left[\frac{N_1 \cos x}{D_N} - \left(\vartheta_1 - \frac{Q_1}{D_Q}\right)\sin x\right]dx}{\displaystyle\int_0^{\alpha} r\left[\frac{\cos^2 x}{D_N} - \left(l_2 - \frac{\sin x}{D_Q}\right)\sin x\right]dx}.
\tag{20.13}
$$

Here,

$$
\begin{aligned}
Q_1 &= \frac{P}{2}\cos\varphi + \cos\varphi \textstyle\int_0^{\varphi} q_1 dx - \sin\varphi \int_0^{\varphi} q_2 dx + (r q_\zeta)\,|_0 \sin\varphi, \\
N_1 &= -\frac{P}{2}\sin\varphi - \sin\varphi \int_0^{\varphi} q_1 dx - \cos\varphi \int_0^{\varphi} q_2 dx + (r q_\zeta)\,|_0 \cos\varphi, \\
M_1 &= M_0 - M_0(\alpha), \quad M_0 = \int_0^{\varphi} r Q_1 dx, \quad \vartheta_1 = \int_0^{\varphi} \frac{r M_1}{D_M} dx, \\
l_1 &= l_0 - l_0(\alpha), \quad l_0 = \int_0^{\varphi} r \sin x \, dx, \quad l_2 = \int_0^{\varphi} \frac{r l_1}{D_M} dx.
\end{aligned}
$$

Note that the proposed semi-analytical approach allows us to solve boundary-value problems for cylindrical shells of a non-circular cross-section of continuously variable or piecewise-continuous thickness, particularly for shells supported by longitudinal overlays or rods (Guz et al. 2009; Maximyuk et al. 2014).

20.4.3 Effect of Changing the Thickness of an Elliptical Cylindrical Shell on Its SSS

Consider the boundary value problem for a closed elliptical cylindrical shell of variable thickness under uniform normal pressure $q = \text{const}$.

The shell thickness varies according to the following law:

$$h = h_0(1 + \beta \cos 2\varphi), \tag{20.14}$$

so that the weight of the shell remains unchanged with variation in the parameter β.

Let us study the effect of changing the shell thickness on the distribution of deflection and stresses along the directrix while its weight remains constant for the shell with the following parameters:

$$r_0/h_0 = 15, \quad r_0 = (a + b)/2, \quad a/b = 2,$$
$$E = 38.4\ \text{GPa}, \quad G_{\varphi\zeta} = 0.005E, \quad \nu = 0.3,$$
$$k = 5/6, \quad \beta = 0, \ \pm 0.187, \ \pm 0.3, \ \pm 0.5,$$

where a ann b are the semi-axes of the ellipse.

Figures 20.7 and 20.8 demonstrate the distribution of the dimensionless deflections $\tilde{w} = wE/10^4 h_0 q$ and stresses $\tilde{\sigma} = \sigma/q$ on the outside ($\tilde{\sigma}^+$) and inside ($\tilde{\sigma}^-$) surfaces of the shell along the directrix ($0 \leqslant \tilde{\varphi} \leqslant 1, \ \tilde{\varphi} = 2\varphi/\pi$) depending on the thickness ($\beta = 0, \ \pm 0.187, \ \pm 0.5$).

Table 20.3 summarizes maximum relative stresses $\tilde{\sigma}_{\max}$ and deflections $\tilde{w}_{\max}$ for specified values of the coefficient β.

The data presented in Fig. 20.7 and Table 20.3 show that the deflections are maximum at the end of the major semi-axis ($\varphi = 0$) for all values of β.

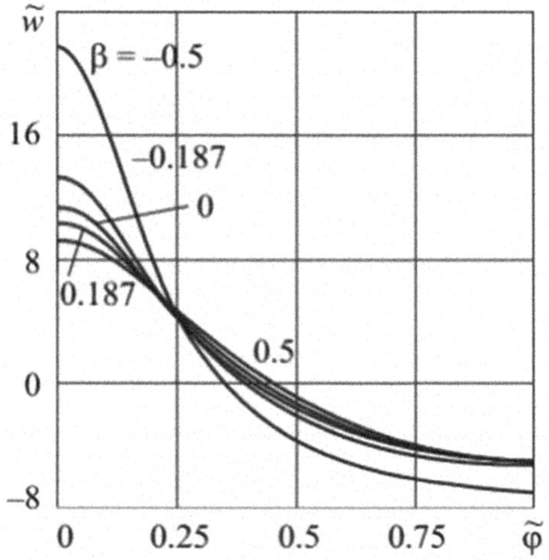

Fig. 20.7 Distribution of deflections along the directrix

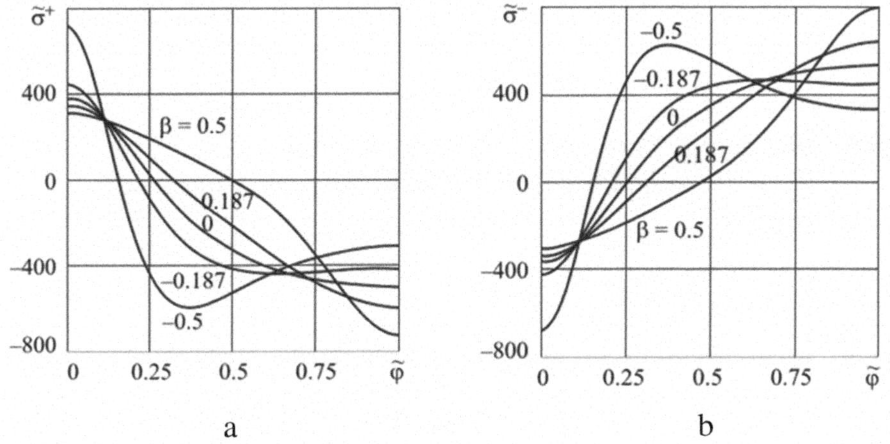

Fig. 20.8 Stress distribution on the outside **a** and inside **b** surfaces of the shell along the directrix

Table 20.3 Dependence of maximum stresses and deflections on β

SSS	β						
	–0.5	–0.3	–0.187	0	0.187	0.3	0.5
$\tilde{\sigma}_{\max}$	721	513	450	542	648	713	807
$\tilde{w}_{\max}$	21.623	15.001	13.059	11.137	10.074	9.638	8.967

The maximum deflection increases with a decrease in the thickness (with a decrease in β) in the vicinity of the apex of the minor axis of the ellipse and at $\beta = -0.5$ exceeds the same value for the shell with a constant thickness ($\beta = 0$) by a factor of 1.94. An increase in the thickness in this region ($\beta = 0.187,\ 0.3,\ 0.5$) results in a negligible decrease in the maximum deflection.

The obtained results (Fig. 20.8 and Table 20.3) indicate that the stress reaches its maximum on the inside surface of the shell in cross-section $\varphi = \pi/2$ at $\beta = 0,\ 0.187,\ 0.3,\ 0.5$, on the outside surface in cross-section $\varphi = 0$ at $\beta = -0.3,\ -0.5$, and at $\beta = -0.187$ on both the inside surface in the section $\varphi = \pi/2$ and the outside surface in the section $\varphi = 0$, i.e., $\tilde{\sigma}_{\max} = \tilde{\sigma}^-(\pi/2) = \tilde{\sigma}^+(0) = 450$. Therefore, at $\beta = 0.187,\ \pm 0.3,\ \pm 0.5$, the stresses are maximum in sections with minimum thickness.

Applying the minimax criterion, which minimizes the maximum stress, to the data in Table 20.3, we can conclude that the thickness variation law (20.14) is optimal at $\beta = -0.187$.

Considering the shell thickness at $\beta = -0.187$ as a reference point, we see that the maximum stress increases with a decrease in the thickness in the vicinity of the major axis ($\beta = 0,\ 0.187,\ 0.3,\ 0.5$) or in the vicinity of the minor semi-axis ($\beta = -0.3,\ -0.5$) and exceeds the stress for the shell with optimal thickness by 60% at $\beta = -0.5$ and by 79% at $\beta = 0.5$.

Thus, by varying the shell thickness, it is possible to control the distribution of the SSS components along the directrix with the weight being unchanged.

20.5 Physically Nonlinear Problems of the Statics of Thin Non-Circular Cylindrical Shells Made of Composite Material

Below, we give the statement of nonlinear problems about the SSS of orthotropic long non-circular cylindrical shells made of nonlinear elastic composite materials. We present an effective numerical method for their solution based on the mixed functional, the variational-difference method (VDM), the successive approximations method (SAM), and the method of Lagrange multipliers. We show the features of deformation of non-circular shells.

20.5.1 Statement of Physically Nonlinear Problems for Orthotropic Shells and Numerical Methods for Their Solution

Let us consider an infinitely long cylindrical shell of arbitrary cross-section, made of nonlinear elastic orthotropic material and loaded with surface forces. The load on the shell can vary along the directrix.

We will write the equation of the midsurface of a long cylindrical shell of an arbitrary non-circular cross-section in the Cartesian coordinate system (X, Y, Z).

For example, the equation of the midsurface of a closed long cylindrical shell of an elliptical cross-section with semi-axes a and b and with a generatrix parallel to OX-axis has (Fig. 20.1) the form (Abrosov et al. 2021):

$$F(Y, Z) = \left(\frac{Y}{a}\right)^2 + \left(\frac{Z}{b}\right)^2 - 1 = 0. \tag{20.15}$$

We attach the curvilinear coordinate system (x, s, ζ) to this surface. In this system, the coordinates x and ζ are the length of the generatrix and the normal to the surface, respectively, and s is the length of the ellipse arc measured from point $(Y = 0, Z = b)$. The orthotropy axes of the nonlinear elastic composite material (Guz et al. 1998) coincide with the axes of principal curvatures of the shell. Obviously, the coefficients of the first quadratic form in this system are equal to unity, and the curvature of the generatrix is zero.

Let small displacements occur in the cross-section due to the application of internal pressure $p(s)$ to an orthotropic homogeneous elastic thin shell of thickness h. Furthermore, there is no movement along the axis. Then the SSS components depend only on the coordinate s. Apparently, the closed shell tends to a nearly circular shape, which causes large deflections and bends near the points of intersections of the ellipse by the symmetry planes. To analyze the SSS under such conditions, we employ the geometrically linear theory of thin shells and a mixed functional to simplify the implementation of the Kirchhoff–Love hypotheses and avoid membrane locking (Maksimyuk et al. 2012).

Nonlinear physical relations for a long shell ($e_{xx} = 0$) under a plane stress state for simple loads are given according to the theory of plasticity of anisotropic media (Guz et al. 1998):

$$\begin{aligned} &\left(\frac{1}{E_{ss}} + \Psi q_{ss}\right)\sigma_{ss} + \left(-\frac{\nu_{sx}}{E_{xx}} + \Psi q_{sx}\right)\sigma_{xx} = e_{ss}, \\ &\left(-\frac{\nu_{xs}}{E_{xx}} + \Psi q_{xs}\right)\sigma_{ss} + \left(\frac{1}{E_{xx}} + \Psi q_{xx}\right)\sigma_{xx} = 0, \end{aligned} \tag{20.16}$$

where σ_{ss} and σ_{xx} are components of circular and longitudinal stresses, respectively; E_{ss}, E_{xx} and ν_{sx}, ν_{xs} are elastic moduli and coefficients of transverse deformation of the orthotropic material, respectively; q_{ss}, q_{xx}, and q_{sx} are components of the tensor that account for the anisotropy of the nonlinear properties. In (20.16), the function Ψ determines the nonlinear properties of the material (Guz et al. 1998), which appear when the quadratic stress function

$$f = \frac{1}{2}\left(q_{ss}\sigma_{ss}^2 + q_{xx}\sigma_{xx}^2 + 2q_{sx}\sigma_{ss}\sigma_{xx}\right)$$

exceeds the value of f_s.

The system of Eq. (20.16) is strongly nonlinear. It can be solved with respect to stresses using, for example, the Newton method. The numerical method is based on variational principles using a mixed functional (Abrosov et al. 2021; Guz et al. 1998; Maksimyuk et al. 2012).

Considering the principle of virtual work and assuming that according to the SAM in the form of additional stresses, the values of the nonlinear components are known from the previous approximation and are not varied, the variational equation can be presented in the form:

$$\delta\Pi = \delta\left(\Pi^L + \Pi^N\right) = 0,$$

where

$$\begin{aligned} \Pi^L = &\frac{1}{2}\iint_\Omega \left(T_{ss}^L \varepsilon_{ss} + M_{ss}^L \mu_{ss}\right) d\Omega - \iint_\Omega p w d\Omega \\ &+ \iint_\Omega T_{s\zeta}^f \varepsilon_{s\zeta} d\Omega - \frac{1}{2}\iint_\Omega C_{ss}(\varepsilon_{ss} - \varepsilon_{ss}^f)^2 d\Omega, \end{aligned} \tag{20.17}$$

$$\Pi^N = \iint_\Omega \left(T_{ss}^{Nf} \varepsilon_{ss} + M_{ss}^N \mu_{ss}\right) d\Omega. \tag{20.18}$$

The functional $\Pi(u, w, \vartheta, T^f_{s\zeta}, \varepsilon^f_{ss})$ depends on four varied functions, including two displacements, angle of rotation, force $T^f_{s\zeta}$ (similar to shearing force), and hoop deformation-function ε^f_{ss}. The advantages of this functional are listed in Maksimyuk et al. (2012). The first term in the linear part of functional (20.17) is the elastic strain energy, the second term is work done by the surface force, the third term represents the geometrical part of the Kirchhoff–Love hypotheses by the method of Lagrange multipliers ($T^f_{s\zeta}$), and the fourth term reduces the membrane locking. In the nonlinear part of the functional (22.18), the nonlinear components of forces depend on the deformation-function ε^f_{ss}. It should be noted that the notation of forces and deformation with the superscript (^{f}) in 22.17 and (22.18) emphasizes the distinction between a "quantity-formula" and a "quantity-function" and has a certain methodological meaning.

The problem linearized in this way is reduced to determining stationary values of the following functional in each approximation:

$$\Pi^{LN} = \Pi^L + \Pi^N. \tag{20.19}$$

The stationarity condition for the functional $\delta\Pi = 0$ leads to the natural static boundary conditions. In the case of symmetry, the principal kinematic conditions are the following (Guz et al. 1998; Maksimyuk et al. 2012):

$$u = 0, \quad \vartheta = 0, \quad T^f_{s\zeta} = 0.$$

To determine the stationary values of functional (20.9), we use the VDM (Guz et al. 1998; Maksimyuk et al. 2012).

20.5.2 *Features of Deformation of Non-Circular Cylindrical Shells. Linear Statement*

The methodology outlined above was used to solve test problems. As a result, the numerical phenomenon of membrane locking is demonstrated, and the way to overcome it is shown. In addition, the features of the deformation of non-circular cylindrical shells were revealed.

Timoshenko's test problem. The SSS of an isotropic linearly elastic infinitely long closed cylindrical shell with an elliptical cross-section of constant thickness under uniform internal pressure was studied in Timoshenko (1941). Compared to (20.17), a linear functional was used, which does not contain the term that reduces the membrane locking. Slow but stable convergence of the calculated VDM of the SSS components to the analytical solution (Timoshenko 1941) and the solution in Storozhuk and Yatsura (2017) was demonstrated by refining the mesh. Accuracy in the maximum values of deflections, stresses, or moments up to the second significant figure is already achieved at $K = 2561$ (K is the number of nodes), and for membrane stresses σ^0 in the middle surface, it is achieved at $K = 5121$. At $K < 321$, the overestimated fictitious stiffness of the shell takes place to such an extent that the sign in the moments calculated near the major semi-axis changes. Such adverse computational effects, caused by large bends at small strains, are called membrane locking or degeneration (Maksimyuk et al. 2012).

Influence of the ellipticity of the cross-section on the linear deformation of the shell. Of particular interest is the occurrence of negative deflections near the major semi-axis of the ellipse. It was found that the region of negative deflections changes nonmonotonically with a decrease in ellipticity. It first slightly increases and then decreases to disappear.

Improvement of convergence in the presence of membrane locking. In the example of the test problem, according to Timoshenko, the improvement of convergence is demonstrated by using the linear functional Π^L, which contains the term (20.17) and reduces the membrane locking. Already at $K = 641$, the maximum deflections and stresses coincide with the analytical solution (Timoshenko 1941) with an accuracy up to the third significant figure, and at $K = 11$, the proposed approach provides an accuracy up to 2% in the maximum stresses, which indicates the effectiveness of improving the convergence by additionally varying the previously small deformation components.

The effect of loading in a linearly elastic shell on the convergence of the numerical method. An almost momentless SSS should appear in the shell if the surface forces will be determined from the equilibrium equations in which the absence of tangential displacement and constant deflection are assumed, i.e., $u = 0$ and $w_0 = \text{const}$. Then the tangential (q) and normal (p) components of the surface load are determined as follows:

$$p = w_0 k_s^2 \frac{Eh}{1 - \nu^2}, \quad q = -w_0 k_s' \frac{Eh}{1 - \nu^2}. \tag{20.20}$$

Since the curvature of the ellipse $k(s)$ and its derivative $k'(s)$ depend on the coordinate s, the load components (20.20) also vary along the arc.

Under such a load, the momentless SSS with strains of the middle surface $\varepsilon_{ss} = k_s w_0$ and stress

$$\sigma_{ss} = w_0 k_s \frac{E}{1 - \nu^2}$$

should be realized in the shell.

Calculations showed fast convergence of stresses. Only six mesh points are enough to achieve accuracy up to the third significant figure. In this case, the deflections range from 9.73 mm near the major semi-axis to 10.3 mm near the minor one. They differ from the exact value $w = h = 10$ mm by 3%. Because of the symmetry at the indicated points, the tangential displacements u are naturally equal to zero; in the intermediate areas, they are almost two orders of magnitude smaller than the deflections, comparable to the deflection error. The ratio of the maximum difference in the stresses on the outer and the inner surfaces of the shell to the stress in the middle surface does not exceed 0.1%, which makes it possible to estimate the degree of SSS momentlessness. At the same time, the convergence in deflections is slower. Thus, the accuracy up to the third significant figure is achieved by dividing the arc into 161 grid points.

Table 20.4 Displacement and stress at the characteristic points of the supercircle

$\tilde{s}$	$\tilde{w}$	σ_s^+	σ_s^0	σ_s^-
0	0.689	5792	100	–5592
0.5	0.289	592	102	–387
1	–0.173	–6614	118	6852

Therefore, a remarkable result is observed when the rate of convergence of the numerical method depends on the nature of the load in the same cylindrical shell of an elliptical cross-section. Moreover, with uniformly distributed pressure, the convergence is slow, and with the selected variable load, the convergence becomes fast. This example and several others (Maksimyuk 2020) confirm that a relation between varied functions in algebraic or differential form unites locking phenomena. Accordingly, a general method exists to overcome locking by additional variation of previously small functions.

Deformation of supercircular cross-section shells. Membrane locking can especially occur in a shell, the section of which has the shape of a supercircle

$$|Y|^n + |Z|^n = R^n, \quad n > 2. \tag{20.21}$$

The SSS of the shell with the following geometric and mechanical parameters was determined: $h = 0.01$ m, $R = 1$ m, $n = 4$, $E = 210$ GPa, $\nu = 0.3$, and $p = 10$ kPa. Furthermore, due to the symmetry of the problem with respect to the $Y = 0$ and $Z = Y$ planes, the one-eighth of the supercircle was considered, i.e., $s \in [0, s_k]$, where $s_k = 0.8772$ m is one-eighth of the supercircle perimeter, which is calculated by the numerical curve discretization algorithm.

For the characteristic points of the supercircle $\tilde{s} = s/s_k$, the deflections ($\tilde{w} = w/h$), pressure-(p) related dimensionless stresses on the outside (σ_s^+), inside (σ_s^-), and middle (σ_s^0) surfaces of the shell are given in Table 20.4.

At the extreme characteristic points of the supercircle $\tilde{s} = s/s_k = 0,\ 0.5,\ 1$, the moments are large and have opposite signs. At the "diameter" point of the supercircle ($\tilde{s} = 0$), the shell bends like a plate. Here, the dimensionless membrane stress in the middle surface is $\sigma_s^0\ (0) = 100$ and coincides with the exact value of $R/h = 100$. At the "diagonal" point ($\tilde{s} = 1$), the shell expands, and the membrane stress $\sigma_s^0\ (s_k) = 118$ becomes slightly larger, since the "diagonal" of the supercircle is greater than the "diameter". At the middle point ($\tilde{s} = 0.5$), the moment is the smallest and an order of magnitude smaller than the maximum.

20.5.3 Physically Nonlinear Deformation of an Orthotropic Cylindrical Shell of Elliptical Cross-Section

The orthotropic shell material, eight-layer organoplastic (Guz et al. 1998), has the following characteristics: $E_{ss} = 26.8$ GPa, $E_{xx} = 46.5$ GPa, $\nu_{sx} = 0.166$, $q_{ss} = 4.32$, $q_{xx} = 2$, $q_{sx} = -0.64$ ($E_{ss} < E_{xx}$). The geometric parameters of the shell: $h = 0.01$ m, $a = 20h$, $b = 15h$. Let us consider the domain $s \in [0, s_k]$, where $s_k = 27.629h$ is one-fourth of the ellipse perimeter. Suppose the orientation of the material orthotropy axes relative to the cylinder axis changes. In that case, the material characteristics change as well: $E_{ss} = 46.5$ GPa, $E_{xx} = 26.8$ GPa, $\nu_{sx} = 0.288$, $q_{ss} = 2$, $q_{xx} = 4.32$, $q_{sx} = -0.64$ ($E_{ss} > E_{xx}$).

Table 20.5 summarizes the results of the SSS analysis of a shell with the orthotropy axes orientation $E_{ss} > E_{xx}$ under loading $p = 3$ MPa. The SAM iteration process was terminated when the relative change in maximum strains in two neighboring approximations became less than 10^{-3}.

As can be seen, the shell cross-section under internal pressure has a nearly circular shape. With the physical nonlinearity taken into account, the stresses decrease, while the deflections and strains near the ellipse poles increase. The maximum stress (σ^-, $\tilde{s} = 1$) decreases by almost 9%.

It should be noted that thin non-circular cylindrical shells demonstrate much greater deflections than circular shells in the same stress state. There may be a situation in which determining the SSS using the geometrically linear shell theory leads to increased virtual deflections while the physical nonlinearity of the material does not yet takes place. For this reason, we considered shells of medium thickness above.

With the orthotropy axes orientation $E_{ss} > E_{xx}$, in contrast to the orientation in Table 20.5, the nonlinear properties of the material appear even at load $p = 2$ MPa (Table 20.6). The effect of nonlinearity is similar here. Since $E_{ss} < E_{xx}$, the deflections under a lower load are greater than those for the orientation $E_{ss} > E_{xx}$ under $p = 3$ MPa.

Let us compare the stress components for the two orientations of the orthotropy axes $E_{ss} < E_{xx}$ and $E_{ss} > E_{xx}$ under $p = 2$ MPa (LP, Table 20.7). Table 20.5 summarizes the hoop (σ_{ss}) and longitudinal (σ_{xx}) stresses on the outer ($\tilde{\zeta} = \zeta/h = 0.5$) and inner surfaces of the shell. As is seen, the hoop stresses in the LP are independent of the material properties, while the longitudinal stresses increase for the orientation

Table 20.5 SSS of the orthotropic nonlinear elastic shell, $E_{ss} > E_{xx}$

Problem	$\tilde{s}$	$\tilde{w}$	$e_{ss}^{+} \cdot 10^2$	$e_{ss}^{-} \cdot 10^2$	σ^+, MPa	σ^-, MPa
LP	0	3.80	1.59	−1.40	776	−686
	1	−2.74	−1.60	1.84	−783	903
NP	0	3.87	1.63	−1.43	732	−667
	1	−2.80	−1.66	1.94	−741	822

$E_{ss} < E_{xx}$ and additionally contribute to the quadratic stress function (Guz et al. 1998). It explains appearing of the nonlinear properties of the material in the shell with the orientation $E_{ss} < E_{xx}$ under lower loads (Table 20.6) than in the shell with the orientation $E_{ss} > E_{xx}$ (Table 20.5).

20.6 Conclusions

In the chapter, based on the refined Timoshenko shell model, a general statement of linear boundary value problems of statics is given, and the basic equations for infinitely long cylindrical shells of the non-circular cross-section made of transversely isotropic material are given. Analytical (exact) expressions for the internal forces, the moment, the rotation angle of the normal, the tangential displacement, and the deflection of long oval cylindrical shells with a continuous and stepped radius of curvature were obtained. Specific numerical results for closed shells under the action of uniformly distributed surface forces were presented. Note that the obtained exact solutions can be used as a reference solution for approximate and numerical methods.

An analytical-numerical approach is developed to solve linear boundary-value problems for long cylindrical shells of arbitrary cross-sections and variable thickness under the action of static surface and linear loads. This approach uses a general solution of static problems for shells of this class, as well as analytical integration and numerical integration with the trapezoidal rule. The proposed approach allows obtaining results with high (almost analytical) accuracy and is free from the so-called membrane and shear locking. Using the developed approach, the effects of the cross-section aspect ratio, the accounting for the transverse shear deformations, and the

Table 20.6 SSS of the orthotropic nonlinear elastic shell, $E_{ss} < E_{xx}$

Problem	$\tilde{s}$	$\tilde{w}$	$e_{ss}^{+} \cdot 10^2$	$e_{ss}^{-} \cdot 10^2$	σ^{+}, MPa	σ^{-}, MPa
LP	0	4.40	1.81	−1.62	517	–457
	1	−3.17	−1.85	2.13	–522	602
NP	0	4.49	1.89	−1.66	485	–444
	1	−3.24	−1.93	2.26	–492	544

Table 20.7 The effect of orthotropy on the SSS in a linearly elastic shell

$\tilde{s}$	ζ	$E_{ss} = 46.5$ GPa		$E_{ss} = 26.8$ GPa	
		σ_{ss}, MPa	σ_{xx}, MPa	σ_{ss}, MPa	σ_{xx}, MPa
0	0.5	517	85	517	149
	–0.5	–457	−75	–457	−131
1	0.5	–522	−86	–522	−150
	–0,5	602	99	602	173

thickness variation law on the SSS of closed elliptical cylindrical shells under the action of uniform pressure were investigated.

Statement of physically nonlinear problems on determining the SSS of orthotropic long non-circular cylindrical shells made of nonlinear elastic composite materials is given within the framework of the classical Kirchhoff–Love model. An effective numerical method of their solution based on mixed functionals, VDM, SAM, and the method of Lagrange multipliers is developed. Mixed functionals, in which the membrane strain is additionally varied, are used to improve convergence and prevent membrane locking. The influence of loads on the numerical method's convergence in linear problems is demonstrated for the first time. Furthermore, the features of nonlinear deformation of orthotropic shells are revealed.

References

Abrosov YY, Maximyuk VA, Chernyshenko IS (2021) Physically nonlinear deformation of a long orthotropic cylindrical shell with elliptic cross-section. Int Appl Mech 57(3):282–289

Boulle A, Dube M, Gosselin FP (2015) Parametric study of an elliptical fuselage made of a sandwich composite structure. Mech Res Comm 69:129–135

Bresse JACH (1866) Cours De Mécanique Appliquée. Première Partie. Résistance Des Matériaux Et Stabilité Des Constructions, Gauthier-Villars, Paris

Chan TM, Gardner L, Law KH (2010) Structural design of elliptical hollow sections: a review. Proc Inst Civil Engrs Struct Build 163(6):391–402

Chen YN, Kempner J (1976) Buckling of oval cylindrical shell under compression and asymmetric bending. AIAA J 14(9):1235–1240

Grigorenko YM, Budak VD, Grigorenko OY (2010) Solving shell problems by discrete-continuum methods. Nikolaev

Grigorenko YM, Kharitonova LV (2007) Deformation of the flexible non-circular cylindrical shells under joint action of two types of loading. Int Appl Mech 43(7):754–760

Grigorenko YM, Zakhriichenko LI (1999) Analysis of the stress-strain state of non-circular cylindrical shells subject to thickness viration and weight retention. Int Appl Mech 35(6):567–576

Guz AN, Kosmodamianskii AS, Shevchenko VP, et al (1998) Stress concentration, of the twelve-volume series Composite mechanics, vol 7. Kyiv

Guz AN, Storozhuk EA, Chernyshenko IS (2009) Nonlinear two-dimensional static problems for thin shells with reinforced curvilinear holes. Int Appl Mech 45(12):1269–1300

Holland M (1976) Pressurized member with elliptic median line: effect of radial thickness function. J Mech Sci 18(5):245–253

Kiseleva TA, Klochkov YV, Nikolaev AP (2015) Comparison of scalar and vector FEM forms in the case of an elliptic cylinder. J Comput Math Math Phys 55(3):422–431

Maksimyuk VA (2020) Locking phenomenon in computational methods of the shell theory. Int Appl Mech 36(3):347–350

Maksimyuk VA, Storozhuk EA, Chernyshenko IS (2012) Variational finite-difference methods in linear and nonlinear problems of the deformation of metallic and composite shells (review). Int Appl Mech 48(6):613–687

Maximyuk VA, Storozhuk EA, Chernyshenko IS (2014) Stress state of flexible composite shells with stiffened holes. Int Appl Mech 50(5):558–565

Meish VF, Kepenach NP (2014) Nonstationary dynamics of longitudinally reinforced elliptic cylindrical shells. Int Appl Mech 50(6):677–682

Novozhilov VV (1964) Thin shell theory. Noordhoff, Groningen

Romano F, Ramlet D (1967) Non-circular rings under shear load. J Frank Instit 284(5):283–299
Sheinman I, Firer M (1994) Buckling analysis of laminated cylindrical shells with arbitrary non-circular cross section. AIAA J 32(3):648–654
Soldatos KP (1999) Mechanics of cylindrical shells with non-circular cross-section: a survey. Appl Mech Rev 52(8):237–274
Storozhuk EA, Yatsura AV (2016) Exact solutions of boundary-value problems for non-circular cylindrical shells. Int Appl Mech 52(4):386–397
Storozhuk EA, Yatsura AV (2017) Analytical-numerical solution of static problems for non-circular cylindrical shells of variable thickness. Int Appl Mech 53(3):313–325
Tennyson RC, Booton M, Caswell RD (1971) Buckling of imperfect elliptical cylindrical shells under axial compression. AIAA J 9(2):250–255
Timoshenko S (1941) Strength of materials. Part II, advanced theory and problems D. Van Nostrand Company, New York
Tornabene F, Fantuzzi N, Bacciocchi M, Dimitri R (2015) Free vibrations of composite oval and elliptic cylinders by the generalized differential quadrature method. Thin Walled Struct 97:114–129
Yamada G, Irie T, Tagawa Y (1984) Free vibration of non-circular cylindrical shells with variable circumferential profile. J Sound Vibr 95(1):117–126
Young WC, Budynas RG (2002) Roark's formulas for stress and strain. McGraw-Hill, New York
Zheleznov LP, Kabanov VV, Boiko DV (2006) Nonlinear deformation and stability of oval cylindrical shells under pure bending and internal pressure. J Appl Mech Tech Phys 47(3):406–411

Chapter 21
Nonlinear Hereditary Creep of Transversely Isotropic Composites of Random Structure

Borys Maslov

21.1 Introduction

Composites with an aluminum matrix reinforced with axially oriented elastic spheroidal inclusions are essential engineering materials. Such materials are widely used, for instance, in aviation and space technology, the shipbuilding industry, and oil refineries (Aboudi et al. 2013; Altenbach and Knape 2020; Christensen 2005; Khoroshun et al. 1993; Lakes 2009). When modeling the long-term response of composites with a metal matrix, it is necessary to consider the effects of creep, which develops even at low enough temperatures. For this purpose, we use the following characteristics to describe the deformation and strength of composites: the creep strain, the rate of creep strain, the relaxation time, the ratio of creep limits, and durability.

One of the urgent problems of the mechanics of hereditary creep of composite materials is to predict the averaged creep properties of the composite based on the properties of its components, their volume content, and reinforcement methods. Many theoretical and experimental papers are devoted to studying various aspects of the creep process of isotropic composite materials. An overview of the obtained results is presented in Aboudi et al. (2013), Altenbach et al. (2018), Christensen (1982), Hashin (1981). The vast majority of the performed studies refer to the linear region of long-term viscoelastic deformation. All solutions are built based on the Boltzmann–Volterra linear theory of viscoelasticity (Christensen 1982, 2005; Hashin 1981; Golub et al. 2016).

It is a well-known fact that for most metal composites, the linearity range is relatively small, and satisfactory results can be obtained at low stress and only for short loading duration (Asaro and Lubarda 2006). Nonlinear equations of viscoelasticity are given in Christensen (2005). However, the representation by infinite series

B. Maslov (✉)
S.P. Timoshenko Institute of Mechanics, National Academy of Science of Ukraine, Kyiv, Ukraine
e-mail: bmaslov@ukr.net

A. N. Guz et al. (eds.), *Advances in Mechanics*, Advanced Structured Materials 191,
https://doi.org/10.1007/978-3-031-37313-8_21

of multiple integrals makes it impossible to identify integral kernels and determine their parameters in typical experiments. An approach based on the similarity of isochronous creep diagrams is more promising for building nonlinear models of hereditary creep (Rabotnov 1969). In Schapery (1990), this algorithm was further developed due to the expansion of the initial condition of similarity, which includes the diagram of instantaneous deformation as an isochrone for the zero moment. The extension of the similarity condition made it possible to build a nonlinear creep model with a time-invariant nature of the nonlinearity, which is determined by the instantaneous deformation diagram. Within the framework of the model, the problem of determining nonlinear creep deformations of polymer with reinforcing fibers was solved (Maslov and Bugai 1992). Nonlinear creep deformations of fibrous unidirectional composites during stretching along the direction of reinforcement are determined. In Aboudi et al. (2013), Maslov (2021), Volterra's theory of hereditary elasticity is used to solve the problems of creep mechanics. The theory describes inversion processes and assumes a linear relationship between stresses and strains. Therefore, it sometimes cannot be used to describe the creep of metals, even in the first approximation. Different generalizations of the theory of hereditary elasticity are possible.

21.2 Problem Statement

The nonlinear creep of isotropic multicomponent composites is considered. The material is stretched for a long time by a uniaxial load along the isotropy axis of the composite sample. A complete adhesion between the fibers and the matrix is assumed.

The external load P does not change over time t, so the tensile stress $\sigma(t)$ is given by the condition $\sigma(t) = \sigma_k H(t)$ $(k = 1, \ldots, n)$, where $\sigma_k = \text{const}$ is the nominal stress value, $H(t)$ is the Heaviside unit function (Christensen 1982). The initial stress state $(t = 0)$ of the composite and its components is elastic $(\sigma_k < \sigma_y)$. Here σ_y is the yield point of the composite (Besson et al. 2010).

The response of the composite to the long-term $(t \gg 0)$ influence of the external load is characterized by the time-dependent value of the total deformation $e(t)$ which includes elastic, e^e, and creep, $e^c(t)$ components: $e(t) = e^e + e^c(t)$. Creep deformation $e^c(t)$ is determined by the change $\Delta l(t)$ of the initial length l_0 of the composite sample (Maslov 2018).

To describe the creep process of the composite, we use the nonlinear creep model of the hereditary type, built based on the hypothesis of the similarity of the isochronous creep diagrams and the instantaneous deformation diagram (Hashin 1981; Golub et al. 2016)

$$\begin{aligned}\frac{de(t)}{dt} &= [E(t)]^{-1}\frac{d}{dt}\left[\int_0^t K_\alpha(t-u)\sigma(u)du\right] \\ &= [E(t)]^{-1}(\sigma * \dot{g})(t).\end{aligned} \tag{21.1}$$

Here $K_\alpha(t-u)$ is the fractional-exponential kernel of hereditary elasticity (it characterizes the influence of stresses at the moment $u \leqslant t$ on creep deformation at the moment of time t); the dot above the symbol means the time derivative. The value $E(t) = d\sigma^e(t)/de(t)$, where $\sigma^e(e)$ determines the instantaneous deformation diagram. Thus, $E(t)$ represents the tangential creep modulus (Golub et al. 2016). The study aims to predict the composite creep behavior when the properties and volume content of its components, as well as the reinforcement method are known.

Deformation in hereditary materials is determined by the stress history $\boldsymbol{\sigma}(u)$ ($u \in [0, t]$), as well as the initial conditions $\boldsymbol{\sigma}(0) = \mathbf{0}$. For a linear medium, after integration in (21.1), it is possible to write down (21.1) in the general tensor form (Schapery 1990, 2021)

$$\mathbf{e}(\mathbf{x}, t) = \int_0^t \mathbf{J}(t-u)\frac{d}{du}\boldsymbol{\sigma}(\mathbf{x}, u)du, \tag{21.2}$$

where $\mathbf{J}(t) = \mathbf{J}^e(\boldsymbol{\sigma})g(t)$ is the nonlinear creep tensor function (retardation). In a short symbolic form (Lakes 2009), we have

$$\mathbf{e}\,(\mathbf{x}, t) = \frac{d}{dt}(\mathbf{J} * \boldsymbol{\sigma})(\mathbf{x}, t) = (\boldsymbol{\sigma} * \dot{\mathbf{J}})(\mathbf{x}, t), \tag{21.3}$$

where the asterisk denotes the integral convolution operation. Thus, the expression $(\boldsymbol{\sigma} * \dot{\mathbf{J}})(t) = (\boldsymbol{\sigma} * d\mathbf{J})(t)$ is a Stieltjes convolution (Christensen 1982).

The following rules for symbolic notation of operations with tensors of the second and fourth ranks are adopted here and further. Dyadic and inner products are introduced as follows

$$(\mathbf{ab})_{ij} = a_{im}b_{mj}, \quad (\mathbf{Ab})_{ij} = A_{ijkl}b_{lk}, \quad (\mathbf{a} \otimes \mathbf{b})_{ijkl} = a_{ij}b_{kl}, \tag{21.4}$$

where summation by repeating indexes is assumed; tensor quantities are marked with bold symbols. Spherical $\mathbf{I}_m$ and deviatoric $\mathbf{I}_d$ operators are given as follows

$$\mathbf{I}_m = \tfrac{1}{3}\mathbf{1} \otimes \mathbf{1}, \quad \mathbf{I}_d = \mathbf{I} - \mathbf{I}_m, \quad (\mathbf{1})_{ij} = \delta_{ij}. \tag{21.5}$$

Here $\mathbf{I}$ is a unit symmetric tensor of the fourth rank, $\mathbf{1}$ is a unit symmetric tensor of the second rank (Maslov 2008). If $f(t)$ is some continuous function on the interval $0 \leqslant t < \infty$ and it behaves as an exponent when $t \to \infty$, then the Laplace–Carson (LC) transformation of the function reads (Christensen 1982):

$$LC\{f(t)\} = s\int_0^\infty f(t)e^{-st}dt. \tag{21.6}$$

Applying the LC transformation to (21.1) and (21.2), we find

$$\begin{aligned}\sigma^e(s) &= (\mathbf{C}\mathbf{e})(s) \\ &= (g\sigma)(s).\end{aligned} \tag{21.7}$$

Here it is assumed that the instantaneous elastic response of the material is physically linear. That is, there is an LC transformation $\mathbf{C}(s)$ from the relaxation function $\mathbf{C}(t)$.

We also apply the law of instantaneous deformation of the second order of deformations (Asaro and Lubarda 2006)

$$\sigma(s) = \mathbf{C}(s)\mathbf{e}(s) - \boldsymbol{\beta}(s), \quad \boldsymbol{\beta}(s) = -\underline{\mathbf{C}}(s)h(s)\mathbf{e}(s), \tag{21.8}$$

where $\boldsymbol{\beta}(\mathbf{e})$ is hardening stress (Altenbach et al. 2018; Rabotnov 1969) (it is the nonlinear function of infinitesimal strain tensor $\mathbf{e}(t)$ or $\mathbf{e}(s)$ in LC domain), $h(t)$ is the reduced relaxation function (Schapery 2022).

In the linear theory of viscoelasticity, the problem's solution can be obtained using the correspondence principle (Christensen 1982; Schapery 1990). It is natural to generalize this principle to problems of hereditary creep. Here we use quasi-linear variant of the hereditary creep equations (Rabotnov 1969), where the concept of so-called modified stresses $\sigma^e(t) = \partial W/\partial \mathbf{e}\,(t)$ and modified (restored elastic) strains $\mathbf{e}^e(t) = \partial U/\partial \sigma\,(t)$ is used. The function $W(\mathbf{e})$ is elastic energy, the function $U(\sigma)$ is complementary elastic energy.

To analyze the problems of deformation prediction within the framework of the hereditary creep model (21.7) and (21.8), we use the generalized Schapery correspondence principle (Schapery 1990) in notations adopted in Lakes (2009), Maslov (2018). Let us denote $\mathbf{x}$ the position of a material point in space; $\mathbf{e}(\mathbf{x}, t)$, $\sigma(\mathbf{x}, t)$, and $\mathbf{u}(\mathbf{x}, t)$ are the current deformation, stress, and displacement, respectively.

Following the accepted hypothesis, we assume that the material exhibits an instantaneous elastic reaction and denote by $\mathbf{e}^e(\mathbf{x}, t)$, $\sigma^e(\mathbf{x}, t)$, and $\mathbf{u}^e(\mathbf{x}, t)$ instantaneous elastic deformation, stress, and displacement, respectively. A material with elastic properties is defined as a medium whose behavior corresponds to the first and the second laws of thermodynamics (Christensen 1982, 2005). From this follows the existence of the stored energy function $W(\mathbf{e})$ and additional energy $U(\sigma)$, which make it possible to find instantaneous deformation (during creep) or instantaneous stress (relaxation process)

$$U = U(\sigma, \mathbf{x}, t), \quad \mathbf{e}^e(t) = \frac{\partial U}{\partial \sigma}(\sigma, \mathbf{x}, t), \tag{21.9}$$

$$W = W(\mathbf{e}, \mathbf{x}, t), \quad \sigma^e(t) = \frac{\partial W}{\partial \mathbf{e}}(\mathbf{e}, \mathbf{x}, t). \tag{21.10}$$

Here, stress and strain are determined in the orthogonal Cartesian coordinate system x_j. The presence of the coordinate $\mathbf{x}$ and time t in the list of arguments indicates the possibility of applicability of (21.9) and (21.10) for the analysis of heterogeneous media and composite materials when the time factor is essential. We write down the constitutive equations in hereditary creep relative to modified (restored) stresses and modified (restored) displacements

$$\mathbf{u}^e(t) = (h * \dot{\mathbf{u}})(t), \quad h(t) = R(t)/R(0). \tag{21.11}$$

These variables are related to physical displacements $u_i = u_i(x_j, \tau)$, in the hereditary media and stresses $\sigma_{ij}(x_j, \tau)$ determined by hereditary integrals of the type (21.3). Therefore, $u_i = u_i(x_j, \tau)$ are physical displacements determined at the time of integration τ. The spatial coordinate x_j is considered to be related to the initially undeformed body ($t = 0$). The $R(t)$ represents the relaxation modulus, a significant characteristic of heredity that affects the material's behavior over time. Thus $h(t)$ can be interpreted as a normalized, dimensionless modulus of hereditary relaxation. We write down the inverse relation to

$$\mathbf{u}(t) = (g * \dot{\mathbf{u}}^e)(t), \quad g(t) = R(0)J(t). \tag{21.12}$$

Here $J(t)$ represents the creep function. At the same time, it is known that $R(t)$ and $J(t)$ satisfy the equation (Christensen 1982; Lakes 2009)

$$\int_{t_0}^{t} J(t-\tau)\dot{R}(\tau - t_0)d\tau = H(t-t_0), \tag{21.13}$$

where $H(t)$ is the Heaviside step function.

21.3 Generalized Schapery Correspondence Principle for Hereditary Creep Problems

The correspondence principle of the linear theory of viscoelasticity is used based on elastic relations, which include Laplace-transformed stresses and displacements. Here we apply the generalized correspondence principle for time-dependent quasi-static solutions of nonlinear elastic and hereditary creep boundary value problems. It makes it possible to analyze the inherited creep problem based on the known nonlinear elastic solution. We write down the equilibrium equation in the form

$$\operatorname{div}\boldsymbol{\sigma} + \mathbf{b} = \mathbf{0}. \tag{21.14}$$

Correlation between Cauchy stresses $\boldsymbol{\sigma}(t)$ and modified (restored) strains

$$\boldsymbol{\sigma}(t) = \left[h * d\left(\frac{\partial W}{\partial \mathbf{e}}\right)\right](t), \quad h(t) = g^{-1}(t). \tag{21.15}$$

Substitution of (21.15) in (21.14) leads to three integro-differential equations for three displacements $u_i(\mathbf{x})$. The solution to the hereditary creep problem that satisfy

(21.14)–(21.15) has the form

$$\boldsymbol{\sigma}(t) = (h * d\boldsymbol{\sigma}^e)(t), \quad \mathbf{u}(t) = (g * d\mathbf{u}^e)(t), \quad h = g^{-1}, \quad g = h^{-1}, \tag{21.16}$$

or

$$\boldsymbol{\sigma}^e(t) = (g * d\boldsymbol{\sigma})(t), \quad \mathbf{u}^e(t) = (h * d\mathbf{u})(t), \quad \mathbf{e}^e(t) = (h * d\mathbf{e})(t), \tag{21.17}$$

where $\boldsymbol{\sigma}^e$ and $\mathbf{u}^e$ satisfy the equation of the basic problem of nonlinear elasticity (21.14) and (21.15) together with the boundary conditions for traction $\mathbf{t}^e = \mathbf{t}$ and the initial condition $\mathbf{u}(0) = \mathbf{u}^e(0) = \mathbf{0}$.

Equation (21.16) determines the dependence of the current stresses $\boldsymbol{\sigma}(t)$ on the restored elastic (modified) stress

$$\begin{aligned}\boldsymbol{\sigma}(t) &= (\boldsymbol{\sigma}^e * dh) \\ &= \left(\frac{\partial W}{\partial \mathbf{e}} * dh\right)(t), \quad h = g^{-1}.\end{aligned} \tag{21.18}$$

And similarly from (21.17) the dependence of the current strain $\mathbf{e}(t)$ on the restored elastic (modified) strain

$$\begin{aligned}\mathbf{e}(t) &= (\mathbf{e}^e * dg)(t) \\ &= \left(\frac{\partial U}{\partial \boldsymbol{\sigma}} * dg\right)(t), \quad g = h^{-1}.\end{aligned} \tag{21.19}$$

If a hereditary body is initially undisturbed, mass forces $\mathbf{b}$ and tractions $\mathbf{t}$ are given, then the solution of the nonlinear problem of hereditary creep (Eqs. (21.14), (21.15), (21.16), and (21.2)) is as follows

$$\boldsymbol{\sigma}(t) = \boldsymbol{\sigma}^e(t), \quad \mathbf{e}(t) = (g * d\mathbf{e}^e)(t), \quad \mathbf{u}(t) = (g * d\mathbf{u}^e)(t), \tag{21.20}$$

where the Stieltjes convolution is defined by (21.3). The field variables $\boldsymbol{\sigma}^e(\mathbf{x})$, $\mathbf{e}^e(\mathbf{x})$, and $\mathbf{u}^e(\mathbf{x})$ satisfy the equation of the corresponding problem of the nonlinear theory of elasticity together with the same boundary conditions (Truesdell and Noll 2004). Thus, $\mathbf{u}^e(\mathbf{x})$ satisfies the equations of the corresponding nonlinear elastic problem together with the same mass forces $\mathbf{b}^e(\mathbf{x}) = \mathbf{b}(\mathbf{x})$ in the body and the boundary conditions $\mathbf{t}^e(\mathbf{x}) = \mathbf{P}(\mathbf{x})$.

It should be emphasized that to describe the instantaneous dependence $\boldsymbol{\sigma}(t) - \mathbf{e}(t)$, it is necessary first to define creep isochrones (Golub et al. 2016). On the one hand, it is related to the approximation of a single integral constitutive equation. On the other hand, in time-dependent stress-strain relations $\boldsymbol{\sigma}(t) - \mathbf{e}(t)$, one of the quantities is initially given, and its current value is equal to the instantaneous value (with index e). The response of the material and its current state includes the hereditary effect of creep.

For example, if in the case of creep problems, the given stress tensor is $\boldsymbol{\sigma}$, then $\boldsymbol{\sigma}(t) = \boldsymbol{\sigma}^e(t)$. So, from (21.14) and (21.15), we see that in this case $g(t) = H(t)$. Since the response of the material to deformations due to creep is characterized by inequalities $\mathbf{e}(t) \neq \mathbf{e}^e(t)$, $g(t) \neq H(t)$, and $h(t) \neq H(t)$ in the equations

$$\begin{aligned} \mathbf{e}(t) &= (\mathbf{e}^e * dg)(t) \\ &= \left(\frac{\partial U}{\partial \boldsymbol{\sigma}} * dg\right)(t), \quad g = 1/h. \end{aligned} \tag{21.21}$$

21.4 A Local Problem of the Mechanics of Hereditary Creep Under a Complex Stress State

Next, we consider composite materials with a metal matrix reinforced with spheroidal inclusions. Materials of this class significantly enhance the operational characteristics of structural members and assemblies operating in extreme conditions (Aboudi et al. 2013; Besson et al. 2010; Lin et al. 2013). Aluminum alloys of several classes are used to produce aluminum-boron and aluminum-silicon oxide composites. As an object for modeling, we refer to the mechanical properties of an aluminum-based alloy of the Al 2024 S type (Lin et al. 2013; Maximov et al. 2014).

Predicting the parameters of hereditary creep of composites with a metal matrix is one of the important factors in the design of modern structural members. It concerns products designed for long-term operation under difficult load conditions. The elastic and viscoelastic properties of reinforcement and matrix can be assumed to be physically linear only in the first approximation. This assumption is valid in most cases for reinforcement. However, the material of the matrix becomes nonlinear, even at sufficiently low stress compared to the corresponding strength. Physical nonlinearity, in combination with the hereditary properties of the matrix, is one of the factors that provide some necessary mechanical properties of the composite, such as low level of concentration of local stresses, high dynamic characteristics, and crack resistance. The ability to determine the deformable properties of the composite in the area of physical nonlinearity based on the specified properties of the reinforcement and the matrix allows us to more accurately establish the actual stress-strain state of structural members (Aboudi et al. 2013; Lakes 2009), as well as to fully use all the resources of the composite in the analysis of problems of long-term strength and optimization of structures with them.

The problem consists in determining the areas of linearity and nonlinearity of the creep process and predicting the creep according to the properties and volume content of components. We consider the material in a steady creep state. We write down the general constitutive equations of hereditary creep in the form (Maslov 2022; Schapery 1990)

$$\sigma^e(t) = (g * \dot{\sigma})(t) = \int_{-\infty}^{t} g(t-s)\dot{\sigma}(s)ds, \quad g(t) = \mathbf{J}(t)/\mathbf{J}(0), \tag{21.22}$$

where $\boldsymbol{\sigma}$ is the Cauchy stress tensor, $\mathbf{e}$ is the tensor of deformations, the superscript e refers to instantaneous (elastic) values of stresses and strains, $\boldsymbol{\sigma}^e(t)$, $\mathbf{e}^e(t)$, and $\boldsymbol{\sigma}(t)$, $\mathbf{e}(t)$ are physical stresses and strains (at the moment of observation). The form of dependence is a consequence of the study of isochronous creep curves (Hashin 1981; Golub et al. 2016). Instantaneous stresses $\boldsymbol{\sigma}^e(t)$ are often called referenced stresses (stresses that are restored from the known stress values $\boldsymbol{\sigma}(t)$) (Schapery 1990).

Relations (21.21) and (21.22) are the constitutive equations of a nonlinear elastic body. To further use the principle of viscoelastic correspondence, let us write down the relationship between the current deformations and the current stresses in the form of the constitutive equations of hereditary creep developed in Maslov (2017, 2019, 2022)

$$\sigma^e(t) = \frac{\partial W}{\partial \mathbf{e}}(\mathbf{e}, \mathbf{x}, t) = \int_{-\infty}^{t} g(t-u)\dot{\sigma}(u)du. \tag{21.23}$$

Hereditary integrals used here are linear functionals with reduced creep functions $g(\mathbf{x}, t)$ depending on the spatial coordinate $\mathbf{x}$. It allows us to consider differentiation concerning coordinate and successive integration over time t as commutative operations. The instantaneous compliance $\mathbf{J}(0) = \mathbf{S}$ and instantaneous relaxation modulus $\mathbf{R}(0) = \mathbf{C}$ are related by $\mathbf{J}(0) = \mathbf{R}^{-1}(0)$, while $g(0) = h(0) = 1$.

Thus, this determines the dependence of the current hereditary creep stresses $\boldsymbol{\sigma}(t)$ on the corresponding restored elastic stresses $\boldsymbol{\sigma}^e(t)$. Similarly, (21.23) is the ratio between the current strain $\mathbf{e}(t)$ and the recovered elastic strain $\mathbf{e}^e(t)$.

Tensor integral identity

$$\frac{\partial W}{\partial \mathbf{e}}(\mathbf{e}, \mathbf{x}, t) = \int_{0}^{t} g(t-u)\frac{\partial \sigma}{\partial u}(\mathbf{x}, u)du \tag{21.24}$$

is a generalization of a known one-dimensional (1D) relation of the correspondence principle (Hashin 1981; Lakes 2009; Maslov 2018) and allows us to transfer the problem with fractional-exponential operators to the domain of symbolic calculations. Here we express the stress in terms of an instantaneous derivative of the deformation energy according to (21.19). Normalized creep function $g(t)$ does not depend on stress, and the normalized relaxation module $h(t)$ does not depend on deformation. Thus, the constitutive equations of hereditary creep (21.24) differ from the equations of linear viscoelasticity (Golub et al. 2016) in form, but their sense is similar (Schapery 1990).

For example, in the case of linear viscoelasticity, if stress is known ($\boldsymbol{\sigma}(t) = \boldsymbol{\sigma}^e(t)$), then after substitution it into the equation $\mathbf{e}^e(t) = g(t)\boldsymbol{\sigma}^e(t) = g(t)\boldsymbol{\sigma}(t)$, we find

$$\mathbf{e}(t) = (\mathbf{J}/\mathbf{J}(0) * d[\mathbf{J}(0)\boldsymbol{\sigma}])(t) = (\boldsymbol{\sigma} * d\mathbf{J})(t). \tag{21.25}$$

Obviously, (21.25) is the usual constitutive equation of linear viscoelasticity (Hashin 1981; Lakes 2009) for the case of a complex stress state.

Based on the integral form of constitutive equations of hereditary creep, the correspondence principle (Maslov 2008; Schapery 1984) allows us to transform the problem with fractional-exponential operators into the field of symbolic calculations. In the case of three-dimensional problems of mechanics of random structure composites under a constant load, the analysis can be transferred to the LC domain (21.6).

Therefore, an anisotropic composite with an aluminum matrix, in a common case, is modeled by a heterogeneous medium occupying a volume V composed of homogeneous phases $V^{(r)}$ ($r \in [0, N]$) with a characteristic volume function $\chi^{(r)}(\mathbf{x})$. In addition, inequality $V^{(r)} \ll V$ holds, and the interphase contact is assumed to be ideal. The creep function of rth phase is denoted by $\mathbf{J}^{(r)}(t)$. Then the creep function of the composite can be represented as piecewise homogeneous:

$$\mathbf{J}(\mathbf{x}, \mathbf{t}) = \sum_{r=1}^{N} \mathbf{J}^{(r)}(t)\chi^{(r)}(\mathbf{x}). \tag{21.26}$$

Here $\chi^{(r)}(\mathbf{x}) = 1\ \forall \mathbf{x} \in V^{(r)}$ and $\chi^{(r)}(\mathbf{x}) = 0,\ \mathbf{x} \notin V^{(r)}$ in another case. Volumetric averaging over V and $V^{(r)}$ is denoted hereafter by $\langle\cdot\rangle$ and $\langle\cdot\rangle^{(r)}$, respectively. Volumetric averaging of the characteristic function by r-phase gives the volume concentration value $c_r = \langle\chi\rangle^{(r)}$. Volumetric averaging of any function f over the representative volume V and over the r-phase,

$$\bar{f} = \langle f \rangle = \frac{1}{V}\int_V f dv,$$

is denoted by $\langle\cdot\rangle^{(r)}$ and

$$\bar{f}^{(r)} = \langle f \rangle^{(r)} = \frac{1}{V^{(r)}}\int_{V^{(r)}} f dv,$$

respectively.

Stochastic equilibrium equations and boundary conditions of the first linear approximation written in the domain of the LC transformations (21.6) can be represented in the form (Khoroshun et al. 1993; Maslov 2017; Rabotnov 1969)

$$\begin{aligned} \mathbf{L}(\nabla)\mathbf{v}(\mathbf{x}, s) &= -\nabla\boldsymbol{\tau}_{(1)}(\mathbf{x}, s), \\ \mathbf{v}(\mathbf{x}, s) &= \mathbf{u}_{(1)}(\mathbf{x}, s) - \bar{\mathbf{u}}(\mathbf{x}, s), \\ \boldsymbol{\tau}_{(1)}(\mathbf{x}, s) &= \mathbf{f}(\mathbf{x}, s)\,\mathbf{e}_{(1)}(\mathbf{x}, s), \\ \mathbf{f}(\mathbf{x}, s) &= \mathbf{C}(\mathbf{x}, s) - \mathbf{L}. \end{aligned} \tag{21.27}$$

The dash bar above indicates the results of statistical averaging in a sample with a random elasticity tensor $\mathbf{C}(\mathbf{x}, s)$, $\mathbf{L}$ is the elastic modulus tensor of a homogeneous body of comparison (Hill 1983). The Fourier transform $\mathbf{L}(\mathbf{k})$ of the differential operator in (21.26) is determined by the relations (Hill 1983; Khoroshun et al. 1993)

$$\begin{aligned} [\mathbf{L}(\mathbf{k})]_{im} &= L_{ijmn} k_j k_m, \\ \mathbf{L} &= \mu_0(\alpha \mathbf{1} \otimes \mathbf{1} + 2\mathbf{I}), \quad \alpha = (\lambda/\mu)_0. \end{aligned} \tag{21.28}$$

That is, μ_0 and λ_0 are the parameters that are constant within the boundaries of the representative volume V; $\mathbf{I}$ is a unit symmetric tensor of the fourth rank; $\boldsymbol{\tau}(\mathbf{x}, s)$ is the polarization stress tensor (Khoroshun et al. 1993). Green's function of the system of equations (21.28) can be found from the condition

$$\mathbf{L}(\nabla)\,\mathbf{G}(\mathbf{x}) + \mathbf{I}\delta(\mathbf{x}) = \mathbf{0}. \tag{21.29}$$

Here $\boldsymbol{\delta}(\mathbf{x})$ is the Dirac delta function, then

$$G_{im}(\mathbf{k}) = [\mathbf{L}(\mathbf{k})]^{-1}_{im}. \tag{21.30}$$

Using the method described in Khoroshun et al. (1993), Maslov (2000, 2018), we write down the solution of the system (21.27) in the form of a convolution type integral over the domain V

$$\mathbf{e}_{(1)}(\mathbf{x}_1, s) = \bar{\mathbf{e}}(\mathbf{x}_1, s) + \boldsymbol{\Gamma}(\mathbf{x}_1, \mathbf{x}_2) * \boldsymbol{\tau}_{(1)}(\mathbf{x}_2, s), \tag{21.31}$$

where $\boldsymbol{\Gamma}(\mathbf{x}_1, \mathbf{x}_2)$ is an operator with a kernel expressed in terms of derivatives of Green's function $\mathbf{G}(\mathbf{x}_1, \mathbf{x}_2)$ from (21.30).

Let us first consider a two-component material with the matrix reinforced with chaotically space oriented spheroidal inclusions. The elastic properties of the inclusions are determined by the potential $W_i(\mathbf{e})$, and the instantaneous properties of the matrix are determined by the potential $W_m(\mathbf{e})$, that is, we take $i = 1$ and $m = 2$. Let us average equation (21.31) under the condition that the coordinate $\mathbf{x}_1$ (the argument of its left part) is located in the volume v_a occupied by inclusions oriented in the $\mathbf{n}_a$-direction, $a \in [1, n]$. As a result, we have

$$\mathbf{e}_a(\mathbf{x}_1, s) = \bar{\mathbf{e}}(s) + \Gamma(\mathbf{x}_1, \mathbf{x}_2) * \left[\mathbf{f}(\mathbf{x}_2) \sum_{b=1}^{n+1} \mathbf{e}_{ba}(\mathbf{x}_2, \mathbf{x}_1, s)\, p_{b|a}(\mathbf{x}_2, \mathbf{x}_1) \right]. \tag{21.32}$$

The following notations are used here for two-point conditional statistical moment functions (Khoroshun et al. 1993; Maslov 2008):

$$p_{b|a}(\mathbf{x}_1, \mathbf{x}_2) = p(\mathbf{x}_1 \in v_b \,|\mathbf{x}_2 \in v_a), \tag{21.33}$$

i.e., the angle brackets, which contain the condition of a point belonging to a particular set, indicate the operation of conditional statistical averaging. Conditional densities of the distribution of transitions from the states $\mathbf{x}_1 \in v_a$ (from the inclusion in $\mathbf{n}_a$-direction) are chosen in the form

$$\begin{aligned} p_{b|a}(\mathbf{x}_1, \mathbf{x}_2) &= p(\mathbf{x}_1, \mathbf{x}_2)\,\delta_{ba} + c_b p^T(\mathbf{x}_1, \mathbf{x}_2)\,(1-\delta_{ba}), \\ p(\mathbf{x}_1, \mathbf{x}_2) &= c_a + c_a^T \varphi(\mathbf{x}_1, \mathbf{x}_2), \quad p^T(\mathbf{x}_1, \mathbf{x}_2) = 1 - p(\mathbf{x}_1, \mathbf{x}_2), \\ p_{m|a}(\mathbf{x}_1, \mathbf{x}_2) &= c_m \varphi^T(\mathbf{x}_1, \mathbf{x}_2), \qquad \varphi^T(\mathbf{x}_1, \mathbf{x}_2) = 1 - \varphi(\mathbf{x}_1, \mathbf{x}_2), \end{aligned} \tag{21.34}$$

where $\varphi(\mathbf{x}_1, \mathbf{x}_2)$ is the two-point correlation function of the field of elastic properties; $c_a^T = 1 - c_a$; c_a is the volume concentration of inclusions oriented in the $\mathbf{n}_a$-direction.

The integration of equations (21.32) is performed according to the method outlined in Maslov (2000, 2008), while the algebraic matrix $\mathbf{g}$ of the operator, which is obtained from the integral $\mathbf{\Gamma}$, has six components (transverse isotropy) (Balbi et al. 2018):

$$\mathbf{g} = (g_1, g_2, g_3, g_4, g_5, g_6). \tag{21.35}$$

Common notations (Maslov 2022; Parnell 2016)

$$g_1 = \tfrac{1}{2}(g_{11} + g_{12}), \; g_2 = g_{13}, \; g_3 = g_{31}, \; g_4 = g_{33}, \; g_5 = g_{66}, \; g_6 = g_{44},$$

are used here for combinations of algebraic matrix elements in the case of transverse isotropy. Each element g_k in (21.35) depends on w-parameter, the ratio of the longitudinal and transverse dimensions of the spheroidal inclusion (Maslov 2008).

Statistical fluctuations of deformations of inclusions $\boldsymbol{\epsilon}_a = \mathbf{e}_a - \bar{\mathbf{e}}$ in the $\mathbf{n}_a$-direction are expressed in terms of average deformations of the matrix material $\mathbf{e}_m$ in the representative volume of the composite V

$$\boldsymbol{\epsilon}_a = c_m \mathbf{z}_a \mathbf{h}_i \mathbf{e}_m, \tag{21.36}$$

where c_m is the volume concentration of the matrix material,

$$\mathbf{h}_i(\mathbf{x}) = \mathbf{C}_i(\mathbf{x}) - \mathbf{C}_m. \tag{21.37}$$

The transversely isotropic tensor $\mathbf{z}_a$ is given by the relations

$$\mathbf{z}_a = (\mathbf{I} - \mathbf{g}_a \mathbf{f}_i)^{-1} \mathbf{g}_a, \quad \mathbf{f}_i = \mathbf{C}_i - \mathbf{L}. \tag{21.38}$$

Let us average equation (21.38) over the set Ω of possible orientations of inclusions. We express the statistical average deformations of inclusions $\mathbf{e}_i$ and matrix material $\mathbf{e}_m$ in terms of macroscopic deformations $\bar{\mathbf{e}}$ of the representative volume V

$$
\begin{aligned}
\mathbf{e}_m &= \mathbf{A}_m\bar{\mathbf{e}}, \quad \mathbf{A}_m = (\mathbf{I} + c_i\mathbf{z}\mathbf{h})^{-1},\\
\mathbf{e}_i &= \mathbf{A}_i\bar{\mathbf{e}}, \quad \mathbf{A}_i = \mathbf{A}_m(\mathbf{I} + \mathbf{z}\mathbf{h}),\\
\mathbf{z} &= \langle \mathbf{z}_a \rangle_\Omega,
\end{aligned}
\tag{21.39}
$$

where angle brackets with an index Ω indicate the operation of statistical averaging over the ensemble of possible orientations of inclusions.

Given the relation (21.39), in the region of the LC transformation, the expressions for the Lame elasticity effective moduli $\lambda(s)$ and $\mu(s)$ for two-component material of random structure can be written in closed form (Maslov 2008).

21.5 Multicomponent Linear Hereditary Materials

In the case of multicomponent nonlinear materials, we use the methodology of conditional statistical moment functions (Khoroshun et al. 1993; Maslov 2008) and the Mori–Tanaka calculation scheme (Mori and Tanaka 1978). To do this, we write down the operator relations that connect the deformation fields of the components and the corresponding macroscopic quantities. We assume that the exact solution for $\mathbf{e}_i$ is known

$$
\mathbf{e}_i = \mathbf{G}_i\mathbf{e}_m, \tag{21.40}
$$

then tensors $\mathbf{A}_i$ and $\mathbf{A}_m$ (where i is the number of inclusions with potential $W_i(\mathbf{e})$, $i \in [1,\ n]$; $m = n + 1$ is the material index of the matrix with elastic potential $W_m(\mathbf{e})$) are determined by the expressions

$$
\mathbf{A}_i = \mathbf{G}_i\mathbf{A}_m, \quad \mathbf{A}_m = \left(c_m\mathbf{I} + \sum_{i=1}^{n} c_i\mathbf{G}_i\right)^{-1}. \tag{21.41}
$$

An approximate solution can be obtained by replacing unknown (in the general case) operator $\mathbf{G}_i$ with a known approximate strain localization operator $\mathbf{T}_i$, which relates the average deformations of the inclusions marked with the number i ($i \in [1, n]$) with the average deformations of the representative volume V, i.e.,

$$
\mathbf{e}_i = \mathbf{T}_i\bar{\mathbf{e}}. \tag{21.42}
$$

Then, we get formulas for finding tensors $\mathbf{A}_i,\ \mathbf{A}_m$:

$$
\mathbf{A}_i = \mathbf{T}_i\mathbf{A}_m, \quad \mathbf{A}_m = \left(c_m\mathbf{I} + \sum_{i=1}^{n} c_i\mathbf{T}_i\right)^{-1}, \quad i \in [1,\ n]. \tag{21.43}
$$

In this work, we define the operator $\mathbf{G}_i$ from the solution obtained on the basis of the one-point approximation for two-component media (21.39), i.e.,

$$\mathbf{G}_i = \underline{\mathbf{1}} + \mathbf{z}_i\mathbf{h}_i, \quad \mathbf{z}_i = \langle \mathbf{z}_a \rangle_\Omega, \quad \mathbf{z}_a = \left(\underline{\mathbf{1}} - \mathbf{g}_a\mathbf{f}_i\right)^{-1}\mathbf{g}_a, \quad \underline{\mathbf{1}} = \mathbf{I}. \tag{21.44}$$

Here, the algebraic matrix $\mathbf{g}_i$ is obtained in (21.36) as a result of the analysis of the stress state in inclusions with the number i. Thus, the tensors $\mathbf{A}_i$ and $\mathbf{A}_m$ are defined as follows

$$\begin{aligned} \mathbf{A}_i &= \mathbf{A}_m(\underline{\mathbf{1}} + \mathbf{zh})_i, \quad i \in [1, n], \\ \mathbf{A}_m &= \left(\underline{\mathbf{1}} + \sum_{i=1}^{n} c_i \mathbf{zh}_i\right)^{-1} \\ &= \left(\underline{\mathbf{1}} + \langle \mathbf{zh} \rangle^{(i)}\right)^{-1}. \end{aligned} \tag{21.45}$$

As in the case of incompressible materials (Maslov 2000), it is possible to verify by direct comparison that, for a two-component material, representations (21.45) are identical to (21.39). From this, it can be concluded, in particular, that the accuracy of (21.45) corresponds to the accuracy of the method of conditional statistical moment functions.

21.6 Nonlinear Solution of Second-Order Hereditary Creep

The equilibrium equations are written in terms of the statistical fluctuations of LC displacements of the second order of smallness $\mathbf{w}(\mathbf{x}, s) = \mathbf{u}_{(2)} - \bar{\mathbf{u}}$. In the coordinates of the reference and for elastic configuration of Maslov (2008), Truesdell and Noll (2004), these equations have the form

$$\begin{aligned} \mathbf{L}(\nabla)\,\mathbf{w}(\mathbf{x}, s) &= -\nabla\boldsymbol{\tau}_{(2)}(\mathbf{x}, s), \quad \forall \mathbf{x} \in V, \\ \mathbf{w}(\mathbf{x}, s) &= 0, \quad \forall \mathbf{x} \in \partial V, \\ \boldsymbol{\tau}_{(2)}(\mathbf{x}, s) &= \mathbf{f}(\mathbf{x}, s)\,\mathbf{e}_{(2)}(\mathbf{x}, s) - \boldsymbol{\beta}_{(1)}(\mathbf{x}, \mathbf{e}, s), \\ \boldsymbol{\beta}_{(1)}(\mathbf{x}, \mathbf{e}, s) &= -\underline{\mathbf{C}}(\mathbf{e}_{(1)})\mathbf{e}_{(1)}(\mathbf{x}, s). \end{aligned} \tag{21.46}$$

The differential operator of system (21.46) coincides in form with the corresponding operator from (21.14). This makes it possible to use Green's function of the linear solution and obtain an integral equation for determining the strain transforms $\mathbf{e}_{(2)}(\mathbf{x}_1, s)$ of the second approximation

$$\mathbf{e}_{(2)}(\mathbf{x}_1, s) = \bar{\mathbf{e}}(s) + \Gamma(\mathbf{x}_1, \mathbf{x}_2) * \boldsymbol{\tau}_{(2)}(\mathbf{x}_2, s). \tag{21.47}$$

Let us perform statistical averaging of (21.47) under the condition that the coordinate (the argument of the left part) is in the volume occupied by spheroidal inclusions, which are oriented in the $\mathbf{n}_a$-direction and have the potential W_i. Then, to determine the average deformations of the inclusions in this direction, we obtain the equation

$$\mathbf{e}_{(2)}^{(a)}(\mathbf{x}_1, s) = \bar{\mathbf{e}}(s) + \Gamma(\mathbf{x}_1, \mathbf{x}_2) * \left[\sum_{b=1}^{n+1} \boldsymbol{\tau}_{ba}(\mathbf{x}_2, \mathbf{x}_1, s)\, p_{b|a}(\mathbf{x}_2, \mathbf{x}_1)\right]_{(1)} . \tag{21.48}$$

Following the proposed algorithm, we find the solution of this equation by performing integration using conditional probability distribution densities of the type (21.34). At the same time, the nonlinear terms in the right-hand part are expressed in terms of macroscopic deformations of the representative volume of the composite material

$$\mathbf{e}_{(1)}^{(r)} = \mathbf{A}^{(r)}\bar{\mathbf{e}}, \quad r \in [1, n+1], \quad n+1 = m. \tag{21.49}$$

Here and further, superscripts in parentheses mean the result of the conditional statistical averaging operation, subscripts—the order of approximation. The first approximation of deformations is $\mathbf{e}_{(1)}{}^{(r)}$. By implementing the procedure defined by (21.48), we obtain

$$\begin{aligned}
&\mathbf{e}_{(2)}^{(r)}(s) = \mathbf{A}^{(r)}(s)\bar{\mathbf{e}}(s) + \mathcal{A}^{(r)}(s)\,(\bar{\mathbf{e}}(s)), \quad r \in [i \cup m], \quad i \in [1, n],\\
&\mathbf{A}_m(s) = (\mathbf{I} + \mathbf{R}(s))^{-1}, \quad \mathbf{R} = \sum_{i=1}^{n} c_i \mathbf{R}_i,\\
&\mathbf{A}_i(s) = (\mathbf{I} + \mathbf{R}_i(s))\,\mathbf{A}_m(s), \quad \mathbf{R}_i = (\mathbf{zh})_i,\\
&\mathcal{A}_m(s) = -\mathbf{A}_m(s)\left(\sum_{i=1}^{n} c_i \mathbf{e}_\beta^i\right), \quad \mathbf{e}_\beta^i = (\mathbf{zh}_\beta)^i, \quad \mathbf{h}_\beta^i = \boldsymbol{\beta}^i - \boldsymbol{\beta}^m,\\
&\mathcal{A}_i(s) = (\mathbf{I} + \mathbf{R}_i(s))\,\mathcal{A}_m(s) + \mathbf{e}_\beta^i(s).
\end{aligned} \tag{21.50}$$

At the same time, the conditions for the normalization of the operators $\mathbf{A}_r$, $\mathcal{A}_r$ are satisfied

$$\sum_{r=1}^{n+1} c_r \mathbf{A}_r = \mathbf{I}, \qquad \sum_{r=1}^{n+1} c_r \mathcal{A}_r = \mathbf{0}. \tag{21.51}$$

Substituting (21.50) into the averaged physically nonlinear relations of the second order from (21.46), we find the dependence of the tensor of macroscopic Cauchy stresses $\bar{\boldsymbol{\sigma}}(t)$ on creep strains $\bar{\mathbf{e}}(t)$ for a multicomponent compressible composite material reinforced with spatially oriented spheroidal inclusions.

21.7 The Function of Stored Strain Energy and Additional Energy

For further analysis of creep processes, we should use the obtained constitutive equations concerning given parameters for the potential of deformations, i.e., additional energy. The energy of the instantaneous elastic deformation per unit volume of the material can be represented by expansion in the Taylor series in the vicinity of the natural unstressed state (Asaro and Lubarda 2006; Biot 2008; Truesdell and Noll 2004)

$$W(\mathbf{e}) = \frac{1}{2!} e_{ij} C_{ijkl} e_{kl} + \frac{1}{3!} e_{ij} (C_{ijklmn} e_{mn}) e_{kl} + \cdots, \tag{21.52}$$

where e_{ij} are the components of the linear tensor of Cauchy deformations; C_{ijkl} are the component of the tensor of linear elasticity constants. The elastic constants of the third order are the components of the tensor of the sixth order.

The symmetric Cauchy stress tensor is the gradient σ_{ij} of elastic energy in relation to elastic strains

$$\begin{aligned}\sigma_{ij} &= \frac{\partial W}{\partial e_{ij}} \\ &= C_{ijkl} e_{kl} + \tfrac{1}{2} C_{ijklmn} e_{mn} e_{kl}, \quad \underline{C}_{ijkl} = C_{ijklmn} e_{mn}.\end{aligned} \tag{21.53}$$

If relations (21.53) are invertible, then we get

$$\begin{aligned}e_{ij} &= \frac{\partial U}{\partial \sigma_{ij}} \\ &= S_{ijkl} \sigma_{kl} + \tfrac{1}{2} S_{ijklmn} \sigma_{mn} \sigma_{kl}.\end{aligned} \tag{21.54}$$

Here

$$\begin{aligned}U(\sigma) &= \sigma_{kl} e_{kl} - W(e_{ij}) \\ &= \tfrac{1}{2} \sigma_{ij} S_{ijkl} \sigma_{kl} + \tfrac{1}{3!} \sigma_{ij} (S_{ijklmn} \sigma_{mn}) \sigma_{kl},\end{aligned} \tag{21.55}$$

is the additional strain energy determined by the Legendre transformation of the function $W(\mathbf{e})$. Since $\partial \sigma_{ij} / \partial \sigma_{kl} = I_{ijkl}$ (unity tensor of the fourth rank) and

$$\partial^2 \sigma_{ij} / \partial \sigma_{kl} \partial \sigma_{mn} = 0,$$

we can obtain the relations connecting the elastic constants of the second C_{ijkl} and third C_{ijklmn} orders with the compliances of the corresponding orders, namely,

$$C_{ijkl} = S^{-1}_{ijkl}, \quad C_{ijklmn} = -C_{ijpq} S_{pqrsuv} C_{rskl} C_{uvmn}. \tag{21.56}$$

As an example, let us further consider the elastic potential of the third order in the form (Asaro and Lubarda 2006; Truesdell and Noll 2004)

$$W = \tfrac{1}{2}(\lambda + 2\mu) I_1^2 + 2\mu I_2 + \frac{l + 2m}{3} 3 I_1^3 + 2m I_1 E_2 + n I_3, \tag{21.57}$$

where I_k are the main invariants of the Cauchy strain tensor. Then the formula for stresses takes the form

$$\boldsymbol{\sigma} = \left[\lambda I_1 + l I_1^2 + (2m - n) I_2\right] \mathbf{1} + [2\mu + (2m - n) I_1] \mathbf{e} + n \mathbf{e}^2. \tag{21.58}$$

Third-order constants l, m and n were introduced by Murnaghan (Besson et al. 2010). Since the elastic constants of the kth order are components of the tensor of rank $2k$, the symmetry property is obtained from the relations

$$C_{ijkl} = \frac{\partial^2 W}{\partial e_{ij} \partial e_{kl}}, \quad C_{ijklmn} = \frac{\partial^3 W}{\partial e_{ij} \partial e_{kl} \partial e_{mn}}. \tag{21.59}$$

The tensor of elastic constants of the third order in a rectangular coordinate system for isotropic materials has the form (Asaro and Lubarda 2006)

$$C_{ijklmn} = \nu_1 \delta_{ij} \delta_{kl} \delta_{mn} + 6\nu_2 \delta_{ij} I_{klmn} + 8\nu_3 I_{ijklmn}. \tag{21.60}$$

At the same time, the Lame constants of the third order ν_k are related to the coefficients l, m and n from (21.57)

$$l = \nu_2 + \tfrac{1}{2}\nu_1, \quad m = \nu_2 + 2\nu_3, \quad n = 4\nu_3.$$

Here, the following relations are used

$$J_{ijklmn} = \tfrac{1}{3}\left(\delta_{ij} I_{klmn} + \delta_{kl} I_{mnij} + \delta_{mn} I_{ijkl}\right), \tag{21.61}$$
$$I_{ijklmn} = \tfrac{1}{4}\left(\delta_{ik} I_{jlmn} + \delta_{il} I_{jkmn} + \delta_{im} I_{klnj} + \delta_{in} I_{klmj}\right). \tag{21.62}$$

In the case of an isotropic material, the elastic compliances of the second order s_{ij} can be expressed in terms of elastic constants

$$s_{11} = \frac{c_{11} + c_{12}}{(c_{11} - c_{12})(c_{11} + 2c_{12})} = -\frac{\nu}{E}, \quad s_{12} = \frac{c_{12}}{(c_{11} - c_{12})(c_{11} + 2c_{12})} = \frac{1}{E}.$$

For isotropic media, we also write down some useful linear combinations of elastic moduli c_{ijk} and third-order compliances s_{ijk}

$$9s_{123} + 18s_{144} + 8s_{456} = -(3s_{12} + 2s_{44})^3 (9c_{123} + 18c_{144} + 8c_{456}), \tag{21.63}$$
$$3s_{144} + 4s_{456} = -4s_{44}^2 (3s_{12} + 2s_{44})(3c_{144} + 4c_{456}), \tag{21.64}$$

with

$$\nu_1 = c_{123}, \quad \nu_2 = c_{144}, \quad \nu_3 = c_{456}. \tag{21.65}$$

Then the expression for the additional strain energy can be written in the form

$$S_{ijklmn} = \eta_1 \delta_{ij} \delta_{kl} \delta_{mn} + 6\eta_2 \delta_{ij} I_{klmn} + 8\eta_3 I_{ijklmn}, \tag{21.66}$$
$$\eta_1 = s_{123}, \quad \eta_2 = s_{144}, \quad \eta_3 = s_{456}. \tag{21.67}$$

In the theory of nonlinear viscoelasticity, the coefficients of proportionality of stress and strain deviators (Hashin 1981) of the second and third order are useful

$$
\begin{aligned}
s_{144} &= -4s_{44}^2(s_{11}c_{144} + 2s_{12}c_{244}),\\
s_{11} + s_{12} &= s_{144}\\
&= -4s_{44}^2[s_{12}c_{144} + (s_{11} + s_{12})c_{244}].
\end{aligned}
\tag{21.68}
$$

Thus, from (21.68), we obtain the inverse relations expressing compliance s_{ijk} in turms of elasticity coefficients c_{ijk}. Nonlinear creep and relaxation parameters of the aluminum matrix may be connected by

$$
\begin{aligned}
9\nu_1 + 18\nu_2 + 8\nu_3 &= -(3\kappa)^3(9\eta_1 + 18\eta_2 + 8\eta_3),\\
3\nu_2 + 4\nu_3 &= -12\mu^2\kappa(3\eta_2 + 4\eta_3), \quad \nu_3 = -8\mu^3\eta_3.
\end{aligned}
\tag{21.69}
$$

In the nonlinear theory of viscoelasticity, data from basic tensile and torsional creep experiments are used (Hashin 1981; Golub et al. 2016) to determine material parameters with (21.68) or (21.69).

One of the functions that monotonically decreases with increasing argument and has a weak singularity at $t = 0$ is the fractional-exponential function (Khoroshun et al. 1993; Maslov 2022). This function is successfully used as the kernel of the linear integral equation (21.69) and describes the behavior of polymers (Balbi et al. 2018) and composite materials within the linear theory of creep. Currently, there is a well-developed mathematical apparatus (Rabotnov 1969) for solving boundary value problems of the theory of linear creep using these kernels as initial ones.

Following (21.19), the constitutive equations of creep, which specify the dependence between the components of the strain tensor, the stress tensor, and the time, can be written in the form

$$
\begin{aligned}
\mathbf{e}_d(t) &= (\mathbf{e}_d^e * dg_d)(t)\\
&= \left(\frac{\partial U}{\partial \boldsymbol{\sigma}_d} * dg_d\right)(t),\\
\mathbf{e}_m(t) &= (\mathbf{e}_m^e * dg_m)(t)\\
&= \left(\frac{\partial U}{\partial \boldsymbol{\sigma}_m} * dg_m\right)(t).
\end{aligned}
\tag{21.70}
$$

Reduced creep functions $g_d(t)$ and $g_m(t)$ are obtained from deviator $J_d(t)$ and volume $J_m(t)$ creep functions of linear viscoelastic material (see Hashin (1981), Golub et al. (2016))

$$
\begin{aligned}
g_d(t) &= 2\mu J_d(t), \quad g_d(t) = 1 + (\xi/\beta)_d(1 - M_d(\alpha_d, \beta_d, t)),\\
g_m(t) &= 3\kappa J_m(t), \quad g_m(t) = 1 + (\xi/\beta)_m(1 - M_m(\alpha_m, \beta_m, t)).
\end{aligned}
\tag{21.71}
$$

Here, $M_q(\alpha, \beta, t)$ is the Mittag-Leffler function

$$
M_q(\alpha, \beta, t) = \sum_{n=0}^{\infty} \frac{(-\beta t^q)^n}{\Gamma(1 + qn)}, \quad q = \alpha + 1; \tag{21.72}
$$

μ and κ are shear and volume deformation moduli, respectively; α, ξ and β are rheological parameters.

The most practically important problem is the identification of creep and relaxation kernels

$$K(\alpha, \beta, t) = -\frac{1}{\beta}\frac{d}{dt}M_{\alpha+1}(\alpha, \beta, t).$$

The identification of creep and relaxation kernels in (21.68) and (21.71) is carried out using two groups of basic experiments on creep under constant stresses (Golub et al. 2016). The first group includes creep tests of solid cylindrical samples under uniaxial tension for determining longitudinal and transverse deformations. In this case, one-dimensional dependences between stresses, strains, and time are given by equations similar to (21.2). The second group of basic experiments considered in Hashin (1981) includes creep tests of solid cylindrical samples under uniaxial tension for determining the longitudinal deformations and creep tests of thin-walled tubular samples under pure torsion for determining the angular deformations.

The creep kernels $K\ (t)$ in (21.1), (21.2) and relaxation $R\ (t)$ in (21.11) are approximated by fractional-exponential functions

$$K(t) = t^{\alpha}\sum_{n=0}^{\infty}\frac{(-\beta)^n(t)^{nq}}{\Gamma\left[(1+n)q\right]}, \qquad R(t) = t^{\alpha}\sum_{n=0}^{\infty}\frac{-(\xi+\beta)^n(t)^{nq}}{\Gamma\left[(1+n)q\right]}, \tag{21.73}$$

where α and β are kernel parameters ($-1 < \alpha < 0, \beta > 0$), $\Gamma[\cdot]$ is Euler's gamma function (Rabotnov 1969).

The relaxation functions corresponding to other homogeneous stress states can also be determined using $\mu(s)$ and $\kappa(s)$. A comparison of the LC transformation of constitutive relations for linear, isotropic elasticity theory shows that the relaxation functions corresponding to other stress states are determined using the relations of the elasticity theory by replacing the elastic moduli with the transformation of the corresponding relaxation functions of the viscoelasticity theory (Christensen 1982; Lakes 2009). So, for example, it follows from the relations of the isotropic elasticity theory that, in the transformation domain, the relaxation function $E(s)$ that characterizes the state of uniaxial tension is determined in terms of transformed functions $\mu(s)$, $\kappa(s)$ as follows

$$E(s) = \frac{9\mu(s)\kappa(s)}{\mu(s)+3\kappa(s)}.$$

It is essential to establish the relationship between the creep and relaxation kernels of nonlinear hereditary materials under a complex stress state and the creep kernels under uniaxial tension and pure torsion. It is the stage of the general procedure for determining the parameters of the kernels given by the fractional-exponential functions (Khoroshun et al. 1993).

For example, consider the hereditary creep of a composite made of a three-component mixture: an aluminum matrix of the Al 2024 S type (Lin et al. 2013;

Maximov et al. 2014), reinforced with oriented along x_3-axis spheroidal elastic inclusions. The elastic characteristics of the components (Aboudi et al. 2013) are given in Table 21.1.

We take into account the hereditary properties of the aluminum matrix. The rheological parameters of the fractional-exponential kernel we obtain by identification of experimental data from Lin et al. (2013), and are given in Table 21.2.

Further, numerical algorithms in the Fortran 95 environment were used for computer simulation (Abate and Valko 2004; Kuhlman 2013; Levesque et al. 2007).

Curves in Fig. 21.1 show the dependence of the normalized transverse creep function $g_T(t)$ over time. The solid line corresponds to the predicted transverse creep function for an aluminum matrix reinforced with axially oriented boron fibers $c_1 = 0.15$. The dashed line illustrates the behavior of the uniaxial creep function $g(t)$ for a pure matrix.

Figure 21.2 shows the dependence curves of the normalized transverse shear creep function $j_T(t)$ over time. The solid line corresponds to the predicted transverse creep function for aluminum matrix reinforced with axially oriented fibers $c_1 = 0.15$. The dashed line illustrates the behavior of the shear creep function $j(t)$ for a pure *Al* 2024 *S* matrix.

Further, Fig. 21.3 shows the varying of transverse relaxation function $E_T(c_1)$ with pure matrix relaxation $E(c_1)$ function under transverse load when volume boron fibers concentration c_1 increases. The dashed curve corresponds to an instantaneous elastic solution, and the solid curve is determined as a result of long-term hereditary creep analysis for $t = T_b = 10\tau$, where τ is the relaxation time.

Figure 21.4 shows the dependence of longitudinal Poisson's ratio $\nu_L(c_1)$ prediction from volume concentration of boron fibers c_1. The dashed curve corresponds to instantaneous elastic solution (Maslov and Bugai 1992; Maslov 1989), and the solid curve is determined as a result of hereditary creep analysis for $T_b = 10\tau$.

Figure 21.5 shows the dependence of transverse Poisson's ratio $\nu_T(c_1)$ from volume concentration of boron fibers c_1. The dashed curve corresponds to the instantaneous elastic solution, and a solid one is determined by the hereditary creep analysis for $T_b = 10\tau$ (Maslov 2018).

Table 21.1 Material constants of metal matrix (Al 2024 S) composite components

Material	E, GPa	ν	ν_1	ν_2	ν_3
Boron	467.3	0.361	−840.0	−420.0	−390.00
SiC	440.3	0.171	−227.2	31.5	−170.75
Al 2024 S	55.8	0.33	−115.0	−160.5	−108.80

Table 21.2 Viscoelastic parameters of aluminum matrix

Material	α	ξ, (hour)$^{-q}$	β, (hour)$^{-q}$
Al 2024 S	0.5	0.33	115.0

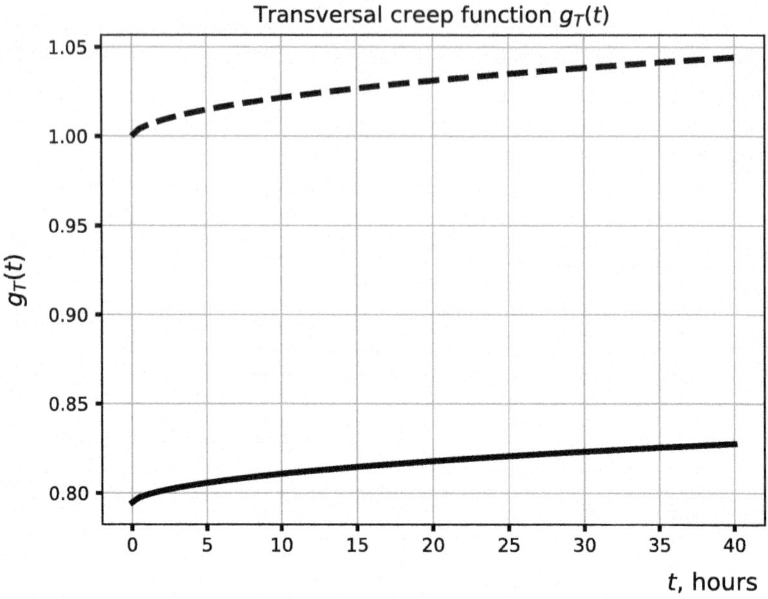

Fig. 21.1 Comparison of the results of normalized transverse creep function $g_T(t)$ prediction for unidirectionally reinforced composite with pure matrix creep function under transverse uniaxial load

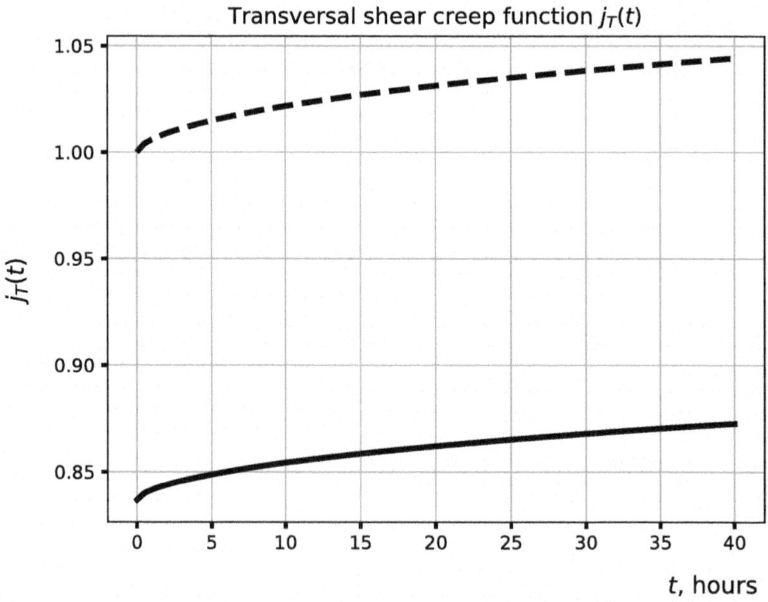

Fig. 21.2 Comparison of the results of normalized transverse shear creep function $j_T(t)$ prediction for unidirectionally reinforced composite with pure matrix creep function under transverse shear load

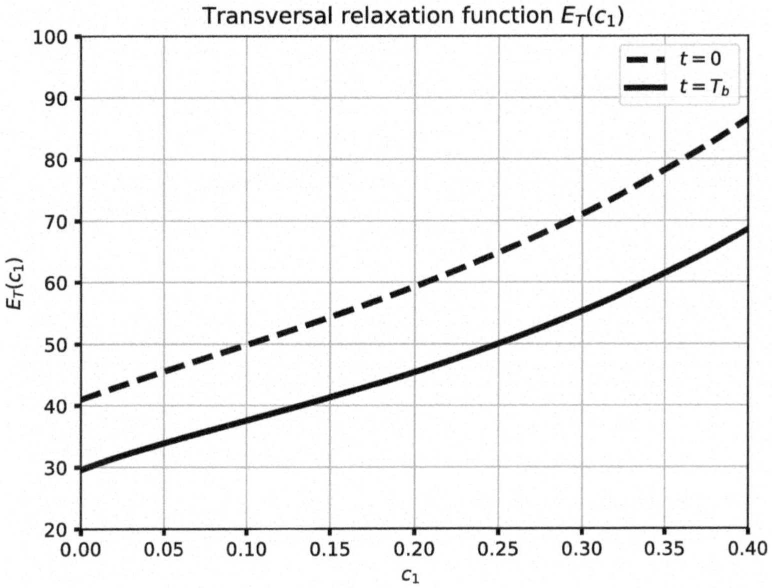

Fig. 21.3 Comparison of the results of transverse relaxation function $E_T(c_1)$ with pure matrix relaxation $E(c_1)$ function under transverse load

Next, Figs. 21.1, 21.2, 21.3, 21.4 and 21.5 let us compare obtained properties of unidirectionally reinforced composite with pure matrix under uniaxial longitudinal and transverse loading.

The calculations are carried out for nonlinear elastic parameters of constituents presented in Table 21.1 and viscoelastic properties of the aluminum matrix presented in Table 21.2.

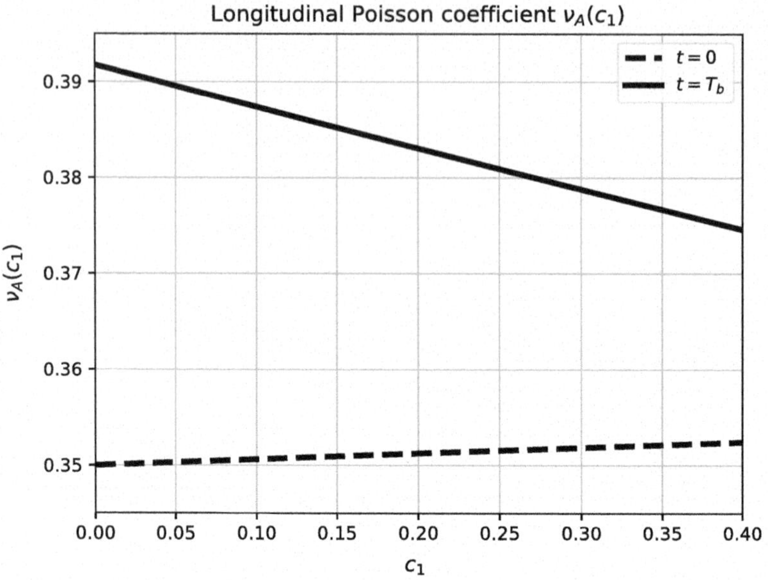

Fig. 21.4 Comparison of the results of longitudinal Poisson's ratio $\nu_A(c_1)$ prediction for unidirectional reinforced composite with pure matrix $\nu(c_1)$ under uniaxial load

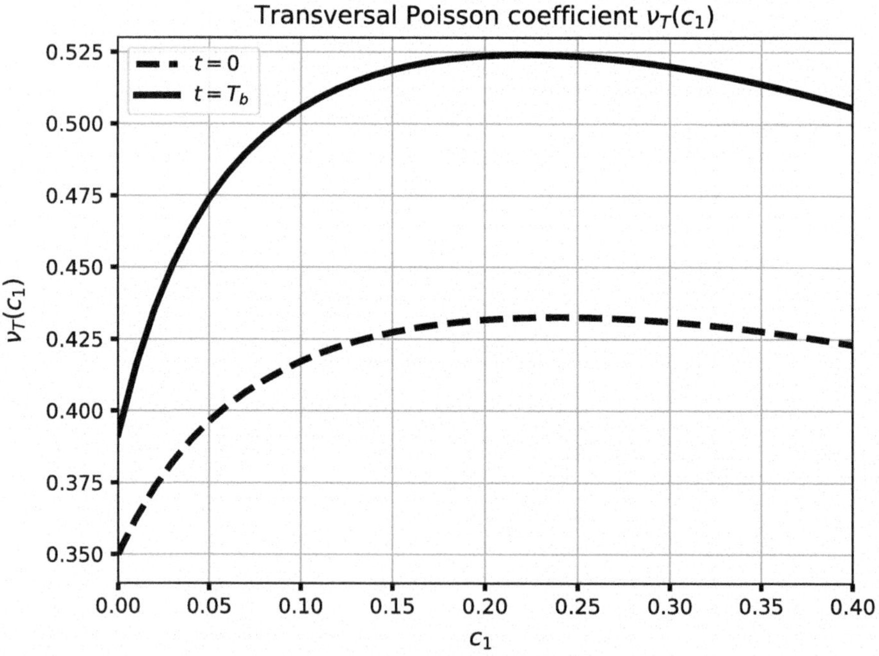

Fig. 21.5 Comparison of the results of transverse Poisson's ratio $\nu_T(c_1)$ prediction for unidirectional reinforced composite with pure matrix creep function $\nu(c_1)$ under uniaxial load

21.8 Conclusion

Composite materials are used in structural elements that are operated both under conditions of long-term, one-time, and cyclic loads. The strength and durability of composites depend on the average or maximum stresses per load cycle in the matrix, inclusions, and their interfaces; there is also dependence on the number of cycles. In this regard, for the design of new materials with given properties, it is of great interest to model and study microstructural stresses and strains that change over time. Therefore, predicting the macroscopic characteristics of nonlinear hereditary creep and determining their dependence on the form of inclusions, the type of spatial reinforcement, and the volume concentration of components are of great practical importance. The article examines a composite material of a random structure and uses an averaging method based on the Schapery-type correspondence principle, Fourier and Laplace–Carson integral transforms. The Fortran 95 environment is used to implement the numerical algorithms. The advantages of the research include obtaining final results in the form of constitutive equations of hereditary creep, namely, the time dependence of deformations on stresses and inverted relaxation relation of stress–deformation.

References

Abate J, Valko P (2004) Multi-precision Laplace transform inversion. Int J Numer Methods Eng 60:979–993

Aboudi J, Arnold S, Bednarcyk B (2013) Micromechanics of composite materials. Elsevier, Boston

Altenbach H et al (2018) Mechanics of composite structural elements mechanics

Altenbach H, Knape K (2020) On the main directions in creep mechanics of metallic materials. Mechanics - Proc NAS Armenia 73:24–43

Altenbach H (2002) Creep analysis of thin-walled structures. Z Angew Math Mech 82(8):507–533

Asaro RJ, Lubarda VA (2006) Mechanics of solids and materials. Cambridge University Press, Cambridge

Balbi V, Shearer T, Parnell WJ (2018) A modified formulation of quasi-linear viscoelasticity for transversely isotropic materials under finite deformation. Proc R Soc A 474:20180231

Besson J, Cailletaud G, Chaboche JL, Forest S (2010) Non-linear mechanics of materials. Springer, New York

Biot MA (2008) Mechanics of incremental deformations. Willey, New York

Christensen RM (2005) Mechanics of composite materials. Dover Publications, New York

Christensen RM (1982) Theory of viscoelasticity. An Introduction. Academic Press, New York

Golub VP, Maslov BP, Fernati PV (2016) Identification of the hereditary kernels of isotropic linear viscoelastic materials in combined stress state. I. Superposition of shear and bulk creep. Int Appl Mech 52(2):165–174

Hashin Z (1981) Fatigue failure criteria for unidirectional fiber composites. Trans ASME 48(12):846–852

Hill R (1983) Interfacial operators in the mechanics of composite media. J Mech Phys Solids 31(4):347–357

Khoroshun LP, Maslov BP, Shikula EN, Nazarenko LV (1993) Statistical mechanics and effective properties of materials, vol. 3 of the 12-volume series mechanics of composite materials. Naukova Dumka, Kyiv [in Russian]

Kuhlman KL (2013) Review of inverse Laplace transform algorithms for Laplace-space numerical approaches. Numer Algorithms 63(2):339–355

Lakes RS (2009) Viscoelastic materials. Cambridge University Press, Cambridge

Levesque M, Gilchris MD, Bouleau N, Derrien K, Baptiste D (2007) Numerical inversion of the Laplace-Carson transform applied to homogenization of randomly reinforced linear viscoelastic media. Comput Mech 40(4):771–789

Lin YC, Xia YC, Chen MS, Jiang YQ, Li LT (2013) Modeling the creep behavior of 2024–T3 Al alloy. Comput Mater Sci 67:243–248

Maslov BP, Bugai SN (1992) Nonlinear deformation and strength of stochastic composites with a damaged matrix. Mech Compos Mater 27:509–515

Maslov BP (2017) Stress concentration in nonlinear viscoelastic composites. J Mech Adv Technol 1(79):5–10

Maslov BP (2019) Application of a quasi-linear visco-elastic model for the creep of a non-heterogeneous geological media. Bull Taras Shevchenko Natl Univ Kyiv Ser Phys Math 1:122–125

Maslov BP (2018) Combined numerical and analytical determination of poissons ratio for viscoelastic isotropic materials. Int Appl Mech 54(2):220–230

Maslov BP (2021) Hereditary creep of isotropic composites of random structure under a complex stress state. Bull Taras Shevchenko Natl Univ Kyiv Ser Phys Math 3:77–80

Maslov BP (2022) Nonlinear hereditary creep of isotropic composites of random structure. Int Appl Mech 58(1):75–90

Maslov BP (2000) Stress concentration in incompressible multicomponent materials. Int Appl Mech 36(3):384–390

Maslov BP (1989) Stress-strain state in matrix of the stochastically reinforced composites. Mech Compos Mater 3:396–402

Maslov BP (2008) Thermal-stress concentration near inclusions in viscoelastic random composites. J Eng Math 61:339–335

Maximov JT et al (2014) Modeling of strain hardening and creep behaviour of 2024T3 aluminium alloy at room and high temperatures. Comput Mater Sci 83:381–393

Mori T, Tanaka K (1978) Average stress in matrix and average elastic energy of materials with miss-fitting inclusions. Acta Metallurgica 21(4):571–574

Parnell WJ (2016) The Eshelby, Hill moment and concentration tensors for ellipsoidal inhomogeneities in the Newtonian potential problem and linear elastostatics. J Elast 125:231–294

Rabotnov YuN (1969) Viscoelastic materials. Creep problems in structural members. North Hollands, Amsterdam

Schapery RA (2021) A theory of viscoelastic crack growth-Revisited (Revised). https://doi.org/10.13140/RG.2.2.10607.36003

Schapery RA (1984) Correspondence principles and a generalized integral for large deformation and fracture analysis of viscoelastic media. Int J Fract 25(3):195–223

Schapery RA (2022) Crack growth in viscoelastic media with large strains: further results and validation of nonlinear theory for rubber (Revised 11/23/22). https://www.researchgate.net/publication/365687544

Schapery RA (1990) On some path independent integrals and their use in fracture of nonlinear viscoelastic media. Int J Fract 42:189–207

Truesdell C, Noll W (2004) The nonlinear field theories of mechanics (ed. Antman, S.). Springer, Berlin

Chapter 22
Boundary Integral Equation Method for 3D Elastodynamic Problems with Chain-Arranged Rigid Disk-Shaped Inclusions

Viktor Mykhas'kiv and Igor Zhbadynskyi

22.1 Introduction

Currently, periodically structured elastic composites or artificial metamaterials are widely applied in modern technical systems because they provide special deformation and strength properties as the consequence of the interaction between ordered constituents. Under wave propagation conditions, their behavior is also distinguished by the multiple scattering and interference phenomena, which can be constructive or destructive in a total sense. In many practical cases, such composites of 3D configuration may contain arranged fillers in the form of thin inclusions (Kadic et al. 2019). The singular character of stresses takes place in the vicinities of inhomogeneities, which should be investigated in the time-harmonic wave loading situation with the consideration of both particle interaction and frequency effects.

Most of the known theoretical results on many thin-walled scatterers in the 3D field of incident elastic waves are related to the elastodynamic problems for solids with systems of crack-like objects. For example, the interaction between the finite number of penny-shaped or square cracks in homogeneous and bi-material elastic solids was considered in Itou (2000), Mykhas'kiv et al. (2010). In addition, physically correct statements of frequency-domain problems, which allow dynamic cracks' closure, are proposed in Guz (2022), Menshykova et al. (2012), Mykhailova et al. (2011). The joint contribution of an infinite number of periodically located penny-shaped or rectangular cracks to a 3D elastic wave field was investigated in Golub and Doroshenko (2019a, b), Mykhas'kiv et al. (2014, 2019), Remizov et al. (2017). Obtained results show that the mutual dynamic influence of cracks can lead to pronounced shielding and amplification in terms of fracture parameters as well as selective wave blocking in terms of transmission coefficients depending on the frequency spectrum.

V. Mykhas'kiv (✉) · I. Zhbadynskyi
Pidstryhach Institute for Applied Problems of Mechanics and Mathematics, Lviv, Ukraine
e-mail: mykhaskiv@gmail.com

A. N. Guz et al. (eds.), *Advances in Mechanics*, Advanced Structured Materials 191,
https://doi.org/10.1007/978-3-031-37313-8_22

Regarding 3D frequency-domain problems with thin-walled inhomogeneities, investigations available in the literature are limited by the time-harmonic elastic wave incidence on single disk-shaped inclusion (Kit et al. 2002). Based on this solution, the effective dynamic properties of a 3D elastic matrix containing a dilute array of disk-shaped inclusions were estimated when the influence of inhomogeneities on each other could be neglected (see Mykhas'kiv et al. (2010, 2018)). However, to the best of the authors' knowledge, the dynamic interaction in the large systems of thin-walled inclusions was not considered. It should be mentioned that the periodic distribution of inhomogeneities is most adapted for such investigation because of the possibility of analysis in a unit cell of ordered structure.

For elastic solids with periodically arranged volumetric inclusions, it was demonstrated that among many analytical and numerical methods such as the plane-wave expansion method, the multiple scattering method, the finite element method (Mahmood et al. 2017; Maslov et al. 2000; Zhang et al. 2003; Zhao et al. 2015), and the boundary integral equation method (BIEM) can be successfully applied for corresponding wave propagation problems. Its efficiency is quite high due to the reduction of the problem dimensionality and the automatic satisfaction of the radiation conditions at infinity through the involvement of appropriate Green's functions. There are usually two ways to analyze the periodic structures by the BIEM. One way is to formulate the boundary integral equations for a unit cell according to the governing wave equations and the conventional Green's functions. Then the periodicity conditions are stated on the boundaries of the unit cell. The other way is that the periodicity conditions are directly substituted into the governing wave equations or integral representations of their solutions. Then the boundary integral equations are formulated on the surface of the representative scatterer with the periodic or quasi-periodic Green's functions in the kernels of corresponding equations. The global effect of the multiple scatterers is taken into account in the periodic Green's functions by the lattice sums, whose effective calculations can be realized by their dual, Ewald's-type or integral representations depending on the problem periodicity (see Bruno (2014), Linton (2010), Martin (2006)). The first way has been successfully applied for periodic problems to simulate the wave fields in 2D elastic solids containing periodic holes and fibers (Li et al. 2013; Wang et al. 2011). The second way was implemented to investigate elastic wave propagation in 3D solids with periodic spherical inclusions (Isakari et al. 2012). In the current chapter, this approach, which does not require the discretization of the boundary of the unit cell, is used to analyze the normal incidence of plane longitudinal elastic wave on an infinite array of one-periodic or chain-arranged coplanar rigid disk-shaped inclusions.

22.2 Boundary Integral Equation in the Domain of Reference Inclusion

Let us consider an infinite isotropic elastic matrix with the mass density ρ, the shear modulus μ, and Poisson's ratio ν containing a one-periodic array of absolutely rigid disk-shaped inclusions, which occupy the domains S_n $(n \in \mathbb{Z})$ as shown in Fig. 22.1, and have the same mass M and radius a. The inclusions are perfectly contacted with the matrix material and lie in the plane $x_3 = 0$ with their centers located on the x_1-axis with the periodic distance $d > 2a$.

The system of inclusions is subjected to an incident longitudinal time-harmonic wave with the amplitude factor u_0 and the circular frequency k, which propagates in the direction of the positive x_3-axis and has the following displacement components:

$$u_1^{in}(\mathbf{x}) = u_2^{in}(\mathbf{x}) = 0, \quad u_3^{in}(\mathbf{x}) = u_0 \exp(ik_1 x_3). \tag{22.1}$$

Here and hereafter, the common steady-state term $\exp(-ikt)$ is omitted, $k_j = k/c_j$ $(j = 1, 2)$ are the wave numbers,

$$c_2 = \sqrt{\mu/\rho}, \quad c_1 = \gamma c_2 \quad \left(\gamma = \sqrt{2(1-\nu)/(1-2\nu)}\right)$$

are the transverse and longitudinal wave velocities, respectively.

By using the superposition principle, the total displacement field $\mathbf{u} = (u_1, u_2, u_3)$ in a matrix with multiple wave scatterers can be written as

$$u_l(\mathbf{x}) = u_l^{in}(\mathbf{x}) + \sum_{n=-\infty}^{\infty} u_{ln}^{sc}(\mathbf{x}), \quad l = 1, 2, 3, \tag{22.2}$$

where $u_{ln}^{sc}(\mathbf{x})$ are the unknown displacements connected with the presence in the system of nth inclusion, which should satisfy the classical differential equation of motion and radiation conditions at infinity (Martin 2006).

In order to obtain the integral representations for the displacement components u_{ln}^{sc}, the Betty–Rayleigh reciprocity theorem is applied in conjunction with the properties

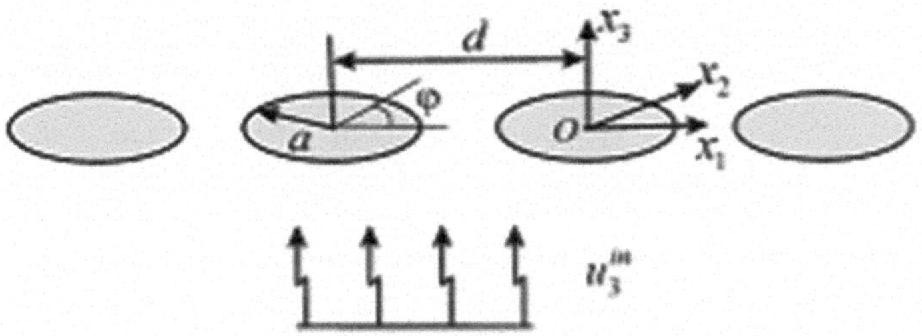

Fig. 22.1 Geometry of the problem

of the elastodynamic fundamental solutions. Then, these representations take a form similar to those for the single inclusion problem (Kit et al. 2002), namely,

$$u_{ln}^{sc}(\mathbf{x}) = \frac{1}{4\pi\mu}\iint_{S_n}\Delta\sigma_n(\xi)\left\{\delta_{3l}\frac{\exp\left(ik_2\left|\mathbf{x}-\boldsymbol{\xi}\right|\right)}{\left|\mathbf{x}-\boldsymbol{\xi}\right|} - \frac{1}{k_2^2}\frac{\partial^2}{\partial x_l \partial x_3}\right.$$
$$\left.\times\left[\frac{\exp\left(ik_1\left|\mathbf{x}-\boldsymbol{\xi}\right|\right)}{\left|\mathbf{x}-\boldsymbol{\xi}\right|} - \frac{\exp\left(ik_2\left|\mathbf{x}-\boldsymbol{\xi}\right|\right)}{\left|\mathbf{x}-\boldsymbol{\xi}\right|}\right]\right\}dS_\xi, \quad l = 1, 2, 3. \quad (22.3)$$

In (22.3), the displacement continuity conditions across the inclusions as absolutely rigid units are accounted for implicitly. In this equation, δ_{jl} is the Kronecker symbol, $|\mathbf{x}-\boldsymbol{\xi}|$ is the distance between the field point $\mathbf{x} = (x_1, x_2, x_3)$ and integration point $\boldsymbol{\xi} = (\xi_1, \xi_2)$, and $\Delta\sigma_n$ is the jump of the normal stresses σ_{33} in the domain S_n of nth inclusion, which is defined by

$$\Delta\sigma_n(\mathbf{x}) = \sigma_{33}^-(\mathbf{x}) - \sigma_{33}^+(\mathbf{x}), \quad \mathbf{x} \in S_n, \quad \sigma_{33}^{\pm}(\mathbf{x}) = \lim_{x_3 \to \pm 0} \sigma_{33}(\mathbf{x}). \quad (22.4)$$

It should be mentioned that the jumps of tangential stresses across the domains S_n are zeros due to the antisymmetry of the scattered wave field with respect to the plane $x_3 = 0$.

The inclusions are considered as rigid units, and their motion under the specified incident wave is described by the translations u_n^* in the direction of the x_3-axis only. Then the following displacement boundary conditions take place in the domains S_n

$$u_3(\mathbf{x}) = u_n^*, \quad (x_1, x_2, 0) \in S_n. \quad (22.5)$$

Equation (22.4) together with the equation of motion of each inclusion as a rigid unit yields the following relation between the translation of the nth inclusion and the stress jump $\Delta\sigma_n$

$$u_n^* = \frac{1}{k^2 M}\iint_{S_n}\Delta\sigma_n(\boldsymbol{\xi})\, dS_\xi\,. \quad (22.6)$$

Due to the periodicity of the problem of the normal incidence of the elastic longitudinal wave on a one-periodic array of inclusions, the following equalities can be postulated for the stress jumps and translation functions

$$\Delta\sigma_n(x_1 + nd, x_2) = \Delta\sigma(x_1, x_2), \quad u_n^* = u^*, \quad n \in \mathbb{Z}. \quad (22.7)$$

In (22.7), $\Delta\sigma = \Delta\sigma_0$ and $u^* = u_0^*$ are the stress jump across the faces of the reference inclusion $S = S_0$ and the translation of the reference inclusion, respectively.

Hence, the components of displacement in the matrix and the kinematical parameters of the inclusions are related to the stress jumps across the inclusions by the relations (22.3) and (22.6). The definition of these functions is based on the substitution of (22.3) into (22.2) and satisfaction of boundary conditions (22.5) with accounting for (22.6) and the periodicity conditions (22.7). In this way, the boundary integral equation for the stress jump $\Delta\sigma$ across the surfaces of reference inclusion is deduced as

$$\iint_S \Delta\sigma\,(\boldsymbol{\xi})\left[R\,(|\mathbf{x}-\boldsymbol{\xi}|)+G\,(\mathbf{x},\boldsymbol{\xi})\right]dS_\xi = -4\pi\mu k_2^2\,u_3^{in}(\mathbf{x}),\quad \mathbf{x}\in S, \tag{22.8}$$

Here the kernel R is the same as for a single rigid disk-shaped inclusion in an infinite elastic matrix (Kit et al. 2002), namely

$$R\,(r)=[1-ik_1r]\,\frac{\exp\,(ik_1r)}{r^3}-\left[1-ik_2r-k_2^2r^2\right]\frac{\exp\,(ik_2r)}{r^3}-\frac{4\pi\rho}{M}, \tag{22.9}$$

the kernel $G(\mathbf{x},\boldsymbol{\xi})$ describes the interaction of the reference inclusion S with the infinite number of inclusions in all replica cells and plays the role of one-periodic Green's function for the considered problem. In addition, this function has the following form of the infinite lattice sum

$$\begin{aligned}
G(\mathbf{x},\boldsymbol{\xi}) &= G_1(\mathbf{x},\boldsymbol{\xi})+G_2(\mathbf{x},\boldsymbol{\xi}),\\
G_1(\mathbf{x},\boldsymbol{\xi}) &= \sum_{n=-b+1}^{b-1}(1-\delta_{n0})R\left[r_n(\mathbf{x},\boldsymbol{\xi})\right],\\
G_2(\mathbf{x},\boldsymbol{\xi}) &= \sum_{n=b}^{\infty}\sum_{j=1}^{2}\mathbf{T}_j^x\left[\frac{\exp\left[ik_jr_n(\mathbf{x},\boldsymbol{\xi})\right]}{r_n(\mathbf{x},\boldsymbol{\xi})}\right]\\
&\quad+\sum_{n=-\infty}^{-b}\sum_{j=1}^{2}\mathbf{T}_j^x\left[\frac{\exp\left[ik_jr_n(\mathbf{x},\boldsymbol{\xi})\right]}{r_n(\mathbf{x},\boldsymbol{\xi})}\right].
\end{aligned} \tag{22.10}$$

Here $r_n(\mathbf{x},\boldsymbol{\xi})=\sqrt{(x_1-\xi_1-nd)^2+(x_2-\xi_2)^2}$, ($b\geqslant 2$) is a fixed integer number, $\mathbf{T}_j^x$ are the differential operators given by

$$\mathbf{T}_1^x=\frac{\partial^2}{\partial x_1^2}+\frac{\partial^2}{\partial x_2^2}+k_1^2,\qquad \mathbf{T}_2^x=-\frac{\partial^2}{\partial x_1^2}-\frac{\partial^2}{\partial x_2^2}. \tag{22.11}$$

From the geometrical point of view, the part of Green's function G, the function G_1 corresponds to the presence of "close" inclusions located in the domains S_n ($n\in[-b+1,b-1]$) on both sides from the reference inclusion, while another part of Green's function G, the function G_2, corresponds to the consideration of rest "far" inclusions. The reason for such a division lies in the fact that the infinite lattice sum G_2 in its direct form (22.10) converges very slowly (conditionally) with increasing frequency due to the presence of the oscillating terms. Therefore, its transformation is needed to provide an exponential convergence. It can be achieved by the special integral representation of the expression $\exp(ik_jr_n)/r_n$. In addition, separating the function G_1 (as the limited sum for the neighboring inclusions) from Green's function G aims to accelerate the convergence of the remaining term. Since $r_n\neq 0$, the function G_1 does not contain singularities and can be evaluated numerically without difficulties.

Thus, to rewrite the function G_2 in a more convenient form, the following integral expression is used (Gradshteyn and Ryzhik 2000):

$$\frac{\exp\left[ik_j r_n(\mathbf{x},\boldsymbol{\xi})\right]}{r_n(\mathbf{x},\boldsymbol{\xi})} = \int_0^\infty t\frac{\exp\left[-\left|x_1-\xi_1-nd\right|\sqrt{t^2-k_j^2}\right]}{\sqrt{t^2-k_j^2}} J_0\left[\left|x_2-\xi_2\right|t\right]dt. \quad (22.12)$$

Here and hereafter, J_n is the Bessel function of the nth order, and the square root is defined according to the radiation conditions at infinity:

$$\sqrt{t^2-k_j^2} = \begin{cases} \sqrt{t^2-k_j^2}, & |t| \geqslant k_j \\ -i\sqrt{k_j^2-t^2}, & |t| < k_j. \end{cases}$$

By using the integral representations (22.12), performing the corresponding summation of the geometrical series, and taking the derivatives present in the differential operators (22.11), the function $G_2(\mathbf{x},\boldsymbol{\xi})$ takes the exponentially convergent form

$$G_2(\mathbf{x},\boldsymbol{\xi}) = \frac{1}{d^3}\sum_{j=1}^{2}\int_0^\infty t\frac{\exp\left(-U_j(t)\right)}{U_j(t)\left[1-\exp\left(-U_j(t)\right)\right]} \times Y_j\left(\frac{x_1-\xi_1}{d},t\right)\Omega_j\left(\left|\frac{x_2-\xi_2}{d}\right|,t\right)dt, \quad (22.13)$$

where

$$U_j(t) = \sqrt{t^2-(k_j d)^2},\quad \Omega_1(r,t) = t\frac{J_1(rt)}{r},\quad \Omega_2(r,t) = k_2^2 d^2 J_0(rt) - t\frac{J_1(rt)}{r},$$
$$Y_j(r,t) = \exp\left[-(b-1-r)\,U_j(t)\right] + \exp\left[-(b-1+r)\,U_j(t)\right].$$

By substituting (22.13) into (22.10) and then into (22.8), we arrive at the boundary integral equation with the frequency-domain Green's function of exponentially convergent form for a one-periodic array of rigid disk-shaped inclusions.

22.3 Regularization and Discretization of Boundary Integral Equation

An efficient numerical solution of boundary integral equation (22.8) requires proper regularization procedures for its integral terms with the weak singularity $|\mathbf{x}-\boldsymbol{\xi}|^{-1}$ of the kernel R at the source point $\boldsymbol{\xi}=\mathbf{x}$ as well as the singularities in the integral representation (22.13) of the kernel G_2 at the points coinciding with the roots of the functions $U_j(t)$ due to the following behavior:

$$\left[U_j(t)\left(1-\exp\left(-U_j(t)\right)\right)\right]^{-1} \quad \to \quad \left[t^2-(k_jd)^2\right]^{-1},$$

if $U_j(t)$ tend to zero.

The singular integrals in the part G_2 of periodic Green's function (22.10) are regularized by the singularity subtraction technique. According to this technique, the exact values of the sums (Gradshteyn and Ryzhik 2000)

$$\sum_{n=1}^{\infty}\frac{\exp\left(ik_jdn\right)}{n}=\int_0^{\infty}t\frac{\exp\left(-U_j(t)\right)}{U_j(t)\left[1-\exp\left(-U_j(t)\right)\right]}dt=A_j+\pi iq_j, \quad (22.14)$$

which play the role of regularizing terms or integrals, are used.

In (22.14), the number $q_j=0,1,2,\ldots$ is determined from the condition $2\pi q_j < k_jd < 2\pi(q_j+1)$ and

$$A_j=\frac{1}{2}\left\{-\ln\left[2-2\cos(k_jd)\right]+i(\pi-k_jd)\right\}.$$

The closed form of the lattice sum (22.14) is valid for all k_jd except the values $k_jd=2\pi n$ $(n=1,2,\ldots)$. These wave numbers correspond to the cut-off wave numbers associated with the n-mode of the emerging transversal and longitudinal waves in the plane of chain-arranged inclusions.

The real part of the representation (22.14) is known as the Clausen function. It is suitable for the computations near the cut-off wave numbers using the rapidly converging series (Linton and McIver 2001)

$$\frac{1}{2}\ln\left[2-2\cos(k_jd)\right]=\ln(k_jd)+\sum_{l=1}^{\infty}\frac{(-1)^l B_{2l}(k_jd)^{2l}}{2l(2l)!},$$

where B_{2l} are the Bernoulli numbers.

On the base of relation (22.14), the regularized form of the lattice sum G_2 can be written as

$$\begin{aligned}G_2(\mathbf{x},\boldsymbol{\xi})=&\frac{1}{d^3}\sum_{j=1}^{2}\int_0^{\infty}t\frac{\exp\left(-U_j(t)\right)}{U_j(t)\left[1-\exp\left(-U_j(t)\right)\right]}\\&\times\left[Y_j\left(\frac{x_1-\xi_1}{d},t\right)\Omega_j\left(\left|\frac{x_2-\xi_2}{d}\right|,t\right)-2\Omega_j\left(\left|\frac{x_2-\xi_2}{d}\right|,k_jd\right)\right]dt\\&+\frac{1}{d^3}\sum_{j=1}^{2}\left[2A_j\Omega_j\left(\left|\frac{x_2-\xi_2}{d}\right|,k_jd\right)+F_j\left(\frac{x_1-\xi_1}{d},\left|\frac{x_2-\xi_2}{d}\right|\right)\right].\end{aligned}$$

(22.15)

Here F_j $(j = 1, 2)$ are the known functions, which depend on the considered range of the wave number parameter $k_j d$. For instance, in the range $0 < k_j d < 4\pi$, they are given by

$$F_j(r,\tau) = \begin{cases} 0, & 0 < k_j d < 2\pi \\ \pi i Y_l\left(r, \sqrt{(k_j d)^2 - 4\pi^2}\right) & \\ \quad \times\, \Omega_j\left(\tau, \sqrt{(k_j d)^2 - 4\pi^2}\right), & 2\pi < k_j d < 4\pi. \end{cases}$$

Due to the fast convergence of the integrand involved in the resulting regularized form of the function (22.15), the corresponding infinite integral can be truncated in the numerical scheme.

To isolate the weak singularity in the kernel R, we similarly rewrite the boundary integral equation (22.8):

$$A \iint_S \frac{\Delta\sigma(\boldsymbol{\xi})}{|\mathbf{x}-\boldsymbol{\xi}|} dS_\xi + \iint_S \Delta\sigma(\boldsymbol{\xi}) \left[\frac{1}{k_2^2} R(|\mathbf{x}-\boldsymbol{\xi}|) - \frac{A}{|\mathbf{x}-\boldsymbol{\xi}|}\right] dS_\xi$$
$$+ \frac{1}{k_2^2} \iint_S \Delta\sigma(\boldsymbol{\xi})\, G(\mathbf{x},\boldsymbol{\xi})\, dS_\xi = -4\pi\mu\, u_3^{in}(\mathbf{x}), \quad \mathbf{x} \in S, \quad (22.16)$$

where $A = (1+\gamma^2)/2$.

In (22.16), the second integral on the left-hand side exists in the ordinary sense. This fact follows from an analysis of the integrand in the limit $\boldsymbol{\xi} \to \mathbf{x}$. Therefore, in the numerical evaluation of this integral, performing the integration over S_x^0 by excluding a small region around the point $\mathbf{x}$ in S is correct.

For a circular rigid disk-shaped inclusion, the stress jump has "square-root" behavior in the vicinity of the inclusion front, which can be expressed as

$$\Delta\sigma(\mathbf{x}) = \frac{\beta(\mathbf{x})}{\sqrt{a^2 - x_1^2 - x_2^2}}, \quad \mathbf{x} \in S, \quad (22.17)$$

where $\beta(\mathbf{x})$ is a sufficiently smooth function.

Substitution of (22.17) into (22.16) results in the boundary integral equation for the new unknown function $\beta(\mathbf{x})$. It has the weak singularity $1/|\mathbf{x}-\boldsymbol{\xi}|$ at the source point $\boldsymbol{\xi} = \mathbf{x}$ and the "square-root" singularity at the front of the inclusion. To regularize the boundary integral equation at the source point, the following interpretation of the involved integral is utilized

$$\iint_S \frac{\beta(\boldsymbol{\xi})}{\sqrt{a^2 - \xi_1^2 - \xi_2^2} \cdot |\mathbf{x}-\boldsymbol{\xi}|} dS_\xi$$
$$= \pi^2 \beta(\mathbf{x}) + \iint_{S_x^0} \frac{\beta(\boldsymbol{\xi}) - \beta(\mathbf{x})}{\sqrt{a^2 - \xi_1^2 - \xi_2^2} \cdot |\mathbf{x}-\boldsymbol{\xi}|} dS_\xi, \quad (22.18)$$

with consideration of the exact value

$$\iint_S \frac{dS_\xi}{\sqrt{a^2 - \xi_1^2 - \xi_2^2}\,|\mathbf{x} - \boldsymbol{\xi}|} = \pi^2.$$

Next, the change of variables $\mathbf{x} = f(\mathbf{y})$, $\boldsymbol{\xi} = f(\boldsymbol{\eta})$ is performed as

$$\begin{cases} x_1 = a \sin y_1 \cos y_2 \\ x_2 = a \sin y_1 \sin y_2 \end{cases}, \qquad \begin{cases} \xi_1 = a \sin \eta_1 \cos \eta_2 \\ \xi_2 = a \sin \eta_1 \sin \eta_2 \end{cases}, \tag{22.19}$$

where $\mathbf{y} = (y_1, y_2)$ and $\boldsymbol{\eta} = (\eta_1, \eta_2)$ are new variables in the rectangular domain $\tilde{S} = \{\mathbf{y}, \boldsymbol{\eta} : 0 \leqslant y_1, \eta_1 \leqslant \pi/2;\ \ 0 \leqslant y_2, \eta_2 \leqslant 2\pi\}$. The transformation (22.19) maps the circular integration domain to a rectangular integration domain and eliminates the "square-root" singularity at the front of the inclusion corresponding to $\eta_1 = \pi/2$.

Application of relations (22.17)–(22.19) in the boundary integral equation (22.16) yields its complete regularized version

$$A\tilde{\beta}(\mathbf{y}) \left[\pi^2 - \iint_{\tilde{S}_y^0} \frac{\sin \eta_1}{L(\mathbf{y}, \boldsymbol{\eta})} dS_\eta \right] + \frac{1}{\vartheta^2} \iint_{\tilde{S}_y^0} \tilde{\beta}(\boldsymbol{\eta})\, \tilde{R}(\mathbf{y}, \boldsymbol{\eta}) \sin \eta_1 dS_\eta$$
$$+ \frac{1}{\vartheta^2} \iint_{\tilde{S}} \tilde{\beta}(\boldsymbol{\eta})\, \tilde{G}(\mathbf{y}, \boldsymbol{\eta}) \sin \eta_1 dS_\eta = -4\pi \mu \tilde{u}_3^{in}(\mathbf{y}), \quad \mathbf{y} \in \tilde{S}, \tag{22.20}$$

where

$$\tilde{\beta}(\mathbf{y}) = \beta_j(f(\mathbf{y})), \quad \tilde{R}(\mathbf{y}, \boldsymbol{\eta}) = a^3 R(f(\mathbf{y}), f(\boldsymbol{\eta})),$$
$$\tilde{G}(\mathbf{y}, \boldsymbol{\eta}) = a^3 G(f(\mathbf{y}), f(\boldsymbol{\eta})), \quad \tilde{u}_3^{in}(\mathbf{y}) = u_3^{in}(f(\mathbf{y})).$$

In (22.20), $\vartheta = k_2 a$ is the normalized wave number, $\tilde{S}_y^0$ is the mapping of the domain S_x^0 due to the transformation (22.19) (in the domain $\tilde{S}_y^0$, the points $\mathbf{y}$ and $\boldsymbol{\eta}$ do not coincide), and

$$L(\mathbf{y}, \boldsymbol{\eta}) = \sqrt{\sin^2 y_1 + \sin^2 \eta_1 - 2 \sin y_1 \sin \eta_1 \cos(y_2 - \eta_2)}.$$

The discretization of the boundary integral equation (22.20) is performed using a collocation scheme, where the boundary element mesh is formed by dividing the domain $\tilde{S}$ into identical rectangular boundary elements. The centroids of each element are chosen as collocation points. This analysis adopts the constant approximation for the function $\beta(\mathbf{y})$ using the standard Gaussian quadratures for the integration. The appropriate choice of the mesh density and truncation limits of the infinite integrals controls the accuracy of the numerical solution.

22.4 Determination of Actual Dynamic Quantities

After the solution of boundary integral equation (22.20), the displacement u^* or the translation of reference inclusion as a rigid unit can be obtained by means of the relations (22.17), (22.7), and (22.6).

For the definition of dynamic fracture parameters or dynamic stress intensity factors in the vicinity of a reference inclusion, let us introduce the local coordinate system $Otqz$ with the center in the inclusion contour point so that the value $z = 0$ corresponds to the inclusion plane, the axes Ot and Oq lie in the normal and tangential directions relative to the inclusion front line, respectively. Then, the corresponding displacement and stress components at the arbitrary point P near the inclusion in the plane $q = 0$ can be approximated as

$$u_z(r,\psi,\varphi) = \frac{\sqrt{2r}}{2\mu}\sin\frac{\psi}{2}\left(1-\sin^2\frac{\psi}{2}\right)K_{\mathrm{I}}(\varphi)$$

$$+\frac{\sqrt{2r}}{2\mu}\cos\frac{\psi}{2}\left(2-4\nu+\cos^2\frac{\psi}{2}\right)K_{\mathrm{II}}(\varphi)+O\left(r^{3/2}\right),$$

$$u_t(r,\psi,\varphi) = -\frac{\sqrt{2r}}{2\mu}\cos\frac{\psi}{2}\left(4-4\nu-\cos^2\frac{\psi}{2}\right)K_{\mathrm{I}}(\varphi)$$

$$-\frac{\sqrt{2r}}{2\mu}\sin\frac{\psi}{2}\left(1-\sin^2\frac{\psi}{2}\right)K_{\mathrm{II}}(\varphi)+O\left(r^{3/2}\right),$$

$$u_q(r,\psi,\varphi) = \frac{\sqrt{2r}}{2\mu}\cos\frac{\psi}{2}K_{\mathrm{III}}(\varphi)+O\left(r^{3/2}\right),$$

$$\sigma_{zz}(r,\psi,\varphi) = -\frac{1}{\sqrt{2r}}\cos\frac{\psi}{2}\left(1-2\nu-\sin\frac{\psi}{2}\sin\frac{3\psi}{2}\right)K_{\mathrm{I}}(\varphi)$$

$$+\frac{1}{\sqrt{2r}}\sin\frac{\psi}{2}\left(2-2\nu-\cos\frac{\psi}{2}\cos\frac{3\psi}{2}\right)K_{\mathrm{II}}(\varphi)+O(1),$$

$$\sigma_{tt}(r,\psi,\varphi) = \frac{1}{\sqrt{2r}}\cos\frac{\psi}{2}\left(3-2\nu-\sin\frac{\psi}{2}\sin\frac{3\psi}{2}\right)K_{\mathrm{I}}(\varphi)$$

$$+\frac{1}{\sqrt{2r}}\sin\frac{\psi}{2}\left(2\nu+\cos\frac{\psi}{2}\cos\frac{3\psi}{2}\right)K_{\mathrm{II}}(\varphi)+O(1),$$

$$\sigma_{qq}(r,\psi,\varphi) = \frac{2\nu}{\sqrt{2r}}\cos\frac{\psi}{2}K_{\mathrm{I}}(\varphi)+\frac{2\nu}{\sqrt{2r}}\sin\frac{\psi}{2}K_{\mathrm{II}}(\varphi)+O(1),$$

$$\sigma_{zq}(r,\psi,\varphi) = \frac{1}{\sqrt{2r}}\sin\frac{\psi}{2}K_{\mathrm{III}}(\varphi)+O(1),$$

$$\sigma_{tq}(r,\psi,\varphi) = -\frac{1}{\sqrt{2r}}\cos\frac{\psi}{2}K_{\mathrm{III}}(\varphi)+O(1),$$

$$\sigma_{zt}(r,\psi,\varphi) = \frac{1}{\sqrt{2r}}\sin\frac{\psi}{2}\left(2-2\nu+\cos\frac{\psi}{2}\cos\frac{3\psi}{2}\right)K_{\mathrm{I}}(\varphi)$$

$$+\frac{1}{\sqrt{2r}}\cos\frac{\psi}{2}\left(1-2\nu+\sin\frac{\psi}{2}\sin\frac{3\psi}{2}\right)K_{\mathrm{II}}(\varphi)+O(1). \tag{22.21}$$

Here r and ψ are the polar coordinates of the point P; φ is the angular coordinate of the inclusion front point; K_{I}, K_{II}, and K_{III} are the modes I, II, and III dynamic stress intensity factors in the inclusion vicinity (Kassir and Sih 1968).

Using (22.21), the K-factors can be defined directly from the function $\beta(\mathbf{x})$ or the solution of boundary integral equation (22.20). For the considered wave scattering problem $K_{\mathrm{I}} = K_{\mathrm{III}} = 0$, and

$$K_{\mathrm{II}}(\varphi) = -\frac{\beta(\mathbf{x})}{4(1-\nu)\sqrt{a}}\bigg|_{\substack{x_1=a\cos\varphi\\ x_2=a\sin\varphi}}, \tag{22.22}$$

where the angular coordinate φ is counted from the Ox_1-axis (see Fig. 22.1).

22.5 Numerical Analysis

For the calculation, 176 rectangular boundary elements with the lengths $\pi/22$ and $\pi/12$ are used on the domain $\tilde{S}$ of the reference inclusion in the chain-arranged massif. The displacement u_0 of the incident wave and the value $K_0 = u_0\mu/\left[4(1-\nu)\sqrt{a}\right]$ are chosen as the normalizing factors for the translation of inclusions as the rigid units and the dynamic stress intensity factor in the inclusion vicinity, respectively, so that $\bar{u} = |u^*|/u_0$ and $\bar{K} = |K_{\mathrm{II}}|/K_0$. Poisson's ratio is taken as $\nu = 0.3$; the normalized inclusion mass is introduced as $\bar{M} = M/(\rho a^3)$ and the periodicity length as $\bar{d} = d/a$.

Testing accuracy and stability of the proposed numerical algorithm is fulfilled for the static analog of the problem, which is achieved in the limit $\vartheta \to 0$ and has trivial solution $u^* = u_0$, $K_{\mathrm{II}} = 0$. A good agreement is observed in this case.

For comparison purposes, the marked curves also show the behavior of pertinent parameters for a single inclusion subjected to the same time-harmonic excitation.

It follows from Figs. 22.2 and 22.3 that as the wave number increases the amplitudes of translation of inclusions $\bar{u}$ increase from unity, when $\vartheta = 0$ (the static value), to the maximums and then decrease and approach their single inclusion counterparts at high frequencies. Therefore, the effect of dynamic interaction in the periodic system of inclusions is leveled in the range of large wave numbers. At low frequencies, the displacements $\bar{u}$ for interacting inclusions are less than for a single inclusion situation what means the exhibition of vibration damping due to the inclusion neighborhood. In the range of intermediate frequencies, the displacements $\bar{u}$ can be both bigger and smaller than the single inclusion counterparts depending on the wave number. The frequencies with the peaks of the displacements $\bar{u}$ are higher for the periodic systems with inclusions of smaller mass (Fig. 22.2) and larger periodicity length (Fig. 22.3).

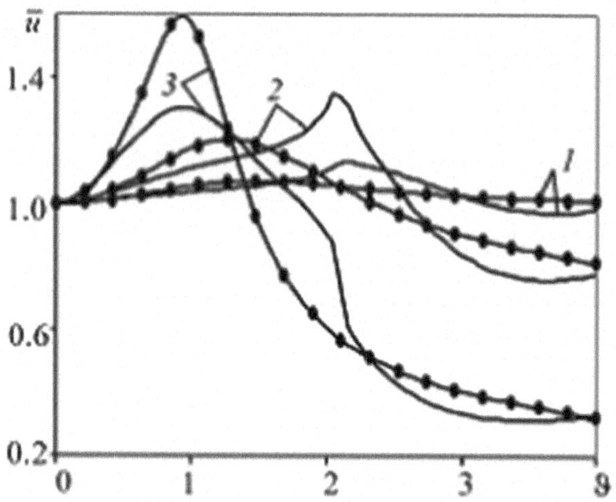

Fig. 22.2 Normalized translation $\bar{u}$ of reference inclusion of different mass in one-periodic array with periodicity length $\bar{d} = 3.0$ as a function of dimensionless wave number ϑ: 1—$\bar{M} = 1$, 2—$\bar{M} = 3$, 3—$\bar{M} = 10$

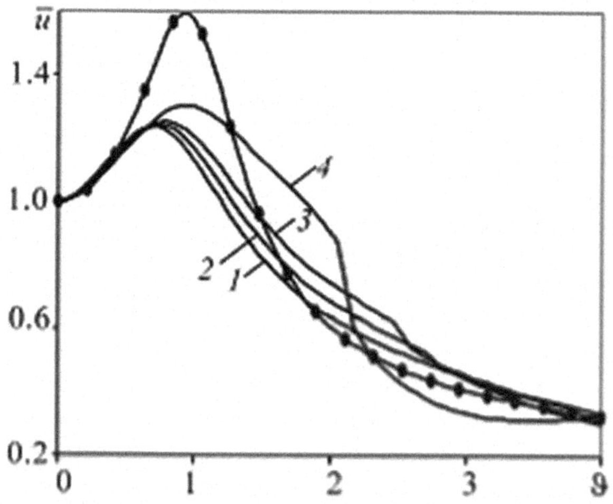

Fig. 22.3 Normalized translation $\bar{u}$ of reference inclusion of mass $\bar{M} = 10$ in one-periodic arrays with different periodicity length as a function of dimensionless wave number ϑ: 1—$\bar{d} = 2.1$, 2—$\bar{d} = 2.3$, 3—$\bar{d} = 2.5$, 4—$\bar{d} = 3.0$

Since involved into analysis problem for the considered configuration and disturbing wave is not axisymmetric relative Ox_3-axis due to the influence of neighboring inclusions on the reference one (unlike a single inclusion case), the dynamic stress intensity factor $\bar{K}$ changes from point to point around the inclusion front while maintaining natural symmetry relative to the periodicity axis Ox_1 and perpendicular direction Ox_2. This fact is demonstrated in Fig. 22.4, where the possibilities of $\bar{K}$-factor maximums both at the nearest and most distant (relative to the neighboring inclusions) front points of the inclusion with $\varphi = 0,\ \pi$ and $\varphi = \pi/2,\ 3\pi/2$, respectively, are detected. However, the closest points are most sensitive to interaction effects since, at these points, the $\bar{K}$-values differ the most from the values of a single inclusion.

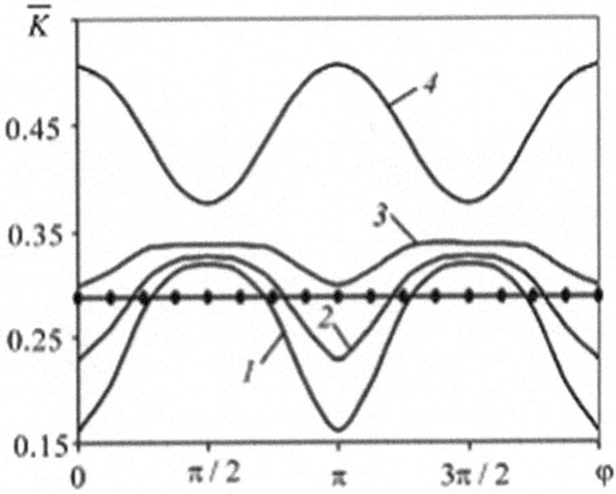

Fig. 22.4 Normalized dynamic stress intensity factor $\bar{K}$ of reference inclusion of mass $\bar{M} = 10$ in one-periodic arrays with different periodicity lengths versus polar angle φ of the inclusion front point: 1—$\bar{d} = 2.1$, 2—$\bar{d} = 2.3$, 3—$\bar{d} = 2.5$, 4—$\bar{d} = 3.0$

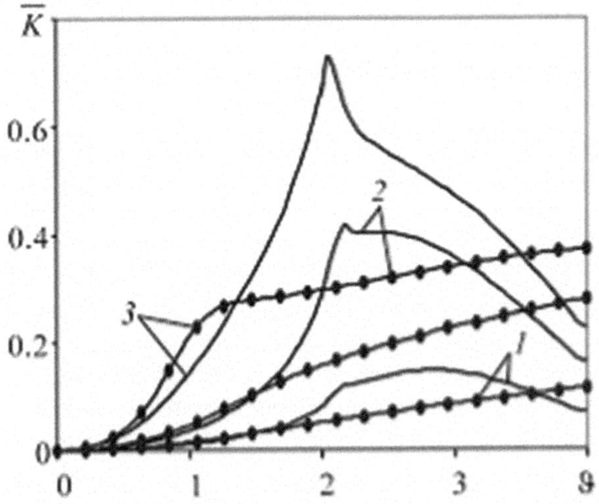

Fig. 22.5 Normalized dynamic stress intensity factor $\bar{K}$ at the front point $\varphi = 0$ of reference inclusion of different mass in a one-periodic array with periodicity length $\bar{d} = 3.0$ versus dimensionless wave number ϑ: 1—$\bar{M} = 1$, 2—$\bar{M} = 3$, 3—$\bar{M} = 10$

In Figs. 22.5 and 22.6, where the frequency dependencies are depicted, $\bar{K}$-factor first increases from zero with the dominance for the periodic inclusions of large mass and the chain arrays with bigger periodicity length. A further increase in ϑ in the case of periodic inclusions with the periodicity length $\bar{d} = 3.0$ leads to a local maximum of $\bar{K}$. At low frequencies, the interaction of inclusions causes the decrease in $\bar{K}$-factor compared to a single inclusion situation. In other words, the periodic system provides a shielding effect on the dynamic stresses in the vicinity of inclusion. However, with increasing the frequency, an opposite tendency for $\bar{K}$-factor takes place, i.e., this effect changes to the amplification one.

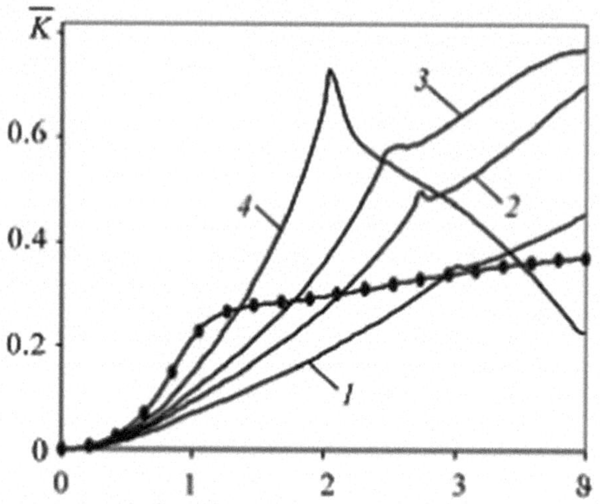

Fig. 22.6 Normalized dynamic stress intensity factor $\bar{K}$ at the front point $\varphi = 0$ of reference inclusion of mass $\bar{M} = 10$ in one-periodic arrays with different periodicity lengths versus dimensionless wave number ϑ: 1—$\bar{d} = 2.1$, 2—$\bar{d} = 2.3$, 3—$\bar{d} = 2.5$, 4—$\bar{d} = 3.0$

22.6 Conclusions

An efficient BIEM for numerical simulation of elastic wave propagation in a 3D infinite elastic matrix with a one-periodic array of coplanar disk-shaped inclusions is presented. The problem with the normal incidence of a plane longitudinal time-harmonic wave on the absolute rigid massive inclusions is considered for convenience. The boundary integral equation formulation for the reference inclusion in a unit cell is improved by presenting the involved periodic Green's function or the infinite lattice sum as exponentially convergent Fourier integrals. In addition, the special properties of the periodic Green's function at the cut-off frequencies are implicitly considered in its equivalent and regularized form. Numerically revealed phenomena confirm that the improved deformability and strength dynamic properties of composite materials can be achieved by using ordered rigid thin-walled inclusions as the structural elements.

The proposed approach provides an efficient way of studying wave fields in 3D elastic solids with periodic thin-walled inclusions for more general cases. They include both longitudinal and transverse elastic waves incident with arbitrary wave vectors or Bloch-type periodical conditions. The present method can be generalized on doubly periodic disk-shaped inclusions. Also, the problems related to phononic crystals and acoustic metamaterials can be analyzed by considering wave penetration through multiple equidistant planes of doubly periodic disk-shaped inclusions or cascade structures. Then, the so-called "wide-spacing approximation" model (Martin 2006) can be used for the determination of the transmission coefficients by correspondingly multiplying the scattering matrix for a single plane of the periodic inclusions.

References

Bruno OP, Delourme B (2014) Rapidly convergent two-dimensional quasi-periodic Green function throughout the spectrum - including Wood anomalies. J Comput Phys 262:262–290

Golub MV, Doroshenko OV (2019a) Boundary integral equation method for simulation scattering of elastic waves obliquely incident to a doubly periodic array of interface delaminations. J Comput Phys 376:675–693

Golub MV, Doroshenko OV (2019b) Wave propagation through an interface between dissimilar media with a doubly periodic array of arbitrary shaped planar delaminations. Math Mech Solids 24:483–498

Gradshteyn IS, Ryzhik IM (2000) Tables of integrals, series and products. Academic Press, New York

Guz AN (2022) Problem 7. Brittle fracture of materials with cracks under action of dynamic loads (with allowance for contact interaction of the crack edges). Adv Struct Mater 159:335–349

Isakari H, Niino K, Yoshikawa H, Nishimura N (2012) Calderon's preconditioning for periodic fast multipole method for elastodynamics in 3D. Int J Numer Methods Eng 90:484–505

Itou S (2000) Three-dimensional dynamic stress intensity factors around two parallel square cracks in an infinite elastic medium subjected to a time-harmonic stress wave. Acta Mech 143:79–90

Kadic M, Milton GW, Van Hecke M (2019) 3D metamaterials. Nat Rev Phys 1:198–210

Kassir MK, Sih GC (1968) Some three-dimensional inclusion problems in elasticity. Int J Solids Struct 4:225–241

Kit HS, Mykhas'skiv VV, Khay OM (2002) Analysis of the steady oscillations of a plane absolutely rigid inclusion in a three-dimensional elastic body by the boundary element method. J Appl Math Mech 66:817–824

Li FL, Wang YS, Zhang Ch, Yu GL (2013) Boundary element method for band gap calculation of two-dimensional solid phononic crystals. Eng Anal Bound Elem 37:225–235

Linton CM (2010) Lattice sums for the Helmholtz equation. SIAM Rev 52:630–674

Linton CM, McIver P (2001) Handbook of mathematical techniques for wave/structure interactions. Chapman & Hall/CRC, London

Mahmood MS, Laghrouche O, Trevelyan J, El Kacimi A (2017) Implementation and computational aspects of a 3D elastic wave modelling by PUFEM. Appl Math Model 49:568–586

Martin PA (2006) Multiple scattering: interaction of time-harmonic waves with N obstacles. Cambridge University Press, Cambridge

Maslov K, Kinra VK, Henderson BK (2000) Elastodynamic response of a coplanar periodic layer of elastic spherical inclusions. Mech Mater 32:785–795

Menshykova MV, Menshykov OV, Mikucka VA, Guz IA (2012) Interface cracks with initial opening under harmonic loading. Compos Sci Technol 72:1057–1063

Mykhailova II, Menshykov OV, Guz IA (2011) Cracks' closure in 3-D fracture dynamics: the effect of relative location of two coplanar cracks. Arch Appl Mech 81:1215–1230

Mykhas'kiv VV, Khay OM, Zhang Ch, Boström A (2010) Effective dynamic properties of 3D composite materials containing rigid penny-shaped inclusions. Waves Random Complex Media 20:491–510

Mykhas'kiv V, Kunets Ya, Matus V, Khay O (2018) Elastic wave dispersion and attenuation caused by multiple types of disc-shaped inclusions. Int J Struct Integr 9:219–232

Mykhas'kiv VV, Zhbadynskyi IYa, Zhang Ch (2014) Dynamic stresses due to time-harmonic elastic wave incidence on doubly periodic array of penny-shaped cracks. J Math Sci 203:114–122

Mykhas'kiv V, Zhbadynskyi I, Zhang Ch (2010) Elastodynamic analysis of multiple crack problem in 3-D bi-materials by a BEM. Int J Numer Methods Biomed Eng 26:1934–1946

Mykhas'kiv VV, Zhbadynskyi IYa, Zhang Ch (2019) On propagation of time-harmonic elastic waves through a double-periodic array of penny-shaped cracks. Eur J Mech Solids 73:306–317

Remizov MYu, Sumbatyan MA (2017) Three-dimensional one-mode penetration of elastic waves through a doubly periodic array of cracks. Math Mech Solids 23:636–650

Wang YF, Wang YS, Su XX (2011) Large bandgaps of two-dimensional phononic crystals with cross-like holes. J Appl Phys 110:113520
Zhang X, Liu Z, Liu Y, Wu F (2003) Elastic wave band gaps for three-dimensional phononic crystals with two structural units. Phys Lett A 313:455–460
Zhao J, Li Y, Liu WK (2015) Predicting band structure of 3D mechanical metamaterials with complex geometry via XFEM. Comput Mech 55:659–672

Chapter 23
Using the Wiener–Hopf Method in Solving the Problem of Elastic Equilibrium of a Bihomogeneous Body with Interfacial Cracks at the Corner Point of a V-Shaped Interface

Volodimir Nazarenko and Alexander Kipnis

23.1 Introduction

In the manufacture of structures from a layered composite material, a typical situation is when the presence of a defect between the layers of the composite leads to the fact that this surface ceases to be smooth but becomes such that it contains a corner point. In the case when this point is a stress concentrator, discontinuity and the initiation of interfacial cracks, including shear cracks with fully contacting edges, are possible in its vicinity. Therefore, the study of the stress state of piecewise homogeneous bodies weakened by interfacial cracks that originated at the corner point of the interface is an actual problem of the mechanics of deformable solids.

In the major part of works devoted to the investigation of two-dimensional static problems of the theory of elasticity for wedge-shaped bodies with cracks at the vertices, the analyzed bodies are considered homogeneous (Bantsuri 1966; Keer et al. 1977; Khrapkov 1971; Nekislykh and Ostrik 2010; Ouchterlony 1977; Stone and Westmann 1981). In the case of piecewise homogeneous bodies, the researchers studied the problems of the lines of discontinuity of displacements at the ends of interfacial cracks and also at the end of a crack appearing on the interface of the media (Kuliev et al. 1978; Loboda and Sheveleva 2003). At the same time, the problems of fracture mechanics devoted to the analysis of the stress state at angular points of piecewise homogeneous elastic bodies were not studied in the presence of interfacial cracks at these points. Among the methods for studying problems of fracture mechanics of materials for bodies with cracks (Bogdanov 2010; Bogdanov et al. 2015; Guz 2014; Guz et al. 2013), one of the most effective is the Wiener–Hopf methods (Noble 1988).

V. Nazarenko · A. Kipnis (✉)
S.P. Timoshenko Institute of Mechanics, National Academy of Science of Ukraine, Kyiv, Ukraine
e-mail: a.l.kipnis@gmail.com

V. Nazarenko
e-mail: v.m.n@inmech.kiev.ua

A. N. Guz et al. (eds.), *Advances in Mechanics*, Advanced Structured Materials 191,
https://doi.org/10.1007/978-3-031-37313-8_23

Below, using the Wiener–Hopf method, an exact solution to the problem of fracture mechanics of small-scale interfacial shear cracks that originate at the corner point of the interface between isotropic elastic media was constructed. It is assumed that there is no friction between the crack faces. The exponent of the stress singularity at the corner point and the stress intensity factor at the end of the interfacial crack are determined.

23.2 Statement of the Problem

Under the conditions of plane deformation, within the framework of a static symmetric problem of the theory of elasticity, we consider a piecewise homogeneous body with a V-shaped interface, formed by isotropic elastic parts with Young's moduli E_1 and E_2 ($E_1 > E_2$), and Poisson's ratios ν_1 and ν_2 (Fig. 23.1).

The corner point O on the interface of media plays the role of stress concentrator with power singularity, and the stresses in the vicinity of this point have the following representation:

$$\begin{aligned} \tau_{r\theta}(r,0) &= Cg_1(\alpha, e_0, \nu_1, \nu_2)r^{\lambda_0} + o(r^{\lambda_0}), \\ \sigma_\theta(r,0) &= Cg_2(\alpha, e_0, \nu_1, \nu_2)r^{\lambda_0} + o(r^{\lambda_0}), \quad r \to 0, \\ e_0 &= E_1/E_2. \end{aligned} \tag{23.1}$$

Here g_1 and g_2 are known functions, and the exponent of the stress singularity λ_0 is a single root in the interval $(-1, 0)$ of the characteristic equation of the corresponding homogeneous problem (*problem K*):

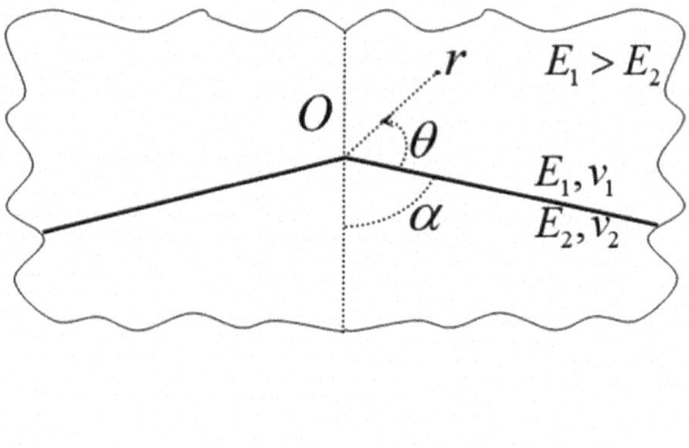

Fig. 23.1 A piecewise homogeneous body with a V-shaped interface

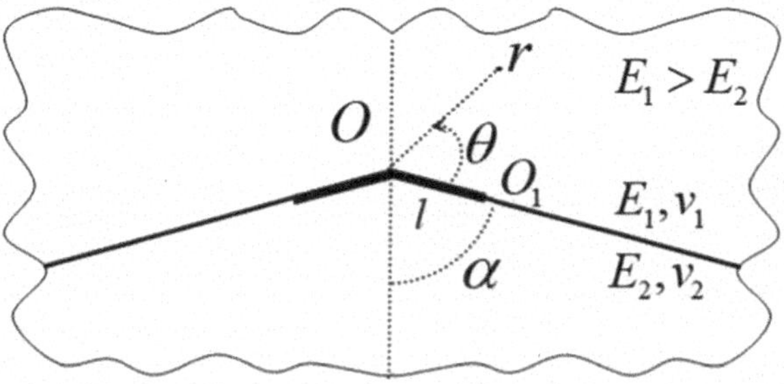

Fig. 23.2 A piecewise homogeneous body in the presence of interfacial cracks

$$
\begin{aligned}
&\Delta(-\lambda - 1) = 0,\\
&\quad \Delta(z) = \delta_0(z) + \delta_1(z)e + \delta_2(z)e^2,\\
&\quad \delta_0(z) = (\sin 2z\alpha + z \sin 2\alpha)[\varkappa_1 \sin 2z(\pi - \alpha) + z \sin 2\alpha],\\
&\quad \delta_1(z) = (1 + \varkappa_1)(1 + \varkappa_2) \sin^2 z\pi\\
&\qquad\quad - (\sin 2z\alpha + z \sin 2\alpha)[\varkappa_1 \sin 2z(\pi - \alpha) + z \sin 2\alpha]\\
&\qquad\quad - [\sin 2z(\pi - \alpha) - z \sin 2\alpha](\varkappa_2 \sin 2z\alpha - z \sin 2\alpha),\\
&\quad \delta_2(z) = [\sin 2z(\pi - \alpha) - z \sin 2\alpha](\varkappa_2 \sin 2z\alpha - z \sin 2\alpha),
\end{aligned}
\tag{23.2}
$$

where $e = (1 + \nu_2)/(1 + \nu_1)e_0$, $\varkappa_{1,2} = 3 - 4\nu_{1,2}$.

The constant C appearing in representations (23.1) is determined from the solution of the corresponding specific problem of the theory of elasticity for a body depicted in Fig. 23.1. This constant depends on the external load and can be regarded as a stress intensity factor at the angular corner on the interface of materials.

A necessary condition for the initiation of interfacial *shear* cracks with fully contacting edges at a corner point (Fig. 23.2) is a high concentration of normal compressive stresses in its vicinity ($\sigma_\theta(r, 0) \to -\infty$ as $r \to 0$). The calculation results show that the specified condition is satisfied if $C < 0$ and $\alpha \in (\alpha_1; \pi/2) \cup (\alpha_2; \pi)$, where α_1 and α_2 are zeros of a function $g_2(\alpha, e_0, \nu_1, \nu_2)$ $(0 < \alpha_1 < \pi/2 < \alpha_2 < \pi)$.

Some values of the exponent of stress singularity are presented in Table 23.1 for different ratios of Young's moduli of materials depending on the angle, and values α_1 and α_2—in Table 23.2.

We further determine the stress intensity factor K_{II} at the tip O_1 of the crack and the exponent of stress singularity at the corner point of the interface (Fig. 23.2).

In view of the small lengths of the cracks, we arrive at a plane static symmetric problem of the theory of elasticity for a piecewise homogeneous isotropic plane with a V-shaped interface containing cuts of finite length originating from the corner point and located on this interface (Fig. 23.3).

At infinity, we specify the asymptotics of the stress field obtained as the solution (23.1) of a similar problem without cuts (*problem K*), i.e., a homogeneous problem of the theory of elasticity for a piecewise homogeneous plane with a V-shaped interface.

Table 23.1 Exponents of stress singularity

α°	e_0			
	2	3	5	10
15	–0.036	–0.068	–0.122	–0.215
30	–0.075	–0.132	–0.232	–0.310
45	–0.112	–0.180	–0.258	–0.332
60	–0.112	–0.184	–0.248	–0.308
75	–0.086	–0.127	–0.167	–0.203
105	–0.025	–0.037	–0.049	–0.059
120	–0.054	–0.081	–0.104	–0.124
135	–0.089	–0.130	–0.168	–0.202
150	–0.117	–0.173	–0.228	–0.278
165	–0.104	–0.168	–0.241	–0.318

Table 23.2 Values of angles α_1 and α_2

e_0	2	3	5	10
α_1°	38.2	34.4	29.3	21.7
α_2°	134.2	133.4	133.1	131.3

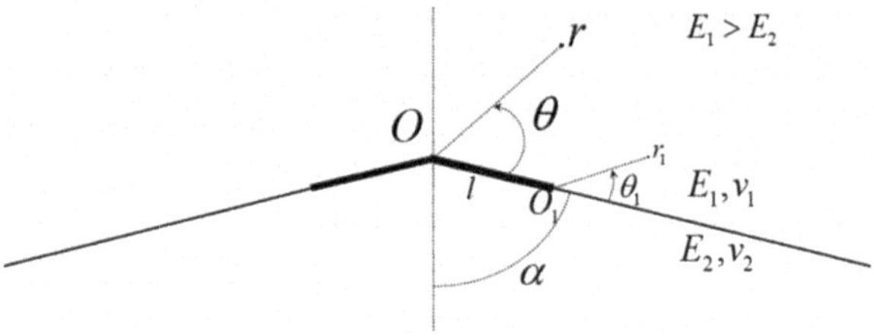

Fig. 23.3 A piecewise homogeneous isotropic plane with a V-shaped interface

The constant C is assumed to be known. It characterizes the intensity of the external load and must be determined from the solution to the external problem.

Taking into account symmetry, the boundary conditions of the problem (Fig. 23.3) are as follows:

$$\begin{gathered}\theta = \pi - \alpha, \quad \tau_{r\theta} = 0, \quad u_\theta = 0, \quad \theta = -\alpha, \quad \tau_{r\theta} = 0, \\ u_\theta = 0, \quad \theta = 0, \quad \langle \sigma_\theta \rangle = \langle \tau_{r\theta} \rangle = 0, \quad \langle u_\theta \rangle = 0,\end{gathered} \tag{23.3}$$

$$\theta = 0, \quad r < l, \quad \tau_{r\theta} = 0, \quad \theta = 0, \quad r > l, \quad \langle u_r \rangle = 0, \tag{23.4}$$

$$\theta = 0, \quad r \to \infty, \quad \tau_{r\theta} = C g_1 r^{\lambda_0} + o\,(1/r) \quad (-\alpha \le \theta \le \pi - \alpha), \tag{23.5}$$

where $\langle a \rangle$ is the jump of a.

The solution of the formulated problem of the theory of elasticity (Fig. 23.3) is the sum of the solutions of two problems. The first problem (*Problem 1*) differs from the original problem by the fact that, instead of the first condition in (23.4), we have

$$\theta = 0, \quad r < l, \quad \tau_{r\theta} = -Cg_1 r^{\lambda_0} \tag{23.6}$$

and the stresses decay at infinity as $o(1/r)$, which is described by condition (23.5) without the first term. The second problem is *Problem K*. Since the solution to the second problem is known, it is sufficient to solve *Problem 1*.

23.3 Wiener–Hopf Functional Equation

In order to construct the exact solution to *Problem 1*, we use the Wiener–Hopf technique together with the method of Mellin integral transform (Ufljand 1967). Consider the mixed boundary-value elasticity problem for a wedge $0 \le \theta \le \pi - \alpha$ and wedge $-\alpha \le \theta \le 0$. Applying the Mellin transform

$$m^*(p) = \int_0^\infty m(r)\, r^p dr,$$

with a complex parameter p ($-\varepsilon < \operatorname{Re} p < 0, \varepsilon$ is a small enough positive number) to the equilibrium equations, consistency conditions for strains, Hooke's law, we introduce into consideration the unknown Mellin transformants of stresses and derivatives of displacements $\sigma_\theta^*(p,\theta)$, $\tau_{r\theta}^*(p,\theta)$, $\sigma_r^*(p,\theta)$, $(\partial u_\theta/\partial r)^*$, and $(\partial u_r/\partial r)^*$. Applying the Mellin transform to the through boundary conditions (23.3), after solving for each of the wedges the corresponding system of linear equations, we obtain a relation connecting the transformants $\tau_{r\theta}^*(p,0)$ and $\langle \partial, u_r/\partial r\rangle^{*\theta=0}$ on the line $\theta = 0$:

$$\tau_{r\theta}{}^*(p,0) = \frac{(1+\varkappa_1)D(p)}{2D_0(p)} \frac{E_1}{4\left(1-\nu_1^2\right)} \left\langle \frac{\partial u_r}{\partial r} \right\rangle^*_{\theta=0},$$

$$\begin{aligned}
D(p) = {} & (1+\varkappa_1)[\cos 2p(\pi-\alpha) - \cos 2\alpha](\sin 2p\alpha + p\sin 2\alpha) \\
& + (1+\varkappa_2)(\cos 2p\alpha - \cos 2\alpha) \\
& \times [\sin 2p(\pi-\alpha) - p\sin 2\alpha]\, e, \\
D_0(p) = {} & (\sin 2p\alpha + p\sin 2\alpha)[\varkappa_1 \sin 2p(\pi-\alpha) + p\sin 2\alpha] \\
& + \Big\{(1+\varkappa_1)(1+\varkappa_2)\sin^2 p\pi - (\sin 2p\alpha + p\sin 2\alpha) \\
& \times [\varkappa_1 \sin 2p(\pi-\alpha) + p\sin 2\alpha] - [\sin 2p(\pi-\alpha) - p\sin 2\alpha] \\
& \times (\varkappa_2 \sin 2p\alpha - p\sin 2\alpha)\Big\}\, e \\
& + [\sin 2p(\pi-\alpha) - p\sin 2\alpha](\varkappa_2 \sin 2p\alpha - p\sin 2\alpha) e^2.
\end{aligned} \tag{23.7}$$

Taking into account the second condition in (23.4) and condition (23.6), we obtain

$$
\begin{aligned}
\tau_{r\theta}^{*}(p,0) &= l^{p+1}\left[\Phi^{+}(p)+\frac{\tau}{p+\lambda_0+1}\right],\\
\frac{E_1}{4\left(1-\nu_1^2\right)}\left\langle\frac{\partial u_r}{\partial r}\right\rangle_{\theta=0}^{*} &= l^{p+1}\Phi^{-}(p)\\
\Phi^{+}(p) &= \int_1^{\infty}\tau_{r\theta}(\rho l,0)\,\rho^{p}d\rho,\\
\Phi^{-}(p) &= \frac{E_1}{4\left(1-\nu_1^2\right)}\int_0^1\left\langle\frac{\partial u_r}{\partial r}\right\rangle\Bigg|_{\substack{r=\rho l\\ \theta=0}}\rho^{p}d\rho\,,\quad \tau=-Cg_1 l^{\lambda_0}.
\end{aligned}
\tag{23.8}
$$

Substituting the expressions (23.8) obtained for the transformants $\tau_{r\theta}^{*}(p,0)$ and $\langle\partial u_r/\partial r\rangle_{\theta=0}^{*}$ into (23.7), we arrive at the following Wiener–Hopf functional equation for unknown functions $\Phi^{+}(p)$ and $\Phi^{-}(p)$ in the strip of the complex plane containing the imaginary axis:

$$
\begin{aligned}
\Phi^{+}(p)+\frac{\tau}{p+\lambda_0+1} &= G_1^0(p)\,\Phi^{-}(p),\\
G_1^0(p) &= \frac{(1+\varkappa_1)\left[a_0(p)+a_1(p)\,e\right]}{2\left[b_0(p)+b_1(p)\,e+b_2(p)\,e^2\right]},\\
a_0(p) &= (1+\varkappa_1)\left[\cos 2p(\pi-\alpha)-\cos 2\alpha\right](\sin 2p\alpha+p\sin 2\alpha),\\
a_1(p) &= (1+\varkappa_2)(\cos 2p\alpha-\cos 2\alpha)\left[\sin 2p(\pi-\alpha)-p\sin 2\alpha\right],\\
b_0(p) &= (\sin 2p\alpha+p\sin 2\alpha)\left[\varkappa_1\sin 2p(\pi-\alpha)+p\sin 2\alpha\right],\\
b_1(p) &= (1+\varkappa_1)(1+\varkappa_2)\sin^2 p\pi-(\sin 2p\alpha+p\sin 2\alpha)\\
&\quad\times\left[\varkappa_1\sin 2p(\pi-\alpha)+p\sin 2\alpha\right]\\
&\quad-\left[\sin 2p(\pi-\alpha)-p\sin 2\alpha\right](\varkappa_2\sin 2p\alpha-p\sin 2\alpha),\\
b_2(p) &= \left[\sin 2p(\pi-\alpha)-p\sin 2\alpha\right](\varkappa_2\sin 2p\alpha-p\sin 2\alpha),
\end{aligned}
\tag{23.9}
$$

$-\varepsilon_1<\mathrm{Re}\,p<\varepsilon_2$, $\varepsilon_{1,2}$ are small enough positive numbers. The function $\Phi^{+}(p)$ is analytic in the half-plane $\mathrm{Re}\,p<\varepsilon_2$, and the function $\Phi^{-}(p)$ is analytic in the half-plane $\mathrm{Re}\,p>-\varepsilon_1$.

Factorization of the coefficient $G_1^0(p)$ of Eq. (23.9) on the imaginary axis is carried out by splitting it into a function factorizable using gamma functions and a function factorizable using the properties of the Cauchy-type integral. Using these factorizations, the principle of analytic continuation, Liouville's theorem, and some other provisions of the theory of a function of a complex variable, an exact solution of the Wiener–Hopf equation is constructed, which is expressed through Cauchy-type integrals and the gamma function (Gakhov 1977; Lavrentev and Shabat 1987; Noble 1988):

$$\Phi^{+}(p) = \frac{\mathrm{K}^{+}(p)\,G^{+}(p)}{p+\lambda_0+1}\left[\frac{\tau}{\mathrm{K}^{+}(-\lambda_0-1)\,G^{+}(-\lambda_0-1)} - \frac{\tau}{\mathrm{K}^{+}(p)\,G^{+}(p)}\right], \quad (\mathrm{Re}\, p<0),$$

$$\Phi^{-}(p) = \frac{\tau p G^{-}(p)}{A\mathrm{K}^{+}(-\lambda_0-1)\,G^{+}(-\lambda_0-1)\,(p+\lambda_0+1)\,\mathrm{K}^{-}(p)} \quad (\mathrm{Re}\, p>0),$$

$$\exp\left[\frac{1}{2\pi i}\int_{-i\infty}^{+i\infty}\frac{\ln G_1^0(z)}{z-p}dz\right] = \begin{cases} G^{+}(p), & \mathrm{Re}\, p<0 \\ G^{-}(p), & \mathrm{Re}\, p>0 \end{cases}$$

$$K^{\pm}(p) = \frac{\Gamma(1\mp p)}{\Gamma(1/2\mp p)}, \quad A = \frac{(1+\varkappa_1)\left[1+\varkappa_1+(1+\varkappa_2)\,e\right]}{2\left[\varkappa_1+(1+\varkappa_1\varkappa_2)\,e+\varkappa_2 e^2\right]} \tag{23.10}$$

($\Gamma(z)$ is gamma function).

23.4 Investigation of the Stress State in the Vicinity of the Corner Point and in the Vicinity of the Tip of the Interfacial Crack

Using the representation of the Mellin transforms of stresses in terms of two analytic functions defined in (23.10) as a solution to the Wiener–Hopf equation and applying the Mellin inversion formula to these representations, we found the stresses in the problem of the theory of elasticity shown in Fig. 23.3. The principal terms of stresses expansion in asymptotic series as $r \to 0$ are determined by formulas of the form

$$\begin{aligned} \sigma_\theta &= r^{\lambda_1} S_1(\theta,\alpha,e_0,\nu_1,\nu_2)C_1 + o(r^{\lambda_1}), \\ \tau_{r\theta} &= r^{\lambda_1} S_2(\theta,\alpha,e_0,\nu_1,\nu_2)C_1 + o(r^{\lambda_1}), \\ \sigma_r &= r^{\lambda_1} S_3(\theta,\alpha,e_0,\nu_1,\nu_2)C_1 + o(r^{\lambda_1}), \\ C_1 &= f_1(\alpha,e_0,\nu_1,\nu_2)Cl^{\lambda_0} - \lambda_1 \quad (-\alpha \le \theta \le \pi-\alpha). \end{aligned} \tag{23.11}$$

In representation (23.11) λ_1 is a single root in the interval $(-1,0)$ of the characteristic equation

$$\begin{aligned} &(1+\varkappa_1)\left[\cos 2(\lambda+1)(\pi-\alpha) - \cos 2\alpha\right]\left[\sin 2(\lambda+1)\alpha + (\lambda+1)\sin 2\alpha\right] \\ &\quad + (1+\varkappa_2)\left[\cos 2(\lambda+1)\alpha - \cos 2\alpha\right] \\ &\quad \times \left[\sin 2(\lambda+1)(\pi-\alpha) - (\lambda+1)\sin 2\alpha\right] e = 0. \end{aligned} \tag{23.12}$$

Some values of λ_1 (the exponent of stress singularity at the corner point) are presented in Table 23.3 for different ratios of Young's moduli $e_0 = E_1/E_2 > 1$ of materials depending on the angle ($\nu_1 = \nu_2 = 0.3$).

Table 23.3 Exponents of stress singularity

α°	e_0			
	2	3	5	10
15	–0.738	–0.702	–0.652	–0.602
30	–0.640	–0.611	–0.577	–0.541
45	–0.547	–0.527	–0.506	–0.484
60	–0.430	–0.419	–0.408	–0.398
75	–0.262	–0.258	–0.255	–0.252
105	–0.274	–0.277	–0.280	–0.283
120	–0.467	–0.476	–0.484	–0.492
135	–0.613	–0.628	–0.641	–0.653
150	–0.731	–0.749	–0.767	–0.782
165	–0.835	–0.855	–0.873	–0.889

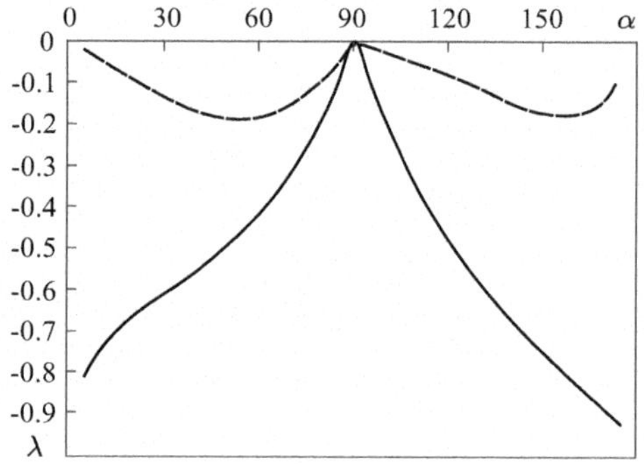

Fig. 23.4 Exponents of stress singularity in the case of the absence (dashed curve) and the presence (solid curve) of interfacial cracks

Figure 23.4 compares the values of the exponent of stress singularity at the corner point O depending on the angle in the case of the absence (λ_0, dashed curve) and the presence (λ_1, solid curve) of interfacial shear cracks that originated at the corner point of the interface O ($e_0 = 3$, $\nu_1 = \nu_2 = 0.3$).

Using the following well-known asymptotic behavior of stresses in the vicinity of the tip of the interfacial crack

$$\theta = 0, \quad r \to l + 0, \quad \tau_{r\theta} \sim \frac{\varkappa_1 + e + 1 + \varkappa_2 e}{2\,(\,1 + \varkappa_2 e)} \frac{K_{II}}{\sqrt{2\pi\,(r - l)}},$$

$$\theta = 0, \quad r \to l - 0, \quad \left\langle \frac{\partial u_r}{\partial r} \right\rangle \sim -\frac{4\left(1 - \nu_1^2\right)}{E_1} \frac{\varkappa_1 + e}{1 + \varkappa_1} \frac{K_{II}}{\sqrt{2\pi\,(l - r)}},$$

$$\theta = 0, \quad r \to l - 0, \quad \sigma_\theta \sim -\frac{1 - \varkappa_1 - (1 - \varkappa_2)e}{2(1 + \varkappa_2 e)} \frac{K_{II}}{\sqrt{2\pi r (l - r)}},$$

based on the solution of the Wiener–Hopf Eq. (23.10), we obtain the following formula for the shear stress intensity factor at the end O_1 of an interfacial crack:

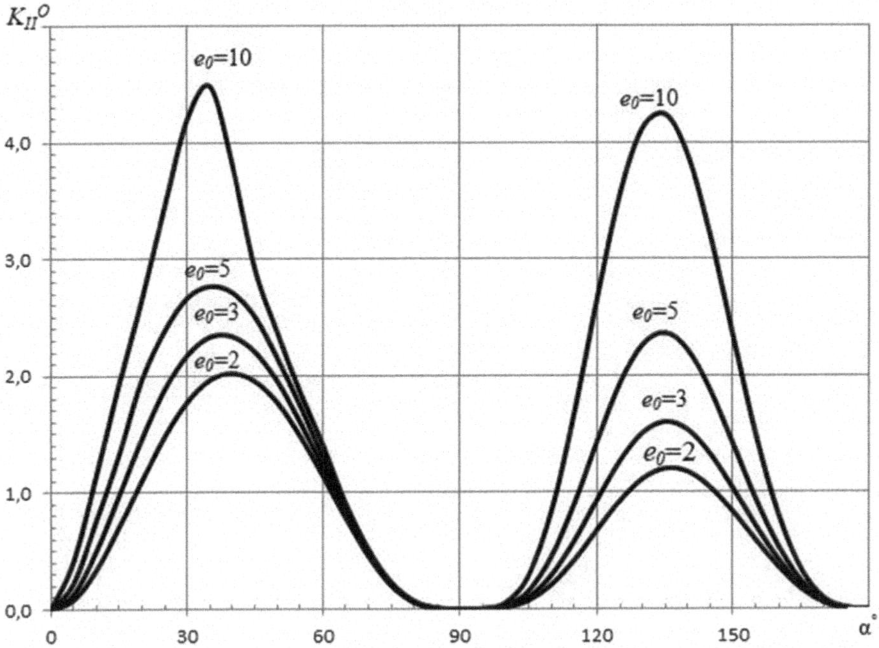

Fig. 23.5 Dimensionless stress intensity factor for different values of the ratio of Young's moduli of the materials

$$K_{II} = g_1(\alpha, e_0, \nu_1, \nu_2)k(\alpha, e_0, \nu_1, \nu_2)Cl^{\lambda_0+1/2},$$

$$k = \frac{2\sqrt{2}(1 + \varkappa_2 e)\Gamma(\lambda_0 + 3/2)}{[1 + \varkappa_1 + (1 + \varkappa_2)e]\,\Gamma(\lambda_0 + 2)G_1^+(-\lambda_0 - 1)}, \tag{23.13}$$

$$G_1^+(-\lambda_0 - 1) = \exp\left[\frac{\lambda_0 + 1}{\pi}\int_0^\infty \frac{\ln G_1(it)}{t^2 + (\lambda_0 + 1)^2}dt\right].$$

The dependences of the dimensionless stress intensity factor $K_{II}^0 = -K_{II}/(Cl^{\lambda_0+1/2})$ on the angle α for different values of the ratio of Young's moduli $e_0 = E_1/E_2 > 1$ of the materials are shown in Fig. 23.5 ($\nu_1 = \nu_2 = 0.3$).

23.5 Conclusions

A plane static symmetric problem of elasticity theory for a bihomogeneous body weakened by interfacial shear cracks at a corner point of a non-smooth interface is considered. The exact analytical solution to the problem was obtained by the Wiener–Hopf method. It is established that the corner point of the interface is a singular point with a power-law singularity, and the exponent of stress singularity at this point is determined. The stress intensity factor at the end of an interfacial crack is determined and its dependence on the elastic constants of materials of the body is studied.

References

Bantsuri RD (1966) The solution of the first fundamental problem in the theory of elasticity for a wedge having a finite slot. Dokl Akad Nauk SSSR 167(6):1256–1259 [in Russian]

Bogdanov VL (2010) Influence of initial stresses on fracture of composite materials containing interacting cracks. J Math Sci 165(3):371–384

Bogdanov VL, Guz AN, Nazarenko VM (2015) Nonclassical problems in the fracture mechanics of composites with interacting cracks. Int Appl Mech 51(1):64–84

Gakhov FD (1977) Bou ndary value problems. Nauka, Moscow [in Russian]

Guz AN (2014) Establishing the foundations of the mechanics of fracture of materials compressed along cracks (Review). Int Appl Mech 50(1):1–57

Guz AN, Nazarenko VM, Bogdanov VL (2013) Combined analysis of fracture under stresses acting along cracks. Arch Appl Mech 83(9):1273–1293

Keer LM, Mendelsohn DA, Achenbach JD (1977) Crack at the apex of a loaded notch. Int J Solids Struct 13(7):615–623

Khrapkov AA (1971) Closed form solutions of problems on the elastic equilibrium of an infinite wedge with nonsymmetric notch at the apex. J Appl Math Mech 35(6):1009–1016

Kuliev VD, Rabotnov YN, Cherepanov GP (1978) Inhibition of cracks at the interface of different elastic media. Izv Akad Nauk SSSR Mekh Tverd Tela No. 4:120–128 [in Russian]

Lavrentev MA, Shabat BV (1987) Methods of the theory of function of complex variable. Nauka, Moscow [in Russian]

Loboda VV, Sheveleva AE (2003) Determining prefracture zones at a crack tip between two elastic orthotropic bodies. Int Appl Mech 39(5):566–572

Nekislykh EM, Ostrik VI (2010) Problems on elastic equilibrium of a wedge with cracks on the axis of symmetry. Mech Solids 45(5):743–756

Noble B (1988) Methods based on the Wiener-Hopf technique. Chelsea, New York

Ouchterlony F (1977) Symmetric cracking of a wedge by concentrated loads. Int J Eng Sci 15(2):109–116

Stone SF, Westmann RA (1981) Stress intensity factors for cracked wedges. Int J Solids Struct 17(3):345–358

Ufljand JS (1967) Integral transforms in problems of elasticity theory. Nauka, Leningrad [in Russian]

Chapter 24
Solving Current Problems in the Dynamics of Space-Rocket Systems

Oleg Pylypenko

24.1 Dynamics of Liquid-Propellant Rocket Propulsion Systems

To get reliable theoretical results on the dynamics of liquid-propellant rocket propulsion systems (LPRPSs), the experiment-and-calculation hydrodynamic model of liquid-propellant engine (LPRE) cavitating pumps (Dolgopolov 2020a) is under continuous refinement. A mathematical model of LPRE pump dynamics was developed to complement the hydrodynamic model of LPRE cavitating pumps by allowing a mathematical simulation of choking regimes (Dolgopolov 2020b, 2021). Theoretical parameters of cavitation self-oscillations in LPRE pump choking regimes were determined using that model. They were shown to be in satisfactory agreement with experimental data. It was found that the computed decrease in the cavitation self-oscillation frequency from the eigenfrequency of the hydraulic system with the cavitating pump under study is close to its experimental value in the case where the choking characteristic intersects the line that connects the values of the pump inlet flow rate and pressure computed at integration steps $i-1$ and i. It was shown (Dolgopolov 2021) that the LPRE pump choking characteristic is a specific nonlinearity associated with the critical cavity flow in the pump and may manifest itself at high parameter oscillation amplitudes. It was found that the choking characteristic of an LPRE pump affects the cavitation oscillation parameters to a greater extent than the cavity volume vs. pump inlet pressure and flow rate relationship does and is the governing nonlinearity in the pump system in choking.

A mathematical model of hard excitation of cavitation oscillations in an LPRE feed system was developed (Pylypenko et al. 2019). Theoretical parameters of cavitation self-oscillations in an LPRE pump's bench feed system in hard excitation regimes

O. Pylypenko (✉)
Institute of Technical Mechanics, National Academy of Sciences of Ukraine and State Space Agency of Ukraine, Dnipro, Ukraine
e-mail: office.itm@nas.gov.ua

A. N. Guz et al. (eds.), *Advances in Mechanics*, Advanced Structured Materials 191,
https://doi.org/10.1007/978-3-031-37313-8_24

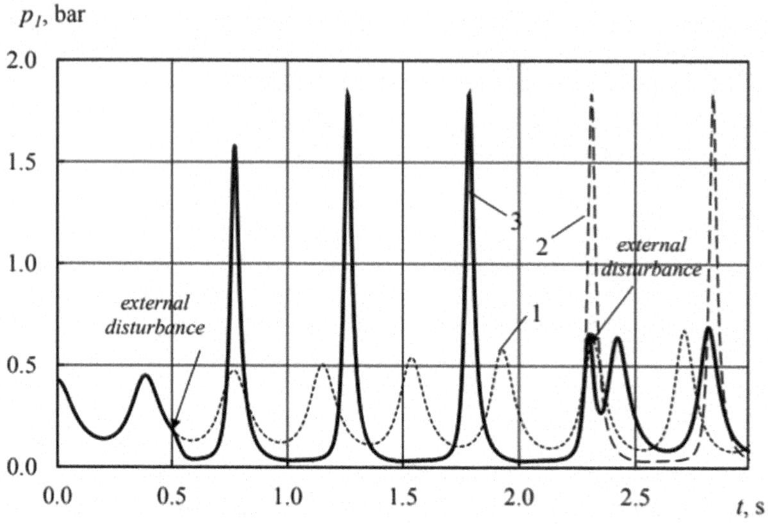

Fig. 24.1 Calculated pump inlet pressure versus time for a hard excitation of cavitation self-oscillations: 1—without external disturbances, 2—an external disturbance at $t = 0.5$ s, 3—external disturbances at $t = 0.5$ s and $t = 2.27$ s

were determined using that model. They were shown to be in satisfactory agreement with experimental data. It was found that if a hydraulic system with a cavitating pump exhibits small-amplitude self-oscillations close to harmonic ones, then, on an external disturbance of sufficient intensity, developed (interrupted) cavitation self-oscillations may be set up in the system. This agrees with known experimental data and indicates a hard excitation regime in a hydraulic system with a cavitating pump (Fig. 24.1). The mathematical model of hard excitation of cavitation self-oscillations made it possible to simulate a case observed in an experiment where it had been possible to stabilize a pump system using an external disturbance (Fig. 24.1).

A mathematical simulation of cavitation self-oscillations in a bench hydraulic system with a cavitating pump was conducted to give cavitation self-oscillation parameters, which made it possible to qualitatively and quantitatively represent complex unsteady dynamic processes that take place in the case of developed cavitation self-oscillations in LPRE pumps: pump choking regimes (flow rate limitation), a significant decrease in the cavitation oscillation frequency in comparison with the hydraulic system eigenfrequencies, and an insignificant effect of the feed line parameters on the parameter oscillation amplitudes and frequencies. The results obtained are a significant development of the most complex area of the theory of cavitation oscillations in pump systems, which is concerned with cavitation self-oscillations in LPRE feed pump systems. In connection with the development of a new Ukrainian launch vehicle (LV), the Cyclone-4M, these results may be used in solving LV longitudinal stability problems and in simulating startup and shutdown transients in the LV propulsion system.

Light was shed on the main current problems in LPRE low-frequency dynamics arising in steady regimes and during an engine startup. A mathematical model of LPRE gas path dynamics was refined as applied to the problem of theoretical prediction of low-frequency process stability in today's staged-combustion LPREs and systems thereof. Methodological recommendations were developed for the numerical study of low-frequency process stability in steady regimes of today's LPREs (in active flight) with account for delays in the equations of gas path low-frequency dynamics.

One of the key problems in LPRE designing is to assure a trouble-free LPRE startup. LPRE bench tryout is highly expensive, and emergency situations may have grave consequences (including engine and bench equipment destruction). Because of this, one of the main tools that allow one to predict the LPRE dynamic characteristics and startup operation features at the design and tryout stages is mathematical simulation. Mathematical models of low-frequency dynamic processes in the hydraulic and gas paths of a staged-combustion LPRE during its startup were refined (Dolgopolov 2017a). A modified mathematical model of cavitating pipe dynamics was developed (Dolgopolov 2017b). The model keeps its structure and operability over a wide cavitation number range and in mutual transitions between the cavitation and the cavitation-free pump operation, which is required for the numerical study of working processes in an LPRE during its startup. An approach to the construction of a nonlinear mathematical model of LPRE hydraulic path filling was developed. The approach allows one, if necessary, to automatically change the scheme of partitioning the hydraulic path into finite hydraulic elements in the process of its filling during engine startup calculations. A scheme of approximate substitution of the delay equations in the mathematical model of LPRE gas path dynamics was proposed. The scheme was constructed with account for the features of calculation of LPRE startup transients, and it allows the simulation accuracy to be improved with the minimum of model complication. The operability of the mathematical models developed was demonstrated by the example of simulating a startup of a staged-combustion oxidizer-rich sustainer LPRE.

An important problem in the mathematical simulation of low-frequency dynamic processes in an LPRE flow controller is the construction of its linear model. The model is a part of a mathematical model of an LPRE as a whole and is used in analyzing LPRE low-frequency dynamics and liquid-propellant rocket longitudinal stability. A methodological approach was developed to construct a linear mathematical model of an LPRE flow controller at different (sizeable) amplitudes of harmonic oscillations of its inlet pressure (Pylypenko et al. 2020a). The approach includes: a numerical determination of equivalent (constructed using the first-harmonic oscillation amplitudes) frequency characteristics of the flow controller by its nonlinear model at different amplitudes of the controller inlet pressure; the derivation of linear-model equations with coefficients that depend on the hydraulic pressure loss in the controller and the dry friction force as a function of the slide valve displacement; and the determination of those coefficients by matching the frequency characteristics found by the nonlinear and the linear model of the low-frequency dynamics of the hydraulic system whose part is the flow controller. Using the proposed approach, the frequency characteristics (the pressure transmission coefficient and the input impedance) of a typical direct-acting flow controller were determined. The results obtained were used to analyze LPRE low-frequency dynamics and provide the longitudinal stability of liquid-propellant rockets.

The need for a lower cost and a shorter time of LPRE development and production often leads to the decision to use clusters of multiple individually developed engines in launch vehicles' sustainer LPRPSs. This opens up prospects for providing a desired thrust by including the necessary number of engines in the cluster. Using sustainer LPRPSs with multiple engines entails additional problems involving nonsimultaneous startups of the engines. This may disrupt the operation of engines that start up with a delay or produce an overturning moment when the rocket detaches from the launcher. A study was conducted into dynamic processes during a startup of a multiengine LPRPS with four staged-combustion oxidizer-rich LPREs with account for the possibility of the engines starting up nonsimultaneously (Pylypenko et al. 2020c). A mathematical model of a startup of the multiengine LPRPS under consideration was developed, and calculations by the model were conducted. It was shown that, as distinct from all the engines starting up simultaneously, their nonsimultaneous startups may result in deep prolonged undershoots in the propellant flow rate accompanied by deep prolonged pressure undershoots at the engine inlets. This may cause a cavitation stall in one or more pumps, which may disrupt the operation of the entire propulsion system and result in an emergency. The results of the mathematical simulation of a four-engine LPRPS startup showed that the character and degree of the effect of possible engine startup delays on transients depend on a variety of factors governed by the LPRPS composition and dynamic performance, startup conditions, etc. Because of this, for multiengine LPRPS startup reliability to be improved in each particular case, i.e., for each new or upgraded LPRPS or launch vehicle, startup transients should be studied numerically with account for nonsimultaneous startups of the LPRPS engines.

The first-stage sustainer LPRPSs of medium, heavy, and superheavy LVs usually include longitudinal dampers, which is a widely used way to eliminate longitudinal oscillations (pogo oscillations) of liquid-propellant LVs. However, until recently, no theoretical studies or analyses of the effect of longitudinal dampers in the feed lines of sustainer LPREs on LPRE startup transients were conducted because of analysis complexity and the lack of reliable nonlinear models of pump cavitation effects. Therefore, a mathematical model was developed to simulate a startup of the multiengine LPRPS of the Cyclone-4M launch vehicle's first stage, which includes four RD 870 sustainer LPREs, with account for the installation of four longitudinal dampers and nonsimultaneous LPRE startups.

A nonlinear mathematical model was developed to simulate low-frequency dynamic processes in longitudinal dampers with bellows gas/liquid separation. It was shown that cavitation in the engine pumps and longitudinal dampers drastically affects the frequency characteristics of the LPRPS oxidizer feed system. The LPRPS startup model made it possible to study dynamic processes during a startup of the multiengine LPRPS of the Cyclone-4M first stage with account for nonsimultaneous LPRE startups and the installation of longitudinal dampers both in the branch lines

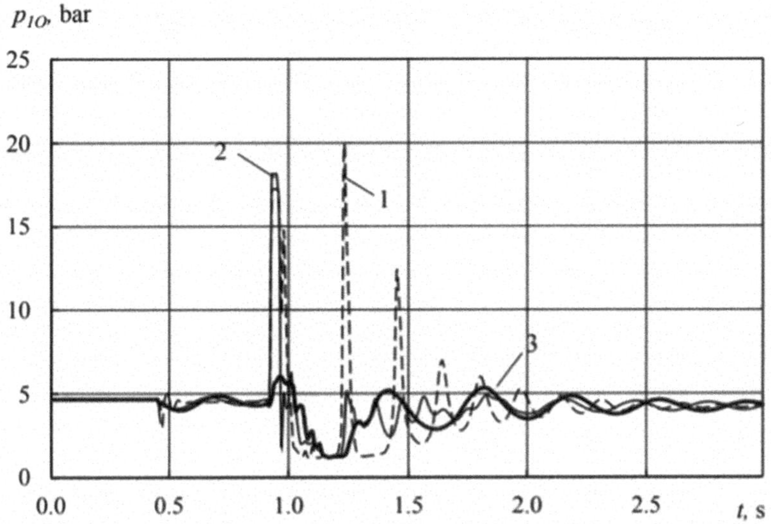

Fig. 24.2 RD 870 LPRE inlet pressure versus time: 1—without dampers, 2—with dampers near the manifold, 3—with dampers at the engine inlets

to the RD 870 LPREs near the manifold (their standard position) and at the LPRE inlets. The analysis of transients in the LPRPS oxidizer feed system and the time dependence of the engine inlet flow rate and pressure showed the following. First, with simultaneous engine startups, the installation of longitudinal dampers near the manifold eliminates nearly all engine inlet pressure overshoots, which occur without dampers. With the dampers at the engine inlets, engine inlet pressure overshoots are virtually absent (Fig. 24.2). Second, nonsimultaneous startups of the RD 870 LPREs adversely affect the time dependence of the engine inlet flow rate and pressure when the dampers are installed near the manifold. Thus, in a time interval from 0.95 s to 1.35 s, some combinations of engine startup delays result in an anomalously high drop in the engine inlet oxidizer flow rate and engine inlet overpressures. With the dampers at the engine inlets, nonsimultaneous RD 870 engine startups affect the LPRPS startup transients only insignificantly in comparison with simultaneous ones. Third, it was thus shown that the installation of longitudinal dampers both at the engine inlets and in the branch lines to the RD 870 LPREs near the manifold has a significant favorable effect on the quality of the transients during a startup of the LPRPS of the Cyclone-4M first stage. Damper installation at the engine inlets is not a standard practice, and it was considered without connection to the LPRPS arrangement. Nevertheless, damper installation at the engine inlets is an efficient way to eliminate under- and overshoots in the LPRPS parameters, particularly in the case of nonsimultaneous startups of the multiengine propulsion system LPREs.

The development of the Cyclone-4M rocket complex at Pivdenne State Design Office with the participation of Pivdennyi Machine-Building Plant, other plants of the State Space Agency of Ukraine, and US and Canadian companies is an important line in the activities of the Ukrainian space industry. The proposed method of feeding the liquid-propellant jet system (LPJS) of the upper stage from the feed lines of the sustainer engine is a pioneering solution in the world's practice. This eliminates the need for a separate system of upper stage LPJS feeding and reduces the upper stage mass, thus allowing one to increase the payload mass. However, this engineering solution entails a number of problems involving the stability of operation of the upper stage motion control system during a startup and a shutdown of the RD 861K sustainer engine of the Cyclone-4M upper stage. Through the common feed system of the sustainer engine and the LPJS, pressure surges and undershoots caused by a sustainer engine startup/shutdown propagate throughout the system without any marked damping and reach the LPJS combustion chambers. In such extreme conditions, the reliability of the upper stage thrust vector control will depend on the LPJS operability. Nonlinear models were developed to simulate the low-frequency dynamics of the oxidizer and fuel feed systems of the Cyclone-4M's upper stage sustainer LPRPS. To test the models, a mathematical simulation of transients in the oxidizer and fuel feed lines was conducted with simulating an engine shutdown using a model liquid. The calculated pressure surge and undershoot magnitudes and durations and the calculated amplitudes and frequencies of decaying processes proved to be in satisfactory agreement with the experimental ones (Fig. 24.3). A nonlinear mathematical model was developed to simulate a startup/shutdown of the RD 861K sustainer engine of the Cyclone-4M upper stage with account for cavitation effects in the oxidizer and fuel pumps over a wide range of their operating conditions. Using the model, a startup of the RD 861K engine was simulated at different combinations of the oxidizer and fuel temperature and pressure at the engine inlet. This made it possible to determine the maximum magnitudes of oxidizer and fuel pressure surges and undershoots at the LPJCS inlet during an engine startup/shutdown, which were used in determining the LPJS operability during a startup/shutdown of the RD 861K sustainer engine.

To execute program motions, the space stages of liquid-propellant launch vehicles are equipped with restartable sustainer engines. After a shutdown of the sustainer engine of a space stage, the liquid propellant in the tank in microgravity conditions moves by inertia upwards as far as possible away from the intake. This results in the potential possibility of the pressurizing gas penetrating into the engine inlet in quantities that make an engine restart impossible. Because of this, motion parameter determination for a liquid moving in propellant tanks in microgravity conditions is a topical problem to be solved in the design of liquid-propellant rocket engines. In Pilipenko and Dolgopolov (1998), Khoriak and Dolhopolov (2017), a methodology was developed for the calculation of the motion parameters of the gas–liquid interface in the propellant tanks of today's liquid-propellant launch vehicles' space stages in microgravity conditions (between a shutdown and a startup of their sustainer engines) taking into account the design features of the in-tank propellant management devices. The methodology uses the finite-element method, the volume-of-fluid method, and up-to-date finite-element CAE systems. It allows one to determine the motion parameters and the shape of the free liquid surface in the tank, the parameters of free gas inclusions formed in the liquid, and the efficiency of the in-tank propel-

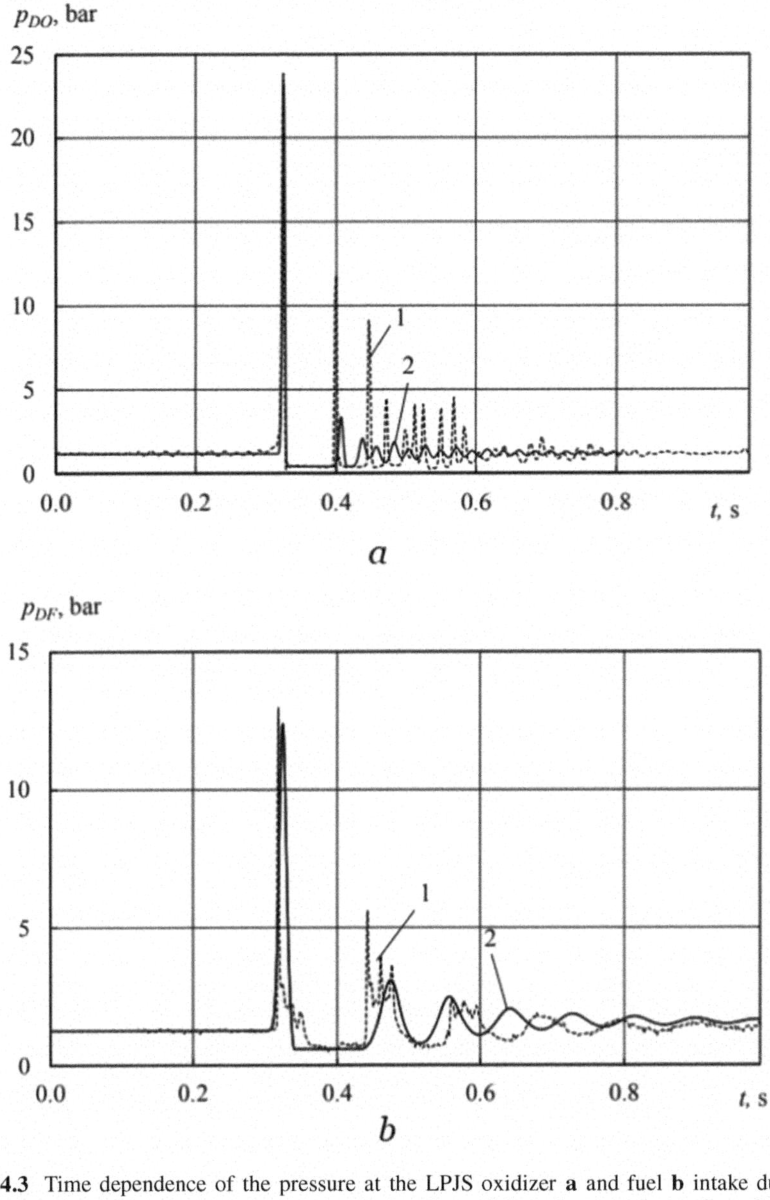

Fig. 24.3 Time dependence of the pressure at the LPJS oxidizer **a** and fuel **b** intake during a shutdown of the RD 861K sustainer engine under bench conditions: 1—experiment, 2—calculation

lant management devices in the passive flight of the launch vehicle for the normal operation of the propulsion system.

For the conditions of motion of a prototype propellant tank with a liquid in a drop tower, which simulates microgravity, the motion of a liquid in a cylindrical tank was simulated numerically with account for the deformation of its free surface. The computed motion parameters of the liquid and the gas–liquid interface are in agreement with experimental data. The methodology developed will allow one to reduce the extent of testing of newly developed and upgraded LV space stages.

24.2 Longitudinal Stability of Liquid-Propellant Launch Vehicles

Low-frequency longitudinal oscillations in active flight are inherent to nearly all liquid-propellant LVs. When LV longitudinal stability is lost, the amplitudes of LV longitudinal oscillations build up and may exceed their acceptable level, thus causing the risk of structural integrity and payload damaging, instrumentation and control system malfunction, unprogrammed engine shutdowns, and other emergencies. Despite the experience in the prevention of liquid-propellant LV longitudinal oscillations gained in Ukraine and abroad, the longitudinal stability problem still remains topical and is among the most important problems in rocket engineering to be solved in developing new launch vehicles and upgrading existing ones (see, for example, (Dolgopolov 2020a; Nikolayev et al. 2018, 2021)). The fact that longitudinal oscillations can be revealed only in flight tests contributes to the complexity of this problem. The elimination of LV longitudinal instability revealed from the results of LV launches calls for too large financial and material expenditures. Because of this, a theoretical analysis of LV longitudinal stability must be conducted at all stages of liquid-propellant LV development.

Longitudinal instability (the pogo effect) is a specific instability form of a self-oscillatory nature. Longitudinal stability is analyzed based on the numerical study of the dynamics of the self-oscillatory system of LPRPS–LV structure, which describes an interaction between the LV structure's elastic longitudinal vibrations and low-frequency processes in the LPRPS. The current requirements for the reliability of theoretical predictions of LV longitudinal stability call for refining mathematical models and methods used in longitudinal stability analysis.

A prediction was made of the longitudinal stability of the Cyclone-4M new two-stage LV during the operation of the first-stage sustainer LPRPS (Nikolayev et al. 2021). The Cyclone-4M launch vehicle, which is under development at Pivdenne State Design Office, has a tandem arrangement of the stages (Fig. 24.4), and the sustainer LPRPS of its first stage comprises four RD 870 staged-combustion oxidizer-rich LPREs. Figure 24.5 shows a simplified flow schematic of the RD 870 engine, which is under development at Pivdenne State Design Office.

To predict longitudinal stability of Cyclone-4M, a mathematical model was developed of the closed-loop dynamic system of first-stage sustainer LPRPS–LV structure. The model describes an interaction between the LV structure's elastic longitudinal vibrations and low-frequency processes in the first-stage LPRPS with account for cavitation effects in the low-pressure and main pumps of the oxidizer and the fuel

Fig. 24.4 Cyclone-4M launch vehicle

(Pylypenko 2020), the injection of gaseous oxygen into the liquid oxygen flow, and delays in the equations of engine gas path dynamics (Dolgopolov 2017a; Kvasha and Zinevych 2018). A theoretical analysis of Cyclone-4M longitudinal stability was made using the Nyquist criterion.

The analysis showed that Cyclone-4M loses longitudinal stability during the operation of the first-stage sustainer LPRPS due to resonance interaction between the LPRPS and the LV structure. The loss of stability occurs early in active flight when

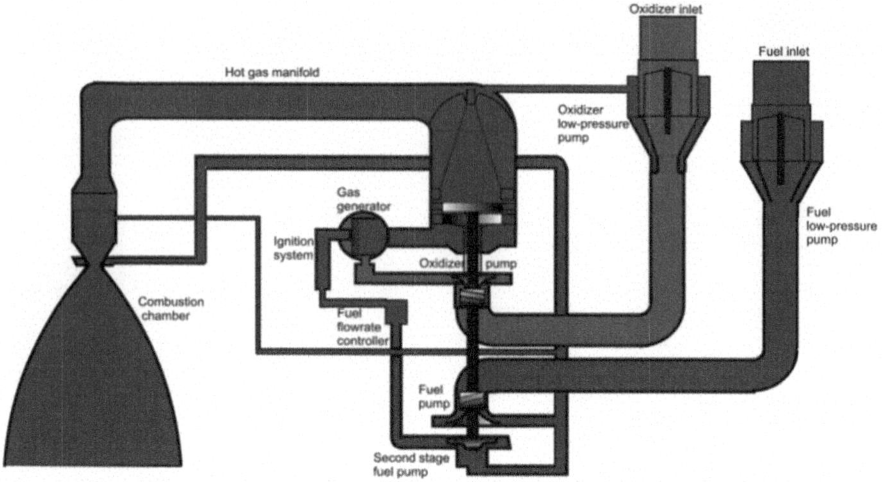

Fig. 24.5 Simplified flow schematic of the RD 870 engine (Nikolayev et al. 2021)

the first frequency of liquid oscillations in the oxidizer feed line approaches the first-tone frequency of the LV structure's longitudinal vibrations. To provide longitudinal stability, it was proposed to install longitudinal dampers in the oxidizer feed lines of the first-stage sustainer LPRPS in order to properly change the dynamic characteristics of the oxidizer feed system and eliminate LPRPS–LV structure resonance interaction.

A mathematical model was developed of the low-frequency dynamics of a damper with bellows gas/liquid separation, and its advisable parameters were determined. It was shown that the installation of dampers of this type in the oxidizer feed lines significantly reduces the liquid oscillation frequencies in the first-stage oxidizer feed system and the maximum modulus of the frequency characteristic of the open-loop system of sustainer LPRE–oxidizer and fuel feed lines–LV structure (tone I), thus providing longitudinal stability of Cyclone-4M throughout its flight at different combinations of the oxidizer and fuel pressures and temperatures at the engine inlets. The calculated tone I and II frequencies of the Cyclone-4M structure's longitudinal vibrations versus the flight time are shown in Fig. 24.6.

The maximum modulus of the frequency characteristic $W_{ol}(j\omega)$ of the open-loop system of sustainer LPRE–oxidizer and fuel feed lines–LV structure (tone I) was calculated as a function of the LV flight time (Fig. 24.7). It was shown that damper installation at the low-pressure pump inlets is more preferable because in this case, the maximum modulus of the frequency characteristic is smaller.

An approach was proposed to the mathematical simulation of liquid-propellant LV longitudinal oscillations (Kvasha et al. 2019) using a 3D finite-element discretization of the LV structure and the liquid propellant in the tanks, representing the LPRPS systems and assemblies as one-dimensional finite elements, and using a nonlinear hydrodynamic model of cavitation effects in the LPRPS pumps. The approach was

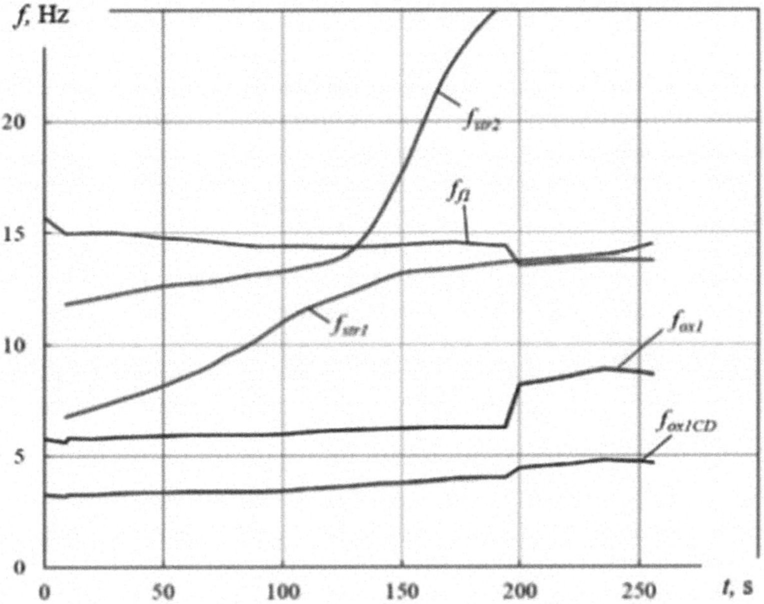

Fig. 24.6 Calculated tone I and II frequencies of the Cyclone-4M structure's longitudinal vibrations (f_{str1}, f_{str2}) and liquid oscillation frequencies in the first-stage LPRPS feed system for the oxidizer (f_{ox1}—without a damper, f_{ox1CD}—with a damper) and the fuel (f_{fl}) versus the flight time t

Fig. 24.7 Calculated maximum modulus of the frequency characteristic of the open-loop system of sustainer LPRE–oxidizer and fuel feed lines– Cyclone-4M structure (tone I) versus the LV flight time with the dampers in the branch lines to the engines (curve 1) and at the low-pressure pump inlets (curve 2)

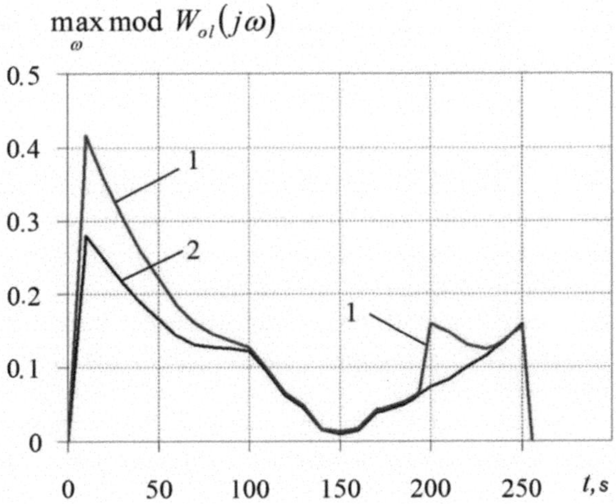

applied to a two-stage LV. A nonlinear mathematical model of the low-frequency dynamics of the LPRPS–LV structure system was developed to simulate the interaction of 3D oscillations of the liquid propellant in the tanks, elastic vibrations of the LV structure, and nonlinear processes in the LPRPS in active flight. The simulation showed that the dynamic system under consideration reaches a limit cycle with a self-oscillation frequency of 15.8 Hz, close to the tone II frequency of the LV structure's longitudinal vibrations. The self-oscillation parameters of this system were determined: the dynamic components of the tank pressures, the displacements and the vibration accelerations of the LV's structural components, and the mass flow rate and the pressure in the LPRPS components.

The proposed approach may be used to study liquid-propellant rocket pogo oscillations (particularly for rockets with an asymmetrical stage arrangement and complex-shaped propellant tanks) and to determine limit cycle parameters in the case of longitudinal instability.

This approach allows one to account for the interaction of 3D oscillations of the liquid propellant in the tanks, elastic vibrations of the LV structure, and nonlinear processes in the LPRPS and for nonlinear dissipative forces in the description of the LV structure's elastic deformations. The approach may also be used in assessing the limit vibration amplitudes and dynamic loads of the structures of prospective liquid-propellant LVs that, according to a preliminary linear analysis, lose longitudinal stability in some portions of their active flight.

24.3 High-Frequency Processes in the LPRE Combustion Chamber

In LPRE tryout, the parameters that govern working processes in the LPRE systems (the pressure, the flow velocity, the gas and liquid temperature, the turbopump speed, etc.) exhibit low- and high-frequency oscillations. High-frequency oscillations in the combustion chamber, which are potentially dangerous to the LPR operational reliability and integrity, are the least understood. The most important tool in the study and development of measures aimed at their elimination in the flight of liquid-propellant LVs is the mathematical simulation of high-frequency processes in the combustion chamber.

The state of the art in the numerical study of high-frequency dynamic processes in LPRE combustion chambers was analyzed with the aim to assess the possibility of using the available numerical methods to simulate the above-mentioned processes in the problem of theoretical prediction of LPRE high-frequency stability and the combustion chamber pressure and flow rate oscillation amplitudes (Kvasha and Zinevych 2019, 2020). Consideration was given to the currently adopted mechanisms of high-amplitude oscillations in the LPRE systems involving the dynamic interaction of physical and chemical processes in the mixing and combustion zone in conditions of periodical heat removal under the action of acoustic oscillations and turbulence in the flow and combustion of the propellant components and combustion products.

The analysis showed that the methods of mathematical simulation of high-frequency acoustic oscillations in an LPRE can be divided into three basic groups: methods for the calculation of the acoustic oscillation parameters in cylindrical chambers based on analytical mathematical models of a relatively low order with the use of the Bessel functions, methods for the study of thermoacoustic effects using approaches of computational fluid dynamics, and hybrid methods in which combustion dynamics is calculated separately from the combustion product acoustic oscillation parameters. The main results obtained in the framework of the above-mentioned groups were discussed. The advantages and drawbacks of the numerical study of combustion product thermoacoustic oscillations in the LPRE combustion chamber were analyzed.

A numerical approach was developed to determine the parameters of acoustic oscillations of the combustion products in LPRE annular combustion chambers, taking into account the features of the configuration of the combustion space and the dependence of the physical properties of the gaseous medium on the axial length of the chamber. The approach is based on a mathematical simulation of the natural oscillations of the closed-loop dynamic system of annular shell structure–gas using the finite-element method.

The approach was tested, and the results were analyzed for an engine with an annular combustion chamber (with an outer-to-inner diameter ratio of 1.5) and the liquid oxygen/methane propellant pair.

A finite-element model was developed to simulate the coupled spatial oscillations of the annular combustion chamber structure and the combustion products. The model allows one to determine the oscillation parameters of the dynamic system under consideration (frequencies, modes, and effective masses) for its dominant acoustic modes together with the vibration amplitudes and vibration accelerations of the combustion chamber walls. Using the model, the LPRE operating parameters potentially dangerous for the development of thermoacoustic instability of the working process in an annular combustion chamber can be identified. To numerically determine the dynamic pressure gains of the combustion chamber under study, a source of harmonic pressure disturbance is introduced into the finite-element model of the dynamic system of annular shell structure–gas (in the elements at the start of the chamber fire space). The shapes and frequencies of the longitudinal, tangential, and radial modes of acoustic oscillations were determined.

24.4 Dynamics of Vibration Protection Systems

The use of a pneumatic springing element can provide a sufficiently small (quasi-zero) stiffness in the working portion of the static characteristic with a progressive stiffness increase in a rebound and a bump stroke. Based on these principles, a new high-efficiency vibration protection system was developed to protect rocket-space hardware and other heavy machinery products in railway, sea, and motor transportation (Kvasha 2020). It was shown that choosing the design and the design parameters allows one to provide a required dynamic performance of the vibration protection system, which is characterized by a superlow natural frequency of the protected object (less than 0.5 Hz) and a required damping coefficient.

A prototype of the proposed vibration protection system was developed, made, and put through dynamic tests, which confirmed its high efficiency in damping impact and harmonic disturbances. The simple design, manufacturability, and high efficiency of the proposed vibration protection system make it possible to recommend it for use both in the transportation of rocket-space hardware by different modes of transport and in the vibration protection of other objects, including new spacecraft.

A new system was proposed to protect the Sich-2M spacecraft from longitudinal vibration loads during an orbital injection. The low-frequency dynamics of the vibration protection system with the spacecraft under the action of an external longitudinal disturbing force was simulated mathematically. It was shown that, when installed between the launch vehicle's upper stage and the spacecraft, the vibration protection system reduces the longitudinal vibration loads on the Sich-2M by a factor of more than two. Equipping launch vehicles with the proposed spacecraft vibration protection system will significantly widen their capabilities of orbiting spacecraft of different purposes.

24.5 Aerodynamic Improvement of Gas Turbine Engine Components

At present, the requirements for the design quality of today's gas turbine engines (GTEs) are ever increasing. Of particular interest is the application of optimization methods to the aerodynamic improvement of GTE components because the degree of their aerodynamic perfection has a significant effect on the power performance of the engine as a whole.

A computational methodology was developed for the aerodynamic improvement of aircraft GTE centrifugal compressor stages using a numerical simulation of 3D turbulent gas flows. An economical (in terms of computational cost) computational procedure was developed using this methodology. The procedure allows one to vary the spatial shape of the guide and diffuser vanes of centrifugal compressor stages in the course of aerodynamic optimization using input sets of vane profile coordinates without any preliminary vane shape approximation. A procedure was developed to vary the spatial shape of centrifugal wheel blades in the course of aerodynamic optimization using a small number of parameters (in comparison with data arrays that specify the blade shape). The procedure also allows the spatial shape of wheel blades to be varied without any preliminary approximation of their initial shape. An algorithm was developed to construct a response surface with a limited number of calls of the objective function calculation procedure in the course of aerodynamic optimization. The algorithm is based on a sequential numerical solution of the Laplace equation and the diffusion equation on uniform meshes (Kvasha and Zinevych 2018). The operability of the proposed response surface construction algorithm was demonstrated for three essentially different test functions. An approach to the aerodynamic optimization of centrifugal compressor stages was proposed. The key feature of the approach is searching for advisable geometrical parameters of guide vanes and centrifugal wheels using uniformly distributed point sequences in the independent-variable region and formulating quality criteria as the mean integral values of the centrifugal stage power characteristics over the operating range of the air flow rate through the stage.

Multiparameter numerical studies made it possible to identify ways to aerodynamically improve centrifugal compressor stage components. By the example of the inlet guide vanes of a centrifugal compressor stage, it was shown by computation that the computational methodology developed does improve the stage power characteristics even in the case of a relatively small number of points uniformly distributed in the variable region (Kvasha et al. 2019). As a result of the aerodynamic improvement of a centrifugal compressor wheel, it was found that varying the shape of the middle part of the centrifugal wheel blades alone, with their inlet and outlet parts remaining unchanged, has rather a strong effect on the flow. This way to vary the blade shape has a significant (in many cases, favorable) effect on the air compression ratio in the wheel while leaving its efficiency almost unaffected. In this case, there is no need for varying the shape of the inlet guide vanes or diffuser vanes, which makes it possible to recommend that in the aerodynamic improvement of a centrifugal wheel, the shape of its blades be varied as indicated above (Kvasha and Zinevych 2019). A numerical simulation of 3D turbulent gas flows showed that in the absence of flow separation in the blade channels of a centrifugal wheel with a given starting shape of the meridional contour, varying that shape has a relatively insignificant effect on the wheel power characteristics. Because of this, in similar cases, it seems to be advisable to aerodynamically improve centrifugal wheels by varying the shape of their blades in the circumferential direction (Kvasha and Zinevych 2020).

Based on a numerical simulation of a 3D turbulent flow in the air duct of one of the air intake configurations for an aircraft turboprop engine, the efficiency of that configuration was assessed. The calculated flow parameter nonuniformity at the engine compressor inlet was compared with that of another air intake configuration for the same engine. It was noted that the second air intake configuration provided a much more uniform flow parameter distribution at the engine compressor inlet (Kvasha 2020).

24.6 Conclusions

Further development has been given to the following research line: the dynamics of liquid-propellant rocket propulsion systems and the longitudinal stability of liquid-propellant launch vehicles with account for cavitation in their inducer-equipped centrifugal pumps.

A new high-efficiency vibration protection system has been developed to protect rocket-space hardware and other heavy machinery products in railway, sea, and motor transportation.

Based on the computational methodology developed, some components of aircraft gas turbine engines have been improved aerodynamically.

References

Dolgopolov SI (2017) Hydrodynamic model of cavitation oscillation for modelling dynamic processes within pump systems at high cavitation number. Teh Meh 2:12–19 [in Russian]

Dolgopolov SI (2020) Verification of a hydrodynamic model of a liquid-propellant rocket engine's cavitating pumps using experimental and theoretical pump transfer matrices. Teh Meh 3:18–29 [in Ukrainian]

Dolgopolov SI (2020) Mathematical simulation of choking under self-oscillations in hydraulic systems with cavitating pumps of liquid-propellant rocket engines. Teh Meh 4:35–42 [in Ukrainian]

Dolgopolov SI (2021) Mathematical simulation of hard excitation of cavitation self-oscillations in a liquid-propellant rocket engine feed system. Teh Meh 1:29–36 [in Ukrainian]

Dolgopolov SI, Nikolaev AD (2017) Mathematical modelling low-frequency dynamics of flow controller at various amplitudes of harmonic disturbance. Teh Meh 1:15–25 [in Russian]

Khoriak NV, Dolhopolov SI (2017) Features of mathematical simulation of gas path dynamics in the problem of the stability of low-frequency processes in liquid-propellant rocket engines. Teh Meh 3:30–44 [in Russian]

Kook JP, JeongUk Y, SiHun L, Jaehyun N, Hyunji K, Juyeon L, Tae-Seong R, Jack JY, Chongam K, SangJoon S (2020) Pogo accumulator optimization based on multiphysics of liquid rockets and neural networks. J Spacecr Rockets 57:809–822

Kvasha YA (2020) Calculation of a 3D turbulent flow in aircraft gas turbine engine ducts. Teh Meh 4:65–71 [in Ukrainian]

Kvasha YA, Zinevych NA (2018) On objective function interpolation in the optimization of engineering systems. Teh Meh 2:71–29 [in Russian]

Kvasha YA, Zinevych NA (2019) Aerodynamic improvement of centrifugal compressor stage impellers. Teh Meh 1:57–67 [in Russian]

Kvasha YA, Zinevych NA (2020) On the effect of the meridional contour shape on the power characteristics of a centrifugal compressor wheel. Teh Meh 3:12–17 [in Ukrainian]

Kvasha YA, Zinevych NA, Petrushenko NV (2019) Aerodynamic improvement of centrifugal compressor stage inlet guide vanes. Teh Meh 3:38–44 [in Russian]

Nikolayev OD, Bashliy ID, Sviridenko NF, Horiak NV (2017) Determination of the parameters of motion of the gas-liquid interface in the fuel tanks of launch vehicle space stages in passive portions of the flight. Teh Meh 4:26–40 [in Russian]

Nikolayev OD, Bashliy ID, Khoryak NV (2018) Computation of the POGO self-oscillation parameters in the dynamic & "propulsion-rocket structure" system by using a 3D structural model. Teh Meh 2:17–29

Nikolayev OD, Bashliy ID, Khoriak NV, Dolgopolov SI (2021) Evaluation of the high-frequency oscillation parameters of a liquid-propellant rocket engine with an annular combustion chamber. Teh Meh 1:16–28

Pilipenko VV, Dolgopolov SI (1998) Experiment-and-calculation determination of the coefficients of the equation of cavity dynamics in inducer-equipped centrifugal pumps of different standard sizes. Teh Meh 8:50–56 [in Russian]

Pilipenko VV, Zadontsev VA, Dovgotko NI, Grigoriev YE, Manko IK (2001) Dynamics of liquid-propellant rocket propulsion systems and liquid-propellant launch vehicle pogo stability. Teh Meh 2:11–37 [in Russian]

Pylypenko MV (2020) System for space hardware vibration protection in transportation. Teh Meh 1:120–130 [in Russian]

Pylypenko OV, Khoriak NV, Dolhopolov SI, Nikolayev OD (2019) Mathematical simulation of dynamic processes in hydraulic and gas paths at the start of a liquid-propellant rocket engine with generator gas after-burning. Teh Meh 4:5–20 [in Russian]

Pylypenko OV, Nikolayev OD, Bashliy D, Dolgopolov SI (2020) Mathematical modeling of dynamic processes in feeding system of space stage main engine of launch vehicle at active and passive flight. Space Sci Technol 1:3–17 [in Russian]

Pylypenko OV, Degtyarev MA, Nikolayev OD, Klimenko DV, Dolgopolov SI, Khoriak NV, Bashliy ID, Silkin LA (2020) Providing of POGO stability of the Cyclone-4M launch vehicle. Space Sci Technol 4:3–20

Pylypenko OV, Nikolayev OD, Bashliy ID, Khoriak NV, Dolgopolov SI (2020) State of the art in the theoretical study of the high-frequency stability of working processes in liquid-propellant rocket combustion chambers. Teh Meh 2:5–21 [in Ukrainian]

Pylypenko OV, Dolhopolov SI, Nikolayev OD, Khoriak NV (2020) Mathematical simulation of the start of a multiengine liquid-propellant rocket propulsion system. Teh Meh 1:5–18 [in Russian]

Chapter 25
Comparison of Evolution of Five Types of Elastic Waves (Harmonic, Gauss, Whittaker, MacDonald, and Friedlander Initial Profiles)

Jeremiah Rushchitsky and Vasyl Yurchuk

25.1 Introduction

The scientific activity of Stepan Timoshenko can be referred to the twentieth century. Most mechanicians around the world know him as a brilliant scientist who worked in the mechanics of materials. His book on mechanics of materials is seemingly one of the most readable books on mechanics. The specificity of the mechanics of materials lies in the fact that it is developing in different directions for many years, both before and after S. Timoshenko. One of the modern directions is the theoretical and numerical analysis of nonlinear waves in materials. This area is being actively researched, which can be confirmed by the publication in recent decades of a series of books (Altenbach et al. 2020; Andrianov et al. 2021; Bedford and Drumheller 1994; Engelbrecht 2015; Fujimoto 2021; Lempriere 2002; Rushchitsky 2011, 2014) in leading publishing houses. The basic wave effect consists of an evolution of the initial profile of waves. It is well-known that the model of linear elastic deformation does not describe an evolution of the waves (waves do not interact) (Andrianov et al. 2021; Rushchitsky 2011). So, nonlinear models should be used.

25.2 Two Basic Nonlinear Equations Describing the Elastic Wave Propagations in Materials

In this chapter, the propagation in the elastic material plane longitudinally polarized and axisymmetric cylindrical waves are studied. The elastic material is assumed to be deformed nonlinearly. The most frequently utilized nonlinear Murnaghan model

J. Rushchitsky (✉) · V. Yurchuk
S.P. Timoshenko Institute of Mechanics, National Academy of Science of Ukraine, Kyiv, Ukraine
e-mail: rushch@inmech.kiev.ua

V. Yurchuk
e-mail: rheol@inmech.kiev.ua

A. N. Guz et al. (eds.), *Advances in Mechanics*, Advanced Structured Materials 191,
https://doi.org/10.1007/978-3-031-37313-8_25

(Lempriere 2002; Rushchitsky 2011), which describes both geometrically and physically nonlinear strains, is chosen. It also describes well the deformation of many engineering materials in a wide range of strains.

25.2.1 The Nonlinear Equation Describing the Propagation of Elastic Plane Longitudinal Waves

Without a doubt, plane waves are the most studied type of classical waves. Two features of the sea waves explain this. First, they have the simplest wavefront—the plane one. Second, many types of waves with curvilinear wavefronts can be considered plane waves when studied at a sufficient distance from the source of wave motion. Owing to the relative simplicity of analysis of plane waves, the nonlinear plane waves in elastic materials are also well studied. Therefore, the choice of the plane wave seems rational in the proposed study of the new kind of nonlinear elastic plane waves—the solitary plane waves.

Note that solitary waves are defined as having a profile that is described by a function given on the finite segment or the function of finite weight. A typical example of a solitary wave is a wave with a profile in the form of a Gauss function (the bell-shaped or hump-shaped wave that is a function of a finite weight). The class of these waves includes solitons, which also have profiles in the form of a hump, but are the exact solutions of certain nonlinear equations. For example, the simplest solitary wave can be described by the classical d'Alembert solution of the linear wave equation, and the soliton is the exact solution of the Korteweg–De Vries equation (Rushchitsky 2011). The essential difference between solitary waves and solitons is that solitary waves interact with each other, while solitons do not.

Note that the studied propagation of plane waves under the condition that they move in the direction of the abscissa axis is described by the nonlinear wave equations of the different order (from 2 to 5) of approximations (Rushchitsky 2011).

The case of the second-order linearity is characterized by the following quadratic nonlinear wave equation (Rushchitsky 2011):

$$u_{1,tt} - (\lambda + 2\mu)\ u_{1,11} = N_1 u_{1,11} u_{1,1} \tag{25.1}$$

$$N_1 = 3\left[(\lambda + 2\mu) + 2\left(A + 3B + C\right)\right], \tag{25.2}$$

where u_1 is the displacement in the direction of the abscissa axis Ox_1, ρ is the density, and λ, μ, A, B, and C are the elastic constants of the five-constant Murnaghan model. The corresponding linear equation has the form

$$u_{1,tt} - (\lambda + 2\mu)\ u_{1,11} = 0. \tag{25.3}$$

25.2.2 The Nonlinear Equation Describing the Propagation of Elastic Cylindrical Wave

The corresponding classical linear statement of the problem on the propagation of the cylindrical radial wave is that it propagates along the radial coordinate in an infinite space with the cylindrical circular cavity (Altenbach et al. 2020; Rushchitsky 2011). The wave motion in the radial direction is excited by the pulse applied to the surface of the cavity.

The cylindrical coordinate system is chosen so that the axis coincides with the axis of the cavity. Then the problem is axisymmetric and depends on two variables—radius and time. The radial displacement and three components of the stress tensor are nonzero. The linear equation of motion has the following form:

$$(c_L)^2\left(u_{r,rr}+\frac{1}{r}u_{r,r}-\frac{u_r}{r^2}\right)-u_{r,tt}=0. \tag{25.4}$$

The corresponding simplest nonlinear equation within the framework of the Murnaghan model is as follows (Rushchitsky 2011):

$$(c_L)^2\left(1-\alpha_1 u_{r,r}\right)\left(u_{r,rr}+\frac{1}{r}u_{r,r}-\frac{u_r}{r^2}\right)-u_{r,tt}=0,$$

$$\alpha_1=3+\frac{2\left(A+3B+C\right)}{\lambda+2\mu}. \tag{25.5}$$

Therefore, the nonlinear equation (25.5), which describes a cylindrical radial wave, has a structure similar to the corresponding Eq. (25.3) for a plane longitudinal wave—a homogeneous linear wave equation with a wave speed that is nonlinearly dependent on the solution. However, the situation here is new compared to the plane wave—the operator of the wave equation is different, the displacement gradient is different, and the wave itself and its phase are different.

25.3 Two Approximate Methods of Solving the Basic Nonlinear Wave Equations

These methods are most frequently used in the study of solitary elastic waves in materials. Both admit the different levels of approximate representation. It was established that the first three approximations (the first one is linear) quite adequately describe the main wave effects.

25.3.1 Method of Successive Approximations (Method 1)

This method (method of successive approximation, the perturbation method, and the method of the small parameter) is based on introducing the small parameter ε.

Let us narrow the description of the method to Eq. (25.3) and consider the displacement $u_1(x_1, t)$, assuming it is sufficiently smooth. According to the method, the function $u_1(x_1, t, \varepsilon)$ is introduced and sought in the form of a convergent series:

$$u_1(x_1, t, \varepsilon) = \sum_{n=1}^{\infty} \varepsilon^{n-1} u^{(n)}(x_1, t)$$
$$= u_1^{(1)}(x_1, t) + \varepsilon u_1^{(2)}(x_1, t) + \varepsilon^2 u_1^{(3)}(x_1, t) + \cdots . \quad (25.6)$$

The first term $u^{(0)}\,(x, t)$ is assumed to be the solution of the linear wave equation (25.1) corresponding to (25.3). Note that the presence of the linear part of the nonlinear equation is a necessary condition of this method.

The solution of the nonlinear equation (25.3) is sought in the form of sequential approximations:

$$u_1(x, t) = u_1(x_1, t, \varepsilon = 1) = \sum_{n=1}^{\infty} u_1^{(n)}(x_1, t)$$
$$= u_1^{(1)}(x_1, t) + u_1^{(2)}(x_1, t) + u_1^{(3)}(x_1, t) + \cdots . \quad (25.7)$$

A feature and characteristic advantage of method 1 is that the arbitrary approximation ${u_1}^{(n)}(x_1, t)$ is found as the solution of the following inhomogeneous linear equation:

$$u_{1,tt}^{(n)} - (c_L)^2\; u_{1,11}^{(n)} = \left(N_1 / \rho\right) u_{1,11}^{(n-1)} u_{1,1}^{(n-1)}. \quad (25.8)$$

Thus, to find the nth approximation, it is necessary to know only the $(n-1)$th approximation and solve only the inhomogeneous linear wave equation. This method works well in the theory of waves in materials when the initial amplitudes do not increase essentially (in some cases, half as much again).

Equation (25.3) is further analyzed under the condition that the initial wave profile is a function that can describe waves with a quite arbitrary initial profile

$$u\,(x_1, t = 0) = F\,(a x_1)\,, \quad (25.9)$$

where a is the arbitrary parameter characterizing the wavelength for the harmonic wave and the wave bottom for the solitary wave.

The wave is assumed to be propagated in the form which is admissible both for the harmonic and solitary wave profiles

$$u\,(x_1, t) = F(\sigma). \quad (25.10)$$

Note that $\sigma = a\,(x_1 - vt)$ is the standard wave phase variable.

25.3.2 Method of Restrictions on the Gradient of Displacement (Method 2)

An important feature and necessary condition of the method of restriction on the gradient of displacement is that the studied nonlinear wave equation must have a particular structure: the right-side part of this equation can be moved to the left-hand-side one, and then formally, the nonlinear equation can be written as the linear wave equation with the variable wave velocity (Rushchitsky 2011). This condition is fulfilled in the case of Eqs. (25.3) and (25.5).

Case of Eq. (25.3). Let us start with the equation:

$$u_{1,tt} - \left\{(c_L)^2 + (N_1/\rho)u_{1,1}\right\} u_{1,11} = 0 \to$$
$$u_{1,tt} - v^2 u_{1,11} = 0, \quad v = c_L\sqrt{1+\alpha u_{1,1}}, \quad \alpha = [N_1/(\lambda+2\mu)]. \quad (25.11)$$

Return now to Eq. (25.11) and introduce the restriction on the gradient of displacement

$$\left|\alpha u_{1,1}\right| \ll 1. \quad (25.12)$$

The condition (25.12) permits to represent approximately the velocity by only two first approximations:

$$v = c_L\sqrt{1+\alpha\, u_{1,1}} \approx c_L\left[1 + {}^1\!/\!_2\,\alpha u_{1,1}\right]. \quad (25.13)$$

Based on (25.13), the phase can also be approximately written in the form of two summands:

$$\sigma \approx \left[a\,(x_1 - c_L t) - (1/2)\,\alpha a c_L u_{1,1} t\right]. \quad (25.14)$$

The representation (25.14) also makes it possible to approximately represent the solution of the nonlinear wave equation (25.11) in the following form:

$$u_1\,(x_1,t) \cong F\left[a\left(x_1 - c_L t - {}^1\!/\!_2\,\alpha c_L u_{1,1} t\right)\right]$$
$$= F\left[a\,(x_1 - c_L t) - {}^1\!/\!_2\,\alpha a c_L u_{1,1} t\right]. \quad (25.15)$$

Finally, we expand the solution (25.15) into a Taylor series with respect to the new small parameter

$$\left|\,\delta = -{}^1\!/\!_2\,\alpha a c_L u_{1,1} t\,\right| \ll 1 \quad (25.16)$$

in the vicinity of the classical constant value $\sigma = a\,(x_1 - c_L t)$ of the wave phase and by saving only two first terms

$$u_1\,(x_1,t) \approx F(\sigma) + F^{/}_{,1}(\sigma) a\delta = F(\sigma) - {}^1\!/\!_2\,\alpha a^2 c_L t\left[F^{/}_{,1}(\sigma)\right]^2. \quad (25.17)$$

Thus, the representation (25.17) of the solution of the nonlinear wave equation (25.11) takes into account the first two approximations and has a general character.

The case of Eq. 25.5. Let us start with a slightly transformed equation (25.5):

$$c_L^2\left(1-\alpha_1 u_{r,r}\right)\left(u_{r,rr}+\frac{1}{r}u_{r,r}-\frac{u_r}{r^2}\right)-u_{r,tt}=0. \qquad (25.18)$$

Assume now that the following function describes the initial wave profile:

$$u_r\left(r,t=0\right)=F\left(r\right). \qquad (25.19)$$

The corresponding wave has the form

$$u_r\left(r,t\right)=u_r^o F\left(a\left(r-vt\right)\right). \qquad (25.20)$$

Here, the unknown phase velocity is defined as follows:

$$v=\sqrt{1-\alpha_1 u_{r,r}}\;c_L \qquad (25.21)$$

and u_r^o is the given constant amplitude coefficient.

Introduce the restriction

$$\left|\alpha_1 u_{r,r}\right|<1 \qquad (25.22)$$

and represent the root approximately by the first three terms of the expansion

$$\sqrt{1-\alpha_1\,u_{r,r}}\;\approx 1-{}^1\!/\!_2\,\alpha_1 u_{r,r}+{}^1\!/\!_8(\alpha_1)^2\left(u_{r,r}\right)^2. \qquad (25.23)$$

This means that the first three approximations will be considered.

It follows from (25.23) that the wave (25.23) can be written approximately as follows:

$$u_r\left(r,t\right)\cong u_r^o F\left[a\left(r-c_L t\right)\right.$$
$$\left.-\,{}^1\!/\!_2\,ac_L\alpha_1 u_{r,r}t+{}^1\!/\!_8\,ac_L\left(\alpha_1\right)^2\left(u_{r,r}\right)^2 t\right]. \qquad (25.24)$$

The next step consists of the representation of the formula (25.24) in the form

$$u_r\left(r,t\right)=u_r^o F\left[a\left(r-vt\right)\right]=u_r^o F\left[a\left(r-c_L t\right)+\delta^*\right]=u_r^o F\left(\sigma+\delta^*\right). \quad (25.25)$$

Here, the new parameter $\delta^*=-{}^1\!/\!_2\,ac_L\alpha_1 u_{r,r}\left[1-{}^1\!/\!_4\,\alpha_1 u_{r,r}\right]t$ is introduced, which should be restricted according to the method

$$\left|\delta^*=-{}^1\!/\!_2\,\alpha_1 u_{r,r}\left[1-{}^1\!/\!_4\,\alpha_1 u_{r,r}\right]t\right|<1. \qquad (25.26)$$

Thus, two restrictions (25.21) and (25.26) are used in this approach. They make it possible to expand the solution (25.25) into the Taylor series with respect to the parameter (25.26) in the vicinity of the classical phase value $\sigma = a\,(r - c_L t)$ with constant wave velocity c_L

$$\begin{aligned} u_r\,(r,t) &= u_r^o a F\left(\sigma + \delta^*\right) \\ &= u_r^o F(\sigma) + u_r^o F^{/}(\sigma)\delta^* + {}^{1}\!/_{2}\, u_r^o F^{//}(\sigma)\delta^{*2} + \cdots . \end{aligned} \tag{25.27}$$

We save only two first terms $u_r\,(r,t) = u_r^o F\,(\sigma + \delta^*) \approx u_r^o F(\sigma) + u_r^o F^{/}(\sigma)\delta^*$. The substitution into the last formula of the expression for the small parameter yields

$$u_r\,(r,t) \approx u_r^o F(\sigma) - {}^{1}\!/_{2}\, u_r^o F^{/}(\sigma) a c_L \alpha_1\, u_{r,r}\left[1 - {}^{1}\!/_{4}\,\alpha_1 u_{r,r}\right] t. \tag{25.28}$$

After some transformations, the final approximate representations can be obtained as follows:

- the solution in the form of the first two approximations

$$u_r\,(r,t) \approx u_r^o F(\sigma) - {}^{1}\!/_{2}\, a^2\left(u_r^o\right)^2 c_L \alpha_1 t\left[F^{/}(\sigma)\right]^2; \tag{25.29}$$

- the solution in the form of the first three approximations

$$\begin{aligned} u_r\,(r,t) &\approx u_r^o F(\sigma) - {}^{1}\!/_{2}\, a^2\left(u_r^o\right)^2 c_L \alpha_1 t\left[F^{/}(\sigma)\right]^2 \\ &\quad + {}^{1}\!/_{8}\, a^3\left(u_r^o\right)^3 (\alpha_1)^2 c_L t\left[F^{/}(\sigma)\right]^3. \end{aligned} \tag{25.30}$$

Solution (25.29) includes two summands. The first corresponds to the linear approach, and the second introduces the quadratic nonlinear correction to the linear approach. Solution (25.30) includes, compared to (25.29), the third summand, which introduces the cubic correction. A difference between the first three approximations, obtained by methods 1 and 2, is described as Comparison 3 in the final part of the article.

25.4 Harmonic Wave as the Classic One (The Symmetric Initial Profile)

The two main features of the harmonic waves are the initial profile in the form of the cosine function and the wave phase in the form $\sigma = a\,(kx_1 - \omega t)$ including the wave frequency ω and wavenumber k. Thus, in the case of plane longitudinal harmonic wave, we have

$$u_1\,(x_1,t) = u_{1o} \cos\left(k_L x_1 - \omega t\right), \tag{25.31}$$

where u_{1o} is the initial wave amplitude. Note that the harmonic wave is periodic. On a single period, this wave has two humps, each of which is symmetrical with respect to the vertical passing through the top of the hump. Only this type of symmetry is considered in this paper.

25.4.1 Method 1. First Two Approximations

Note once again that when analyzing Eq. (25.3) by method 1, the first approximation $u_1^{(1)}(x_1, t)$ is assumed to be the solution of the corresponding linear wave equation (25.1) and coincides with (25.31).

The second approximation $u_1^{(2)}(x_1, t)$ is a solution of the inhomogeneous linear wave equation

$$u_{1,tt}^{(2)} - (v_L)^2 u_{1,11}^{(2)} = (N_1/\rho) u_{1,11}^{(1)} u_{1,1}^{(1)} \tag{25.32}$$

and is characterized by the second harmonic (Engelbrecht 2015)

$$u_1^{(2)}(x_1, t) = x_1 \left[\frac{N_1}{8(\lambda + 2\mu)} (u_{1o})^2 k_L^2 \right] \cos 2(k_L x_1 - \omega t). \tag{25.33}$$

Thus, the approximate solution corresponding to method 1 consists of the sum of two harmonics (Engelbrecht 2015)

$$\begin{aligned} u_1^{(1+2)}(x_1, t) &= u_1^{(1)}(x_1, t) + u_1^{(2)}(x_1, t) = u_{1o} \cos(k_L x_1 - \omega t) \\ &+ x_1 \left[\frac{N_1}{8(\lambda + 2\mu)} (u_{1o})^2 k_L^2 \right] \cos 2(k_L x_1 - \omega t). \end{aligned} \tag{25.34}$$

Solution (25.34) confirms a generation of the second harmonic. This classical nonlinear wave effect is formed in a few stages (Engelbrecht 2015). As a result, the effect of the second harmonic increases, and it becomes the dominant one. Thus, the evolution of the wave consists of a gradual transition from the profile of the first harmonic to the profile of the second harmonic.

All the results in this study are obtained for metallic materials, small strains, and ultrasound frequencies. The 2D picture shows an evolution in coordinates "distance of propagation–displacement" and the 3D picture—in coordinates "time of propagation–distance of propagation–displacement."

Figure 25.1 (Rushchitsky 2011) presents the exact transition from the first harmonic to the second one. Here, two different (initial and final) parts of the same plot with $\omega = 100$ KHz, $u_{1o} = 0.1$ mm are superposed (see two rows of distance values on the horizontal axis).

The process of transition can be divided into a few stages, one of which is shown in Figs. 25.2 and 25.3. As seen from the figures, the initial symmetric (in the above sense) profile is distorted symmetrically.

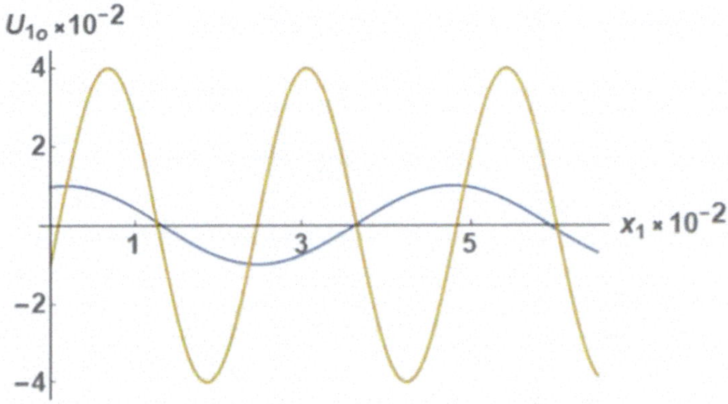

Fig. 25.1 The transformation of the initially excited first harmonic into the second one

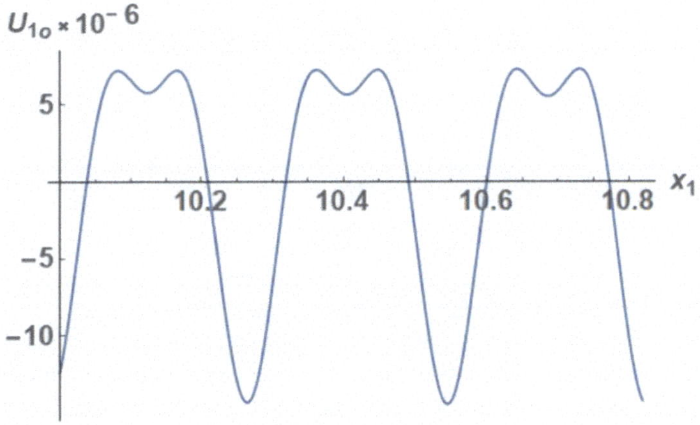

Fig. 25.2 Evolution of the initial harmonic profile (2D picture)

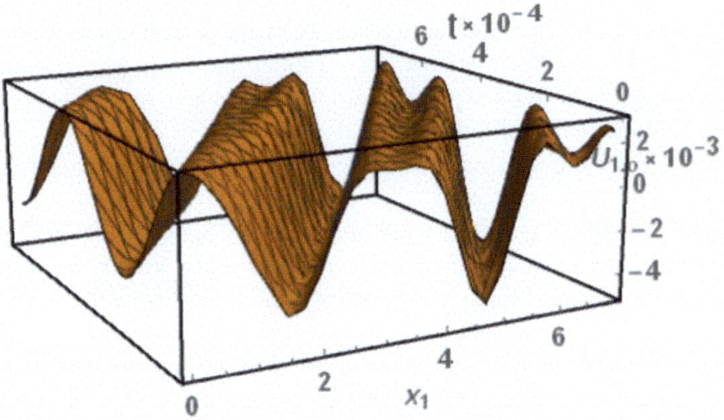

Fig. 25.3 Evolution of the initial harmonic profile (3D picture)

25.4.2 Method 1. First Three Approximations

The third approximation is found as the solution of the following wave equation:

$$u_{1,tt}^{(2)} - (A_L)^2 u_{1,11}^{(2)} = \left(N_1/\rho\right) u_{1,11}^{(1)} u_{1,1}^{(1)}. \tag{25.35}$$

Then the approximate solution of the basic nonlinear wave equation (25.3) within the framework of the first three approximations is formed of the sum of the first, second, and fourth harmonics, having the form (Rushchitsky 2011):

$$u_2^{(1+2+3)}(x_1,t) = u_1^{(1)}(x_1,t) + u_1^{(2)}(x_1,t) + u_1^{(3)}(x_1,t) = \tag{25.36}$$

$$= u_{1o}\cos\sigma + u_{1o} M_L x_1 \cos 2\sigma + u_{1o} M_L^3 x_1^3 \left[-\frac{8}{3} + \frac{5}{2k_L x_1} \sin 4\sigma + \left(-\frac{4}{3} + \frac{11}{8k_L^2 x_1^2} \right) \cos 4\sigma \right], \tag{25.37}$$

$$M = \frac{N_1}{8(\lambda + 2\mu)} u_{1o} k_L^2 = \frac{1}{8\rho} N_1 u_{1o} \frac{k_L^2}{v_L^2} = \frac{1}{8\rho} N_1 u_{1o} \frac{\omega^2}{v_L^4}.$$

Based on (25.37), hundreds of plots were built (Engelbrecht 2015), showing the evolution for different values of three basic parameters—the initial maximal amplitude, wave frequency, and mechanical characteristics of the material.

The main conclusion from the obtained graphs is a general scheme of changing the wave profile (evolution scenario). First, when propagating in a nonlinearly elastic material, the harmonic wave interacts with itself and generates new waves, creating a picture of evolution. The dependence of the wave amplitude on the time and distance of the wave propagation is shown in the following figures. Figure 25.4 corresponds to the initial stage of the wave motion, taking into account the first+second+third approximations. Figure 25.5 shows two stages of the longer wave motion (about ten wavelengths) and the first approximation (upper line for positive amplitude values), first+second (lower line for positive amplitude values), and first+second+third (middle line for positive amplitude values).

Figure 25.6 shows the 3D image of the evolution in the coordinates "amplitude–distance–time."

Note that the nonlinear wave equation of type (25.3) appears not only in mechanics. For example, the corresponding optical waves were studied by Blombergen (Bloembergen 1965; Rushchitsky 2011; Yariv 1967), and his results were awarded the Nobel Prize in Physics in 1980 (for his contribution to the development of laser spectroscopy). These results include a description of the transformation of the first harmonic (red light from a ruby laser with a wavelength of 6940 Å) to the second harmonic (blue light with a wavelength of 3470 Å) when the light propagates through a potassium dihydrogen phosphate crystal KH_2PO_4 [28].

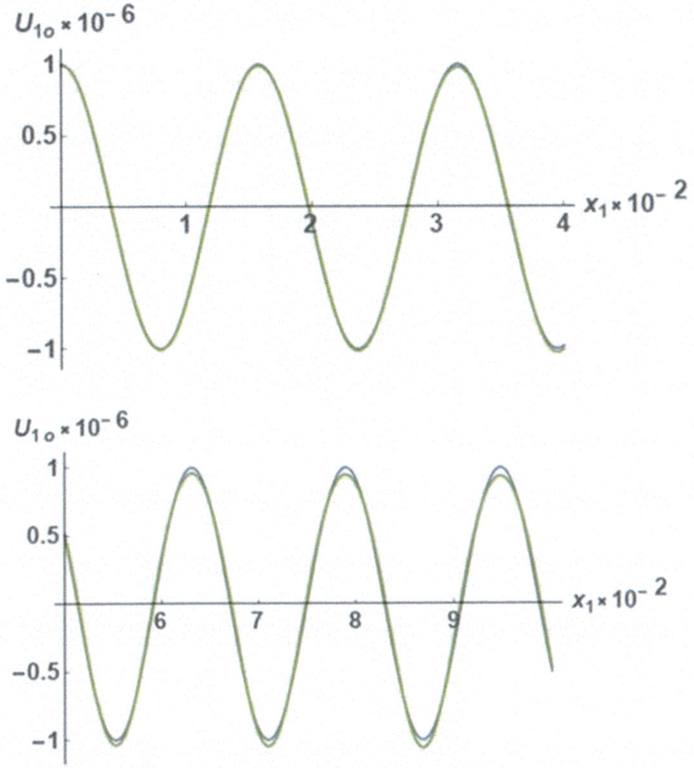

Fig. 25.4 Evolution of the initial harmonic profile (2D picture)

The phenomenon of the transformation of the first harmonic into the second one can be seen in Fig. 25.5—the bottom line for positive amplitude values shows the gradual transformation of one hump of the periodic profile into two in the period of one hump.

Figure 25.6 also shows that taking into account the third approximation reveals another wave effect—a wave with a periodic system of humps in each period turns one hump into three in one period during its motion. At that, formula (25.22) shows the existence of the third term with the fourth harmonic.

Thus, the evolution of harmonic waves is that with an increase in the time or distance of wave propagation, its profile in one period forms several humps from one. This nonlinear wave effect is accompanied by the constancy of the period—the period (sole) of the wave does not change during its evolution.

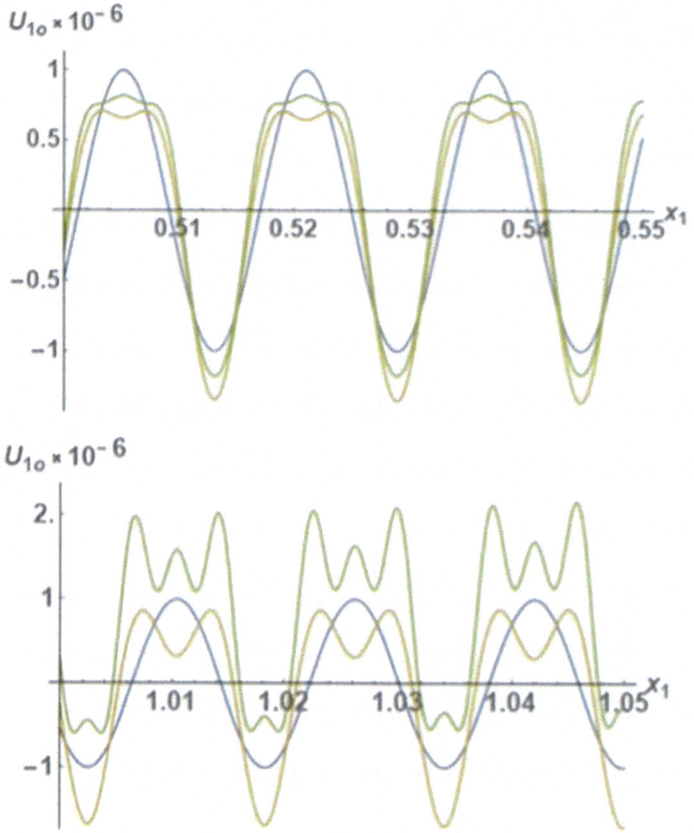

Fig. 25.5 Evolution of the initial harmonic profile (2D picture)

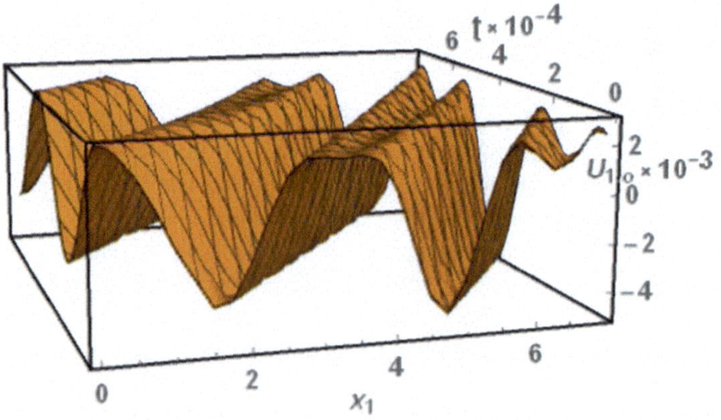

Fig. 25.6 Evolution of the initial harmonic profile (3D picture)

25.4.3 Method 2. First Two Approximations

This method assumes that the initial profile has the form

$$F\left(x_1\right) = \mathrm{e}^{-ik_L x_1}. \tag{25.38}$$

The formula for the solution that includes only the first and second approximations is shown above as formula (25.17):

$$u_1\left(x_1, t\right) \approx F(\sigma) + F'_{,1}(\sigma) a\delta = F(\sigma) - {}^1\!/_2\, \alpha a^2 c_L t \left[F'_{,1}(\sigma)\right]^2.$$

When expression (25.38) is substituted into (25.17), the following new formula gives the theoretical solution for the problem indicated in the title of this subsection:

$$u_1\left(x_1, t\right) = a^o \mathrm{e}^{-ik_L(x_1 - c_L t)} - {}^1\!/_2\, \alpha k_L^2 \left(a^o\right)^2 \mathrm{e}^{-2ik_L(x_1 - c_L t)}. \tag{25.39}$$

Comparison of expression (25.39) with expression (25.34) for the corresponding solution obtained by method 1 shows the identity of these two representations, which does not change the qualitative picture of evolution. This identity is observed only for the final stage of evolution. The comparison of all stages shows some distinctions, highlighted at the end of this chapter as Comparison 1.

25.5 Wave with the Symmetric Initial Profile in the Form of Gauss Function

25.5.1 Gauss Profile (Method 1, First Two Approximations)

The wave with this profile (bell-shaped, hump-shaped) has the form (Rushchitsky 2011; Weisstein 2013)

$$u_1\left(x_1, t\right) = u_{1o} \mathrm{e}^{-{}^1\!/_2 \sigma^2}. \tag{25.40}$$

According to the method, the first approximation is the solution of the corresponding linear equation (25.5) and has the form (25.24). The second approximation can be sought as a solution to the inhomogeneous equation

$$u_{1,tt}^{(2)} - v_L^2 u_{1,11}^{(2)} = \left(N_1/\rho\right) u_{1,11}^{(1)} u_{1,1}^{(1)}$$

or

$$u_{1,tt}^{(2)} - v_L^2\, u_{1,11}^{(2)} = \left(N_1/\rho\right) \left(u_{1o}\right)^2 a^3 \sigma \left(1 - \sigma^2\right) \mathrm{e}^{-\sigma^2}. \tag{25.41}$$

If the solution of linear equation (25.41) is found in the form $u_1^{(2)} = A(\sigma)e^{-\sigma^2}$, then this form is the solution of a corresponding homogeneous equation and can be treated as d'Alembert's wave. Therefore, the solution should be complicated to the form $u_1^{(2)} = tA(\sigma)e^{-\sigma^2}$, and, being substituted into the left part of (25.27) with the notation $B(\sigma) = tA(\sigma)e^{-\sigma^2}$, the differential equation relative to function $B(\sigma)$ can be obtained as

$$\left[B''(\sigma) + 3\left(\sigma^2 - 1\right)B(\sigma)\right] = (u_{1o})^2 \alpha a\sigma\left(1 - \sigma^2\right)e^{1/2\,\sigma^2}. \tag{25.42}$$

The homogeneous equation $B''(\sigma) + 3(\sigma^2 - 1)B(\sigma) = 0$ corresponds to Eq. (11) from Sect. 2.173 of Kamke (1977) with $a = 0$, $b = 3$, $c = \sqrt{3}$ and can be reduced to the so-called Whittaker equation (Gradstein and Ryzhik 2007; Press et al. 1992). So, the solution looks pretty complicated owing to the factor $\sigma\left(\sigma^2 - 1\right)$.

Thus, the second approximation will have a complex mathematical form that has yet to be found, and, accordingly, the analytical and numerical analysis of wave evolution within the framework of the first two approximations looks very unpromising for the profile in the form of the bell-shaped function. In this situation, method 2 has some advantages.

25.5.2 *Gauss Profile (Method 2, First Two Approximations)*

We still consider the initial profile in the form of (25.26), and the basic formula for the first two approximations is (25.17). Therefore, the starting formula for the case of the first two approximations has the following form:

$$u_1(x_1, t) = u_{1o}e^{-1/2\,\sigma^2} - 1/2\, t\alpha c_L a^2\sigma^2 (u_{1o})^2 e^{-\sigma^2}. \tag{25.43}$$

First, expression (25.43) testifies that the Gauss wave evolves and the profile is distorted symmetrically due to the appearance of a "second harmonic" whose amplitude increases nonlinearly with the propagation time of the wave.

Note that the term "harmonic" should be conditionally used for this wave. If to assume that the subsequent harmonics after the first one $\exp\{-1/2\,\sigma^2\}$ are distinguished only by the coefficient (simple number) before the phase, and this number gives the number for harmonic, then $\exp\{-\sigma^2\}$ can be called the second harmonic. However, the notion "harmonic" is nevertheless used in the harmonic analysis based on the completeness of the functions-harmonics. This completeness is not valid for the Gauss function. Therefore, here the term "second harmonic" should, strictly speaking, be written with quotation marks.

We also note that the simplicity of obtaining the second harmonic is due to utilizing only the first derivative of the Gaussian function.

The evolution of the wave in time and distance of wave propagation is shown in Figs. 25.7 and 25.8. In Fig. 25.7, the initial profile is superimposed on the distorted

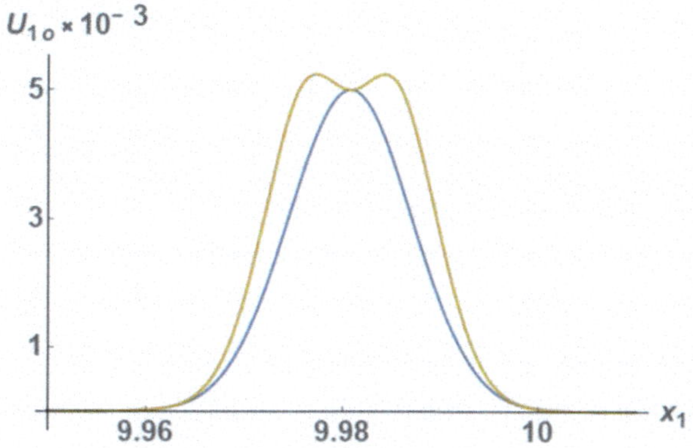

Fig. 25.7 Evolution of the initial Gauss profile (2D picture)

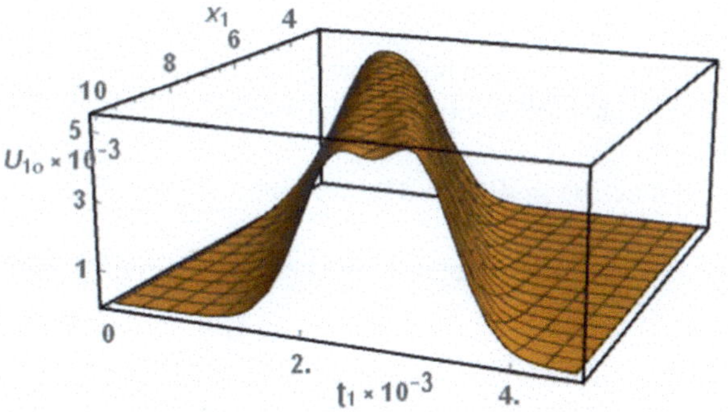

Fig. 25.8 Evolution of the initial Gauss profile (3D picture)

one and shows evolution with distance. Figure 25.8 is 3D and shows the evolution in time and space.

A comparison with the change in the harmonic symmetric wave profile shows that the symmetric Gauss wave changes its profile symmetrically in a slightly different way. The second harmonic always gives a negative term. So, the slopes of the distorted "bell or hump" become steeper. The upper part of the "bell" falls and forms two "bells." Since the wave is elastic and there can be no energy loss, the profile change is consistent with the law of conservation of wave energy during wave propagation. This fact is described in the Results of comparisons as Comparison 3.

25.5.3 Gauss Profile (Method 2, First Three Approximations)

First of all, method 2 proposed above should be extended to the case when the first three approximations are taken into account. In this case, three terms should be kept in the approximate representation of the root instead of two:

$$\sqrt{1+\alpha u_{1,1}} \approx 1 + {}^{1}\!/_{2}\,\alpha u_{1,1} - {}^{1}\!/_{8}\,\alpha^2 \left(u_{1,1}\right)^2 . \tag{25.44}$$

The wave velocity is now written in the more complicated form

$$v = c_L \left[1 + {}^1\!/_2 \alpha u_{1,1} - {}^1\!/_8 \alpha^2 \left(u_{1,1} \right)^2 \right] \tag{25.45}$$

and the approximate solution can be represented in a form that generalizes the corresponding representation (25.43)

$$u_1 (x_1, t) \cong F \left\{ a (x_1 - c_L t) - {}^1\!/_2 t \alpha c_L u_{1,1} \left[1 - {}^1\!/_4 \alpha u_{1,1} \right] \right\}. \tag{25.46}$$

The following procedure is similar to the previous one—the smallness of the parameter δ^* should be assumed:

$$\left| \delta^* = - {}^1\!/_2 t \alpha v_L u_{1,1} \left[1 - {}^1\!/_4 \alpha u_{1,1} \right] \right| \ll 1, \tag{25.47}$$

and then the solution is expanded into a Taylor series with respect to a small parameter δ^* in the neighborhood of a phase value $\sigma = a (x_1 - c_L t)$

$$u (x_1, t) = F(\sigma + \delta^*) = F(\sigma) + F^{/}(\sigma)\delta^* + {}^1\!/_2 F^{//}(\sigma)\delta^{*2} + \cdots . \tag{25.48}$$

The final representation is obtained by saving only the first two terms:

$$\begin{aligned} u_1^{1+2+3}(x_1, t) &\approx F(\sigma) - F^{/}_{,1}(\sigma) a^2 \left\{ {}^1\!/_2 t \alpha c_L F^{/}_{,1}(\sigma) \left[1 - {}^1\!/_4 \alpha a F^{/}_{,1}(\sigma) \right] \right\} \\ &= F(\sigma) - {}^1\!/_2 \alpha a^2 c_L t \left[F^{/}_{,1}(\sigma) \right]^2 \left[1 - {}^1\!/_4 \alpha a F^{/}_{,1}(\sigma) \right]. \end{aligned} \tag{25.49}$$

The corresponding to the Gauss profile solution is as follows:

$$\begin{aligned} u (x_1, t) = u_{1o} e^{-{}^1\!/_2 \sigma^2} &- {}^1\!/_2 \alpha v_L a^2 t \left(u_{1o} \right)^2 \sigma^2 e^{-\sigma^2} \\ &- {}^1\!/_8 \alpha^2 v_L a^3 t \left(u_{1o} \right)^3 \sigma^3 e^{-{}^3\!/_2 \sigma^2}. \end{aligned} \tag{25.50}$$

Thus, taking into account the third approximation introduces new features in the evolution. First, the representation (25.32) (first+second+third) introduces the asymmetric changes in the form of an initial profile contrary to the case first+second, which introduces the symmetric changes. Besides that, the representation (25.32) contains the cubic nonlinearity in the third approximation, which in the case of a harmonic profile means the presence of the third harmonic in contrast to the effect of the classical case, where after the second harmonic the fourth one is generated. This fact forms an essence of Comparison 4 from the Results of comparisons.

Note that the expression for the wave amplitude (25.50) includes the squared phase in the second term and the cube in the third term. This fact is characteristic of solitary waves and is not presented in the description of the evolution of a harmonic wave.

Figures 25.9, 25.10, 25.11, and 25.12 show the 2D plots—the dependence of the wave amplitude on the wave propagation distance. All graphs are obtained accord-

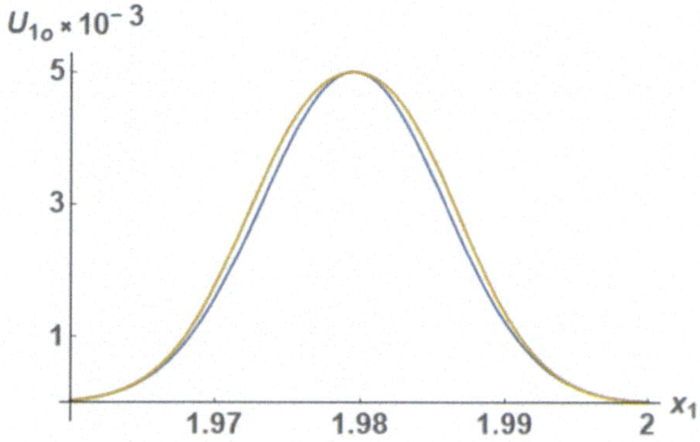

Fig. 25.9 Evolution of the initial Gauss profile (2D picture, stage 1)

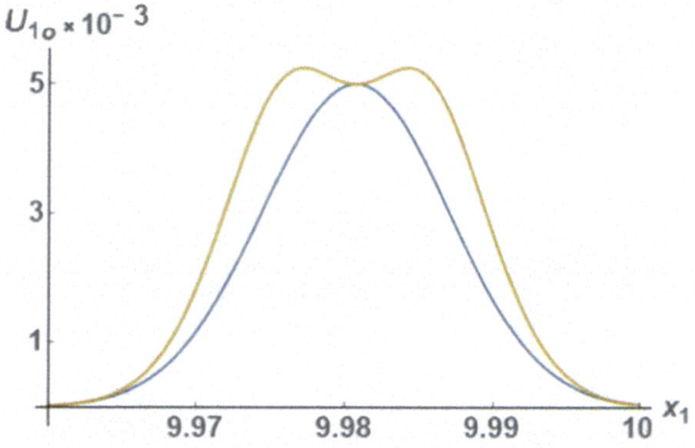

Fig. 25.10 Evolution of the initial Gauss profile (2D picture, stage 2)

ing to (25.32). Figure 25.9 corresponds to the initial stage of evolution. The lower graph shows the initial profile, and the upper one shows the profile for the two first approximations.

Figure 25.10 corresponds to the advanced stage of evolution and differs from the previous one only in the upper graph, which is obtained taking into account the first three approximations.

Figures 25.11 and 25.12 show two profiles ($u_1^{(1+2)}$ and $u_1^{(1+2+3)}$), each of which have two humps, but the plots differ in the distance passed by the wave (the second one corresponds to a fairly formed evolution).

Figure 25.13 shows the 3D picture of the Gauss wave evolution.

So, in the evolution of the Gauss wave, some features are observed:

A. The second harmonic always gives a negative addition. Therefore, the bell in its upper part becomes thinner. That is, the slopes of the bell become steeper.
B. The upper part of the bell collapses, and two bells are formed.
C. The evolution of the right and left bells is different—it shows an increase in the left and a decrease in the right bells. This effect is unexpected, like several other third approximation nonlinear wave effects.

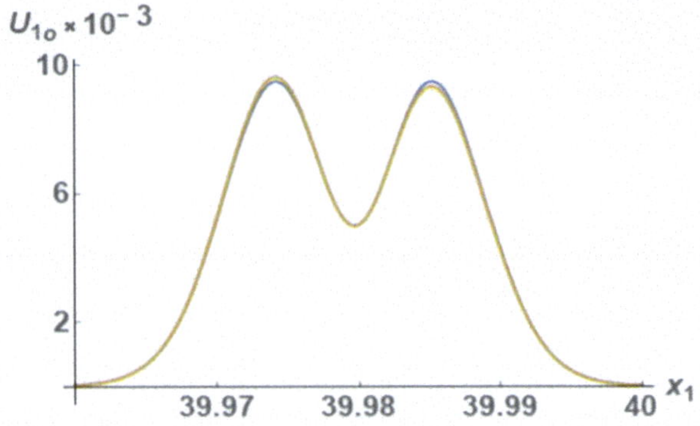

Fig. 25.11 Evolution of the initial Gauss profile (2D picture, stage 3)

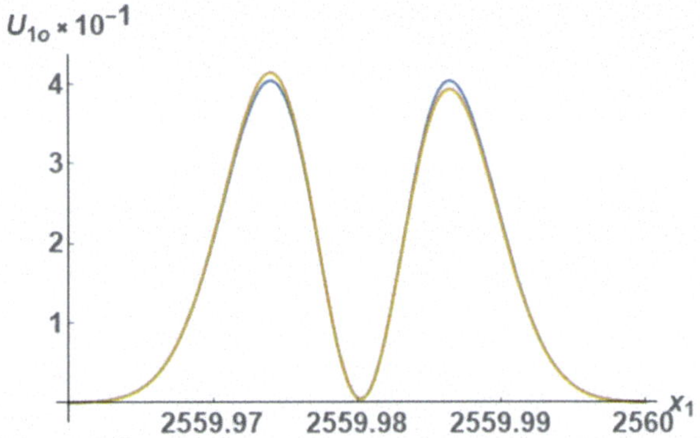

Fig. 25.12 Evolution of the initial Gauss profile (2D picture, stage 4)

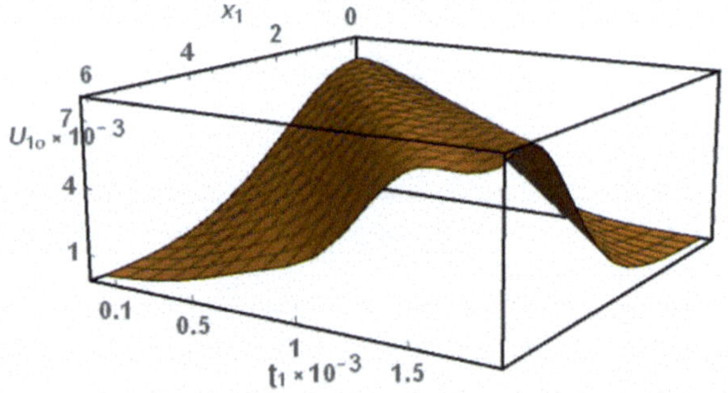

Fig. 25.13 Evolution of the initial Gauss profile (3D picture)

25.6 Wave with the Asymmetric Initial Profile in the Form of Whittaker Function

The set of Whittaker functions (Gradstein and Ryzhik 2007; Olde Daalhuis 2010) has the common characteristic property—their plots have the form of one hump, which is asymmetric relative to the vertical straight line passing through the hump top and corresponds to some experimentally observed profiles of excited waves (Rushchitsky

2011). These functions depend on two parameters and satisfy some special ordinary differential equation of the second order. Further, the Whittaker function is considered to describe the initial profile of some solitary wave:

$$F(x_1) = W_{1/4\,;1/4}(ax_1), \tag{25.51}$$

where a is the arbitrary constant parameter defining the size of the wave bottom.

Then the approximate representation of the solitary wave with the initial profile in the form (25.51), according to the basic formula (25.31), has the form:

- in the case of the first two approximations

$$u_1^{1+2}(x_1,t) = u_{1o}W_{1/4\,;3/4}(\alpha\sigma) - {}^1\!/_2\, t\alpha c_L a^2 (u_{1o})^2 \left[W'_{1/4\,;1/4}(\alpha\sigma)\right]^2; \tag{25.52}$$

- in the case of the first three approximations

$$\begin{aligned} u_1^{1+2+3}(x_1,t) &= u_{1o}W_{1/4\,;3/4}(\alpha\sigma) \\ &- {}^1\!/_2\, t\alpha c_L a^2 (u_{1o})^2 \left[W'_{1/4\,;1/4}(\alpha\sigma)\right]^2 + \\ &+ {}^1\!/_8\, t\alpha^2 c_L a^3 (u_{1o})^3 \left[W'_{1/4\,;1/4}(\alpha\sigma)\right]^3. \end{aligned} \tag{25.53}$$

Now, the derivative of the function $W_{\kappa,\mu}(z)$ must be calculated by the general formula (Chandra et al. 2012):

$$\frac{d}{dz}W_{\lambda,\mu}(z) = \left(\frac{\lambda}{z} - \frac{1}{2}\right)W_{\lambda,\mu}(z) - \frac{1}{z}\left[\mu^2 - \left(\lambda - \frac{1}{2}\right)^2\right]W_{\lambda-1,\mu}(z),$$

according to which the following formula is valid:

$$\left(W_{3/4;\,1/4}(\sigma)\right)' = \left(\frac{3}{4\sigma} - \frac{1}{2}\right)W_{3/4,\,1/4}(\sigma). \tag{25.54}$$

Then, in the case of the first two approximations, the solutions (25.52) and (25.53) have the form:

$$\begin{aligned} u_1^{1+2}(x_1,t) &= u_{1o}W_{3/4;\,1/4}(a\sigma) - \\ &- {}^1\!/_2\, t\alpha c_L a^2 (u_{1o})^2 \left(\left(\frac{3}{4\sigma} - \frac{1}{2}\right)W_{3/4,\,1/4}(a\sigma)\right)^2; \end{aligned} \tag{25.55}$$

in the case of the first three approximations, we have

Fig. 25.14 Evolution of the initial Whittaker profile (2D picture, 1 and 1+2 approximations, stage 1)

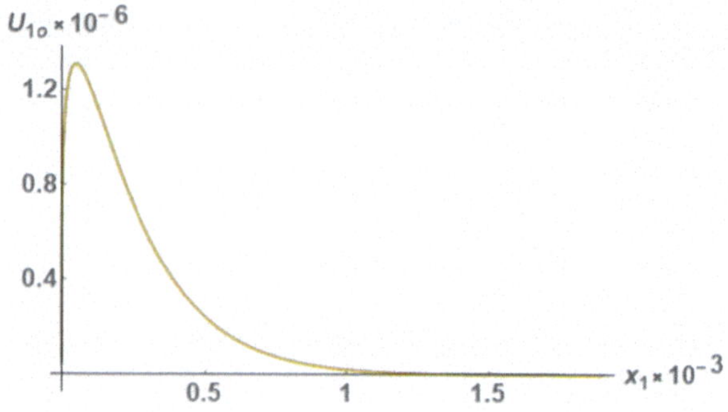

Fig. 25.15 Evolution of the initial Whittaker profile (2D picture, 1 and 1+2 approximations, stage 2)

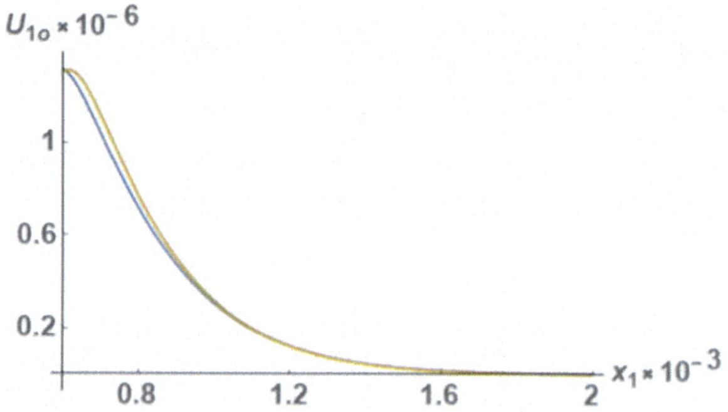

$$\begin{aligned} u_1^{1+2+3}(x_1,t) &= u_{1o} W_{3/4;1/4}(a\sigma) - \\ &- {}^1\!/_2\, t\alpha c_L a^2 (u_{1o})^2 \left(\left(\frac{3}{4\sigma} - \frac{1}{2} \right) W_{3/4,1/4}(a\sigma) \right)^2 + \\ &+ {}^1\!/_8\, t\alpha^2 c_L a^3 (u_{1o})^3 \left(\left(\frac{3}{4\sigma} - \frac{1}{2} \right) W_{3/4,1/4}(a\sigma) \right)^3 . \end{aligned} \tag{25.56}$$

Two features follow from the form of solutions (25.55) and (25.56): they describe the distortion in the initial profile of the wave due to the direct dependence of the nonlinear component on time and the "spreading" of the initial profile due to the presence of nonlinear components.

Figures 25.14, 25.15, and 25.16 show three 2D plots with two profiles in each in coordinates "displacement u_1—passing by the wave distance x_1" which are built by formulas (25.55). Figure 25.14 presents the initial stage with two practically identical profiles. Figure 25.15 presents the next stage with two profiles, which correspond to the first (linear) approximation (upper line) and the first two approximations (lower line). Figure 25.16 presents stage 3, where the difference between profiles is significant due to the top hump shift to the left. Figure 25.17 shows stage 3 in the 3D format (displacement–distance–time).

The subsequent two figures (Figs. 25.18 and 25.19) show curves corresponding to the 1+2 approximations and 1+2+3 approximations for different distances passed

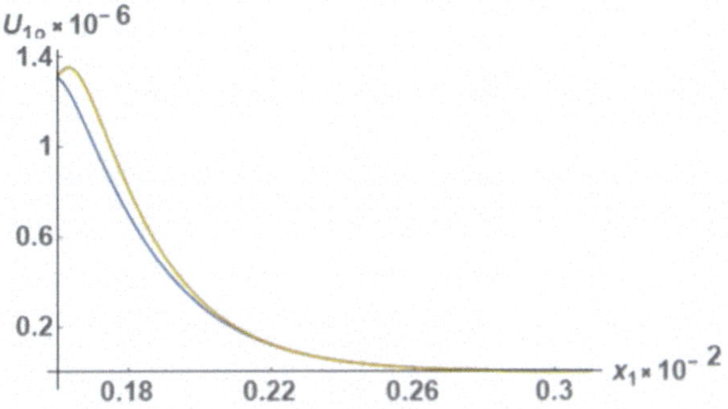

Fig. 25.16 Evolution of the initial Whittaker profile (2D picture, 1 and 1+2 approximations, stage 3)

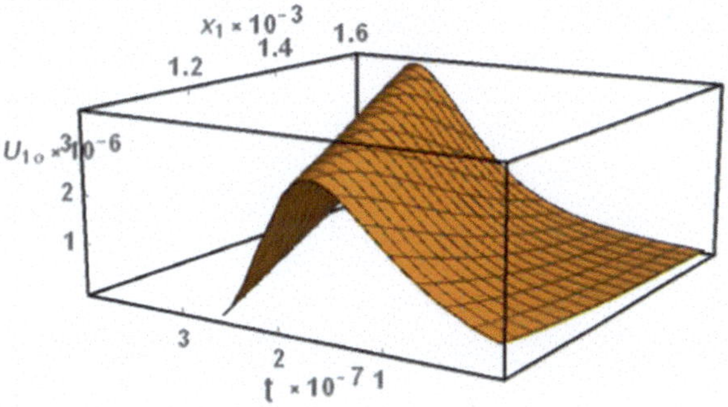

Fig. 25.17 Evolution of the initial Whittaker profile (3D picture, 1 and 1+2 approximations, stage 3)

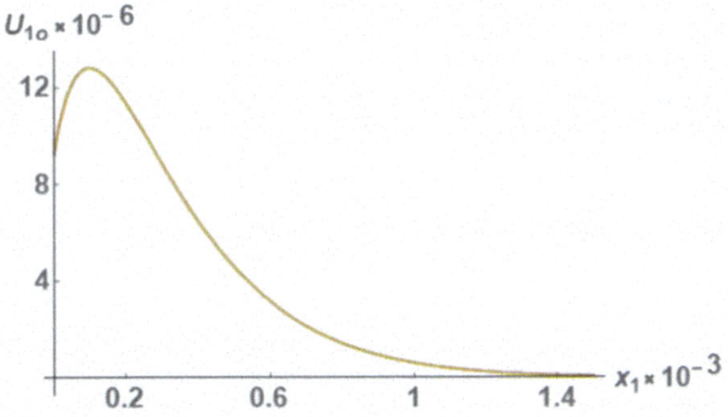

Fig. 25.18 Evolution of the initial Whittaker profile (2D picture, 1+2 and 1+2+3 approximations, stage 1)

by the wave. Figure 25.18 shows the initial stage 1 when the curves are practically identical. Figure 25.19 corresponds to stage 2 when the curves are slightly different (here and further, the upper curve corresponds to 1+2 approximations and lower—to 1+2+3 approximations).

Figures 25.20, 25.21, and 25.22 are similar to the previous ones but correspond to the longer propagation distances. Figure 25.20 shows only a reduction in the hump height, whereas Fig. 25.21 demonstrates the forming of two humps instead the initial hump. Figure 25.23 shows stage 5 in the 3D format (displacement–distance–time).

The presented plots show a series of features of the evolution of the Whittaker wave:

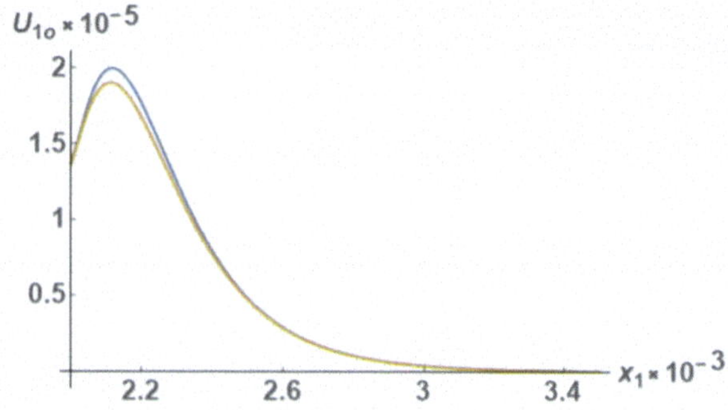

Fig. 25.19 Evolution of the initial Whittaker profile (2D picture, 1+2 and 1+2+3 approximations, stage 2)

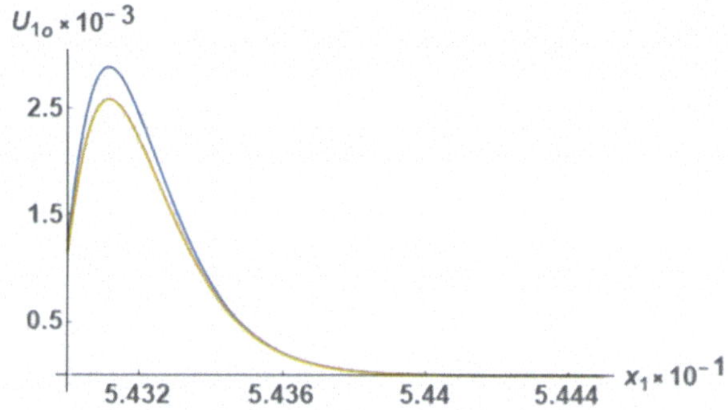

Fig. 25.20 Evolution of the initial Whittaker profile (2D picture, 1+2 and 1+2+3 approximations, stage 3)

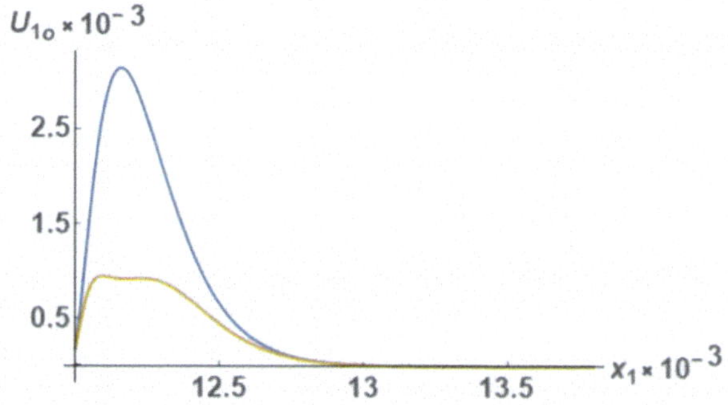

Fig. 25.21 Evolution of the initial Whittaker profile (2D picture, 1+2 and 1+2+3 approximations, stage 4)

A. The evolution of nonsymmetric profile occurs nonsymmetrically.
B. The maximal value of amplitude increases.
C. The wave bottom is not changed.
D. Accounting for the third approximation introduces essential changes to the profile distortion. First, the case 1+2+3 approximations demonstrate some tendency to form two nonsymmetric humps.

Of course, the evolution of the Whittaker wave gives new possibilities for comparing the evolution of the Gauss (symmetric profile) and Whittaker (nonsymmetric profile) waves. Such a comparison is given in Sect. 25.9 as Comparisons 9–11.

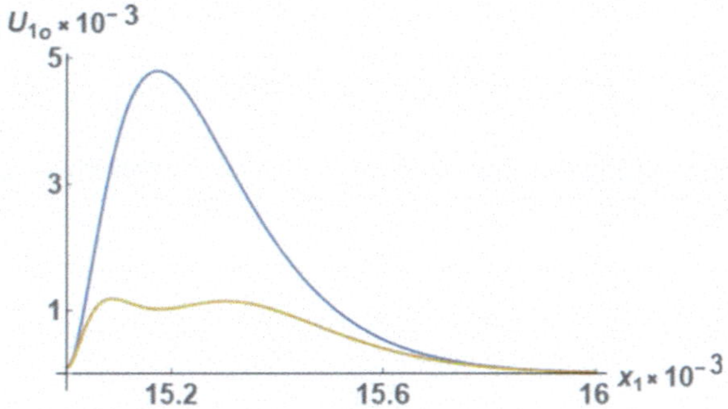

Fig. 25.22 Evolution of the initial Whittaker profile (2D picture, 1+2 and 1+2+3 approximations, stage 3)

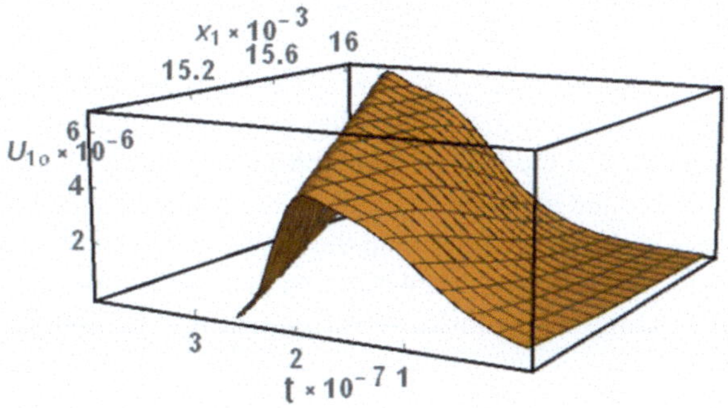

Fig. 25.23 Evolution of the initial Whittaker profile (3D picture, first three approximations, stage 4)

25.7 Wave with the Asymmetric Initial Profile in the Form of the MacDonald Function

In the prior sections, three types of initial profiles are considered—the harmonic profile with the symmetric periodic humps, the solitary Gauss profile with one symmetric hump, and the solitary Whittaker profile with one nonsymmetric hump. Now we consider the fourth type—the solitary MacDonald profile (Olver and Maximon 2010; Press et al. 1992), which is nonsymmetric and has no hump. This profile is close to the wave profiles experimentally observed in the case of blast loading. Besides that, the wave is already not plane and is related to the elastic cylindrical radial waves of displacement. It is described by the nonlinear wave equation (25.5).

Recall here the classical linear description of these waves (Altenbach et al. 2020; Rushchitsky 2011). This wave propagates in an infinite space with a cylindrical circular cavity. The wave motion is plane axisymmetric in the radial direction and excited by a harmonic in-time pulse applied to the boundary surface.

The cylindrical system of coordinates $Or\vartheta z$ is used, and the z-axis coincides with the axis of the cavity. Due to the axisymmetry, the motion depends only on the radius r and time t. Only the radial displacements and σ_{rr}, $\sigma_{\vartheta\vartheta}$, and σ_{zz} are nonzero. The linear and nonlinear motion equations have the forms (25.4) and (25.5), respectively.

Let us start with the case when the solution of (25.4) is sought as the harmonic in the time wave of displacement. When the wave propagates from the cylindrical cavity of the radius r_o and is excited by the harmonic in time displacement $u_r(r_o, t) = u_{ro}e^{i\omega t}$ or load $\sigma^{rr}(r_o, t) = p_o e^{i\omega t}$, the nonlinear problem is solved in Engelbrecht (2015) by method 1 taking into account the first two approximations. The linear approximation has a form

$$u_r^{(1)}(r, t) = u_{ro} H_1^{(1)}(k_L r) e^{i\omega t}, \tag{25.57}$$

where the Hankel function of the first kind is used, and u_{ro} is the amplitude factor

$$u_{ro} = -\frac{p_o k_L}{k_L(\lambda + 2\mu) H_0^{(1)}(k_L r_o) - \frac{2\mu}{r_o} H_1^{(1)}(k_L r_o)}. \tag{25.58}$$

The solution (25.57) shows that the wave is harmonic only asymptotically with respect to the spatial coordinate. The wave intensity decreases over time due to the properties of the Hankel function $H_1^{(1)}$.

Consider now the case when the function $F[a(r - vt)]$ describes the solitary wave and is not harmonic in the time wave. First, the linear wave equation should be analyzed. The cylindrical function of a real argument—the Hankel function $H_\lambda(r)$—is no longer the solution of this equation. This linear equation has a solution in the form of the cylindrical function—the MacDonald function $K_\lambda(r)$.

Let us use method 2 and (25.30). Choosing the initial profile in the form of the function $F(\sigma) = K_0(\sigma)$ and substituting this expression into (25.30), we obtain the representation of the solution in the form of the first three approximations

$$\begin{aligned} u_r(r, t) \approx u_{ro} a K_0(\sigma) - {}^1\!/\!_2\, a^3 (u_{ro})^2 c_L \alpha_1 t \left[K_0^{/}(\sigma)\right]^2 + \\ + {}^1\!/\!_8\, a^5 (u_{ro})^3 c_L (\alpha_1)^2 t \left[K_0^{/}(\sigma)\right]^3. \end{aligned} \tag{25.59}$$

Further, the derivative in (25.59) should be rewritten using the known formula $K_0^{/}(\sigma) = -K_1(\sigma)$, transforming the approximate representation of the nonlinear wave into the new form

$$\begin{aligned} u_r(r, t) \approx {} & u_{ro} K_0(a(r - c_L t)) \\ & - {}^1\!/\!_2\, a^3 (u_{ro})^2 \alpha_1 a c_L t \left[K_1(a(r - c_L t))\right]^2 + \\ & + {}^1\!/\!_8\, a^5 (u_{ro})^3 (\alpha_1)^2 c_L t \left[K_1(a(r - c_L t))\right]^3. \end{aligned} \tag{25.60}$$

The form of solution (25.60) shows two evident features:

A. The MacDonald wave changes its initial profile since the nonlinear summands are directly dependent on time.
B. Because the MacDonald wave profile has no humps and likes the hyperbola, it does not evolve into the profile with the humps. The distorted profile will be smooth and repeat the initial profile shape.

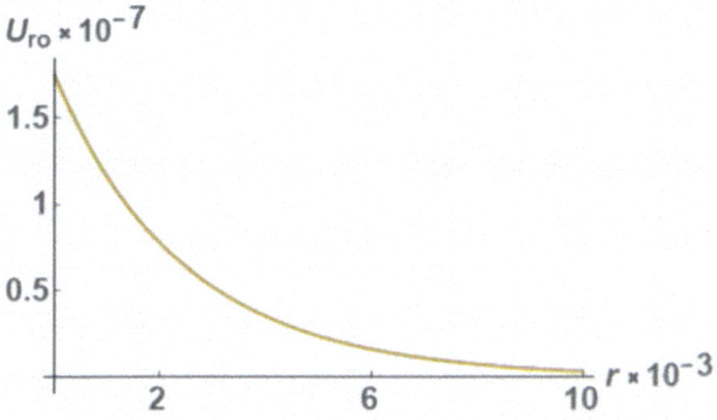

Fig. 25.24 Evolution of MacDonald wave (1+2, stage 1)

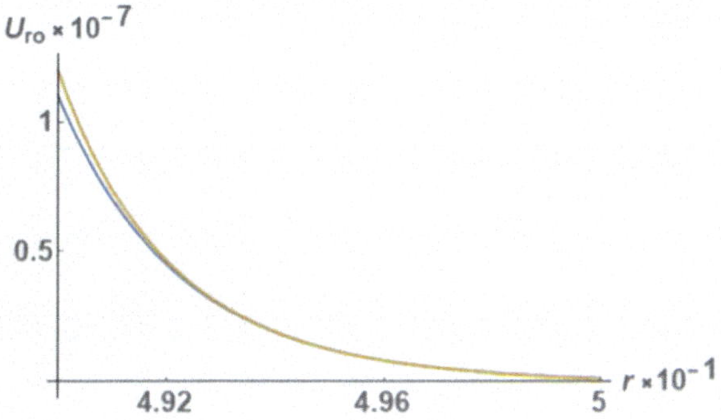

Fig. 25.25 Evolution of MacDonald wave (1+2, stage 2)

The formula (25.60) is used for the numerical simulation of wave evolution. The next group of the 2D plots is built in coordinates "displacement u_r—distance r" for the cases of 1+2 and 1+2+3 approximations.

Figure 25.24 corresponds to stage 1 and shows the small effect of nonlinearity on the evolution. The left curve corresponds to the linear approximation, and the right one—to the 1+2 approximation. Figure 25.25 corresponds to stage 2 and shows the more developed evolution of the initial profile. It saves the displayed in stage 1 features. The distorted profile becomes steeper and shifts more to the right side from the initial profile. The wave bottom is still shortened.

Figure 25.26 corresponds to stage 3 and shows the more developed evolution. It also saves the displayed in stages 1 and 2 features. The distorted profile becomes steeper and more shifted to the right side from the initial profile. The wave bottom continued to shorten. Figure 25.27 proposes a new 3D picture. Thus, the distorted profile becomes gently steeper and is shifted to the right side from the initial profile. The bottom of the distorted profile is shortened.

So, the shown plots for the MacDonald wave demonstrate that this new wave with a profile not similar to the previous three wave profiles is distorted in a new way. This way is characterized by some features differing from the basic features:

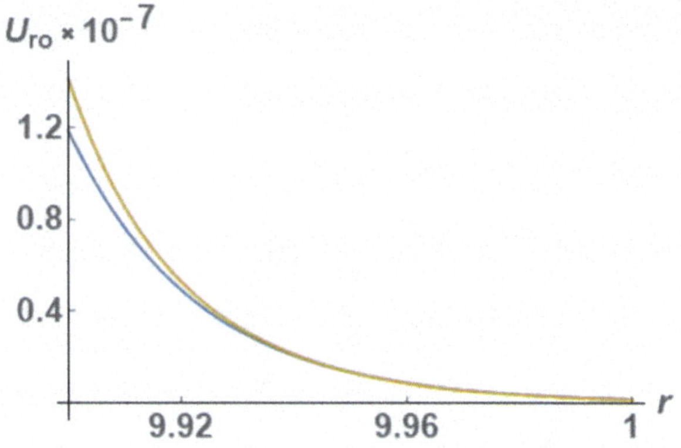

Fig. 25.26 Evolution of MacDonald wave (1+2, stage 3)

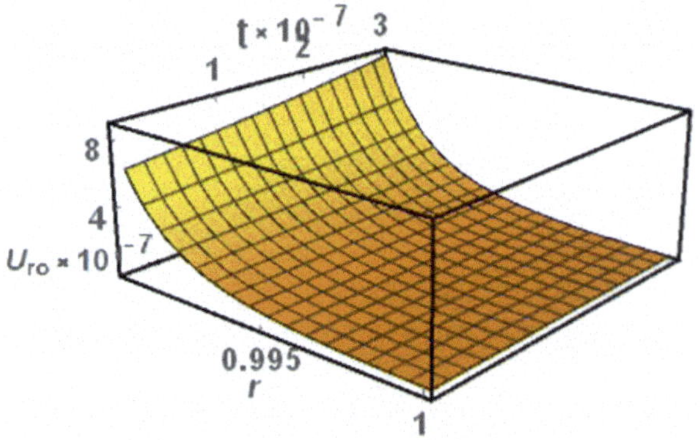

Fig. 25.27 Evolution of MacDonald wave (1+2, 3D picture)

A. Its profile without the hump does not generate humps.
B. A non-symmetry of the profile is saved, and the distorted profile becomes steeper and steeper.
C. The wave profile bottom becomes shorter and shorter. These facts allow comparisons of the harmonic, Gauss, Whittaker, and MacDonald waves. Such comparisons are given in Sect. 25.9, Comparisons 12–14.

25.8 Wave with the Asymmetric Initial Profile in the Form of Friedlander Function

The Friedlander function (initial wave profile) is less known in the mechanics of materials than the four profiles considered above. This initial profile, known in the physical theory of an explosion (originally a nuclear explosion), has been studied for almost a hundred years, and these studies continue to this day. These studies are the experimental ones that deal with the modern problems of science, not related to nuclear explosions, but, in particular, other military and medical problems (Kuriakose

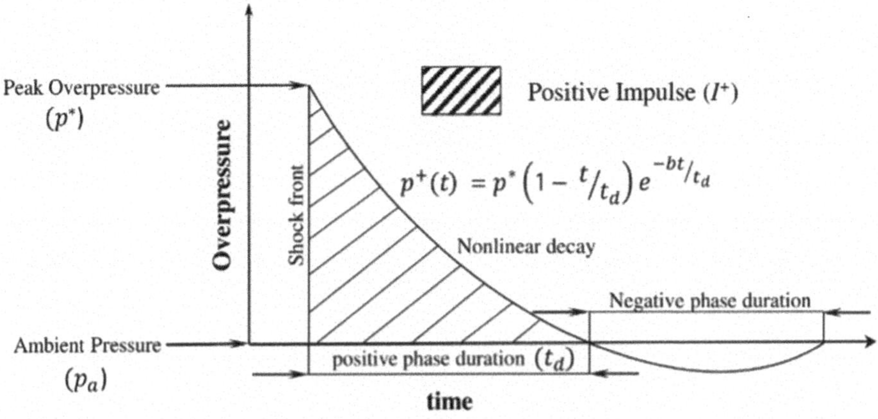

Fig. 25.28 The Friedlander function (basic scheme) generated by the explosion. The authors of Press et al. (1992) consider the Friedlander profile acceptable for interpreting the experiment

et al. 2016; Press et al. 1992). These studies are experimental and deal with modern scientific problems unrelated to nuclear explosions, particularly other military and medical problems (Kuriakose et al. 2016; Press et al. 1992).

So, choose the Friedlander profile (Gradstein and Ryzhik 2007; Kuriakose et al. 2016; Press et al. 1992) and write it in the form

$$F(r) = u_{ro} e^{-br/r_{att}} \left(1 - r/r_{att}\right), \tag{25.61}$$

where u_{ro} is the amplitude factor, and the constant b defines the profile bottom.

Figures 25.27 and 25.28 show the shape of the Friedlander initial profile, which is given in recent experimental works (Kuriakose et al. 2016; Press et al. 1992).

Note that profile (25.12) is considered one of the simplest and is often used in the interpretation of experiments. Figure 25.28 shows the experimentally investigated evolution of the initial impulse

When substituting (25.61) into (25.17), the following approximate formula is obtained for the numerical modeling of the distortion of the Friedlander profile

$$\begin{aligned} u_1(r,t) = {} & u_{ro} e^{-ba(r - v_L t)/r_{att}} \left[1 - a\left(r - v_L t\right)/r_{att}\right] - \\ & - {}^1\!/\!_2\, \alpha a^4 v_L t \left(1/r_{att}\right)^2 \left(u_{ro}\right)^2 \\ & \times \{1 + b\left[1 - a(r - v_L t)/r_{att}\right]\}^2 e^{-2ba(r - v_L t)/r_{att}}. \end{aligned} \tag{25.62}$$

Equation (25.62) describes the change in the initial wave profile due to the direct dependence of the nonlinear term on time. Figures 25.29, 25.30, 25.31, and 25.32 show the evolution for the first two and the first three approximations. The left curve corresponds to the first two approximations, and the right—to the first three.

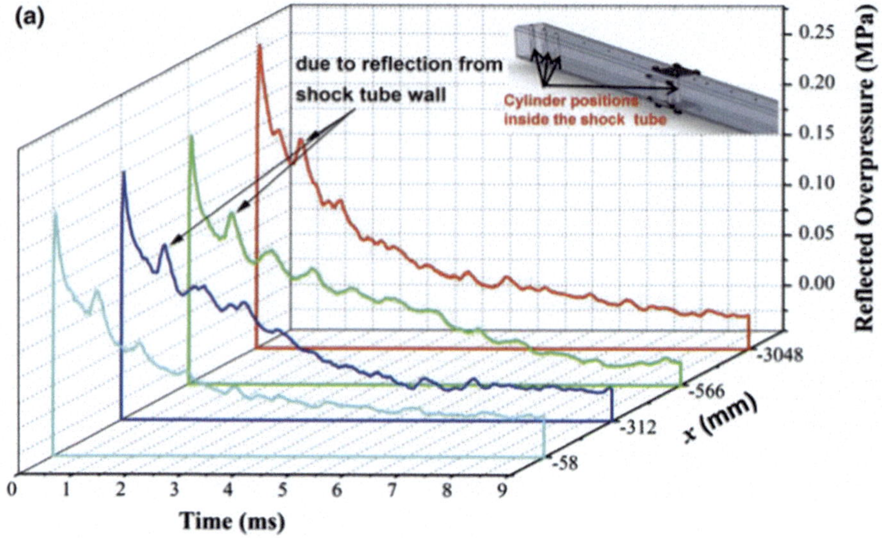

Fig. 25.29 The evolution of the Friedlander function (experiment)

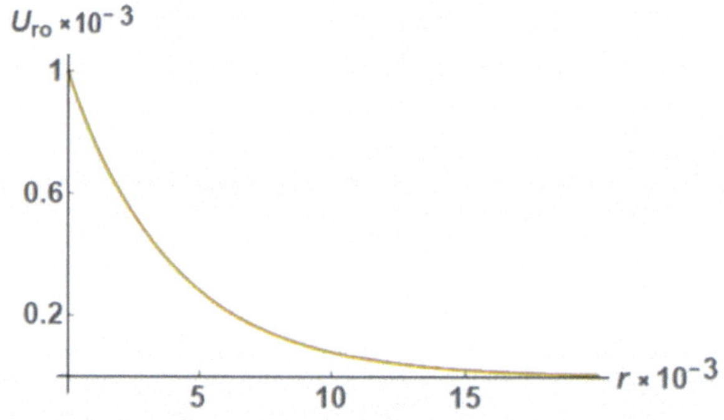

Fig. 25.30 Evolution of Friedlander wave (stage 1)

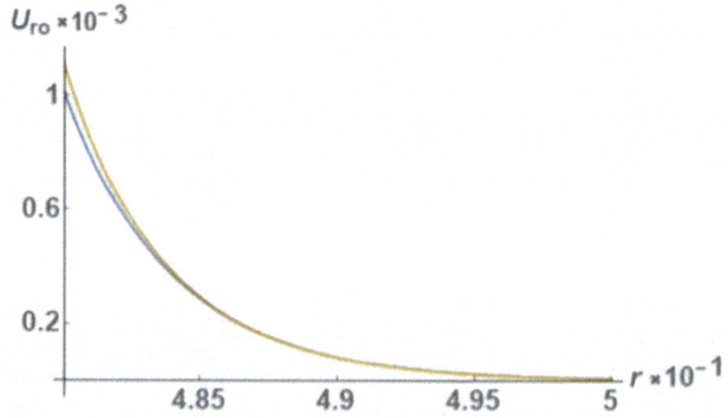

Fig. 25.31 Evolution of Friedlander wave (stage 2)

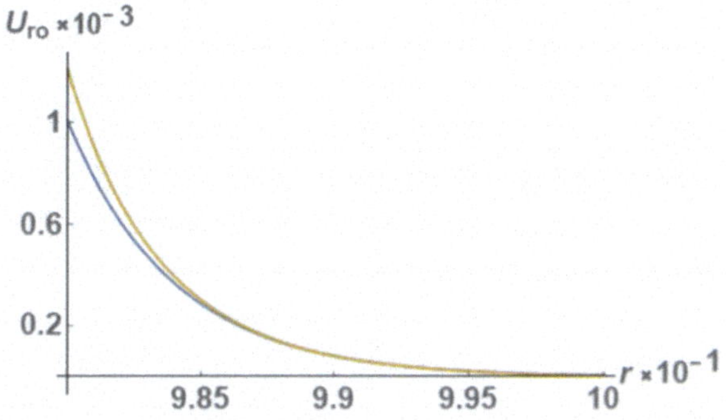

Fig. 25.32 Evolution of Friedlander wave (stage 3)

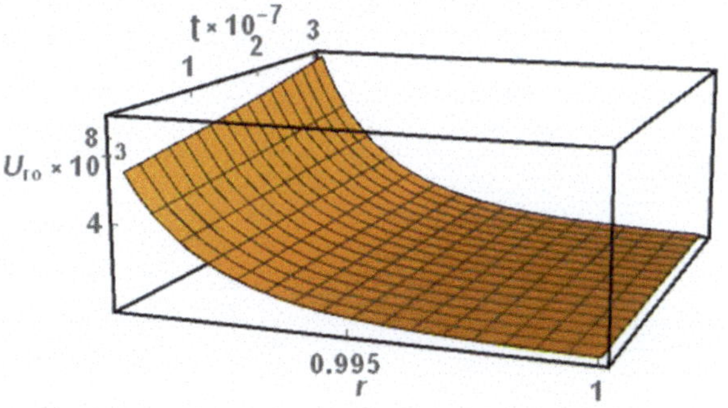

Fig. 25.33 Evolution of Friedlander wave (3D picture)

Figures 25.30, 25.31, 25.32, and 25.33 are very similar to Figs. 25.24, 25.25, 25.26, and 25.27. Therefore, it seems worth concentrating on comparing the evolution of the MacDonald and Friedlander waves.

Note that a feature common to the initial profiles of both waves (MacDonald and Friedlander) is that their graphic representation is almost identical. However, the mathematical expressions of these profiles are different—the usual function gives the Friedlander profile, whereas the special function expresses the MacDonald profile. Therefore, the question arises of how the difference in the mathematical representation of the profiles will affect the description of the evolution of the wave.

Further, the basic formulas (25.60) (MacDonald wave) and (25.62) (Friedlander wave) are used in the computational modeling of evolution. The obtained plots show the results of comparing the evolution of the two waves.

Figures 25.34, 25.35, and 25.36 represent the plots obtained within the framework of the first three approximations (two curves correspond to the studied two waves). All plots indicate that from the beginning of the wave motion, an evolutionary change occurs in the wave profile caused by the nonlinearity of the propagation medium: the convexity of the wave becomes larger. The plots show the above nonlinear effect of increasing the convexity. They also show that the distortion of both profiles at the initial stage of wave propagation is significant, and the difference between the compared plots is small. At the developed stages of evolution, the increase in the

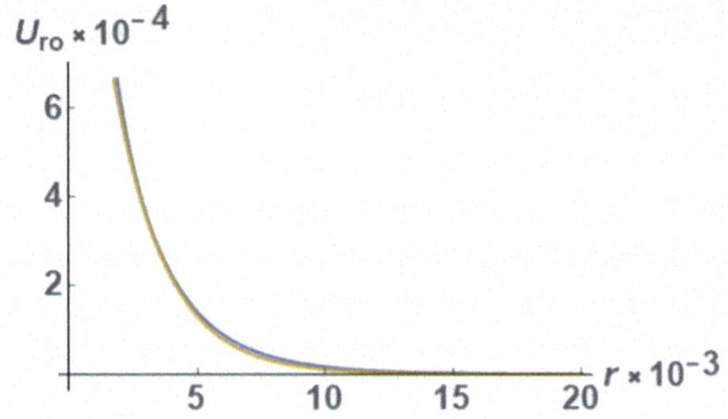

Fig. 25.34 Comparison of the evolution of the Friedlander wave

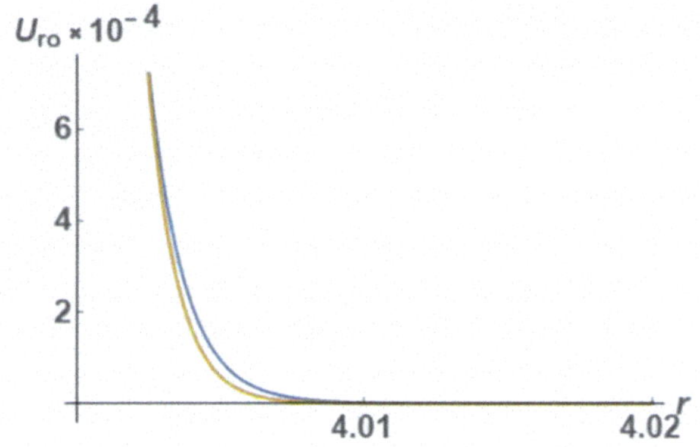

Fig. 25.35 Comparison of the evolution of Friedlander and MacDonald waves (stage 1) and MacDonald waves (stage 2)

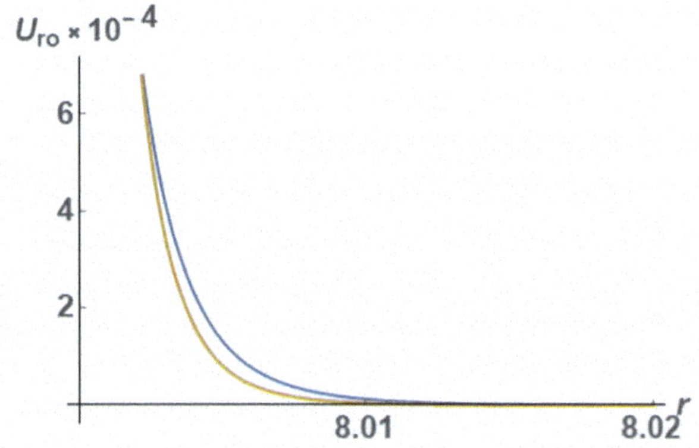

Fig. 25.36 Comparison of the evolution of Friedlander and MacDonald waves (stage 3)

convexity of both plots continues, but the difference between them is slight and remains unchanged.

So, a theoretical study and numerical modeling showed that the evolution of profiles is very similar. Both profiles are interchangeable in the framework of evolutionary analysis, although mathematically, they are represented differently.

A comparison of MacDonald and Friedlander profiles as solutions of the linear wave equation showed that if the MacDonald function is considered as an exact solution to the equation, then the Friedlander function can be interpreted as an approximate solution of the same equation. Furthermore, the presented results show that it is possible to represent the solution in the form of a single wave for the Friedländer function as an approximation of the Macdonald function.

25.9 Results of Comparisons

Comparison 1 (comparison of methods 1 and 2 as applied to the evolution of harmonic wave within the framework of the first two approximations). When comparing the plots obtained by the two methods, it can be seen that the evolution of the initial profile takes place according to the same scheme but at different speeds. In the case of method 2, the speed of distortion of the profile is significantly higher than in the case of method 1. Therefore, the distortion depends on the choice of the method.

Let us pay attention to the differences in the restrictions that are used for the two methods. Method 1 includes only one restriction in its mathematical formulation—the series composed of successive approximations $\sum_n u^n$ must be convergent. Note that, at that, the general term of the series must converge to zero $\lim\limits_{n\to\infty} u_n = 0$. In engineering calculations, with a few constraints, each approximation must be smaller than the previous one. Method 2 is based on the restriction on the displacement gradient $\left|\alpha u_{1,1}\right| \ll 1$ and $\left|\delta^* = -{}^1\!/_2\, t\alpha v_L u_{1,1}\left[1 - {}^1\!/_4\, \alpha u_{1,1}\right]\right| \ll 1$. Hence, the constraints used in both methods are different. Therefore, the evolution of the harmonic wave obtained above differs in specific parameters. However, at the same time, they are qualitatively close since they take into account only the second approximation.

Note also that the evolution occurs asymmetrically. The positive part of the profile would remain in the form of a profile of the first harmonic, while the negative part shows a tendency to transform into a profile of the second harmonic.

Comparison 2 (comparison of methods 1 and 2 as applied to the evolution of solitary wave within the framework of the first two approximations).

It is shown in Sect. 25.5.1 that the solution for the second approximation has a complicated mathematical form, and analytical and numerical analysis of wave evolution within the framework of the first two approximations looks very unpromising for the profile in the form of the Gauss function. Clearly, the situation is the same for any other solitary profile. In this case, some advantage has method 2.

Comparison 3 (comparison of methods 1 and 2 as applied to the new harmonics within the framework of the first three approximations). The application of method 1 introduces the fourth-order correction as the third approximation, and method 2 introduces the third-order correction. Thus, method 1 doubles the orders of harmonics (1+2+4+6+…), and method 2 takes into account all harmonics (1+2+3+4+…).

Comparison 4 (comparison of harmonic and Gauss wave evolutions within the framework of the first two approximations).

The comparison of the evolution of the harmonic symmetric wave profile with the symmetric Gauss wave shows that the Gauss profile distorts symmetrically in a slightly different way. The second harmonic always gives a negative addition. So, the slopes of the distorted "bell or hump" become steeper. The upper part of the "bell" falls and forms two "bells." Since the wave is elastic and there can be no energy loss, the profile change is consistent with the law of conservation of wave energy during wave propagation.

Comparison 5 (comparison of harmonic and Gauss wave evolutions in the framework of the first three approximations relative to the presented harmonics). The representation of the Gauss wave evolution within the framework of the first three approximations contains the term with the cubic nonlinearity in the third approximation, which can be commented on as the appearance of the third harmonic after the second one. In the case of the harmonic profile, the corresponding solution includes the fourth harmonic after the second one. This difference is significant.

Comparison 6 (comparison of harmonic and Gauss wave evolutions in the framework of the first three approximations relative to the symmetry of profiles). Thus, considering the third approximation introduces new features in the evolution. First, the representation (25.31) (first+second+third) leads to asymmetric changes in the form of an initial profile, unlike the first+second case, which leads to symmetric changes.

Comparison 7 (similarity of harmonic and Gauss wave evolutions in the framework of the first three approximations). Two effects in the evolution of harmonic and Gauss waves are important. The first effect is that both waves with the same bottom turn one hump into two and then into three. Another effect is that the bottom is stable during evolution (it does not change).

Comparison 8 (comparison of harmonic, Gauss, and Whittaker waves relative to the symmetry of profile distortion). Here, symmetry is considered relative to the vertical line passing through the hump top. In this case, the symmetric profiles are distorted symmetrically, and the nonsymmetric—nonsymmetrically.

Comparison 9 (comparison of changes of bottoms of harmonic, Gauss, and Whittaker waves). The obtained plots indicate that the wave bottom is unchanged in all three wave types. So, these waves are similar relative to the bottom size changes.

Comparison 10 (comparison of hump heights of harmonic, Gauss, and Whittaker waves). Harmonic and Gaussian waves are very different from Whittaker waves in this respect. So the first two do not change the maximal value of amplitude, whereas the third increases this value.

Comparison 11 (ability to transform the profile from one-humped to two-humped). The considered three one-humped waves show the tendency to form the two-humped profiles. This nonlinear wave effect is illustrated using the first three approximations. At that, the symmetric profiles form the symmetric two humps, whereas the nonsymmetric profile forms two nonsymmetric humps.

Comparison 12 (comparison by the presence of humps). Unlike the three previous profiles, the MacDonald profile has no hump and does not generate a hump.

Comparison 13 (comparison by the ability to maintain non-symmetry). Like the Whittaker profile, the MacDonald profile maintains non-symmetry, and the distorted profile becomes steeper and steeper.

Comparison 14 (comparison by the profile bottom). Unlike the other discussed profiles, the bottom of the MacDonald profile is getting shorter and shorter.

Comparison 15 (comparison of MacDonald and Friedlander waves by similarity attribute). A theoretical study and numerical modeling of MacDonald and Friedlander waves show that the evolution of profiles is very similar. It follows from this that both profiles are interchangeable in the framework of evolutionary analysis, although, mathematically, they are represented differently.

Comparison 16 (comparison of MacDonald and Friedlander profiles as solutions of the nonlinear wave equation). It shows that if the MacDonald function is considered an exact solution to the equation, then the Friedlander function can be interpreted as an approximate solution of the same equation. It follows from the plots that the representation of a solution as a solitary wave is possible for the Friedlander function as an approximation of the MacDonald function.

25.10 Final Conclusions

Five kinds of nonlinear elastic waves of initial profiles (harmonic-cosinusoidal, Gauss, Whittaker, MacDonald, and Friedlander), different in mathematical form, are considered. A nonlinearity of elastic deformation is described by the Murnaghan five-constant model. The study's main goal was to compare the similarities and differences between the studied waves. The evolution of these waves is studied theoretically and numerically. The approximate methods of solving are utilized, and for comparison of the features of evolution, the first three approximations are used.

The main similarity for all the profiles is that they distort the form (evolute) during the wave's motion due to the wave's nonlinear interaction with itself. However, distortion occurs for each profile in its way.

The comparison results are listed in 16 positions, where the observed similarities and differences are specified.

References

Altenbach H, Eremeyev VA, Pavlov IS, Porubov AV (2020) Nonlinear wave dynamics of materials and structures. Springer, Cham

Andrianov IV, Awrejcewicz J, Danishevskyy V (2021) Linear and nonlinear waves in microstructured solids: homogenization and asymptotic approaches. CRC Press, Boca Raton

Bedford A, Drumheller D (1994) Introduction to elastic wave propagation. Wiley, New York

Bloembergen N (1965) Nonlinear optics. (A lecture note, W.A. Benjamin Inc.), New York–Amsterdam

Chandra N, Ganpule S, Kleinschmit NN, Feng R, Holmberg AD, Sundaramurthy A, Selvan V, Alai A (2012) Evolution of blast wave profiles in simulated air blasts: experiment and computational modeling. Shock Waves 22:403–415

Engelbrecht J (2015) Questions about elastic waves. Springer, Berlin

Freidlander FG (1946) The diffraction of sound pulses. (I. Diffraction by a semi-infinite plate). Proc Roy Soc Lond A 186:322–344

Fujimoto M (2021) Introduction to the mathematical physics of nonlinear waves, 2nd edn. IOP Publishing Ltd, Bristol

Gradstein I, Ryzhik I (2007) Table of integrals, series, and products. (7th revised edition, edited by Jeffrey, A. and Zwillinger, D.). Academic Press Inc., New York

Kamke E (1977) Differentialgleichungen. Lösungmethoden und Lösungen. (Differential equations. Methods of solving and solutions). Vieweg + Teubner Verlag, Wiesbaden. Springer Fachmedien Wiesbaden GmbH

Kuriakose M, Skotak M, Misistia A, Kahali S, Sundaramurthy A, Chandra N (2016) Tailoring the blast exposure conditions in the shock tube for generating pure, (primary shock waves: the end plate facilitates elimination of secondary loading of the specimen). PLoS ONE 11(9)

Lempriere B (2002) Ultrasound and elastic waves: frequently asked questions. Academic Press, New York

Murnaghan F (1985) Finite deformation in an elastic solid, 3rd edn. Gloucester. Peter Smith Publisher Inc., MA, USA

Olde Daalhuis AB (2010) Confluent hypergeometric functions. Chapter 13, 383–402. Whittaker functions. 13.14-13.26. In: Olver FWJ, Lozier DW, Bousvert RF, Clark CW (eds) NIST (National Institute of Standards and Technology) Handbook of mathematical functions. Cambridge University Press, Cambridge

Olver FWJ, Maximon LC (2010) Bessel functions. Chapter 10, 215–286. In: Olver FWJ, Lozier DW, Bousvert RF, Clark CW (eds) NIST (National Institute of Standards and Technology) Handbook of mathematical functions. Cambridge University Press, Cambridge

Press WH, Flannery BP, Teukolsky SA, Vetterling WT (1992) Bessel functions of fractional order, airy functions, spherical bessel functions. (§6.7 in Numerical recipes in FORTRAN: the art of scientific computing, second ed). Cambridge University Press, Cambridge, pp 234–245

Rushchitsky JJ (2011) Theory of waves in materials. Ventus Publishing ApS, Copenhagen

Rushchitsky JJ (2014) Nonlinear elastic waves in materials. Springer, Heidelberg

Weisstein EW (2013) Fourier transform–Gaussian. MathWorld

Yariv A (1967) Quantum electronics. Wiley, New York

Chapter 26
Stress–Strain State of Sheet Elements at One- and Two-Layer Build-Up with Liquid Metal

Ihor Senchenkov, Ihor Ryabtsev, Olha Chervinko, and Nina Yakovenko

26.1 Introduction

Surfacing of metal members is a technological process that provides the working surfaces of structural elements with special properties: wear resistance, heat resistance, corrosion resistance, etc. (Senchenkov and Oksenchuk 2013; Senchenkov et al. 2005).

The modeling of processes during the surfacing of thin-walled elements, such as pipes or sheets, when the thickness of the surfaced metal is commensurate with the thickness of the parts, is one of the build-up classes of tasks. The specificity of the process modeling is to estimate the distortion as well as to determine the stress–strain state.

It should be noted that when building up, especially multilayer, thin-walled elements, there is a significant variability of thermomechanical fields in thickness. Therefore, models based on the hypotheses of plates and shells have questionable reliability. The results obtained within the spatial formulation of the problem are more reliable.

The objects of mathematical modeling are the current and residual thermomechanical state of parts during the build-up process. The results of investigations on this problem are summarized in Motovilovets and Kozlov (1987), Popov and Popova (1961), Ryabtsev and Senchenkov (2013), Senchenkov and Oksenchuk (2013), Senchenkov et al. (2002, 2005). The literature review shows that these surfacing problems need to be studied more.

I. Senchenkov · O. Chervinko (✉) · N. Yakovenko
S.P. Timoshenko Institute of Mechanics, National Academy of Science of Ukraine, Kyiv, Ukraine
e-mail: chop497@gmail.com

I. Senchenkov
e-mail: term@inmech.kiev.ua

I. Ryabtsev
E.O. Paton Institute of Electric Welding, National Academy of Science of Ukraine, Kyiv, Ukraine

A. N. Guz et al. (eds.), *Advances in Mechanics*, Advanced Structured Materials 191,
https://doi.org/10.1007/978-3-031-37313-8_26

The chapter presents the results of modeling thermomechanical processes during the growth sheet elements of rectangular cross sections. St0, St1, St2, and St3 denote the used steel grades. The following chemical compositions of steel are used in the current study:

- St0—0.3% C;
- St1—0.08% C;
- St2—0.25% C, Cr 5.0%, V 1.0%, Mo 1.0%, Si 1.0%;
- St3—0.10% C, Cr 18.0%, Ni 9%, Ti 1.0%.

A basic rectangular element made of St0 steel is built up with one or two layers of liquid metal made of St1, St2, and St3 steels using a growing body model (Arutyunyan et al. 1987; Lindgren 2007; Senchenkov and Tabieva 1996; Senchenkov et al. 1998; Shorshorov and Belov 1972).

The Bodner–Partom inelastic flow equations describe the thermomechanical behavior of materials (Bodner 2000; Shevchenko et al. 2012). Microstructural transformations are governed by means of thermokinetic diagrams of austenite decomposition during cooling (Ryabtsev et al. 2015; Zubchenko 2003). The problem is solved using the finite element technique (Radaj 2003).

26.2 Object of Study

A sheet (plate) element made of steel St0 with a rectangular (in the plane Oxy) cross section is considered. The shape and size designation of the plate and deposited beads are shown in Fig. 26.1a. In Fig. 26.1b, c, the scheme of smooth support and support with a gap (free support) between the plate and the base and fixing along the side edges at a length l_c is shown.

The scheme of simultaneous (instantaneous) imposition of the bead in the direction of the z-axis is applied to reduce the three-dimensional problem to a two-dimensional one. Thus, the problem is simplified to the plane strain (PDS) or plane stress (PSS) problem in the plane Oxy depending on the conditions of fixing the ends $z = 0$ and $z = l_z$.

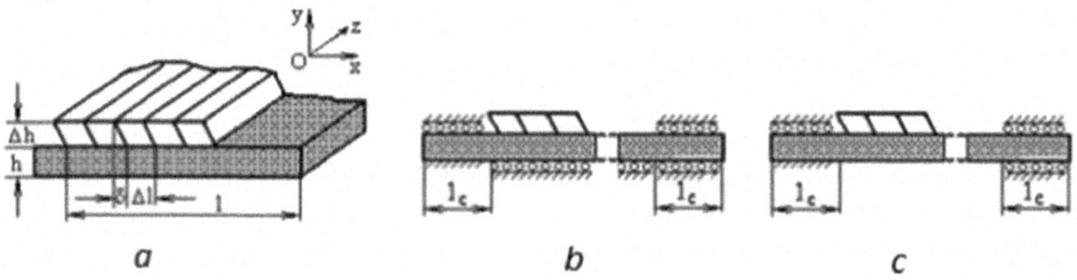

Fig. 26.1 Object of study

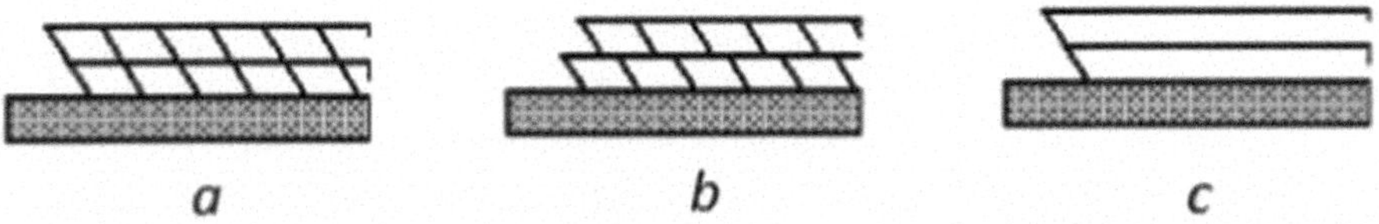

Fig. 26.2 The schemes of layer build-up

In the case of two-layer surfacing, we consider the schemes of beads build-up with (Fig. 26.2b) and without (Fig. 26.2a) shift, as well as a simplified scheme of the sequential instantaneous build-up of the first and then the second layers (Fig. 26.2c).

26.3 Model of Microstructural Transformations

In the thermomechanical processes of build-up, allotropic transformations in a solid body due to the decomposition of austenite ($\xi = A$) in steels into the phases of ferrite ($\xi = F$), pearlite ($\xi = P$), bainite ($\xi = B$), and martensite ($\xi = M$) during cooling are taken into account. The thermokinetic decomposition diagram (CCT) describes the transformation of austenite. Figure 26.3 presents such diagrams for steel St0 and steel St2 (Ryabtsev et al. 2015; Zubchenko 2003). Bold lines show the boundaries of the transformation region, and thin lines correspond to the cooling curves. Figures show the volume percentage of austenite decomposition at the outlet of the transformation region.

The law of new phase ξ accumulation in the appropriate regions along the cooling trajectory for martensitic transformations is given by the phenomenological

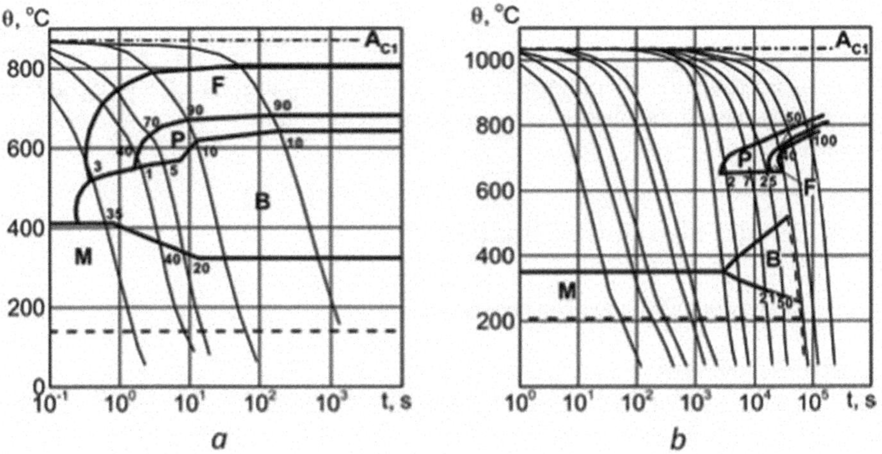

Fig. 26.3 CCT diagrams for steel St0 **a** and steel St2 **b**

Coistinen–Marburger equation (Makhnenko 2006). Diffusive transformations such as the decomposition of austenite into ferrite, pearlite, or bainite, which depend on temperature and time, are described using the method (Klingbeil et al. 2002) based on the JMAK (Johnson–Mehl–Avrami–Kolmogorov) equation.

Thermomechanical properties of each phase Y_ξ are calculated taking into account the dependence on temperature $Y_\xi = Y_\xi(\theta)$. The linear rule of mixtures is used to determine the macro-characteristics $\bar{Y}$ for an arbitrary phase composition. The general formula is written in the form

$$\bar{Y}(\theta, t) = C_\xi(\theta, t)\, Y_\xi(\theta), \quad 0 \leqslant C_\xi \leqslant 1, \quad \sum_\xi C_\xi = 1, \tag{26.1}$$

where C_ξ is the volume fraction of the phase.

The rule of mixtures can calculate the following physical quantities: heat capacity (c_υ), thermal conductivity (k), Young's modulus (E), coefficient of linear thermal expansion (α), Poisson's ratio (ν), and parameters of the flow model. In (26.1) and below, repeated indices imply the summation is to be done. The rule of mixtures is used in a wide temperature range, which is consistent with the experiments in Lindgren (2007), Popov and Popova (1961), Ryabtsev and Senchenkov (2013).

Thermophase deformation ${\varepsilon_{ij}}^{\theta ph}$ is determined through the specific volumes of phases V_ξ by the formula

$$\varepsilon_{ij}^{\theta ph}\left(\theta, \theta_r, C_\xi\right) = \frac{V_\xi(\theta)\, C_\xi(\theta) - V_\xi(\theta_r)\, C_\xi(\theta_r)}{3V_\xi(\theta_r)\, C_\xi(\theta_r)} \delta_{ij}. \tag{26.2}$$

Here θ is the current temperature and θ_r is some reference temperature.

Temperature dependences of specific volumes for the phase ξ in m^3/kg, referred to $\theta_r = 20°$ C, are taken as follows (Yuryev 1950):

$$V_\xi(\theta) = V_{\xi 0}[1 + 3\alpha_\xi(\theta - \theta_r)],$$

where α_ξ is the coefficient of linear thermal expansion of the phase ξ and $V_{\xi 0}$ are the specific volumes of the phase ξ at the reference temperature.

26.4 Model of Growing Bodies

The peculiarity of the model of growing bodies is that all components of the stress tensor are designated on the surface of the growth (Arutyunyan et al. 1987; Senchenkov et al. 2002; Shorshorov and Belov 1972). In the case under consideration, taking into account the schemes of Figs. 26.1 and 26.2, the following conditions must be satisfied in the built-up element:

$$\sigma_{ij} = 0, \quad t = t^*, \tag{26.3}$$

where t^* is the moment of building up the element (bead).

Let the empty mesh element $\Delta V(t^*)$ have deformation ε_{ij}^* at the moment t^* of filling, and let it be filled with material of temperature θ^*. Thus, conditions (26.1), in fact, mean that at $t = t^*$

$$\sigma_{ij}(\varepsilon_{ij}^*, \theta^*) = 0, \quad \text{in } \Delta V(t^*). \tag{26.4}$$

It is assumed that the inelastic deformation of the layer element that is built up at $t = t^*$ is equal to zero

$$\varepsilon_{ij}^p(t^*) = 0, \quad \text{in } \Delta V(t^*). \tag{26.5}$$

The process of filling the elements with reference to the technological parameters of surfacing is clarified in more detail in Senchenkov and Oksenchuk (2013), Senchenkov et al. (2005).

26.5 Flow Model

The thermomechanical behavior of the material is described by the equations of the Bodner–Partom model (Bodner 2000)

- equations of the modified Hooke's law

$$\varepsilon_{ij} = \varepsilon_{ij}^e + \varepsilon_{ij}^p + \varepsilon_{ij}^{\theta ph} + \varepsilon_{ij}^*, \quad \varepsilon_{kk}^p = 0, \tag{26.6}$$

$$s_{ij} = 2G(e_{ij} - \varepsilon_{ij}^p - e_{ij}^{\theta ph} - e_{ij}^*), \quad \sigma_{kk} = 3K_V(\varepsilon_{kk} - \varepsilon_{kk}^{\theta ph} - \varepsilon_{kk}^*), \tag{26.7}$$

  for the basic metal St0, we assume $\varepsilon_{ij}^*(0) = 0$;
- the flow equations

$$\dot{\varepsilon}_{ij}^p = D_0 \exp\left\{-\frac{1}{2}\left[\frac{(\bar{K}_0 + K)^2}{3s_i}\right]^n\right\}\frac{s_{ij}}{s_i} = 0, \quad \varepsilon_{ij}^p(0) = 0; \tag{26.8}$$

- the evolution equation for the isotropic strengthening parameter

$$\dot{K} = m_1\left(\bar{K}_1 - K\right)\dot{w}_p, \quad K(0) = 0, \tag{26.9}$$

  where G, G_f, K_V, and K_{Vf} are the shear and bulk compression moduli; $\bar{K}_0$ and $\bar{K}_1$ are determined by (26.1); m_1, n, and D_0 are model parameters; $\dot{w}_p$ is the plastic power, s_i is the second invariant of the stress tensor, $\dot{w}_p = \sigma_{ij}\dot{\varepsilon}_{ij}{}^p$ and $s_i{}^2 = {}^1/_2 s_{ij}s_{ij}$.

The averaged values $\bar{K}_0$ and $\bar{K}_1$ for the current phase state are calculated according to Zubchenko (2003) using the technique described in Shevchenko et al. (2012).

For materials with dissimilar diagrams (Bezukhov et al. 1965), modifications of the standard model (26.6)–(26.9), which are proposed in Bodner (2000), Senchenkov et al. (2013), Yuryev (1950), are used.

26.6 Problem Statement

In addition to the presented Eqs. (26.6)–(26.9), the problem statement includes equations of equilibrium and heat conduction

$$\sigma_{ij,j} = 0, \quad \bar{c}_v \dot{\theta} = (\bar{k}\,\theta_{,i}\,)_{,i} + Q \tag{26.10}$$

with the following boundary and initial conditions

$$-\,\bar{k}\,n_i \theta_{,i} = -q + \gamma(\theta - \theta_0) + \sigma\,\varepsilon\,(\theta^4 - \theta_0^4), \quad \theta(0) = \theta_0, \tag{26.11}$$

where Q is the heat source, $\bar{k}$ and $\bar{c}_v$ are averaged coefficients of thermal conductivity and volumetric heat capacity of the mixture of phases, q is the heat flux, γ is the heat transfer coefficient, σ is the Stefan–Boltzmann constant, and ε is the degree of blackness.

The conditions of growth and fixation of the element specify the mechanical boundary conditions.

26.7 Thermomechanical Characteristics of the Material

To calculate the parameters of the flow model, uniaxial tensile diagrams at constant strain rate ($\dot{\varepsilon} = 5 \cdot 10^{-4} \mathrm{s}^{-1}$) and different temperatures are used. Such diagrams for steels St0–St3, taken from (Bezukhov et al. 1965), are shown in Fig. 26.4. Dashed lines correspond to experimental data and solid lines—to calculated data. Numbers on the right side of the figures indicate the temperature.

Temperature dependences of the Bodner–Partom model parameters for these materials are given in Senchenkov (2005). Temperature dependences of thermophysical characteristics of materials are taken from Bezukhov et al. (1965).

26.8 Method of Numerical Simulation

The problem is solved by the method developed in Radaj (2003), Senchenkov and Oksenchuk (2013), Senchenkov et al. (2005). Equations with time derivatives are integrated using an implicit scheme with a variable integration step. The nonlinear boundary value problem of thermomechanics is solved at each step by a simple

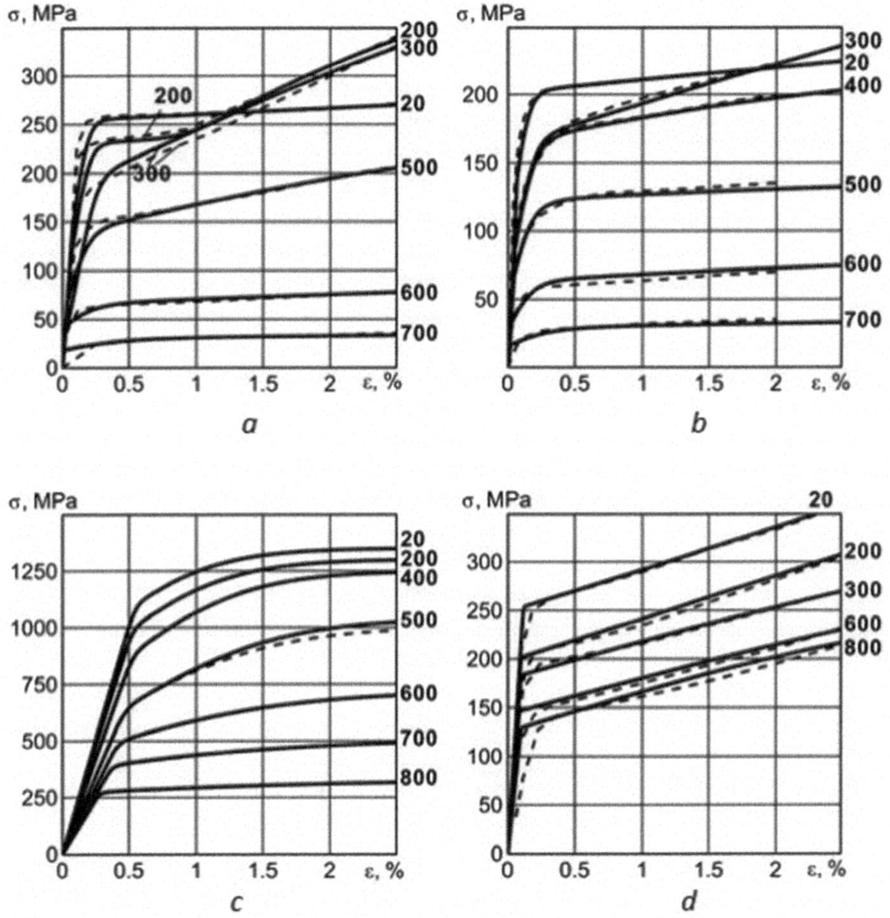

Fig. 26.4 Uniaxial tensile diagrams at constant strain rate for some temperatures: **a**—steel of base layer St0; **b**, **c**, and **d** corresponds to steel of built-up layers St1, St2, and St3, respectively

iteration method with accelerated convergence according to the Stephensen–Eitken procedure. The linearized problem at each iteration is solved by the finite element method using quadrilateral isoparametric elements.

26.9 Results of Calculations of Deflections of Sheet Elements at Surfacing

The superposition of beads was modeled according to the scheme shown in Figs. 26.1 and 26.2. We chose the following values of geometrical parameters: $l = 100$ mm, $h = 3$ mm, $\Delta h = 2.3$ mm, $l = 5$ mm, $l_c = 25$ mm, $l_z = 200$ mm (the length of the element in the direction of the z-axis), $v_z = 8.7$ mm/s (the deposition speed in the direction of the z-axis). The heat transfer coefficient from the surface of the plate γ is chosen to be 30 W/m^2°C and from the surface of the base—100 W/m^2°C.

The bead and the adjacent strips of material are preheated during the time $t_q = 9$ s by volumetric sources. Their power corresponds to the enthalpy of the melt that is put into it, as well as heating by an electric arc, and provides heating up to $\theta \approx 1550$°C.

When this temperature level is reached, the actual temperature θ^* and deformation ε^*_{ij} of the empty elements of the bead are fixed. Then they are filled with the build-up material, the mechanical properties of which are described by Eqs. (26.6)–(26.9).

After cooling for the time $t_c = 23$ s and $t_c = l_z/v_z$, the surfacing cycles are repeated until the end of the process of building up the sheet element. A total of 20 beads are applied. After cooling, the element is released from the fixtures shown in Fig. 26.1b, c, and its residual deflection is determined.

The effect of the martensitic transformation on displacement kinetics is illustrated at the point in the base material under the eleventh bead ($x = 55$ mm, $y = 2.5$ mm), which is built up in the time interval according to the scheme in Fig. 26.1c under conditions of free lower boundary. The deflection curves marked with 1 and 2 correspond to the build-up of steels St1 and St2, respectively.

Steel St2 undergoes phase transformations during cooling. As a result, there is a significant increase in deflection in the region of martensitic transformation. The points on the time–temperature curve indicate the moments of entry into and exit from the austenite–martensite transformation region. Vertical dashed lines delimit this region on the θ and t-axes. The instantaneous change in deflection after cooling corresponds to the release of the plate edges from fixation.

Figure 26.5b shows a comparison of residual deflections at the bead surfacing of the layer (Fig. 26.1c) when surfacing with steels St1 and St2, respectively, and the free lower boundary. When building up with St2 steel, as a result of martensitic transformations, we have a higher level of deflection, especially under the conditions

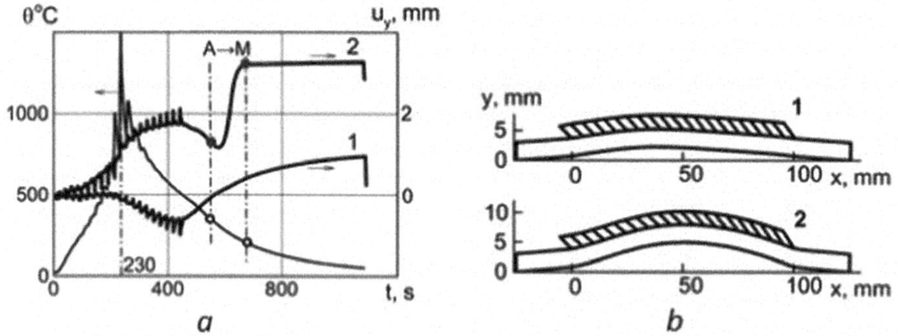

Fig. 26.5 Comparison of deflections under the eleventh bead for steel without (1—St1) and with (2—St2) martensite transformations and the free lower boundary; **a**—dependence of deflections on time, **b**—residual forms of build-up plates

Table 26.1 Comparison of calculated and experimental values of residual deflections

Conditions support	PDS/PSS	Material for building-up								
		St3			St2			St1		
		Deflections: calculated (1, 2) and experimental (3) data, mm								
		1	2	3	1	2	3	1	2	3
Smooth	PDS	1.45	1.61	0.7	1.52	1.69	1.7	1.21	1.22	0.7
	PSS	1.34	1.38	0.7	1.93	1.72	1.7	1.13	1.12	0.7
With gap	PDS	2.77	2.18	2.0	2.81	4.96	2.5	2.84	2.27	1.2
	PSS	2.66	1.68	2.0	2.55	4.22	2.5	2.35	1.69	1.2

of a free lower boundary. On the other hand, residual deflections for similar problems with smooth support at the lower boundary are significantly smaller.

In the case of simultaneous surfacing, we have the following results. Although the displacement kinetics is different in the case of phase transformation, the residual deflection values do not differ significantly. The bead surfacing model predicts higher deflection values for both the smooth support and free boundary cases than the simultaneous surfacing model.

For the three grades of cladding steel (St1, St2, and St3), experimental and calculated data for single-layer surfacing under the conditions of smooth support and support with a gap are given in Table 26.1.

The characteristic deflections were determined as follows. After cooling and releasing the fixed edges, the sample was placed on a smooth plate, and the normal displacement of the upper surface of the sample from the plate was measured with an indicator. For the characteristic deflection given in the table, the maximum local value above the defined deflection in the section $z = l_z/2$ of the welded part of the plate was taken. With this definition, the characteristic deflection is always positive. Columns 1 and 2 correspond to the calculated data for simultaneous and bead surfacing. The PDS and PSS terms correspond to the calculation models. Column 3 shows the experimental data.

As can be seen from the above data, the PDS model predicts larger deflections compared to the PSS model, except for the material with martensitic transformation (St2) under smooth support conditions. For this material, larger deflections occur under both bearing conditions due to the volume effects of transformation. It should also be noted that more significant deflections are observed for surfacing with a gap between the plate and the welding table than in the case of tight pressing of the plate to the table.

Similar calculations and experiments were carried out for the case of two-layer surfacing. For two-layer surfacing with steel St1, under conditions of smooth support, the schemes without overlapping beads (Fig. 26.2a) give deflections of 1.02 mm (PDS) and 0.91 mm (PSS). The scheme of the sequential simultaneous surfacing of layers gives a deflection of 1.00 mm (PDS) and 0.91 mm (PSS). The experimental

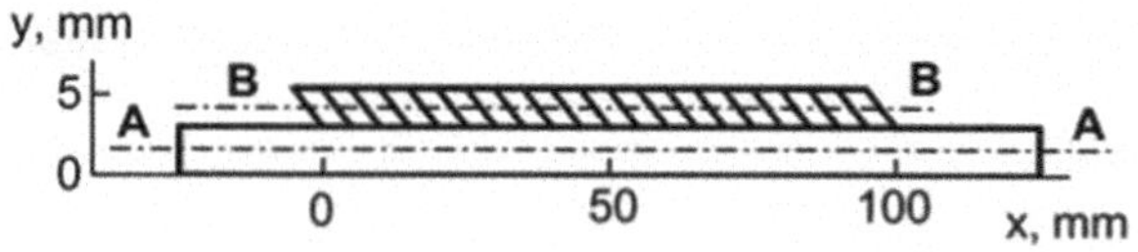

Fig. 26.6 Scheme of a single-layer surfacing

deflection value is 0.8 mm. The results of the calculation of deflections for the case of surfacing beads with (Fig. 26.2b) and without (Fig. 26.2a) overlapping have a difference of less than 10%.

The discrepancy between the above-calculated and experimental results is determined, on the one hand, by the inaccuracy of the mathematical model, which does not take into account the contact interaction of the plate with the support surface under conditions of smooth support, and, on the other hand, by the technical difficulties in providing all the conditions for fixing the edges of the plates and measuring deflections during experimental studies, as well as by the insufficient number of samples used.

26.10 Results of Calculations of Stresses and Plastic Deformations

Consider a single-layer surfacing according to the scheme shown in Fig. 26.6. Cross sections A–A and B–B are shown here with the stress and plastic deformation distributions.

Let us study the influence of support conditions for the case of the plane strain and St1 steel. For the support with a gap (Fig. 26.1b), Figs. 26.7a (section A–A) and 26.7b (section B–B) show the distributions of residual stresses σ_{xx}, σ_{zz} and stress intensities s_i at the zone of surfacing. Figure 26.7c shows the distributions of plastic deformation intensity e_i^p and $e_i^p = (2e_{ij}^p e_{ij}^p/3)^{1/2}$ in both sections.

Similar results for smooth support conditions are shown in Fig. 26.8. Again, it can be seen that the stress–strain state levels in both cases of support differ insignificantly, except for the growth of plastic deformations in the last beads when building up with a gap (Fig. 26.8c). The largest are tensile stresses σ_{zz}.

Stress distributions at section D–D (Fig. 26.9a) for the conditions of smooth support are shown in Fig. 26.9b, c. The numbers in Fig. 26.9a indicate the numbers of the built-up beads. The interval $0 \leqslant y \leqslant 3$ mm corresponds to the base layer, and $3\ \text{mm} < y \leqslant 5.3$ mm to the built-up layer.

The most stressed is the base metal, and the most plastically deformed is the built-up layer near the boundary with the base metal. However, distributions for build-up with a gap differ insignificantly.

Let us consider the influence of the build-up material on the example of the PSS model and smooth support conditions. Residual distributions of parameters s_i and

e_i^p at sections A–A and B–B for surfacing with steel grades St1 and St2 are shown in Figs. 26.10a, b and 26.11a, b, respectively.

The stress level corresponding to surfacing with steel St1 (Fig. 26.10a) is significantly lower than that for St2 (Fig. 26.11a) due to strengthening of the latter after martensitic transformations. Accordingly, the plastic deformations in the martensitic phase of the deposited layer (Fig. 26.11b) are much smaller than in the layer of steel St1. At the same time, the stresses in the St2 steel layer are more significant than the stresses in the base metal (Fig. 26.10a), and the plastic deformations are comparable (Fig. 26.11b).

The above results are confirmed by comparing the distributions of stresses at D–D section, intensity s_i, and deformations e_i^p shown in Figs. 26.12a, b and 26.12c, d for steels St1 and St2, respectively.

Let us compare the results obtained within the framework of the hypotheses of the PDS and the PSS on the example of building up by steel St2 under conditions of smooth support. The corresponding distributions of the characteristics of the stress–

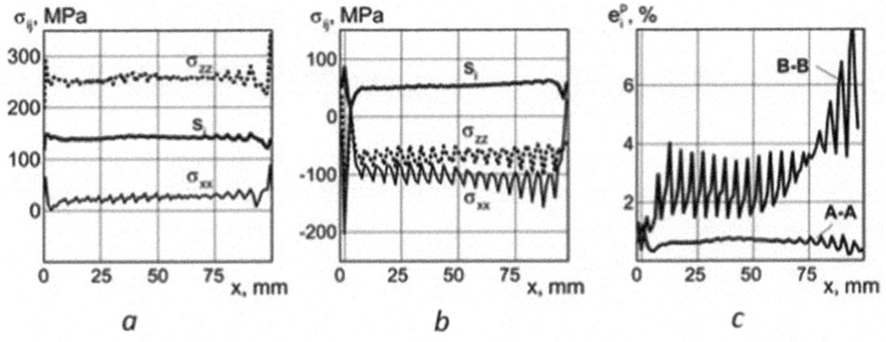

Fig. 26.7 Stress–strain state for the case of build-up with St1 steel, the free lower boundary, and PDS

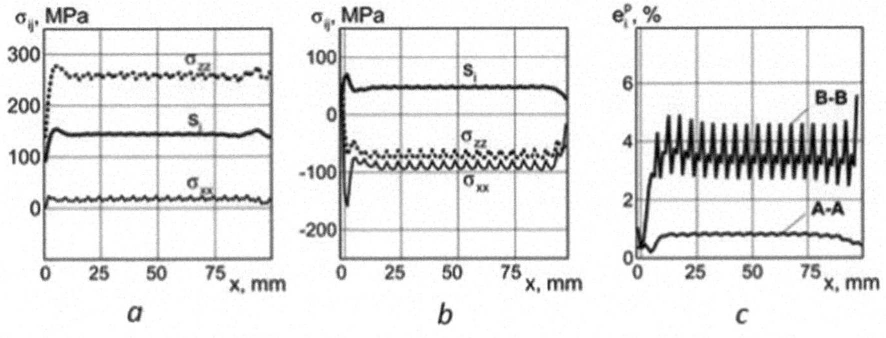

Fig. 26.8 Stress–strain state for the case of build-up with St1 steel, the smooth lower boundary, and PDS

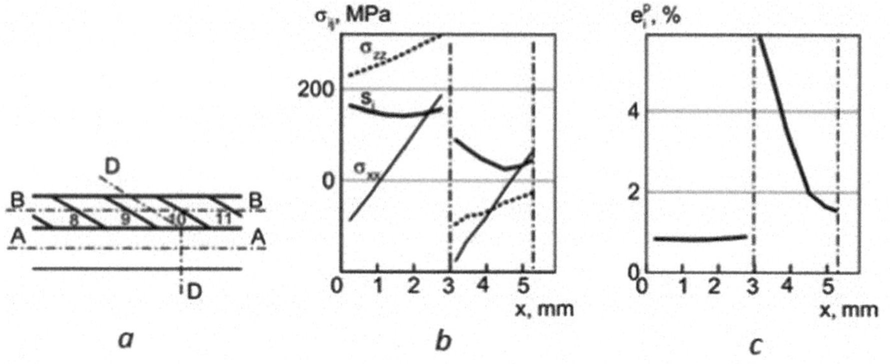

Fig. 26.9 Stress–strain state for the case of build-up with St1 steel in cross section D–D, the smooth lower boundary, and PDS

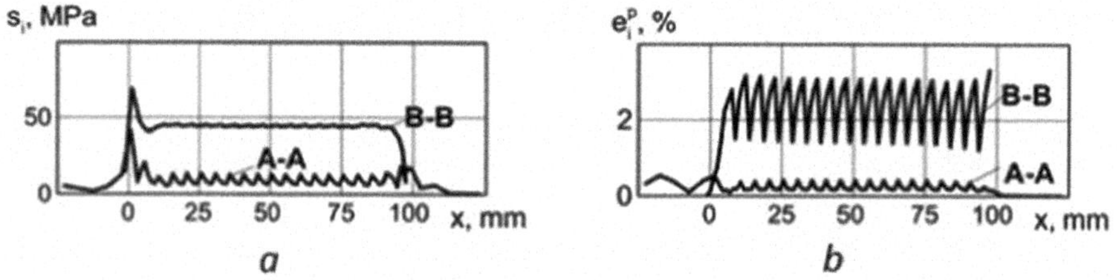

Fig. 26.10 Stress–strain state for the case of build-up with St1 steel, the smooth lower boundary, and PSS

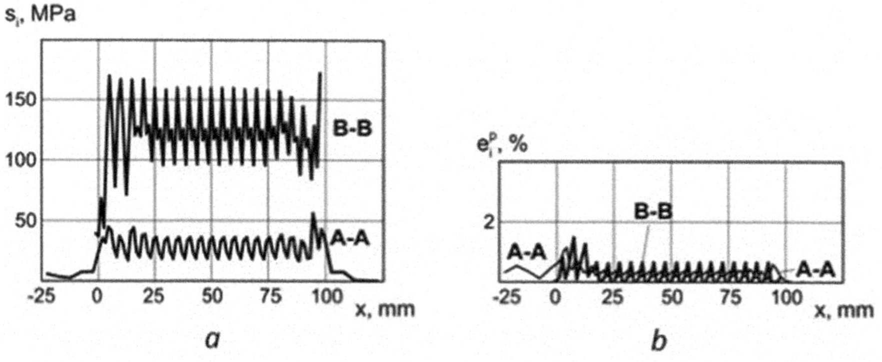

Fig. 26.11 Stress–strain state for the case of build-up with St2 steel, the smooth lower boundary, and PSS

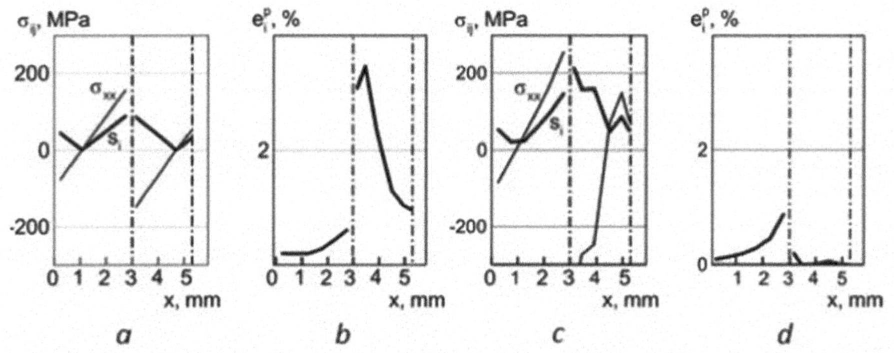

Fig. 26.12 Stress–strain state in cross section D–D for the case of build-up with St1 steel **a**, **b** and St2 steel **c**, **d**, smooth lower boundary, and PSS

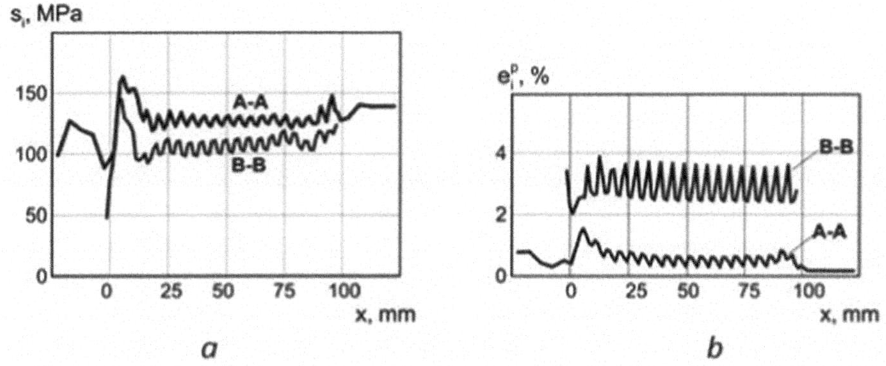

Fig. 26.13 Stress–strain state in cross sections A–A and B–B for the case of build-up with St2, smooth lower boundary, and PDS

strain state for the PDS and the PSS are shown in Figs. 26.13a, b and 26.11a, b, respectively.

The PDS model predicts approximately the same level of stresses in the built-up layer and the base metal (Fig. 26.13a) and a significantly higher level of plastic deformation in the deposited layer (Fig. 26.13b).

The PSS model predicts significantly higher stresses in the deposited layer compared to the base metal (Fig. 26.11a) and significantly lower plastic deformations in the deposited layer (Fig. 26.11b) compared to the PDS. At the same time, both models show similar levels of plastic deformation in the base metal.

These results show that the conditions of fixing the edges of the plate $z = 0$ and $z = l_z$ significantly affect the residual stress–strain state of the built-up element.

Consider the case of a two-layer build-up with steel St1 under conditions of smooth support. The second layer is built up after cooling the first layer to $\theta \approx 300°C$.

The results of calculations of residual stress–strain state under the PDS conditions for two-layer surfacing with steel St1 (the scheme in Fig. 26.14) are shown in

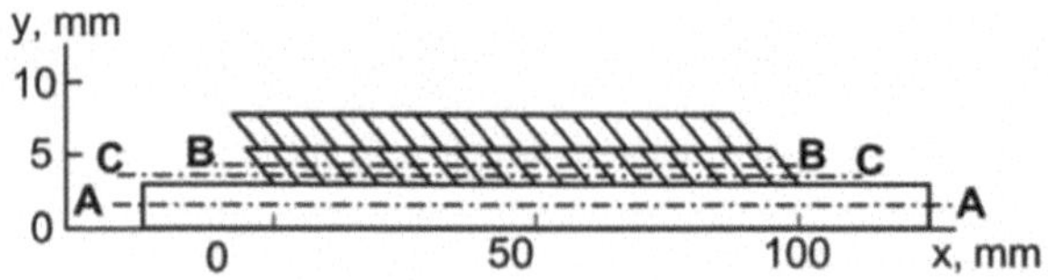

Fig. 26.14 Scheme of two-layer surfacing

Fig. 26.15. Figure 26.15a corresponds to section A–A at the zone of surfacing, and Fig. 26.15b—section B–B.

The maximum tensile stresses σ_{zz} are achieved in the base metal (section A–A), and the maximum compressive stresses σ_{xx} are formed in the first layer (section C–C) near the boundary with the base metal. In this section, there are maximum values of the intensity of residual plastic deformations, which have an oscillatory character with a period Δl (Fig. 26.15c).

The distribution of stresses and plastic deformation intensities along the y-axis in middle section D–D in the tenth bead (the scheme in Fig. 26.16a) is shown in Fig. 26.16b, c.

There is a discontinuity of stresses and plastic deformations at the point of transition from the base metal to the build-up metal. The discontinuity is caused by different properties of metal grades and conditions of layers deposition. In this case, the gap at the boundary of layers is practically absent.

Calculations based on a simplified scheme of the sequential instantaneous build-up of the first and second layers with St1 steel under the PDS conditions show that the results are in good agreement in terms of stresses. However, the distribution and level of plastic deformations differ significantly, both quantitatively and qualitatively. The approximate scheme, with the exception of the edge zones, predicts much smaller plastic deformations in the deposited metal.

The effect of fixing the plate on the edges perpendicular to the z-axis is illustrated by comparing the above results with similar results under the PSS conditions (Figs. 26.17 and 26.18). Comparing with the data in Figs. 26.15 and 26.16, we can

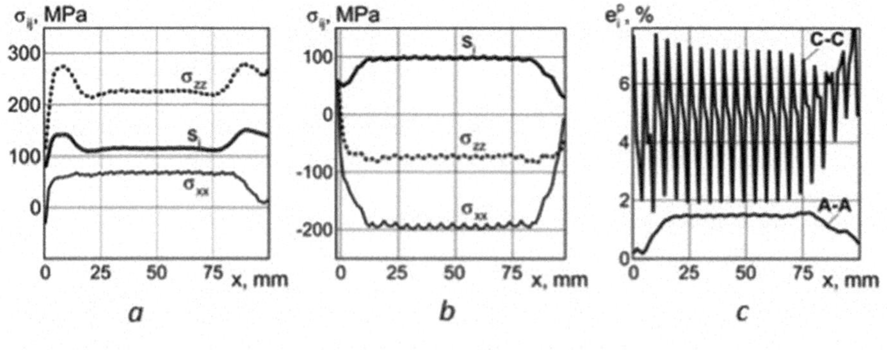

Fig. 26.15 Stress–strain state for the case of two-layer build-up with St1 steel, smooth lower boundary, and PDS

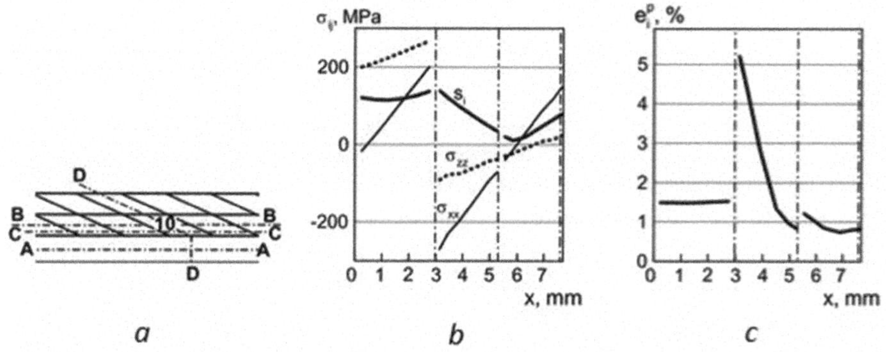

Fig. 26.16 Stress–strain state in cross section D–D for the case of two-layer build-up with St1 steel, smooth lower boundary, and PDS

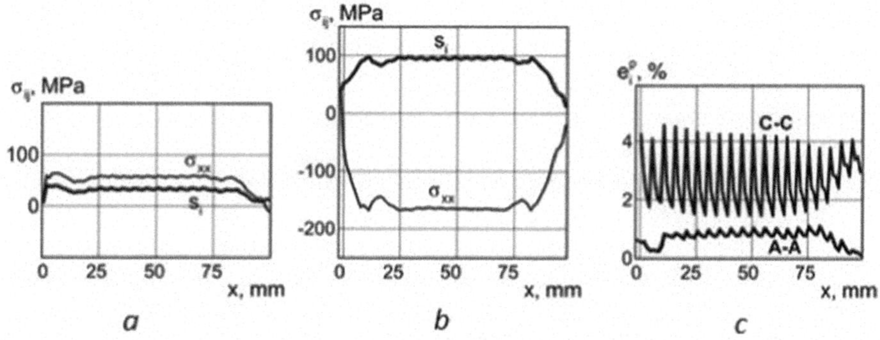

Fig. 26.17 Stress–strain state for the case of two-layer build-up with St1 steel, smooth lower boundary, and PSS

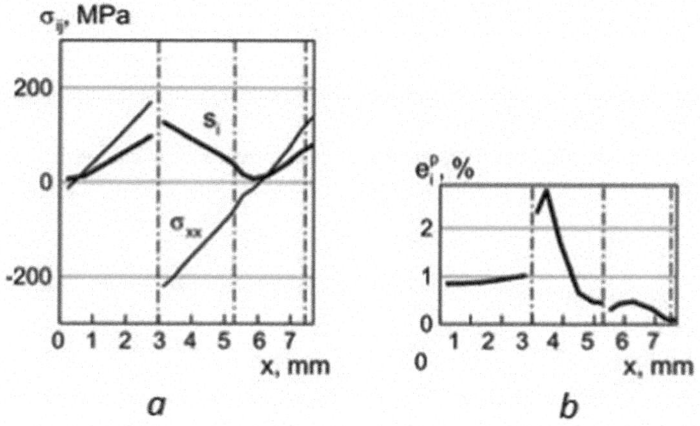

Fig. 26.18 Stress–strain state in cross section D–D for the case of two-layer build-up with St1 steel, smooth lower boundary, and PSS

see that under the PSS conditions, we have a significantly lower level of stresses and plastic deformations in the base metal and almost the same level of stresses in the built-up metal.

The plastic deformations in the case of the PSS due to the absence of the component σ_{zz} are much smaller than in the case of the PDS.

The results of calculations carried out for the cases of not overlapping and overlapping beads (Fig. 26.2a, b) differ insignificantly.

26.11 Conclusions

1. The calculations of the stress–strain state of sheet elements made of steel St0, when built up with steels St1, St2, and St3 under conditions of free boundary and smooth support on the lower surface of the element, were carried out. It is established that there is a smaller residual deflection under smooth conditions compared to the free boundary conditions.
2. When building up with materials with martensitic transformations, more considerable deflections take place.
3. The conditions of smooth support and smooth fixing of edges provide minimal deflections.
4. The conditions of element fixation at the boundaries $z = 0$ and $z = l_z$ significantly affect the residual stress–strain state of the built-up element. Therefore, the results obtained under the PDS/PNS conditions can be considered as limit estimates of the characteristics of the stress–strain state.
5. In the case of two-layer build-up, the results for the cases with and without overlapping beads differ insignificantly.
6. A simplified scheme of the sequential instantaneous build-up of the first and second layers shows that the stress results agree with the scheme of bead build-up in the build-up area. However, the plastic strain distributions differ significantly both quantitatively and qualitatively. The approximate scheme predicts much smaller strains in the built-up layer.
7. The results of calculations of deflections satisfactorily correlate with experimental data.

References

Arutyunyan NK, Drozdo AD, Naumov VE (1987) Mechanics of growing visco-elastoplastic bodies. Nauka, Moscow

Bezukhov NI, Bazhanov VL, Goldenblat II, Nikolaenko NA, Sinyukov AM (1965) Calculations of strength. Stability and vibrations in high temperature conditions, Machine Engineering, Moscow

Bodner SR (2000) Unified plastisity-an engineering approach. Final Rep. Technion-Israel Inst. of Tech, Haifa

Carlone P, Palazzo G (2011) Development and validation of thermos-mechanical finite element model of steel quenching process including solid-solid phase changes. Int Appl Mech 46:955–971

Klingbeil NW, Beuth JL, Chin RK, Amon CH (2002) Residual stress-induced warping in direct metal solid freeform fabrication. Int J Mech Sci 44(1):57–77

Koistinen D, Marburger R (1959) A general equation prescribing the extent of the austenite-martensite transformation in pure-carbon alloys and plain carbon steels. Acta Metallica 7:59–60

Lindgren l-E, (2007) Computational welding mechanics: Thermo mechanical and microstructural simulations. Woodhead Publishing Limited, CRC Press, Cambridge

Makhnenko VI (2006) Resource of safe operation of welded joints and joints of modern structures: Monogr. E.O. Paton institute of welding of NAS of placecountry-regionUkraine, Naukova Dumka, Kyiv

Motovilovets IA, Kozlov VI (1987) Thermoelasticity. Mechanics of coupled fields in materials and structural elements in 5 vols, vol 1. Naukova Dumka, Kyiv

Popov AA, Popova AE (1961) Isothermal and thermokinetic diagrams of overcooled austenite decomposition. Handbook of the Terminist, GSTI Mashlit, Moscow-Sverdlovsk

Radaj D (2003) Welding residual stresses and distortion. DVS Verlag GmbH, Düsseldorf, Calculation and measurement

Ryabtsev IA, Senchenkov IK, Turyk EV (2015) Surfacing. Materials, technologies, mathematical modeling. Ed. of the Silesian Polytechnic Institute (Wydawnictwo Politechniki Slaskiej). Gliwice

Ryabtsev IA, Senchenkov IK (2013) Theory and practice of surfacing works. Ekotehnologiya, Kyiv

Senchenkov IK, Oksenchuk ND (2013) Modeling of a nonisothermal flow with regard for the dependence of plastic properties on the microstructure of a material. J Math Sci 190:796–803

Senchenkov IK (2005) Thermomechanical model of growing cylindrical bodies made of physically nonlinear materials. Int Appl Mech 41(9):1059–1065

Senchenkov IK, Oksenchuk ND (2013) Modeling of a nonisothermal flow with regard for the dependence of plastic properties on the microstructure of a material. J Math Sci 190:796–803

Senchenkov IK, Tabieva GA (1996) Determination of the parameters of the Bodner-Partom model for thermoviscoplastic deformation of materials. Int Appl Mech 32(2):132–139

Senchenkov IK, Lobanov LM, Chervinko OP, Paschin NA (1998) Laws of relative longitudinal displacements of plates during butt electric welding. Dop NAN Ukrainy 6:66–70

Senchenkov IK, Ryabtsev IO, Chervinko OP, Babinets AA (2022) Calculation of deflections during the build-up of sheet elements with liquid metal. Appl Mech 58(5):97–108

Senchenkov IK, Ryabtsev IA, Turyk E, Tabieva GA (2005) Calculation of residual stresses under multilayer spiral surfacing of cylindrical parts on the basis of the theory of viscoplastic body building up. Welding Prod 9:18–25

Shevchenko AY, Banyas MV, Senchenkov IK (2012) A variant of the equations of nonisothermal plastic flow. Int Appl Mech 48(5):602–607

Shorshorov MK, Belov VV (1972) Phase transformations and change of steel properties at welding. Atlas, Nauka, Moscow

Yuryev SF (1950) Specific volumes of phases in martensite transformation of austenite. Metallurgizdat, Moscow

Zubchenko AS (2003) Handbook of steels and alloys, 2nd edn supplementary and revised. Mashinostroenie, Moscow

Chapter 27
Nonstationary Thermoelastic Problem for a Multilayer Coating/Half-Space Assembly at Radiative and Convective Loading

Victor Shevchuk

27.1 Introduction

In actual operating conditions, non-uniform temperature fields give rise to significant thermal strains and stresses in the elements of constructions and equipment with one- or multilayer coatings of various functionality. As a result, this may cause cracking, exfoliation of coatings, and loss of bearing capacity of such systems.

The problems of determining the thermal stressed state of multilayered structures under radiative–convective heat exchange with the environment were considered in Jendrzejewski et al. (2006), Jiguang and Fei (2015), Kushnir et al. (2002), Protsyuk and Horun (2017), Zhao et al. (2017), where existing mathematical models are used, and corresponding boundary value problems of heat conduction and thermoelasticity are solved based on analytical–numerical and various numerical methods.

One of the problems arising in the study and calculation of the temperature fields of structural elements under radiative and convective heating or cooling is the choice of an efficient method for solving appropriate nonlinear nonstationary heat conduction boundary value problems and elaboration of their approximate solutions. The well-known ways of solving nonlinear boundary value heat conduction problems are their linearization, use of numerical methods, construction of iteration schemes, application of variational or integral methods, and the perturbation technique (Aziz and Na 1984; Hogan and Gartling 2008; Kushnir and Popovych 2011). Recently, the following new approaches have become widespread: the Adomian decomposition method (Oguntala et al. 2019), the homotopy analysis method (Liao 2003; Liao et al. 2006), the homotopy perturbation method (Ganji 2006; Torabi et al. 2011), and the differential transform method (Bervillier 2012; Yaghoobi and Torabi 2011).

Among the approaches used to solve thermal problems with a nonlinear boundary condition is the reduction to the equivalent nonlinear integral equation of the Volterra

V. Shevchuk (✉)
Pidstryhach Institute for Applied Problems of Mechanics and Mathematics, Lviv, Ukraine
e-mail: shevchuk@iapmm.lviv.ua

A. N. Guz et al. (eds.), *Advances in Mechanics*, Advanced Structured Materials 191,
https://doi.org/10.1007/978-3-031-37313-8_27

type using Green functions (Kushnir et al. 2002) and Laplace integral transformation (Villasenor 1994). To solve this nonlinear integral equation, the implicit Runge–Kutta method is applied in Abdalkhani (1992), linear splines—in Kushnir et al. (2002), and the reduction to a nonlinear algebraic equation—in Villasenor (1994). In Onur and Sivrioglu (1993), Sundén (1989), the finite-difference computing technique and in Fetni et al. (2021), Guo et al. (2016), the finite element method are used for solving such heat conduction problems. A numerical–analytical procedure for determining the nonstationary temperature field in the half-space, taking into account thermal radiation, is proposed in Protsyuk (2021). The nonlinear heat conduction problem is reduced to the recurrent system of nonlinear algebraic equations using the Green function and linear splines. The method of successive approximations (Popovych et al. 2012) and the lumped parameter model approach (An and Su 2011; Tan et al. 2009) are applied to determine the temperature field in bodies under radiative–convective heat exchange with the environment. In Choudhury (2006), Fakoor-Pakdaman et al. (2015), the fourth-order nonlinear effect of radiative heat transfer in the boundary condition is linearized, assuming that the difference between the ambient and initial solid temperature is small.

To solve the nonlinear boundary value problem of nonstationary heat conduction and investigate the influence of radiation on heat transfer in plates with thin coatings, the quasi-linearization of boundary conditions and the finite-difference method are used in Terlets'kyi and Turii (2013). By solving the boundary value heat conduction problem for a half-space under radiative cooling in the environment, the substantial advantage of the method of quasi-linearization, which has a high rate of convergence compared with other iterative methods, has been shown (Shevchuk and Havrys 2017). For solving boundary value problems of heat conduction for solids with thin coatings, the application of generalized linear boundary conditions (Shevchuk 2012, 2013, 2014; Shevchuk and Gavris 2020) (for solids with one- or multilayer coatings) and generalized nonlinear boundary conditions (Terlets'kyi and Turii 2013) (for bodies with one-layer coatings) appears to be the most efficient approach. In Shevchuk (2022), generalized nonlinear boundary conditions have been obtained for describing radiative–convective heat transfer of the bodies with the environment via multilayer coating.

In this study, using a nonlinear boundary condition (Shevchuk 2022), the results of determining the temperature field in a half-space during convective interaction with the environment through a multilayer coating (Shevchuk 2013) are generalized for the case of radiative–convective heating or cooling of such a system. Based on the method of quasi-linearization (Bellman and Kalaba 1965), an iterative scheme for determining the temperature in the half-space is built. Using the Laplace transform (Luikov 1968), the approximate analytical solution of the linearized problem is obtained for each iteration.

For a half-space/two-layer wear-resistant coating assembly, the influence of thermophysical and thermomechanical characteristics of the system and the conditions of interaction with the ambient medium on the distribution of temperature and stresses has been studied.

27.2 Statement of the Heat Conduction Problem

We study the radiative–convective heat exchange for a half-space with an ambient medium via an n-layer coating of thickness $\delta = \sum_{i=1}^{n} \delta_i$, where δ_i is the thickness of the ith coating layer. The rectangular coordinates (x, y, z) are chosen so that plane $z = 0$ coincides with the coating/substrate interface, and the z-axis is directed toward the system.

The one-dimensional nonstationary boundary value problem of heat conduction is formulated as follows:

- heat equation and initial condition

$$\frac{\partial T_j}{\partial \tau} = a_j \frac{\partial^2 T_j}{\partial z^2}, \quad j = \mathrm{b}, 1, 2, \ldots, n, \tag{27.1}$$

$$T_j\big|_{\tau=0} = T_0 = \text{const}, \tag{27.2}$$

- boundary condition of radiative–convective heat transfer between the coating and the ambient medium

$$\lambda_n \frac{\partial T_n}{\partial z} = \varepsilon \sigma_0 \left(T_n^4 - T_\mathrm{a}^4\right) + \mu \left(T_n - T_\mathrm{a}\right) \quad \text{at} \quad z = z_n = -\delta, \tag{27.3}$$

- conditions of perfect thermal contact on the interfaces of coating layers and coating/half-space

$$\begin{aligned} &T_i = T_{i-1}, \ \lambda_i \frac{\partial T_i}{\partial z} = \lambda_{i-1} \frac{\partial T_{i-1}}{\partial z} && \text{at } z = z_{i-1} = -\sum_{m=1}^{i-1} \delta_m, \ i = 2, \ldots, n, \\ &1 = T_\mathrm{b}, \quad \lambda_1 \frac{\partial T_1}{\partial z} = \lambda_\mathrm{b} \frac{\partial T_\mathrm{b}}{\partial z} && \text{at } z = z_0 = 0, \end{aligned} \tag{27.4}$$

- condition at infinity

$$T_\mathrm{b} \to T_0 \quad \text{at} \quad z \to \infty. \tag{27.5}$$

Here, the following notations are introduced: T is the absolute temperature, $a = \lambda/(\omega\rho)$ is the thermal diffusivity, λ is the thermal conductivity, ω is the specific heat, ρ is the density, τ is the time, μ is the heat transfer coefficient for the coating surface, ε is the emissivity factor of the coating surface, σ_0 is the Stefan–Boltzmann constant.

Indices i, b, and a refer to the number of a coating layer, body (half-space), and ambient medium, respectively.

To solve the problem, we use an approach based on the simulation of the influence of the coating on the heat conduction process in the system by generalized boundary conditions. Under this approach, the temperature is determined by the solution of Eq. (27.1) for a half-space with initial condition (27.2), condition at infinity (27.5), and the generalized boundary condition, which has the following form for this case (Shevchuk 2022)

$$\lambda_b\left(1+\frac{\mu}{H}\right)\frac{\partial T_b}{\partial z}+\mu\left(T_a-T_b\right)-\Omega\frac{\partial T_b}{\partial \tau}+\varepsilon\sigma_0 T_a^4$$
$$-\varepsilon\sigma_0\sum_{m=0}^{4}(-1)^m C_4^m\left(\frac{\lambda_b}{H}\right)^m T_b^{4-m}\left(\frac{\partial T_b}{\partial z}\right)^m=0,$$
$$T_b\Big|_{\tau=0}=T_0 \quad \text{at} \quad z=0, \tag{27.6}$$

where $\Omega=\sum_{i=1}^{n}\omega_i\rho_i\delta_i$ and $H^{-1}=\sum_{i=1}^{n}\delta_i/\lambda_i$ are the reduced heat capacity and thermal resistance of the whole coating, and $C_4^m=4!/(m!(4-m)!)$ are the binomial coefficients.

We introduce the dimensionless temperatures

$$\theta_b=T_b/T_*,\quad \theta_0=T_0/T_*,\quad \theta_a=T_a/T_*, \tag{27.7}$$

where the reference temperature

$$T_*=\begin{cases} T_a & \text{during heating of the system}\\ T_0 & \text{during its cooling.}\end{cases} \tag{27.8}$$

Here $\bar{z}=z/z_*$ is the dimensionless coordinate; $\mathrm{Fo}=a_b\tau/z_*^2$ is the Fourier number; $\mathrm{Bi}=\mu z_*/\lambda_b$ is the Biot criterion; $\mathrm{Sk}=\varepsilon\sigma_0 T_*^3 z_*/\lambda_b$ is the Stark criterion; $\xi=\lambda_b/(Hz_*)$ is the relative effective thermal resistance of the coating; $\eta=\Omega/(\omega_b\rho_b z_*)$ is the relative effective heat capacity of the coating (z_* is a selected scale unit).

Then the nonlinear boundary value problem of heat conduction (27.1), (27.2), (27.5), and (27.6) for a half-space is given in dimensionless quantities as follows:

$$\frac{\partial^2\theta_b(\bar{z},\mathrm{Fo})}{\partial\bar{z}^2}=\frac{\partial\theta_b(\bar{z},\mathrm{Fo})}{\partial\mathrm{Fo}},\quad \bar{z}>0,\quad \mathrm{Fo}>0, \tag{27.9}$$

$$\theta_b(\bar{z},0)=\theta_0, \tag{27.10}$$

$$\lim_{\bar{z}\to\infty}\theta_b(\bar{z},\mathrm{Fo})=\theta_0, \tag{27.11}$$

$$(1+\xi\mathrm{Bi})\frac{\partial\theta_b}{\partial\bar{z}}+\mathrm{Bi}\left(\theta_a-\theta_b\right)-\eta\frac{\partial\theta_b}{\partial\mathrm{Fo}}$$
$$-\mathrm{Sk}\left[\sum_{m=0}^{4}(-1)^m C_4^m\xi^m\theta_b^{4-m}\left(\frac{\partial\theta_b}{\partial\bar{z}}\right)^m-\theta_a^4\right]=0,$$
$$\theta_b\Big|_{\mathrm{Fo}=0}=\theta_0 \quad \text{at} \quad \bar{z}=0. \tag{27.12}$$

It may be noted that the introduction of dimensionless temperature θ_b using (27.7) and (27.8), as in Campo (1977), allows formulating the nonlinear heat conduction problem of radiative–convective heating or cooling, including cooling in the ambient medium with zero temperature, by one system of relations.

27.3 Solution of the Heat Conduction Problem With a Generalized Boundary Condition

27.3.1 Scheme of the Quasi-linearization Method

We apply the iterative procedure based on the quasi-linearization method (Bellman and Kalaba 1965; Campo 1977) to the solution of the boundary value problem (27.9)–(27.12). We represent the boundary value problem for the kth approximation as follows:

$$\frac{\partial^2 \theta_{\mathrm{b}}^{(k)}(\bar{z}, \mathrm{Fo})}{\partial \bar{z}^2} = \frac{\partial \theta_{\mathrm{b}}^{(k)}(\bar{z}, \mathrm{Fo})}{\partial \mathrm{Fo}}, \quad \bar{z} > 0, \quad \mathrm{Fo} > 0, \tag{27.13}$$

$$\theta_{\mathrm{b}}^{(k)}(\bar{z}, 0) = \theta_0, \tag{27.14}$$

$$\lim_{x \to \infty} \theta_{\mathrm{b}}^{(k)}(\bar{z}, \mathrm{Fo}) = \theta_0, \tag{27.15}$$

$$(1 + \xi \mathrm{Bi}) \frac{\partial \theta_{\mathrm{b}}^{(k)}}{\partial \bar{z}} + \mathrm{Bi}\left(\theta_{\mathrm{a}} - \theta_{\mathrm{b}}^{(k)}\right) - \eta \frac{\partial \theta_{\mathrm{b}}^{(k)}}{\partial \mathrm{Fo}}$$
$$- \mathrm{Sk}\left[\sum_{m=0}^{4} (-1)^m C_4^m \xi^m \left(\theta_{\mathrm{b}}^{(k)}\right)^{4-m} \left(\frac{\partial \theta_{\mathrm{b}}^{(k)}}{\partial \bar{z}}\right)^m - \theta_{\mathrm{a}}^4\right] = 0 \text{ at } \bar{z} = 0,$$
$$\theta_{\mathrm{b}}^{(k)}(0, 0) = \theta_0. \tag{27.16}$$

To linearize the nonlinear term in the boundary condition (27.16), we use the following formula (Bellman and Kalaba 1965; Campo 1977):

$$f\left(\theta_{\mathrm{b}}^{(k)}, \frac{\partial \theta_{\mathrm{b}}^{(k)}}{\partial \bar{z}}\right) = f\left(\theta_{\mathrm{b}}^{(k-1)}, \frac{\partial \theta_{\mathrm{b}}^{(k-1)}}{\partial \bar{z}}\right)$$
$$+ f'_{\theta_{\mathrm{b}}}\left(\theta_{\mathrm{b}}^{(k-1)}, \frac{\partial \theta_{\mathrm{b}}^{(k-1)}}{\partial \bar{z}}\right)\left(\theta_{\mathrm{b}}^{(k)} - \theta_{\mathrm{b}}^{(k-1)}\right)$$
$$+ f'_{\partial\theta_{\mathrm{b}}/\partial\bar{z}}\left(\theta_{\mathrm{b}}^{(k-1)}, \frac{\partial \theta_{\mathrm{b}}^{(k-1)}}{\partial \bar{z}}\right)\left(\frac{\partial \theta_{\mathrm{b}}^{(k)}}{\partial \bar{z}} - \frac{\partial \theta_{\mathrm{b}}^{(k-1)}}{\partial \bar{z}}\right), \tag{27.17}$$

where

$$f\left(\theta_{\mathrm{b}}, \frac{\partial \theta_{\mathrm{b}}}{\partial \bar{z}}\right) = \sum_{m=0}^{4} (-1)^m C_4^m \xi^m (\theta_{\mathrm{b}})^{4-m} \left(\frac{\partial \theta_{\mathrm{b}}}{\partial \bar{z}}\right)^m .$$

Then the linearized condition (27.16) has the following form:

$$\frac{\partial \theta_{\mathrm{b}}^{(k)} (0, \mathrm{Fo})}{\partial \bar{z}} - \mathrm{M}^{(k)}(\mathrm{Fo}) \frac{\partial \theta_{\mathrm{b}}^{(k)} (0, \mathrm{Fo})}{\partial \mathrm{Fo}}$$
$$= \mathrm{Bi}_*^{(k)}(\mathrm{Fo})\theta_{\mathrm{b}}^{(k)} (0, \mathrm{Fo}) + Q^{(k)}(\mathrm{Fo}), \quad \theta_{\mathrm{b}}^{(k)}(0, 0) = \theta_0, \tag{27.18}$$

where

$$\mathrm{Bi}_*^{(k)}(\mathrm{Fo}) = \frac{\mathrm{Bi} + \mathrm{Sk}\alpha_2^{(k-1)}(\mathrm{Fo})}{1 + \xi\mathrm{Bi} - \mathrm{Sk}\alpha_1^{(k-1)}(\mathrm{Fo})},$$
$$\mathrm{M}^{(k)}(\mathrm{Fo}) = \frac{\eta}{1 + \xi\mathrm{Bi} - \mathrm{Sk}\alpha_1^{(k-1)}(\mathrm{Fo})}, \tag{27.19}$$
$$Q^{(k)}(\mathrm{Fo}) = -\frac{\mathrm{Bi} + \left(1+\alpha_3^{(k-1)}(\mathrm{Fo})\right)\mathrm{Sk}}{1 + \xi\mathrm{Bi} - \mathrm{Sk}\alpha_1^{(k-1)}(\mathrm{Fo})},$$

$$\alpha_1^{(k-1)} = \sum_{m=1}^{4} (-1)^m \, m C_4^m \xi^m \left(\theta_{\mathrm{b}}^{(k-1)}\right)^{4-m} \left(\frac{\partial \theta_{\mathrm{b}}^{(k-1)}}{\partial \bar{z}}\right)^{m-1},$$
$$\alpha_2^{(k-1)} = \sum_{m=1}^{4} (-1)^{m-1} (5-m) C_4^{m-1} \xi^{m-1} \left(\theta_{\mathrm{b}}^{(k-1)}\right)^{4-m} \left(\frac{\partial \theta_{\mathrm{b}}^{(k-1)}}{\partial \bar{z}}\right)^{m-1}, \tag{27.20}$$
$$\alpha_3^{(k-1)} = 3 \sum_{m=0}^{4} (-1)^m C_4^m \xi^m \left(\theta_{\mathrm{b}}^{(k-1)}\right)^{4-m} \left(\frac{\partial \theta_{\mathrm{b}}^{(k-1)}}{\partial \bar{z}}\right)^{m}.$$

27.3.2 Construction of an Approximate Solution of the Linearized Problem for the kth Approximation

To solve the problem, we use the integral Laplace transform (Luikov 1968)

$$\tilde{\theta}_{\mathrm{b}}^{(k)}(\bar{z}, s) = \mathscr{L}\left[\theta_{\mathrm{b}}^{(k)}(\bar{z}, \mathrm{Fo})\right] = \int_0^\infty \theta_{\mathrm{b}}^{(k)}(\bar{z}, \mathrm{Fo}) \exp(-s\mathrm{Fo}) d\mathrm{Fo}.$$

To obtain an analytical solution, we consider the quantities $\mathrm{Bi}_*^{(k)}$, $Q^{(k)}$, and $\mathrm{M}^{(k)}$ constant at each iteration. Then the solution of (27.13) in the half-space in the transform domain, taking into account the conditions (27.14), (27.15), and (27.18), has the following form:

$$\tilde{\theta}_{\mathrm{b}}^{(k)}(\bar{z}, s) = \frac{\theta_0}{s} - \frac{g^{(k)} \exp(-q\bar{z})}{s\left(\psi^{(k)} q^2 + \left(\mathrm{Bi}_*^{(k)}\right)^{-1} q + 1\right)}, \tag{27.21}$$

where $q = \sqrt{s}$, $\psi^{(k)} = \mathrm{M}^{(k)}/\mathrm{Bi}_*^{(k)}$, and $g^{(k)} = \theta_0 + Q^{(k)}/\mathrm{Bi}_*^{(k)}$.

In obtaining the inverse transform of (27.21), some cases of representation are possible depending on the character of the roots of the equation

$$\psi^{(k)} q^2 + \left(\mathrm{Bi}_*^{(k)}\right)^{-1} q + 1 = 0: \tag{27.22}$$

1. For $\eta \neq 0$, when $(\mathrm{Bi}_*^{(k)})^{-2} - 4\psi^{(k)} \neq 0$, Eq. (27.22) has two different roots:

$$q_{1,2}^{(k)} = \frac{-1 \pm \sqrt{1 - 4\psi^{(k)}\left(\mathrm{Bi}_*^{(k)}\right)^2}}{2\psi^{(k)}\mathrm{Bi}_*^{(k)}}. \tag{27.23}$$

Accordingly, transform (27.21) can be represented in the form

$$\tilde{\theta}_{\mathrm{b}}^{(k)}(\bar{z}, s) = \frac{\theta_0}{s} - \frac{g^{(k)} \exp\left(-\bar{z}\sqrt{s}\right)}{\psi^{(k)}\left(q_1^{(k)} - q_2^{(k)}\right) s}\left(\frac{1}{\sqrt{s} - q_1^{(k)}} - \frac{1}{\sqrt{s} - q_2^{(k)}}\right). \tag{27.24}$$

Using the inversion formula of the Laplace transform (Luikov 1968, p. 664)

$$\mathscr{L}^{-1}\left[\frac{\exp\left(-l\sqrt{s}\right)}{s\left(\sqrt{s} + b\right)}\right] = \frac{1}{b}\left[\operatorname{erfc}\left(\frac{l}{2\sqrt{\tau}}\right) - \exp\left(bk + b^2\tau\right)\operatorname{erfc}\left(\frac{l}{2\sqrt{\tau}} + b\sqrt{\tau}\right)\right], \tag{27.25}$$

we finally obtain the expression to determine the kth temperature approximation:

$$\theta_{\mathrm{b}}^{(k)}(\bar{z}, \mathrm{Fo}) = \theta_0 - g^{(k)}\Big\{\operatorname{erfc}\varphi + \frac{1}{2\Delta^{(k)}}\Big[\left(1 + \Delta^{(k)}\right) F_1^{(k)}(\bar{z}, \mathrm{Fo}) - \left(1 - \Delta^{(k)}\right) F_2^{(k)}(\bar{z}, \mathrm{Fo})\Big]\Big\}, \quad 0 \leqslant \bar{z} < \infty. \tag{27.26}$$

Here

$$\varphi = \frac{\bar{z}}{2\sqrt{\mathrm{Fo}}}, \quad \Delta^{(k)} = \sqrt{1 - 4\mathrm{M}^{(k)}\mathrm{Bi}_*^{(k)}},$$
$$F_m^{(k)}(\bar{z}, \mathrm{Fo}) = \exp\left(-q_m^{(k)}\bar{z} + \left(q_m^{(k)}\right)^2 \mathrm{Fo}\right)\operatorname{erfc}\left(\varphi - q_m^{(k)}\sqrt{\mathrm{Fo}}\right), \quad m = 1, 2, 3, 4. \tag{27.27}$$

2. For $\eta \neq 0$, when $(\mathrm{Bi}_*^{(k)})^{-2} - 4\psi^{(k)} = 0$, Eq. (27.22) has one real double root

$$q_3^{(k)} = -0.5/\mathrm{M}^{(k)}. \tag{27.28}$$

Then transform (27.21) has the form

$$\tilde{\theta}_{\mathrm{b}}^{(k)}(\bar{z}, s) = \frac{\theta_0}{s} - \frac{g^{(k)} \exp\left(-\bar{z}\sqrt{s}\right)}{s\psi^{(k)}\left(\sqrt{s} - q_3^{(k)}\right)^2}. \tag{27.29}$$

Using the inversion formula of the Laplace transform (Luikov 1968, p. 665)

$$\mathscr{L}^{-1}\left[\frac{\exp\left(-l\sqrt{s}\right)}{s\left(\sqrt{s}+b\right)^2}\right] = \frac{1}{b^2}\operatorname{erfc}\left(\frac{l}{2\sqrt{\tau}}\right) - \frac{2}{b}\sqrt{\frac{\tau}{\pi}}\exp\left(-\frac{l^2}{4\tau}\right)$$
$$+\left(2\tau + \frac{l}{b} - \frac{1}{b^2}\right)\exp\left(bl + b^2\tau\right)\operatorname{erfc}\left(\frac{l}{2\sqrt{\tau}} + b\sqrt{\tau}\right),$$

we obtain the inverse transform on the kth iteration of finding the temperature

$$\theta_{\mathrm{b}}^{(k)}(\bar{z}, \mathrm{Fo}) = \theta_0 - g^{(k)}\left\{\operatorname{erfc}\varphi + 2q_3^{(k)}\sqrt{\frac{\mathrm{Fo}}{\pi}}\exp\left(-\varphi^2\right)\right.$$
$$\left.+\left[2\left(q_3^{(k)}\right)^2\mathrm{Fo} - q_3^{(k)}\bar{z} - 1\right]F_3^{(k)}(\bar{z}, \mathrm{Fo})\right\}, \quad 0 \leqslant \bar{z} < \infty. \tag{27.30}$$

3. For $\eta = 0$, Eq. (27.22) has one real root

$$q_4^{(k)} = -\mathrm{Bi}_*^{(k)}, \tag{27.31}$$

and transform (27.21) can be written as

$$\tilde{\theta}_{\mathrm{b}}^{(k)}(\bar{z}, s) = \frac{\theta_0}{s} - \frac{\mathrm{Bi}_*^{(k)}\theta_0 + Q^{(k)}}{s\left(\sqrt{s} + \mathrm{Bi}_*^{(k)}\right)}\exp\left(-\bar{z}\sqrt{s}\right). \tag{27.32}$$

Using inversion formula (27.25), we write the expression for the kth temperature approximation:

$$\theta_{\mathrm{b}}^{(k)}(\bar{z}, \mathrm{Fo}) = \theta_0 - g^{(k)}\left[\operatorname{erfc}\varphi - F_4^{(k)}(\bar{z}, \mathrm{Fo})\right], \quad 0 \leqslant \bar{z} < \infty. \tag{27.33}$$

27.3.3 Iterative Scheme for Determining Temperature in the Half-Space

We present the computational algorithm as follows:

$$\theta_{\mathrm{b}}^{(0)} = \theta_0, \quad \theta_{\mathrm{b}}^{(k)}(\bar{z}, \mathrm{Fo}) = \Pi\left(\theta_{\mathrm{b}}^{(k-1)}, \frac{\partial\theta_{\mathrm{b}}^{(k-1)}}{\partial\bar{z}}\right), \quad k = 1, 2, \ldots \tag{27.34}$$

where the form of the operator Π depending on the character of the roots of Eq. (27.22) is given by (27.26), (27.30), or (27.33).

27.4 Relations for Determining the Coating Temperature

The temperature in the multilayer coating is determined via the boundary values of the temperature and of its derivative using the recovery formula (Shevchuk 2013, 2014, 2022)

$$\theta_i(\bar{z}, \mathrm{Fo}) = \theta_\mathrm{b}(0, \mathrm{Fo}) + r_i(\bar{z})\frac{\partial \theta_\mathrm{b}}{\partial x}(0, \mathrm{Fo}), \quad \bar{z}_i \leqslant \bar{z} \leqslant \bar{z}_{i-1}, \quad i = 1, \ldots, n, \tag{27.35}$$

where

$$r_i(\bar{z}) = \lambda_\mathrm{b}\left(-\sum_{j=1}^{i-1}\frac{\delta_j}{z_*\lambda_j} + \frac{\bar{z} - \bar{z}_{i-1}}{\lambda_i}\right), \quad \theta_i = \frac{T_i}{T_*}, \quad \bar{z}_i = \frac{z_i}{z_*}, \quad \bar{z}_0 = 0.$$

We obtain expressions for the temperature in the ith coating layer by substituting approximations (27.26), (27.30), and (27.33) into recovery formula (27.35) on the last iteration at $k = \tilde{k}$. Here $\tilde{k}$ is the number of iterations required to achieve the specified accuracy of calculations.

1. For $\eta \neq 0$, when $(\mathrm{Bi}_*^{(\tilde{k})})^{-2} - 4\psi^{(\tilde{k})} \neq 0$

$$\theta_i\,(\bar{z}, \mathrm{Fo}) = \theta_0 + g^{(\tilde{k})}\left\{1 - \frac{1}{2\Delta^{(\tilde{k})}}\right.$$
$$\left.\sum_{m=1}^{2}\left((-1)^{m+1} + \Delta^{(\tilde{k})}\right)\left(1 + r_i\,(\bar{z})\,q_m^{(\tilde{k})}\right)F_m^{(\tilde{k})}\,(0, \mathrm{Fo})\right\}. \tag{27.36}$$

2. For $\eta \neq 0$, when $(\mathrm{Bi}_*^{(\tilde{k})})^{-2} - 4\psi^{(\tilde{k})} = 0$

$$\theta_i\,(\bar{z}, \mathrm{Fo}) = \theta_0 - g^{(\tilde{k})}\left\{1 + 2\left(1 + q_3^{(\tilde{k})}r_j\,(\bar{z})\right)q_3^{(\tilde{k})}\sqrt{\mathrm{Fo}/\pi}\right.$$
$$\left.-\left[1 - 2q_3^{(\tilde{k})}\left(1 + q_3^{(\tilde{k})}r_j\,(\bar{z})\right)\mathrm{Fo}\right]F_3^{(\tilde{k})}\,(0, \mathrm{Fo})\right\}. \tag{27.37}$$

3. For $\eta = 0$

$$\theta_i\,(\bar{z}, \mathrm{Fo}) = \theta_0 - g^{(\tilde{k})}\left[1 - \left(1 + \mathrm{Bi}_*^{(\tilde{k})}r_i\,(\bar{z})\right)F_4^{(\tilde{k})}\,(0, \mathrm{Fo})\right]. \tag{27.38}$$

27.5 Solution of the Problem of Thermoelasticity

The non-uniform distribution of temperature in the system causes temperature stresses in it, which can be determined by the relations (Shevchuk 2017)

$$\begin{aligned}\sigma_{xx}^{\mathrm{b}} &= \sigma_{yy}^{\mathrm{b}} = \sigma^{\mathrm{b}}(z,\tau) = -\frac{E_{\mathrm{b}}}{1-\nu_{\mathrm{b}}}\beta_{\mathrm{b}}\left[T_{\mathrm{b}}(z,\tau)-T_0\right], \; 0 \le z < \infty, \\ \sigma_{xx}^{i} &= \sigma_{yy}^{i} = \sigma^{i}(z,\tau) = -\frac{E_i}{1-\nu_i}\beta_i\left[T_i(z,\tau)-T_0\right], \\ & i = 1,2,\ldots,n, \quad z_i \leqslant z \leqslant z_{i-1},\end{aligned} \tag{27.39}$$

where E_j are Young's moduli, β_j are the coefficients of linear thermal expansion, and ν_j are Poisson's ratios.

Having introduced dimensionless stresses by formula

$$\tilde{\sigma}^j(\bar{z},\mathrm{Fo}) = \sigma^j(z,\tau)\left(\frac{E_{\mathrm{b}}\beta_{\mathrm{b}}(T_{\mathrm{a}}-T_0)}{1-\nu_{\mathrm{b}}}\right)^{-1}, \quad j = \mathrm{b},1,2,\ldots,n, \tag{27.40}$$

we write the relations to determine the thermal stresses in the system:

- during heating

$$\tilde{\sigma}^j(\bar{z},\mathrm{Fo}) = \frac{E_j\beta_j}{1-\nu_j}\left(\frac{E_{\mathrm{b}}\beta_{\mathrm{b}}}{1-\nu_{\mathrm{b}}}\right)^{-1}\frac{\theta_0-\theta_j(\bar{z},\mathrm{Fo})}{1-\theta_0}; \tag{27.41}$$

- during cooling

$$\tilde{\sigma}^j(\bar{z},\mathrm{Fo}) = \frac{E_j\beta_j}{1-\nu_j}\left(\frac{E_{\mathrm{b}}\beta_{\mathrm{b}}}{1-\nu_{\mathrm{b}}}\right)^{-1}\frac{\theta_j(\bar{z},\mathrm{Fo})-1}{1-\theta_{\mathrm{a}}}. \tag{27.42}$$

27.6 Numerical Results and Their Analysis

Consider the radiative–convective heating of a half-space made of stainless steel 316L with a two-layer wear-resistant coating made of nichrome Cr–Ni (inner layer is $\delta_1 = 0.1$ mm in thickness) and WC–Co alloy (outer layer is $\delta_2 = 0.3$ mm in thickness).

Computations are performed at $T_{\mathrm{a}} = 1073$ K, $T_0 = 293$ K, $\varepsilon = 0.5$, and $z_* = 1$ m. The input parameters of the problem are selected as some average values in the considered temperature range (Sayman et al. 2009); they are presented in Table 27.1 (with such parameters $\mathrm{Sk} = 2.06$).

Figure 27.1 illustrates the effect of increasing the intensity of convective (solid curves) and radiative–convective (dashed curves) heating on the change in time of the contact temperature $\theta_{\mathrm{b}}(0,\mathrm{Fo})$.

Table 27.1 Properties of materials

Thermomechanical parameters	316L steel	Cr–Ni	WC–Co
Thermal conductivity λ, W/(m · K)	17	13	24
Specific heat, ω, J/(kg·K)	535	530	166
Density, ρ, kg/m^3	8031	8050	13900
Elasticity modulus, E, GPa	170	200	367
Poisson's ratio, ν	0.29	0.29	0.29
Thermal expansion coefficient, β, K^{-1} × 10^{-6}	18	14	6.5

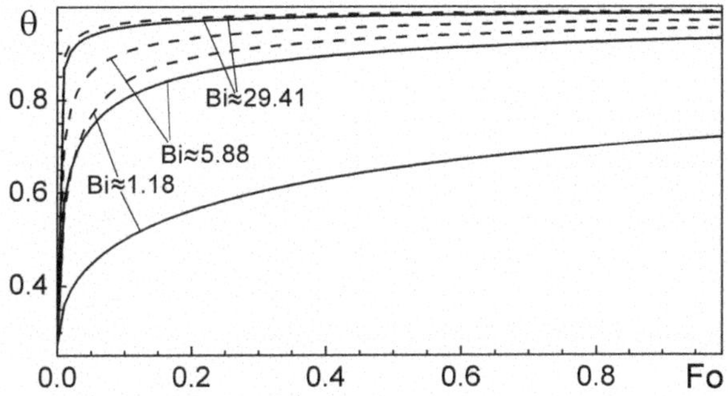

Fig. 27.1 Influence of heating intensity on variation of contact temperature $\theta = \theta_b$ (0, Fo): solid curves—Sk = 0; dashed curves—2.06

As we can see, if at $\mu = 20$ W/(m^2 · K) (Bi ≈ 1.18) and $\mu = 100$ W/(m^2 · K) (Bi ≈ 5.88), the radiative component (Sk = 2.06) significantly increases the contact temperature, then at $\mu \geqslant 500$ W/(m^2 · K) (respectively Bi ⩾ 29.41), its influence on heating of the system is insignificant (the upper dashed and solid curves practically merge).

The temperature change in the half-space/two-layer coating system over the spatial coordinate for some moments of time, depending on the intensity of convective and radiative–convective heating, is shown in Fig. 27.2.

Intensification of heating causes a significant increase in temperature drop along the depth of the base. Note that, as for the case of the contact temperature (Fig. 27.1), the influence of the radiative component on the temperature distribution in the system with an increase in the heat transfer coefficient up to 500 W/(m^2 · K) (Bi ≈ 29.41) appears to be insignificant. It should be noted that with such relations of the characteristics of the coating and the base, there is practically no temperature drop over the thickness of the coating layers in the considered range of μ. However, for another half-space/multilayer coating assembly, the influence of the intensity of heat exchange can cause a noticeable temperature drop over the thickness of the coating layers.

Figure 27.3 shows the change in time of the dimensionless contact stresses $\tilde{\sigma}^b$ (0, Fo) in a half-space on the coating/half-space interface depending on heating intensity. As previously for the case of Fig. 27.1, this causes an increase in compres-

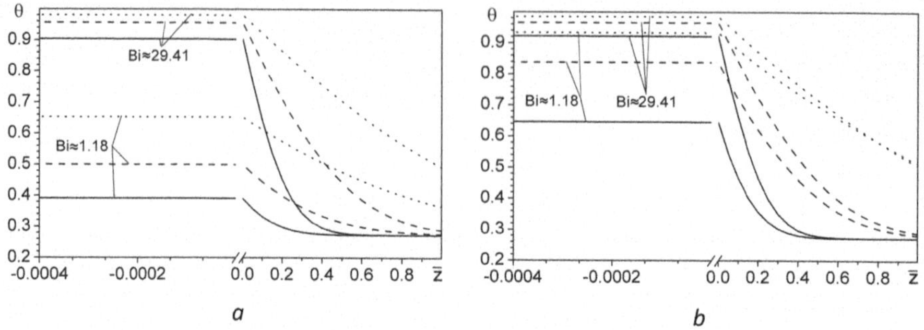

Fig. 27.2 Temperature distribution $\theta = \theta_j\ (\bar{z}, \mathrm{Fo})$ in the half-space/two-layer coating assembly for some time moments: **a**—convective heating (Sk = 0), **b**—radiative–convective heating (Sk = 2.06), solid curves—Fo = 0.02, dashed curves—0.1, and dotted curves—0.5

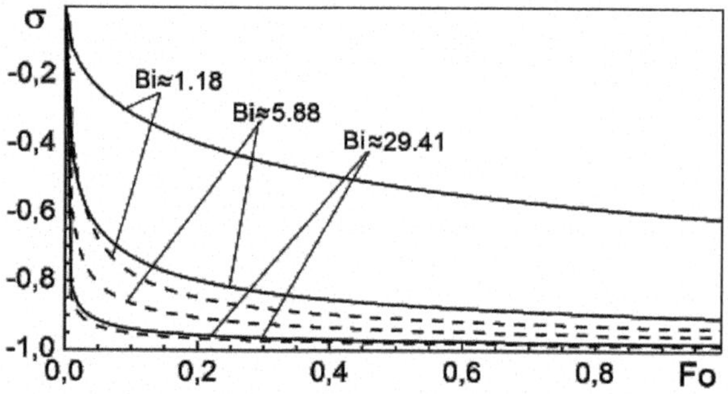

Fig. 27.3 Influence of heating intensity on contact stresses $\sigma = \tilde{\sigma}^{\mathrm{b}}(0, \mathrm{Fo})$: solid curves—Sk = 0, dashed curves—2.06

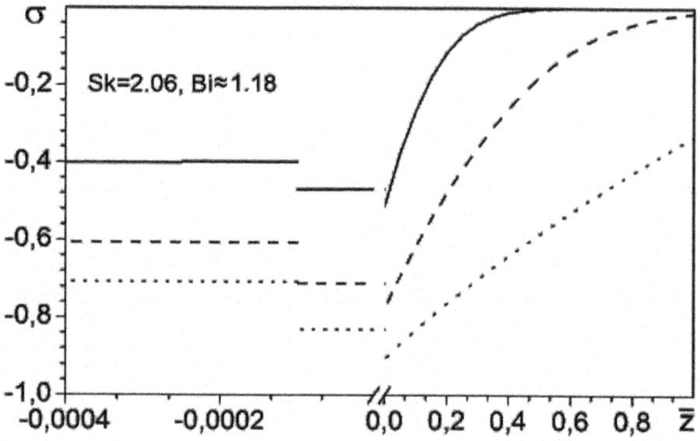

Fig. 27.4 Thermal stress distribution $\sigma = \tilde{\sigma}^{j}\ (\bar{z}, \mathrm{Fo})$ in the half-space/two-layer coating assembly for some time moments: solid curve—Fo = 0.02, dashed curve—0.1, and dotted curve—0.5

sive stresses on the substrate surface, and if $\mathrm{Bi} \geqslant 29.41$ ($\mu \geqslant 500\ \mathrm{W/(m^2 \cdot K)}$), the effect of the radiative component on the contact stresses is insignificant.

In Figs. 27.4 and 27.5, the distribution of stresses over spatial coordinate in the half-space/two-layer coating assembly is shown for some time moments under radiative–convective heating at $\mathrm{Bi} \approx 1.18$, $\mathrm{Sk} = 2.06$ (Fig. 27.4) and for some values of Bi and Sk at $\mathrm{Fo} = 0.02$ (Fig. 27.5).

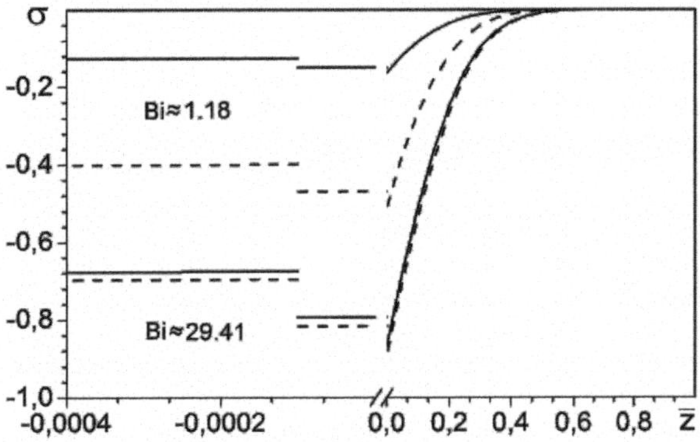

Fig. 27.5 Thermal stress distribution $\sigma = \tilde{\sigma}^j(\bar{z}, \text{Fo})$ in the half-space/two-layer coating assembly at fixed time moment Fo = 0.02: solid curves—Sk = 0, dashed curves—2.06

Hence, compressive stresses in the coating have a discontinuous character and increase with time. The relation

$$\frac{E_2\beta_2}{1-\nu_2}\left(\frac{E_1\beta_1}{1-\nu_1}\right)^{-1}$$

determines their jump on the coating layers' interface and

$$\frac{E_1\beta_1}{1-\nu_1}\left(\frac{E_b\beta_b}{1-\nu_b}\right)^{-1},$$

on the coating/substrate interface.

As for the temperature field, if $\mu \geqslant 500\ \text{W}/(\text{m}^2 \cdot \text{K})$ (Bi $\geqslant$ 29.41), the radiative component has a negligible effect on the thermal stress state of the system, which is particularly noticeable in the character of the distribution of compressive stresses in the depth of the half-space, and somewhat less—on their values in the coating layers (Fig. 27.5).

27.7 Conclusions

Based on generalized nonlinear boundary conditions, the method of quasi-linearization, and the integral Laplace transform, an effective approach has been developed for determining the thermal stress state of a half-space with a multilayer thin coating during radiative–convective heat exchange with the environment. The cooling and heating processes of the system are combined and described by unified relations. For the half-space/two-layer wear-resistant coating assembly, the effect of the radiative–convective heat exchange parameters on the nonstationary temperature field and the thermal stress state caused by it has been investigated. The range of variation in the heat transfer coefficient, in which the influence of the radiative component upon the thermal stressed state of the system is non-essential, has been revealed.

References

Abdalkhani J (1992) The nonlinear cooling of a semi-infinite solid—implicit Runge-Kutta (IRK) methods. Appl Math Comput 52:233–237

An C, Su J (2011) Improved lumped models for transient combined convective and radiative cooling of multi-layer composite slabs. Appl Therm Eng 31:2508–2517

Aziz A, Na TY (1984) Perturbation method in heat transfer. Hemisphere Publishing Corporation, Washington DC

Bellman RE, Kalaba RE (1965) Quasilinearization and nonlinear boundary-value problems. American Elsevier Publishing Company, New York

Bervillier C (2012) Status of the differential transformation method. Appl Math Comput 218:10158–10170

Campo A (1977) A quasilinearization approach for the transient response of bodies with surface radiation. Lett Heat Mass Transf 4:291–298

Choudhury BK (2006) Transient temperature and elastic response of a space-based mirror in the radiation-conduction environment. Quart Appl Math 64:201–228

Fakoor-Pakdaman M, Ahmadi M, Bagheri F, Bahrami M (2015) Optimal time-varying heat transfer in multilayered packages with arbitrary heat generations and contact resistance. J Heat Transf 137:081401

Fetni S, Enrici TM, Niccolini T, Tran HS, Dedry O, Duchêne L, Mertens A, Habraken AM (2021) Thermal model for the directed energy deposition of composite coatings of 316L stainless steel enriched with tungsten carbides. Mater Des 204:109661

Ganji DD (2006) The application of He's homotopy perturbation method to nonlinear equations arising in heat transfer. Phys Lett A 355:337–341

Guo HF, Tian ZJ, Huang YH (2016) Laser surface remelting of WC-12Co coating: finite element simulations and experimental analyses. Mater Sci Technol 32:813–822

Hogan RE, Gartling DK (2008) Solution strategies for coupled conduction/radiation problems. Commun Numer Meth Eng 24:523–542

Jendrzejewski R, Sliwinski G, Krawczuk M, Ostachowicz W (2006) Temperature and stress during laser cladding of double-layer coatings. Surf Coat Technol 201:3328–3334

Jiguang C, Fei G (2015) Temperature field and thermal stress analyses of high-speed train brake disc under pad variations. Open Mech Eng J 9:371–378

Kushnir RM, Popovych VS (2011) Heat conduction problems of thermosensitive solids under complex heat exchange. In: Vikhrenko V (ed) Heat conduction–basic research. In Tech, Croatia, pp 131–154

Kushnir RM, Protsyuk BV, Synyuta VM (2002) Temperature stresses and displacements in a multilayer plate with nonlinear conditions of heat exchange. Mater Sci 38:798–808

Liao S (2003) Beyond perturbation: Introduction to the homotopy analysis method. Chapman & Hall/CRC Press, Boca Raton

Liao S, Su J, Chwang AT (2006) Series solutions for a nonlinear model of combined convective and radiative cooling of a spherical body. Int J Heat Mass Transf 49:2437–2445

Luikov AV (1968) Analytical heat diffusion theory. Academic Press, New York, NY

Oguntala G, Sobamowo G, Ahmed Y, Abd-Alhameed R (2019) Thermal prediction of convective-radiative porous fin heatsink of functionally graded material using Adomian decomposition method. Comput 7:19

Onur N, Sivrioglu M (1993) Transient heat conduction with uniform heat generation in a slab subjected to convection and radiation cooling. Heat Mass Transf 28:345–349

Popovych VS, Vovk OM, Harmatii HY (2012) Investigation of the static thermoelastic state of a thermosensitive hollow cylinder under convective-radiant heat exchange with environment. J Math Sci 187:726–736

Protsyuk BV (2021) Nonstationary nonlinear problems of heat conduction for a half space. J Math Sci 256:551–566

Protsyuk BV, Horun OP (2017) Quasistatic thermoelastic state of a heat-sensitive three-component layer under the conditions of convective-radiative heat exchange. J Math Sci 223:117–131

Sayman O, Sen F, Celik E, Arman Y (2009) Thermal stress analysis of Wc-Co/Cr-Ni multilayer coatings on 316L steel substrate during cooling process. Mater Des 30:770–774

Shevchuk VA (2012) Nonstationary one-dimensional problem of heat conduction for a cylinder with a thin multilayer coating. J Math Sci 184:215–223

Shevchuk VA (2013) Analytical solution of nonstationary heat conduction problem for a half-space with a multilayer coating. J Eng Phys Thermophys 86:450–459

Shevchuk VA (2014) Generalized boundary conditions to solving thermal stress problems for bodies with thin coatings. In: Hetnarski RB (ed) Encyclopedia of thermal stresses, vol 4. Springer, Dordrecht, pp 1942–1953

Shevchuk VA (2017) Thermoelasticity problem for a multilayer coating/half-space assembly with undercoat crack subjected to convective thermal loading. J Therm Stress 40:1215–1230

Shevchuk VA (2022) Generalized boundary conditions of radiant-convection heat exchange of bodies with ambient medium through multilayer nonplanar coating. J Math Sci 261:95–114

Shevchuk VA, Gavris AP (2020) Nonstationary heat-conduction problem for a half-space with a multilayer coating upon cyclic change in the ambient temperature. J Eng Phys Thermophys 93:1489–1497

Shevchuk VA, Havrys' OP (2017) Choice of the iterative method for the solution of nonlinear nonstationary problem of heat conduction for a half space in the course of radiative cooling. J Math Sci 220:226–234

Sundén B (1989) Transient conduction in a cylindrical shell with a time-varying incident surface heat flux and convective and radiative surface cooling. Int J Heat Mass Transf 32:575–584

Tan Z, Su G, Su J (2009) Improved lumped models for combined convective and radiative cooling of a wall. Appl Therm Eng 29:2439–2443

Terlets'kyi RF, Turii OP (2013) Modeling and investigation of heat transfer in plates with thin coatings with regard for the influence of radiation. J Math Sci 192:703–722

Torabi M, Yaghoobi H, Saedodin S (2011) Assessment of homotopy perturbation method in nonlinear convective-radiative non-Fourier conduction heat transfer equation with variable coefficient. Therm Sci 15(2):263–274

Villasenor R (1994) A comparative study between an integral equation approach and a finite difference formulation for heat diffusion with nonlinear boundary conditions. Appl Math Model 18:321–327

Yaghoobi H, Torabi M (2011) The application of differential transformation method to nonlinear equations arising in heat transfer. Int Commun Heat Mass Transf 38:815–820

Zhao X, Iyer A, Promoppatum P, Yao S-C (2017) Numerical modeling of the thermal behavior and residual stress in the direct metal laser sintering process of titanium alloy products. Addit Manuf 14:126–136

Chapter 28
Model of Nonlinear Deformation of Layered-Fibrous Materials with Physically Nonlinear Components

Elena Shikula and Viktor Vyshnivskyi

28.1 Introduction

When the load increases, many homogeneous and composite materials exhibit a nonlinear nature of the dependences between macrostresses and macrostrains. It may be due to the physical nonlinearity of component deformation (Kauderer 1958). This type of nonlinearity is typical for composites with plastic metal matrix or polymer matrix at elevated temperatures. Experimental studies show (Wolf and Romadin 1967) that high-modulus materials such as organic glass fibers exhibit nonlinearity at sufficiently high temperatures. Figure 28.1 shows the experimental dependence of stress on strain for organic glass at some temperatures. As we can see, the relations between stress and strain have a parabolic character at sufficiently high temperatures. Therefore, studying the physically nonlinear deformation of composite materials with nonlinear deformation of both matrix and fibers is essential.

Many authors studied nonlinear fibrous and layered-fibrous composite materials theoretically and experimentally. For example, the effective properties of materials within the elastic range were studied by Choy et al. (1992), Geiger and Weich (1997) to determine their ultimate load-bearing capacity. Damage models of fibrous composites were built by Voyiadjis and Kattan (2006). The effective properties of fibrous composite materials were determined by Chen et al. (2007), Duan et al. (2007), Gray (2009), Hashin (1990), Hashin (2002), Liu et al. (2004), Khoroshun et al. (2013), Nazarenko and Stolarski (2016), Nazarenko et al. (2017a, b, 2018a, b, 2019), Maslov (2022) and others. However, the authors considered materials with a nonlinear or elastoplastic matrix but fibers were assumed to be elastic. The deforma-

E. Shikula (✉)
S.P. Timoshenko Institute of Mechanics, National Academy of Science of Ukraine, Kyiv, Ukraine
e-mail: ensh@ukr.net

V. Vyshnivskyi
State University of Telecommunications, Kyiv, Ukraine
e-mail: vish_vv@ukr.net

A. N. Guz et al. (eds.), *Advances in Mechanics*, Advanced Structured Materials 191,
https://doi.org/10.1007/978-3-031-37313-8_28

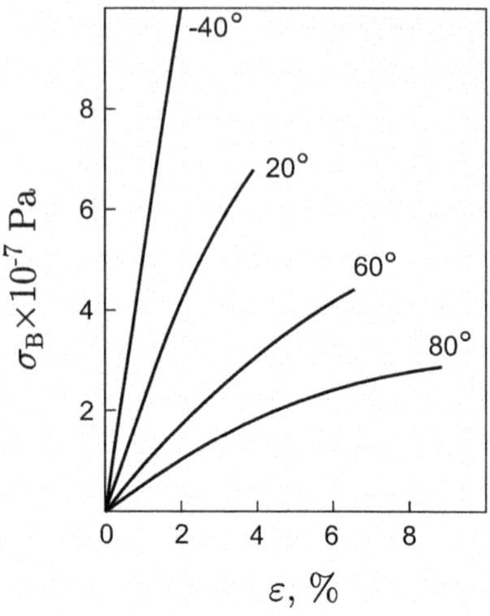

Fig. 28.1 Experimental dependence of stress on strain for organic glass for some temperatures

tion of fibrous and layered-fibrous composites for the case of nonlinear deformation of both matrix and fibers has yet to be investigated. For fibrous materials, this problem was solved by Khoroshun and Shikula (2016). Therefore, constructing a model of nonlinear deformation of layered-fibrous composite materials with a nonlinearly deformable matrix and fibers is hugely important.

28.2 Nonlinear Equations for Determination of the Effective Deformative Properties and Stress-Strain State of the Composite

Let us consider a layered-fibrous composite material of oblique winding. It is a layered material, in which each layer is a unidirectional fibrous composite material. Two types of layers with different orientation angles are possible, while the volume content of layers with different fiber orientations is the same. In addition, the elastic characteristics and volume contents of fibers and matrix are the same in layers with different fiber orientations. Finally, we consider a composite with isotropic, physically nonlinear fibers and a matrix.

We choose a global coordinate system for the layered-fibrous material so that the fibers in the layers lie in the plane x_1x_2, and the x_3-axis is normal to the planes of all layers. The angle between the direction of the fibers in the layer and the x_1-axis equals γ. Two orientation angles are possible for layered-fibrous materials with oblique winding: $\gamma_1 = \gamma$ and $\gamma_2 = -\gamma$. Let us choose the local coordinate system of the layer so that the axis x_1' coincides with the direction of the fibers in this layer, and the axis x_3' coincides with the axis x_3, which is normal to the planes of all layers.

Determining the stress-strain state and effective deformative characteristics of layered-fibrous materials is reduced to two consecutive problems. The first one is the determination of the stress-strain state of the fibers ($\langle {\sigma_{pq}}^1 \rangle$ and $\langle {\varepsilon_{pq}}^1 \rangle$) and the matrix ($\langle {\sigma_{pq}}^2 \rangle$ and $\langle {\varepsilon_{pq}}^2 \rangle$) of the layer, as well as the effective properties λ_{11}^*, λ_{22}^*, λ_{12}^*, λ_{23}^*, and λ_{55}^* of the unidirectional fibrous material of the layer under given deformations of the layer $\langle \varepsilon_{pq} \rangle$. The second problem is to determine the stress-strain state of the layer ($\langle \sigma_{pq} \rangle$ and $\langle \varepsilon_{pq} \rangle$) and the effective properties of the layered-fibrous material l_{11}^*, l_{22}^*, l_{33}^*, l_{12}^*, l_{13}^*, l_{23}^*, l_{44}^*, l_{55}^*, and l_{66}^* under given macrostrains of the composite $\langle\langle \varepsilon_{pq} \rangle\rangle$.

Consider the unidirectional fibrous material of the layer. We denote the volume contents of fibers and matrix by c_1 and c_2; K_1 and μ_1 are the bulk and shear moduli of the fibers, respectively, K_2 and μ_2—of the matrix. Assume that the volume deformations of the components are linear, i.e., the bulk moduli K_1 and K_2 do not depend on strains. Nonlinear diagrams describe the shear deformations, that is, $\mu_1(\langle \varepsilon_1 \rangle)$ and $\mu_2(\langle \varepsilon_2 \rangle)$. Then the dependence between the average stresses $\langle \sigma_{pq} \rangle_l$ and strains $\langle \varepsilon_{pq} \rangle_l$ in the local coordinate system of the layer can be written in the form

$$\begin{aligned}
\langle \sigma_{11} \rangle_l &= \lambda_{11}^* \langle \varepsilon_{11} \rangle_l + \lambda_{12}^* \langle \varepsilon_{rr} \rangle_l, \\
\langle \sigma_{km} \rangle_l &= (\lambda_{22}^* - \lambda_{23}^*) \langle \varepsilon_{km} \rangle_l + (\lambda_{12}^* \langle \varepsilon_{11} \rangle_l + \lambda_{23}^* \langle \varepsilon_{rr} \rangle_l)\, \delta_{km}, \\
\langle \sigma_{1k} \rangle_l &= 2\lambda_{55}^* \langle \varepsilon_{1k} \rangle_l \quad (k, m, r = 2, 3),
\end{aligned} \tag{28.1}$$

where the effective deformative characteristics of the layer λ_{11}^*, λ_{22}^*, λ_{12}^*, λ_{23}^*, and λ_{55}^* depend on the strains in the layer $\langle \varepsilon_{pq} \rangle_l$. In the local coordinate system of this layer, the effective characteristics are determined by the bulk and shear moduli of the fibers and the matrix of this layer as follows (Khoroshun and Shikula 1996, 2002, 2012)

$$\begin{aligned}
\lambda_{11}^* &= c_1(\lambda_1 + 2\mu_1) + c_2(\lambda_2 + 2\mu_2) \\
&\quad - \frac{c_1 c_2 (\lambda_1 - \lambda_2)^2}{c_1(\lambda_2 + \mu_2) + c_2(\lambda_1 + \mu_1) + m}, \\
\lambda_{12}^* &= c_1\lambda_1 + c_2\lambda_2 - \frac{c_1 c_2\, (\lambda_1 + \mu_1 - \lambda_2 - \mu_2)(\lambda_1 - \lambda_2)}{c_1(\lambda_2 + \mu_2) + c_2(\lambda_1 + \mu_1) + m}, \\
\lambda_{22}^* + \lambda_{23}^* &= 2c_1(\lambda_1 + \mu_1) + 2c_2(\lambda_2 + \mu_2) \\
&\quad - \frac{2c_1 c_2\, (\lambda_1 + \mu_1 - \lambda_2 - \mu_2)^2}{c_1(\lambda_2 + \mu_2) + c_2(\lambda_1 + \mu_1) + m}, \\
\lambda_{22}^* - \lambda_{23}^* &= 2c_1\mu_1 + 2c_2\mu_2 - \frac{2c_1 c_2 (\mu_1 - \mu_2)^2}{c_1\mu_2 + c_2\mu_1 + \dfrac{mn}{n + 2m}}, \\
\lambda_{55}^* &= c_1\mu_1 + c_2\mu_2 - \frac{c_1 c_2 (\mu_1 - \mu_2)^2}{c_1\mu_2 + c_2\mu_1 + m},
\end{aligned} \tag{28.2}$$

where

$$\lambda_\chi = K_\chi - \tfrac{2}{3}\mu_\chi \quad (\chi = 1,\, 2),$$

$$m = c_1\mu_1 + c_2\mu_2; \quad n = c_1(\lambda_1 + \mu_1) + c_2(\lambda_2 + \mu_2) \tag{28.3}$$

if the stiffness of the matrix is greater than the stiffness of the fibers in this layer, and

$$m = \left(\frac{c_1}{\mu_1} + \frac{c_2}{\mu_2}\right)^{-1}, \quad n = \left(\frac{c_1}{\lambda_1 + \mu_1} + \frac{c_2}{\lambda_2 + \mu_2}\right)^{-1} \tag{28.4}$$

otherwise.

In the local coordinate system, for the layer, the average strains in the fibers $\langle \varepsilon_{pq}{}^1 \rangle$ and in the matrix $\langle \varepsilon_{pq}{}^2 \rangle$ are related to average strains in this layer $\langle \varepsilon_{pq} \rangle_l$ by the following dependence (Khoroshun and Shikula 2002, 2008, 2012):

$$\begin{aligned}
\langle \varepsilon_{11}^1 \rangle &= \frac{1}{\Delta_1}\{[(\lambda_{11}^* - \lambda_2 - 2\mu_2)a_1 - 2(\lambda_{12}^* - \lambda_2)a_2]\langle \varepsilon_{11} \rangle_l \\
&\quad + [(\lambda_{12}^* - \lambda_2)a_1 - (\lambda_{22}^* + \lambda_{23}^* - 2\lambda_2 - 2\mu_2)a_2] \\
&\quad \times (\langle \varepsilon_{22} \rangle_l + \langle \varepsilon_{33} \rangle_l)\}, \\
\langle \varepsilon_{pq}^1 \rangle &= \frac{\lambda_{22}^* - \lambda_{23}^* - 2\mu_2}{2c_1(\mu_1 - \mu_2)}\langle \varepsilon_{pq} \rangle_l \\
&\quad + \frac{1}{\Delta_1}\{[(\lambda_{12}^* - \lambda_2)(a_1 - a_2) - (\lambda_{11}^* - \lambda_2 - 2\mu_2)a_2]\langle \varepsilon_{11} \rangle_l \\
&\quad + [-(\lambda_{22}^* - \lambda_2 - 2\mu_2)a_2 + (\lambda_{23}^* - \lambda_2)a_1 - (\lambda_{12}^* - \lambda_2)a_2] \\
&\quad \times (\langle \varepsilon_{22} \rangle_l + \langle \varepsilon_{33} \rangle_l)\}, \\
\langle \varepsilon_{1q}^1 \rangle &= \frac{\lambda_{55}^* - \mu_2}{c_1(\mu_1 - \mu_2)}\langle \varepsilon_{1q} \rangle_l \quad (p, q = 2, 3),
\end{aligned} \tag{28.5}$$

$$\begin{aligned}
\langle \varepsilon_{11}^2 \rangle &= \frac{1}{\Delta_2}\{[(\lambda_{11}^* - \lambda_1 - 2\mu_1)a_1 - 2(\lambda_{12}^* - \lambda_1)a_2]\langle \varepsilon_{11} \rangle_l \\
&\quad + [(\lambda_{12}^* - \lambda_1)a_1 - (\lambda_{22}^* + \lambda_{23}^* - 2\lambda_1 - 2\mu_1)a_2] \\
&\quad \times (\langle \varepsilon_{22} \rangle_l + \langle \varepsilon_{33} \rangle_l)\}, \\
\langle \varepsilon_{pq}^2 \rangle &= \frac{\lambda_{22}^* - \lambda_{23}^* - 2\mu_1}{2c_2(\mu_2 - \mu_1)}\langle \varepsilon_{pq} \rangle_l \\
&\quad + \frac{1}{\Delta_2}\{[(\lambda_{12}^* - \lambda_1)(a_1 - a_2) - (\lambda_{11}^* - \lambda_1 - 2\mu_1)a_2]\langle \varepsilon_{11} \rangle_l \\
&\quad + [-(\lambda_{22}^* - \lambda_1 - 2\mu_1)a_2 + (\lambda_{23}^* - \lambda_1)a_1 - (\lambda_{12}^* - \lambda_1)a_2] \\
&\quad \times (\langle \varepsilon_{22} \rangle_l + \langle \varepsilon_{33} \rangle_l)\}, \\
\langle \varepsilon_{1q}^2 \rangle &= \frac{\lambda_{55}^* - \mu_1}{c_2(\mu_2 - \mu_1)}\langle \varepsilon_{1q} \rangle_l \quad (p, q = 2, 3),
\end{aligned} \tag{28.6}$$

where

$$\begin{aligned}
\Delta_1 &= 2c_1(\mu_1 - \mu_2)(3\lambda_1 + 2\mu_1 - 3\lambda_2 - 2\mu_2), \\
\Delta_2 &= 2c_2(\mu_2 - \mu_1)(3\lambda_1 + 2\mu_1 - 3\lambda_2 - 2\mu_2), \\
a_1 &= 2(\lambda_1 + \mu_1 - \lambda_2 - \mu_2), \quad a_2 = (\lambda_1 - \lambda_2).
\end{aligned} \tag{28.7}$$

Average strains in the layer in the local $\langle \varepsilon_{pq} \rangle_l$ and global $\langle \varepsilon_{pq} \rangle$ coordinate systems of this layer are related as follows

$$\begin{aligned}
\langle \varepsilon_{11} \rangle_l &= \langle \varepsilon_{11} \rangle \cos^2 \gamma + \langle \varepsilon_{22} \rangle \sin^2 \gamma + \langle \varepsilon_{12} \rangle \sin 2\gamma, \\
\langle \varepsilon_{22} \rangle_l &= \langle \varepsilon_{11} \rangle \sin^2 \gamma + \langle \varepsilon_{22} \rangle \cos^2 \gamma - \langle \varepsilon_{12} \rangle \sin 2\gamma, \\
\langle \varepsilon_{12} \rangle_l &= \langle \varepsilon_{12} \rangle \cos 2\gamma + \tfrac{1}{2}(\langle \varepsilon_{22} \rangle - \langle \varepsilon_{11} \rangle) \sin 2\gamma, \\
\langle \varepsilon_{33} \rangle_l &= \langle \varepsilon_{33} \rangle, \\
\langle \varepsilon_{13} \rangle_l &= \langle \varepsilon_{13} \rangle \cos \gamma + \langle \varepsilon_{23} \rangle \sin \gamma, \\
\langle \varepsilon_{23} \rangle_l &= -\langle \varepsilon_{13} \rangle \sin \gamma + \langle \varepsilon_{23} \rangle \cos \gamma.
\end{aligned} \tag{28.8}$$

Consider the layered-fibrous material of oblique winding. We assume that the macrostrains in the composite $\langle\langle \varepsilon_{pq} \rangle\rangle$ are given. Then the macrostresses $\langle\langle \sigma_{pq} \rangle\rangle$ are depend on them by formulas

$$\begin{aligned}
\langle\langle \sigma_{11} \rangle\rangle &= l^*_{11} \langle\langle \varepsilon_{11} \rangle\rangle + l^*_{12} \langle\langle \varepsilon_{22} \rangle\rangle + l^*_{13} \langle\langle \varepsilon_{33} \rangle\rangle, \\
\langle\langle \sigma_{22} \rangle\rangle &= l^*_{12} \langle\langle \varepsilon_{11} \rangle\rangle + l^*_{22} \langle\langle \varepsilon_{22} \rangle\rangle + l^*_{23} \langle\langle \varepsilon_{33} \rangle\rangle, \\
\langle\langle \sigma_{33} \rangle\rangle &= l^*_{31} \langle\langle \varepsilon_{11} \rangle\rangle + l^*_{32} \langle\langle \varepsilon_{22} \rangle\rangle + l^*_{33} \langle\langle \varepsilon_{33} \rangle\rangle, \\
\langle\langle \sigma_{13} \rangle\rangle &= l^*_{44} \langle\langle \varepsilon_{13} \rangle\rangle, \\
\langle\langle \sigma_{23} \rangle\rangle &= l^*_{55} \langle\langle \varepsilon_{23} \rangle\rangle, \\
\langle\langle \sigma_{12} \rangle\rangle &= l^*_{66} \langle\langle \varepsilon_{12} \rangle\rangle,
\end{aligned} \tag{28.9}$$

where the effective deformative characteristics of the composite material l^*_{11}, l^*_{22}, l^*_{33}, l^*_{12}, l^*_{13}, l^*_{23}, l^*_{44}, l^*_{55}, and l^*_{66} are functions of macrostrains $\langle\langle \varepsilon_{pq} \rangle\rangle$.

The effective deformative modulus $l^*_{pq} l^*{}_{pq}$ of the layered-fibrous composite are determined by the formulas (Khoroshun and Shikula 1996, 2002)

$$\begin{aligned}
l^*_{11} &= \lambda^*_{12} + (\lambda^*_{22} - \lambda^*_{12}) \sin^4 \gamma + (\lambda^*_{11} - \lambda^*_{12}) \cos^4 \gamma + \lambda^*_{55} \sin^2 2\gamma, \\
l^*_{22} &= \lambda^*_{12} + (\lambda^*_{22} - \lambda^*_{12}) \cos^4 \gamma + (\lambda^*_{11} - \lambda^*_{12}) \sin^4 \gamma + \lambda^*_{55} \sin^2 2\gamma, \\
l^*_{33} &= \lambda^*_{22}, \quad l^*_{12} = \lambda^*_{12} + \tfrac{1}{4}(\lambda^*_{11} + \lambda^*_{22} - 2\lambda^*_{12} - 4\lambda^*_{55}) \sin^2 2\gamma, \\
l^*_{13} &= \lambda^*_{12} + \lambda^*_{23} \sin^2 \gamma, \quad l^*_{23} = \lambda^*_{12} + \lambda^*_{23} \cos^2 \gamma, \\
l^*_{66} &= l_{66} - l^2_{36}/l^*_{33}, \quad l^*_{44} = l_{44} - l^2_{45}/l_{55}, \quad l^*_{55} = l_{55} - l^2_{45}/l_{44},
\end{aligned} \tag{28.10}$$

where

$$\begin{aligned}
l_{36} &= \tfrac{1}{2}(\lambda^*_{12} - \lambda^*_{23}) \sin 2\gamma, \\
l_{66} &= \lambda^*_{55} + \tfrac{1}{4}(\lambda^*_{11} + \lambda^*_{22} - 2\lambda^*_{12} - 4\lambda^*_{55}) \sin^2 2\gamma, \\
l_{44} &= \tfrac{1}{2}(\lambda^*_{22} - \lambda^*_{23}) \sin^2 \gamma + \lambda^*_{55} \cos^2 \gamma, \quad l_{45} = \tfrac{1}{4}(2\lambda^*_{55} - \lambda^*_{22} + \lambda^*_{23}) \sin 2\gamma, \\
l_{55} &= \tfrac{1}{2}(\lambda^*_{22} - \lambda^*_{23}) \cos^2 \gamma + \lambda^*_{55} \sin^2 \gamma.
\end{aligned} \tag{28.11}$$

In the global coordinate system, the average strains in the layer $\langle \varepsilon_{pq} \rangle$ are related to macrostrains $\langle\langle \varepsilon_{pq} \rangle\rangle$ by the following relations

$$
\begin{aligned}
\langle \varepsilon_{km} \rangle &= \langle\!\langle \varepsilon_{km} \rangle\!\rangle \quad (k, m = 1, 2), \\
\langle \varepsilon_{33} \rangle &= \langle\!\langle \varepsilon_{33} \rangle\!\rangle \mp \frac{2l_{36}}{l_{33}^{*}} \langle\!\langle \varepsilon_{12} \rangle\!\rangle, \\
\langle \varepsilon_{13} \rangle &= \langle\!\langle \varepsilon_{13} \rangle\!\rangle \mp \frac{2l_{45}}{l_{55}} \langle\!\langle \varepsilon_{23} \rangle\!\rangle, \\
\langle \varepsilon_{23} \rangle &= \langle\!\langle \varepsilon_{23} \rangle\!\rangle \mp \frac{2l_{45}}{l_{44}} \langle\!\langle \varepsilon_{13} \rangle\!\rangle.
\end{aligned}
\tag{28.12}
$$

Here the sign "–" refers to layers with fiber orientation γ, and the sign "+"—to $-\gamma$.

Equations (28.2)–(28.8) and (28.10)–(28.12) are the system of nonlinear equations for determining the effective deformative characteristics of the composite and the average strains in the fibers and the matrix of the layer.

28.3 Algorithm for Determination of the Effective Deformative Properties and Stress-Strain State of the Composite

The solution of the system of nonlinear Eqs. (28.2)–(28.8), (28.10)–(28.12) can be found by the iterative method according to the following scheme. We assume that the volume strains and stresses in the fibers and the matrix are linearly related, i.e., their bulk moduli

$$K_\chi = \lambda_\chi + \tfrac{2}{3}\mu_\chi \quad (\chi = 1, 2)$$

do not depend on the strains, and the nonlinear law

$$\langle \sigma_{pq}^{\chi} \rangle = 2\mu_\chi (J_\varepsilon^{\chi}) \langle \varepsilon_{pq}^{\chi} \rangle, \quad J_\varepsilon^{\chi} = \left(\langle \varepsilon_{pq}^{\chi} \rangle' \langle \varepsilon_{pq}^{\chi} \rangle' \right)^{12} \quad (\chi = 1, 2) \tag{28.13}$$

relates the deviators of stresses and strains. At the same time, we assume that the nonlinear deformation diagrams of the fibers and the matrix under small deformations have linear regions corresponding to the shear moduli $\mu_1(0)$ and $\mu_2(0)$, respectively.

In the local coordinate system of the layer, the effective deformative characteristics of the fibrous composite that forms the layer are determined in the nth approximation by the following formulas

$$
\begin{aligned}
\lambda_{11}^{*(n)} &= c_1 \left[\lambda_1 \left(J_\varepsilon^{1(n)} \right) + 2\mu_1 \left(J_\varepsilon^{1(n)} \right) \right] + c_2 \left[\lambda_2 \left(J_\varepsilon^{2(n)} \right) + 2\mu_2 \left(J_\varepsilon^{2(n)} \right) \right] \\
&\quad - \frac{c_1 c_2 \left[\lambda_1 \left(J_\varepsilon^{1(n)} \right) - \lambda_2 \left(J_\varepsilon^{2(n)} \right) \right]^2}{c_1 \left[\lambda_2 \left(J_\varepsilon^{2(n)} \right) + \mu_2 \left(J_\varepsilon^{2(n)} \right) \right] + c_2 \left[\lambda_1 \left(J_\varepsilon^{1(n)} \right) + \mu_1 \left(J_\varepsilon^{1(n)} \right) \right] + m},
\end{aligned}
$$

$$
\begin{aligned}
\lambda_{12}^{*(n)} &= c_1 \lambda_1 \left(J_\varepsilon^{1(n)} \right) + c_2 \lambda_2 \left(J_\varepsilon^{2(n)} \right) \\
&\quad - c_1 c_2 \left[\lambda_1 \left(J_\varepsilon^{1(n)} \right) + \mu_1 \left(J_\varepsilon^{1(n)} \right) - \lambda_2 \left(J_\varepsilon^{2(n)} \right) - \mu_2 \left(J_\varepsilon^{2(n)} \right) \right]
\end{aligned}
$$

$$\times \frac{\lambda_1\left(J_\varepsilon^{1(n)}\right)-\lambda_2\left(J_\varepsilon^{2(n)}\right)}{c_1\left[\lambda_2\left(J_\varepsilon^{2(n)}\right)+\mu_2\left(J_\varepsilon^{2(n)}\right)\right]+c_2\left[\lambda_1\left(J_\varepsilon^{1(n)}\right)+\mu_1\left(J_\varepsilon^{1(n)}\right)\right]+m},$$

$$\lambda_{22}^{*(n)}+\lambda_{23}^{*(n)}=2c_1\left[\lambda_1\left(J_\varepsilon^{1(n)}\right)+\mu_1\left(J_\varepsilon^{1(n)}\right)\right]+2c_2\left[\lambda_2\left(J_\varepsilon^{2(n)}\right)+\mu_2\left(J_\varepsilon^{2(n)}\right)\right]$$
$$-\frac{2c_1c_2\left[\lambda_1\left(J_\varepsilon^{1(n)}\right)+\mu_1\left(J_\varepsilon^{1(n)}\right)-\lambda_2\left(J_\varepsilon^{2(n)}\right)-\mu_2\left(J_\varepsilon^{2(n)}\right)\right]^2}{c_1\left[\lambda_2\left(J_\varepsilon^{2(n)}\right)+\mu_2\left(J_\varepsilon^{2(n)}\right)\right]+c_2\left[\lambda_1\left(J_\varepsilon^{1(n)}\right)+\mu_1\left(J_\varepsilon^{1(n)}\right)\right]+m},$$

$$\lambda_{22}^{*(n)}-\lambda_{23}^{*(n)}=2c_1\mu_1\left(J_\varepsilon^{1(n)}\right)+2c_2\mu_2\left(J_\varepsilon^{2(n)}\right)$$
$$-\frac{2c_1c_2\left[\mu_1\left(J_\varepsilon^{1(n)}\right)-\mu_2\left(J_\varepsilon^{2(n)}\right)\right]^2}{c_1\mu_2\left(J_\varepsilon^{2(n)}\right)+c_2\mu_1\left(J_\varepsilon^{1(n)}\right)+\dfrac{mn}{n+2m}},$$

$$\lambda_{55}^{*(n)}=c_1\mu_1\left(J_\varepsilon^{1(n)}\right)+c_2\mu_2\left(J_\varepsilon^{2(n)}\right)-\frac{c_1c_2\left[\mu_1\left(J_\varepsilon^{1(n)}\right)-\mu_2\left(J_\varepsilon^{2(n)}\right)\right]^2}{c_1\mu_2\left(J_\varepsilon^{2(n)}\right)+c_2\mu_1\left(J_\varepsilon^{1(n)}\right)+m},\tag{28.14}$$

$$\begin{gathered}\lambda_\chi\left(J_\varepsilon^{\chi(n)}\right)=K_\chi-\tfrac{2}{3}\mu_\chi\left(J_\varepsilon^{\chi(n)}\right)\quad(\chi=1,2),\\ m=c_1\mu_1(0)+c_2\mu_2(0),\quad n=c_1[\lambda_1(0)+\mu_1(0)]+c_2[\lambda_2(0)+\mu_2(0)],\end{gathered}\tag{28.15}$$

where if the stiffness of the matrix is greater than the stiffness of the fibers in this layer

$$m=\left[\frac{c_1}{\mu_1(0)}+\frac{c_2}{\mu_2(0)}\right]^{-1},\quad n=\left[\frac{c_1}{\lambda_1(0)+\mu_1(0)}+\frac{c_2}{\lambda_2(0)+\mu_2(0)}\right]^{-1}.\tag{28.16}$$

otherwise

In the global coordinate system, for the nth approximation, the effective deformative characteristics of the layered-fibrous composite material of oblique winding are related to the effective characteristics of the fibrous composite (28.14)–(28.16) as follows

$$l_{11}^{*(n)}=\lambda_{12}^{*(n)}+\left(\lambda_{22}^{*(n)}-\lambda_{12}^{*(n)}\right)\sin^4\gamma$$
$$+\left(\lambda_{11}^{*(n)}-\lambda_{12}^{*(n)}\right)\cos^4\gamma+\lambda_{55}^{*(n)}\sin^2 2\gamma,$$
$$l_{12}^{*(n)}=\lambda_{12}^{*(n)}+\tfrac{1}{4}\left(\lambda_{11}^{*(n)}+\lambda_{22}^{*(n)}-2\lambda_{12}^{*(n)}-4\lambda_{55}^{*(n)}\right)\sin^2 2\gamma,$$

$$
\begin{aligned}
l_{13}^{*(n)} &= \lambda_{12}^{*(n)} + \lambda_{23}^{*(n)} \sin^2 \gamma, \\
l_{23}^{*(n)} &= \lambda_{12}^{*(n)} + \lambda_{23}^{*(n)} \cos^2 \gamma, \\
l_{22}^{*(n)} &= \lambda_{12}^{*(n)} + \left(\lambda_{22}^{*(n)} - \lambda_{12}^{*(n)}\right) \cos^4 \gamma \\
&\quad + \left(\lambda_{11}^{*(n)} - \lambda_{12}^{*(n)}\right) \sin^4 \gamma + \lambda_{55}^{*(n)} \sin^2 2\gamma, \\
l_{33}^{*(n)} &= \lambda_{22}^{*(n)}, \\
l_{66}^{*(n)} &= l_{66}^{(n)} - \left[l_{36}^{(n)}\right]^2 / l_{33}^{*(n)}, \\
l_{44}^{*(n)} &= l_{44}^{(n)} - \left[l_{45}^{(n)}\right]^2 / l_{55}^{*(n)}, \\
l_{55}^{*(n)} &= l_{55}^{(n)} - \left[l_{45}^{(n)}\right]^2 / l_{44}^{(n)},
\end{aligned}
\tag{28.17}
$$

where

$$
\begin{aligned}
l_{36}^{(n)} &= \tfrac{1}{2}\left(\lambda_{12}^{*(n)} - \lambda_{23}^{*(n)}\right) \sin 2\gamma, \\
l_{66}^{(n)} &= \lambda_{55}^{*(n)} + \tfrac{1}{4}\left(\lambda_{11}^{*(n)} + \lambda_{22}^{*(n)} - 2\lambda_{12}^{*(n)} - 4\lambda_{55}^{*(n)}\right) \sin^2 2\gamma, \\
l_{44}^{(n)} &= \tfrac{1}{2}\left(\lambda_{22}^{*(n)} - \lambda_{23}^{*(n)}\right) \sin^2 \gamma + \lambda_{55}^{*(n)} \cos^2 \gamma, \\
l_{45}^{(n)} &= \tfrac{1}{4}\left(2\lambda_{55}^{*(n)} - \lambda_{22}^{*(n)} + \lambda_{23}^{*(n)}\right) \sin 2\gamma, \\
l_{55}^{(n)} &= \tfrac{1}{2}\left(\lambda_{22}^{*(n)} - \lambda_{23}^{*(n)}\right) \cos^2 \gamma + \lambda_{55}^{*(n)} \sin^2 \gamma.
\end{aligned}
\tag{28.18}
$$

For the $(n+1)$th approximation, average strains in layers $\langle \varepsilon_{pq} \rangle^{(n+1)}$ depends on macrostrains $\langle\langle \varepsilon_{pq} \rangle\rangle$ according to the following relations

$$
\begin{aligned}
\langle \varepsilon_{km} \rangle^{(n+1)} &= \langle\langle \varepsilon_{km} \rangle\rangle \quad (k, m = 1, 2), \\
\langle \varepsilon_{33} \rangle^{(n+1)} &= \langle\langle \varepsilon_{33} \rangle\rangle \mp 2 l_{36}^{(n)} / l_{33}^{*(n)} \langle\langle \varepsilon_{12} \rangle\rangle, \\
\langle \varepsilon_{13} \rangle^{(n+1)} &= \langle\langle \varepsilon_{13} \rangle\rangle \mp 2 l_{45}^{(n)} / l_{55}^{(n)} \langle\langle \varepsilon_{23} \rangle\rangle, \\
\langle \varepsilon_{23} \rangle^{(n+1)} &= \langle\langle \varepsilon_{23} \rangle\rangle \mp 2 l_{45}^{(n)} / l_{44}^{(n)} \langle\langle \varepsilon_{13} \rangle\rangle.
\end{aligned}
\tag{28.19}
$$

In the $(n+1)$th approximation, average strains in the layer relative to its local (with subscript l) and the global (without subscript) coordinate systems are related by dependence

$$
\begin{aligned}
\langle \varepsilon_{11} \rangle_l^{(n+1)} &= \langle \varepsilon_{11} \rangle^{(n+1)} \cos^2 \gamma + \langle \varepsilon_{22} \rangle^{(n+1)} \sin^2 \gamma + \langle \varepsilon_{12} \rangle^{(n+1)} \sin 2\gamma, \\
\langle \varepsilon_{22} \rangle_l^{(n+1)} &= \langle \varepsilon_{11} \rangle^{(n+1)} \sin^2 \gamma + \langle \varepsilon_{22} \rangle^{(n+1)} \cos^2 \gamma - \langle \varepsilon_{12} \rangle^{(n+1)} \sin 2\gamma, \\
\langle \varepsilon_{33} \rangle_l^{(n+1)} &= \langle \varepsilon_{33} \rangle^{(n+1)}, \\
\langle \varepsilon_{12} \rangle_l^{(n+1)} &= \langle \varepsilon_{12} \rangle^{(n+1)} \cos 2\gamma + \tfrac{1}{2}\left(\langle \varepsilon_{22} \rangle^{(n+1)} - \langle \varepsilon_{11} \rangle^{(n+1)}\right) \sin 2\gamma, \\
\langle \varepsilon_{13} \rangle_l^{(n+1)} &= \langle \varepsilon_{13} \rangle^{(n+1)} \cos \gamma + \langle \varepsilon_{23} \rangle^{(n+1)} \sin \gamma, \\
\langle \varepsilon_{23} \rangle_l^{(n+1)} &= -\langle \varepsilon_{13} \rangle^{(n+1)} \sin \gamma + \langle \varepsilon_{23} \rangle^{(n+1)} \cos \gamma.
\end{aligned}
\tag{28.20}
$$

For the $(n+1)$th approximation, in the local coordinate system, the average strains in the fibers (with superscript 1) and the matrix (with superscript 2) are related to the average strains in the layer $\langle \varepsilon_{pq} \rangle_l{}^{(n+1)}$ by the following dependence

$$
\begin{aligned}
\langle \varepsilon^1_{11} \rangle^{(n+1)} = {} & \frac{1}{\Delta_1^{(n+1)}} \Big\{ \Big[\Big(\lambda_{11}^{*(n)} - \lambda_2 \left(J_\varepsilon^{2(n)} \right) - 2\mu_2 \left(J_\varepsilon^{2(n)} \right) \Big) a_1^{(n+1)} \\
& - 2 \Big(\lambda_{12}^{*(n)} - \lambda_2 \left(J_\varepsilon^{2(n)} \right) \Big) a_2^{(n+1)} \Big] \langle \varepsilon_{11} \rangle_l^{(n+1)} \\
& + \Big[\Big(\lambda_{12}^{*(n)} - \lambda_2 \left(J_\varepsilon^{2(n)} \right) \Big) a_1^{(n+1)} - \Big(\lambda_{22}^{*(n)} + \lambda_{23}^{*(n)'} - 2\lambda_2 \left(J_\varepsilon^{2(n)} \right) \\
& - 2\mu_2 \left(J_\varepsilon^{2(n)} \right) \Big) a_2^{(n+1)} \Big] \Big(\langle \varepsilon_{22} \rangle_l^{(n+1)} + \langle \varepsilon_{33} \rangle_l^{(n+1)} \Big) \Big\},
\end{aligned}
$$

$$
\begin{aligned}
\langle \varepsilon^1_{pq} \rangle^{(n+1)} = {} & \frac{\lambda_{22}^{*(n)} - \lambda_{23}^{*(n)} - 2\mu_2 \left(J_\varepsilon^{2(n)} \right)}{2c_1 \left[\mu_1 \left(J_\varepsilon^{1(n)} \right) - \mu_2 \left(J_\varepsilon^{2(n)} \right) \right]} \langle \varepsilon_{pq} \rangle_l^{(n+1)} \\
& + \frac{1}{\Delta_1^{(n+1)}} \Big\{ \Big[\Big(\lambda_{12}^{*(n)} - \lambda_2 \left(J_\varepsilon^{2(n)} \right) \Big) \Big(a_1^{(n+1)} - a_2^{(n+1)} \Big) \\
& - \Big(\lambda_{11}^{*(n)} - \lambda_2 \left(J_\varepsilon^{2(n)} \right) - 2\mu_2 \left(J_\varepsilon^{2(n)} \right) \Big) a_2^{(n+1)} \Big] \langle \varepsilon_{11} \rangle_l^{(n+1)} \\
& + \Big[- \Big(\lambda_{22}^{*(n)} - \lambda_2 \left(J_\varepsilon^{2(n)} \right) - 2\mu_2 \left(J_\varepsilon^{2(n)} \right) \Big) a_2^{(n+1)} + \Big(\lambda_{23}^{*(n)} - \lambda_2 \left(J_\varepsilon^{2(n)} \right) \Big) a_1^{(n+1)} \\
& - \Big(\lambda_{12}^{*(n)} - \lambda_2 \left(J_\varepsilon^{2(n)} \right) \Big) a_2^{(n+1)} \Big] \Big(\langle \varepsilon_{22} \rangle_l^{(n+1)} + \langle \varepsilon_{33} \rangle_l^{(n+1)} \Big) \Big\},
\end{aligned}
$$

$$
\langle \varepsilon^1_{1q} \rangle^{(n+1)} = \frac{\lambda_{55}^{*(n)} - \mu_2 \left(J_\varepsilon^{2(n)} \right)}{c_1 \left[\mu_1 \left(J_\varepsilon^{1(n)} \right) - \mu_2 \left(J_\varepsilon^{2(n)} \right) \right]} \langle \varepsilon_{1q} \rangle_l^{(n+1)} \quad (p, q = 2, 3), \tag{28.21}
$$

$$
\begin{aligned}
\langle \varepsilon^2_{11} \rangle^{(n+1)} = {} & \frac{1}{\Delta_2^{(n+1)}} \Big\{ \Big[\Big(\lambda_{11}^{*(n)} - \lambda_1 \left(J_\varepsilon^{1(n)} \right) - 2\mu_1 \left(J_\varepsilon^{1(n)} \right) \Big) a_1^{(n+1)} \\
& -2 \Big(\lambda_{12}^{*(n)} - \lambda_1 \left(J_\varepsilon^{1(n)} \right) \Big) a_2^{(n+1)} \Big] \langle \varepsilon_{11} \rangle_l^{(n+1)} \\
& + \Big[\Big(\lambda_{12}^{*(n)} - \lambda_1 \left(J_\varepsilon^{1(n)} \right) \Big) a_1^{(n+1)} - \Big(\lambda_{22}^{*(n)} + \lambda_{23}^{*(n)} - 2\lambda_1 \left(J_\varepsilon^{1(n)} \right) \\
& - 2\mu_1 \left(J_\varepsilon^{1(n)} \right) \Big) a_2^{(n+1)} \Big] \Big(\langle \varepsilon_{22} \rangle_l^{(n+1)} + \langle \varepsilon_{33} \rangle_l^{(n+1)} \Big) \Big\},
\end{aligned}
$$

$$
\begin{aligned}
\langle \varepsilon^2_{pq} \rangle^{(n+1)} = {} & \frac{\lambda_{22}^{*(n)} - \lambda_{23}^{*(n)} - 2\mu_1 \left(J_\varepsilon^{1(n)} \right)}{2c_2 \left[\mu_2 (J_\varepsilon^{2(n)}) - \mu_1 \left(J_\varepsilon^{1(n)} \right) \right]} \langle \varepsilon_{pq} \rangle_l^{(n+1)} \\
& + \frac{1}{\Delta_2^{(n+1)}} \Big\{ \Big[\Big(\lambda_{12}^{*(n)} - \lambda_1 \left(J_\varepsilon^{1(n)} \right) \Big) \Big(a_1^{(n+1)} - a_2^{(n+1)} \Big)
\end{aligned}
$$

$$
\left.- \left(\lambda_{11}^{*(n)} - \lambda_1\left(J_\varepsilon^{1(n)}\right) - 2\mu_1\left(J_\varepsilon^{1(n)}\right)\right) a_2^{(n+1)}\right] \langle\varepsilon_{11}\rangle_l^{(n+1)}
$$
$$
+ \left[-\left(\lambda_{22}^{*(n)} - \lambda_1\left(J_\varepsilon^{1(n)}\right) - 2\mu_1\left(J_\varepsilon^{1(n)}\right)\right) a_2^{(n+1)} + \left(\lambda_{23}^{*(n)} - \lambda_1\left(J_\varepsilon^{1(n)}\right)\right) a_1^{(n+1)}\right.
$$
$$
\left. - \left(\lambda_{12}^{*(n)} - \lambda_1\left(J_\varepsilon^{1(n)}\right)\right) a_2^{(n+1)}\right]\left(\langle\varepsilon_{22}\rangle_l^{(n+1)} + \langle\varepsilon_{33}\rangle_l^{(n+1)}\right)\Big\},
$$

$$
\langle\varepsilon_{1q}^2\rangle^{(n+1)} = \frac{\lambda_{55}^{*(n)} - \mu_1\left(J_\varepsilon^{1(n)}\right)}{c_2\left[\mu_2\left(J_\varepsilon^{2(n)}\right) - \mu_1\left(J_\varepsilon^{1(n)}\right)\right]}\langle\varepsilon_{1q}\rangle_l^{(n+1)} \quad (p, q = 2, 3), \tag{28.22}
$$

where

$$
\begin{aligned}
\Delta_1^{(n+1)} &= 2c_1[\mu_1(J_\varepsilon^{1(n)}) - \mu_2(J_\varepsilon^{2(n)})] \\
&\quad \times \left[3\lambda_1\left(J_\varepsilon^{1(n)}\right) + 2\mu_1\left(J_\varepsilon^{1(n)}\right) - 3\lambda_2\left(J_\varepsilon^{2(n)}\right) - 2\mu_2\left(J_\varepsilon^{2(n)}\right)\right], \\
\Delta_2^{(n+1)} &= 2c_2\left[\mu_2\left(J_\varepsilon^{2(n)}\right) - \mu_1\left(J_\varepsilon^{1(n)}\right)\right] \\
&\quad \times \left[3\lambda_1\left(J_\varepsilon^{1(n)}\right) + 2\mu_1\left(J_\varepsilon^{1(n)}\right) - 3\lambda_2\left(J_\varepsilon^{2(n)}\right) - 2\mu_2\left(J_\varepsilon^{2(n)}\right)\right], \\
a_1^{(n+1)} &= 2\left[\lambda_1\left(J_\varepsilon^{1(n)}\right) + \mu_1\left(J_\varepsilon^{1(n)}\right) - \lambda_2\left(J_\varepsilon^{2(n)}\right) - \mu_2\left(J_\varepsilon^{2(n)}\right)\right], \\
a_2^{(n+1)} &= \left[\lambda_1\left(J_\varepsilon^{1(n)}\right) - \lambda_2\left(J_\varepsilon^{2(n)}\right)\right].
\end{aligned} \tag{28.23}
$$

The linear deformation of a layered-fibrous composite material is associated with a zero approximation.

28.4 Influence of Component Nonlinearity on the Deformation of the Composite

As an example, we study the nonlinear deformation of a layered-fibrous composite material, in which the bulk moduli of the fibers and the matrix are constant for the layers, and the following functions give the shear moduli

$$
\mu_1\left(J_\varepsilon^1\right) = \begin{cases} \mu_{01}, & J_\varepsilon^1 < \dfrac{k_1}{2\mu_{01}} \\ \mu_{01} - \mu_1'\left(1 - \dfrac{k_1}{2J_\varepsilon^1}\right), & J_\varepsilon^1 \geqslant \dfrac{k_1}{2\mu_{01}} \end{cases}
$$
$$
\mu_2\left(J_\varepsilon^2\right) = \begin{cases} \mu_{02}, & J_\varepsilon^2 < \dfrac{k_2}{2\mu_{02}} \\ \mu_2' + \left(1 - \dfrac{\mu_2'}{\mu_{02}}\right)\dfrac{k_{i2}}{2J_\varepsilon^{i2}}, & J_\varepsilon^2 \geqslant \dfrac{k_2}{2\mu_{02}}, \end{cases} \tag{28.24}
$$

where $\mu_{0\chi}, \mu_\chi', k_\chi = \sigma_{0\chi}\sqrt{2/3}$ are fiber (at $\chi = 1$) and matrix (at $\chi = 2$) constants, $\sigma_{0\chi}$ are their yield strength, $J_\varepsilon^\chi = \left(\langle\varepsilon_{pq}^\chi\rangle'\langle\varepsilon_{pq}^\chi\rangle'\right)^{1/2}$, $\langle\varepsilon_{pq}^\chi\rangle'$ are the deviator of the average strains in fibers (at $\chi = 1$) and matrix (at $\chi = 2$).

For numerical solutions, diagrams of nonlinear deformation (28.24) for glass fibers with the following constants (Khoroshun and Shikula 2016; Kregers 1988; Mikheev et al. 2002; Wolf and Romadin 1967) are used

$$K_1 = 27.78 \text{ GPa}, \quad \mu_{01} = 20.83 \text{ GPa},$$
$$\mu_1' = 0.184 \text{ GPa}, \quad \sigma_{01} = 1.8 \text{ GPa};$$

volume contents $c_1 = \{0, 0.2, 0.4, 0.6, 1.0\}$ are chosen for the solutions; the epoxy matrix has a diagram of linear strengthening (28.24) with the following constants (Khoroshun and Shikula 2016; Kregers 1988; Mikheev et al. 2002; Wolf and Romadin 1967)

$$K_2 = 3.33 \text{ GPa}, \ \mu_{02} = 1.11 \text{ GPa},$$
$$\mu_2' = 0.02 \text{ GPa}, \ \sigma_{02} = 0.12 \text{ GPa}.$$

Based on the obtained dependence, the effective diagrams of nonlinear deformation of the layered-fibrous composite material at specified volume concentrations of the components in the layers and some values of the winding angle are studied.

In the case when the uniaxial tension of the composite is given by

$$\langle\langle \varepsilon_{11} \rangle\rangle \neq 0; \quad \langle\langle \sigma_{22} \rangle\rangle = \langle\langle \sigma_{33} \rangle\rangle = 0, \tag{28.25}$$

the macrostress $\langle\langle \sigma_{11} \rangle\rangle$ in the composite depends on the macrostrain $\langle\langle \varepsilon_{11} \rangle\rangle$ as follows

$$\langle\langle \sigma_{11} \rangle\rangle = \frac{1}{l_{22}^* l_{33}^* - (l_{23}^*)^2} \big[l_{11}^* l_{22}^* l_{33}^* + 2 l_{12}^* l_{13}^* l_{23}^* - (l_{12}^*)^2 l_{33}^* - (l_{13}^*)^2 l_{22}^* - (l_{23}^*)^2 l_{11}^* \big] \langle\langle \varepsilon_{11} \rangle\rangle. \tag{28.26}$$

At that

$$\langle\langle \varepsilon_{22} \rangle\rangle = \frac{l_{13}^* l_{23}^* - l_{12}^* l_{33}^*}{l_{22}^* l_{33}^* - (l_{23}^*)^2} \langle\langle \varepsilon_{11} \rangle\rangle, \quad \langle\langle \varepsilon_{33} \rangle\rangle = \frac{l_{12}^* l_{23}^* - l_{13}^* l_{22}^*}{l_{22}^* l_{33}^* - (l_{23}^*)^2} \langle\langle \varepsilon_{11} \rangle\rangle. \tag{28.27}$$

Figure 28.2 shows the dependence of $\langle\langle \sigma_{11} \rangle\rangle / \mu_2$ on $\langle\langle \varepsilon_{11} \rangle\rangle$ for some winding angles $\gamma = \{\pi/12, \pi/6, \pi/4\}$. The solid curves correspond to $\gamma = \pi/12$, dashed curves—$\gamma = \pi/6$, and dotted curves—$\gamma = \pi/4$.

In the case when the uniaxial tension of the composite is given by

$$\langle\langle \varepsilon_{22} \rangle\rangle \neq 0, \quad \langle\langle \sigma_{11} \rangle\rangle = \langle\langle \sigma_{33} \rangle\rangle = 0, \tag{28.28}$$

the macrostress $\langle\langle \sigma_{22} \rangle\rangle$ of composite depends on the macrostrain $\langle\langle \varepsilon_{22} \rangle\rangle$ according to

$$\langle\langle \sigma_{22} \rangle\rangle = \frac{1}{l_{11}^* l_{33}^* - (l_{13}^*)^2} [l_{11}^* l_{22}^* l_{33}^* + 2 l_{12}^* l_{13}^* l_{23}^* - (l_{12}^*)^2 l_{33}^* - (l_{13}^*)^2 l_{22}^* - (l_{23}^*)^2 l_{11}^*] \langle\langle \varepsilon_{22} \rangle\rangle. \tag{28.29}$$

At that

$$\langle\langle \varepsilon_{11} \rangle\rangle = \frac{l^*_{13} l^*_{23} - l^*_{12} l^*_{33}}{l^*_{11} l^*_{33} - (l^*_{13})^2} \langle\langle \varepsilon_{22} \rangle\rangle, \quad \langle\langle \varepsilon_{33} \rangle\rangle = \frac{l^*_{12} l^*_{13} - l^*_{23} l^*_{11}}{l^*_{11} l^*_{33} - (l^*_{13})^2} \langle\langle \varepsilon_{22} \rangle\rangle. \qquad (28.30)$$

Figure 28.3 shows the dependence of $\langle\langle \sigma_{22} \rangle\rangle / \mu_2$ on $\langle\langle \varepsilon_{22} \rangle\rangle$ for some winding angle $\gamma = \{\pi/12, \pi/6, \pi/4\}$. The same notations as in Fig. 28.2 are used.

The figures give evidence of the significant influence of the nonlinearity of the deformative properties of the components on the deformation of the composite.

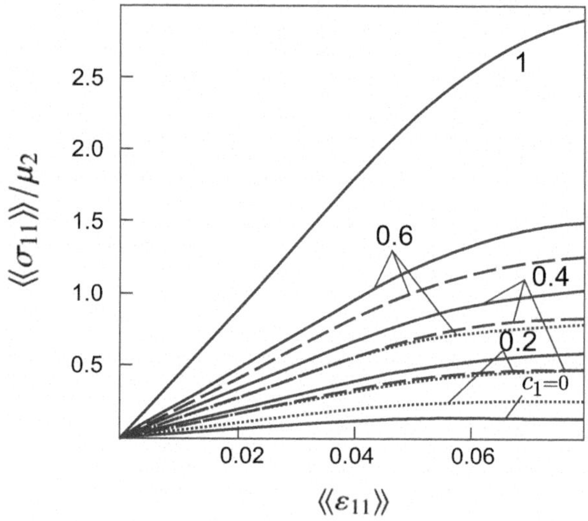

Fig. 28.2 Dependence of the macrostress $\langle\langle \sigma_{11} \rangle\rangle$ on the macrostrain $\langle\langle \varepsilon_{11} \rangle\rangle$ for some winding angles

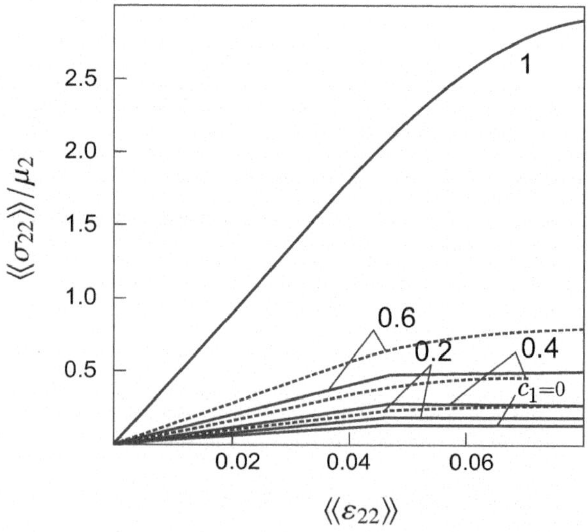

Fig. 28.3 Dependence of the macrostress $\langle\langle \sigma_{22} \rangle\rangle$ on the macrostrain $\langle\langle \varepsilon_{22} \rangle\rangle$ for some winding angles

As we can see, when $c_1 > 0$, the dependence $\langle\langle\sigma_{11}\rangle\rangle/\mu_2$ on $\langle\langle\varepsilon_{11}\rangle\rangle$ is parabolic; the dependence $\langle\langle\sigma_{22}\rangle\rangle/\mu_2$ on $\langle\langle\varepsilon_{22}\rangle\rangle$ consists of two linear sections for $0 < c_1 < 0.4$.

The winding angle γ significantly affects the dependence $\langle\langle\sigma_{11}\rangle\rangle/\mu_2$ on $\langle\langle\varepsilon_{11}\rangle\rangle$. For a fixed strain $\langle\varepsilon_{11}\rangle$, the stress $\langle\langle\sigma_{11}\rangle\rangle$ increases with the angle γ. For a fixed strain $\langle\langle\varepsilon_{22}\rangle\rangle$, according to the dependence of $\langle\langle\sigma_{22}\rangle\rangle/\mu_2$ on $\langle\langle\varepsilon_{22}\rangle\rangle$, the stress $\langle\langle\sigma_{22}\rangle\rangle$ decreases with an increase in the winding angle. Its influence on the deformation curves is insignificant for sufficiently large winding angles ($\gamma = \pi/6 \div \pi/4$)).

28.5 Conclusions

The process of nonlinear deformation of layered-fibrous composite materials of oblique winding with physically nonlinear components has been modeled. For the mentioned materials, in the framework of the proposed model, the algorithms for determining the effective deformative properties and stress-strain state are built. Furthermore, the effect of nonlinearity on the deformation of the composite is investigated.

References

Chen T, Dvorak GJ, Yu CC (2007) Size-dependent elastic properties of unidirectional nano-composites with interface stresses. Acta Mech 188:39–54

Choy CL, Leung WP, Kowk KW, Lau FP (1992) Elastic moduli and thermal conductivity of injection molded short fiber reinforced thermoplastics. Polym Compos 13:69–80

Duan HL, Yi X, Huang ZP, Wang J (2007) A unified scheme for prediction of effective moduli of multiphase composites with interface effects. Mech Mater 39:81–103

Geiger AL, Weich P (1997) Tensile properties and thermal expansion of discontinuously reinforced aluminum composites at subambient temperatures. J Math Sci 32:2611–2616

Gray RM (2009) Probability, random processes, and ergodic properties. Springer-Verlag, New York

Hashin Z (1990) Thermoelastic properties of fiber composites with imperfect interface. Mech Mater 8:333–348

Hashin Z (2002) Thin interphase/imperfect interface in elasticity with application to coated fiber composites. J Mech Phys Solid 50:2509–2537

Kauderer H (1958) Nichlineare mechanik. Springer-Verlag, Berlin

Khoroshun LP, Nazarenko LV (2013) Deformation and damage of composite materials with anisotropic components (Review). Int Appl Mech 49(4):388–455

Khoroshun LP, Shikula EN (1996) Nonlinear deformation of fiber-reinforced composite laminates with microscopic fractures in the binder. Int Appl Mech 32(11):858–864

Khoroshun LP, Shikula EN (2002) Nonlinear deformative properties of dispersion-reinforced materials. Mech Compos Mater 38(4):473–486

Khoroshun LP, Shikula EN (2002) Short-term damage micromechanics of laminated fibrous composites under thermal actions. Int Appl Mech 37(9):1083–1093

Khoroshun LP, Shikula EN (2008) Deformation of physically nonlinear stochastic composites. Int Appl Mech 44(12):1325–1351

Khoroshun LP, Shikula EN (2012) Deformation and long-term damage of homogeneous and composite materials of stochastic structure. Int Appl Mech 48(1):7–55

Khoroshun LP, Shikula EN (2016) Effective deformative properties of fibrous composite materials under nonlinear deformation of components. Dopov Nac akad nauk Ukr 6:47–55 [in Russian]
Kregers AF (1988) Mathematical modeling of the thermal expansion of spatially reinforced composites. Mech Compos Mater 3:316–325
Liu HT, Sun LZ, Ju JW (2004) An interfacial debonding model for particle-reinforced composites. Int J Damage Mech 13:163–185
Maslov BP (2022) Nonlinear hereditary creep of isotropic composites of random structure. Int Appl Mech 58(1):75–90
Mikheev SV, Stroganov GB, Romashin AG (2002) Ceramic and composite materials in aviation technology. Altex, Moscow [in Russian]
Nazarenko L, Stolarski H (2016) Energy-based definition of equivalent inhomogeneity for various interphase models and analysis of effective properties of particulate composites. Compos B 94:82–94
Nazarenko L, Stolarski H, Altenbach H (2017) Thermo-elastic properties of random particulate nano-materials for various models of interphase. IJMS 126:130–141
Nazarenko L, Stolarski H, Altenbach H (2017) A model of cylindrical inhomogeneity with spring layer interphase and its application to analysis of short-fiber composites. Compos Struct 160:635–652
Nazarenko L, Stolarski H, Altenbach H (2018) On modeling and analysis of effective properties of carbon nanotubes reinforced materials. Compos Struct 189:718–727
Nazarenko L, Stolarski H, Altenbach H (2018) Effective properties of particulate composites with surface-varying interphases. Compos B 149:268–284
Nazarenko L, Stolarski H, Altenbach H (2019) A statistical interphase damage model of random particulate composites. Int J Plast 116:118–142
Voyiadjis GZ, Kattan PI (2006) Advances in damage mechanics: Metals and metal matrix composites with an introduction to fabric tensors. Elsevier, Oxford
Wolf BK, Romadin KP (1967) Aviation material science. Mechanical engineering, Moscow [in Russian]

Chapter 29
Contact Problems for Cylindrical Stamps and Elastic Bodies with Initial (Residual) Stresses

Natalia Yarets'ka

29.1 Introduction

Studying the problems of contact mechanics, which is the primary purpose of this chapter, is a fundamental issue since contact interaction is one of the most common ways of transmitting external loads in practice. Contact mechanics allows us to find die-pressure distribution, study its concentration, and develop ways to reduce it. The significance of this problem has no doubts from the point of view of the development of fundamental achievements in contact mechanics as from the point of view of the applied branches of modern technology. Furthermore, contact mechanics theory is critical in mechanical engineering since the contact of structural members with each other carries out the force transfer in machines and columns of buildings. Also, similar problems can occur when calculating the critical characteristics of foundations of building columns, chimneys, cooling towers, water towers, and other high-rise structures for wind load or load from their weight.

The circle of problems about the contact of elastic, viscoelastic, and plastic bodies without initial stresses is quite broad (Grilitskiy and Kizyima 1981). Nevertheless, modern engineering practice demands have presented some new problems to researchers. They require the use of more complex continuous medium models. These are problems with complicated physical and mechanical properties. Models of contact interaction must consider the following factors: heat generation, the effect of friction, stiffness, surface properties of the material, and wear resistance of the surface.

Another fundamental problem of contact mechanics is the consideration of initial stresses, which need to be studied more intensely. Various factors, such as technological operations, manufacturing processes, and assembly of structures, cause initial stresses. When applying geostatic and geodynamic forces, the initial stresses arise in the earth's crust. We consider initial stresses when solving problems about

N. Yarets'ka (✉)
Khmelnytskyi National University, Khmelnytskyi, Ukraine
e-mail: yaretskano@khmnu.edu.ua

A. N. Guz et al. (eds.), *Advances in Mechanics*, Advanced Structured Materials 191,
https://doi.org/10.1007/978-3-031-37313-8_29

the deformation of soils (especially frozen ones). They arise due to technological processes in the creation of composite materials. Initial stresses are present in the blood vessels of living organisms. Internal residual stresses may exist in elastoplastic bodies after the removal of the load. Sometimes it is appropriate to deliberately create initial stresses (residual and technological) to compensate for those that arise in structural members. Such an approach increases the strength characteristics of the structure. Of particular interest is the study of contact problems for prestressed bodies due to the manufacturing of new artificial materials that can withstand large initial deformations. Considering the initial stresses in the approbation of critical structural members makes it possible to use strength resources more effectively, significantly reducing material consumption.

Quite often, in order to increase the strength of the structure, there is a need to strengthen some of its load-bearing elements with elastic fasteners (stringers). The results of research that accounted for initial stresses in the structure were carried out in Babich and Dikhtyaruk (2001), Babich et al. (2019). As in the mentioned papers, the current study is characterized by all considered punches being elastic, and the bases are prestressed.

Two approaches have historically been developed while studying problems of contact interaction of bodies with initial stresses. The first one is related to studying bodies with a specific form of elastic potential. The paper (Kurashige 1969) became pioneering in this field of research. It deals with the problem of a coin-shaped crack in the case of an incompressible elastic body with initial stresses for the Treloar potential (a body of the neo-Hookean type). Studies related to this approach are reviewed in Aleksandrov and Arutyunyan (1969), Dhaliwal et al. (1980) for other potentials.

The second approach (Guz 2003, 2019, 2022) is developed in parallel with the first one. It is related to the study of problems formulated for elastic bodies with initial stresses with an arbitrary structure of elastic potential. In the mentioned papers, the problems are solved in a general way for compressible and incompressible materials. For this purpose, the theories of large (finite) initial strains and two versions of the theory of small initial strains for equal and unequal roots of the characteristic equation are used (Guz et al. 2015; Guz 2022). All the results presented in this chapter are obtained within the framework of the second approach. The authors believe it has several advantages compared to the first approach.

Thus, the same problem (contact or crack problem) for prestressed bodies was considered by some authors, for example, for the Treloar potential and by other authors for the Mooney potential, that is, for a specific form of elastic potential. This chapter presents the research results in a single general form for compressible and incompressible prestressed bodies for an arbitrary structure of the elastic potential. Specific elastic potentials were used only at the final research stage (obtaining numerical results).

Contact problems for rigid and elastic punches within the framework of the second approach in Ukraine were solved in the works of Guz (2022), Guz et al. (2015), Guz et al. (2021), Yaretskaya (2014), Yaretskaya (2018), Babych and Yarets'ka (2021), Hutchinson and Thompson (2017), Guz et al. (1998). In the current chapter, using the relations of the linearized theory of elasticity, the main results of research on contact interaction are presented, namely: the 3D problem on a prestressed cylindrical punch and an elastic layer with initial stresses (Hutchinson and Thompson 2017; Yaretskaya 2014); problems dealing with the pressure of two prestressed half-spaces on an elastic cylindrical punch with initial stresses (Babych et al. 2022); problems on a rigid ring-shaped punch with a half-space with initial stresses (Yaretskaya 2018); problems dealing with the pressure of a prestressed elastic ring-shaped punch with a flat base on a half-space with initial stresses (Babych and Yarets'ka 2021). The results of the study are expressed in a general form without taking into account the friction for compressible and incompressible bodies. We obtained the solutions within the theory of large (finite) initial deformations and two theories of small initial deformations with an arbitrary form of the elastic potential (within the framework of the second approach).

As the initial stresses cannot be taken into account in the linear elastic mechanics of materials, the general nonlinear theory of elasticity (Guz 2019, 2022; Hutchinson and Thompson 2017) can be used. However, it is challenging to get a solution directly in this case. Therefore, with sufficiently large initial stresses, it is better to use its linearized option (Guz et al. 2015; Guz 2022; Hutchinson and Thompson 2017).

Thus, let us assume that the following fundamental conditions of linearized elasticity theory hold (Guz et al. 2015):

1. Contact interaction of an elastic finite cylindrical or ring-shaped punch with initial stresses (or without it) with a prestressed elastic body (layer or half-space) occurs after appearing of the initial stress state.
2. Additional (relative to the initial state) external loading causes much less stress-strain disturbance in prestressed bodies compared to the corresponding values of the initial stress state.
3. The initial stress-strain state in the area of contact interaction can be approximately considered homogeneous.
4. The solution of linearized problems of the elasticity theory of the contact interaction of prestressed bodies and punches (cylindrical or circular) is unique (Guz et al. 2015). Thus, condition (2.23) in Guz et al. (2015) is satisfied.

The second condition can be violated in the points where boundary conditions change (Guz et al. 2015); at these points, the contact stresses are unbounded. A detailed discussion of this phenomenon in the theory of contact problems of the linear and linearized theory of elasticity is given in Grilitskiy and Kizyima (1981), Guz et al. (2015). Based on the results of mentioned papers, the following can be concluded. There are power singularities in the solutions of contact problems for elastic and rigid bodies. The contact stresses are $O(\rho^{1-\gamma})$, where ρ is the distance from a point to the point where boundary conditions change, γ is a parameter that is determined by a transcendental equation (Grilitskiy and Kizyima 1981) and depends on the elastic constants of contacting bodies, as well as on the elastic potential. Thus, at the mentioned points, the stresses have no physical sense and do not affect determining the integral characteristics of the contact problems.

29.2 Main Relations

Let elastic solids interact with elastic or rigid punches. The surfaces outside the contact boundary are assumed to remain free from the influence of external forces, and displacements and stresses are continuous at the contact boundary.

We use the coordinates Oy_i $(i = 1, 2, 3)$, which are associated with the Lagrangian coordinates x_i $(i = 1, 2, 3)$ by the relations: $Oy_i = \lambda_i x_i$ $(i = 1, 2, 3)$, where λ_i $(i = 1, 2, 3)$ are the elongation factors determining displacements of the initial state, $\lambda_i = \text{const}$. The y_3-axis is normal to the contact boundary.

Consider elastic isotropic bodies (compressible or incompressible) with an arbitrary form of the elastic potential (Guz et al. 2015). For orthotropic bodies, the elastic-equivalent directions are assumed to coincide with the coordinate axes in the deformed state y_i $(i = 1, 2, 3)$.

We assume that the initial stress state is homogeneous, the contact boundary of elastic bodies is in the plane $y_3 = \text{const}$, and the initial stresses act along the contact boundary (Guz et al. 2015; Hutchinson and Thompson 2017)

$$y_m = x_m + U_m^0, \quad U_m^0 = \delta_{mi}(\lambda_m - 1)\lambda_i^{-1} y_i \quad (i = 1, 2, 3),$$

where δ_{mi} is the Kronecker symbol.

Then, for compressible bodies, the basic equation in terms of displacements (Guz et al. 2015; Hutchinson and Thompson 2017) is the following

$$L'_{m\alpha} U_\alpha = 0, \quad L'_{m\alpha} = \frac{\omega'_{ij\alpha\beta}\partial^2}{\partial y_i \partial y_{\beta_i}} \quad (i, m, \alpha, \beta = 1, 2, 3); \tag{29.1}$$

for non-rigid bodies, the incompressibility condition is satisfied

$$\begin{aligned} &L'_{m\alpha} U_\alpha + \frac{q'_{\alpha m}\partial p'}{\partial y_\alpha} = 0, \quad L'_{m\alpha} = \frac{\kappa'_{ij\alpha\beta}\partial^2}{\partial y_i \partial y_{\beta_i}}, \\ &q'_{ij}\frac{\partial U_j}{\partial y_i} = 0, \quad q'_{ij} = \lambda_i q_{ij} \quad (i, j, m, \alpha, \beta = 1, 2, 3). \end{aligned} \tag{29.2}$$

Stress tensor components for compressible bodies at $y_i = \text{const}$ $(i = 1, 2, 3)$ reeds

$$Q'_{ij} = \omega'_{ij\alpha\beta}\frac{\partial U_\alpha}{\partial y_\beta}, \quad \omega'_{ij\alpha\beta} = \frac{\lambda_i \lambda_\beta}{\lambda_1 \lambda_2 \lambda_3}\omega_{ij\alpha\beta},$$

and for non-compressible bodies

$$Q'_{ij} = \kappa'_{ij\alpha\beta}\frac{\partial U_\alpha}{\partial y_\beta} + q'_{ij} p, \quad \kappa'_{ij\alpha\beta} = \frac{\lambda_i \lambda_\beta}{\lambda_1 \lambda_2 \lambda_3}\kappa_{ij\alpha\beta},$$

where $\omega'_{im\alpha\beta} = \omega'_{im\alpha\beta}\left(S_{11}{}^0,\ S_{22}{}^0,\ S_{33}{}^0\right)$ and $\kappa'_{im\alpha\beta} = \kappa'_{im\alpha\beta}\left(S_{11}{}^0,\ S_{22}{}^0,\ S_{33}{}^0\right)$ are the components of the fourth-order tensor of elasticity modules.

The initial stresses being homogeneous, the following condition takes place

$$S_0^{11} = S_0^{22} \neq 0, \quad S_0^{33} = 0, \quad \lambda_1 = \lambda_2 \neq \lambda_3. \tag{29.3}$$

Taking into account (29.3), the solution of (29.1) and (29.2) can be represented using function $\tilde{\chi}$ that, in cylindrical coordinates (r, θ, y_3), satisfies the characteristic equation

$$\left(\Delta_1 + \xi_2'^2 \frac{\partial^2}{\partial y_3^2}\right)\left(\Delta_1 + \xi_3'^2 \frac{\partial^2}{\partial y_3^2}\right)\tilde{\chi} = 0, \tag{29.4}$$

where $\Delta_1 = \partial^2/\partial r^2 + \partial/\partial r$.

We take into account the uniqueness of the solution of the linearized theory of elasticity for compressible and incompressible bodies and present two possible variants of the general solution of (29.4):

- the case of equal roots ($\xi_2'^2 = \xi_3'^2$):

$$\tilde{\chi} = \tilde{\chi}_1 + y_3\tilde{\chi}_2, \quad \left(\Delta_1 + \xi_2'^2 \frac{\partial^2}{\partial y_3^2}\right)\tilde{\chi}_1 = 0, \quad \left(\Delta_1 + \xi_2'^2 \frac{\partial^2}{\partial y_3^2}\right)\tilde{\chi}_2 = 0; \tag{29.5}$$

- case of unequal roots ($\xi_2'^2 \neq \xi_3'^2$):

$$\tilde{\chi} = \tilde{\chi}_1 + \tilde{\chi}_2, \quad \left(\Delta_1 + \xi_2'^2 \frac{\partial^2}{\partial y_3^2}\right)\tilde{\chi}_1 = 0, \quad \left(\Delta_1 + \xi_3'^2 \frac{\partial^2}{\partial y_3^2}\right)\tilde{\chi}_2 = 0. \tag{29.6}$$

In circular cylindrical coordinates (r, θ, z_i) ($z_i = v_i^{-1} y_3$, $v_i = \sqrt{n_i}$ $(i = 1, 2)$, $n_1 = \xi_2{}'^2$ and $n_2 = \xi_3{}'^2$) we obtained the solutions for finite cylindrical (circular) punches with initial stresses using variable separation methods (Fourier method) in the form of infinite system for constants. The stress-strain state in elastic bodies with initial stresses for cases (29.5) and (29.6) is determined in terms of harmonic functions in the form of Hankel integrals.

Further, we consider the solved problems in detail.

29.3 Spatial Contact Problem for Prestressed Cylindrical Punch and Elastic Layer With Initial (Residual) Stresses

The section presents the problem statement, boundary conditions, method of solution, and numerical results.

29.3.1 Problem Statement and Boundary Conditions

Let the finite elastic cylindrical punch (with the radius of the base R and height H) with initial stresses be pressed into the elastic layer under the action of force P (Fig. 29.1). We use the following notations in Fig. 29.1: h_1 is the thickness of the layer in the initial stress state, which is related to the thickness h_2 in the undeformed state by the ratio $\lambda_3 = h_2/h_1$. We assume that P is applied only to the free end of the elastic punch. All the points of the end of the punch move in the direction of the symmetry axis y_3 by the same value ε. In the case of a prestressed layer, three types of contact interaction are considered: (1) the layer is located on a rigid foundation without friction; (2) the layer with initial stresses is rigidly fixed on an undeformed foundation; (3a) the layer with the initial stresses lays without friction on the foundation with the initial stresses; (3b) the layer with initial stresses lays without friction on a foundation without initial stresses.

In a cylindrical coordinate system (r, θ, z_i), the following boundary conditions correspond to the problem statement:

1. At the end of the elastic punch $z_i = Hv_i^{-1}$:

$$u_3^{\prime(1)} = -\varepsilon, \quad Q_{3r}^{\prime(1)} = 0 \quad (0 \leqslant r \leqslant R). \tag{29.7}$$

2. On the edge of the elastic layer in the contact area:

$$u_3^{\prime(1)} = u_3^{\prime(2)}, \quad Q_{33}^{\prime(1)} = Q_{33}^{\prime(2)}, \quad Q_{3r}^{\prime(1)} = Q_{3r}^{\prime(2)} = 0$$
$$(z_i = 0, \quad i = 1, 2; \quad 0 \leqslant r \leqslant R). \tag{29.8}$$

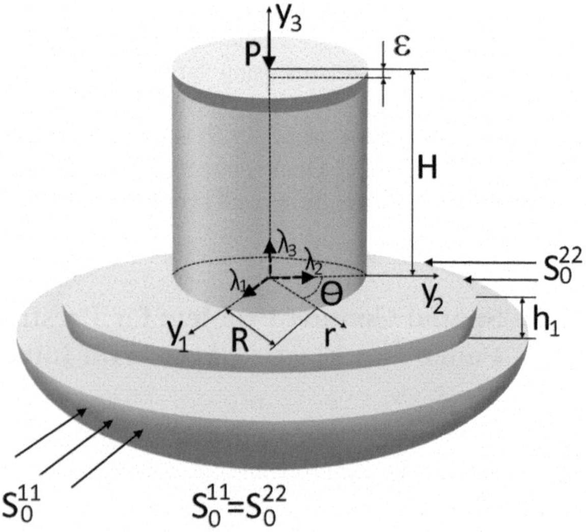

Fig. 29.1 Prestressed cylindrical punch on the elastic layer with initial stresses

3. On the edge of the elastic layer outside the contact area:

$$Q'^{(2)}_{33} = 0, \quad Q'^{(2)}_{3r} = 0 \quad (z_i = 0, \quad i = 1, 2; \quad R \leqslant r < \infty). \tag{29.9}$$

4. On the side surface of the elastic punch:

$$Q'^{(1)}_{rr} = 0, \quad Q'^{(1)}_{3r} = 0 \quad (0 \leqslant z_i \leqslant H v_i^{-1}; \quad r = R). \tag{29.10}$$

5. On the lower surface of the layer $z_i = -\lambda_3 h_2 v_i^{-1} = -H v_i^{-1}$ $(i = 1, 2)$:
   a. for a layer with initial stresses lying without friction on an undeformed foundation:
   $$u'^{(2)}_3 = 0, \quad Q'^{(2)}_{3r} = 0 \quad (0 \leqslant r < \infty); \tag{29.11}$$
   b. for a layer with initial stresses rigidly fixed to an undeformed foundation:
   $$u'^{(2)}_3 = 0, \quad u'^{(2)}_r = 0 \quad (0 \leqslant r < \infty); \tag{29.12}$$
   c. for a layer with initial stresses lying without friction on an elastic basis with initial stresses:
   $$u'^{(2)}_3 = u'^{(3)}_3, \quad Q'^{(2)}_{3r} = Q'^{(3)}_{3r} = 0 \quad (0 \leqslant r < \infty). \tag{29.13}$$

29.3.2 Method of Solution

Solutions are found in the case of roots (29.5) and (29.6) of (29.4). For example, the solution in general form for a cylindrical punch reads

- for $n_1 = n_2$:

$$\tilde{\chi} = \varepsilon \Bigg\{ v_1 z_1 \left[(m_2 - 1)^{-1} + \chi_0 \left((1 - m_2)^{-1} - 2E(3H\theta_2)^{-1}(3r^2 - 2z_1^2) \right) \right]$$

$$+ R \sum_{k=1}^{\infty} \chi_k \Bigg[R(2\gamma_k)^{-1} b_1^{(k)} \left(H \left(1 + \frac{s_0(1 - I_0(v_1 \gamma_k R))}{v_1 \gamma_k R I_1(v_1 \gamma_k R)} \right) + z_1 \right) I_0(\gamma_k z_1 v_1)$$

$$\times \sin(\gamma_k z_1 v_1) + J_0(\alpha_k r) \mu_k^{-1} (\tilde{S}_2(\alpha_k z_1) + z_1 \tilde{S}_3(\alpha_k z_1)) \Bigg] \Bigg\};$$

- for $n_1 \neq n_2$:

$$\tilde{\chi} = \frac{\varepsilon}{2}\left\{\frac{1}{\theta_8}(r^2 - z_1^2 - z_2^2) - \chi_0\left[r^2\left(\frac{1}{\theta_8} + \frac{z_1+z_2}{2H\theta_6}\right) - \frac{z_1^2+z_2^2}{\theta_8} - \frac{z_1^3+z_2^3}{3H\theta_6}\right]\right\}$$
$$-\sum_{k=1}^{\infty}\left\{b_3^{(k)}\left[s_0\frac{I_1(\gamma_k v_2 R)}{I_1(\gamma_k v_1 R)} I_0(\gamma_k v_1 r)\sin(\gamma_k z_1 v_1) + I_0(\gamma_k v_2 r)\sin(\gamma_k z_2 v_2)\right]\right.$$
$$\left. - J_0(\alpha_k r)\left[\tilde{S}_2(\alpha_k z_1) + \tilde{S}_3(\alpha_k z_2)\right]\right\}\chi_k,$$

where $J_\upsilon(x)$, $I_\upsilon(x)$ are Bessel functions of real and imaginary arguments; for $n_1 = n_2$, $\gamma_k = 2\pi k H^{-1}$; for $n_1 \neq n_2$, $\gamma_k = \pi(2k+1)H^{-1}$, $(k = 0, 1, 2, \ldots)$;

$$s_0 = (1+m_2)(1+m_1)^{-1}, \quad \theta_8 = m_1 n_1^{-1} + m_2 n_2^{-1}, \quad \theta_6 = m_1 v_1^{-3} + m_2 v_2^{-3},$$
$$b_3^{(k)} = 4\varepsilon R^2 J_0(\nu_k)\left[\frac{\tilde{c}_1 - \tilde{c}_0}{\nu_k^2 + (\gamma_k v_1 R)^2} - \frac{v_2}{v_1 s_0}\frac{\tilde{c}_2 - \tilde{c}_0}{\nu_k^2 + (\gamma_k v_2 R)^2}\right]$$
$$\times \frac{v_1 H \gamma_k^3 I_1(\gamma_k v_2 R)}{v_2 W_k(2) - v_1 s_0 W_k(1)}, \quad W(j) = \frac{(\tilde{c}_0 - \tilde{c}_j) I_0(\gamma_k v_j R)}{I_1(\gamma_k v_j R)} + \frac{1-\tilde{c}_0}{\gamma_k v_j R},$$

$$m_1 = \begin{cases} (\omega'_{1111} n_1 - \omega'_{1331})(\omega'_{1133}\omega'_{1313})^{-1}, & \text{for compressible bodies} \\ \lambda_1 q_1 n_1 (\lambda_3 q_3)^{-1}, & \text{for non-compressible bodies,} \end{cases}$$

$$\tilde{c}_0 = \begin{cases} \omega'_{1111}\omega'^{-1}_{1122}, & \text{for compressible bodies} \\ \lambda_1 q_1 (\lambda_3 q_3)^{-1}(\kappa'_{1133} + \kappa'_{1313})\kappa'_{1122}, & \text{for non-compressible bodies,} \end{cases}$$

$$m_2 = \begin{cases} (\omega'_{1133} - \omega'_{1313})(\omega'_{1133} + \omega'_{1313})^{-1}, & \text{for compressible bodies,} \\ 1, & \text{for non-compressible bodies,} \end{cases}$$

$$\tilde{c}_i = \begin{cases} \lambda_3 \omega'_{1133} m_i \omega'^{-1}_{1122} n_i^{-1}, & \text{for compressible bodies;} \\ (\kappa'_{1133} m_i - \kappa'_{3113})\kappa'^{-1}_{1122} n_i^{-1} (i = 1, 2), & \text{for non-compressible bodies,} \end{cases}$$

$$b_1^{(k)} = \frac{J_0(\nu_k)\gamma_k t_{00}\left[t_{14}\sinh^2(\alpha_k H v_1^{-1})\cosh(\alpha_k H v_1^{-1}) + t_{11}t_{29}\right]}{(I_0(\alpha_k v_1 R) - 1)t_{33} + c_1(\cosh(\alpha_k H v_1^{-1}) - 1)t_{34} + \sinh(\alpha_k H v_1^{-1})t_{35}},$$

$$M^{(k)} = M_k N_k^{-1}, \quad E^{(k)} = -E_k N_k^{-1},$$

$$\tilde{S}_2(\alpha_k z_1) = R s_0 \mu_k^{-1}\cosh(\alpha_k z_1) + E^{(k)}\sinh(\alpha_k z_1),$$
$$\tilde{S}_3(\alpha_k z_2) = -\sinh(\alpha_k z_1) - M^{(k)}\cosh(\alpha_k z_1),$$

$$
\begin{aligned}
&t_{00} = \mu_k^2(\mu_k^2 + R^2 v_1^2 \gamma_k^2), \quad t_{01} = (1 + m_2 t_{23} + t_{10}\tilde{c}_1 s_0), \quad t_{02} = 2(H\tilde{c}_1\alpha_k - \tilde{c}_0 s_0),\\
&t_{03} = 2(m_1 + 1)(m_2 - 1), \quad t_{10} = \alpha_k H(m_1 + 1),\\
&t_{14} = \alpha_k H\tilde{c}_1\left(v_1(m_1 + 1)(m_2 - 1) + m_1(1 + m_2)\right), \quad t_{12} = \alpha_k H\tilde{c}_0 - \tilde{c}_1 s_0,\\
&t_{13} = v_1(\tilde{c}_2 - \tilde{c}_1 + \tilde{c}_0) + \alpha_k H\tilde{c}_1 - \tilde{c}_0 s_0, \quad t_{11} = v_1(1 - m_2) + s_0 m_1,\\
&t_{15} = v_1(\tilde{c}_0 - \tilde{c}_1 + \tilde{c}_2)t_{10} - \tilde{c}_1 s_0(m_2 + 1), \quad t_{23} = v_1(\tilde{c}_2 - \tilde{c}_1 + \tilde{c}_0) - \tilde{c}_0 s_0,\\
&t_{21} = H(H(3 + 2\tilde{c}_2 - 4\tilde{c}_1 + \tilde{c}_0) + 1 + 2\tilde{c}_1 + \tilde{c}_0), \quad t_{22} = 2v_1^4\gamma_k^2 R^2(m_2 - 1),\\
&t_{24} = v_1(1 + m_1)(\tilde{c}_2 - \tilde{c}_1 + \tilde{c}_0) + \tilde{c}_0(1 + m_2),\\
&\quad J_1(\alpha_k R) = 0, \quad \alpha_k = \mu_k R^{-1},\\
&t_{27} = c_0\sinh(\alpha_k H v_1^{-1}) + c_1(1 - \cosh(\alpha_k H v_1^{-1})), \quad t_{26} = t_{12}\sinh(\alpha_k H v_1^{-1}) + t_{13},\\
&t_{28} = t_{10}\cosh(\alpha_k H v_1^{-1})t_{27} + c_1(1 + m_2)\sinh(\alpha_k H v_1^{-1})\left(1 - \cosh(\alpha_k H v_1^{-1})\right),\\
&t_{29} = (1 + m_1)\sinh^2(\alpha_k H v_1^{-1})t_{26} + \cosh(\alpha_k H v_1^{-1})t_{28},\\
&t_{30} = t_{12}\cosh(\alpha_k H v_1^{-1}) + c_1 s_0 + t_{23}\sinh(\alpha_k H v_1^{-1}),\\
&t_{31} = c_1 t_{10}\cosh(\alpha_k H v_1^{-1})(1 + \cosh(\alpha_k H v_1^{-1})),\\
&t_{32} = t_{31} + (1 + m_1)t_{30}\sinh(\alpha_k H v_1^{-1}), \quad t_{33} = t_{11}t_{21}t_{00}\sinh^2(\alpha_k H v_1^{-1}) + t_{22}t_{32},\\
&t_{34} = t_{10}\cosh(\alpha_k H v_1^{-1}) + (1 + m_2)\sinh(\alpha_k H v_1^{-1}),\\
&t_{35} = c_0 t_{10}\cosh(\alpha_k H v_1^{-1}) - t_{24}\sinh(\alpha_k H v_1^{-1}).
\end{aligned}
$$

Applying (29.7)–(29.13), the stress-strain state in the prestressed layer is defined for equal roots (29.5) (29.1) by

$$
\begin{aligned}
u_3^{\prime(2)} &= \theta_3\left(\int_0^\infty \frac{F(\eta)}{\eta}J_0(\eta\rho)d\eta - \int_0^\infty \frac{F(\eta)}{\eta}G(\eta h)J_0(\eta\rho)d\eta\right),\\
Q_{33}^{\prime(2)} &= \theta_1\int_0^\infty F(\eta)J_0(\eta\rho)d\eta, \quad Q_{3r}^{\prime(2)} = 0,
\end{aligned}
\tag{29.14}
$$

where

$$
\theta_1 = C_{44}l_1(1 + m_1)\kappa, \quad h = \frac{h_1}{R}, \quad \theta_3 = \frac{m_1}{v_1}(s_1 - s_0), \quad s_1 = \frac{m_1 - 1}{m_1},
$$

$$
C_{44} = \begin{cases} \omega'_{1313}, & \text{for compressible bodies}\\ \kappa'_{1313}, & \text{for non-compressible bodies,}\end{cases}
$$

$$
m_1 = \begin{cases} \dfrac{\omega'_{1111}n_i - \omega'_{3113}}{\omega'_{1133} + \omega'_{1313}}, & \text{for compressible bodies}\\ \dfrac{\lambda_1 q_1}{\lambda_3 q_3}n_i, & \text{for non-compressible bodies,}\end{cases}
$$

$$
l_i = \begin{cases} \dfrac{\omega'_{1331}}{\omega'_{1313}} + \dfrac{\omega'_{1313} - \omega'_{1331}}{\omega'_{1313}}\dfrac{\omega'_{1133} + \omega'_{1313}}{\omega'_{1111}n_i + \omega'_{1133}}, & \text{for compressible bodies}\\ \dfrac{\kappa'_{1331}}{\kappa'_{1313}} + \dfrac{\kappa'_{1313} - \kappa'_{1331}}{\kappa'_{1313}}\dfrac{\lambda_3 q_3}{\lambda_3 q_3 + \lambda_1 q_1 n_i}, & \text{for non-compressible bodies,}\end{cases}
$$

In (29.14), we introduces the notation $F(\eta) = \eta^3 B_2 R^{-3}(1 - G(\eta))^{-1}$; the function $G(\eta)$ is determined by (29.7)–(29.13).

Next, we introduce variables χ_k $(k = 0, 1, 2, \dots)$, which define the stress state in the layer, punch, and foundation for equal (29.5) and unequal roots (29.6); these variables depend on the elastic potential.

From the conditions of continuity of stresses and displacements (29.8) at the contact boundary and outside it, we write down the dual integral equations with the unknown function $F(\eta)$:

$$\int_0^\infty F(\eta)\eta^{-1} J_0(\eta\rho)d\eta = f(\rho) \quad (\rho < 1),$$
$$\int_0^\infty F(\eta) J_0(\eta\rho)d\eta = 0 \quad (\rho > 1),$$

where, in the case of unequal roots, we have

$$f(\rho) = \frac{\varepsilon}{\theta_3}\left(\chi_0 - 1 - \theta_4 \sum_{k=1}^{\infty} \chi_k J_0(\mu_k \rho) + \frac{\theta_3}{\varepsilon}\int_0^\infty \frac{\eta}{F(\eta)} G(\eta h) J_1(\eta\rho)d\eta\right),$$

$\theta_4 = n_1^{-1}(v_1(m_2 - 1) - m_1 s_0)$.

Using the inversion formula (Guz et al. 2015), we obtain integral Fredholm equations of the second kind with the unknown function $F(\eta)$ determined by (29.5):

$$\begin{aligned}\frac{F(\eta)}{\eta} &= -\frac{2\varepsilon}{\pi\theta_3}\left((1 - \chi_0)\Psi_0(\eta, 0) - 2(m_2 - 1)\frac{R^2}{\theta_2}\chi_0\Psi_1(\eta, 0)\right)\\ &\quad - \frac{2\varepsilon}{\pi\theta_3}\left(\theta_4 \sum_{k=1}^{\infty} \chi_k \Psi_0(\eta, \mu_k) + \frac{m_2 - 1}{2} R^2 \sum_{k=1}^{\infty} b_1^{(k)} \chi_k \Psi_0(\eta, i\gamma_k v_1 R)\right)\\ &\quad + \frac{2}{\pi}\int_0^\infty \frac{F(u)}{u} G(uh)\Psi_0(\eta, u)du \end{aligned} \quad (29.15)$$

for equal roots (29.5), and

$$\begin{aligned}\frac{F(\eta)}{\eta} &= \frac{2\varepsilon}{\pi\theta_3}\left((\chi_0 - 1)\Psi_0(\eta, 0) - \theta_4 \sum_{k=1}^{\infty} \chi_k \Psi_0(\eta, \mu_k)\right)\\ &\quad + \frac{2}{\pi}\int_0^\infty \frac{F(u)}{u} G(uh)\Psi_0(\eta, u)du \end{aligned} \quad (29.16)$$

for unequal roots (29.6). In (29.15) and (29.16)

$$\Psi_n(x, y) = \int_0^1 t^n \cos(xt)\cos(y)t dt.$$

We use the method of successive approximations to find the solution of (29.15) and (29.16) in the form

$$F(\eta) = \sum_{k=1}^{\infty} F^{(k)}(\eta). \tag{29.17}$$

The convergence of the method of successive approximations is studied based on the principle of compressed mappings. Note that the process of successive approximations (29.17) is convergent for $\lambda_1 > \lambda_{kp}$. This method was used when solving Eqs. (29.14) and (29.15) under the condition

$$h > v_1\sqrt{D_1(2\pi)^{-1}}, \quad D_n = \int_0^{\infty} t^n G(t)dt.$$

In the case of harmonic potential, the minimum layer thickness values h are given in Hutchinson and Thompson (2017). In the case of the absence of initial stresses in the cylinder, the thickness of the layer is given for comparison. It was shown that the initial stresses affect the implementation of the method of successive approximations.

Using the boundary conditions (29.7)–(29.13) and the orthogonality of the Bessel functions, we obtain the infinite quasi-regular system of linear equations

$$\vartheta_k \chi_k + \sum_{n=0}^{\infty} \vartheta_{kn} \chi_n = \bar{\omega}_k \qquad (k = 0, 1, 2, \dots). \tag{29.18}$$

The quasi-regularity of (29.18) can be established using asymptotic representations for the Bessel functions, and μ_k and boundedness of integrals $\Psi(\mu_k, \mu_n)$ for $\lambda_1 > \lambda_{kp}$ (Hutchinson and Thompson 2017). Thus, the problem is reduced to determining the constants χ_k $(k = 0, 1, 2, \dots)$, by which the characteristics of the stress-strain state of the elastic layer, punch, and foundation with initial stresses are expressed. The relation between the punch base displacements and the load P for equal (29.5) and unequal roots (29.6) of (29.4) are the following

$$P = 8\pi\varepsilon E\theta_1(\kappa\theta_2 I R)^{-1}\chi_0, \quad P = 2\pi\varepsilon E\theta_5\theta_1(\kappa H)^{-1}\chi_0,$$

where $I = H/R$, $\theta_5 = (v_2 + v_1 s)n_1 n_2((m_1 v_2^3 + m_2 v_1^3)E)^{-1}$.

When determining the stress state of the layer and foundation, most of the integrals cannot be found analytically due to the complexity of the function $G(t)$. Therefore, starting from the second approximation of the function (29.17), we expand the expressions under integrals into power series h^{-i} $(i = 1, 2, \dots)$; it allows us to calculate the coefficients of (29.18) approximately.

29.3.3 Numerical Results

Here, the impact of initial stresses on the contact interaction of the elastic cylinder and layer is studied for the potentials of Bartenev–Khazanovich, Treloar, and harmonic potential. We find a numerical solution of the quasi-regular system (29.18) at $k = 32$. The algorithm based on the reduction method was tested on some reference problems and showed its efficiency. Numerical analysis is presented in dimensionless coordinates (Figs. 29.2, 29.3, 29.4, 29.5 and 29.6). In the figures, the dashed line corresponds to a stress state without initial stresses.

Figures 29.2 and 29.3 show the die pressure for the cylinder with initial stresses at $h = 1.6$. The impact of initial stresses on contact displacements in the cylinder and layer is illustrated in Figs. 29.4 and 29.5. Figure 29.6 shows that tangential stresses are the most concentrated near the contact zone.

By comparing the stress state of bodies with initial stresses and the corresponding state for an isotropic body without initial stresses, the following equation is obtained at $z_i = 0$

$$U_3(r, 0) = kU_3^0(r, 0), \quad Q_{33}(r, 0) = k_s Q_{33}^0(r, 0), \tag{29.19}$$

where $U_3(r, 0)$ and $Q_{33}(r, 0)$ are displacement and stress under the punch pressed into the layer with initial stresses; $U_3{}^0(r, 0)$ and $Q_{33}{}^0(r, 0)$ are displacement and stress under the punch pressed into the layer without initial stresses; k and k_s are coefficients that illustrate the effect of initial stresses on contact stresses and displacement of elastic cylinder and layer.

The dependence of coefficients changes k and k_s from (29.19) is presented in Hutchinson and Thompson (2017). When the elongation coefficient approaches the

Fig. 29.2 Harmonic potential

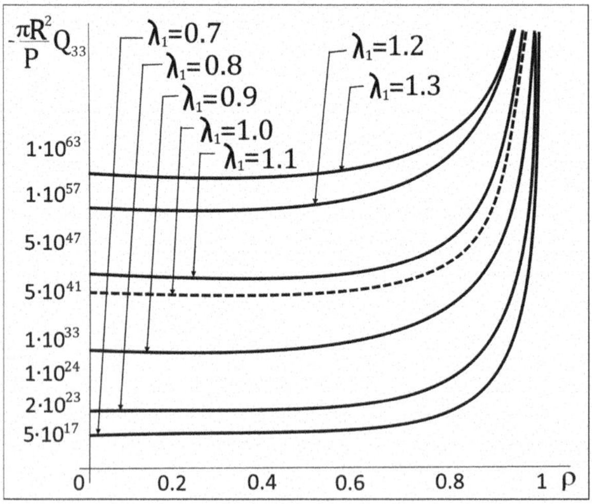

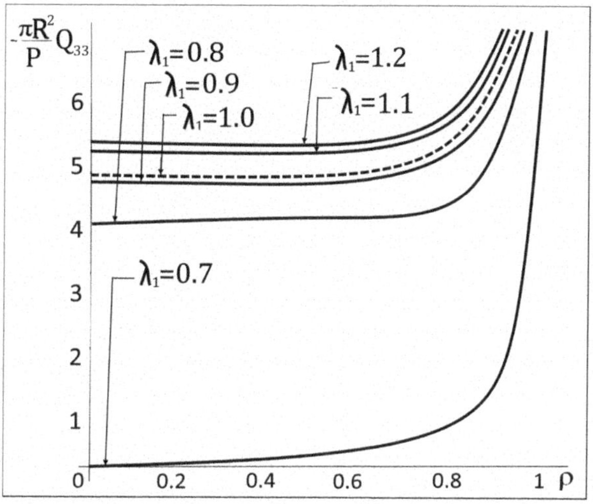

Fig. 29.3 Treloar potential

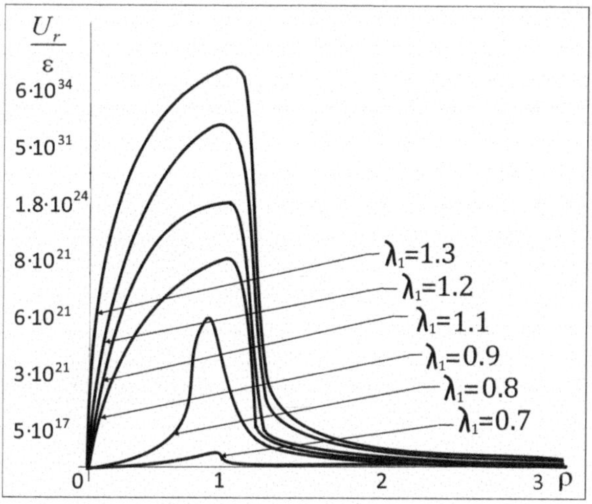

Fig. 29.4 Bartenev–Khazanovich potential

value of the surface instability of the material, displacements increase unboundedly, and stresses tend to zero.

29.4 Pressure of Two Prestressed Half-Spaces on Elastic Cylindrical Punch With Initial (Residual) Stresses

The section presents the problem statement, boundary conditions, solution method, and numerical results for this problem.

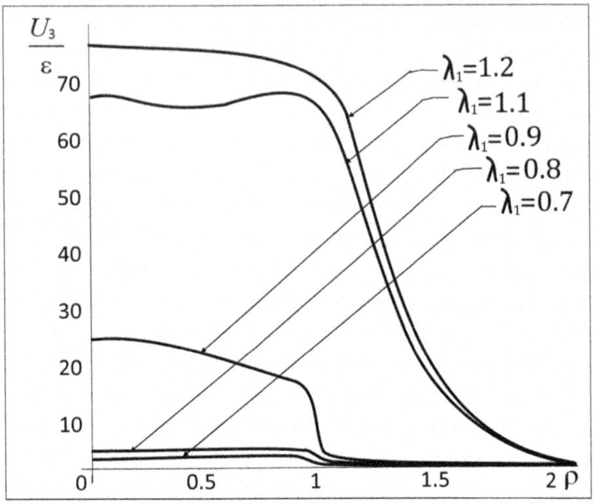

Fig. 29.5 Treolar potential

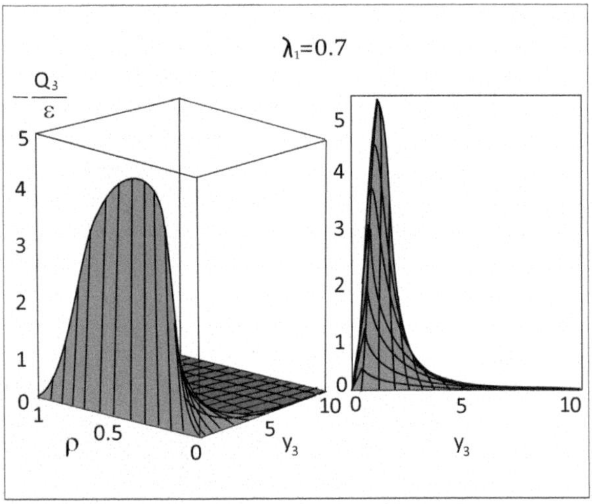

Fig. 29.6 Treolar potential

29.4.1 Problem Statement and Boundary Conditions

Let the finite elastic cylindrical punch with initial stresses is compressed (stretched) by two identical prestressed half-spaces reduced to equal and oppositely directed forces P (Fig. 29.7). The geometrical axis of symmetry of the punch coincides with the y_3-axis of the cylindrical coordinate system (r, θ, y_3). The external load is applied so that the points of the unloaded surfaces of the bases move in the direction of $y_3 = 0$ by ε. Let $h = 0.5H$.

The following boundary conditions correspond to the problem statement

1. at the base of the elastic punch

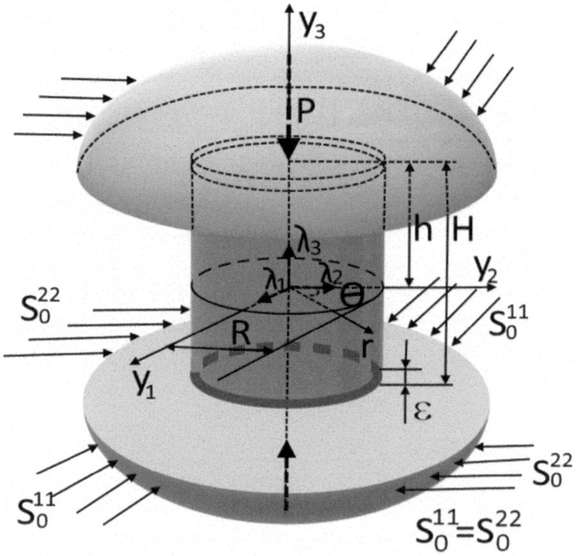

Fig. 29.7 Two prestressed half-spaces and an elastic cylindrical punch with initial stresses

$$u_3^{\prime(i)} - u_3^{\prime(3)} = \varepsilon, \quad Q_{33}^{\prime(3)} = Q_{33}^{\prime(i)}, \quad Q_{3r}^{\prime(3)} = 0, \quad Q_{3r}^{\prime(i)} = 0$$
$$(z_i = \pm h/v_i, \quad i = 1, 2; \quad 0 \leqslant r \leqslant R), \qquad (29.20)$$

2. on the boundaries of elastic half-spaces outside the contact area

$$Q_{33}^{\prime(3)} = 0, \quad Q_{3r}^{\prime(i)} = 0, \quad u_3^{\prime(i)} = 0, \quad (z_i = \pm h/v_i, \quad i = 1, 2; \quad r > R), \qquad (29.21)$$

3. on the side surface of the elastic punch

$$Q_{rr}^{\prime(3)} = 0, \quad Q_{3r}^{\prime(3)} = 0 \quad (|z_i| \leqslant h/v_i, \quad i = 1, 2; \quad r = R). \qquad (29.22)$$

The equilibrium condition that establishes the dependence of load P on displacements of punch bases is as follows:

$$P = -2\pi \int_0^R r \left| Q_{33}^{\prime(i)} \right| dr, \quad \left| Q_{33}^{\prime(i)} \right| = \left| Q_{3r}^{\prime(i)} \right|_{z_i = \pm H/v_i} \quad (i = 1, 2). \qquad (29.23)$$

29.4.2 Method of Solution

In the case of unequal roots ($\xi_2^{\prime 2} \neq \xi_3^{\prime 2}$) of the characteristic Eq. (29.4), the solution for a cylindrical elastic punch with initial stresses reads

$$\tilde{\chi} = \frac{\varepsilon \chi_0}{2h\theta_6}\left(\frac{r^2}{2}(z_1+z_2) - \frac{1}{3}(z_1^3+z_2^3)\right)$$
$$+ \frac{\varepsilon\omega_2}{Rn_1}\sum_{k=1}^{\infty}\mu_k\chi_k\left\{\left[\frac{s_0 I_1(\gamma_k v_2 R)}{I_1(\gamma_k v_1 R)} I_0(\gamma_k v_1 r)\sin(\gamma_k v_1 z_1) - I_0(\gamma_k v_2 r)\sin(\gamma_k v_2 z_1)\right]\right.$$
$$-J_0(\alpha_k r)F_k^*\left[\frac{v_2(\tilde{c}_0-\tilde{c}_2)\left(\cot\left(\frac{\alpha_k h}{v_2}\right)\sinh\left(\frac{\alpha_k h}{v_1}\right) - \cosh\left(\frac{\alpha_k h}{v_1}\right)\right)\sinh(\alpha_k z_1)}{\sinh(\alpha_k h v_1^{-1})\,(v_2(\tilde{c}_0-\tilde{c}_2)+v_1 s_0(\tilde{c}_1-\tilde{c}_0))}\right.$$
$$+\cosh(\alpha_k z_1) + \frac{n_2(\tilde{c}_1-\tilde{c}_0)\left(\cot\left(\frac{\alpha_k h}{v_2}\right)\sinh\left(\frac{\alpha_k h}{v_1}\right) - \cosh\left(\frac{\alpha_k h}{v_2}\right)\right)\sinh(\alpha_k z_2)}{v_1\sinh(\alpha_k h v_1^{-1})\,(v_2(\tilde{c}_0-\tilde{c}_2)+v_1 s_0(\tilde{c}_1-\tilde{c}_0))}$$
$$\left.\left. - \frac{n_2\sinh(\alpha_k h v_1^{-1})\cosh(\alpha_k z_2)}{n_1 s_0\sinh(\alpha_k h v_2^{-1})}\right]\right\}, \qquad (29.24)$$

where

$$\omega_2 = v_1^3 m_1^{-1}(s_3-s_2)^{-1},\quad \gamma_k = \pi k h^{-1},\quad \alpha_k = \mu_k R^{-1}\ (J_1(\alpha_k R)=0),$$

$$F_k^* = (1+\alpha_k^3)\gamma_k^3\left[\frac{n_1\tilde{A}_k^* G_k(1;R)}{1-\gamma_k^2 v_1^2}\left(\gamma_k v_1\sin(\gamma_k h)\cos\left(\frac{h}{v_1}\right) - \cos(\gamma_k h)\sin\left(\frac{h}{v_1}\right)\right)\right.$$
$$\left.+\frac{n_2 G_k(2;R)}{1-\gamma_k^2 v_2^2}\left(\gamma_k v_2\sin(\gamma_k h)\cos\left(\frac{h}{v_2}\right) - \cos(\gamma_k h)\sin\left(\frac{h}{v_2}\right)\right)\right]$$
$$\times\left(\alpha_k^3 J_0(\alpha_k R)\left[\frac{\tilde{c}_1-\tilde{c}_0}{v_1}E_k^*\left(\alpha_k\sinh\left(\frac{\alpha_k h}{v_1}\right)\cos\left(\frac{h}{v_1}\right) + \cosh\left(\frac{\alpha_k h}{v_1}\right)\sin\left(\frac{h}{v_1}\right)\right)\right.\right.$$
$$\left.\left.+\frac{\tilde{c}_2-\tilde{c}_0}{v_2}N_k^*\left(\alpha_k\sinh\left(\frac{\alpha_k h}{v_2}\right)\cos\left(\frac{h}{v_2}\right) + \cosh\left(\frac{\alpha_k h}{v_2}\right)\sin\left(\frac{h}{v_2}\right)\right)\right]\right).$$

The stress-strain state in prestressed half-spaces is defined by the following linearized equations (Guz et al. 2015)

$$Q_{33}^{\prime(i)}(\rho;\zeta_i) = \frac{C_{44}(1+m_1)l_1}{R}\int_0^\infty F(\eta)(se^{\eta\zeta_2} - s_3 e^{\eta\zeta_1})J_0(\eta\rho)d\eta,$$
$$Q_{3r}^{\prime(i)}(\rho;\zeta_i) = -\frac{C_{44}(1+m_1)}{v_1 R}\int_0^\infty F(\eta)(e^{\eta\zeta_2} - e^{\eta\zeta_1})J_1(\eta\rho)d\eta,$$
$$U_3^{\prime(i)}(\rho;\zeta_i) = -\frac{m_1}{v_1}\int_0^\infty\frac{F(\eta)}{\eta}(s_2 e^{\eta\zeta_2} - s_3 e^{\eta\zeta_1})J_0(\eta\rho)d\eta, \qquad (29.25)$$
$$U_r^{\prime(i)}(\rho;\zeta_i) = -\int_0^\infty\frac{F(\eta)}{\eta}(e^{\eta\zeta_2} - s_3 e^{\eta\zeta_1})J_1(\eta\rho)d\eta,$$

where $\xi = z_i v_i R^{-1}, \zeta_i = \xi v_i^{-1} = z_i R^{-1}, \eta = \xi R\,(i=1,2), s = s_0 l_2 l_1^{-1}, s_1 = (m_1 - 1)m_1^{-1}, s_2 = v_1 m_2 (v_2 m_1)^{-1}, s_3 = s_0 v_1 v_2^{-1}$ and $F(\eta)$ is the unknown function.

From the first boundary conditions in (29.20)–(29.21), the unknown function $F(\eta)$ can be defined from the dual integral equations

$$\begin{aligned} &\int_0^\infty \frac{F(\eta)}{\eta} J_0(\eta\rho)d\eta = q(\rho) \ \ (0 < \rho < 1), \\ &\int_0^\infty F(\eta) J_0(\eta\rho)d\eta = 0 \ \ (\rho > 1), \end{aligned} \tag{29.26}$$

where

$$q(\rho) = \varepsilon \left\{ (1 - \chi_0)\frac{\omega_2}{n_1} + \sum_{k=1}^{\infty} \alpha_k J_0(\mu_k \rho) F_k^* \left[\frac{m_1}{n_1} \left(E_k^* \sinh\left(\frac{\mu_k h}{v_1 R}\right) \right.\right.\right.$$
$$\left.\left.\left. + \cosh\left(\frac{\mu_k h}{v_1 R}\right)\right) + \frac{m_2}{n_2}\left(N_k^* \sinh\left(\frac{\mu_k h}{R v_2}\right) + M_k^* \cosh\left(\frac{\mu_k h}{R v_2}\right)\right)\right] \chi_k \right\}.$$

Using the inversion formula (Guz et al. 2015), we obtain

$$\frac{F(\eta)}{\eta} = \frac{2\varepsilon}{\pi}\left((1 - \chi_0)\frac{\omega_2}{n_1}\Psi(\eta, 0) \right.$$
$$+ \sum_{k=1}^{\infty} \frac{\mu_k}{R} F_k^* \left[\frac{m_1}{n_1}\left(E_k^* \sinh\left(\frac{\mu_k h}{v_1 R}\right) + \cosh\left(\frac{\mu_k h}{v_1 R}\right)\right)\right.$$
$$\left.\left. + \frac{m_2}{n_2}\left(N_k^* \sinh\left(\frac{\mu_k h}{R v_2}\right) + M_k^* \cosh\left(\frac{\mu_k h}{R v_2}\right)\right)\right] \chi_k \Psi(\eta, \mu_k) \right). \tag{29.27}$$

To determine the coefficients $N_k{}^*$, $E_k{}^*$, $M_k{}^*$ and the function $F(\eta)$, new variables are introduced:

$$\begin{aligned} &\int_0^\infty \frac{F(\eta)}{\eta} J_1(\eta\rho)d\rho = -\frac{\varepsilon R(v_2 + v_1 s)}{2 v_1 v_2 h (s - s_3)\theta_6}\chi_0, \quad \chi_k = -\frac{R n_1}{\varepsilon \mu_k \omega_2}\tilde{B}_k, \\ &\int_0^\infty \eta \Psi(\eta, \mu_k) \int_0^1 \rho J_0(\mu_n \rho) J_0(\eta\rho) d\rho d\eta = \Psi(\mu_n, \mu_k). \end{aligned} \tag{29.28}$$

The infinite system, which is similar to (29.18) for the unknowns χ_k $(k = 0, 1, \ldots)$ in (29.24)–(29.27), is obtained. Thus, this problem is also reduced to the determination of constants χ_k $(k = 0, 1, 2, \ldots)$ that determine the stress state of the elastic punch and two half-spaces with initial stresses. Using the equilibrium condition (29.23), the formula for the load P is found as $P = \pi\varepsilon R^2 C_{44}(1 + m_1) l_1 (v_2 + s v_1)(v_2 v_1 h \theta_6)^{-1}$.

After defining the unknowns χ_k $(k = 0, 1, 2, \ldots)$ from a system of linear algebraic equations similar to (29.18), it is possible to obtain displacements and stresses in both elastic half-spaces and elastic punch (Babych et al. 2022). Problem solutions are also represented as series containing an infinite system of constants χ_k $(k = 0, 1, \ldots)$.

29.4.3 *Numerical Results*

Numerical analysis is presented for the Treloar potential (for Neo-Hookean bodies). The distribution of normal stresses $Q_{33}{}'^{(3)}/P$ in the contact zone (at $z_i = h/v_i$) and along the elastic cylinder is shown in Fig. 29.7. Displacement $U_3{}'^{(3)}/\varepsilon$ in the contact zone (at $z_i = h/v_i$) and along the cross section of the elastic cylinder (at $z_i = 0$) is shown in Fig. 29.8, where the dashed curve corresponds to contact without initial stresses ($\lambda_1 = 1$), and the solid curve—with initial stresses. All quantities in Figs. 29.9 and 29.10 are presented in dimensionless coordinates. When initial (residual) stresses are absent ($\lambda_1 = 1$), the graph of contact stresses distribution corresponds to the known solutions of the contact problem about the pressure of two half-spaces on a cylinder without accounting for initial stresses (Grilitskiy and Kizyima 1981). The dependence of the equivalent load P on the elongation coefficient λ_1 is illustrated in Fig. 29.10 for some argument values.

29.5 Contact Problem for a Rigid Ring Punch With Half-Space With Initial (Residual) Stresses

The section presents the problem formulation, boundary conditions, solution method, and numerical results for the problem.

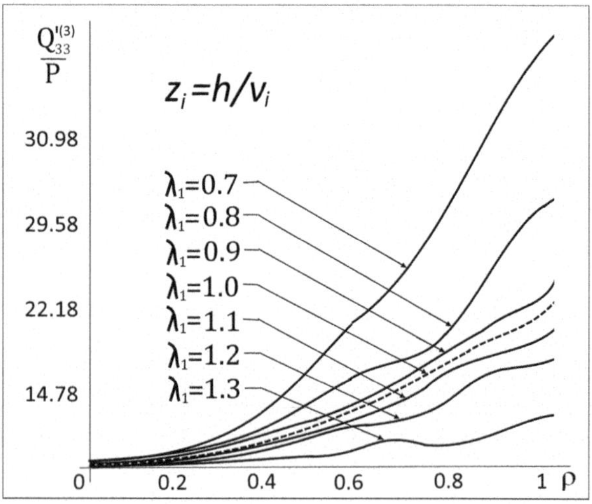

Fig. 29.8 Contact stresses $Q_{33}'^{(3)}/P$

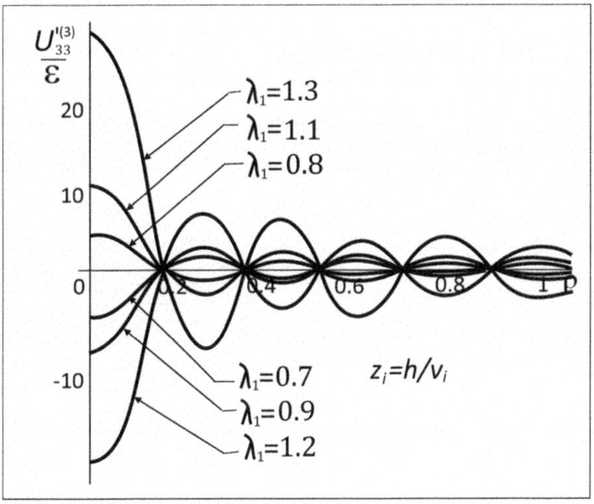

Fig. 29.9 Contact displacement $U_3'^{(3)}/\varepsilon$ in the contact zone

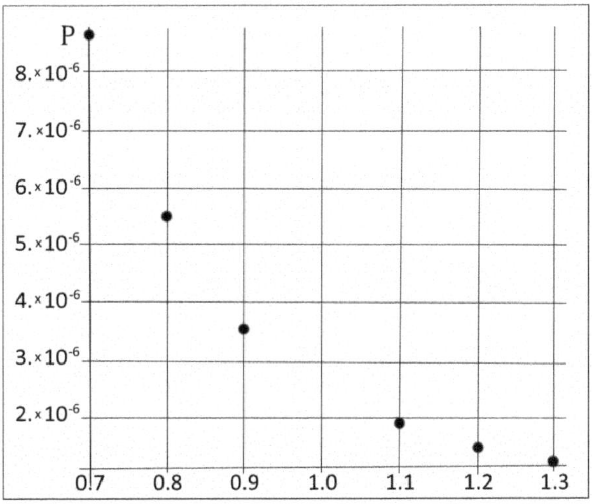

Fig. 29.10 Dependence of equivalent force P on the elongation coefficient λ_1

29.5.1 Problem Statement and Boundary Conditions

Let a finite rigid ring-shaped punch (with a flat base) is pressed into a prestressed half-space with force P (Fig. 29.11). The geometric symmetry axis of the punch coincides with the y_3-axis (in the cylindrical coordinate system) directed into the half-space. The geometry parameters R_1 and R_2 are the inner and outer radii of the punch, respectively. The external load is assumed to be applied only to the free end of the elastic punch. All the points of the punch base move to the direction of the symmetry axis y_3 by the same value ε.

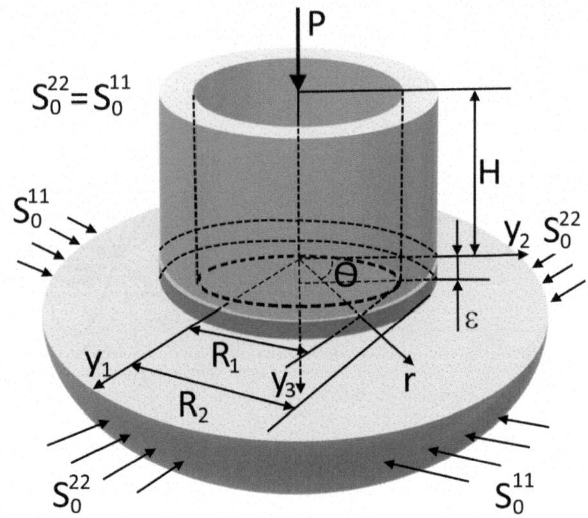

Fig. 29.11 Rigid ring-shaped punch on elastic half-space with initial stresses

In the circular cylindrical coordinate system (r, θ, z_i), the following boundary conditions correspond to the given above problem statement:

$$U_z = \varepsilon \ (R_1 < r < R_2), \tag{29.29}$$

$$Q_{zz} = 0 \ (0 < r < R_1 \ R_2 < r < \infty), \tag{29.30}$$

$$Q_{rz} = 0 \ (0 < r < \infty), \tag{29.31}$$

$$U_z = 0 \ (0 < r < R_1, \ R_2 < r < \infty), \tag{29.32}$$

$$Q_{zz} = -\sigma_z^0 \ (R_1 < r < R_2), \tag{29.33}$$

where $\sigma_z^0 = 8\varepsilon\omega_3(\pi\sqrt{1-r^2})^{-1}$ for $R_1 < r < R_2$, and $\sigma_z^0 = 0$ for $0 < r < R_1, R_2 < r < \infty$; $\omega_3 = C_{44}(1+m_1)l_1(s-s_0)$.

The equilibrium condition that establishes the relation between the punch indentation and the equivalent load P is as follows

$$P = -2\pi \int_{R_1}^{R_2} r Q_{zz}(0, r)dr. \tag{29.34}$$

29.5.2 *Method of Solution*

The stress-strain state in prestressed half-space is determined based on conditions (29.29)–(29.33) and $z_1 = 0$ for unequal roots by the following quantities (Habrusieva and Shelestovskyi 2011; Yaretskaya 2018)

$$
\begin{aligned}
Q_{33} &= \frac{\omega_4}{R_0}\int_0^{\infty} F(\eta)J_0(\eta r)d\eta, \\
U_3 &= -\frac{1}{\omega_5}\int_0^{\infty} \frac{F(\eta)}{\eta}J_0(\eta r)d\eta, \\
U_r &= \omega_6\int_0^{\infty} \frac{F(\eta)}{\eta}J_1(\eta r)d\eta,
\end{aligned}
\tag{29.35}
$$

where

$$
R_0 = \frac{R_1}{R_2}, \quad \omega_4 = C_{44}(1+m_1)l_1(s-s_3), \quad \omega_5 = \frac{v_1}{m_1(s_2-s_3)}, \quad \omega_6 = s_3 - 1.
$$

Taking into account condition (29.33), we obtain:

$$
\frac{\omega_4}{R_0}\int_0^{\infty} F(\eta)J_0(\eta r)d\eta = 0 \quad (0 < r < R_1, \ \ R_2 < r < \infty) \tag{29.36}
$$

Introducing the unknown continuous function $f(r)$ $(R_1 < r < R_2)$ defining the distribution of die pressure, and extending (29.36) to $0 < r < \infty$, we get (Habrusieva and Shelestovskyi 2011)

$$
\frac{\omega_4}{R_0}\int_0^{\infty} F(\eta)J_0(\eta r)d\eta = f(r)\left(\delta(r-R_1) - \delta(r-R_2)\right) \quad (0 < r < \infty) \tag{29.37}
$$

where $\delta(r)$ is the delta function.

Since the function $f(r)$ ($f(r) = 0$ at $r \leqslant R_1$ and $r \geqslant R_2$) is continuous, we present it by a segment of the generalized Furrier series (Habrusieva and Shelestovskyi 2011; Yaretskaya 2018):

$$
L_n(r) = J_0\left(\frac{\gamma_n}{R_1}r\right)Y_0(\gamma_n) - J_0(\gamma_n)Y_0\left(\frac{\gamma_n}{R_1}r\right),
$$

where γ_n are the positive roots of the equation

$$
J_0\left(\frac{\gamma_n}{R_1}R_2\right)Y_0(\gamma_n) - J_0(\gamma_n)Y_0\left(\frac{R_2}{R_1}\gamma_n\right) = 0,
$$

$Y_0(x)$ is the Weber function. Thus,

$$
f(r) = \sum_{n=1}^{\infty} a_n L_n(r),
$$

where a_n are unknown coefficients.

Applying the inversion formula of the integral Hankel transform to (29.37), we get

$$\frac{F(\eta)}{\eta} = \frac{R_0}{\omega_4} \sum_{n=1}^{\infty} a_n \Phi_n(\eta) \quad (0 < \eta < \infty), \tag{29.38}$$

where

$$\Phi_n(\eta) = \int_{R_1}^{R_2} r L_n(r) J_0(\eta r) dr$$
$$= \frac{\gamma_n \eta^2}{\gamma_n^2 - (\eta R_1)^2} \left\{ \frac{R_2}{R_1} \left[J_1\left(\frac{R_2}{R_1}\gamma_n\right) Y_0(\gamma_n) - Y_1\left(\frac{R_2}{R_1}\gamma_n\right) J_0(\gamma_n) \right] \right.$$
$$\left. \times J_0(\eta R_2) - \left[J_1(\gamma_n) Y_0(\gamma_n) - Y_1(\gamma_n) J_0(\gamma_n) \right] J_0(\eta R_1) \right\}.$$

Using the second expression in (29.35), Eqs. (29.38) and (29.40), we obtain

$$\sum_{n=1}^{\infty} a_n \phi_n(r) = \frac{\varepsilon}{k_1} \quad (R_1 < r < R_2), \tag{29.39}$$

where

$$k_1 = -\frac{R_0}{\omega_4 \omega_5}, \quad \phi_n(r) = \int_0^{\infty} \Phi_n(\eta) J_0(\eta r) d\eta.$$

After determining a_n from (29.39), it is possible to obtain the components of the stress-strain state in the elastic half-space using (29.38) and (29.35). The relation between the indentation displacement and the equivalent force P, according to (29.34), reads $P = 16\omega_3\varepsilon(1 - \sqrt{1 - R_0})$.

29.5.3 Numerical Results

This section presents a numerical analysis corresponding to the Treloar potential (Neo-Hookean bodies). The normal contact stress $\varepsilon^{-1} Q_{33}$ and radial displacement $\varepsilon^{-1} U_r$ is shown in Figs. 29.12 and 29.13 in dimensionless coordinates. The larger the elongation coefficients λ_1, the higher the curve for mentioned relations. Dotted curves correspond to half-space without initial stresses ($\lambda_1 = 1$), and solid curves—with initial stresses.

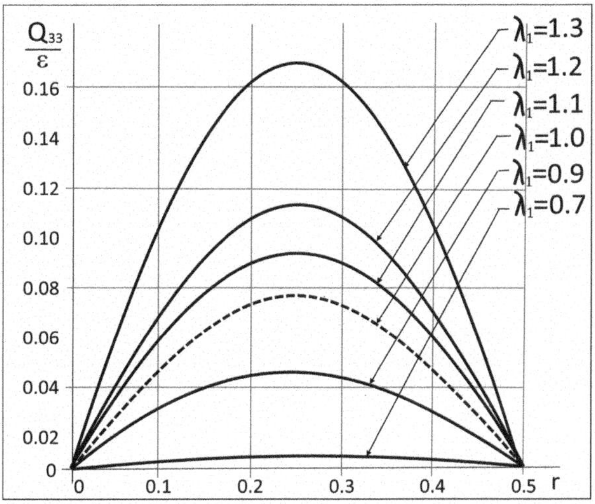

Fig. 29.12 Contact stresses $Q_{33}\varepsilon^{-1}$. The potential of Treloar

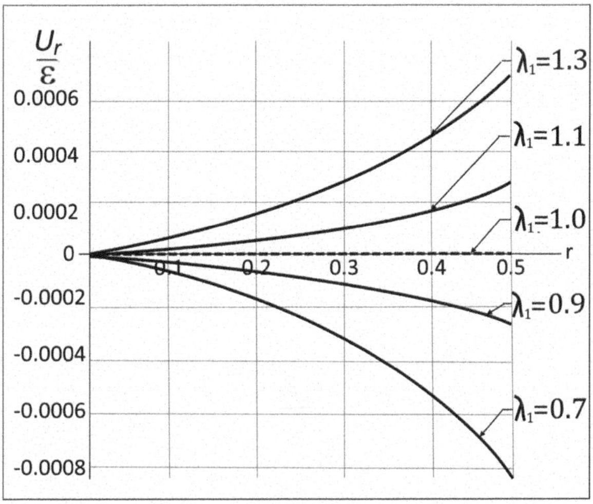

Fig. 29.13 Contact displacement $U_r\varepsilon^{-1}$. The potential of Treloar

29.6 Pressure of Elastic Ring Punch With Initial Stresses on Prestressed Half-Space

The section presents the formulation of the problem, boundary conditions, solution method, and numerical results of the problem on the indentation of an elastic ring-shaped punch with initial stresses into a prestressed half-space.

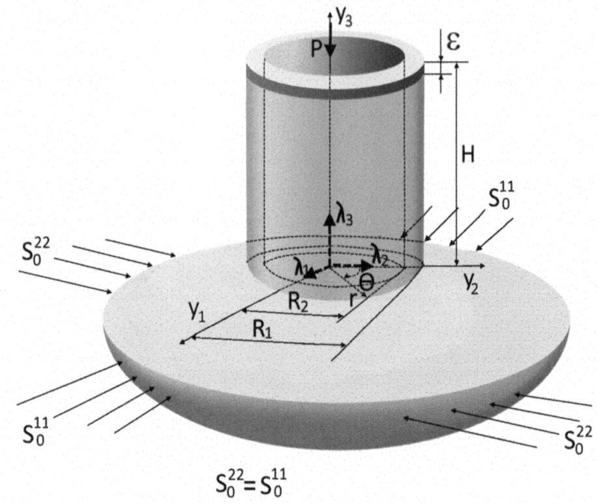

Fig. 29.14 Elastic ring-shaped punch and elastic half-space with initial stresses

29.6.1 Problem Statement and Boundary Conditions

Let a finite elastic ring-shaped punch of height H is pressed into a half-space with force P. The geometric axis of symmetry coincides with the axis y_3 of the cylindrical coordinate system (r, θ, y_3) and is directed inside the half-space (Fig. 29.14). Geometry parameters R_1 and R_2 are the inner and outer radii of the punch, respectively. The external load is assumed to be applied only to the free end of the punch, where all points move along the axis of symmetry y_3 by the same value ε. The punch is assumed to be prestressed.

In the system of cylindrical coordinates (r, θ, y_3), the following equations correspond to the boundary conditions:

$$U_3^{\prime(1)} = -\varepsilon, \quad Q_{3r}^{\prime(1)} = 0 \quad (z_i = H v_i^{-1}, \quad i = 1, 2; \quad R_1 < r < R_2), \tag{29.40}$$

$$U_3^{\prime(1)} = U_3^{\prime(2)}, \quad Q_{33}^{\prime(1)} = Q_{33}^{\prime(2)}, \quad Q_{3r}^{\prime(1)} = Q_{3r}^{\prime(2)} = 0$$
$$(z_i = 0, \quad i = 1, 2; \quad R_1 < r < R_2), \tag{29.41}$$

$$Q_{33}^{\prime(2)} = 0, \quad Q_{3r}^{\prime(2)} = 0 \quad (z_i = 0, \quad i = 1, 2; \quad 0 < r < R_1, \quad R_2 < r < \infty), \tag{29.42}$$

$$Q_{rr}^{\prime(1)} = 0, \quad Q_{3r}^{\prime(1)} = 0 \quad (0 \leqslant z_i \leqslant H v_i^{-1}, \quad r = R_1, \quad r = R_2). \tag{29.43}$$

The equilibrium condition that establishes the relation between the indentation displacement and the equivalent force P is as follows

$$P = -2\pi \int_{R_1}^{R_2} r Q_{33}^{\prime(2)}(0, r)dr. \tag{29.44}$$

29.6.2 Method of Solution

To determine the stress-strain state in an elastic ring-shaped punch with initial stresses, was use the linearized equations (see p. 78 in Babych and Yarets'ka (2021)). These equations give the displacement and the stress for compressible and non-compressible bodies. The general solution $\chi = \chi_1 + \chi_2$ for the case of unequal roots of the characteristic equation (Babych and Yarets'ka 2021) is obtained in the form

$$\begin{aligned}
\tilde{\chi} &= 2A_0(r^2 - z_1^2 - z_2^2) + 3r^2 C_0(z_1^3 + z_2^3) \\
&+ \sum_{k=1}^{\infty} \Big\langle C_k \Big\{ \Big[A_k^{(1)} I_0(\gamma_k \upsilon_1 r) + A_k^{(2)} K_0(\gamma_k \upsilon_1 r) \Big] \sin(\gamma_k \upsilon_1 z_1) \\
&\quad + \Big[A_k^{(1)} I_0(\gamma_k \upsilon_1 r) + A_k^{(2)} K_0(\gamma_k \upsilon_1 r) \Big] \sin(\gamma_k \upsilon_1 z_1) \Big\} \\
&\quad + M_k \Big[T_k^{(1)} J_0(\alpha_k r) + T_k^{(2)} Y_0(\alpha_k r) \Big] \Big(\tilde{S}_2(\alpha_k z_1) + \tilde{S}_3(\alpha_k z_2) \Big) \Big\rangle,
\end{aligned}$$

where $A_k^{(2)}$, $B_k^{(2)}$, $T_k^{(2)}$, C_0, C_k, $\tilde{M}_k$ = const, and M_k are the unknowns,

$$\begin{aligned}
&\tilde{S}(x) = \tilde{E}_k \sinh(x) + \tilde{F}_k \cosh(x), \quad \tilde{S}_3(x) = \tilde{N}_k \sinh(x) + \cosh(x), \\
&T_k^{(1)} = -Y_1(\alpha_k R_1 R_2^{-1})(J_1(\alpha_k R_1 R_2^{-1}))^{-1} T_k^{(2)}, \\
&A_k^{(1)} = K_1(\gamma_k \upsilon_1 R_1)(I_1(\gamma_k \upsilon_1 R_1))^{-1} A_k^{(2)}, \\
&B_k^{(1)} = K_1(\gamma_k \upsilon_2 R_1)(I_1(\gamma_k \upsilon_2 R_1))^{-1} B_k^{(2)}, \\
&\tilde{N}_k = -\coth(\alpha_k H \upsilon_2^{-1}), \quad \tilde{S}_4(x) = \tilde{E}_k \cosh(x) + \tilde{F}_k \sinh(x), \\
&A_0 = (3C_0 H(\upsilon_1)^{-1} - \varepsilon n_1 n_2 (4(m_1 n_2 + m_2 n_1))^{-1}), \\
&M_k = \tilde{M}_k T_k^{(2)}, \quad \tilde{E}_k = (1 + m_2) n_1 ((1 + m_1) n_2)^{-1} \coth(\alpha_k H \upsilon_1^{-1}), \\
&\tilde{F}_k = -(1 + m_2) n_1 ((1 + m_1) n_2)^{-1}, \quad \gamma_k = \pi k H^{-1}, \\
&\alpha_k = \mu_k R_2 R_1^{-1}, \quad J_1(\mu_k) Y_1(\mu_k R_2 R_1^{-1}) - Y_1(\mu_k) J_1(\mu_k R_2 R_1^{-1}) = 0.
\end{aligned}$$

Basing on (29.40)–(29.43) and $z_1 = 0$, the stress-strain state in the prestressed half-space is obtained for unequal roots in the form (Babych and Yarets'ka 2021; Guz et al. 2015; Guz 2022):

$$
\begin{aligned}
Q_{33}^{\prime(2)} &= \frac{\omega_3}{R_2 - R_1}\int_0^{\infty} F(\eta) J_0(\eta r) d\eta, \\
u_3^{\prime(2)} &= -\frac{1}{\omega_2}\int_0^{\infty} \frac{F(\eta)}{\eta} J_0(\eta r) d\eta, \\
u_r^{\prime(2)} &= \omega_1 \int_0^{\infty} \frac{F(\eta)}{\eta} J_1(\eta r) d\eta,
\end{aligned}
\qquad (29.45)
$$

where $\omega_2 = v_1\left(m_1(s_3 - s_2)\right)^{-1}$, $\omega_1 = s_0 - 1$.

Satisfying the first condition (29.41), we define the unknown function $F(\eta)$ in (29.45) from of the triple integral equations

$$
\begin{aligned}
&\int_0^{\infty} F(\eta) J_0(\eta r) d\eta = 0 \qquad (R_2 < r < \infty), \\
&\int_0^{\infty} \frac{F(\eta)}{\eta} J_0(\eta r) d\eta = f(r) \ (R_1 < r < R_2), \\
&\int_0^{\infty} F(\eta) J_0(\eta r) d\eta = 0 \qquad (0 < r < R_1),
\end{aligned}
\qquad (29.46)
$$

where

$$
\begin{aligned}
f(r) &= \frac{\omega_2}{R_2}\left(\varepsilon + t_1 \sum_{k=1}^{\infty} \alpha_k^2 \left(\frac{Y_1(\alpha_k R_1 R_2^{-1})}{J_1(\alpha_k R_1 R_2^{-1})} J_0(\alpha_k R_2^{-1} r) - Y_0\left(\alpha_k R_2^{-1} r\right)\right) M_k\right), \\
t_1 &= \frac{m_1 - m_2}{n_2(1 + m_1)}.
\end{aligned}
$$

Further, we reduce the integral Eq. (29.46) to a single equation (Bosakov 2018). The function $F(\eta)$ then can be sought in the form (Bosakov 2018):

$$
F(\eta) = R_2 \sum_{n=0}^{\infty} W_{2n} J_{2n}\left(\tfrac{1}{2}\eta(R_2 - R_1)\right) J_{2n}\left(\tfrac{1}{2}\eta(R_2 + R_1)\right), \qquad (29.47)
$$

where W_{2n} are unknown constants. Substituting (29.47) in (29.46), we get the integral equation

$$
\sum_{n=0}^{\infty} W_{2n} \int_0^{\infty} J_{2n}\left(\tfrac{1}{2}\eta(R_2 - R_1)\right) J_{2n}\left(\tfrac{1}{2}\eta(R_2 + R_1)\right) J_0(\eta r) d\eta = f(r). \qquad (29.48)
$$

From the second boundary condition (29.41), we obtain

$$\sum_{n=0}^{\infty} W_{2n} \int_0^{\infty} J_{2n}\left(\tfrac{1}{2}\eta(R_2 - R_1)\right)$$
$$\times J_{2n}\left(\tfrac{1}{2}\eta(R_2 + R_1)\right) \int_{R_1}^{R_2} r J_0(\mu_k r) J_0(\eta r) dr d\eta$$
$$= \frac{\hat{N}_{44}(R_2 - m_1)(1 + m_2)}{\omega_3 v_2} \alpha_k^2 t_1 \left[\frac{Y_1(\alpha_k R_1 R_2^{-1})}{J_1(\alpha_k R_1 R_2^{-1})} \tilde{O}_k^{(1)} - \tilde{O}_k^{(2)} \right] M_k. \quad (29.49)$$

where

$$\tilde{O}_k^{(1)} = \frac{R_2}{\alpha_k} \left[R_1 J_1\left(\alpha_k R_1 R_2^{-1}\right) - R_2 J_1(\alpha_k) \right],$$
$$\tilde{O}_k^{(2)} = \frac{R_2}{\alpha_k} \left[R_1 Y_1\left(\alpha_k R_1 R_2^{-1}\right) - R_2 Y_1(\alpha_k) \right].$$

To determine the constants M_i and W_{2i} $(i = 0, 1, \dots)$ from (29.45) and (29.46), we use an infinite system of linear algebraic equations consisting of (29.48) and (29.49). This system is solved by the reduction method, taking into account that $W_0 = \omega_2 \varepsilon \pi (R_2 - R_1)(8\omega_3 R_2)^{-1}$. Using the equilibrium condition (29.44), we establish the relation between the indention displacement and the equivalent force P in the following form

$$P = \frac{2\omega_2 \omega_3 \varepsilon}{\pi (R_2 - R_1)}.$$

Having determined the unknown constants M_i and W_{2i} $(i = 0, 1, \dots)$, we can obtain displacement and stress in the elastic half-space as well as in the punch. It should be noted that the coefficients depend on the elastic potential and the height of the elastic punch H.

29.6.3 Numerical Results

In this section, numerical solutions corresponding to the Treloar potential are presented. Figures 29.15 and 29.16 show the normal contact stress $P^{-1} Q_{33}{}^{(1)}$ distribution and displacement $-(\varepsilon R_2)^{-1} U_3{}^{(1)}$ under the ring-shaped punch at the contact area boundary in dimensionless coordinates. The dashed curve corresponds to the half-space without initial stresses ($\lambda_1 = 1$), and the solid curve—with initial stresses.

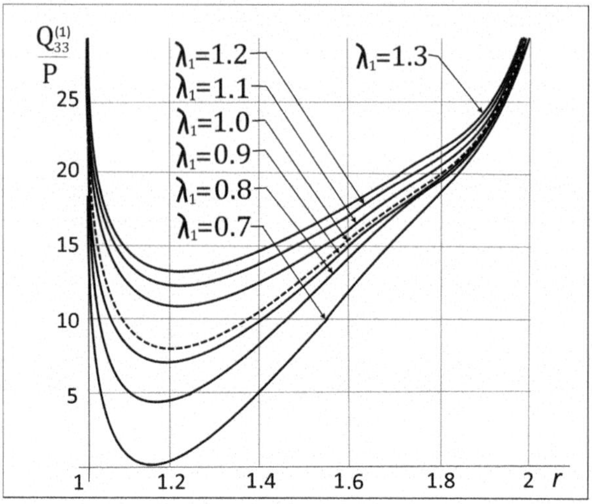

Fig. 29.15 Contact stress

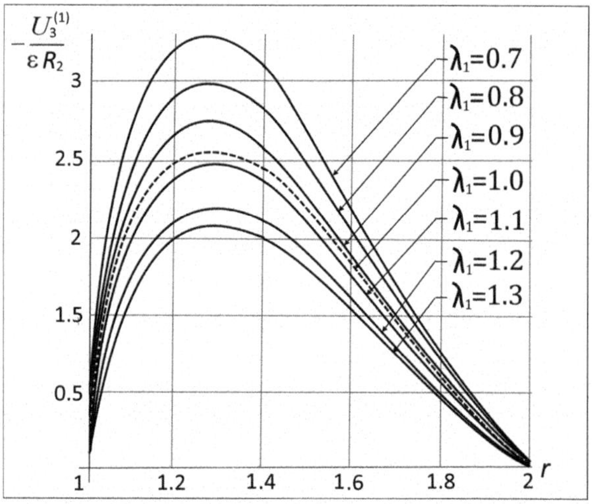

Fig. 29.16 Contact displacement

29.7 Conclusion

The general engineering conclusions for the considered problems are as follows:

1. initial stresses in bodies lead to a decrease in stresses and an increase in the absolute values of displacements in compression ($\lambda_1 < 1$) an increase in stresses and a decrease in the absolute values of displacements in tension ($\lambda_1 > 1$). Thus, a prestressed state during the contact interaction of elastic bodies makes it possible to adjust contact stress and displacement when determining the strength of structures and their members. Moreover, for contact stresses, initial stresses are

dangerous in case of tension, and for displacements, initial stresses are dangerous in case of compression;
2. more significantly, in quantitative terms, initial stresses act in highly elastic materials in comparison with more rigid ones, but their influence remains qualitatively;
3. in cases of absence of initial stresses, the obtained results coincide with the classical ones.

In the case of the pressure of two prestressed half-spaces on the elastic cylindrical punch and the pressure of the cylindrical punch on the layer, it was established that the most significant effect of initial stresses is observed on the side surface of the punches. The thickness of the layer does not affect the nature of the initial stress and affects only their value. The situation is dangerous when initial stresses approach the values of surface instability since contact stresses and displacement dramatically change their values.

Thus, from the study of the contact interaction of two prestressed half-spaces and an elastic cylindrical punch, it can be concluded that the closer to the central cross section of the cylindrical punch, the faster the normal stresses tend to zero. Furthermore, displacement $U_3{'^{(3)}}/\varepsilon$ takes significantly higher values closer to the axis of the cylindrical punch than to its side surface. The values of the equivalent force P decrease with an increase in the elongation coefficient λ_1. Thus, the force P takes greater values during tension than during compression.

The research made it possible to

1. obtain analytical and numerical dependencies that determine the behavior of stresses and displacements in the contact zone;
2. develop algorithms for numerical determination of contact characteristics;
3. use the proposed research principle and solution algorithm when designing technological types of equipment, buildings, and other structures.

Thus, this chapter presents research on contact problems for prestressed bodies with elastic or rigid punches. We believe that further progress in the development of the mechanics of contact interaction of prestressed bodies (both in the case of rigid and elastic punches) is determined by the research of more complex classes of problems, for example, taking into account friction, conducting experimental researches on the effect of initial stresses on the main characteristics of contact interaction of structural materials.

References

Aleksandrov VM, Arutyunyan NK (1969) Contact problems for prestressed deformed bodies. Sov Appl Mech 20(3):209–215

Babich SY, Dikhtyaruk NN (2001) Load transfer from an infinite inhomogeneous stringer to an elastic strip clamped by one face with initial stresses. Int Appl Mech 55(6):346–356

Babich SY, Dikhtyaruk NN, Degtyar SV (2019) Contact problem for two identical strips reinforced by periodically arranged fasteners with initial stresses. Int Appl Mech 55(6):629–635

Babych SY, Yarets'ka NO (2021) Contact problem for an elastic ring punch and a half-space with initial (residual) stresses. Int Appl Mech 57(3):297–305

Babych SY, Yaretska NO, Lazar VF, Shchekan NP (2022) Analytical solutions of the static problem about the pressure of prestressed half-spaces and an elastic cylinder with initial stresses. Sci Bull Uzhhorod Univ 41(2):91–102 [in Ukrainian]

Bosakov SV (2018) Two contact problems on the indentation of an annular punch into an elastic half-space. Sci Technol 6(17):458–464 [in Russian]

Dhaliwal RS, Singh BM, Rokne JG (1980) Axisymmetric contact and crack problems for a initially stressed Neo-Hooken elastic layer. Int J Eng Sci 18(1):169–179

Grilitskiy DV, Kizyima YM (1981) Axisymmetric contact problems of the theory of elasticity and thermoelasticity. Higher School, Lviv [in Russian]

Guz AN, Babich SY, Glukhov YP (2015) Mixed problems for an elastic foundation with initial stresses. In: Mixed problems for prestressed elastic foundation. Lap Lambert Academic Publishing, Germany [in Russian]

Guz AN (2003) Establishing the fundamentals of the theory of stability of mine workings. Int Appl Mech 39(1):20–48

Guz AN (2019) Nonclassical problems of fracture?: Failure mechanics: on the occasion of the 50th anniversary of research (review). III. Int Appl Mech 55(4):343–415

Guz AN (2022) On general list of references to the monograph "Eight non-classical problems of fracture mechanics.âŁž Int Appl Mech 58(1):1–29

Guz OM, Babych SY, Rudnitskii VB (1998) Contact problems for elastic bodies with initial stresses: Focus on Ukrainian research. Int Appl Mech Rew 51(5):343–371

Guz OM, Babych SY, Glukhov AY (2021) Axisymmetric waves in prestressed highly elastic composite material. Long wave approximation. Int Appl Mech 57(2):134–147

Habrusieva TY, Shelestovskyi BH (2011) Contact interaction of a ring die with a prestressed isotropic layer. MMFMP 54(3):138–146 [in Ukrainian]

Hutchinson JW, Thompson JMT (2017) Nonlinear buckling interaction for spherical shells subject to pressure and probing forces. J Appl Mech 84(6):061001

Kurashige M (1969) Circular crack problem for initially stressed neo-Hookean solid. ZAMM 49(8):671–678

Yaretskaya NA (2014) Three-dimensional contact problem for an elastic layer and a cylindrical punch with prestresses. Int Appl Mech 50(4):378–388

Yaretskaya NA (2018) Contact problem for the rigid ring punch and the half-space with initial (residual) stresses. Int Appl Mech 54(5):539–543

Chapter 30
Express Diagnostics of Material Strength Properties Under Shock-Wave Destruction

Yaroslav Zhuk, Mykola Melnichenko, Volodymyr Andruschenko, and Mykhailo Vodotovka

30.1 Introduction

The development of aerospace and military equipment, energy, and other branches of industry calls for a detailed experimental study of the reaction of various structural materials to shock-wave loading. The analysis of shock-wave phenomena in solids allows us to establish the peculiarities of the course of deformation processes at high-impact velocities. Currently, there is still no unified theory that would explain the key effects that determine the elastic-plastic properties and dynamic strength of materials under intense loadings. Peculiarities of the behavior of metals and ceramics under conditions of dynamic and shock-wave loading caused by the development of defects are the subject of modern active research, which consists of the experimental study of the behavior of materials and the development of wide-range modeling methods.

For a long time, obtaining planar non-stationary shock waves to study the thermodynamic and mechanical properties of solid and liquid continua under the influence of impulse pressure was carried out by traditional methods, such as the detonation of a charge of a chemical explosive located on the surface of a sample, or by the impact of a thin plate previously accelerated by the products of the explosion, or electric gun (Saxena et al. 2010; Zirnheld et al. 2009). The use of these methods of studying the mechanical properties of materials under conditions of high deformation rates at low impulse loads is limited either by the critical thickness of the explosive substance or by the stability of the acceleration of thin plates.

Y. Zhuk (✉) · M. Melnichenko · V. Andruschenko · M. Vodotovka
Faculty of Mechanics and Mathematics, Taras Shevchenko National University of Kyiv, Kyiv, Ukraine
e-mail: yaroslavzhuk@knu.ua

V. Andruschenko
e-mail: realcrystallab@univ.kiev.ua

M. Vodotovka
e-mail: nanocrystallab@univ.kiev.ua

A. N. Guz et al. (eds.), *Advances in Mechanics*, Advanced Structured Materials 191,
https://doi.org/10.1007/978-3-031-37313-8_30

Therefore, for these purposes, an electric explosion of a metal foil placed on the surface of the tested samples is currently used. The electric explosion method provides for the possibility of rapid conversion of electromagnetic energy into thermal energy of the foil material and ensures the production of fading shock waves with an amplitude ranging from several units to tens of GPa (Grigor'ev and Pavlenko 2009; Muffoletto et al. 2018). At the same time, it was experimentally proven that the method of an electrical explosion of the foil allows for studying the mechanical properties under planar-impulse loading. Shock waves initiated by the electrical explosion of conductors in various environments are widely used due to the ability to control their characteristics in wide ranges by changing the parameters of the discharge RLC circuit, the size, and the material of the exploding conductor. Studies of the dynamic destruction of solids under shock-wave loading demonstrate a number of effects (Kanel et al. 2013; Stepanov and Makovei 1993) that contradict classical models of strength and crack resistance (Kundu 2008). The latter is very important since dynamic destruction is decisive in armor-piercing and projectile fragmentation. At the same time, the complexity of dynamic research consists of the need to record the process parameters in the microsecond time range and to increase the informativeness; it is necessary to evaluate the behavior of materials under various types of loading and destruction: in plane load waves, during the development of a single crack, and when breaking through an obstacle.

The specificity of shock-wave effects is that due to the short duration of the loading time, there is practically no mutual influence of individual sections of the loaded body containing any large defects, and, accordingly, their contribution to the destruction process. In addition, at the same time, extremely high rates of deformation develop, heating of the substance occurs, and the mechanism of plastic deformation itself changes (for example, additional sliding planes are activated in metals, the contribution of twinning to deformation increases even for those materials where deformation twins do not form under normal loading conditions).

The relevance of research in this field is also due to the need to develop mathematical models capable of describing the connection between elastic-plastic behavior and the destruction of materials with the kinetics of defect accumulation. At the same time, further studies of changes in strength at the micro level and mechanisms of plastic deformation will contribute to the design of new high-strength materials and the improvement of their processing technology.

30.2 Problem Statement and Experimental Technique

The main goal of the work was to develop a method of express diagnostics for the experimental determination of the strength properties of ceramic, glass, and polymer materials during their shock-wave destruction.

For this purpose, the authors developed a shock-wave experimental device based on the use of the specifics of shock-wave effects. It was designed to test the dynamic strength of ceramics, glass, and Plexiglas materials of various thicknesses. The shock-wave device allows for generating an elastic compression pulse up to 5 GPa for less than 0.4 microseconds on an area of several tens of square millimeters. In this case, a thin metal foil is clamped between two smooth plates (for example, made of glass) and explodes (melts and evaporates) from the discharge of the storage capacitor. The compression wave reaches the upper free boundary of the tested plate and, reflecting from it, forms a wave of destruction, which ensures conical spalling destruction. With the help of such a device, it is possible to accurately determine the value of dynamic strength at the beginning of spalling failure and, thus, to modify the technological process to obtain the required strength parameters or to compare and select materials for specific purposes.

30.3 Experimental Results

Using the shock-wave experimental device designed by the authors, an express diagnosis of the strength properties of the studied plates made of ceramics, artificial stone, and composite material during their shock-wave destruction was carried out.

30.3.1 Investigation of Ceramic Plate Strength Properties

The peculiarities of the properties of ceramics as a material are determined both by the raw materials (clays and their mixtures with mineral additives, as well as oxides and their compounds) and by the production technology (sintering of components with subsequent cooling). Therefore, the shock-wave strength characteristics can differ significantly for different compositions. In practice, it is important to know at least two experimental parameters. The first is the magnitude of the shock pulse when quasi-static failure begins. In this case, the plate is destroyed only with the formation of radial cracks. The second is the value of the shock pulse when the cone spalling failure begins. Figure 30.1 presents photos of the destruction of ceramic plates measuring 100 × 100 × 10 mm from household ceramics.

The test at the minimum power of the electric explosion leads to quasi-static destruction with the formation of radial cracks Fig. 30.1a. With an increase in the power of the explosion, not only quasi-static destruction is observed but also spalling destruction due to the formation of one cone, Fig. 30.1b, c.

A similar pattern of simultaneous quasi-static destruction and spalling destruction was observed for armor plates made of silicon carbide (Fig. 30.2a). Figure 30.2b shows the spalling cone that was formed after the shock-wave destruction.

A further increase in the power of the explosion and a decrease in the duration of the front leads to the appearance of multi-conical spalling destruction. It should be noted that spalling damage is possible with the formation of two (Fig. 30.3a, b, c), three, or four cones with impressive damage patterns. The analysis of ceramic samples obtained in this work after quasi-static destruction makes it possible to

a *b* *c*

Fig. 30.1 Destruction patterns of 100 × 100 × 10 mm ceramic plates: quasi-static loading **a**, spalling cone formation **b**, **c** at high power loading

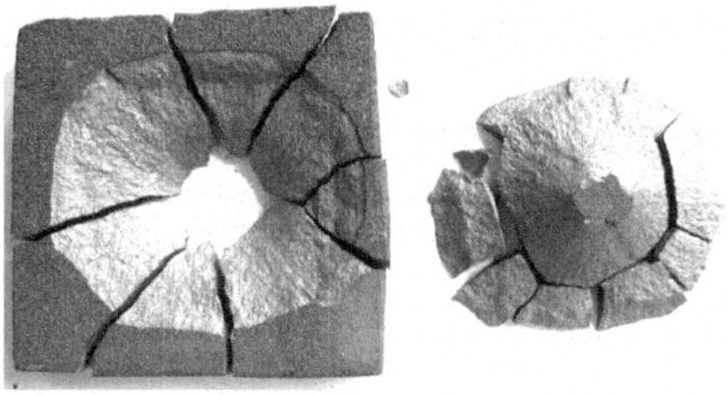

a *b*

Fig. 30.2 Mixed quasi-static and spalling destruction pattern of armor plate made of silicon carbide **a**; spalling cone formed as a result of the shock-wave destruction **b**

a *b* *c*

Fig. 30.3 Multi-conical spalling destruction image at elevated power: without cone **a**, one- **b**, and two-cone **c** patterns

estimate the speed of crack growth in ceramics which in this case is, on average, equal to an impressive 180 km/s.

For comparison, Fig. 30.4a, b shows a picture of the characteristics of the destruction of ceramic plates under the action of penetration of high-velocity impact elements. For shock tests, a table-top acceleration stand was used in which a ball with a diameter of 8 mm made of SHX15 steel is accelerated by the energy of the gases of

Fig. 30.4 Patterns of the ceramic plate destruction as a result of high-velocity impact element penetration at the velocity of 400 m/s **a** and 600 m/s **b**

the mounting cartridge while moving along a channel 570 mm long at a controlled velocity, which makes it possible to form shock waves in a wide range of pressures.

As can be seen from Fig. 30.4, destruction processes are very different depending on the velocity of the impact element (Fig. 30.4a—the velocity of the impact element is 400 m/s, Fig. 30.4b—the velocity of the impact element is 600 m/s). At the same time, quasi-static failure, spalling failure, and specific edge failure in the form of fragmentation into a large number of particles of different configurations and sizes are observed.

30.3.2 Strength Properties of Artificial Stone

In the work, systematic studies of the dynamics of destruction processes were carried out for samples of artificial stone manufactured according to factory requirements. Artificial stone is created from natural stone chips, primarily quartzite. The strength characteristics of quartzite are very high. Artificial stone surpasses not only marble but also granite in all indicators of strength: impact, bending, compression, and compression during cooling. The size of the sample was 60 mm by 60 mm and 3 mm thick. During the experimental study of artificial stone samples after shock-wave loading, no sample destruction was observed. At the same time, only the formation of radial cracks is observed (see Fig. 30.5a, b). Analysis of Fig. 30.5 demonstrates the need for metallographic studies of radial cracks to illuminate samples at different angles.

The obtained results of the study of the formation and movement of cracks after shock-wave loading of artificial stone samples make it possible to evaluate the rate of crack growth in such samples. Unlike ceramics, under the same load conditions,

a *b*

Fig. 30.5 Radial cracks formation in artificial stone samples 60 × 60 × 3 mm under different angles of illumination **a**, **b**

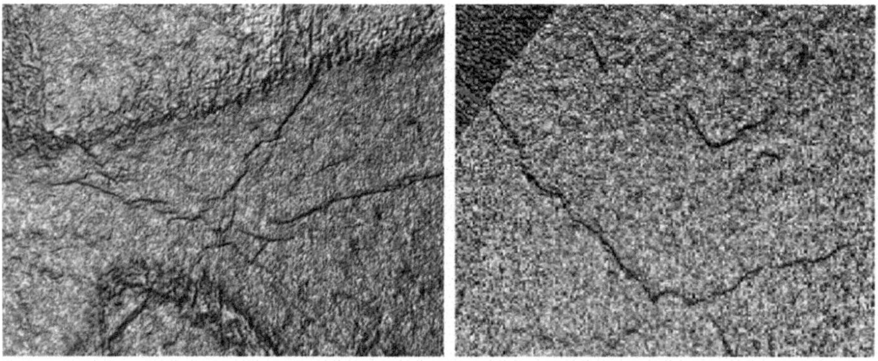

a *b*

Fig. 30.6 3D surface of the central part **a** of the artificial stone sample and its periphery **b**

the rate of crack growth in artificial stone samples is, on average, 75 km/s, which is almost two times less than in ceramics.

For a detailed study of the nucleation and movement of cracks after shock-wave loading, software tools for 3D reconstruction of the surface with high spatial resolution were used in the work. Figure 30.6a, b shows the obtained 3D surface of the central part of the artificial stone sample and its periphery. 3D reconstruction of the surface makes it possible to observe and study the entire pattern of crack growth from the center to the periphery of the sample.

In this work, the study and analysis of individual parts of the crack were also carried out, in particular, in the areas where their bifurcation and further growth occur (Fig. 30.7a, b). Analysis of Fig. 30.6b indicates the possibility of studying in detail

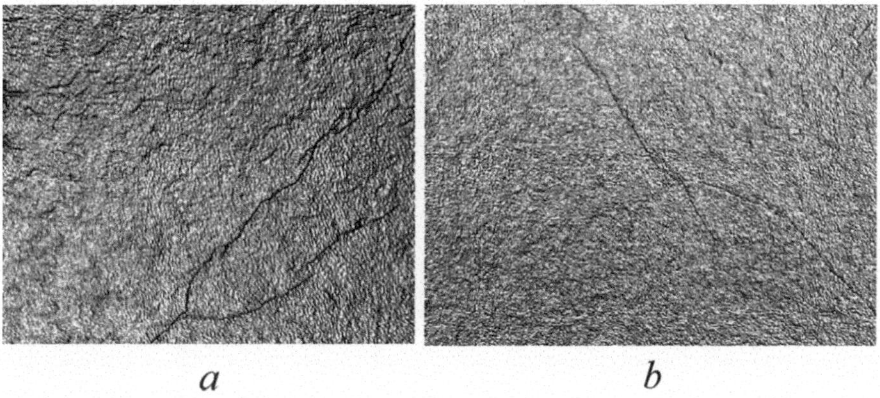

Fig. 30.7 Images of different parts of the crack: bifurcation area **a** and further growth area **b**

the movement of cracks to the edge of the samples under the action of high-velocity impact load, regardless of the angles of illumination necessary for metallographic studies. At the same time, the nucleation and movement of local cracks in different places of the samples are also observed. The latter makes it possible to assert the existence of complex multi-level destruction of samples under the action of shock-wave loading. It is very important to be able to study the growth of the crack with the bifurcation of the last feature (Fig. 30.7a, b). After the bifurcation of the crack, the energy for growth is usually enough for only one branch of the bifurcated crack. All obtained experimental results are necessary and very important for the mathematical modeling of these processes in order to obtain new materials with high-strength values necessary to resist high-velocity shock loads.

30.3.3 Investigation of Composite Material Strength Properties

Plates made of the composite material were also a subject of research in this work. Conductive plates were made of thermally expanded graphite in a rubber matrix. Samples were obtained from one plate by cutting it into four parts. The sample size was 70 mm by 70 mm and 10 mm thick. When investigating the impact of high-velocity shock-wave loading, it was established that a single shock-wave load with the parameters indicated above did not lead to the destruction of the samples or the appearance of cracks on it. Therefore, in this work, experiments were carried out on two-fold shock-wave loading of samples. In this case, after loading, the metallographic method revealed the appearance of radial cracks (Fig. 30.8a, b). Figure 30.8 shows the view of the surface of the sample at different magnifications (initial—Fig. 30.8a, ten times optical magnification—Fig. 30.8b). Analysis of the samples obtained in the work after loading makes it possible to estimate the speed of crack

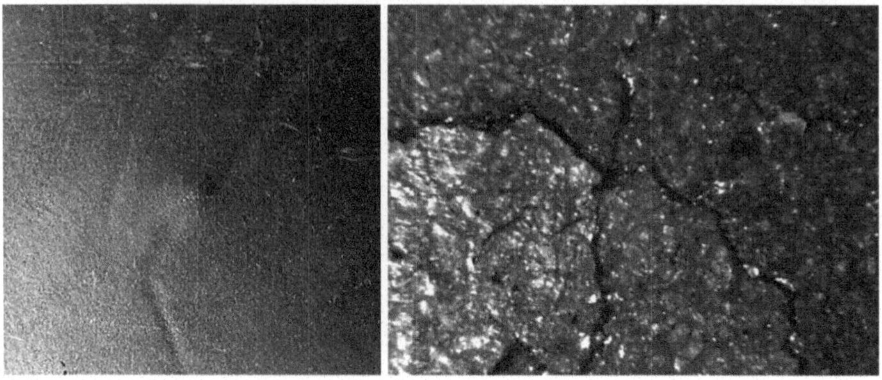

a *b*

Fig. 30.8 Crack formation in the 70 × 70 × 10 mm composite plate(thermally expanded graphite in a rubber matrix) under two-fold shock-wave loading: normal view **a**; ten times magnified image **b**

Fig. 30.9 Surface structure changes from the impact site to the periphery of the sample

growth in the samples, which in this case is, on average, equal to a velocity of 25 km/s, which is much less than in the ceramics and artificial stone samples. The latter indicates significant resistance to high-velocity shock-wave loading.

The study of the surface structure by the metallographic method (at a small magnification of 5 times) from the side of the applied load indicates significant changes in the surface structure from the center of the impact to the periphery of the samples (Fig. 30.9).

30.4 Conclusions

The analysis of shock-wave phenomena in solids allows for studying the peculiarities of deformation processes at high-impact velocities. The work used a developed and manufactured compact shock-wave device for rapid testing of the dynamic strength of such materials as ceramics, glass, and Plexiglas of various thicknesses, as well as an experimental study of the physical and mechanical parameters of the studied materials after exposure to high-velocity shock loads.

The developed design of the experimental complex is capable of forming a pressure pulse with a maximally planar front and with the possibility of increasing the power of the explosion of a thin metal foil to achieve spalling destruction. The shock-wave device allows for generating an elastic compression pulse up to 5 GPa for less than 0.4 microseconds on an area of several tens of square millimeters.

It was demonstrated in the example of ceramic samples that at the minimum value of the shock-wave load, the plate is destroyed only with the formation of radial cracks. When the power of the explosion increases (the shock-wave load increases), not only quasi-static destruction is observed, but also spalling destruction due to the formation of one cone. A further increase in the power of the explosion leads to the appearance of multi-cone spalling destruction. At the same time, the rate of growth of cracks in ceramics is, on average, equal to an impressive velocity of 180 km/s. The use of such a device allows you to accurately determine the value of dynamic strength at the beginning of spalling failure and thus modify the technological process to obtain the necessary strength parameters or compare and select materials for specific purposes.

In this work, systematic studies of the dynamics of destruction processes and the evolution of defects are carried out on the example of artificial stone samples. During the experimental evaluation of samples made of artificial stone after shock-wave loading, no destruction of the samples was observed. At the same time, only the formation of radial cracks is observed. In contrast to ceramics, under the same loading conditions, the rate of crack growth in artificial stone samples is, on average, 75 km/s. Software for 3D surface reconstruction with high spatial resolution were used to study the initiation and movement of cracks after shock-wave loading. 3D reconstruction of the surface makes it possible to observe and study the entire pattern of crack growth from the center to the periphery of the sample. The study and analysis of individual parts of the crack were carried out, particularly in the areas where their bifurcation and further growth occur. The nucleation and movement of local cracks in different places of the samples are also observed. The latter makes it possible to assert the existence of complex multi-level destruction of samples under the effect of shock-wave loading.

The work also investigated plates made of composite material—electrically conductive plates made of thermally expanded graphite in a rubber matrix. When studying the effect of high-velocity shock-wave loading, it was established that a single shock-wave load with the parameters indicated for previous samples did not lead to the destruction of the composite material samples or the appearance of cracks on them. Therefore, in this work, experiments were carried out on two-fold shock-wave loading. The appearance of radial cracks was detected by the metallographic method after two-fold shock-wave loading. Analysis of the samples obtained in the work after loading makes it possible to estimate the speed of crack growth in the samples, which in this case is, on average, equal to a velocity of 25 km/s, which is much less than in the previous ceramics and artificial stone samples. The latter indicates significant resistance to high-velocity shock-wave loading. The study of the surface structure by the metallographic method from the side of the applied load shows substantial changes in the surface structure from the impact center to the periphery of the samples.

It should be noted that the developed method of shock-wave testing allows for the investigation of a whole complex of effects associated with high-velocity destruction. The importance of the technique is due to the fact that there is no single theory that explains the crucial effects that determine the elastic-plastic properties and dynamic strength of the latest materials under intense loads. Therefore, the development of relevant theoretical models is an urgent problem. From this point of view, in addition to the direct value of the obtained experimental data for applied applications, it is necessary and important for the mathematical modeling of these processes in order to obtain new materials with high-strength values necessary to resist high-velocity shock loads. The use of computer modeling will allow for obtaining the necessary information at the selected point of interaction and evaluating the degree of deformation and destruction of the studied samples.

References

Grigor'ev AN, Pavlenko AV (2009) Pressure generated by the electric explosion of metal foils. Tech Phys Lett 35(9):865–868

Kanel GI, Razorenov SV, Fortov VE (2013) Shock-wave phenomena and the properties of condensed matter. Springer, New York

Kundu T (2008) Fundamentals of fracture mechanics. CRC Press

Muffoletto DP, Burke KM, Zirnheld JL, Olabisi SO (2018) Effects of inductance on the pressure producedfrom exploding aluminum metallized capacitorGrade polypropylene films. IEEE Trans Radiat Plasma Med Sci 2:624–628

Saxena AK, Kaushik TC, Gupta SC (2010) Shock experiments and numerical simulations on low energy portable electrically exploding foil accelerators. Rev Sci Instrum 81:033508

Stepanov GV, Makovei VA (1993) Dynamic crack resistance of structural materials. Naukova Dumka, Kiev (in Russian)

Surkaev AL, Usachev VI, Kumysh MM (2011) Millisecond electric explosion of metal conductors. Tech Phys Lett 37(12):1135–1138

Zirnheld JL, Olabisi S, Burke K, Disanto TM, Moore HL Jr, Singh H (2009) Electric explosion of aluminum metallized film. IEEE Trans Plasma Sci 37(12):2378–2384

The manufacturer's authorised representative in the EU is Springer Nature Customer Service Centre GmbH, Europaplatz 3, 69115 Heidelberg, Germany. If you have any concerns regarding our products, please contact ProductSafety@springernature.com

Printed and bound by CPI Group (UK) Ltd, Croydon, CR0 4YY
01/10/2026
02225855-0016